smartwork

wwnorton.com/physics

SmartWork offers ready-made assignments for each chapter in *Physics for Engineers and Scientists*. Questions are drawn from the end-of-chapter problems and review problems. Instructors may use these assignments as is, select alternate questions from the question bank, or use the SmartWork authoring tools to develop new content.

Students and instructors access SmartWork through the StudySpace website for *Physics*, students' one-stop resource for online learning materials — at wwnorton.com/physics.

SMARTWORK HIGHLIGHTS
- an intuitive and easy-to-use interface
- extensive hinting and answer feedback— including multi-step guided tutorial problems
- a wide range of supported question types: numeric answer, equation entry, and vector diagramming

OTHER HELPFUL FEATURES
- an easy-to-use tool for composing mathematical expressions
- algorithmically generated variables
- fully worked solutions to which instructors may restrict or allow access
- adjustable tolerance for numeric answers
- vector diagramming functionality
- gradable units of measurement or value
- a full complement of instructor tools for managing homework assignments and grades
- a fully integrated ebook

Two types of problems expand upon the exposition of concepts in the text. **Simple Feedback Problems** anticipate common misconceptions and offer prompts at just the right moment to help students discover the correct solution. Students receive instant scoring, helpful hints, and answer feedback that offers the help they need, when they need it.

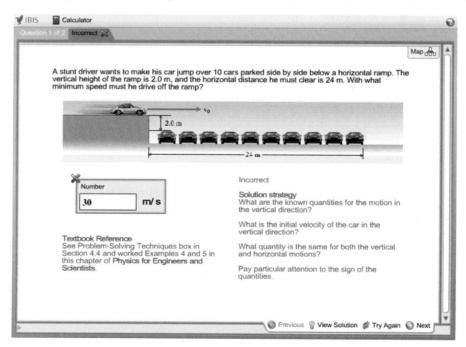

Guided Tutorial Problems address more challenging topics. If a student answers such a problem incorrectly, SmartWork guides the student through a series of discrete tutorial steps that lead to a general solution. The student then returns to the original problem ready to apply this newly obtained knowledge.

Shown here is a "parent" question and three tutorial steps that offer help.

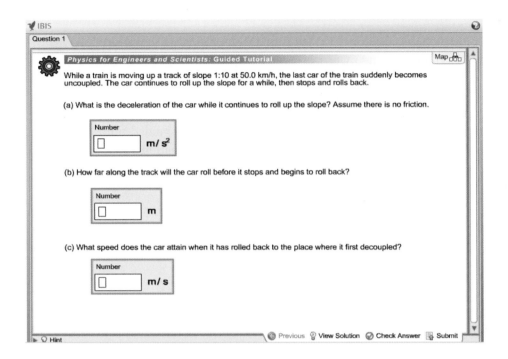

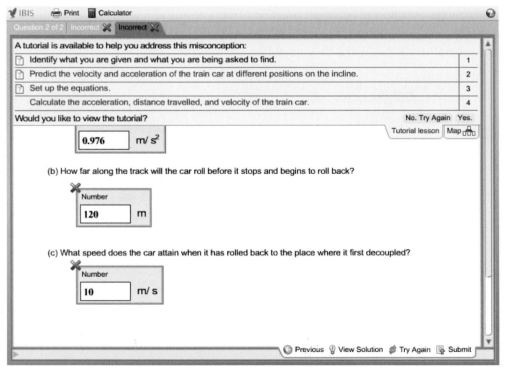

Each tutorial step is a simple feedback question that the student answers, with hints if necessary.

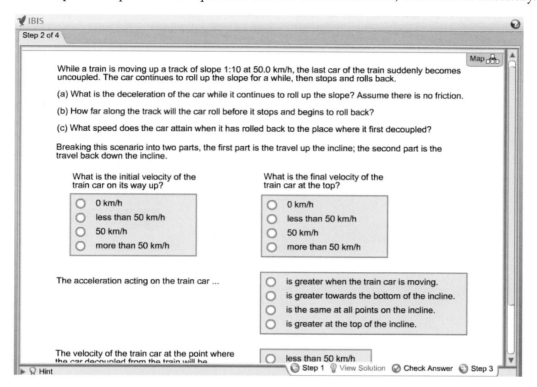

An intuitive click-to-select equation editor makes it easy for students to compose mathematical expressions.

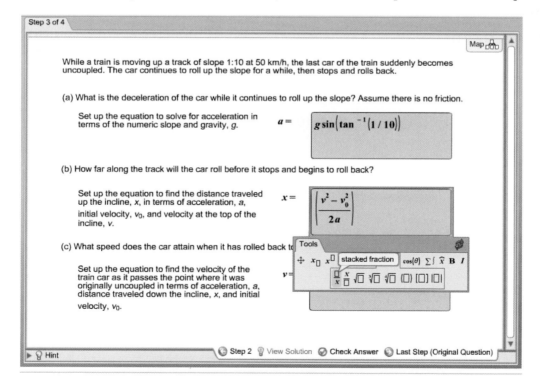

Numeric answers that fall within a predefined range of the right answer are scored as correct so that students are not penalized for small rounding errors or mistakes in identifying the significant digits. Instructors can easily customize pre-set tolerance settings. A calculator is built in.

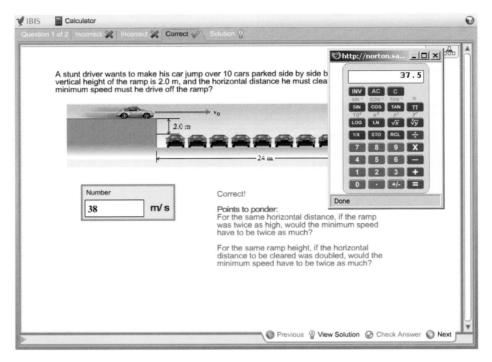

After a student has answered the question correctly, she may view a fully worked solution. Instructors have full control over when—or if—students may view the worked solution.

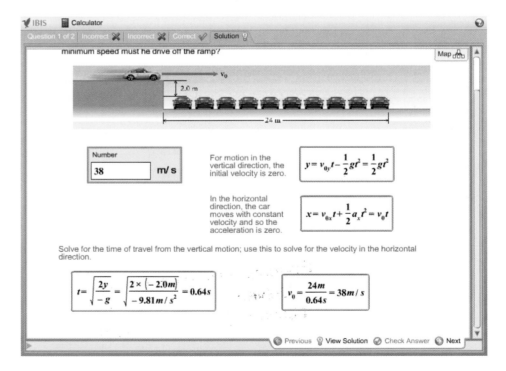

Physics for Engineers and Scientists

Third Edition

Volume 2 (Chapters 22–36)

ELECTRICITY AND MAGNETISM

WAVES AND OPTICS

THEORY OF SPECIAL RELATIVITY

W · W · NORTON & COMPANY NEW YORK · LONDON

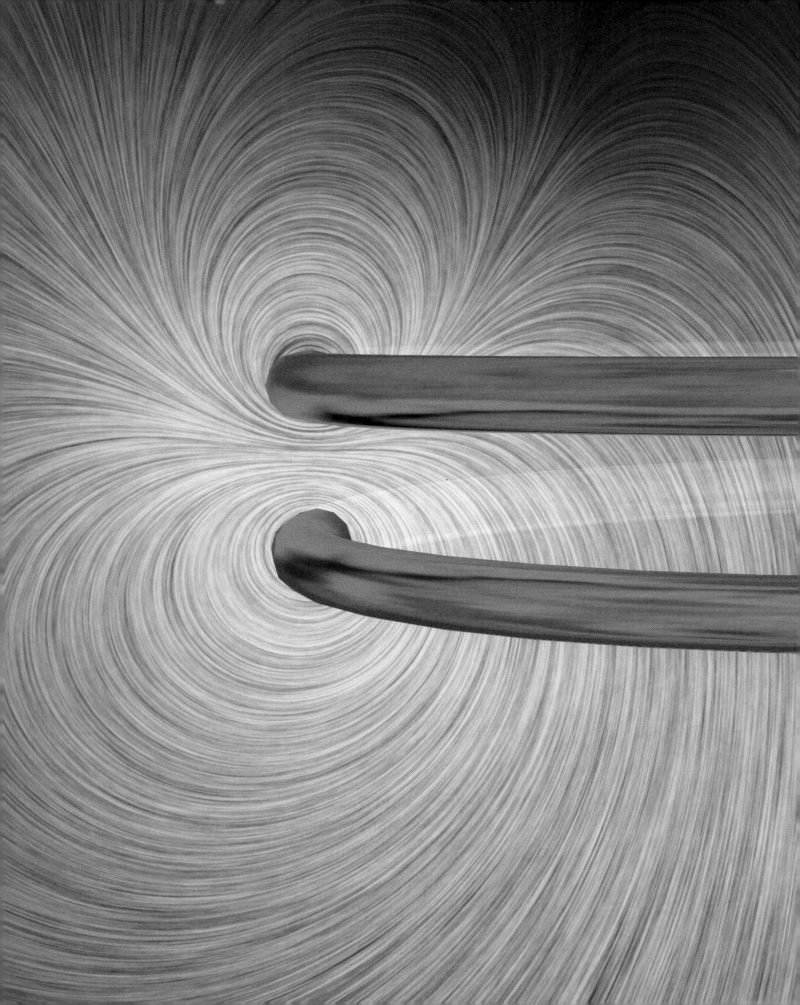

Physics for Engineers and Scientists Third Edition

Volume 2 (Chapters 22–36)

ELECTRICITY AND MAGNETISM

WAVES AND OPTICS

THE THEORY OF SPECIAL RELATIVITY

HANS C. OHANIAN,
UNIVERSITY OF VERMONT

JOHN T. MARKERT,
UNIVERSITY OF TEXAS AT AUSTIN

W · W · NORTON & COMPANY   NEW YORK · LONDON

To Susan Ohanian, writer, who gently tried to teach me some of her craft.—H.C.O.

To Frank D. Markert, a printer by trade; to Christiana Park, for her thirst for new knowledge; and to Erin, Ryan, Sean, and Gwen, for their wonder and clarity.—J.T.M.

Copyright © 2007 by W.W. Norton & Company, Inc.

Composition: Techbooks
Manufacturing: RR Donnelley & Sons Company
Editor: Leo A. W. Wiegman
Media Editor: April E. Lange
Director of Manufacturing—College: Roy Tedoff
Senior Project Editor: Christopher Granville
Photo Researcher: Kelly Mitchell
Editorial Assistant: Lisa Rand, Sarah L. Mann
Copy Editor: Richard K. Mickey
Book designer: Sandy Watanabe
Layout artist: Paul Lacy
Illustration Studio: Penumbra Design, Inc.
Cover Illustration: John Belcher, inter alia.
Cover Design: Joan Greenfield

Library of Congress Cataloging-in-Publication Data has been applied for.
ISBN 0-393-93004-1
ISBN 978-0-393-93004-7

W. W. Norton & Company, Inc., 500 Fifth Avenue, New York, N.Y. 10110
www.wwnorton.com

W. W. Norton & Company Ltd., Castle House, 75/76 Wells Street, London W1T 3QT

1234567890

W. W. Norton & Company has been independent since its founding in 1923, when William Warder Norton and Mary D. Herter Norton first published lectures delivered at the People's Institute, the adult education division of New York City's Cooper Union. The Nortons soon expanded their program beyond the Institute, publishing books by celebrated academics from America and abroad. By mid-century, the two major pillars of Norton's publishing program—trade books and college texts— were firmly established. In the 1950s, the Norton family transferred control of the company to its employees, and today—with a staff of four hundred and a comparable number of trade, college, and professional titles published each year—W. W. Norton & Company stands as the largest and oldest publishing house owned wholly by its employees.

Brief Contents

Table of Contents

APPENDICES

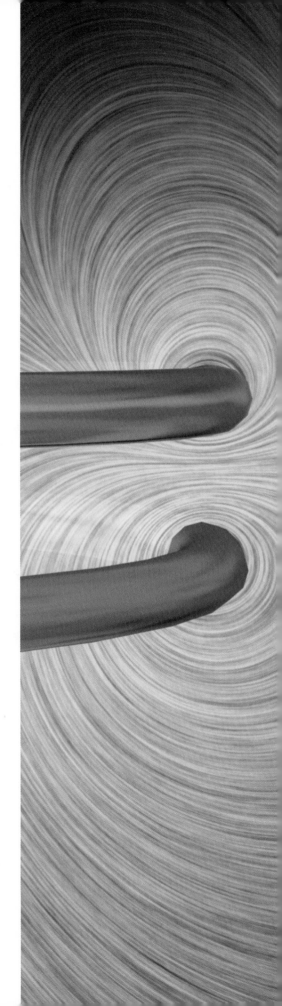

Preface

Our aim in *Physics for Engineers and Scientists,* Third Edition, is to present a modern view of classical mechanics and electromagnetism, including some optics and quantum physics. We also want to offer students a glimpse of the practical applications of physics in science, engineering, and everyday life.

The book and its learning package emerged from a collaborative effort that began more than six years ago. We adapted the core of Ohanian's earlier *Physics* (Second Edition, 1989) and combined it with relevant findings from recent physics education research on how students learn most effectively. The result is a text that presents a clear, uncluttered explication of the core concepts in physics, well suited to the needs of undergraduate engineering and science students.

Organization of Topics

The 41 chapters of the book cover the essential topics of introductory physics: mechanics of particles, rigid bodies, and fluids; oscillations, wave motion, heat and thermodynamics; electricity and magnetism; optics; special relativity; and atomic and subatomic physics.

Our arrangement and treatment of topics are fairly traditional with a few deliberate distinctions. We introduce the principle of superposition of forces early in Chapter 5 on Newton's laws of motion, and we give the students considerable exposure to the vector superposition of gravitational forces in Chapter 9. This leaves the students well prepared for the later application of vector superposition of electric and magnetic forces generated by charge or current distributions. We place gravitation in Chapter 9 immediately after the chapters on work and energy, because we regard gravitation as a direct application of these concepts (instructors who prefer to postpone gravitation can, of course, do so). We introduce forces on stationary electric charges in a detailed, complete exposition in Chapter 22, before proceeding to the less obvious concept of the electric field in Chapter 23. We start the study of magnetism in Chapter 29 with the force on a moving charged particle near a current, instead of the more common practice of starting with a postulate about the magnetic field in the abstract. With our approach, the observed magnetic forces on moving charges lead naturally to the magnetic field, and this progression from magnetic force to magnetic field will remind students of the closely parallel progression from electric force to electric field. For efficiency and brevity, we sometimes combine in one chapter closely related topics that other authors elect to spread over more than one chapter. Thus, we cover induction and inductance together in Chapter 31 and interference and diffraction together in Chapter 35.

Concise Writing with Sharp Focus on Core Concepts

Our goal is concise exposition with a sharp focus on core concepts. Brevity is desirable because long chapters with a large number of topics and excessive verbiage are confusing and tedious for the student. In our writing, we obey the admonitions of Strunk and White's *Elements of Style*: use the active voice; make statements in positive form; use definite, specific, concrete language; omit needless words.

We strove for simplicity in organizing the content. Each chapter covers a small set of core topics—rarely more than five or six—and we usually place each core topic in a section of its own. This divides the content into manageable segments and gives the chapter a clear and clean outline. Transitional sentences at the beginning or end of sections spell out the logical connections between each section and the next. Within each section, we strove for a seamless narrative leading from the discussions of concepts to their applications in Example problems. We sought to avoid the patchy, cobbled structure of many texts in which the discussions appear to serve as filler between one equation and the next.

Emphasis on the Atomic Structure of Matter

Throughout the book, we encourage students to keep in mind the atomic structure of matter and to think of the material world as a multitude of restless electrons, protons, and neutrons. For instance, in the mechanics chapters, we emphasize that all macroscopic bodies are systems of particles and that the equations of motion for macroscopic bodies emerge from the equations of motion of the individual particles. We emphasize that macroscopic forces are the result of a superposition of the forces among the particles of the system, and we consider atoms and their bonds in the qualitative discussions of elasticity, thermal expansion, and changes of state. By exposing students to the atomic structure of matter in the first semester, we help them to grasp the nature of the charged particles that play a central role in the treatment of electricity and magnetism in the second semester. Thus, in the electricity chapters, we introduce the concepts of positive and negative charge by referring to protons and electrons, not by referring to the antiquated procedure of rubbing glass rods with silk rags.

We try to make sure that students are always aware of the limitations of the nineteenth-century fiction that matter and electric charge are continua. Blind reliance on this old fiction has often been justified by the claim that, although engineering students need physics as a problem-solving tool, the atomic structure of matter is of little concern to them. This supposition may be adequate for a superficial treatment of mechanical engineering. Yet much of modern engineering—from materials science to electronics—hinges on understanding the atomic structure of matter. For this purpose, engineers need a physicist's view of physics.

Real-World Examples Begin Each Chapter

Each chapter opens with a "Concepts in Context" photograph illustrating a practical application of physics. The caption for this photo explores various core concepts in a concrete real-world context. The questions included in the caption are linked to several solved Examples or discussions later in the chapter. Such revisiting of the chapter-opening application provides layers of learning, as new concepts are carefully built upon a foundation firmly planted in the real world. The emphasis on real-world data is also evident throughout other Examples and in the end-of-chapter problems. By exposing students to realistic data, we give them confidence to apply physics in their later science or engineering courses.

Conceptual Discussions Precede and Motivate the Math

Only after a careful exposition of the conceptual foundations in a qualitative physical context does each section proceed to the mathematical treatment. Thus, we ensure that the mathematical formulas and their consequences and variations are rooted in a firm conceptual foundation. We were very careful to provide clear, thorough, and accurate explanations and derivations of all mathematical statements, to ensure that students acquire a good intuition about why particular equations are applied. Immediately after such derivations, we provide solved Examples to establish a firm connection between theory and concrete practical applications.

Examples Enliven the Text

We devote significant portions of each chapter to carefully selected Examples of solved problems—about 390 altogether or 9 on average per chapter. These Examples are concrete illustrations of the preceding conceptual discussions. They build cumulatively upon each other, from simple to more complicated as the chapter progresses. To enliven the text, we employ realistic data in the Examples, such as students would actually encounter outside the classroom. The solved Examples are designed to cover most variations of possible problems, with solutions that include both general approaches and specific details on how to extract the important information for the given problem. For instance, when such keywords as *initially* or *at rest* occur in a solved Example, we are careful to point out their importance in the problem-solving process. Comments appended to some Examples draw attention to limitations in the solution or to wider implications.

Checkup Questions Implement Active Learning

We conclude each section of a chapter with a series of brief *Checkup* questions. These permit students to test their mastery of core concepts, and they can be of great help in clearing up common misconceptions. Checkup questions include variations and "flip sides" of simple concepts that often occur to students but are rarely addressed. We give detailed answers to each Checkup question at the back of the chapter. The entire book contains roughly 5 Checkup questions per section—comprising a total of about 800 Checkup questions.

The final Checkup question of each section is always in multiple-choice forrmat— specifically designed for interactive teaching. At the University of Texas, instructors use such multiple-choice questions as classroom concept quizzes for welcome breaks in conventional lecturing. When more than one answer is popular, the instructor and class immediately know that more discussion or more examples are needed. Such occasions lend themselves well to peer instruction, in which the students explain to one another their reasoning before responding. This pedagogy implements an active, participatory alternative to the traditional lecture format. In addition, several supplements to the textbook, including the Student Activity Workbook, Online Concept Tutorials, Smartwork online homework, and PhysiQuizzes also implement active learning and a mastery-based approach.

Problem-Solving Techniques

Many chapters have inserts in the form of boxes devoted to Problem-Solving Techniques. These 39 skill boxes summarize the main steps or approaches for the solution of common classes of problems. Often deployed after several seemingly disparate Examples, the Problem-Solving Techniques boxes underscore the unity and generality of the techniques used in the Examples. The boxes list the steps or approaches to be taken, providing a handy reference and review.

Math Help

We have placed a Math Help box wherever students encounter a mathematical concept or technique that may be difficult or unfamiliar. These 6 skill boxes briefly review and summarize such topics as trigonometry, derivatives, integrals, and ellipses. Students can find more detailed help in Appendix 2 on basic algebra, 3 on trigonometry and geometry, 4 on calculus, and 5 on propagation of uncertainties.

Physics in Practice

Many chapters have a short essay on Physics in Practice that illustrates an application of physics in engineering and everyday life. These 27 essay boxes discuss practical topics, such as ultracentrifuges, communication and weather satellites, magnetic levitation, etc. Each of these essays provides a wealth of interesting detail and offers a practical supplement to some of the chapter topics. They have been designed to be engaging, yet sufficiently qualitative to provide some respite from the more analytical discussions, Examples, and Questions.

Figures and Balloon Captions

Over 1,500 figures illustrate the text. We made every effort to assemble a visual narrative as clear as the verbal narrative. Each figure in a sequence carefully builds upon the visual information in the figure that precedes it. Many figures in the text contain a caption in "balloon" that points to important features within the figure. The balloon caption is a concise and informative supplement to the conventional figure caption. The balloons make immediately obvious some details that would require a long, wordy explanation in the conventional caption. Often the balloon captions are arranged so that some cause-effect or other sequential thought process becomes immediately evident. All drawn figures are available to instructors in digital form for use in the course.

End–of–Chapter Summary

Each chapter narrative closes with several support elements, starting with a brief Summary. The Summary contains the essential physical laws, quantities, definitions, and key equations introduced in the chapter. A page reference, key equation number, and often a thumbnail figure accompany these laws, definitions, and equations. The Summary does not include repetition of the detailed explanations of the chapter. The Summary is followed by Questions for Discussion, Problems, Review Problems, and Answers to Checkups.

Questions for Discussion

After the chapter's Summary, we include a large selection of qualitative Questions for Discussion — about 700 in the entire book or roughly 17 per chapter. We intend these qualitative end-of-chapter Questions to stimulate student thinking. Some of these questions are deliberately formulated so as to have no unique answer, which is intended to promote class discussion.

Problems

After the chapter's qualitative Questions, we include computational Problems grouped by chapter section — about 3000 in the entire book, or roughly 73 per chapter. Each problem's level of difficulty is indicated by no asterisk, one asterisk (*), or two asterisks (**). Most no-asterisk Problems are easy and straightforward, only requiring students to "plug in" the correct values to compute answers or to retrace the steps of an Example. One-asterisk Problems are of medium difficulty. They contain a few complications requiring the combination of several concepts or the manipulation of several formulas. Two-asterisk Problems are difficult and challenging. They demand considerable thought and perhaps some insight, and occasionally require appreciable mathematical skill. When an Online Concept Tutorial (see below) is available for help in mastering the concepts in a given section, a dagger footnote (†) tells students where to find the tutorial.

We tried to make the Problems interesting for students by drawing on realistic examples from technology, science, sports, and everyday life. Many of the Problems are based on data extracted from engineering handbooks, car repair manuals, *Jane's Book of Aircraft*, *The Guinness Book of World Records*, newspaper reports, research and industrial instrumentation manuals, etc. Many other Problems deal with atoms and subatomic particles. These Problems are intended to reinforce the atomistic view of the material world. In some cases, experts will perhaps consider the use of classical physics somewhat objectionable in a problem that really ought to be handled by quantum mechanics. But we believe that the advantages of familiarization with atomic quantities and magnitudes outweigh the disadvantages of an occasional naive use of classical mechanics.

Among the Problems are a smaller number of somewhat contrived, artificial Problems that make no pretense of realism (for example, "A block slides on an inclined plane tied by string..."). Such unrealistic Problems are sometimes the best way to bring an important concept into sharp focus. Some Problems are formulated as guided problems, with a series of questions that take the student through an important problem-solving procedure, step by step.

Review Problems

After the Problems section of each chapter, we offer an extra selection of Review Problems — about 600 in the entire book or roughly 15 per chapter. We wrote these Review Problems specifically to help students prepare for examinations. Hence, Review Problems often test comprehension by requiring students to apply concepts from more than one section of the chapter and occasionally from prior, related chapters. Answers to all odd-numbered Problems and Review Problems are given in Appendix 11.

Units and Significant Figures

We use the SI system of units exclusively throughout the text. In the abbreviations for the units we follow the recommendations of the International Committee for Weights and Measures (CIPM), although we retain some traditional units, such as revolution and calorie that have been discontinued by the CIPM. In addition, for the sake of clarity we spell out the name of the unit in full whenever the abbreviation is likely to lead to ambiguity and confusion (for instance, in the case of V for volt, which is easily confused with V for potential; or in the case of C for coulomb, which might be confused with C for capacitance). We try to use realistic numbers of significant figures, with most Examples and Problems using two or three. In cases where it is natural to employ some data with two significant figures and some with three, we have been careful to propagate the appropriate number of significant figures to the result.

For reference purposes, we give the definitions of the British units. Currently only the United States, Bangladesh, and Liberia still adhere to these units. In the United States, automobile manufacturers have already switched to metric units for design and construction. The U. S. Army has also switched to metric units, so soldiers give distances in meters and kilometers (in army slang, the kilometer is called a "klick," a usage that is commendable itself for its brevity). British units are not used in examples or in problems, with the exception of a handful of problems in the early chapters. In the definitions of the British units, the pound (lb) is taken to be the unit of mass, and the pound force (lbf) is taken to be the unit of force. This is in accord with the practice approved by the American National Standards Institute (ANSI), the Institute of Electrical and Electronics Engineers (IEEE), and the U. S. Department of Defense.

Optional Sections and Chapters

We recognize course content varies from institution to institution. Some sections and some chapters can be regarded as optional and can be omitted without loss of continuity. These optional sections are marked by asterisks in the Table of Contents.

Mathematical Prerequisites

In order to accommodate students who are taking an introductory calculus course concurrently, derivatives are used slowly at first (Chapter 2), and routinely later on. Likewise, the use of integrals is postponed as long as possible (Chapter 7), and they come into heavy use only in the second volume (after Chapter 21). For students who need a review of calculus, Appendix 4 contains a concise primer on derivatives and integrals.

Acknowledgments

We have had the benefit of a talented author team for our support resources. In addition to their primary role in the assembly of the learning package, they all have also made substantial contributions to the accuracy and clarity of the text.

Stiliana Antonova, Barnard College
Charles Chiu, University of Texas-Austin
William J. Ellis, University of California-Davis
Mirela Fetea, University of Richmond
Rebecca Grossman, Columbia University
David Harrison, University of Toronto
Prabha Ramakrishnan, North Carolina State University
Hang Deng-Luzader, Frostburg State University
Stephen Luzader, Frostburg State University
Kevin Martus, William Paterson University
David Marx, Illinois State University
Jason Stevens, Deerfield Academy
Brian Woodahl, Indiana University–Purdue University-Indianapolis
Raymond Zich, Illinois State University

And at Sapling Systems and Science Technologies in Austin, Texas, for content, James Caras, Ph.D.; Jon Harmon, B.S.; Kevin Nelson, Ph.D.; John A. Underwood, Ph.D.; and Jason Vestuto, M.S. and for animation and programming, Jeff Sims and Nathan Wheeler.

Our manuscript was subjected to many rounds of peer review. The reviewers were instrumental in identifying myriad improvements, for which we are grateful:

Yildirim Aktas   University of North Carolina–Charlotte
Patricia E. Allen   Appalachian State University
Steven M. Anlage   University of Maryland
B. Antanaitis   Lafayette College
Laszlo Baksay   Florida Institute of Technology
Marco Battaglia   University of California-Berkeley
Lowell Boone   University of Evansville
Marc Borowczak   Walsh University
Amit Chakrabarti   Kansas State University
D. Cornelison   Northern Arizona University
Corbin Covault   Case Western Reserve University
Kaushik De   University of Texas at Arlington
William E. Dieterle   California University of Pennsylvania
James Dunne   Mississippi State University
R. Eagleton   California Polytechnic University-Pomona
Gregory Earle   University of Texas-Dallas
William Ellis   University of California-Davis
Mark Eriksson   University of Wisconsin-Madison
Morten Eskildsen   University of Notre Dame
Bernard Feldman   University of Missouri–St. Louis
Mirela Fetea   University of Richmond
J. D. Garcia   University of Arizona
U. Garg   University of Notre Dame
Michael Gurvitch   State University of New York at Stony Brook
David Harrison   University of Toronto
John Hernandez   University of North Carolina–Chapel Hill
L. Hodges   Iowa State University
Jean-Pierre Jouas   United Nations International School
Kevin Kimberlin   Bradley University
Sebastian Kuhn   Old Dominion University
Tiffany Landry   Folsom Lake College
Dean Lee   North Carolina State University
Frank Lee   George Washington University
Stephen Luzader   Frostburg State University
Kevin Martus   William Paterson University
M. Matkovich   Oakton Community College
David McIntyre   Oregon State University
Rahul Mehta   University of Central Arkansas
Kenneth Mendelson   Marquette University
Laszlo Mihaly   State University of New York at Stony Brook
Richard Mistrick   Pennsylvania State University
Rabindra Mohapatra   University of Maryland
Philip P. J. Morrison   University of Texas at Austin
Greg Mowry   University of Saint Thomas
David Murdock   Tennessee Technological University
Anthony J. Nicastro   West Chester University
Scott Nutter   Northern Kentucky University
Robert Oerter   George Mason University
Ray H. O'Neal, Jr.   Florida A & M University
Frederick Oho,   Winona State University
Paul Parris   University of Missouri–Rolla
Ashok Puri   University of New Orleans
Michael Richmond   Rochester Institute of Technology
John Rollino   Rutgers University–Newark
David Schaefer   Towson State University
Joseph Serene   Georgetown University
H. Shenton   University of Delaware
Jason Stevens   Deerfield Academy

Jay Strieb	Villanova University
John Swez	Indiana State University
Devki N. Talwar	Indiana University of Pennsylvania
Chin-Che Tin	Auburn University
Tim Usher	California State University-San Bernardino
Andrew Wallace	Angelo State University
Barrett Wells	University of Connecticut
Edward A.P. Whittaker	Stevens Institute of Technology
David Wick	Clarkson University
Don Wieber	Contra Costa College
J. William Gary	University of California-Riverside
Suzanne Willis	Northern Illinois University
Thomas Wilson	Marshall University
William. J. F. Wilson	University of Calgary
Brian Woodahl	Indiana University–Purdue University-Indianapolis
Hai-Sheng Wu	Mankato State University

We thank John Belcher, Michael Danziger, and Mark Bessette of the Massachusetts Institute of Technology for creating the cover image. It illustrates the magnetic field generated by two currents in two copper rings. This is one frame of a continuous animation; at the instant shown, the current in the upper ring is opposite to that in the lower ring and is of smaller magnitude. The magnetic field structure shown in this picture was calculated using a modified intregration technique. This image was created as part of the Technology Enabled Active Learning (TEAL) program in introductory physics at MIT, which teaches physics interactively, combining desktop experiments with visualizations of those experiments to "make the unseen seen."

We thank the several editors that supervised this project: first Stephen Mosberg, then Richard Mixter, John Byram, and finally Leo Wiegman, who had the largest share in the development of the text, and also gave us the benefit of his incisive line-by-line editing of the proofs, catching many slips and suggesting many improvements. We also thank the editorial staff at W. W. Norton & Co., including Chris Granville, April Lange, Roy Tedoff, Rubina Yeh, Rob Bellinger, Kelly Mitchell, Neil Hoos, Lisa Rand, and Sarah Mann, as well as the publishing professionals whom Norton engaged, such as Paul Lacy, Richard K. Mickey, Susan McLaughlin, and John B. Woolsey for their enthusiasm and their patience in dealing with the interminable revisions and corrections of the text and its support package. In addition, JTM is grateful to Robert W. Christy of Dartmouth University for various pointers on textbook writing.

HANS C. OHANIAN
Burlington, Vermont
hohanian@uvm.edu

JOHN T. MARKERT
Austin, Texas
jmarkert@physics.utexas.edu

Publication Formats

Physics for Engineers and Scientists comprises six parts. The text is published in two hardcover versions and several paperback versions.

Hardcover Versions

Third Extended Edition, Parts I–VI, 1450 pages, ISBN 0-393-92631-1
    (Chapters 1–41 including Relativity, Quanta and Particles)
Third Edition, Parts I–V, 1282 pages, ISBN 0-393-97422-7
    (Chapters 1–36, including Special Relativity)

Paperback Versions

Volume 1, (Chapters 1–21) 778 pages, ISBN 0-393-93003-3
    Part I Motion, Force, and Energy (Chapters 1–14)
    Part II Oscillations, Waves, and Fluids (Chapters 15–18)
    Part III Temperature, Heat, and Thermodynamics (Chapter 19–21)
Volume 2, (Chapters 22–36) 568 pages, ISBN 0-393-93004-1
    Part IV Electricity and Magnetism (Chapters 22–32)
    Part V Waves and Optics (Chapters 33–35 and Chapter 36 on Special Relativity)
Volume 3, (Chapters 36–41) 250 pages, ISBN 0-393-92969-8
    Part VI Relativity, Quanta, and Particles
In addition, to explore customized versions, please contact your Norton representative.

Two Norton ebook Options

nortonebooks.com

Physics for Engineers and Scientists is available in a Norton ebook format that retains the content of the print book. The ebook offers a variety of tools for study and review, including sticky notes, highlighters, zoomable images, links to Online Concept Tutorials, and a search function. Purchased together, the SmartWork with integrated ebook bundle makes it easy for students to check text references when completing online homework assignments.

    The ebook may also be purchased as a standalone item. The downloadable PDF version is available for purchase from Powells.com.

Package Options

Each version of the text purchased from Norton—with or without SmartWork—will come with free access to our website at Norton's StudySpace that includes the valuable Online Concept Tutorials. Each version of the text may be purchased as a stand-alone book or as a package that includes—each for a fee—Norton's new SmartWork online homework system or the Student Activity Workbook by David Harrison and William Ellis. Hence, several optinal packages are available to instructors:
• Textbook–StudySpace–Online Concept Tutorials + Student Activity Workbook
• Textbook–StudySpace–Online Concept Tutorials + SmartWork/ebook
• Textbook–StudySpace–Online Concept Tutorials + SmartWork/ebook + Student Activity Workbook

The Support Program

To enhance individual learning and also peer instruction, a carefully integrated support program accompanies the text. Each element of the support program has two goals. First, each support resource mirrors the text's emphasis on sharply focused core concepts. Second, treatment of a core concept in a support resource offers a perspective that is different from but compatible with that of the text. If a student needs help beyond the text, he or she would more likely benefit from a fresh presentation on the same concept rather than from one that simply repeats the text presentation.

Hence, the text and its support package offers three or more different approaches to the core concepts. For example, Newton's First and Second Laws are rendered with interactive animations in the Online Concept Tutorial "Forces," with pencil-and-paper exercises in Chapter 5 of the Student Activity Workbook crafted by David Harrison and William Ellis, and with concept test inquiries in PhysiQuiz questions written by Charles Chiu and edited by Jason Stevens.

Both printed and digital resources are offered within the support program. Outstanding web-based resources for both instructors and students include tutorials and a homework system.

SmartWork Online Homework System

smartwork
www.wwnorton.com/physics

SmartWork—Norton's online homework management system—provides ready-made automatically graded assignments, including guided problems, simple feedback questions, and animated tutorials—all specifically designed to extend the text's emphasis on core concepts and problem-solving skills.

Developed in collaboration with Sapling Systems, SmartWork features an intuitive, easy-to-use interface that offers instructors flexible tools to manage assignments, while making it easy for students to compose mathematical expressions, draw vectors and graphs, and receive helpful and immediate feedback. Two different types of questions expand upon the exposition of concepts in the text:

Simple Feedback Problems present students with problems that anticipate common misconceptions and offer prompts at just the right moment to help them discover the correct solution.

Guided Tutorial Problems addresses more challenging topics. If a student answers a problem incorrectly, SmartWork guides the student through a series of discrete tutorial steps that lead to a general solution. Each step is a simple feedback question that the student answers, with hints if necessary. After completing all of the tutorial steps, the student returns to the original problem ready to apply this newly-obtained knowledge.

SmartWork problems use algorithmic variables so two students are unlikely to see exactly the same problem. Instructors can use the problem sets provided, or can customize these ready-made questions and assignments, or use SmartWork to create their own.

SmartWork is available bundled with the Norton ebook of *Physics for Engineers and Scientists*. Where appropriate, SmartWork prompts students to review relevant sections in the textbook. Links to the **ebook** make it easy for students to consult the text while working through problems online.

Online Concept Tutorials

Online
Concept
Tutorial
www.wwnorton.com/physics

Developed in collaboration with Science Technologies specifically for this course, these 45 tutorials feature interactive animations that reinforce conceptual understanding and develop students' quantitative skills. In-text icons alert students to the availability of a tutorial. All Online Concept Tutorials are available on the free StudySpace web site and are integrated into SmartWork. Tutorials can also be accessed from a CD-ROM that requires no installation, browser tune-ups, or plug-ins.

StudySpace Website

www.wwnorton.com/physics

STUDYSPACE www.wwnorton.com/physics. This free and open website is the portal for both public and premium content. Free content at StudySpace includes the Online Concept Tutorials and a Study Plan for each chapter in *Physics for Engineers and Scientists*. Premium content at StudySpace includes links to the online ebook and to SmartWork.

WebAssign

Selected end-of-chapter problems from Physics for Engineers and Scientists are available in WebAssign, with additional problems available to adopting instructors by request to WebAssign.

Additional Instructor Resources

TEST BANK by Mirela Fetea, University of Richmond; Kevin Martus, William Paterson University; and Brian Woodahl, Indiana University-Purdue University-Indianapolis. The Test Bank offers approximately 2000 multiple-choice questions, available in ExamView, WebCT, BlackBoard, rich-text, and printed format.

INSTRUCTOR SOLUTIONS MANUAL by Stephen Luzader and Hang-Deng Luzader, both of Frostburg State University, and David Marx of Illinois State University. The Instructor Solution Manual offers solutions to all end-of-chapter Problems and Review Problems, checked for accuracy and clarity.

PHYSIQUIZ "CLICKER" QUESTIONS by Charles Chiu, University of Texas at Austin, with Jason Stevens, Deerfield Academy. The PhysiQuiz multiple-choice questions are designed for use with classroom response, or "clicker", systems. The 300 PhysiQuiz questions are available as PowerPoint slides, in printed format, and as transparency masters.

NORTON MEDIA LIBRARY INSTRUCTOR CD-ROM The Media Library for instrutors includes selected figures, tables, and equations from the text in JPEG and PowerPoint formats, PhysiQuiz "clicker" questions, and PowerPoint-ready offline versions of the Online Concept Tutorials.

INSTRUCTOR RESOURCE MANUAL offers a guide to the support package with descriptions of the Online Concept Tutorials, information about the SmartWork homework problems available for each chapter, printed PhysiQuiz "clicker" questions, and instructor notes for the workshop activities in the Student Activity Workbook.

TRANSPARENCY ACETATES Approximately 200 printed color acetates of key figures from the text.

BLACKBOARD AND WEBCT COURSE CARTRIDGES Course Cartridges for BlackBoard and WebCT include access to the Online Concept Tutorials, a Study Plan for each chapter, multiple-choice tests, plus links to the premium, password-protected contents of the Norton ebook and SmartWork.

Additional Student Resources

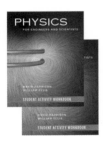

STUDENT ACTIVITY WORKBOOK by David Harrison, University of Toronto, and William Ellis, University of California Davis. The *Student Activity Workbook* is an important part of the learning package. For each chapter of *Physics for Engineers and Scientists*, the Workbook's Activities break down a physical condition into constituent parts. The Activities are pencil and paper exercises well suited to either individual or small group collaboration. The Activities include both conceptual and quantitative exercises. Some Activities are guided problems that pose a question and present a solution scheme via follow up questions. The Workbook is available in two paperback volumes: Volume 1 comprises Chapters 1–21 and Volume 2 comprises Chapters 22–41.

STUDENT SOLUTIONS MANUAL by Stephen Luzader and Hang-Deng Luzader, both of Frostburg State University, and David Marx of Illinois State University. The Student Solutions Manual contains detailed solutions to approximately 25% of the problems in the book, chosen from the odd-numbered problems whose answers appear in the back of the book. The Manual is available in two paperback volumes: Volume 1 comprises Chapters 1–21 and Volume 2 comprises Chapters 22–41.

ONLINE CONCEPT TUTORIALS CD-ROM The 45 Online Concept Tutorials (see above) can also be accessed from an optional CD-ROM that requires no installation, browser tune-up, or plug-in.

About the Authors

Hans C. Ohanian received his B.S. from the University of California, Berkeley, and his Ph.D from Princeton University, where he worked with John A. Wheeler. He has taught at Rensselaer Polytechnic Institute, Union College, and the University of Vermont. He is the author of several textbooks spanning all undergraduate levels: *Physics*, *Principles of Physics*, *Relativity: A Modern Introduction*, *Modern Physics*, *Principles of Quantum Mechanics*, *Classical Electrodynamics*, and, with Remo Ruffini, *Gravitation and Spacetime*. He is also the author of dozens of articles dealing with gravitation, relativity, and quantum theory, including many articles on fundamental physics published in the *American Journal of Physics*, where he served as associate editor for some years. He lives in Vermont. hohanian@uvm.edu

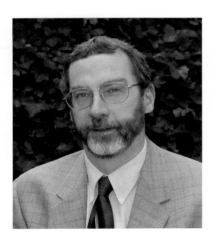

John T. Markert received his B.A. in physics and mathematics from Bowdoin College (1979), and his M.S. (1984) and Ph.D. (1987) in physics from Cornell University, where he was recipient of the *Clark Award for Excellence in Teaching*. After postdoctoral research at the University of California, San Diego, he joined the faculty at the University of Texas at Austin in 1990, where he has received the *College of Natural Sciences Teaching Excellence Award* and is currently Professor of Physics and Department Chair. His introductory physics teaching methods emphasize context-based approaches, interactive techniques, and peer instruction. He is author or coauthor of over 120 journal articles, including experimental condensed-matter physics research in superconductivity, magnetism, and nanoscience. He lives in Austin, Texas, with his spouse and four children. jmarkert@physics.utexas.edu

Owner's Manual for *Physics for Engineers and Scientists*

These pages give a brief tour of the features of *Physics for Engineers and Scientists* and its study resources. Some resources are found within the book. Others are located in accompanying paperback publications or at the StudySpace web portal. Features on the text pages shown here come chiefly from the discussion of friction in Chapter 6, but are common in other chapters.

The learning resources listed below help students study by offering alternative explanations of the core concepts found in the text. These student resources are briefly described at the end of this owner's manual:

- Online Concept Tutorials
- SmartWork Online Homework
- StudySpace
- Student Activity Workbook
- Student Solutions Manual

Each chapter of the textbook starts with a real-world example of a core concept. Chapter 6 opens with the concept of friction and uses automobile tires as an example of friction that is revisited in several different conditions. The opening photograph, it's caption and the caption's closing questions all discuss this example.

CHAPTER 6

Further Applications of Newton's Laws

CONCEPTS IN CONTEXT

Concepts — in — Context

Automobiles rely on the friction between the road and the tires to accelerate and to stop. We will see that one of two types of contact friction, kinetic or static, is involved. To see how these friction forces affect linear and circular motion, we ask:

? In an emergency, an automobile brakes with locked and skidding wheels. What deceleration can be achieved? (Example 1, page 176)

? What is the steepest slope of a street on which an automobile can rest without slipping? (Example 4, page 179)

? When braking without skidding, what maximum deceleration can be achieved? (Example 5, page 180)

? How quickly can a racing car round a curve without skidding sideways? (Example 10, page 186)

? How does a banked curve help to avoid skidding? (Example 11, page 186)

6.1 Friction

6.2 Restoring Force of a Spring: Hooke's Law

6.3 Force for Uniform Circular Motion

6.4 The Four Fundamental Forces

173

Most chapters have six or fewer sections. Most sections are four or five pages in length and cover one major topic.

In this chapter, the rubber tires of an automobile are revisited to explore concepts in friction on pages 176, 179, 180, and 186, as indicated.

The icon indicates an **Online Concept Tutorial** is available for a key concept. Each such icon includes the identification number of the tutorial—8, in this case. These tutorials offer a visual guide and self-quiz for the concept at hand. Find all the Tutorials at www.wwnorton.com/physics.

In mathematical expressions, such as $m\mathbf{a}=\mathbf{F}$, the **bold type** indicates a **vector** and *italic* indicates variables that are not vectors.

Text in *italic type* indicates major definitions of laws or statements of general principles.

Text in **bold type** highlights the first use of a key term and is generally accompanied by an explanation.

Key concepts or important variants of these concepts have a key-term label in the margin.

Short **biographical sketches** appear in the margins of this text. Each offer a brief glimpse into the life of some major contributor to our knowledge about the physical world—in this case, Italian artist and engineer Leonardo da Vinci.

Highlighted equations are key equations that express central physics concepts mathematically.

---

174   **CHAPTER 6** Further Applications of Newton's Laws

To find a solution of the equation of motion means to find a force **F** and a corresponding acceleration **a** such that Newton's equation $m\mathbf{a} = \mathbf{F}$ is satisfied. For a physicist, the typical problem involves a known force and an unknown motion; for example, the physicist knows the forces between the planets and the Sun, and she seeks ... for an engineer, the reverse problem with ... often of practical importance; for example, ... a given curve at 60 km/h, and he seeks to ... wheels must withstand. A special problem ... tics; here we know that the body is at rest ... wish to compute the forces that will main- ... depending on the circumstances, we can ... f the equation $m\mathbf{a} = \mathbf{F}$ as an unknown that ... t the other side.

... ome solutions of the equation of motion ... ight and constant pushes or pulls. In this ... s of the equation of motion, and we will examine other, more complicated forces, such as friction and the forces exerted by springs.

Online Concept Tutorial

LEONARDO da VINCI (1452–1519)
Italian artist, engineer, and scientist. Famous for his brilliant achievements in painting, sculpture, and architecture. Leonardo also made pioneering contributions to science. But Leonardo's investigations of friction were forgotten, and the laws of friction were rediscovered 200 years later by Guillaume Amontons, a French physicist.

6.1 FRICTION

Friction forces, which we have ignored up to now, play an important role in our environment and provide us with many interesting examples of motion with constant force. For instance, if the driver of ... wheels will lock and begin t... an (approximately) constan... the automobile at an (appro... the friction force depends on... the heavy friction of rubber... of the wheels, which introdu...

For the sake of simplicit... solid block of metal sliding... in the shape of a brick, slidi... velocity and then let it coast... are the weight **w**, the normal... ward with a magnitude mg. ... upward; the magnitude of th... The friction force **f** exerted... tabletop, in a direction oppo... contact force which acts ove... 6.1 it is shown as though ac...

The friction force arises... in the block form bonds wi... these bonds are continually... represents the effort require... scopic level the phenomeno... the resulting friction force... law, first enunciated by Leo...

6.1   Friction

The magnitude of the friction force between unlubricated, dry surfaces sliding one over the other is proportional to the magnitude of the normal force acting on the surfaces and is independent of the area of contact and of the relative speed.

Friction involving surfaces in relative motion is called **sliding friction**, or **kinetic friction**. According to the above law, the magnitude of the force of kinetic friction can be written mathematically as

$$f_k = \mu_k N \qquad (6.1)$$

where μ_k is the **coefficient of kinetic friction**, a constant characteristic of the material involved. Table 6.1 lists typical friction coefficients for various materials.

Note that Eq. (6.1) states that the magnitudes of the friction force and the normal force are proportional. The *directions* of these forces are, however, quite different: the normal fo... ... orce $\mathbf{f}_k$ is parallel... The a... laws. It is only a ... merely a descriptio... detailed theoretica... m this simple law... devi- ations in many everyday engineering problems in which the speeds are not extreme. The simple friction law is then a reasonably good approximation for a wide range of materials, and it is at its best for metals sliding on metals.

The fact that the friction force is independent of the area of contact means that the friction force of the block sliding on the tabletop is the same whether the block slides on a large face or on one of the small faces (see Fig. 6.2). This may seem surprising at first—we might expect the friction force to be larger when the block slides on the larger face, with more area in contact with the tabletop. However, the normal force is then distributed over a larger area, and is therefore less effective in pressing the atoms together; and the net result is that the friction force is independent of the area of contact.

force of kinetic friction

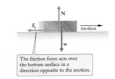

The friction force acts over the bottom surface in a direction opposite to the motion.

FIGURE 6.1 Forces on a block sliding on a plate.

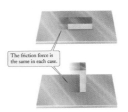

The friction force is the same in each case.

FIGURE 6.2 Steel block on a steel plate, sliding on a large face or on a small face.

TABLE 6.1	KINETIC AND STATIC FRICTION COEFFICIENTS[a]	
MATERIALS	μ_k	μ_s
Steel on steel	0.6	0.7
Steel on lead	0.9	0.9
Steel on copper	0.4	0.5
Copper on cast iron	0.3	1.1
Copper on glass	0.5	0.7
Waxed ski on snow		
at $-10°C$	0.2	—
at $0°C$	0.05	—
Rubber on concrete	≈ 1	≈ 1

[a] The friction coefficient depends on the condition of the surfaces. The values in this table are typical for dry surfaces but not entirely reliable.

Examples are a critical part of each chapter.
• Examples provide concrete illustrations of the concepts being discussed.
• As the chapter unfolds, Examples progress from simple to more complex.

Throughout the text, **figures** often build on each other with a new layer of information.
• **Balloon comments** often point out components of special note in the figure.

The **Concept in Context** icon here indicates the chapter-opening example —automobile tires—is being revisited. In this Example, we explore the slowing down of a skidding automobile with a specific coefficient of kinetic friction for a rubber tire.

---

178   **CHAPTER 6**   Further Applications of Newton's Laws

EXAMPLE 3   A man pushes a heavy crate over a floor. The man pushes downward and forward, so his push makes an angle of 30° with the horizontal (Fig. 6.5a). The mass of the crate is 60 kg, and the coefficient of sliding friction is $\mu_k = 0.50$. What force must the man exert to keep the crate moving at uniform velocity?

(a)   (b)   A push at an angle has both horizontal and vertical components.

FIGURE 6.5 (a) Man pushing a crate. (b) "Free-body" diagram for the crate.

SOLUTION: Figure 6.5b is a "free-body" diagram for the crate. The forces on the crate are the push **P** of the man, the weight **w**, the normal force **N**, and the friction force $\mathbf{f}_k$. Note that because the man pushes the crate down against the floor, the magnitude of the normal force is not equal to mg; we will have to treat the magnitude of the normal force as unknown. Taking the x axis horizontal and the y axis vertical, we see from Fig. 6.5b that the x and y components of the forces are (see also Fig. 5.37)

$$P_x = P\cos 30° \qquad P_y = -P\sin 30°$$
$$w_x = 0 \qquad w_y = -mg$$
$$N_x = 0 \qquad N_y = N$$
$$f_{k,x} = -\mu_k N \qquad f_{k,y} = 0$$

[the] acceleration of the crate is zero in both the x and the y directions, the [sum i]n each of these directions must be zero:

$$P\cos 30° + 0 + 0 - \mu_k N = 0$$
$$-P\sin 30° - mg + N + 0 = 0$$

[T]wo equations for the two unknowns P and N. By multiplying the second [b]y μ_k and then adding the resulting equation to the first, we can elimi-[nate N] and we find an equation for P:

$$P\cos 30° - \mu_k P\sin 30° - \mu_k mg = 0$$

[Solving t]his for P, we find

$$P = \frac{\mu_k mg}{\cos 30° - \mu_k \sin 30°} = \frac{0.50 \times 60 \text{ kg} \times 9.81 \text{ m/s}^2}{\cos 30° - 0.50 \times \sin 30°} \qquad (6.4)$$
$$= 4.8 \times 10^2 \text{ N}$$

Solutions in Examples may cover both general approaches and specific details on how to extract the information from the problem statement.

---

176   **CHAPTER 6**   Further Applications of Newton's Laws

Concepts —in— Context

A kinetic friction force acts on each wheel, but diagram shows these forces combined in a single force $\mathbf{f}_k$.

Skidding motion is opposed by kinetic friction.

FIGURE 6.3 "Free-body" diagram for an automobile skidding with locked wheels.

EXAMPLE 1   Suppose that the coefficient of kinetic friction of the hard rubber of an automobile tire sliding on the pavement of a street is $\mu_k = 0.8$. What is the deceleration of an automobile on a flat street if the driver brakes sharply, so all the wheels are locked and skidding? (Assume the vehicle is an economy model without an antilock braking system.)

SOLUTION: Figure 6.3 shows the "free-body" diagram with all the forces on the automobile. These forces are the weight **w**, the normal force **N** exerted by the street, and the friction force $\mathbf{f}_k$. The normal force must balance the weight; hence the magnitude of the normal force is the same as the magnitude of the weight, or $N = w = mg$. According to Eq. (6.1), the magnitude of the friction force is then

$$f_k = \mu_k N = 0.8 \times mg$$

Since this friction force is the only horizontal force on the automobile, the deceleration of the automobile along the street is

$$a_x = -\frac{f_k}{m} = -\frac{0.8 \times mg}{m} = -0.8 \times g = -0.8 \times 9.8 \text{ m/s}^2$$
$$= -8 \text{ m/s}^2$$

COMMENT: The normal forces and the friction forces act on all the four wheels of the automobile; but in Fig. 6.3 (and in other "free-body" diagrams in this chapter) these forces have been combined into a net force **N** and a net friction force $\mathbf{f}_k$, which, for convenience, are shown as though acting at the center of the automobile. To the extent that the motion is treated as purely translational motion (that is, particle motion), it makes no difference at what point of the automobile the forces act. Later, in Chapter 13, we will study how forces affect the rotational motion of bodies, and it will then become important to keep track of the exact point at which each force acts.

Comments occasionally close an Example to point out the particular limitations and broader implications of a Solution.

EXAMPLE 2   A ship is launched toward the water on a slipway making an angle of 5° with the horizontal direction (see Fig. 6.4). The coefficient of kinetic friction between the bottom of the ship and the slipway is $\mu_k = 0.08$. What is the acceleration of the ship along the slipway? What is the speed of the ship after accelerating from rest through a distance of 120 m down the slipway to the water?

SOLUTION: Figure 6.4b is the free-body" diagram for the ship. The forces shown are the weight **w**, the normal force exerted by the slipway **N**, and the friction force $\mathbf{f}_k$. The magnitude of the weight is $w = mg$.

Since there is no motion in the direction perpendicular to the slipway, we find, as in Eq. (5.36), that the normal force is

$$N = mg\cos\theta$$

and the magnitude of the friction force is

$$f_k = \mu_k N = \mu_k mg\cos\theta \qquad (6.2)$$

With the x axis parallel to the slipway, the x component of the weight is (see Fig. 6.4c)

$$w_x = mg\sin\theta$$

A **Checkup** appears at the end of each section within a chapter.
• Each Checkup is a self-quiz to test the reader's mastery of the concepts in the preceding section.
• Each Checkup has an answer (see below).

Problem-Solving Techniques boxes appear in relevant places throughout the book and offers tips on how to approach problems of a particular kind—in this case, problems involving the use of friction or centripetal force.

Answers to Checkups appear at the very back of each chapter, after the Review Problems.

---

190   **CHAPTER 6**   Further Applications of Newton's Laws

 Checkup 6.3

QUESTION 1: A stone is being whirled around a circle at the end of a string when the string suddenly breaks. Describe the motion of the stone after the string breaks; ignore gravity.

QUESTION 2: At an intersection, a motorcycle makes a right turn at constant speed. During this turn the motorcycle travels along a 90° arc of a circle. What is the direction of the acceleration of the motorcycle during this turn?

QUESTION 3: A car moves at constant speed along a road leading over a small hill with a spherical top. What is the direction of the acceleration of the car when at the top of this hill?

QUESTION 4: In Example 12, for the aircraft looping the loop, does the chair exert a centripetal or a centrifugal force on the pilot? Does the pilot exert a centripetal or a centrifugal force on the chair? What is the direction of the pilot's apparent, increased weight at the instant the aircraft passes through the bottom of the loop? Does the direction of the apparent weight change as the aircraft climbs up the loop?

QUESTION 5: Two cars travel around a traffic circle in adjacent (outer and inner) lanes. If the two cars travel at the same constant speed, which completes the circle first? Which has the larger acceleration?

(A) Outer; outer.   (B) Inner; outer.
(C) Outer; inner.   (D) Inner; inner.

PROBLEM-SOLVING TECHNIQUES	FRICTION FORCES AND CENTRIPETAL FORCES

The problems involving applications of Newton's laws in this chapter can be solved by the techniques discussed in the preceding chapter. In dealing with friction forces and with the centripetal force for uniform circular motion, pay special attention to the directions of the forces.

1   The magnitude of the sliding friction force is proportional to the magnitude of the normal force, but the direction is not the direction of the normal force. Instead, the sliding friction force is always parallel to the sliding surfaces, opposite to the direction of motion.

2   The static friction force is also always parallel to the sliding surfaces, opposite to the direction in which the body tends to move. If you have any doubts about the direction of the static friction force, pretend that the friction is absent, and ask yourself in what direction the body would then move; the static friction force is in the opposite direction.

3   Uniform circular motion requires a force toward the center of the circle, that is, a centripetal force. When preparing a "free-body" diagram for a body in uniform circular motion, include all the pushes and pulls acting on the moving body, but do *not* include a "centripetal mv^2/r force." This would be a mistake, like including an "ma force" in the "free-body" diagram for a body with some kind of translational motion. The quantity mv^2/r is not a force; it is merely the product of mass and centripetal acceleration. This acceleration is caused by one force or by the resultant of several forces already included among the pushes and pulls displayed in the "free-body" diagram. For instance, in Example 11 the resultant force is $w \tan\theta$, in Example 12 the resultant force is $N - mg$, and these resultants equal mv^2/r by Newton's Second Law [see Eqs. (6.17) and (6.18)]. To prevent confusion, do not include the resultant in the "free-body" diagram for a body in uniform circular motion. Instead, draw the resultant on a separate diagram (see Fig. 6.21b).

---

202   CHAPTER

*85. Two springs of constants 2.0×10^3 N/m and $3.0 \times$ are connected in tandem, and a mass of 5.0 kg hang from this spring. By what amount does the mass st combined spring? Each individual spring?

*86. A block of mass 1.5 kg is placed on a flat surface, a being pulled horizontally by a spring with a spring 1.2×10^3 N/m (see Fig. 6.48). The coefficient of st between the block and the table is $\mu_s = 0.60$, and th cient of sliding friction is $\mu_k = 0.40$.

(a) By what amount must the spring be stretched t block moving?

(b) What is the acceleration of the block if the str spring is maintained at a constant value equal t required to start the motion?

(c) By what amount must the spring be stretched to keep the mass moving at constant speed?

FIGURE 6.48 Mass pulled by spring.

*87. A block of mass 1.5 kg is placed on a plane inclined at 30°, and it is being pulled upward by a spring with a spring constant 1.2×10^3 N/m (see Fig. 6.49). The direction of pull of the spring is parallel to the inclined plane. The coefficient of static ... lined plane is $\mu_s = 0.60$, ... $\mu_k = 0.40$.

... be stretched to start the

... ck if the stretch of the
... value equal to that

(c) By what amount must the spring be stretched to keep the mass moving at constant speed?

*88. A mass m_1 slides on a smooth, frictionless table. The mass is constrained to move in a circle by a string that passes through a hole in the center of the table and is attached to a second mass m_2 hanging vertically below the table (Fig. 6.50). If the radius of the circular motion of the first mass is r, what must be its speed?

FIGURE 6.50 Mass in circular motion and hanging mass.

89. An automobile enters a curve of radius 45 m at 70 km/h. Will the automobile skid? The curve is not banked, and the coefficient of static friction between the wheels and the road is 0.80.

*90. A stone of 0.90 kg attached to a rod is being whirled around a vertical circle of radius 0.92 m. Assume that during this motion the speed of the stone is constant. If at the top of the circle the tension in the rod is (just about) zero, what is the tension in the rod at the bottom of the circle?

Answers to Checkups

Checkup 6.1

1. The weight of the second book results in a normal force between the first book and the table that is twice as large, so the friction force, and thus the horizontal push to overcome it, will be twice as large, or 20 N. If the first book pushes the second, then the friction force of the second book on the first adds to the friction force of the first to require a push also twice as large as the original, or 20 N.

2. While the block coasts up the incline, the friction, which always opposes the *motion*, is directed down the plane (the corresponding "free-body" diagram would have the weight

---

9.4   Elliptical Orbits; Kepler's Laws   **283**

MATH HELP   ELLIPSES

An ellipse is defined geometrically by the condition that the sum of the distance from one focus of the ellipse and the distance from the other focus is the same for all points on the ellipse. This geometrical condition leads to a simple method for the construction of an ellipse: Stick pins into the two foci and tie a length of string to these points. Stretch the string taut to the tip of a pencil, and move this pencil around the foci while keeping the string taut (see Fig. 1a).

An ellipse can also be constructed by slicing a cone obliquely (see Fig. 1b). Because of this, an ellipse is said to be a conic section.

The largest diameter of the ellipse is called the major axis, and the smallest diameter is called the minor axis. The semimajor axis and the semiminor axis are one-half of these diameters, respectively (see Fig. 1c).

If the semimajor axis of length a is along the x axis and the semiminor axis of length b is along the y axis, then the x and y coordinates of an ellipse centered on the origin satisfy

$$\frac{x^2}{a^2} + \frac{y^2}{b^2} = 1$$

The foci are on the major axis at a distance f from the origin given by

$$f = \sqrt{a^2 - b^2}$$

The separation between a planet and the Sun is $a - f$ at perihelion and is $a + f$ at aphelion.

(a)   (b)   (c)

focus   focus

semiminor axis — Sun — f — a — semimajor axis

FIGURE 1 (a) Constructing an ellipse. (b) Ellipse as a conic section. (c) Focal distance f, semimajor axis a, and semiminor axis b of an ellipse.

Figure 9.10 illustrates this law. The two colored areas are equal, and the planet takes equal times to move from P to P' and from Q to Q'. According to Fig. 9.10, the speed of the planet is larger when it is near the Sun (at Q) than when it is far from the Sun (at P).

Kepler's Second Law, also called the law of areas, is a direct consequence of the central direction of the gravitational force. We can prove this law by a simple geometrical argument. Consider three successive positions P, P', P'' on the orbit, separated by a relatively small distance. Suppose that the time intervals between P, P' and between P', P'' are equal—say, each of the two intervals is one second. Figure 9.11 shows the positions P, P', P''. Between these positions the curved orbit can be approximated by straight line segments PP' and $P'P''$. Since the time intervals are one unit of time (1 second), the lengths of the segments PP' and $P'P''$ are in proportion to the

---

> **Math Help** boxes offer specific mathematical guidance at the initial location in the text where that technique is most relevant.
> • In this case in Chapter 9, ellipses are important in studying orbits.
> • Additional math help is available in Appendices 2, 3, 4, and 5 at the back of the textbook.

> Throughout the text, **Physics in Practice** boxes offer specific details on a real-world application of the concept under discussion—in this case, forces at work in automobile collisions in Chapter 11.

> The text frequently offers **tables of typical values** of physical quantities.
> • Such tables usually are labeled "**Some ...**," as in this case, from Chapter 5.
> • These tables give some impression of the magnitudes encountered in the real world.

11.1   Impulsive Forces   **343**

PHYSICS IN PRACTICE   AUTOMOBILE COLLISIONS

Concepts in Context

We can fully appreciate the effects of the secondary impact on the human body if we compare the impact speeds of a human body on the dashboard or the windshield with the speed attained by a body in free fall from some height. The impact of the head on the windshield at 15 m/s is equivalent to falling four floors down from an apartment building and landing headfirst on a hard surface. Our intuition tells us that this is likely to be fatal. Since our intuition about the dangers of heights is much better than our intuition about the dangers of speeds, it is often instructive to compare impact speeds with equivalent heights of fall. The table lists impact speeds and equivalent heights, expressed as the number of floors the body has to fall down to acquire the same speed.

The number of fatalities in automobile collisions has been reduced by the use of air bags. The air bag helps by cushioning the impact over a longer time, reducing the time-average force. To be effective, the air bag must inflate quickly, before the passenger reaches it, typically in about 10 milliseconds. Because of this, a passenger, especially a child, too near an air bag prior to inflation can be injured or killed by the impulse from the inflation. But for a properly seated adult passenger, the inflated air bag cushions the passenger, reducing the severity of injuries.

However, the impact can still be fatal—you wouldn't expect to survive a jump from an 11-floor building onto an air mattress.

For maximum protection, a seat belt should always be worn even in vehicles equipped with air bags. In lateral collisions, in repeated collisions (such as in car pileups), and in rollovers, an air bag is of little help, and a seat belt is essential. The effectiveness of seat belts is well demonstrated by the experiences of race car drivers. Race car drivers wear lap belts and crossed shoulder belts. Even in spectacular crashes at very high speeds (see the figure), the drivers rarely suffer severe injuries.

COMPARISON OF IMPACT SPEEDS AND HEIGHTS OF FALL

SPEED	SPEED	EQUIVALENT HEIGHT (NUMBER OF FLOORS)[a]
15 km/h	9 mi/h	$\frac{1}{3}$
30	19	1
45	28	3
60	37	5
75	47	8
90	56	11
105	65	15

[a]Each floor is 2.9 m.

In a race at the California Speedway in October 2000, a car flips over and breaks in half after a crash, but the driver, Luis Diaz, walks away from the wreck.

SOLUTION: The only horizontal force on the ball is the normal force exerted by the wall; this force reverses the motion of the ball (see Fig. 11.3). Since the wall is very massive, the reaction force of the ball on the wall will not give the wall any appreciable velocity. Hence the kinetic energy of the system, both before and after the collision, is merely the kinetic energy of the ball. Conservation of this

---

TABLE 5.1	SOME FORCES	
Gravitational pull of Sun on Earth	3.5×10^{22} N	
Thrust of Saturn V rocket engines (a)	3.3×10^{7} N	
Pull of large tugboat	1×10^{6} N	
Thrust of jet engines (Boeing 747)	7.7×10^{5} N	
Pull of large locomotive	5×10^{5} N	
Decelerating force on automobile during braking	1×10^{4} N	
Force between two protons in a nucleus	$\approx 10^{4}$ N	(a)
Accelerating force on automobile	7×10^{3} N	
Gravitational pull of Earth on man	7.3×10^{2} N	
Maximum upward force exerted by forearm (isometric)	2.7×10^{2} N	
Gravitational pull of Earth on apple (b)	2 N	
Gravitational pull of Earth on 5¢ coin	5.1×10^{-2} N	
Force between electron and nucleus of atom (hydrogen)	8×10^{-8} N	
Force on atomic-force microscope tip	10^{-12} N	
Smallest force detected (mechanical oscillator)	10^{-19} N	(b)

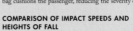

Each chapter closes with a Summary followed by Questions for Discussion, Problems, Review Problems, and Answers to Checkups.

A **Summary** lists the subjects and page references for any special content in this chapter—such as Math Help, Problem-Solving Techniques, or Physics in Practice boxes.
• Next the Summary lists the chapter's core concepts in the order they are treated. The concept appears on the left in bold.
• The mathematical expression for the concept appears in the middle column with an equation number on the far right.

About 15 or more **Questions for Discussion** follow the Summary in each chapter.
• These questions require thought, but not calculation; e.g. "Why are wet streets slippery?"
• Some of these questions are intended as brain teasers that have no unique answer, but will lead to provocative discussions.

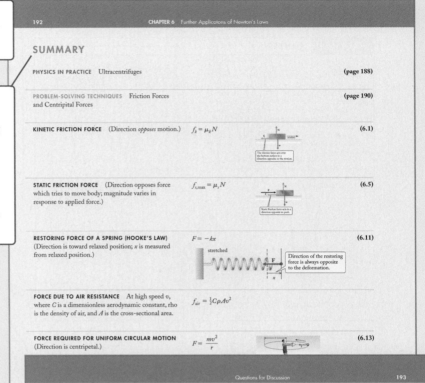

192   CHAPTER 6 Further Applications of Newton's Laws

SUMMARY

PHYSICS IN PRACTICE Ultracentrifuges		**(page 188)**
PROBLEM-SOLVING TECHNIQUES Friction Forces and Centripetal Forces		**(page 190)**
KINETIC FRICTION FORCE (Direction *opposes* motion.)	$f_k = \mu_k N$	**(6.1)**
STATIC FRICTION FORCE (Direction opposes force which tries to move body; magnitude varies in response to applied force.)	$f_{s,max} = \mu_s N$	**(6.5)**
RESTORING FORCE OF A SPRING (HOOKE'S LAW) (Direction is toward relaxed position; x is measured from relaxed position.)	$F = -kx$	**(6.11)**
FORCE DUE TO AIR RESISTANCE At high speed v, where C is a dimensionless aerodynamic constant, rho is the density of air, and A is the cross-sectional area.	$f_{air} = \frac{1}{2}C\rho A v^2$	
FORCE REQUIRED FOR UNIFORM CIRCULAR MOTION (Direction is centripetal.)	$F = \dfrac{mv^2}{r}$	**(6.13)**

Direction of the restoring force is always opposite to the deformation.

Questions for Discussion   193

QUESTIONS FOR DISCUSSION

1. According to the adherents of parapsychology, some people are endowed with the supernormal power of psychokinesis, e.g., spoon-bending-at-a-distance via mysterious psychic forces emanating from the brain. Physicists are confident that the only forces acting between pieces of matter are those listed in Section 6.4, none of which are implicated in psychokinesis. Given that the brain is nothing but a (very complicated) piece of matter, what conclusions can a physicist draw about psychokinesis?

2. If you carry a spring balance from London to Hong Kong, do you have to recalibrate it? If you carry a beam balance?

3. When you stretch a rope horizontally between two fixed points, it always sags a little, no matter how great the tension. Why?

4. What are the forces on a soaring bird? How can the bird gain altitude without flapping its wings?

5. How could you use a pendulum suspended from the roof of your automobile to measure its acceleration?

6. When an airplane flies along a parabolic path similar to that of a projectile, the passengers experience a sensation of weightlessness. How would the airplane have to fly to give the passengers a sensation of enhanced weight?

7. A frictionless chain hangs over two adjoining inclined planes (Fig. 6.24a). Prove the chain is in equilibrium, i.e., the chain will not slip to the left or to the right. [Hint: One method of proof, due to the seventeenth-century engineer and mathematician Simon Stevin, asks you to pretend that an extra piece of chain is hung from the ends of the original chain (Fig. 6.24b). This makes it possible to conclude that the original chain cannot slip.]

FIGURE 6.24 Frictionless chain over two inclines.

8. Seen from a reference frame moving with the wave, the motion of a surfer is analogous to the motion of a skier down a mountain.[2] If the wave were to last forever, could the surfer ride it forever? In order to stay on the wave as long as possible, in what direction should the surfer ski the wave?

9. Excessive polishing of the surfaces of a block of metal increases its friction. Explain.

10. Some drivers like to spin the wheels of their automobiles for a quick start. Does this give them greater acceleration? (Hint: $\mu_s > \mu_k$.)

[2] There is, however, one complication: surf waves grow higher as they approach the beach. Ignore this complication.

11. Cross-country skiers like to use a ski wax that gives their skis a large coefficient of static friction, but a low coefficient of kinetic friction. Why is this useful? How do "waxless" skis achieve the same effect?

12. Designers of locomotives usually reckon that the maximum force available for moving the train ("tractive force") is one-fourth or one-fifth of the weight resting on the drive wheels of the locomotive. What value of the friction coefficient between the wheels and the track does this implicitly assume?

13. When an automobile with rear-wheel drive accelerates from rest, the maximum acceleration that it can attain is less than the maximum deceleration that it can attain while braking. Why? (Hint: Which wheels of the automobile are involved in acceleration? In braking?)

14. Can you think of some materials with $\mu_k > 1$?

15. For a given initial speed, the stopping distance of a train is much longer than that of a truck. Why?

16. Why does the traction on snow or ice of an automobile with rear-wheel drive improve when you place extra weight over the rear wheels?

17. Why are wet streets slippery?

18. In order to stop an automobile on a slippery street in the shortest distance, it is best to brake as hard as possible without initiating a skid. Why does skidding lengthen the stopping distance? (Hint: $\mu_s > \mu_k$.)

19. Suppose that in a panic stop, a driver locks the wheels of his automobile and leaves skid marks on the pavement. How can you deduce his initial speed from the length of the skid marks?

20. Hot-rod drivers in drag races find it advantageous to spin their wheels very fast at the start so as to burn and melt the rubber on their tires (Fig. 6.25). How does this help them to attain a larger acceleration than expected from the static coefficient of friction?

FIGURE 6.25 Drag racer at the start of the race.

About 70 **Problems** and 15 **Review Problems** follow each chapter's Questions for Discussion.
• The Problem's statement contains data and conditions upon which a solution will hinge.
• Problems are grouped by chapter section and proceed from simple to more complex within each section.
• Many Problems employ real-world data and occasionally may introduce applications beyond those treated in the chapter.

Review Problems are specifically designed to help students prepare for examinations.
• Review Problems often test comprehension of concepts from more than one section within the chapter.
• Review Problems often take a guided approach by posing series of questions that build on each other.

---

194   CHAPTER 6 Further Applications of Newton's Laws

21. A curve on a highway consists of a quarter circle connecting two straight segments. If this curve is banked perfectly for motion at some given speed, can it be joined to the straight segments without a bump? How could you design a curve that is banked perfectly along its entire length and merges smoothly into straight segments without any bump?

22. Automobiles with rear engines (such as the old VW "Beetle") tend to oversteer; that is, in a curve the rear end tends to swing toward the outside of the curve, turning the car excessively into the curve. Explain.

23. When rounding a curve in your automobile, you get the impression that a force tries to pull you toward the outside of the curve. Is there such a force?

24. If the Earth were to stop spinning (other things remaining equal), the value of g at all points of the surface except the poles would become slightly larger. Why?

25. (a) If a pilot in a fast aircraft very suddenly pulls out of a dive (Fig. 6.26a), he will suffer blackout caused by loss of blood pressure in the brain. If he suddenly begins a dive while climbing (Fig. 6.26b), he will suffer *redout* caused by excessive blood pressure in the brain. Explain.

   (b) A pilot wearing a G suit—a tightly fitting garment that squeezes the tissues of the legs and abdomen—can tolerate $8g$ while pulling out of a dive (Fig. 6.26c). How does this G suit prevent blackout? A pilot can tolerate no more

than $-2g$ while beginning a dive. Why does the G suit not help against redout?

26. While rounding a curve at high speed, a motorcycle rider leans the motorcycle toward the center of the curve. Why?

(a)

(b)

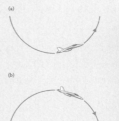

FIGURE 6.26 (a) Aircraft pulling out of a dive. (b) Aircraft beginning a dive. (c) Pilot wearing a G suit.

PROBLEMS

6.1 Friction†

1. The ancient Egyptians moved large stones by dragging them across the sand in sleds. How many Egyptians were needed to drag an obelisk of 700 metric tons? Assume that $\mu_s = 0.30$ for the sled on sand and that each Egyptian exerted a horizontal force of 360 N.

2. The base of a winch is bolted to a mounting plate with four bolts. The base and the mounting plate are flat surfaces made of steel; the friction coefficient of these surfaces in contact is $\mu_s = 0.40$. The bolts provide a normal force of 2700 N each. What maximum static friction force will act between the steel surfaces and help oppose lateral slippage of the winch on its base?

3. According to tests performed by the manufacturer, an automobile with an initial speed of 65 km/h has a stopping distance of 20 m on a level road. Assuming that no skidding occurs during braking, what is the value of μ_s between the wheels and the road required to achieve this stopping distance?

4. A crate sits on the load platform of a truck. The coefficient of friction between the crate and the platform is $\mu_s = 0.40$. If the truck stops suddenly, the crate will slide forward and crash into the cab of the truck. What is the maximum braking deceleration that the truck may have if the crate is to stay put?

5. When braking (without skidding) on a dry road, the stopping distance of a sports car with a high initial speed is 38 m. What would have been the stopping distance of the same car with the same initial speed on an icy road? Assume that $\mu_s = 0.85$ for the dry road and $\mu_s = 0.20$ for the icy road.

6. In a remarkable accident on motorway M1 (in England), a Jaguar car initially speeding "in excess of 100 mph" skidded 290 m before coming to a rest. Assuming that the wheels were completely locked during the skid and that the coefficient of kinetic friction between the wheels and the road was 0.80, find the initial speed.

† For help, see Online Concept Tutorial 8 at www.wwnorton.com/physics

---

Review Problems

REVIEW PROBLEMS

76. At liftoff, the Saturn V rocket used for the Apollo missions has a mass of 2.45×10^6 kg.

   (a) What is the minimum thrust that the rocket engines must develop to achieve liftoff?

   (b) The actual thrust that the engines develop is 3.3×10^7 N. What is the vertical acceleration of the rocket at liftoff?

   (c) At burnout, the rocket has spent its fuel, and its remaining mass is 0.75×10^6 kg. What is the acceleration just before burnout? Assume that the motion is still vertical and that the strength of gravity is the same as when the rocket is on the ground.

77. If the coefficient of static friction between the tires of an automobile and the road is $\mu_s = 0.80$, what is the minimum distance the automobile needs in order to stop without skidding from an initial speed of 90 km/h? How long does it take to stop?

78. Suppose that the last car of a train becomes uncoupled while the train is moving upward on a slope of 1:6 at a speed of 48 km/h.

   (a) What is the deceleration of the car? Ignore friction.

   (b) How far does the car coast up the slope before it stops?

79. A 40-kg crate falls off a truck traveling at 80 km/h on a level road. The crate slides along the road and gradually comes to a halt. The coefficient of kinetic friction between the crate and the road is 0.80.

   (a) Draw a "free-body" diagram for the crate sliding on the road.

   (b) What is the normal force the road exerts on the crate?

   (c) What is the friction force the road exerts on the crate?

   (d) What is the weight force on the crate? What is the net force on the crate?

   (e) What is the deceleration of the crate? How far does the crate slide before coming to a halt?

80. A 2.0-kg box rests on an inclined plane which makes an angle of 30° with the horizontal. The coefficient of static friction between the box and the plane is 0.90.

   (a) Draw a "free-body" diagram for the box.

   (b) What is the normal force the inclined plane exerts on the box?

   (c) What is the friction force the inclined plane exerts on the box?

   (d) What is the net force the inclined plane exerts on the box? What is the direction of this force?

81. The body of an automobile is held above the axles of the wheels by means of four springs, one near each wheel. Assume

that the springs are ver[...]
springs are the same. Th[...]
is 1200 kg, and the spri[...]
N/m. When the autom[...]
far are the springs comp[...]

*82. A block of wood rests o[...]
The coefficient of static[...]
paper is $\mu_s = 0.70$, and [...]
$\mu_k = 0.50$. If you tilt the [...]
begin to move?

*83. Two blocks of masses m_1 and m_2 are connected by a string. One block slides on a table, and the other hangs from the string, which passes over a pulley (see Fig. 6.46). The coefficient of sliding friction between the first block and the table is $\mu_k = 0.20$. What is the acceleration of the blocks?

m_1

m_2

FIGURE 6.46 Mass on table, pulley, and hanging mass.

*84. A man of mass 75 kg is pushing a heavy box on a flat floor. The coefficient of sliding friction between the floor and the box is 0.20, and the coefficient of static friction between the man's shoes and the floor is 0.80. If the man pushes horizontally (see Fig. 6.47), what is the maximum mass of the box he can move?

FIGURE 6.47 Pushing a box.

The **dagger footnote** (†) that accompanies a Problem heading—in this case, "6.1 Friction"—indicates the availability of an Online Concept Tutorial on this specific topic and states its web address.

Problems and Review Problems are marked by **level of difficulty**:
• Those without an asterisk are the most common and require very little manipulation of existing equations; or they may merely require retracing the steps of a worked Example.
• Problems marked with one asterisk (*) are of medium difficulty and may require use of several concepts and manipulation more than one equation to isolate and solve for the unknown variable.
• Problems marked with two asterisks (**) are challenging, demand considerable thought, may require significant mathematical skill, and are the least common.

Online Concept Tutorials

www.wwnorton.com/physics

An **Online Concept Tutorial** accompanies many central topics in this textbook. When a Tutorial is available, its numbered icon appears at section heading within the chapter and a dagger footnote appears in the end-of-chapter Problems section as reminder. These Tutorials are digitally delivered, either via the Internet or via a CD-ROM for those without Internet access.

Online *Concept* Tutorial 8

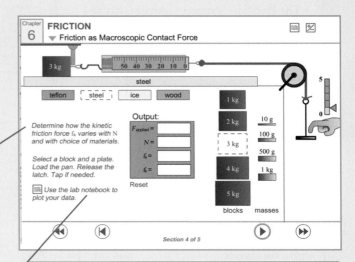

Many Tutorials contain **online experiments**—in this case, determining how the kinetic friction force varies with the normal force and with the choice of materials.

The online experiments allow students to change independent variables—in this case, mass and material.
- Students may collect and display data in a built-in **lab notebook**.
- Each Tutorial includes an interactive **self-quiz**.

The Online Concept Tutorials listed here indicate each textbook section supported by the tutorial (in paratheses).

www.wwnorton.com/physics

SmartWork is a subscription-based online homework-management system that makes it easy for instructors to assign, collect and grade end-of-chapter problems from *Physics for Engineers and Scientists*. Built-in hinting and feedback address common misperceptions and help students get the maximum benefit from these assignments.

Simple Feedback Problems anticipate common misconceptions and offer prompts at just the right moment to help students reach the correct solution.

Guided Tutorial Problems address challenging topics.
 • If a student solves one of these problems incorrectly, she is presented with a series of discrete tutorial steps that lead to a general solution.
 • Each step includes hinting and feedback. After working through these remedial steps, the student returns to a restatement of the original problem, ready to apply this newly obtained knowledge.

SmartWork is available as a stand-alone purchase, or with an integrated ebook version of *Physics for Engineers and Scientists*.
 • Where appropriate, SmartWork prompts students to review relevant sections of the text.
 • Links to the ebook make it easy for students consult the text while working through problems.

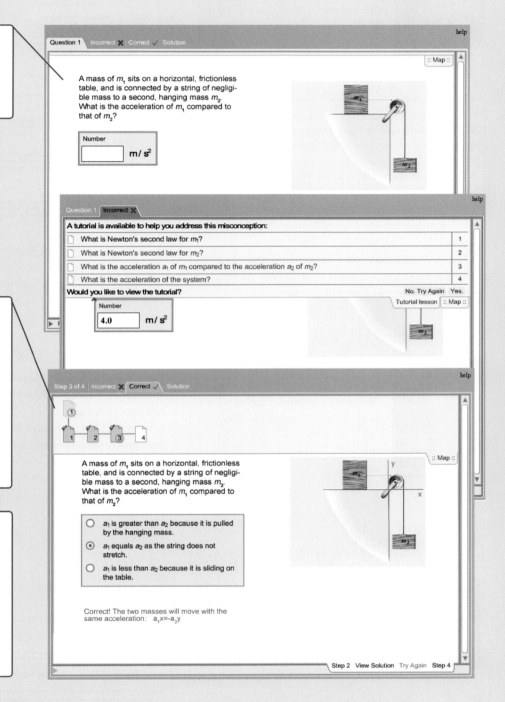

Student Activity Workbook

The Student Activity Workbook is available in two paperback volumes: Volume 1 comprises Chapters 1–21 and Volume 2 comprises Chapters 22–41.

For each chapter of the textbook, the **Student Activity Workbook** offers Activities designed to break down a physical condition into constituent parts.
• The Activities are unique to the Workbook and not found in the textbook.
• The Activities are pencil and paper exercises well suited to either individual or small group collaboration.
• The Activities include both conceptual and quantitative questions.

ACTIVITY 7

Joe is standing on the ground, Pete is standing on a 10m high cliff, and Amanda is at the bottom of a 20m deep pit, as shown. All three are using coordinate systems with the vertical axis directed up.

Joe's coordinate system has the zero of the vertical axis and zero of gravitational potential at ground level.

Peter's coordinate system has the zero of the vertical axis and zero of gravitational potential at the height of the cliff.

Amanda's coordinate system has the zero of the vertical axis and zero of gravitational potential at the bottom of the pit.

A ball of mass m is initially at rest at ground level, Position A.

a) What is the gravitational potential energy of the ball as measured by Joe, by Peter, and by Amanda?

b) The ball is then raised to the height of the cliff, Position B, and is held at rest. What is the gravitational potential energy of the ball at Position B as measured by Joe, by Peter, and by Amanda?

c) The ball is then released from rest and strikes the ground at the bottom of the pit, Position C. What is the gravitational potential energy of the ball at Position C as measured by Joe, by Peter, and by Amanda?

d) What is this change in gravitational potential energy measured by J, P, and A between positions A and C?

Student Solution Manual

The Student Solution Manual is available in two paperback volumes: Volume 1 comprises Chapters 1–21 and Volume 2 comprises Chapters 22–41.

The **Student Solutions Manual** contains worked solutions for about 50% of the odd-numbered Problems and Review Problems in the text.
• Appendix 11 in the back of the textbook contains only the final answer for odd-numbered problems in the chapters, not the intermediate steps of the solutions.

StudySpace   www.wwnorton.com/physics

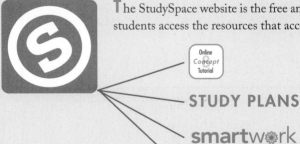

The StudySpace website is the free and open portal through which students access the resources that accompany this text.

• 45 Online Concept Tutorials—at no additional cost.

• 41 Study Plans, one for each chapter—at no additional cost.

• Smartwork online homework system—a subscription service.

• ebook links to textbook chapters—as part of subscription service.

Prelude

The World of Physics

Physics is the study of matter. In a quite literal sense, physics is the greatest of all natural sciences: it encompasses the smallest particles, such as electrons and quarks; and it also encompasses the largest bodies, such as galaxies and the entire Universe. The smallest particles and the largest bodies differ in size by a factor of more than ten thousand billion billion billion billion! In the pictures on the following pages we will survey the world of physics and attempt to develop some rough feeling for the sizes of things in this world. This preliminary survey sets the stage for our explanations of the mechanisms that make things behave in the way they do. Such explanations are at the heart of physics, and they are the concern of the later chapters of this book.

Since the numbers we will be dealing with in this prelude and in the later chapters are often very large or very small, we will find it convenient to employ the **scientific notation** for these numbers. In this notation, numbers are written with powers of 10; thus, hundred is written as 10^2, thousand is written as 10^3, ten thousand is written as 10^4, and so on. A tenth is written as 10^{-1}, a hundredth is written as 10^{-2}, a thousandth is written as 10^{-3}, and so on. The following table lists some powers of ten:

$10 = 10^1$	$0.1 = 1/10 = 10^{-1}$
$100 = 10^2$	$0.01 = 1/100 = 10^{-2}$
$1000 = 10^3$	$0.001 = 1/1000 = 10^{-3}$
$10\,000 = 10^4$	$0.0001 = 1/10\,000 = 10^{-4}$
$100\,000 = 10^5$	$0.00001 = 1/100\,000 = 10^{-5}$
$1\,000\,000 = 10^6$	$0.000001 = 1/1\,000\,000 = 10^{-6}$ etc.

Note that the power of 10, or the exponent on the 10, simply tells us how many zeros follow the 1 in the number (if the power of 10 is positive) or how many zeros follow the 1 in the denominator of the fraction (if the power of 10 is negative).

In scientific notation, a number that does not coincide with one of the powers of 10 is written as a product of a decimal number and a power of 10. For example, in this notation, $1\,500\,000\,000$ is written as 1.5×10^9. Alternatively, this number could be written as 15×10^8 or as 0.15×10^{10}; but in scientific notation it is customary to place the decimal point immediately after the first nonzero digit. The same rule applies to numbers smaller than 1; thus, 0.000015 is written as 1.5×10^{-5}.

The pictures on the following pages fall into two sequences. In the first sequence we zoom out: we begin with a picture of a woman's face and proceed step by step to pictures of the entire Earth, the Solar System, the Galaxy, and the Universe. This ascending sequence contains 27 pictures, with the scale decreasing in steps of factors of 10.

Most of our pictures are photographs. Many of these have become available only in recent years; they were taken by high-flying aircraft, Landsat satellites, astronauts, or sophisticated electron microscopes. For some of our pictures no photographs are available and we have to rely, instead, on drawings.

PART I: THE LARGE-SCALE WORLD

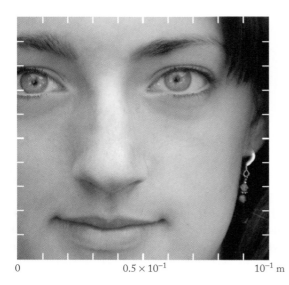

0   0.5×10^{-1} 10^{-1} m

Fig. P1 SCALE 1:1.5   This is Erin, an intelligent biped of the planet Earth, Solar System, Orion Spiral Arm, Milky Way Galaxy, Local Group, Local Supercluster. Erin belongs to the phylum Chordata, class Mammalia, order Primates, family Hominidae, genus *Homo*, species *sapiens*. She is made of 5.4×10^{27} atoms, with 1.9×10^{28} electrons, the same number of protons, and 1.5×10^{28} neutrons.

0   0.5×10^{0} 10^{0} m

Fig. P2 SCALE 1:1.5 × 10   Erin has a height of 1.7 meters and a mass of 57 kilograms. Her chemical composition (by mass) is 65% oxygen, 18.5% carbon, 9.5% hydrogen, 3.3% nitrogen, 1.5% calcium, 1% phosphorus, and 1.2% other elements.

The matter in Erin's body and the matter in her immediate environment occur in three states of aggregation: **solid**, **liquid**, and **gas**. All these forms of matter are made of atoms and molecules, but solid, liquid, and gas are qualitatively different because the arrangements of the atomic and molecular building blocks are different.

In a solid, each building block occupies a definite place. When a solid is assembled out of molecular or atomic building blocks, these blocks are locked in place once and for all, and they cannot move or drift about except with great difficulty. This rigidity of the arrangement is what makes the aggregate hard—it makes the solid "solid." In a liquid, the molecular or atomic building blocks are not rigidly connected. They are thrown together at random and they move about fairly freely, but there is enough adhesion between neighboring blocks to prevent the liquid from dispersing. Finally, in a gas, the molecules or atoms are almost completely independent of one another. They are distributed at random over the volume of the gas and are separated by appreciable distances, coming in touch only occasionally during collisions. A gas will disperse spontaneously if it is not held in confinement by a container or by some restraining force.

Fig. P3 SCALE 1:1.5 × 10^2   The building behind Erin is the New York Public Library, one of the largest libraries on Earth. This library holds 1.4×10^{10} volumes, containing roughly 10% of the total accumulated knowledge of our terrestrial civilization.

0   0.5×10^1 10^1 m

Fig. P4 SCALE 1:1.5 × 10^3   The New York Public Library is located at the corner of Fifth Avenue and 42nd Street, in the middle of New York City, with Bryant Park immediately behind it.

0   0.5×10^2 10^2 m

Fig. P5 SCALE 1:1.5 × 10^4   This aerial photograph shows an area of 1 kilometer × 1 kilometer in the vicinity of the New York Public Library. The streets in this part of the city are laid out in a regular rectangular pattern. The library is the building in the park in the middle of the picture. The photograph was taken early in the morning, and the high buildings typical of New York cast long shadows.

The photograph was taken from an airplane flying at an altitude of a few thousand meters. North is at the top of the photograph.

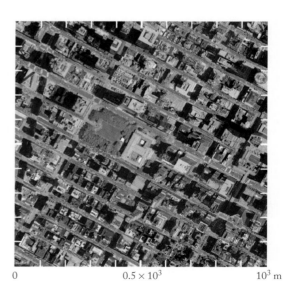

0   0.5×10^3 10^3 m

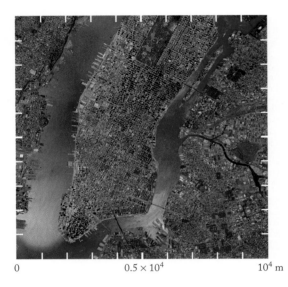

0   0.5×10^4 10^4 m

Fig. P6 SCALE 1:1.5 × 10⁵   This photograph shows a large portion of New York City. We can barely recognize the library and its park as a small rectangular patch slightly above the center of the picture. The central mass of land is the island of Manhattan, with the Hudson River on the left and the East River on the right.

This photograph and the next two were taken by satellites orbiting the Earth at an altitude of about 700 kilometers.

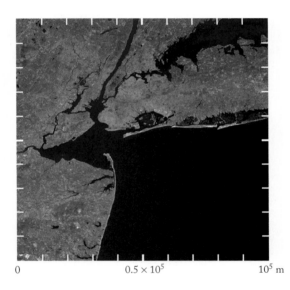

0   0.5×10^5 10^5 m

Fig. P7 SCALE 1:1.5 × 10⁶   In this photograph, Manhattan is in the upper middle. On this scale, we can no longer distinguish the pattern of streets in the city. The vast expanse of water in the lower right of the picture is part of the Atlantic Ocean. The mass of land in the upper right is Long Island. Parallel to the south shore of Long Island we can see a string of very narrow islands; they almost look man-made. These are barrier islands; they are heaps of sand piled up by ocean waves in the course of thousands of years.

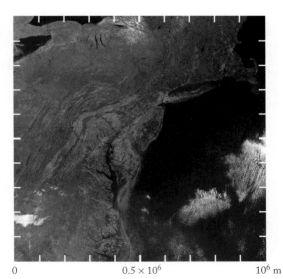

0   0.5×10^6 10^6 m

Fig. P8 SCALE 1:1.5 × 10⁷   Here we see the eastern coast of the United States, from Cape Cod to Cape Fear. Cape Cod is the hook near the northern end of the coastline, and Cape Fear is the promontory near the southern end of the coastine. Note that on this scale no signs of human habitation are visible. However, at night the lights of large cities would stand out clearly.

This photograph was taken in the fall, when leaves had brilliant colors. Streaks of orange trace out the spine of the Appalachian mountains.

Fig. P9 SCALE 1:1.5 × 10^8 In this photograph, taken by the Apollo 16 astronauts during their trip to the Moon, we see a large part of the Earth. Through the gap in the clouds in the lower middle of the picture, we can see the coast of California and Mexico. We can recognize the peninsula of Baja California and the Gulf of California. Erin's location, the East Coast of the United States, is covered by a big system of swirling clouds on the right of the photograph.

Note that a large part of the area visible in this photograph is ocean. About 71% of the surface of the Earth is ocean; only 29% is land. The atmosphere covering this surface is about 100 kilometers thick; on the scale of this photograph, its thickness is about 0.7 millimeter. Seen from a large distance, the predominant colors of the planet Earth are blue (oceans) and white (clouds).

0   0.5 × 10^7 10^7 m

Fig. P10 SCALE 1:1.5 × 10^9 This photograph of the Earth was taken by the Apollo 16 astronauts standing on the surface of the Moon. Sunlight is striking the Earth from the top of the picture.

As is obvious from this and from the preceding photograph, the Earth is a sphere. Its radius is 6.37 × 10^6 meters and its mass is 5.98 × 10^{24} kilograms.

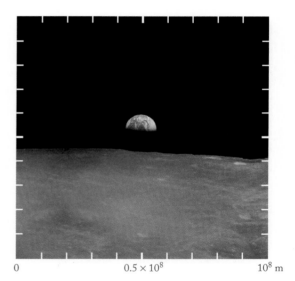

0   0.5 × 10^8 10^8 m

Fig. P11 SCALE 1:1.5 × 10^{10} In this drawing, the dot at the center represents the Earth, and the solid line indicates the orbit of the Moon around the Earth (many of the pictures on the following pages are also drawings). As in the preceding picture, the Sun is far below the bottom of the picture. The position of the Moon is that of January 1, 2000.

The orbit of the Moon around the Earth is an **ellipse**, but an ellipse that is very close to a circle. The solid red curve in the drawing is the orbit of the Moon, and the dashed green curve is a circle; by comparing these two curves we can see how little the ellipse deviates from a circle centered on the Earth. The point on the ellipse closest to the Earth is called the **perigee**, and the point farthest from the Earth is called the **apogee**. The distance between the Moon and the Earth is roughly 30 times the diameter of the Earth. The Moon takes 27.3 days to travel once around the Earth.

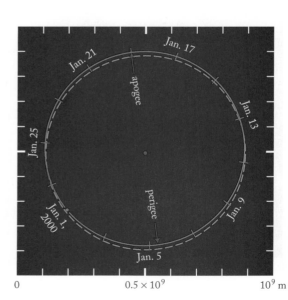

0   0.5 × 10^9 10^9 m

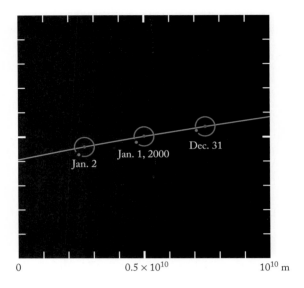

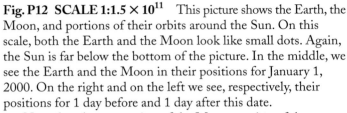

Fig. P12 SCALE 1:1.5 × 10¹¹   This picture shows the Earth, the Moon, and portions of their orbits around the Sun. On this scale, both the Earth and the Moon look like small dots. Again, the Sun is far below the bottom of the picture. In the middle, we see the Earth and the Moon in their positions for January 1, 2000. On the right and on the left we see, respectively, their positions for 1 day before and 1 day after this date.

Note that the net motion of the Moon consists of the combination of two simultaneous motions: the Moon orbits around the Earth, which in turn orbits around the Sun.

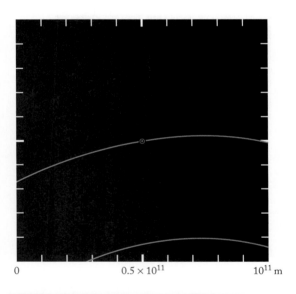

Fig. P13 SCALE 1:1.5 × 10¹²   Here we see the orbits of the Earth and of Venus. However, Venus itself is beyond the edge of the picture. The small circle is the orbit of the Moon. The dot representing the Earth is much larger than what it should be, although the artist has drawn it as minuscule as possible. On this scale, even the Sun is quite small; if it were included in this picture, it would be only 1 millimeter across.

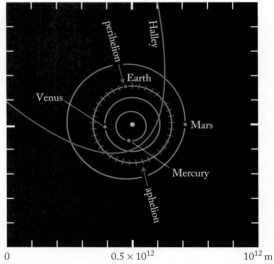

Fig. P14 SCALE 1:1.5 × 10¹³   This drawing shows the positions of the Sun and the inner planets: Mercury, Venus, Earth, and Mars. The positions of the planets are those of January 1, 2000. The orbits of all these planets are ellipses, but they are close to circles. The point of the orbit nearest to the Sun is called the **perihelion** and the point farthest from the Sun is called the **aphelion**. The Earth reaches perihelion about January 3 and aphelion about July 6 of each year.

All the planets travel around their orbits in the same direction: counterclockwise in our picture. The marks along the orbit of the Earth indicate the successive positions at intervals of 10 days.

Beyond the orbit of Mars, a large number of asteroids orbit around the Sun; these have been omitted to prevent excessive clutter. Furthermore, a large number of comets orbit around the Sun. Most of these have pronounced elliptical orbits. The comet Halley has been included in our drawing.

The Sun is a sphere of radius 6.96×10^8 meters. On the scale of the picture, the Sun looks like a very small dot, even smaller than the dot drawn here. The mass of the Sun is 1.99×10^{30} kilograms.

The matter in the Sun is in the **plasma** state, sometimes called the fourth state of matter. Plasma is a very hot gas in which violent collisions between the atoms in their random thermal motion have fragmented the atoms, ripping electrons off them. An atom that has lost one or more electrons is called an **ion**. Thus, plasma consists of a mixture of electrons and ions engaging in frequent collisions. These collisions are accompanied by the emission of light, making the plasma luminous.

Fig. P15 SCALE 1:1.5 × 10¹⁴ This picture shows the positions of the outer planets of the Solar System: Jupiter, Saturn, Uranus, Neptune, and Pluto. On this scale, the orbits of the inner planets are barely visible. As in our other pictures, the positions of the planets are those of January 1, 2000.

The outer planets move slowly and their orbits are very large; thus they take a long time to go once around their orbit. The extreme case is that of Pluto, which takes 248 years to complete one orbit.

Uranus, Neptune, and Pluto are so far away and so faint that their discovery became possible only through the use of telescopes. Uranus was discovered in 1781, Neptune in 1846, and the tiny Pluto in 1930. Pluto is now known as one of several dwarf planets.

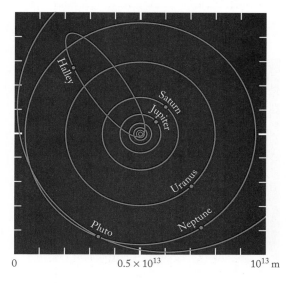

Fig. P16 SCALE 1:1.5 × 10¹⁵ We now see that the Solar System is surrounded by a vast expanse of space. Although this space is shown empty in the picture, the Solar System is encircled by a large cloud of millions of comets whose orbits crisscross the sky in all directions. Furthermore, the interstellar space in this picture and in the succeeding pictures contains traces of gas and of dust. The interstellar gas is mainly hydrogen; its density is typically 1 atom per cubic centimeter.

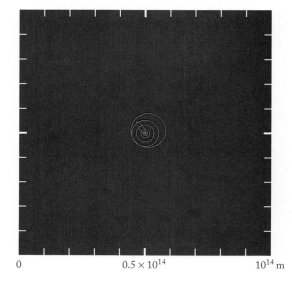

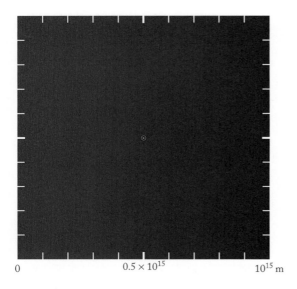

0   0.5×10^{15} 10^{15} m

Fig. P17 SCALE 1:1.5 $\times$ 10^{16}   More interstellar space. The small circle is the orbit of Pluto.

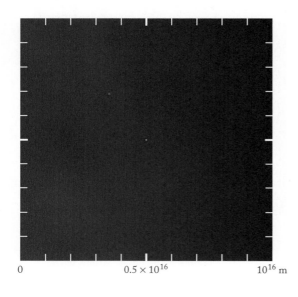

0   0.5×10^{16} 10^{16} m

Fig. P18 SCALE 1:1.5 $\times$ 10^{17}   And more interstellar space. On this scale, the Solar System looks like a minuscule dot, 0.1 millimeter across.

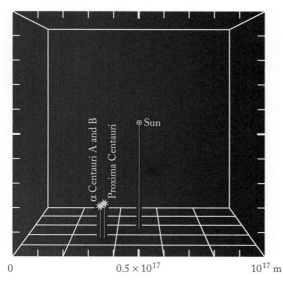

0   0.5×10^{17} 10^{17} m

Fig. P19 SCALE 1:1.5 $\times$ 10^{18}   Here, at last, we see the stars nearest to the Sun. The picture shows all the stars within a cubical box 10^{17} meters $\times$ 10^{17} meters $\times$ 10^{17} meters centered on the Sun: Alpha Centauri A, Alpha Centauri B, and Proxima Centauri. All three are in the constellation Centaurus, in the southern sky.

The star closest to the Sun is Proxima Centauri. This is a very faint, reddish star (a "red dwarf"), at a distance of 4.0 $\times$ 10^{16} meters from the Sun. Astronomers like to express stellar distances in light-years: Proxima Centauri is 4.2 light-years from the Sun, which means light takes 4.2 years to travel from this star to the Sun.

Fig. P20 SCALE 1:1.5 × 10^{19} This picture displays the brightest stars within a cubical box 10^{18} meters $\times$ 10^{18} meters $\times$ 10^{18} meters centered on the Sun. There are many more stars in this box besides those shown—the total number of stars in this box is close to 2000.

Sirius is the brightest of all the stars in the night sky. If it were at the same distance from the Earth as the Sun, it would be 28 times brighter than the Sun.

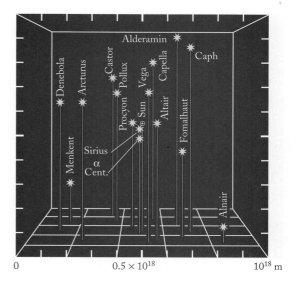

Fig. P21 SCALE 1:1.5 × 10^{20} Here we expand our box to 10^{19} meters $\times$ 10^{19} meters $\times$ 10^{19} meters, again showing only the brightest stars and omitting many others. The total number of stars within this box is about 2 million. We recognize several clusters of stars in this picture: the Pleiades Cluster, the Hyades Cluster, the Coma Berenices Cluster, and the Perseus Cluster. Each of these has hundreds of stars crowded into a fairly small patch of sky. In this diagram, Starbursts signify single stars, circles with starbursts indicate star clusters, and a circle with a single star indicate a star cluster with its brightest star.

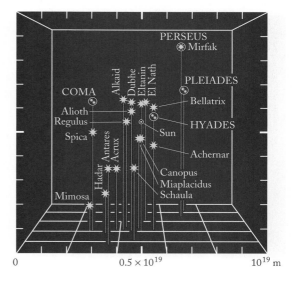

Fig. P22 SCALE 1:1.5 × 10^{21} This photograph shows a view of the Milky Way in the direction of the constellation Sagittarius. Now there are so many stars in our field of view that they appear to form clouds of stars. There are about a million stars in this photograph, and there are many more stars too faint to show up distinctly. Although this photograph is not centered on the Sun, it is similar to what we would see if we could look toward the Solar System from very far away.

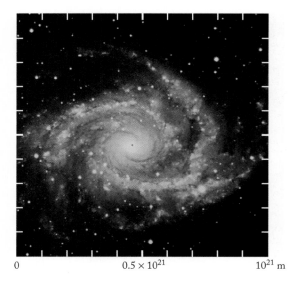

0   0.5×10^{21} 10^{21} m

Fig. P23 SCALE 1:1.5 × 10²² This is the spiral galaxy NGC 2997. Its clouds of stars are arranged in spiral arms wound around a central bulge. The bright central bulge is the nucleus of the galaxy; it has a more or less spherical shape. The surrounding region, with the spiral arms, is the disk of the galaxy. This disk is quite thin; it has a thickness of only about 3% of its diameter. The stars making up the disk circle around the galactic center in a clockwise direction.

Our Sun is in a spiral galaxy of roughly similar shape and size: the **Milky Way Galaxy**. The total number of stars in this galaxy is about 10^{11}. The Sun is in one of the spiral arms, roughly one-third inward from the edge of the disk toward the center.

0   0.5×10^{22} 10^{22} m

Fig. P24 SCALE 1:1.5 × 10²³ Galaxies are often found in clusters of several galaxies. Some of these clusters consist of just a few galaxies, others of hundreds or even thousands. The photograph shows a cluster, or group, of galaxies beyond the constellation Fornax. The group contains an elliptical galaxy like a luminous yellow egg (center), three large spiral galaxies (left), and a spiral with a bar (bottom left).

Our Galaxy is part of a modest cluster, the **Local Group**, consisting of our own Galaxy, the great Andromeda Galaxy, the Triangulum Galaxy, the Large Magellanic Cloud, plus 16 other small galaxies.

According to recent investigations, the dark, apparently empty, space near galaxies contains some form of distributed matter, with a total mass 20 or 30 times as large as the mass in the luminous, visible galaxies. But the composition of this invisible, extragalactic **dark matter** is not known.

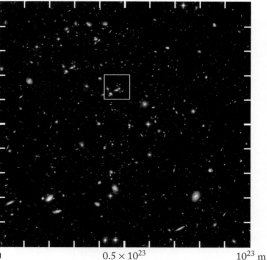

0   0.5×10^{23} 10^{23} m

Fig. P25 SCALE 1:1.5 × 10²⁴ The Local Group lies on the fringes of a very large cluster of galaxies, called the **Local Supercluster**. This is a cluster of clusters of galaxies. At the center of the Local Supercluster is the **Virgo Cluster** with several thousand galaxies. Seen from a large distance, our supercluster would present a view comparable with this photograph, which shows a multitude of galaxies beyond the constellation Fornax, all at a very large distance from us. The photograph was taken with the Hubble Space Telescope coupled to two very sensitive cameras using an exposure time of almost 300 hours.

All these distant galaxies are moving away from us and away from each other. The very distant galaxies in the photo are moving away from us at speeds almost equal to the speed of light. This motion of recession of the galaxies is analogous to the outward motion of, say, the fragments of a grenade after its explosion. The motion of the galaxies suggests that the Universe began with a big explosion, the **Big Bang**, that launched the galaxies away from each other.

Fig. P26 SCALE 1:1.5 × 10²⁵ On this scale a galaxy equal in size to our own Galaxy would look like a fuzzy dot, 0.1 millimeter across. Thus, the galaxies are too small to show up clearly on a photograph. Instead we must rely on a plot of the positions of the galaxies. The plot shows the positions of about 200 galaxies. The dense cluster of galaxies in the lower half of the plot is the Virgo Cluster.

Since we are looking into a volume of space, some of the galaxies are in the foreground, some are in the background; but our plot takes no account of perspective.

The luminous stars in the galaxies constitute only a small fraction of the total mass of the Universe. The space around the galaxies and the clusters of galaxies contains dark matter, and the space between the clusters contains **dark energy**, a strange form of matter that causes an acceleration of the expansion of the Universe.

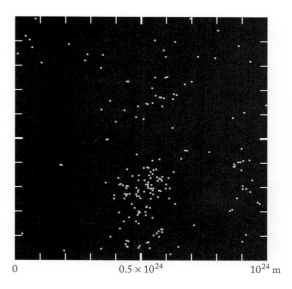

0   0.5×10^{24} 10^{24} m

Fig. P27 SCALE 1:1.5 × 10²⁶ This plot shows the positions of about 100,000 galaxies in a patch of the sky at distances of up to 1×10^9 light years from the Earth. The false color in this image indicates the distance–red for shorter distances, blue for larger distances.

The visible galaxies plotted here contribute only about 5% of the total mass in the universe. The dark matter near the galaxies contribute another 25%. The remaining 70% of the total mass in the universe is in the form of dark energy, which is uniformly distributed over the vast reaches of intergalactic space.

This is the last of our pictures in the ascending series. We have reached the limits of our zoom out. If we wanted to draw another picture, 10 times larger than this, we would need to know the shape and size of the entire Universe. We do not yet know that.

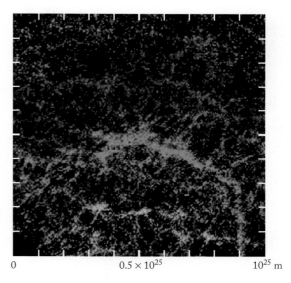

0   0.5×10^{25} 10^{25} m

PART II: THE SMALL-SCALE WORLD

MAGNIFICATION 0.67 ×

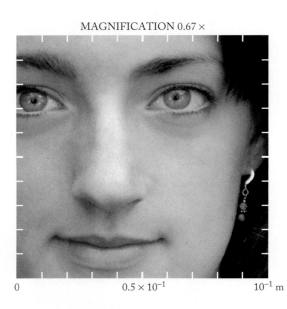

0   0.5×10^{-1} 10^{-1} m

Fig. P28 SCALE 1:1.5   We now return to Erin and zoom in on her eye. The surface of her skin appears smooth and firm. But this is an illusion. Matter appears continuous because the number of atoms in each cubic centimeter is extremely large. In a cubic centimeter of human tissue there are about 10^{23} atoms. This large number creates the illusion that matter is continuously distributed—we see only the forest and not the individual trees. The solidity of matter is also an illusion. The atoms in our bodies are mostly vacuum. As we will discover in the following pictures, within each atom the volume actually occupied by subatomic particles is only a very small fraction of the total volume.

MAGNIFICATION 6.7 ×

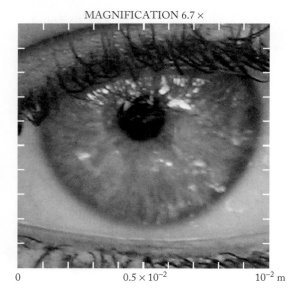

0   0.5×10^{-2} 10^{-2} m

Fig. P29 SCALE 1:1.5 × 10^{-1}   Our eyes are very sophisticated sense organs; they collect more information than all our other sense organs taken together. The photograph shows the pupil and the iris of Erin's eye. Annular muscles in the iris change the size of the pupil and thereby control the amount of light that enters the eye. In strong light the pupil automatically shrinks to about 2 millimeters; in very weak light it expands to as much as 7 millimeters.

Fig. P30 SCALE 1:1.5 × 10⁻²   This false-color photograph shows the delicate network of blood vessels on the front surface of the retina, the light-sensitive membrane lining the interior of the eyeball. The rear surface of the retina is densely packed with two kinds of cells that sense light: cone cells and rod cells. In a human retina there are about 6 million cone cells and 120 million rod cells. The cone cells distinguish colors; the rod cells distinguish only brightness and darkness, but they are more sensitive than the cone cells and therefore give us vision in faint light ("night vision").

This and the following photographs were made with various kinds of **electron microscopes**. An ordinary microscope uses a beam of light to illuminate the object; an electron microscope uses a beam of electrons. Electron microscopes can achieve much sharper contrast and much higher magnification than ordinary microscopes.

MAGNIFICATION 6.7 × 10 ×

0   0.5 × 10⁻³   10⁻³ m

Fig. P31 SCALE 1:1.5 × 10⁻³   Here we have a false-color photograph of rod cells prepared with a scanning electron microscope (SEM). For this photograph, the retina was cut apart and the microscope was aimed at the edge of the cut. In the top half of the picture we see tightly packed rods. Each rod is connected to the main body of a cell containing the nucleus. In the bottom part of the picture we can distinguish tightly packed cell bodies of the cell.

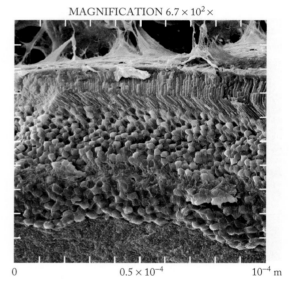

MAGNIFICATION 6.7 × 10² ×

0   0.5 × 10⁻⁴   10⁻⁴ m

Fig. P32 SCALE 1:1.5 × 10⁻⁴   This is a close-up view of a few rods cells. The upper portions of the rods contain a special pigment—visual purple—which is very sensitive to light. The absorption of light by this pigment initiates a chain of chemical reactions that finally trigger nerve pulses from the eye to the brain.

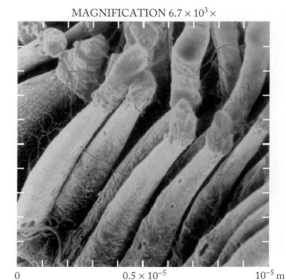

MAGNIFICATION 6.7 × 10³ ×

0   0.5 × 10⁻⁵   10⁻⁵ m

MAGNIFICATION $6.7 \times 10^4 \times$

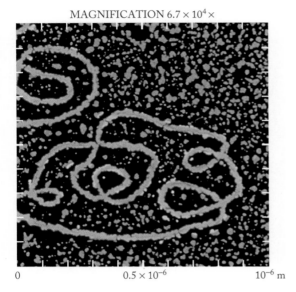

0   0.5×10^{-6} 10^{-6} m

Fig. P33 SCALE $1{:}1.5 \times 10^{-5}$ These are strands of DNA, or deoxyribonucleic acid, as seen with a transmission electron microscope (TEM) at very high magnification. DNA is found in the nuclei of cells. It is a long molecule made by stringing together a large number of nitrogenous base molecules on a backbone of sugar and phosphate molecules. The base molecules are of four kinds, the same in all living organisms. But the sequence in which they are strung together varies from one organism to another. This sequence spells out a message—the base molecules are the "letters" in this message. The message contains all the genetic instructions governing the metabolism, growth, and reproduction of the cell.

The strands of DNA in the photograph are encrusted with a variety of small protein molecules. At intervals, the strands of DNA are wrapped around larger protein molecules that form lumps looking like the beads of a necklace.

MAGNIFICATION $6.7 \times 10^5 \times$

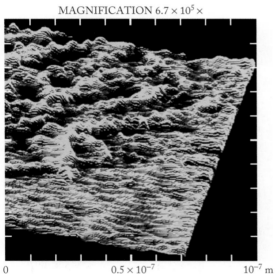

0   0.5×10^{-7} 10^{-7} m

Fig. P34 SCALE $1{:}1.5 \times 10^{-6}$ The highest magnifications are attained by a newer kind of electron microscope, the scanning tunneling microscope (STM). This picture was prepared with such a microscope. The picture shows strands of DNA deposited on a substrate of graphite. In contrast to the strands of the preceding picture, these strands are uncoated; that is, they are without protein encrustations.

MAGNIFICATION $6.7 \times 10^6 \times$

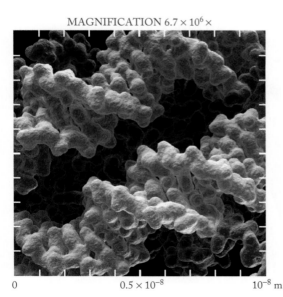

0   0.5×10^{-8} 10^{-8} m

Fig. P35 SCALE $1{:}1.5 \times 10^{-7}$ This close-up picture of strands of DNA reveals the helical structure of this molecule. The strand consists of a pair of helical coils wrapped around each other. This picture was generated by a computer from data obtained by illuminating DNA samples with X rays (X-ray scattering).

Fig. P36 SCALE 1:1.5 × 10⁻⁸   This picture shows a layer of palladium atoms on surface of graphite as seen with an STM. Here we have visual evidence of the atomic structure of matter. The palladium atoms are arranged in a symmetric, repetitive hexagonal pattern. Materials with such regular arrangements of atoms are called **crystals**.

Each of the palladium atoms is approximately a sphere, about 3×10^{-10} meter across. However, the atom does not have a sharply defined boundary; its surface is somewhat fuzzy. Atoms of other elements are also approximately spheres, with sizes that range from 2×10^{-10} to 4×10^{-10} meter across.

At present we know of more than 100 kinds of atoms or chemical elements. The lightest atom is hydrogen, with a mass of 1.67×10^{-27} kilogram; the heaviest is element 114, ununquadium, with a mass about 289 times as large.

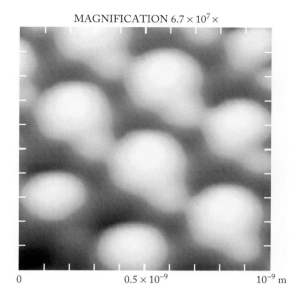

MAGNIFICATION $6.7 \times 10^7 \times$

0   0.5×10^{-9} 10^{-9} m

Fig. P37 SCALE 1:1.5 × 10⁻⁹   The drawing shows the interior of an atom of neon. This atom consists of 10 electrons orbiting around a nucleus. In the drawing, the electrons have been indicated by small dots, and the nucleus by a slightly larger dot at the center of the picture. These dots have been drawn as small as possible, but even so the size of these dots does not give a correct impression of the actual sizes of the electrons and of the nucleus. The electron is smaller than any other particle we know; maybe the electron is truly pointlike and has no size at all. The nucleus has a finite size, but this size is much too small to show up on the drawing. Note that the electrons tend to cluster near the center of the atom. However, the overall size of the atom depends on the distance to the outermost electron; this electron defines the outer edge of the atom.

The electrons move around the nucleus in a very complicated motion, and so the resulting electron distribution resembles a fuzzy cloud, similar to the STM image of the previous picture. This drawing, however, shows the electrons as they would be seen at one instant of time with a hypothetical microscope that employs gamma rays instead of light rays to illuminate an object; no such microscope has yet been built.

The mass of each electron is 9.11×10^{-31} kilogram, but most of the mass of the atom is in the nucleus; the 10 electrons of the neon atom have only 0.03% of the total mass of the atom.

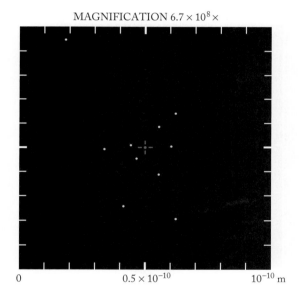

MAGNIFICATION $6.7 \times 10^8 \times$

0   0.5×10^{-10} 10^{-10} m

MAGNIFICATION $6.7 \times 10^9 \times$

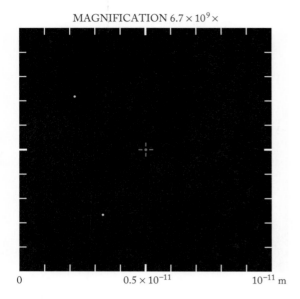

0   0.5×10^{-11} 10^{-11} m

Fig. P38 SCALE 1:1.5 $\times$ 10^{-10} Here we are closing in on the nucleus. We are seeing the central part of the atom. Only two electrons are in our field of view; the others are beyond the margin of the drawing. The size of the nucleus is still much smaller than the size of the dot at the center of the drawing.

MAGNIFICATION $6.7 \times 10^{10} \times$

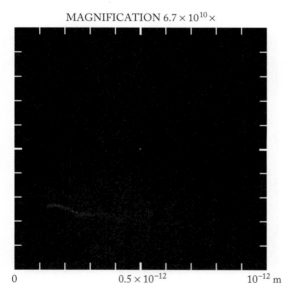

0   0.5×10^{-12} 10^{-12} m

Fig. P39 SCALE 1:1.5 $\times$ 10^{-11} In this drawing we finally see the nucleus in its true size. At this magnification, the nucleus of the neon atom looks like a small dot, 0.5 millimeter in diameter. Since the nucleus is extremely small and yet contains most of the mass of the atom, the density of the nuclear material is enormous. If we could assemble a drop of pure nuclear material of a volume of 1 cubic centimeter, it would have a mass of 2.3×10^{11} kilograms, or 230 million metric tons!

Our drawings show clearly that most of the volume within the atom is empty space. The nucleus occupies only a very small fraction of this volume.

MAGNIFICATION $6.7 \times 10^{11} \times$

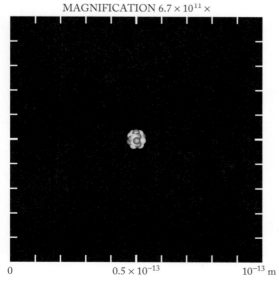

0   0.5×10^{-13} 10^{-13} m

Fig. P40 SCALE 1:1.5 $\times$ 10^{-12} We can now begin to distinguish the nuclear structure. The nucleus has a nearly spherical shape, but its surface is slightly fuzzy.

Fig. P41 SCALE 1:1.5 × 10⁻¹³ At this extreme magnification we can see the details of the nuclear structure. The nucleus of the neon atom is made up of 10 protons (white balls) and 10 neutrons (red balls). Each proton and each neutron is a sphere with a diameter of about 2×10^{-15} meter, and a mass of 1.67×10^{-27} kilogram. In the nucleus, these protons and neutrons are tightly packed together, so tightly that they almost touch. The protons and neutrons move around the volume of the nucleus at high speed in a complicated motion.

Magnification 6.7 × 10¹² ×

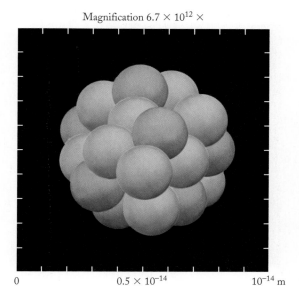

0   0.5 × 10⁻¹⁴   10⁻¹⁴ m

Fig. P42 SCALE 1:1.5 × 10⁻¹⁴ This final picture shows three pointlike bodies within a proton. These pointlike bodies are **quarks**—each proton and each neutron is made of three quarks. Recent experiments have told us that the quarks are much smaller than protons, but we do not yet know their precise size. Hence the dots in the drawing probably do not give a fair description of the size of the quarks. The quarks within protons and neutrons are of two kinds, called **up** and **down**. The proton consists of two *up* quarks and one *down* quark joined together; the neutron consists of one *up* quark and two *down* quarks joined together.

This final picture takes us to the limits of our knowledge of the subatomic world. As a next step we would like to zoom in on the quarks and show what they are made of. According to a speculative theory, they are made of small snippets or loops of strings, 10^{-35} m long. But we do not yet have any evidence for this theory.

Magnification 6.7 × 10¹³ ×

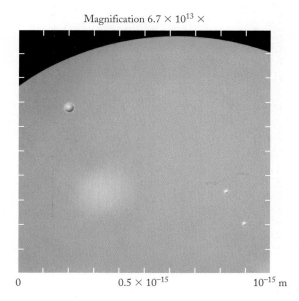

0   0.5 × 10⁻¹⁵   10⁻¹⁵ m

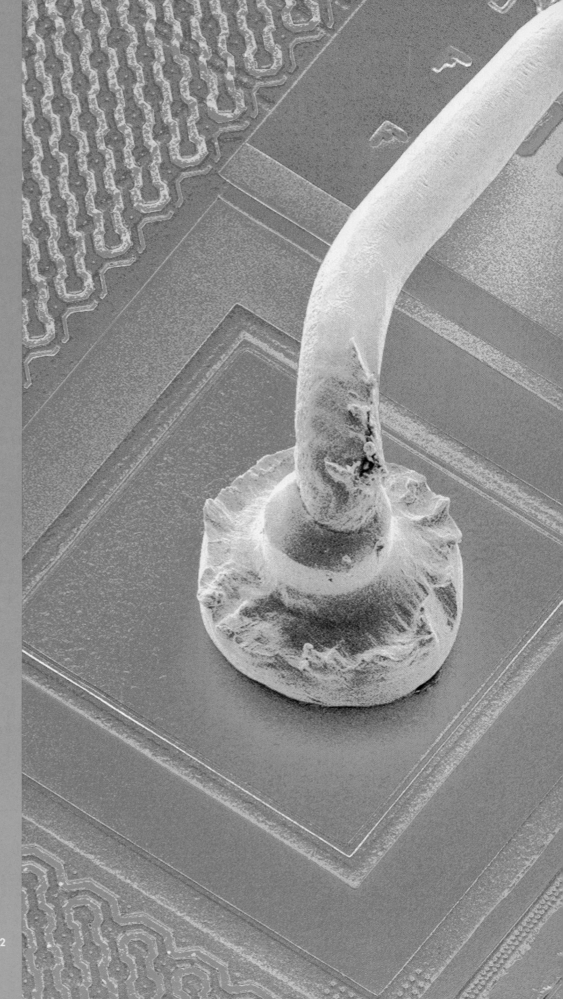

4

Electricity and Magnetism

CONTENTS

A microchip contains precision-engineered circuits packed into a very small space. This gold micro-wire is one of several connections to the microchip, and it leads to larger pins that are used to plug the microchip into a circuit board, such as found inside a computer.

Electric Force and Electric Charge

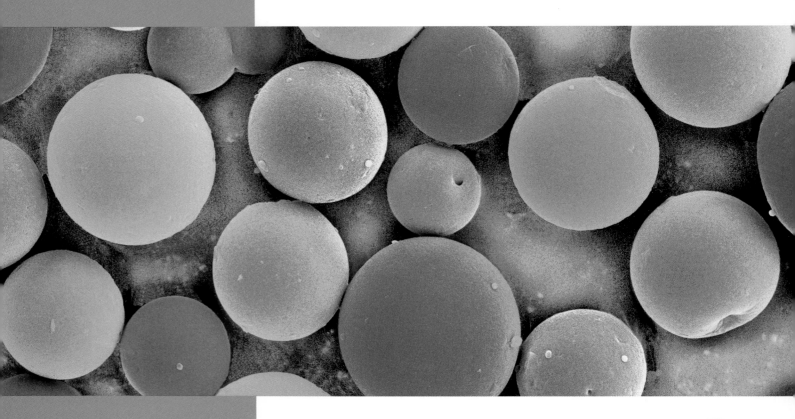

CONCEPTS IN CONTEXT

Concepts
— in —
Context

Photocopiers and laser printers use toner particles, such as these magenta, cyan, yellow, and black modern toner particles made spherical with a polymer coating. An attractive electric force holds the particles to the plate beneath them; the particles also exert repulsive electric forces on one another.

With the properties of electric force and electric charge introduced in this chapter, we can answer such questions as:

? What is the force that one toner particle exerts on another? (Example 4, page 701)

? What is the total force on a particle surrounded by several other particles? (Example 7, page 704)

? How are the toner particles transferred to form an image? (Physics in Practice: Xerography, page 709)

Our society is dependent on electricity. An electric power failure demonstrates our dependence—subways and elevators stop; traffic lights, streetlights, and the lights in our homes go out; refrigerators fail; food can't be cooked; homes can't be heated; radios, TVs, and computers can't be operated. But our dependence on electricity runs even deeper than our reliance on electrical machinery and gadgetry would suggest. Electricity is an essential ingredient in all the atoms in our bodies and in our environment. The forces that hold the parts of an atom together are electric forces. Furthermore, so are the forces that bind atoms in a molecule and hold these building blocks together in large-scale macroscopic structures, such as a rock, a tree, a human body, a skyscraper, or a supertanker. All the mechanical "contact" forces of everyday experience—the push of a hand against a door, the pull of an elevator cable, the pressure of water against the hull of a ship—are nothing but the combined electric forces of many atoms. Thus, our immediate environment is dominated by **electric forces**.

electric force

In the following chapters, we will study electric forces and their effects. For a start (Chapters 22–28), we will assume that the particles exerting these forces are at rest or moving only very slowly. The electric forces exerted under these conditions are called **electrostatic forces**. Later on (Chapters 29–31), we will consider the electric forces when the particles are moving with uniform velocity or nearly uniform velocity. Under these conditions the electric forces are modified—besides the electrostatic force there arises a **magnetic force**, which depends on the velocities of the particles. The combined electrostatic and magnetic forces are called **electromagnetic forces**. Finally, we will consider the forces exerted when the particles are moving with accelerated motion (Chapter 33). The electromagnetic forces are then further modified with a drastic consequence, that is, the emission of electromagnetic waves, such as light and radio waves.

magnetic force

Electricity was first discovered through friction. The ancient Greeks noticed that rods of amber (*elektron*, in Greek), when rubbed with a cloth or with fur, gave off sparks and attracted small bits of straw or feathers. You can easily duplicate this ancient discovery by rubbing a plastic comb on a shirt or a sweater; in the dark, you can then see a multitude of small sparks produced by this rubbing process, and the electrified comb will attract small bits of paper or lint. In the nineteenth century, practical applications of electricity were gradually developed, but it was only in the twentieth century that the pervasive presence of electric forces holding together all the matter of our environment was recognized.

22.1 THE ELECTROSTATIC FORCE

Online
Concept
Tutorial

Ordinary matter—solids, liquids, and gases—consists of atoms, each with a nucleus surrounded by a swarm of electrons. For example, Fig. 22.1 shows the structure of an atom of neon. At the center of this atom there is a nucleus made of ten protons and ten neutrons packed very tightly together—the diameter of the nucleus is only about 6×10^{-15} m. Moving around this nucleus are ten electrons; these electrons are confined to a roughly spherical region about 3×10^{-10} m across.

The atom somewhat resembles the Solar System, with the nucleus as Sun and the electrons as planets. In the Solar System, the force that holds a planet near the Sun is the gravitational force. In the atom the force that holds an electron near the nucleus is the electric force of attraction between the electron and the protons in the nucleus. This electric force resembles gravitation in that it decreases in proportion to the inverse square of the distance. But the electric force is much stronger than the gravitational force.

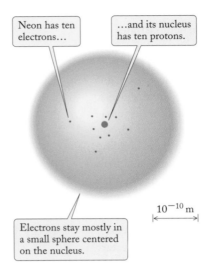

Neon has ten electrons…

…and its nucleus has ten protons.

Electrons stay mostly in a small sphere centered on the nucleus.

10^{-10} m

FIGURE 22.1 Neon atom. This drawing shows the electrons at one instant of time, as they would appear in a (hypothetical) extremely powerful microscope.

The electric attraction between an electron and a proton (at any given distance) is about 2×10^{39} times as strong as the gravitational attraction. Thus, the electric force is by far the strongest force felt by an electron in an atom.

The other great difference between the gravitational force and the electric force is that the gravitational force between two particles is always attractive, whereas electric forces can be attractive, repulsive, or zero, depending on what two particles we consider. Table 22.1 gives a qualitative summary of the electric forces between fundamental particles.

electric charge   Particles that exert electric forces are said to have an **electric charge**; particles that do not exert electric forces are said to have no electric charge. Thus, *electric charge is thought of as the source of electric force*, just as mass is the source of gravitational force. Electrons and protons have electric charge, but neutrons have no electric charge. The electron–proton force, the electron–electron force, and the proton–proton force all have the same magnitudes (for a given distance). Thus, the strengths of the electric forces associated with electrons and with protons are of equal magnitudes; that is, their electric charges are of equal magnitudes. For the mathematical formulation of the law of electric force, we assign a positive charge to the proton and a negative charge (of equal magnitude) to the electron. We designate these charges of the proton and the electron by $+e$ and $-e$, respectively. Table 22.2 summarizes these values of the charges.

In terms of these electric charges, we can then state that *the electric force between charges of like sign is repulsive and the electric force between charges of unlike sign is attractive.*

coulomb (C)   The numerical value of the charge e of the proton depends on the system of units. In the SI system of units, the electric charge is measured in **coulombs** (C), and the corresponding numerical values of the fundamental charges of the proton and of the electron are[1]

charge of proton and electron

$$e = 1.60 \times 10^{-19} \text{ C for proton}$$

$$-e = -1.60 \times 10^{-19} \text{ C for electron} \tag{22.1}$$

TABLE 22.1	
ELECTRIC FORCES (QUALITATIVE)	
PARTICLES	**FORCE**
Electron and proton	Attractive
Electron and electron	Repulsive
Proton and proton	Repulsive
Neutron and anything	Zero

TABLE 22.2	
ELECTRIC CHARGES OF PROTONS, ELECTRONS, AND NEUTRONS	
PARTICLE	**CHARGE**
Proton, p	$+e$
Electron, e	$-e$
Neutron, n	0

[1]As with all physical constants, we have rounded these values to three significant figures. A more exact value can be found in Appendix 6 or inside the book covers.

In the SI system of units, the coulomb is defined in terms of a standard electric current, that is, a standard rate of flow of charge: one coulomb is the amount of electric charge that a current of one ampere delivers in one second. Unfortunately, the definition of the standard current involves the use of magnetic fields, and we will therefore have to postpone the question of the precise definition of ampere and coulomb to a later chapter.

One coulomb of charge represents a large number of fundamental charges; this is easily seen by taking the inverse of Eq. (22.1),

$$1 \text{ C} = \frac{1}{1.60 \times 10^{-19}} e = 6.25 \times 10^{18} \times e$$

Thus, a coulomb is more than 6 billion billion fundamental charges. Since the fundamental charge is so small, we can often ignore the discrete character of charge in practical and engineering applications of electricity, and we can treat macroscopic charge distributions as continuous. This is analogous to treating macroscopic mass distributions as continuous, even though, on a microscopic scale, the mass consists of discrete atoms.

The net electric charge of a body containing some number of electrons and protons is the (algebraic) sum of the electron and proton charges. For instance, the net electric charge of an atom containing equal numbers of electrons and protons is zero; that is, the atom is electrically neutral. Sometimes atoms lose an electron, and sometimes they gain an extra electron. *Such atoms with missing electrons or with extra electrons are called* **ions**. They have a net positive charge if they have lost electrons, and a net negative charge if they have gained electrons. The positive or negative charge on a macroscopic body—such as on a plastic comb electrified by rubbing—arises in the same way, from a deficiency or an excess of electrons.

The electric forces between two neutral atoms tend to cancel; each electron in one atom is attracted by the protons in the nucleus of the other atom, and simultaneously it is repelled by the equal number of electrons of that atom. However, the cancellation of these electric attractive and repulsive forces among the electrons and the protons in the two atoms is sometimes not complete. For instance, the "contact" force between two atoms close together arises from an incomplete cancellation of the attractive and repulsive forces. The force between the atoms depends on the relative locations of the electrons and the nuclei. If the distributions of the electrons are somewhat distorted so, on the average, the electrons in one atom are closer to the electrons of the neighboring atom than to its nucleus, then the net force between these atoms will be repulsive. Figure 22.2a shows such a distortion that leads to a repulsive net force; the distortion may either be intrinsic to the structure of the atom or induced by the presence of the neighboring atom. Figure 22.2b shows a distortion that leads to an attractive force.

Likewise, the electric forces between two neutral macroscopic bodies separated by some appreciable distance tend to cancel. For example, if the macroscopic bodies are a baseball and a tennis ball separated by a distance of 2 m, then each electron of the baseball is attracted by the protons of the tennis ball, but simultaneously it is repelled by the electrons of the tennis ball; and these forces cancel each other. Only when the surfaces of the two balls are very near one another ("touching") will the atoms in one surface exert a net electric force on those in the other surface.

This cancellation of the electric forces between neutral macroscopic bodies explains why we do not see large electric attractions or repulsions between the macroscopic bodies in our environment, even though the electric forces between individual electrons and protons are much stronger than the gravitational forces. Most such macroscopic bodies are electrically neutral, and they therefore will exert no net electric forces on each other, except for contact forces.

ion

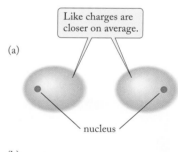

(a)

Like charges are closer on average.

nucleus

(b)

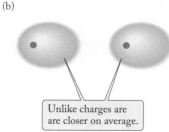

Unlike charges are are closer on average.

FIGURE 22.2 (a) Two neighboring distorted atoms. The colored regions represent the average distribution of the electrons. The electrons of the left atom are closer to most of the electrons of the right atom than to its nucleus. This results in a net repulsive force between the atoms. (b) The electrons of the left atom are closer to the nucleus of the right atom than to most of its electrons. This results in a net attractive force between the atoms.

CHARLES AUGUSTIN de COULOMB
(1736–1806) *French physicist, shown here with the torsion balance, which he invented. He established that the electric force between small charged balls obeys an inverse-square law.*

Online
Concept
Tutorial

24

FIGURE 22.3 Coulomb's torsion balance. The beam carries a small charged ball (a) at one end, and a counterweight on the other. A second small charged ball (b) is brought near the first ball. If the balls carry charges of equal signs, they repel each other, and the beam of the balance rotates.

 Checkup 22.1

QUESTION 1: The planets in the Solar System exert large gravitational forces on each other, but only insignificant electric forces. The electrons in an atom exert large electric forces on each other, but only insignificant gravitational forces. Explain this difference.

QUESTION 2: A stone of mass 1.0 kg rests on the ground. What is the net electric force that the ground exerts on the stone?

QUESTION 3: Suppose that in Fig. 22.2a, the atomic nuclei were displaced toward each other, instead of away from each other. Would the net electric force between the atoms be attractive or repulsive?

QUESTION 4: Six electrons are added to 1.0 coulomb of positive charge. The net charge is approximately

(A) 7.0 C   (B) −5.0 C   (C) 1.0 C   (D) −6e   (E) −5e

22.2 COULOMB'S LAW

As already mentioned above, the electric force between two particles decreases with the inverse square of the distance, just as does the gravitational force. The dependence of the electric force on distance was discovered through experiments by Charles Augustin de Coulomb, who investigated the repulsion between small balls that he charged by a rubbing process. To measure the force between the balls, he used a delicate torsion balance (see Fig. 22.3) similar to the torsion balance later used by Henry Cavendish to measure gravitational forces. His experimental results are summarized in **Coulomb's Law**:

> *The magnitude of the electric force that a particle exerts on another particle is directly proportional to the product of their charges and inversely proportional to the square of the distance between them. The direction of the force is along the line joining the particles.*

Mathematically, the electric force F that a particle of charge q' exerts on a particle of charge q at a distance r is given by the formula

$$F = k \times \frac{q'q}{r^2} \tag{22.2}$$

where k is a constant of proportionality. This formula not only gives the magnitude of the force, but also the direction, if we *interpret a positive value of the force F as repulsive and a negative value as attractive*. For instance, in the case of the force exerted by a proton on an electron, the charges are $q' = e$ and $q = -e$, and the formula (22.2) yields

$$F = k \times \frac{e \times (-e)}{r^2} = -k \frac{e^2}{r^2} \tag{22.3}$$

which is negative, indicating attraction.

The electric force that the particle of charge q exerts on the particle of charge q' has the same magnitude as the force exerted by q' on q, but the opposite direction. These mutual forces are an action–reaction pair (see Fig. 22.4 for two examples).

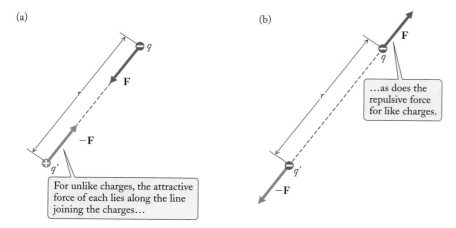

(a)

(b)

...as does the repulsive force for like charges.

For unlike charges, the attractive force of each lies along the line joining the charges...

FIGURE 22.4 (a) Two charged particles q and q', one of which is positive and one of which is negative. (b) Two other particles q and q'; both are negative.

In the SI system of units, the constant of proportionality k in Coulomb's Law, called the **Coulomb constant**, or the electric force constant, has the value

$$k = 8.99 \times 10^9 \ \text{N·m}^2/\text{C}^2 \qquad (22.4)$$

Coulomb constant

This constant is traditionally written in the more complicated but equivalent form

$$k = \frac{1}{4\pi\epsilon_0} \qquad (22.5)$$

with

$$\epsilon_0 = 8.85 \times 10^{-12} \ \text{C}^2/(\text{N·m}^2) \qquad (22.6)$$

permittivity constant

The quantity ϵ_0 ("epsilon nought") is called the **electric constant** or the **permittivity constant**. In terms of the permittivity constant, Coulomb's Law for the force that a particle of charge q' exerts on a particle of charge q becomes

$$F = \frac{1}{4\pi\epsilon_0} \frac{q'q}{r^2} \qquad (22.7)$$

Coulomb's Law

Remember that a positive value of F indicates a repulsive force directed along the line joining the charges, and a negative value indicates an attractive force. Although the expression (22.2), with the value of the constant given in Eq. (22.4), is most convenient for the numerical calculation of the Coulomb force, the somewhat more complicated expression (22.7) is widely used in physics and engineering. Of course, the two expressions are mathematically equivalent, and they give the same results.

Coulomb's Law applies to particles—electrons and protons—and also to any small charged bodies, provided that the sizes of these bodies are much smaller than the distances between them; such bodies are called **point charges**. Equation (22.2) obviously resembles Newton's Law for the gravitational force (see Section 9.1); the constant k is analogous to the gravitational constant G, and the electric charges are analogous to the gravitating masses.

point charge

EXAMPLE 1 Compare the magnitudes of the gravitational force of attraction and of the electric force of attraction between the electron and the proton in a hydrogen atom. According to Newtonian mechanics, what is the acceleration of the electron? Assume that the distance between these particles in a hydrogen atom is 5.3×10^{-11} m.

SOLUTION: From Chapter 9, the magnitude of the gravitational force is

$$F_g = G\frac{mM}{r^2}$$

$$= (6.67 \times 10^{-11}\,\text{N·m}^2/\text{kg}^2) \times \frac{(9.11 \times 10^{-31}\,\text{kg})(1.67 \times 10^{-27}\,\text{kg})}{(5.3 \times 10^{-11}\,\text{m})^2}$$

$$= 3.6 \times 10^{-47}\,\text{N}$$

The magnitude of the electric force is

$$F_e = \frac{1}{4\pi\epsilon_0}\frac{e \times e}{r^2} = (8.99 \times 10^9\,\text{N·m}^2/\text{C}^2) \times \frac{(1.60 \times 10^{-19}\,\text{C})^2}{(5.3 \times 10^{-11}\,\text{m})^2}$$

$$= 8.2 \times 10^{-8}\,\text{N}$$

The ratio of these forces is $(8.2 \times 10^{-8}\,\text{N})/(3.6 \times 10^{-47}\,\text{N}) = 2.3 \times 10^{39}$. Thus the electric force overwhelms the gravitational force.

Since the gravitational force is insignificant compared with the electric force, it can be neglected. The acceleration of the electron is then

$$a = \frac{F}{m} = \frac{8.2 \times 10^{-8}\,\text{N}}{9.11 \times 10^{-31}\,\text{kg}} = 9.0 \times 10^{22}\,\text{m/s}^2$$

This is a gigantic acceleration. If it occurred along the electron's motion instead of centripetally, such an acceleration could boost the electron's velocity close to one-third of the speed of light in only a femtosecond (10^{-15} s)!

COMMENTS: Note that for the ratio of the electric force and the gravitational force between the proton and electron, we would obtain the same immense value 2.3×10^{39} whatever the separation between the two particles, since both are inverse-square forces. Also notice that for the given atomic-scale distance, the electric force has a measurable value, the same as weighing an 8-microgram mass, whereas the gravitational force is far below the current limits of detection (the highest sensitivity attained by a measurement of force is near 10^{-20} N).

EXAMPLE 2 How much negative charge and how much positive charge are there on the electrons and the protons in a cup of water (0.25 kg)?

SOLUTION: The "molecular mass" of water is 18 g; hence, 250 g of water amounts to 250/18 moles. Each mole has 6.02×10^{23} molecules, giving $(250/18) \times 6.0 \times 10^{23}$ molecules in the cup. Each molecule consists of two hydrogen atoms (one electron apiece) and one oxygen atom (eight electrons). Thus, there are 10 electrons in each molecule, and the total negative charge on all the electrons together is

$$(250\text{ g})(1\text{ mole}/18\text{ g})(6.02 \times 10^{23}\text{ molecules/mole})$$

$$\times (10\text{ electrons/molecule})(-1.60 \times 10^{-19}\text{ C/electron})$$

$$= -1.3 \times 10^7 \, \text{C}$$

The positive charge on the protons is the opposite of this.

EXAMPLE 3   What is the magnitude of the attractive force exerted by the electrons in a cup of water on the protons in a second cup of water at a distance of 10 m?

SOLUTION: According to the preceding example, the charge on the electrons in the cup is -1.3×10^7 C and the charge on the protons is $+1.3 \times 10^7$ C. If we treat both of these charges as point charges, the force on the protons is

$$F = \frac{1}{4\pi\epsilon_0} \frac{qq'}{r^2}$$

$$= (8.99 \times 10^9 \, \text{N·m}^2/\text{C}^2) \frac{(-1.3 \times 10^7 \, \text{C})(1.3 \times 10^7 \, \text{C})}{(10 \, \text{m})^2}$$

$$= -1.5 \times 10^{22} \, \text{N}$$

This is approximately the weight of a billion billion tons! This enormous attractive force on the protons is precisely canceled by an equally large repulsive force exerted by the protons in one cup on the protons in the other cup. Thus, the cups exert no net forces on each other.

EXAMPLE 4   Consider two toner particles separated by 1.2×10^{-5} m; each of the two particles has a negative charge of -3.0×10^{-14} C. What is the electric force that one particle exerts on the other? Treat the toner particles approximately as point particles.

Concepts
— *in* —
Context

SOLUTION: Since we may treat the particles approximately as point charges, the force that one particle exerts on the other is

$$F = \frac{1}{4\pi\epsilon_0} \frac{qq'}{r^2}$$

$$= (8.99 \times 10^9 \, \text{N·m}^2/\text{C}^2) \frac{(-3.0 \times 10^{-14} \, \text{C})^2}{(1.2 \times 10^{-5} \, \text{m})^2}$$

$$= +5.6 \times 10^{-8} \, \text{N}$$

The positive sign reminds us that the force is repulsive, tending to push each particle directly away from the other, along the line joining them (see Fig. 22.4b). This mutual repulsion helps to keep the toner particles dispersed, so they do not clump up in one region.

EXAMPLE 5   A simple electroscope for the detection and measurement of electric charge consists of two small foil-covered cork balls of 1.5×10^{-4} kg each suspended by threads 10 cm long (see Fig. 22.5). When equal electric charges are placed on the balls, the electric repulsive force pushes them apart, and the angle between the threads indicates the magnitude of the electric

(a)

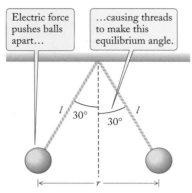

(b)

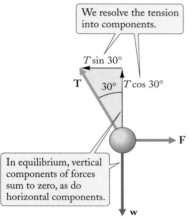

FIGURE 22.5 (a) Two equal charged balls suspended by threads. (b) "Free-body" diagram for the right ball.

charge. If the equilibrium angle between the threads is 60°, what is the magnitude of the charge?

SOLUTION: Figure 22.5b shows a "free-body" diagram for one of the balls. The electric force **F** acts along the line joining the two charges and is thus horizontal. In equilibrium, the vector sum of the electric repulsion **F**, the weight **w**, and the tension **T** of the thread must be zero. Accordingly, the horizontal component of the tension must balance the electric repulsion, and the vertical component of the tension must balance the weight:

$$F = T \sin 30°$$

$$mg = T \cos 30°$$

We can eliminate the tension from the problem by taking the ratio of these equations, yielding

$$F = mg \tan 30°$$

From Fig. 22.5a we see that the distance between the balls is $r = 2l \sin 30°$, so Coulomb's Law tells us

$$F = \frac{1}{4\pi\epsilon_0} \frac{q^2}{(2l \sin 30°)^2}$$

Equating these two expressions for F, we find

$$mg \tan 30° = \frac{1}{4\pi\epsilon_0} \frac{q^2}{(2l \sin 30°)^2}$$

and

$$q = \sqrt{4\pi\epsilon_0 mg \tan 30°} \times 2l \sin 30°$$

$$= \sqrt{(4\pi)(8.85 \times 10^{-12} \text{ C}^2/\text{N·m}^2)(1.5 \times 10^{-4} \text{ kg})(9.81 \text{ m/s}^2)(\tan 30°)}$$

$$\times (2)(0.10 \text{ m})(\sin 30°)$$

$$= 3.1 \times 10^{-8} \text{ C}$$

 Checkup 22.2

QUESTION 1: Suppose that the electric force between two charges is attractive. What can you conclude about the signs of these charges?

QUESTION 2: Suppose that the electric force between two charges separated by a distance of 1 m is 1×10^{-4} N. What will be the electric force if we increase the distance to 10 m? To 100 m?

QUESTION 3: Two balls, separated by some distance, carry equal electric charges and exert a repulsive electric force on each other. If we transfer a fraction of the electric charge of one ball to the other, will the electric force increase or decrease?

QUESTION 4: Two particles are separated by a distance of 3.0 m; each exerts an electric force of 1.0 N on the other. If one particle carries 10 times as much electric charge as the other, what is the magnitude of the smaller charge?

(A) 10 pC   (B) 10 μC   (C) 10 C   (D) 10 kC

22.3 THE SUPERPOSITION OF ELECTRIC FORCES

Online
Concept
Tutorial

The electric force, like any other force, has a magnitude and a direction; that is, the electric force is a vector. According to Coulomb's Law, the *magnitude* of the electric force exerted by a point charge q' on a point charge q is

$$F = \frac{1}{4\pi\epsilon_0} \frac{qq'}{r^2} \tag{22.8}$$

The *direction* of this force is along the line from one charge to the other. As illustrated in Fig. 22.6, this force can be represented by a vector **F** pointing along the line from one charge to the other.

If several point charges q_1, q_2, q_3, . . . simultaneously exert electric forces on the charge q, then *the net force on q is obtained by taking the vector sum of the individual forces* (see Fig. 22.7). Thus, if the vectors representing the individual electric forces produced by q_1, q_2, q_3, . . . are **F**$_1$, **F**$_2$, **F**$_3$, . . ., respectively, then the net force is

$$\mathbf{F} = \mathbf{F}_1 + \mathbf{F}_2 + \mathbf{F}_3 + \cdots \tag{22.9}$$

Superposition Principle of electric forces

Equation (22.9) expresses the **Superposition Principle of electric forces**. According to Eq. (22.9), the force contributed by each charge is independent of the presence of the other charges. For instance, the charge q_2 does not affect the interaction of charge q_1 with q; it merely adds its own interaction with q. This simple combination law is an important empirical fact about electric forces. Since the contact forces of everyday experience, such as the normal force and the friction force, arise from electric forces between the atoms, they will likewise obey the Superposition Principle, and they can be combined with simple vector addition. Incidentally: The gravitational forces on the Earth and within the Solar System also obey the Superposition Principle. Thus, all the forces in our immediate environment obey this principle. We have already made much use of the Superposition Principle in our study of mechanics; now we recognize that *the superposition of mechanical forces, such as "contact" forces, hinges on the superposition of electric forces.*

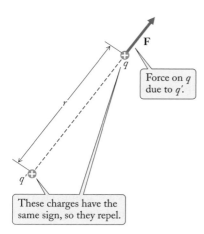

FIGURE 22.6 A charge q' exerts an electric force **F** on the charge q.

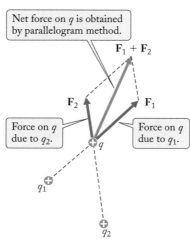

FIGURE 22.7 Two point charges q_1 and q_2 exert electric forces **F**$_1$ and **F**$_2$ on the point charge q. The net force on q is the vector sum of these forces.

(a)

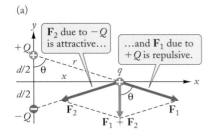

(b)

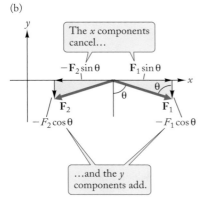

FIGURE 22.8 (a) The charges $+Q$ and $-Q$ exert forces $\mathbf{F}_1$ and $\mathbf{F}_2$ on the charge q. The net force is the vector sum $\mathbf{F}_1 + \mathbf{F}_2$. (b) The x and y components of $\mathbf{F}_1$ and $\mathbf{F}_2$.

EXAMPLE 6   Two point charges $+Q$ and $-Q$ are separated by a distance d, as shown in Fig. 22.8. A positive point charge q is equidistant from these charges, at a distance x from their midpoint. What is the electric force $\mathbf{F}$ on q?

SOLUTION: As illustrated in Fig. 22.8a, the charge $+Q$ produces a repulsive force on the charge q, and the charge $-Q$ produces an attractive force. Thus, the vector $\mathbf{F}_1$ points away from $+Q$, and the vector $\mathbf{F}_2$ points toward $-Q$. From the geometry of Fig. 22.8a, the distance from each of the charges $+Q$ and $-Q$ to the charge q is $r = \sqrt{x^2 + d^2/4}$. Hence, the *magnitudes* of the individual Coulomb forces exerted by $+Q$ and $-Q$ are

$$F_1 = F_2 = \frac{1}{4\pi\epsilon_0}\frac{qQ}{r^2} = \frac{1}{4\pi\epsilon_0}\frac{qQ}{x^2 + d^2/4} \quad \text{(magnitudes)} \quad (22.10)$$

From Fig. 22.8b we see that in the vector sum $\mathbf{F} = \mathbf{F}_1 + \mathbf{F}_2$, the horizontal components (x components) of $\mathbf{F}_1$ and $\mathbf{F}_2$ cancel, and the vertical components (y components) add, giving a net vertical component twice as large as each individual vertical component. Thus in this case the net force $\mathbf{F}$ has a y component but no x component. In terms of the angle θ shown in Fig. 22.8b, the y component of $\mathbf{F}_1$ is $-F_1 \cos\theta$, and the y component of $\mathbf{F}_2$ is $-F_2 \cos\theta$. Since these are equal, the net force is then

$$F = F_y = -F_1 \cos\theta - F_2 \cos\theta = -2F_1 \cos\theta$$

$$= -2\frac{1}{4\pi\epsilon_0}\frac{qQ}{x^2 + d^2/4}\cos\theta \quad (22.11)$$

From Fig. 22.8a, we see that $\cos\theta = \frac{1}{2}d/(x^2 + d^2/4)^{1/2}$, and therefore

$$F = F_y = -\frac{1}{4\pi\epsilon_0}\frac{qQd}{(x^2 + d^2/4)^{3/2}} \quad (22.12)$$

COMMENTS: Note that if the charge q is at a large distance from the two charges $\pm Q$, then d^2 can be neglected compared with x^2, so $(x^2 + d^2/4)^{3/2} \approx (x^2)^{3/2} = x^3$. The force F is then proportional to $1/x^3$; that is, the force decreases in proportion to the inverse cube of the distance. Thus, although the force contributed by each point charge $\pm Q$ is an inverse-square force, the net force has a different behavior, because at large distances the force contributed by one charge tends to cancel the force contributed by the other. We will further discuss this $1/x^3$ behavior far from a pair of equal and opposite charges in Chapter 23.

If an arrangement of point charges is symmetric in some way, the calculation of the net force is often simplified. In the previous example, the position of the charge q, equidistant from the charges $+Q$ and $-Q$, resulted in a cancellation of one component of the force. For a more symmetric arrangement, the result can be even simpler, as in the following example.

Concepts — in — Context

EXAMPLE 7   Similar to the chapter photo, Fig. 22.9a is a scanning electron micrograph of toner particles. We see that the arrangement of toner particles is nearly such that one is at the center of a hexagon and the other six are at the vertices of the hexagon. If the toner particles carry equal charges and are arranged precisely at the center and vertices of a hexagon, what is the net force on the central charge? Assume for simplicity that each particle acts as a point charge located at the center of the particle.

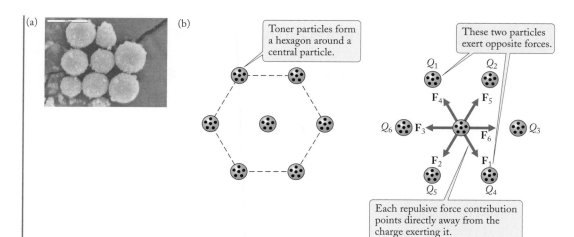

FIGURE 22.9 (a) Scanning electron micrograph of laser printer toner particles. (b) Electric forces on the central particle due to each of the other six.

SOLUTION: Like charges repel, so the force on the central particle due to any one of the other charges will point directly away from that charge. Since all charges are equal and the distance from the center to each vertex is the same, each of the force contributions from the outer six charges will have the same magnitude. Figure 22.9b shows the force vectors for the electric forces on the central particle due to each of the six surrounding toner particles. Inspecting this diagram, we see that the force vectors cancel pairwise; for example, F_1 exactly balances F_4. The net force on the central particle is zero.

Thus the mutual repulsion of the particles not only keeps them dispersed, as mentioned in Example 4, but also tends to keep them in equilibrium. Repulsive forces between particles, atoms, or other entities often result in stable, hexagonal structures, such as the examples shown in Fig. 22.10.

FIGURE 22.10 Because of mutual repulsion, many types of objects tend to form hexagonal arrangements: (a) billiard balls, (b) microscopic polystyrene beads, (c) superconducting vortices, and (d) atoms in an oxide crystal.

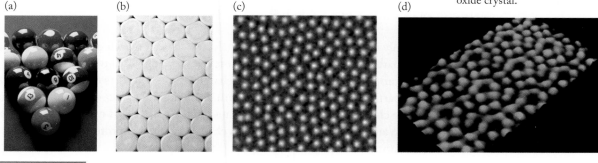

✔ Checkup 22.3

QUESTION 1: Suppose in Example 6 that instead of charges $\pm Q$, both charges Q were positive. What is the direction of the electric force F on the positive charge q in this case?

QUESTION 2: Three identical point charges are at the vertices of an equilateral triangle. A fourth, identical point charge is placed at the midpoint of one side of the triangle. As a result of the three electric force contributions from the vertex charges, the fourth charge

(A) Is in equilibrium and remains at rest
(B) Is pushed toward the center of the triangle
(C) Is pushed outside the triangle

| **PROBLEM-SOLVING TECHNIQUES** | **SUPERPOSITION OF ELECTRIC FORCES** |

- To find the total electric force on one charge due to an arrangement of several other point charges (such as the point charges $\pm Q$ of Example 6), you need to calculate the vector sum of the individual electric forces. The magnitude of the electric force on a point charge q due to each individual point charge q' is given by Eq. 22.7, $F = (1/4\pi\epsilon_0) \, qq'/r^2$; and the direction of this electric force is toward q' if q and q' have unlike signs, and away from q' if q and q' have like signs. The techniques for evaluating the vector sum of the electric force vectors of several charges are the same as for the sum of any other kind of vector.

- Recall the basic techniques for calculating vector sums: First, draw a careful diagram for the forces *on a given charge* q_1 *due to several other* charges q_2, q_3, . . ., with each electric force vector along the line through the

charge exerting the force and the charge acted upon. It may be easiest to put the tail of each contributing vector at the charge acted upon (as in Example 6), pointing away from the charge exerting the force for charges with like signs, and toward the charge exerting the force for unlike signs.

- Use geometry to decompose each vector into its x and y (and, if necessary, z) components. You can then obtain the total electric force vector by separately summing the x, y, and z components of the contributing vectors. You can obtain the magnitude of the total electric force in the usual way, from $F = \sqrt{F_x^2 + F_y^2 + F_z^2}$. In many examples of calculations of electric forces, some components of the force will cancel. If you select the x, y, and z axes judiciously, you can often achieve cancellation of all components except one (see Example 6).

22.4 CHARGE QUANTIZATION AND CONSERVATION

Not only electrons and protons exert electric forces on each other, but so do many other particles. The magnitudes of these electric forces are given by Eq. (22.7) with the appropriate values of the electric charges. Table 22.3 lists the electric charges of some particles; a more complete list will be found in Chapter 41. Antiparticles have electric charges that are opposite to those of the corresponding particles; for example, the antielectron (or positron) has charge $+e$, the antiproton has charge $-e$, the antineutron has charge 0, and so on.

All the known particles have charges that are some integer multiple of the fundamental charge; that is, the charges are always 0, $\pm e$, $\pm 2e$, $\pm 3e$, etc. Why no other charges exist is a mystery for which classical physics offers no explanation. Since charges exist in discrete packets, we say that charge is **quantized**—the fundamental charge e is called the quantum of charge. Thus any amount of charge that is ever encountered is an integer multiple of the fundamental charge e. Of course, as discussed in Section 22.1, the extreme smallness of the fundamental charge permits us to treat macroscopic charge distributions as continuous.

charge quantization

charge conservation

The electric charge is a **conserved quantity**: *in any reaction involving charged particles, the total charges before and after the reaction are always the same.* For instance, here is an example of a reaction in which particles are destroyed, yet the net electric charge remains constant:

matter–antimatter annihilation:

$$[\text{electron}] + [\text{antielectron}] \rightarrow 2[\text{photons}]$$

charges: $-e$ + e $\rightarrow$ 0   (22.13)

The same charge conservation holds for any reaction. No reaction that creates or destroys any net electric charge has ever been discovered.

Charge is of course also conserved in chemical reactions. For instance, in a lead–acid battery (automobile battery), plates of lead and of lead dioxide are immersed in an electrolytic solution of sulfuric acid (Fig. 22.11). The reactions that take place on these plates involve sulfate ions (SO_4^{2-}, where the superscript 2– indicates an ion with two extra electrons) and hydrogen ions (H^+); the reactions release electrons at the lead plate, and they absorb electrons at the lead dioxide plate:

at lead plate:

$$Pb + SO_4^{2-} \rightarrow PbSO_4 + 2[\text{electrons}]$$

$$\text{charges:} \quad 0 + (-2e) \rightarrow \quad 0 \quad + \quad (-2e)$$

(22.14)

at lead dioxide plate:

$$PbO_2 + 4H^+ + SO_4^{2-} + 2[\text{electrons}] \rightarrow PbSO_4 + 2H_2O$$

$$\text{charges:} \quad 0 \; + \; 4e \; + (-2e) + \quad (-2e) \quad \rightarrow \quad 0 \; + \; 0$$

(22.15)

The plates of such a battery are connected by an external circuit (a wire), and the electrons released by the reaction (22.14) travel from one plate to the other, forming an electric current (see Fig. 22.11).

TABLE 22.3	
ELECTRIC CHARGES OF SOME PARTICLES	
PARTICLE	**CHARGE**
Electron, e	$-e$
Muon, μ	$-e$
Pion, π^0	0
Pion, π^+	$+e$
Pion, π^-	$-e$
Proton, p	$+e$
Neutron, n	0
Neutrino, ν	0
Photon, γ	0
Delta, Δ^+	$+e$
Delta, Δ^{++}	$+2e$

EXAMPLE 8   A fully "charged" battery contains a large amount of sulfuric acid in the electrolytic solution (H_2SO_4 in the form of SO_4^{2-} ions and H^+ ions). As the battery delivers electric charge to the external circuit connecting its terminals, the amount of sulfuric acid in solution gradually decreases. Suppose that while discharging completely, the positive terminal of an automobile battery delivers an electric charge of 1.8×10^5 C through the external circuit. How many grams of sulfuric acid will be used up in this process?

SOLUTION: Let us first consider the number of electrons transferred, then the number of sulfate ions used up, and finally obtain the amount of sulfuric acid needed. Since the charge per electron is -1.6×10^{-19} C/electron, the number of electrons in -1.8×10^5 C is $(-1.8 \times 10^5 \text{ C})/(-1.6 \times 10^{-19} \text{ C/electron}) = 1.1 \times 10^{24}$ electrons. According to the reactions (22.14) and (22.15), whenever two electrons are transferred from the lead to the lead dioxide plate, two sulfate ions are absorbed (one at each plate). Thus, 1.1×10^{24} sulfate ions will be used up, which means 1.1×10^{24} molecules of sulfuric acid will be used up, since sulfuric acid has one sulfate ion per molecule (H_2SO_4). The required number of moles of sulfuric acid is therefore $(1.1 \times 10^{24} \text{ molecules})/(6.02 \times 10^{23} \text{ molecules/mole}) = 1.9$ moles, and, since the molecular mass of sulfuric acid is 98 grams per mole, the required mass of sulfuric acid is $(1.9 \text{ mole}) \times (98 \text{ g/mole}) = 183$ g.

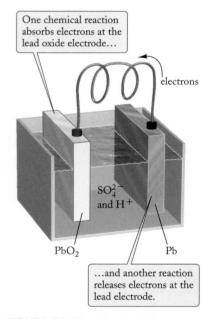

One chemical reaction absorbs electrons at the lead oxide electrode…

electrons

SO_4^{2-} and H^+

PbO_2

Pb

…and another reaction releases electrons at the lead electrode.

FIGURE 22.11 A lead–acid battery.

The conservation of electric charge in chemical reactions, such as the reactions (22.14) and (22.15), is a trivial consequence of the conservation of electrons and protons. All such reactions involve nothing but a rearrangement of the electrons and protons in the molecules; during this rearrangement, the numbers of electrons and of protons remain constant. Obviously, the net electric charge must then also remain constant.

The same argument applies to all macroscopic electric processes, such as the operation of electrostatic machines and generators, the flow of currents on wires, the storage of charge in capacitors, the electric discharge of thunderclouds, etc. All such processes involve nothing but a rearrangement of electrons and protons. Consequently, the net electric charge must remain constant.

✔ Checkup 22.4

QUESTION 1: If an atom loses two electrons, what is the electric charge of the resulting ion? If an atom loses three electrons?

QUESTION 2: Is it possible for a body to have an electric charge of 2.0×10^{-19} C? 3.2×10^{-19} C?

(A) Yes; yes   (B) Yes; no   (C) No; yes   (D) No; no

Online
Concept
Tutorial

conductor

insulator

22.5 CONDUCTORS AND INSULATORS; CHARGING BY FRICTION OR BY INDUCTION

A **conductor**—such as copper, aluminum, or iron—*is a material that permits the motion of electric charges through its volume.* An **insulator**—such as glass, porcelain, rubber, or nylon—*is a material that does not readily permit the motion of electric charges.* Thus, when we place some electric charge on one end of a conductor, it immediately spreads out over the entire conductor until it finds an equilibrium distribution. (We will study the conditions for the equilibrium of electric charge on a conductor in Chapter 24. It turns out that when the charges finally reach equilibrium, they will all be located on the surface of the conductor.) In contrast, when we place some charge on one end of an insulator, it stays in place (see Fig. 22.12).

All metals are good conductors. They readily permit the motion of electric charges, and therefore metallic wires are widely used in electric circuits, such as the circuits in your home, where copper wires serve as conduits for the flow of electric charge. The motion of charge in a metal is due to the motion of electrons. *In a metal, some of the electrons of each atom are free,* that is, they are not bound to any particular atom although they are bound to the metal as a whole. The free electrons come from the outer parts of the atoms. The outer electrons of the atom are not very strongly attached and readily come loose; the inner electrons are firmly bound to the nucleus of the atom and are likely to stay put. The free electrons wander through the entire volume of the metal, suffering occasional collisions, but they experience a restraining force only when they encounter the surface of the metal. The electrons are held inside the metal in much the same way as particles of a gas are held inside a container—the particles of gas can wander through the volume of the container, but they are restrained by the walls. In view of this analogy, electrons in a metallic conductor are often said to form a free-electron gas. If one end of a metallic conductor has an excess or deficit of electrons, the motion of the free-electron gas will quickly distribute this excess or deficit to other parts of the metallic conductor.

The charging of a body of metal is usually accomplished by the removal or the addition of electrons. A body will acquire a net positive charge if electrons are removed, and a net negative charge if electrons are added. Thus, positive charge on a body of metal is simply a deficit of electrons, and negative charge an excess of electrons.

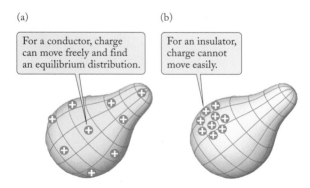

(a)

For a conductor, charge can move freely and find an equilibrium distribution.

(b)

For an insulator, charge cannot move easily.

FIGURE 22.12 (a) Charge placed on a conductor spreads out over the entire conductor. (b) Charge placed on an insulator stays at its original place.

PHYSICS IN PRACTICE   XEROGRAPHY

The process of xerography (from Greek words meaning "dry writing"), commonly used in photocopiers and laser printers, exploits electric forces to put marks on paper. The figures illustrate the process. First, in Fig. 1a, the surface of an insulating cylindrical drum is positively charged, either by contact with a positively charged rod or by proximity to an electron-attracting wire ("corona wire"). The drum has a special surface containing selenium or some other photoconductive material having the unusual property that it becomes a conductor when exposed to light. This property permits the creation of a temporary "charge image" on the drum by exposure to light from an original document in a photocopier, or from microprocessor-controlled laser light in a laser printer. The charge from the exposed, conductive regions is neutralized by electrons from the metal drum below the photoconductive surface layer, and only the dark, insulating areas maintain their layer of positive charge (Fig. 1b). Next, small glass or plastic carrier particles, each with many negatively charged toner particles attached, are sprinkled onto the drum; the toner particles are attracted and stick to the regions of the drum where the positively charged image resides (Fig. 1c). We have seen in Examples 4 and 7 that the mutual repulsion of the toner particles maintains a more or less uniform dispersion throughout the dark image areas. The paper receiving the final image is positively charged and attracts the toner particles upon contact, creating a delicate powder-on-paper dry image (Fig. 1d). Finally, the toner particles are "fixed" (Fig. 1e) by means of a hot roller that melts the black plastic toner particles so they bond with the paper.

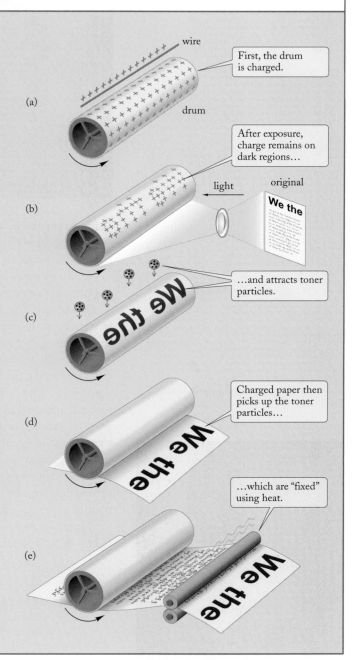

FIGURE 1 The xerographic process.

Liquids containing ions (atoms or molecules with missing electrons or with excess electrons) also are good conductors. For instance, a solution of common salt in water contains ions of Na^+ and Cl^-. The motion of charges through the liquid is due to the motion of these ions. *Liquid conductors with an abundance of ions are called* **electrolytes**.

    Incidentally: Very pure distilled water is a poor conductor because it lacks ions. But ordinary water is a good conductor because it contains some ions contributed by dissolved impurities. The ubiquitous water in our environment makes many substances into conductors. For example, earth (ground) is a reasonably good conductor, mainly because of the presence of the water. Furthermore, on a humid day many insulators

electrolyte

FIGURE 22.13 Lightning.

acquire a microscopic surface film of water, and this permits electric charge to leak away along the surface of the insulator; thus, on humid days, it is difficult to store electric charge on bodies supported by insulators.

Ordinary gases are insulators, but ionized gases are good conductors. For example, ordinary air is an insulator, but the ionized air found in a lightning bolt is a good conductor (see Fig. 22.13). This ionized air contains a mixture of positive ions and free electrons; the motion of charge in such a mixture is due mainly to the motion of the electrons. *Such an ionized gas is called a* **plasma**.

Although lightning gives us the most spectacular evidence of electrical activity in a thunderstorm, the largest part of the electrical activity in a thunderstorm proceeds silently and almost imperceptibly in the form of **corona discharge**. Charge on the ground, attracted to opposite charge in a thundercloud, concentrates at any sharp points—such as the tips of the leaves of trees. The concentration of charge causes a steady, nearly imperceptible ionization of the air near the sharp points. The charge from the ground can then leak into this ionized air and flow toward the thundercloud, whose electric charge it gradually neutralizes. This kind of ionization of air is called corona discharge because it can sometimes be seen as a glowing halo surrounding the pointed object (see Fig. 22.14).

Lightning rods, invented by Benjamin Franklin, were originally intended to inhibit lightning by facilitating corona discharge. However, the traditional lightning rod, with

plasma

corona discharge

FIGURE 22.14 Corona discharge from power lines.

BENJAMIN FRANKLIN (1706–1790)
American scientist, statesman, and inventor. He is most often remembered for his hazardous experiments with a kite in a thunderstorm, which demonstrated that lightning is an electric phenomenon, and for his invention of lightning rods. Franklin also made other significant contributions to the experimental and theoretical studies of electricity, and he was admired and honored by the leading scientific associations in Europe. Among these contributions were his formulation of the Law of Conservation of Electric Charge and his introduction of the modern notation for plus and minus charges, which he regarded as an excess or deficiency of "electric fluid."

a single sharp point, cannot produce a sufficient amount of corona discharge; its main benefit is to conduct the electric charge of a lightning stroke harmlessly to the ground. Some lightning rods of modern design are much better at producing corona discharge. These dissipative lightning rods end in a multitude of sharp points (see Fig. 22.15), which can dissipate a substantial fraction of the total charge of a thundercloud passing overhead.

We will end this chapter with a brief discussion of two different ways in which a body may be charged. The first, frictional electricity, has already been mentioned in the introduction. The second, charging by induction, is a simple application of the Coulomb force that utilizes the properties of a conductor. We discuss each in turn.

Frictional electricity is quite common. It is easy to accumulate electric charge on a glass rod merely by rubbing it with a piece of silk. The silk becomes negatively charged, and the glass positively charged—the rubbing motion between the surfaces of the silk and the glass rips charges off one of these surfaces and makes them stick to the other, but the detailed mechanism is not well understood. It is believed that what is usually involved is a transfer of ions from one surface to the other. Contaminants residing on the rubbed surfaces play a crucial role in frictional electricity. If glass is rubbed with an absolutely clean piece of silk or other textile material, the glass becomes negatively charged rather than positively charged. Ordinary pieces of silk apparently have such a large amount of dirt on their surfaces that the charging process is dominated by the dirt rather than by the silk. Even air can act as contaminant for some surfaces; for instance, careful experiments on the rubbing of platinum with silk show that in vacuum the platinum becomes negatively charged, but in air it becomes positively charged.

The electric charge that can be accumulated on the surface of a body of ordinary size (a centimeter or more) by rubbing may be as much as 10^{-9} or 10^{-8} coulomb per square centimeter. If the charge concentration on a body is higher than that, it will cause an electric discharge into the surrounding air (a corona discharge).

A second, quite different way to charge an object is called **charging by induction**. Once we have accumulated some charge, say, positive charge on a rod of glass, we can produce charges on other bodies by the induction process, as follows: First we bring the glass rod near a metallic body supported on an insulating stand (see Fig. 22.16a). The positive charge on the rod will then attract free electrons to the near side of the body and leave a deficit of free electrons on the far side; thus, the near side will acquire negative charge and the far side positive charge. If we next connect the far side to the ground with a wire, electrons will flow from the ground, attracted by the positive charge, which they neutralize (Fig. 22.16b). We then disconnect the wire; this leaves the metallic body with a net negative charge (Fig. 22.16c). When we finally withdraw the glass rod, the induced charge will remain on the metallic body, distributing itself over the entire surface, as in Fig. 22.16d. Thus the positive glass rod "induces" a charge distribution (and, ultimately, a net negative charge) on the metal sphere without actually making contact with the sphere.

FIGURE 22.15 Modern dissipative lightning rod. The rod ends in a tassel of fine wires, which provide many sharp points for corona discharge.

charging by induction

FIGURE 22.16 Charging by induction. (a) The positively charged glass rod induces a charge distribution on a neutral metallic sphere. (b) When the far side of the sphere is connected to the ground by means of a wire, electrons move from ground to the sphere and neutralize the positive charge. (c) When the grounding wire is removed, a net negative charge remains. (d) When the glass rod is moved away, the net negative charge spreads over the sphere.

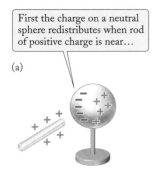

First the charge on a neutral sphere redistributes when rod of positive charge is near…

(a)

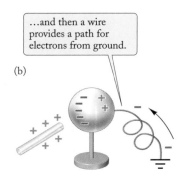

…and then a wire provides a path for electrons from ground.

(b)

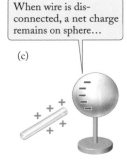

When wire is disconnected, a net charge remains on sphere…

(c)

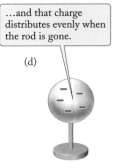

…and that charge distributes evenly when the rod is gone.

(d)

Checkup 22.5

QUESTION 1: What is the difference between an ordinary liquid and an electrolyte? What is the difference between an ordinary gas and a plasma?

QUESTION 2: Figure 22.16 shows a conducting body mounted on an insulating stand. Suggest some material for making the conducting body and some material for the insulating stand.

QUESTION 3: Suppose you have two metallic balls of equal size, both like the one illustrated in Fig. 22.16d, one with a charge of $+1 \times 10^{-7}$ C, and the other with a charge of -3×10^{-7} C. If you touch them together, what will be the resulting charge on each ball?

(A) There will remain $+1 \times 10^{-7}$ C on one and -3×10^{-7} C on the other
(B) There will be 0 C on one and -2×10^{-7} C on the other
(C) There will be -1×10^{-7} C on each

SUMMARY

PROBLEM-SOLVING TECHNIQUES Superposition of Electric Forces	**(page 706)**
PHYSICS IN PRACTICE Xerography	**(page 709)**
ELECTRIC CHARGES May be positive, negative, or zero; like charges repel, unlike charges attract.	
SI UNIT OF CHARGE	1 coulomb = 1 C
FUNDAMENTAL CHARGE, OR CHARGE OF PROTON	$e = 1.60 \times 10^{-19}$ C **(22.1)**
CHARGE OF ELECTRON	$-e = -1.60 \times 10^{-19}$ C **(22.1)**

COULOMB'S LAW   Direction of Coulomb force is along the line joining the particles.

$$F = \frac{1}{4\pi\epsilon_0} \frac{qq'}{r^2}$$ **(22.7)**

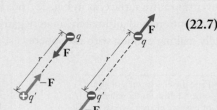

PERMITTIVITY CONSTANT (Electric constant)	$\epsilon_0 = 8.85 \times 10^{-12}$ C^2/N·m^2 **(22.6)**
COULOMB CONSTANT	$k = \dfrac{1}{4\pi\epsilon_0} = 8.99 \times 10^9$ N·m^2/C^2 **(22.4, 22.5)**

SUPERPOSITION PRINCIPLE   Net force is vector sum of individual forces.

$$\mathbf{F} = \mathbf{F}_1 + \mathbf{F}_2 + \mathbf{F}_3 + \cdots$$ (22.9)

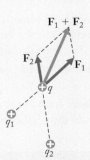

CHARGE CONSERVATION   In any reaction or process, the net electric charge remains constant.

CHARGE QUANTIZATION   Any charge is an integer multiple of the fundamental charge.

$$q = 0, \pm e, \pm 2e, \pm 3e, \cdots$$

ION   An atom with net charge (missing or extra electrons).

ELECTROLYTE   A liquid with many dissolved ions.

PLASMA   A gas with many ionized atoms and free electrons.

CONDUCTOR   Permits the motion of charge.

INSULATOR   Does not permit the motion of charge.

conductor   insulator

QUESTIONS FOR DISCUSSION

1. Suppose that the Sun has a positive electric charge and each of the planets a negative electric charge, and suppose that there is no gravitational force. In what way will the motions of the planets predicted by this "electric" model of the Solar System differ from the observed motions?

2. Describe how you would set up an experiment to determine whether the electric charges on an electron and a proton are exactly the same.

3. Assume that neutrons have a small amount of electric charge, say, positive charge. Discuss some of the consequences of this assumption for the behavior of matter.

4. If we were to assign a positive charge to the electron and a negative charge to the proton, would this affect the mathematical statement [Eq. (22.7)] of Coulomb's Law?

5. In the cgs, or Gaussian, system of units, Coulomb's Law is written as $F = qq'/r^2$. In terms of grams, centimeters, and seconds, what are the units of electric charge in this system?

6. Could we use the electric charge of an electron as an atomic standard of electric charge to define the coulomb? What would be the advantages and disadvantages of such a standard?

7. Besides electric charge, what other physical quantities are conserved in reactions among particles? Which of these quantities are quantized?

8. Since the free electrons in a piece of metal are free to move any which way, why don't they all fall to the bottom of the piece of metal under the influence of the pull of gravity?

9. If the surface of a piece of metal acts like a container in confining the free electrons, why can't we cause these electrons to spill out by drilling holes in the surface?

10. Water is a conductor, but (dry) snow is an insulator. How can this be?

11. If you rub a plastic comb, it will attract hairs or bits of paper (Fig. 22.17), even though they have no net electric charge. Explain.

FIGURE 22.17 The electric charge on this plastic comb attracts small bits of paper.

12. Some old-fashioned physics textbooks define positive electric charge as the kind of charge that accumulates on a glass rod when rubbed with silk. What is wrong with this definition?

13. When you rub your shoes on a carpet, you sometimes pick up enough electric charge to feel an electric shock if you subsequently touch a radiator or some other metallic body connected to the ground. Why is this more likely to happen in winter than in summer?

14. Some automobile operators hang a conducting strap on the underside of their automobiles, so that this strap drags on the street. What is the purpose of this arrangement?

15. An amount of electric charge has been deposited on a Ping-Pong ball. How can you find out whether the charge is positive or negative?

16. Two aluminum spheres of equal radii hang from the ceiling on insulating threads. You have a glass rod and a piece of silk. How can you give these two spheres exactly equal amounts of electric charge?

PROBLEMS

22.1 The Electrostatic Force[†]
22.2 Coulomb's Law[‡]

1. Within a typical thundercloud there are electric charges of -40 C and $+40$ C separated by a vertical distance of 5.0 km (Fig. 22.18). Treating these charges as pointlike, find the magnitude of the electric force of attraction between them.

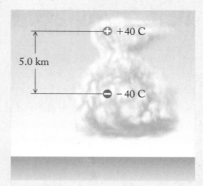

FIGURE 22.18 Charges in a thundercloud.

2. A crystal of NaCl (common salt) consists of a regular arrangement of ions of Na^+ and Cl^-. The distance from one ion to its neighbor is 2.82×10^{-10} m. What is the magnitude of the electric force of attraction between the two ions? Treat the ions as point charges.

3. Suppose that the two protons in the nucleus of a helium atom are at a distance of 2.0×10^{-15} m from each other. What is the magnitude of the electric force of repulsion that they exert on each other? What would be the acceleration of each if this were the only force acting on them? Treat the protons as point charges.

4. An alpha particle (a helium nucleus with charge $+2e$) is launched at high speed toward a nucleus of uranium (charge $+92e$). What is the magnitude of the electric force on the alpha particle when it is at a distance of 5.0×10^{-14} m from the nucleus? What is the corresponding instantaneous acceleration of the alpha particle? Treat the alpha particle and the nucleus as point charges.

5. According to recent theoretical and experimental investigations, the subnuclear particles are made of quarks and of antiquarks (see Chapter 41). For example, a positive pion is made of a u quark and a d antiquark. The electric charge on the u quark is $\frac{2}{3} e$ and that on the d antiquark is $\frac{1}{3} e$. Treating the quarks as classical particles, calculate the electric force of repulsion between the quarks in the pion if the distance between them is 1.0×10^{-15} m.

[†]For help, see Online Concept Tutorial 25 at www.wwnorton.com/physics
[‡]For help, see Online Concept Tutorial 24 at www.wwnorton.com/physics

6. How many electrons do you need to remove from an initially neutral bowling ball to give it a positive electric charge of 1.0×10^{-6} C?

7. A lightning stroke typically deposits -25 C on the ground. How many electrons is this?

8. The mass of the electron cannot be measured directly, since macroscopic amounts of mass always contain a combination of electrons, protons, and neutrons, never pure electrons. Instead, the mass is calculated from a measurement of the electric charge $-e$ of the electron and a measurement of the charge-to-mass ratio $-e/m_e$. The best values for these quantities are $-e = -1.602\,177 \times 10^{-19}$ C and $-e/m_e = -1.758\,820 \times 10^{11}$ C/kg. What best value of the mass of the electron can you deduce from this?

9. In an HCl molecule, the nuclei of the H and the Cl atoms, with charges $+e$ and $+17e$, respectively, are separated by a distance of 1.28×10^{-10} m. What is the electric force of repulsion between these nuclei?

10. Consider two protons separated by a distance of 1.0×10^{-12} m.

    (a) What is the gravitational force of attraction between the protons?

    (b) What is the electric force of repulsion between these protons? What is the ratio of the electric force and the gravitational force?

    (c) Consider a second pair of protons separated by a larger distance, so their electric repulsion matches the gravitational attraction of the other, closer pair calculated in part (a). How far apart would these protons have to be?

11. According to recent theoretical speculations, there might exist an elementary particle of mass 1.0×10^{-11} kg. If such a particle carries a charge e, what is the gravitational force and what is the electric force on an identical particle placed at a distance of 1.0×10^{-10} m?

12. In the lead atom, the nucleus has an electric charge $82e$. The innermost electron in this atom is typically at a distance of 6.5×10^{-13} m from the nucleus. What is the electric force that the nucleus exerts on such an electron? What is the acceleration that this force produces on the electron? Treat the electron as a classical particle.

13. The electric charge in one mole of protons is called **Faraday's constant**. What is its numerical value?

14. The electric charge flowing through an ordinary 115-volt, 150-watt lightbulb is 1.3 C/s. How many electrons per second does this amount to?

15. A maximum electric charge of 7.5×10^{-6} C can be placed on a metallic sphere of radius 15 cm before the surrounding air suffers electric breakdown (sparks). How many excess electrons (or missing electrons) does the sphere have when breakdown is about to occur?

16. How many electrons are in a paper clip of iron of mass 0.30 g?

17. Suppose that you remove all the electrons in a copper penny of mass 2.7 g and place them at a distance of 2.0 m from the remaining copper nuclei. What is the electric force of attraction on the electrons?

18. What is the number of electrons and of protons in a human body of mass 73 kg? The chemical composition of the body is roughly 70% oxygen, 20% carbon, and 10% hydrogen (by mass).

19. It is possible to dissolve 36 g of sodium chloride (table salt) in 100 g of water. By what factor does the number of electrons (or protons) in the solution exceed that of the water alone?

20. An introductory physics laboratory experiment uses two small spheres, each charged with -2.0×10^{-6} C of charge. What is the electric force between the spheres when they are 1.0 m apart?

21. The value of the Coulomb constant is defined to be exactly $k = 1/4\pi\epsilon_0 = 8.987\,551\,787 \times 10^9$ N·m^2/C^2. What is the approximate difference (in percent) between the simple number 9.0×10^9 N·m^2/C^2 and the exact value of the Coulomb constant? (For most calculations, the simple value suffices.)

22. From far away, any charge distribution with a net charge behaves more or less like a point charge. Consider two thin disks, each of radius 1.0 cm. Each disk has a charge per unit area of 2.5×10^{-8} C/m^2. What is the electric force between the disks when they are separated by 2.0 m?

*23. A long, linear organic molecule is initially 1.9 μm in length. An atom at each end is then singly ionized; overall, the molecule remains neutral. The two ionizations result in a length change of -1.2%. What is the effective spring constant for this molecule?

*24. Deimos is a small moon of Mars, with a mass of 2.0×10^{15} kg. Suppose that an electron is at a distance of 100 km from Deimos. What is the gravitational attraction acting on the electron? What negative electric charge would have to be placed on Deimos to balance this gravitational attraction? How many electron charges does this amount to? Treat the mass and the charge distribution as pointlike in your calculations.

*25. A small charge of -2.0×10^{-8} C is at the point $x = 2.0$ m, $y = 0$ on the x axis. A second small charge of -3.0×10^{-6} C is at the point $x = 0, y = -3.0$ m on the y axis (see Fig. 22.19). What is the electric force that the first charge exerts on the second? What is the force that the second charge exerts on the first? Express your answers as vectors, with x and y components.

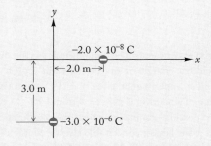

FIGURE 22.19 Two pointlike charges.

*26. Two tiny chips of plastic of masses 5.0×10^{-5} g are separated by a distance of 1.0 mm. Suppose that they carry equal and opposite electrostatic charges. What must the magnitude of the charge be if the electric attraction between them is to equal their weight?

*27. How many extra electrons would we have to place on the Earth and on the Moon so that the electric repulsion between these bodies cancels their gravitational attraction? Assume that the numbers of extra electrons on the Earth and on the Moon are in the same proportion as the radial dimensions of these bodies (6.38:1.74).

*28. At a place directly below a thundercloud, the induced electric charge on the surface of the Earth is $+1.0 \times 10^{-7}$ coulomb per square meter of surface. How many singly charged positive ions per square meter does this represent? The number of atoms on the surface of a solid is typically 2.0×10^{19} per square meter. What fraction of these atoms must be ions to account for the above electric charge?

*29. Although the best available experimental data are consistent with Coulomb's Law, they are also consistent with a modified Coulomb's Law of the form

$$F = \frac{1}{4\pi\epsilon_0} \frac{q_1 q_2}{r^2} e^{-r/r_0}$$

where r_0 is a constant with the dimensions of length and a numerical value that is known to be no less than 10^9 m and is probably much larger. Here, e is the base of the natural logarithms. Assuming that $r_0 = 1.0 \times 10^9$ m, what is the fractional deviation between Coulomb's Law and the modified Coulomb's Law for $r = 1.0$ m? For $r = 1.0 \times 10^4$ m? (Hint: Use the approximation $e^x \approx 1 + x$ for small x.)

*30. A proton is at the origin of coordinates. An electron is at the point $x = 4.0 \times 10^{-11}$ m, $y = 2.0 \times 10^{-11}$ m in the x–y plane (see Fig. 22.20). What are the x and y components of the electric force that the proton exerts on the electron? That the electron exerts on the proton?

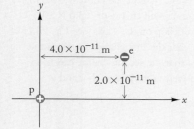

FIGURE 22.20
A proton and an electron.

*31. Precise experiments have established that the magnitudes of the electric charges of an electron and a proton are equal to within an experimental error of $\pm 10^{-21}e$ and that the electric charge of a neutron is zero to within $\pm 10^{-21}e$. Making the worst possible assumption about the combination of errors, what is the largest conceivable electric charge of an oxygen atom consisting of 8 electrons, 8 protons, and 8 neutrons? Treating the atoms as point particles, compare the electric

force between two such oxygen atoms with the gravitational force between these atoms. Is the net force attractive or repulsive?

*32. Under the influence of the electric force of attraction, the electron in a hydrogen atom orbits around the proton on a circle of radius 5.3×10^{-11} m. What is the orbital speed? What is the orbital period?

22.3 The Superposition of Electric Forces[†]

33. Suppose that in Example 6 both charges Q are positive. What are the magnitude and direction of the electric force on the electric charge q in this case?

34. The distribution of electric charges in a thundercloud can be approximated by several pointlike charges placed at different heights. Suppose that a thundercloud has electric charges of +10 C, −40 C, and +40 C at altitudes of 2.0 km, 5.0 km, and 10 km, respectively (see Fig. 22.21). Treating these charges as pointlike, find the net electric force that the two charges of ±40 C exert on the charge of +10 C.

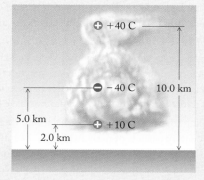

FIGURE 22.21 Charges in a thundercloud.

35. Figure 22.22 shows the arrangement of nuclear charges (positive charges) in an HCl molecule. The magnitudes of the H and Cl nuclear charges are e and $17e$, respectively, and the distance between them is 1.28×10^{-10} m. What is the net electric force that these charges exert on an electron placed 5.0×10^{-11} m above the H nucleus?

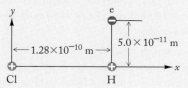

FIGURE 22.22 The positive nuclear charges in the chlorine and hydrogen atoms exert electric forces on an electron.

[†]For help, see Online Concept Tutorial 26 at www.wwnorton.com/physics

36. Five identical charges $+Q$ are at the vertices of a pentagon. What is the net electric force due to those five charges on a sixth charge $+q$ at the center of the pentagon?

37. Suppose that two balls of mass 2.5×10^{-4} kg each carry equal charges and are suspended by identical threads of length 10 cm, similar to Example 5; however, these threads are anchored at points 25 cm apart. If each thread makes an angle of 20° with the vertical, what is the charge on each ball?

*38. Point charges $+Q$ and $-2Q$ are separated by a distance d. A point charge q is equidistant from these charges, at a distance x from their midpoint (see Fig. 22.23). What is the electric force on q?

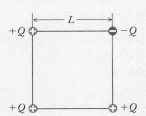

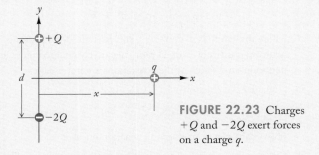

FIGURE 22.23 Charges $+Q$ and $-2Q$ exert forces on a charge q.

*39. Three positive point charges $+Q$ are placed at three corners of a square, and a negative point charge $-Q$ is placed at the fourth corner (see Fig. 22.24). The side of the square is L. Calculate the net electric force that the positive charges exert on the negative charge.

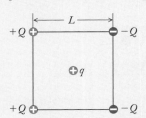

FIGURE 22.24 Three positive point charges and one negative point charge.

*40. Four point charges of $\pm Q$ are arranged on the corners of a square of side L as illustrated in Fig. 22.25. What is the net electric force that these charges exert on a point charge q placed at the center of the square?

FIGURE 22.25 Five point charges.

*41. Figure 22.26 shows the approximate charge distribution in a thundercloud consisting of a pointlike charge of $+40$ C at a height of 10.0 km and a pointlike charge of -30 C at a height of 4.0 km. What is the force that these two charges exert on

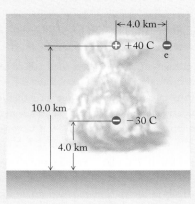

FIGURE 22.26 Charges in a thundercloud.

an electron located at a height of 10.0 km and at a horizontal distance of 4.0 km to the right of the charges?

*42. Repeat the preceding problem but with the electron at a height of 7.0 km and at a horizontal distance of 4.0 km to the right.

*43. Suppose that two balls are suspended by identical threads of length 10 cm anchored at the same point, similar to Example 5, except that these balls have different masses and charges. When one ball carries a charge of $+2.0 \times 10^{-7}$ C and the other a charge of $+6.0 \times 10^{-8}$ C, the threads each make the same equilibrium angle of 25° with the vertical. What is the mass of each ball?

*44. Three charges ($+q$, $+q$, and $-q$) have equal magnitudes and are located at the vertices of an equilateral triangle. Find the magnitude of the total force on one of the positive charges due to the other two charges.

*45. Two equal charges $+Q$ are at two vertices of an equilateral triangle of side a; a third charge $-q$ is at the other vertex. A fourth charge q_0, located a distance $a/2$ outside the triangle along the perpendicular bisector of the $+Q$ charges (see Fig. 22.27), experiences zero net force. Find the value of the ratio q/Q.

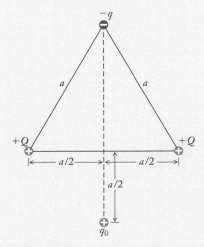

FIGURE 22.27 Three charges $+Q$, $+Q$, and $-q$ exert forces on a fourth charge q_0.

*46. Eight equal charges q are located at the corners of a cube of side a. Find the magnitude of the total force on one of the charges due to the other seven charges.

*47. Two point charges $+Q$ and $-Q$, separated a distance d (an **electric dipole**), are on the x axis at $x = +d/2$ and $x = -d/2$, respectively. Find an expression for the net force on a third charge $+q$ also on the x axis at $x > d/2$. Simplify your result and obtain the form of the approximate net force for $x \gg d$. Compare your result with that discussed in the Comment of Example 6.

*48. A thin rod of length L is placed near a point charge q (see Fig. 22.28), with the nearest end a distance d from the charge and oriented radially as shown. The rod carries a uniform distribution of charge λ coulombs per meter. Find the electric force that acts on the rod. (Hint: Sum the force contributions $dF = kq\,dq'/r^2$ due to each small charge $dq' = \lambda\,dx$ on the rod to obtain the total force $F = \int dF$.)

FIGURE 22.28 A charged rod and a point charge q.

**49. Four equal point charges $+Q$ are located at the vertices of a regular tetrahedron of side a. What is the force on one of the charges due to the other three?

**50. Two thin rods of length L carry equal charges Q uniformly distributed over their lengths. The rods are aligned, and their nearest ends are separated by a distance x (see Fig. 22.29). What is the electric force of repulsion between these two rods? (Hint: Sum the force contributions $dF = kdq_1\,dq_2/r^2$ due to each pair of small charges $dq_1 = (Q/L)\,dx_1$ and $dq_2 = (Q/L)\,dx_2$ to obtain the total force $F = \iint dF$.)

FIGURE 22.29 Two aligned charged rods.

**51. Suppose that in Example 6 both charges Q are positive. Find the value of x for which the force on q has its maximum value. (Hint: Be careful to put all varying quantities in terms of x before maximizing the force.)

**52. Suppose that in Example 6 both charges Q are positive but the other charge is negative, $-q$.

(a) What are the magnitude and direction of the electric force on $-q$ in this case?

(b) For small values of x, show that this force is proportional to x.

(c) For such a force, one expects simple harmonic motion. What is the period of such motion? (Assume that the charge $-q$ has a mass m.)

22.4 Charge Quantization and Conservation

53. Consider the following hypothetical reactions involving the collision between a high-energy proton (from an accelerator) and a stationary proton (in the nucleus of a hydrogen atom serving as target):

$$p + p \rightarrow n + n + \pi^+$$
$$p + p \rightarrow n + p + \pi^0$$
$$p + p \rightarrow n + p + \pi^+$$
$$p + p \rightarrow p + p + \pi^0 + \pi^0$$
$$p + p \rightarrow n + p + \pi^0 + \pi^-$$

where the symbols p, n, π^+, π^-, and π^0 stand for a proton, neutron, positively charged pion, negatively charged pion, and neutral pion. Which of these reactions are impossible?

54. Consider the reaction

$$Ni^{2+} + 4H_2O \rightarrow NiO_4^{2-} + 8H^+ + [\text{electrons}]$$

How many electrons does this reaction release?

55. Lithium ions are often dissolved for use in electrolytes. The reactions in a rechargeable lithium–cobalt (Li–Co) battery can be represented as

$$Li \rightarrow Li^+ + 1[\text{electron}]$$

at the electron-releasing lithium plate and

$$Co^{4+} + N[\text{electrons}] \rightarrow Co^{3+}$$

at the electron-absorbing cobalt-based plate. Use charge balance to determine the number N of electrons absorbed per cobalt atom during the reaction.

56. A spherical shell has net charge only on its inner and outer surfaces. The total charge on the entire shell is $Q_\text{total} = -1.0 \times 10^{-8}$ C. The charge on the inner surface of the shell is $Q_\text{inner} = +2.0 \times 10^{-8}$ C. What charge is on the outer surface of the shell?

*57. We can silver-plate a metallic object, such as a spoon, by immersing the spoon and a bar of silver (Ag) in a solution of silver nitrate ($AgNO_3$). If we then connect the spoon and the silver

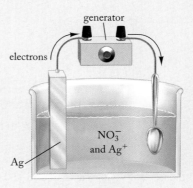

FIGURE 22.30 Silver-plating a spoon.

bar to an electric generator and make a current flow from one to the other, the following reactions will occur at the immersed surfaces (Fig. 22.30):

$$Ag^+ + [electron] \rightarrow Ag_{(metal)}$$

$$Ag_{(metal)} \rightarrow Ag^+ + [electron]$$

The first reaction deposits silver on the spoon, and the second removes silver from the silver bar. How many electrons must we make flow from the silver bar to the spoon in order to deposit 1.0 g of silver on the spoon?

REVIEW PROBLEMS

58. By rubbing a small glass ball against a small nylon ball, you deposit a charge of 6.0×10^{-11} C on the glass ball and a charge of -6.0×10^{-11} C on the nylon ball. If you then separate the balls to a distance of 20 cm, what is the attractive electric force between them?

59. Suppose that two grains of dust of equal masses each have a single electron charge. What must be the masses of the grains if their gravitational attraction is to balance their electric repulsion?

60. The arm in Coulomb's torsion balance was a rod with a charged ball at one end and a counterweight on the other (see Fig. 22.31). Suppose that the length of the arm in such a balance is 15 cm. Suppose that the probe is at a distance of 3.0 cm in a direction perpendicular to the arm. If the charged ball and the probe both carry $+2.0 \times 10^{-9}$ C, what is the torque exerted on the balance arm?

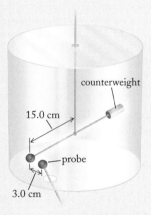

FIGURE 22.31 The arm in Coulomb's balance.

61. Three identical positive point charges $+Q$ are at the vertices of an equilateral triangle. A negative point charge is at the center of the triangle. The four charges are in equilibrium. What is the value of the negative charge?

62. Two small balls, each of mass 2.0×10^{-4} kg, carry opposite charges of equal magnitude. The balls are suspended by identical threads of length 10 cm, similar to Example 5;

however, these threads are anchored at points 20 cm apart. If the equilibrium separation of the balls is 10 cm, what is the magnitude of the charge on each ball?

63. Water drops in thunderclouds carry electric charges. Suppose that two such drops are falling side by side separated by a horizontal distance of 1.0 cm. Each drop has a radius of 0.5 mm and carries a charge of 2.0×10^{-11} C. What is the electric repulsive force on each drop? What is the instantaneous horizontal acceleration of each?

64. Suppose that during a thunderstorm, the corona discharge from a dissipative lightning rod into the surrounding air amounts to 1.0×10^{-4} C of positive charge per second. If this discharge goes on more or less steadily for an hour, how much electric charge flows out of the lightning rod? How many electrons flow into the lightning rod from the surrounding air?

65. Two small balls of plastic carry equal charges of opposite signs and of unknown magnitudes. When the balls are separated by a distance of 18 cm, the attractive force between them is 0.30 N. What is the excess of electrons on one ball and the deficit of electrons on the other?

66. A different version of the electroscope discussed in Example 5 uses a fixed cork ball and a suspended cork ball (see Fig. 22.32). The mass of the suspended ball is 1.5×10^{-4} kg, and the length of the suspension thread is 10 cm. The fixed ball is

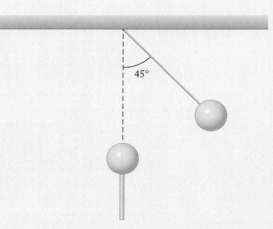

FIGURE 22.32 A charge suspended by a thread and a fixed charge.

located 10 cm directly below the point of suspension if the suspended ball. Assume that when equal electric charges are placed on the two balls, the electric repulsive force pushes the suspended ball up so its thread makes an angle of 45° with the vertical. What is the magnitude of the electric charge?

67. In each of the following decay reactions of elementary particles there is a missing particle. What is its electric charge?

$$n \rightarrow p + e + ?$$
$$\Delta^{++} \rightarrow p + \pi^0 + ?$$
$$\Delta^+ \rightarrow n + ?$$
$$\pi^- \rightarrow \mu^- + ?$$

Answers to Checkups

Checkup 22.1

1. Planets have large masses, and so exert large gravitational forces. However, planets have negligible net charge (if any); overall, they are electrically neutral. In an atom, on the other hand, the electrons have (relatively) large charges and (relatively) small masses.

2. The ground exerts the normal "contact" force on the stone, which is electric in origin; for a stone at rest, the magnitude of the contact force is the same as the gravitational force it cancels: $F = mg = (1.0 \text{ kg})(9.8 \text{ m/s}^2) = 9.8 \text{ N}$.

3. The signs of the charges that are, on average, closest to each other determine the direction of the net force. Thus with two nuclei (like charges) closest to each other, the net electric force is repulsive. The overall effect is somewhat subtle, as it depends on summing four forces.

4. (C) 1.0 C. The addition of a few elementary charges will cause only negligible change (here, in the eighteenth decimal place) in the net charge of any macroscopic quantity of charge, such as one coulomb.

Checkup 22.2

1. For point charges, the two charges must have opposite signs.

2. Since the electric force varies in proportion to the inverse square of distance, $1/r^2$, the increase from 1 m to 10 m will decrease the force by a factor of 100 to 1×10^{-6} N, and the increase from 1 m to 100 m will result in a force of 1×10^{-8} N.

3. Consider any two equal charges q, with an electric force $F = kq^2/r^2$. If we transfer a charge δ from one to the other, the charges become $(q + \delta)$ and $(q - \delta)$, so the force becomes $F' = k(q + \delta)(q - \delta)/r^2$. But $(q + \delta)(q - \delta) = q^2 - \delta^2$, which is less than q^2, so the force decreases.

4. (B) 10 μC. For the given force and distance, and with $q' = 10q$, Coulomb's Law gives $1.0 \text{ N} = F = kqq'/r^2 = (9.0 \times 10^9 \text{ N·m}^2/\text{C}^2) \times 10q^2/(3.0 \text{ m})^2 = 1.0 \times 10^{10} \text{ N/C}^2 \times q^2$. Thus $q = \sqrt{1.0 \times 10^{-10} \text{ C}^2} = 1.0 \times 10^{-5} \text{ C} = 10 \ \mu\text{C}$.

Checkup 22.3

1. In this case, the two forces on q are both away from the Q charges, so now the $+y, -y$ contributions cancel and the two $+x$ contributions add. Thus the total force is to the right, in the $+x$ direction (sketch a quick vector diagram to convince yourself of this).

2. (C) Is pushed outside the triangle. Like charges repel. The forces from the two charges at vertices adjacent to the fourth charge are equal but opposite and cancel. The net force is thus the same as the contribution from the charge at the opposite vertex; that repulsion will push the fourth charge away, out of the triangle.

Checkup 22.4

1. If an atom (which is neutral, in contrast to an ion) loses two electrons, it leaves behind an ion with a net positive charge of $+2e = +3.2 \times 10^{-19}$ C. If three electrons are lost, the remaining ion has a charge $+3e = +4.8 \times 10^{-19}$ C.

2. (C) No; yes. The charge of any body must be an integral multiple of the fundamental charge, such as $e, 2e, 3e = 1.6 \times 10^{-19}$ C, 3.2×10^{-19} C, 4.8×10^{-19} C. Thus the first charge given is impossible, while the second charge given, twice the fundamental charge, is possible—indeed, it is common.

Checkup 22.5

1. An electrolyte has an abundance of ions and is a good conductor; an ordinary liquid has few or no ions and is a poor conductor or an insulator. Similarly, a plasma has an abundance of ions and is a conductor, while an ordinary gas does not.

2. A good material for the conductor would be a metal such as aluminum or copper; for the stand, glass, as well as most rubber or plastic, would be a good insulator.

3. (C) There will be -1×10^{-7} C on each. Since the balls are metal, charge is free to move. When touching, the arrangement is symmetric (balls of equal size), and so the net charge will distribute evenly between the two balls. The net charge is $(+1 - 3) \times 10^{-7}$ C $= -2 \times 10^{-7}$ C, so the charge on each will be -1×10^{-7} C.

The Electric Field

CONCEPTS IN CONTEXT

Lightning is a spectacular visual result of the electric forces generated by the charge distribution in a thundercloud. Electric forces accelerate electrons to high speeds; the electrons ionize atoms, resulting in an avalanche of moving charge, a lightning bolt. In this chapter, we introduce the *electric field,* which is the electric force per unit charge.

With this and other concepts developed in this chapter, we can investigate questions such as

? What is the approximate distribution of charge in a thundercloud, and what electric field does it produce some distance from the cloud? (Example 2, page 725)

? How does the charge distribution in a thundercloud affect the arrangement of charges on the (conducting) ground below the cloud? What is the resulting electric field at ground level? (Example 3, page 727)

? What is the electric force on an electron at ground level? What is the resulting acceleration of the electron? (Example 4, page 728)

U p to here we have taken the view that the gravitational forces and the electric forces between particles are **action-at-a-distance**, that is, a particle exerts a direct gravitational or electric force on another particle even when the particles are widely separated. Such an interpretation of gravitational and electric forces as a ghostly tug-of-war between distant bodies is suggested by Newton's Law of Gravitation and Coulomb's Law. However, according to the modern view, there is a physical entity that acts as mediator of force, conveying the force over the distance from one body to another. This entity is the **field**. A gravitating or electrically charged body generates a gravitational or electric field which permeates the (apparently) empty space around the body, and this field exerts pushes or pulls whenever it comes in contact with another body. Thus, fields convey forces from one body to another through **local action**, or **action-by-contact**.

In the present chapter, we will become acquainted with the electric field that conveys the electric force from one body to another body. We will first examine the electric field due to a point charge. Then, using a superposition of electric fields similar to the superposition of electric forces, we will consider the electric field generated by several point charges, and by continuous distributions of charges. We will also introduce the useful concept of lines of electric field. Finally, we will consider what motion occurs when the electric field acts on a charge.

action-at-a-distance

field

action-by-contact

23.1 THE ELECTRIC FIELD OF POINT CHARGES

Online *Concept* Tutorial **27**

According to the naive interpretation of Coulomb's Law, the electric forces between charges are action-at-a-distance, that is, a charge q' exerts a direct force on a charge q even though these charges are separated by a large distance and are not touching. However, such an action-at-a-distance interpretation of electric forces leads to serious difficulties in the case of moving charges. Suppose that we suddenly move the charge q' somewhat nearer to the charge q; then the electric force on q has to increase, according to the inverse-square law. But the required increase cannot occur instantaneously—the increase can be regarded as a signal from q' to q, and it is a fundamental principle of physics, based on the theory of relativity, that no signal can propagate faster than the speed of light. This suggests that, when we suddenly move the charge q', some kind of physical disturbance propagates through space from q' to q and adjusts the electric force to the new increased value (see Fig. 23.1). Thus, *charges exert forces on one another by means of disturbances that they generate in the space surrounding them. These disturbances are called* **electric fields**.

Fields are a form of matter—they are endowed with energy and momentum (as we shall find in Chapter 33), and they therefore exist in a material sense. In the context of the above example, it is easy to see why the disturbance, or field, generated by the sudden displacement of q' must carry momentum: when we suddenly move q' toward q, the increase in the force that q' exerts on q will be delayed until a signal has had time to propagate from q' carrying the information regarding the changed position of q'. But the increase in the force that q exerts on q' does not suffer a similar delay. Since we have not moved q, we have not disturbed the way in which q causes electric forces on other charges, according to Coulomb's Law. This temporary deviation between the

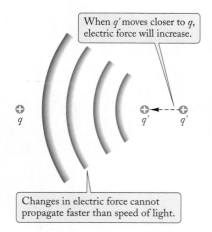

When q' moves closer to q, electric force will increase.

Changes in electric force cannot propagate faster than speed of light.

FIGURE 23.1 A disturbance emanates from the charge q' and reaches the charge q.

electric field

forces exerted by q' on q and by q on q' means that, when only the charges are considered, Newton's Third Law on the balance of action and reaction fails. Thus an extra entity, such as the field, is needed to take up the momentum and energy missing from the particles.

Although the above argument for the existence of the electric field arose by considering charges in motion, we will now adopt the very natural view that the forces on charges at rest involve the same mechanism. We suppose that each charge at rest generates a permanent, static disturbance in the space surrounding it, and that this disturbance exerts forces on other charges when they come in contact with the disturbance. Thus, we take the view that *the electric interaction between charges is action-by-contact: a charge q' generates an electric field which permeates the surrounding space and exerts forces on any other charges that it touches.* The electric field serves as the mediator of forces according to the scheme

charge $q' \Rightarrow$ electric field of charge $q' \Rightarrow$ force on charge q

To translate this conceptual scheme into mathematical language, we start with Coulomb's Law for the force exerted by the charge q' on the charge q,

$$F = \frac{1}{4\pi\epsilon_0}\frac{qq'}{r^2}$$

and we separate this expression into a product of two factors: a factor q characteristic of the point charge on which the force is being exerted, and a factor q'/r^2 characteristic of the point charge that exerts the force at the distance r. We also include the constant of proportionality $1/4\pi\epsilon_0$ in the second factor, so the expression for the force becomes

$$F = q \times \left(\frac{1}{4\pi\epsilon_0}\frac{q'}{r^2}\right) \tag{23.1}$$

The second factor is defined to be the electric field generated by the point charge q'; we designate this electric field by E:

$$E = \frac{1}{4\pi\epsilon_0}\frac{q'}{r^2} \tag{23.2}$$

This says that the magnitude of the electric field of the point charge q' is directly proportional to the magnitude of this charge and inversely proportional to the square of the distance. The force that this electric field exerts on the charge q is then simply

$$F = qE \tag{23.3}$$

Like the force, the electric field is a vector, and it can be represented by an arrow (see Fig. 23.2). The direction of the electric field depends on the sign of the charge q'. *The electric field is directed radially outward if q' is positive, and radially inward if q' is negative.* In vector language, the force on the charge q becomes

$$\mathbf{F} = q\mathbf{E} \tag{23.4}$$

With this special case of the electric field of a point charge as guidance, we can proceed to the definition of the electric field for the general case of an arbitrary charge distribution, such as the charge distribution in a thundercloud, or the charge distribution on an electric power line. To find the electric field that the charge distribution generates

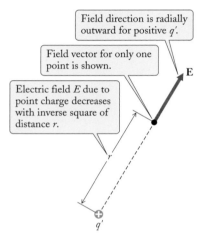

Field direction is radially outward for positive q'.

Field vector for only one point is shown.

Electric field E due to point charge decreases with inverse square of distance r.

FIGURE 23.2 A charge q' generates an electric field **E** at some distance r. The direction of this electric field is along the radial line.

electric field of point charge

TABLE 23.1	SOME ELECTRIC FIELDS
At surface of pulsar	$\approx 10^{14}$ N/C
At orbit of electron in hydrogen atom	6×10^{11}
In X-ray tube	5×10^6
Electrical breakdown of air	3×10^6
In Van de Graaff accelerator	2×10^6
Within lightning bolt	$\approx 10^4$
Under thundercloud	1×10^4
Near radar transmitter (FPS-6)	7×10^3
In sunlight (rms)	1×10^3
In atmosphere (fair weather)	1×10^2
In beam of small laser (rms)	1×10^2
In fluorescent lighting tube	10
In radio wave	$\approx 10^{-1}$
Within household wiring	$\approx 3 \times 10^{-2}$

at a given position, we take a point charge q (a "test charge") and place it at that position. The charge q will then experience an electric force F. By dividing out the charge q, we isolate the factor characteristic of the charge distribution and we eliminate the factor characteristic of the test charge. We define the electric field **E** as the force **F** divided by the magnitude q of the charge:

electric field and electric force

$$\mathbf{E} = \frac{\mathbf{F}}{q} \tag{23.5}$$

This means that the electric field is the force per unit charge. Note that with this definition, the electric field is independent of the magnitude of the test charge. The electric field is a very useful concept: for a given charge distribution, instead of calculating the force that the charges exert on another charge q at some point (which would be different for different values of q), the electric field tells us the force per unit charge on *any* charge placed at that point. Thus, the electric field **E** depends on the magnitudes and positions of the charges that produce the electric field, but it does not depend on the magnitude of the test charge q used to detect it.

newton per coulomb (N/C)

The SI unit of electric field is the **newton per coulomb** (N/C).[1] Table 23.1 gives the magnitudes of some typical electric fields.

The net electric field generated by any distribution of point charges can be calculated by forming the vector sum of the individual electric fields due to the point charges, where each individual electric field is given by Eq. (23.2). This procedure is justified by the Principle of Superposition for electric forces stated in Section 22.3. Thus, to find the total electric field at some point in space, you simply sum the individual vectors [with magnitude given by Eq. (23.2), and direction away from positive charges or toward negative charges].

[1] As we shall see in Chapter 25, newton/coulomb is the same thing as volt/meter.

EXAMPLE 1   Consider two charges $\pm Q$ of equal magnitudes and opposite signs, separated by a distance d. Such an arrangement of charges, similar to the ones exerting the force in Example 6 of Chapter 22, is called an **electric dipole**. Find the electric field at a point equidistant from the two charges, a distance x from their midpoint (see Fig. 23.3). What is the dependence on distance of this "dipole field" for $x \gg d$?

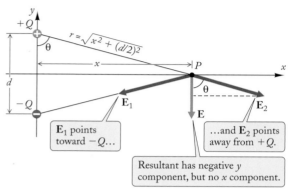

SOLUTION: The *magnitude* of each of the two individual electric fields is $E_1 = E_2 = (1/4\pi\epsilon_0)Q/r^2$, where $r = \sqrt{x^2 + (d/2)^2}$. At the point P, the electric field due to the $+Q$ charge points away from this charge, and the electric field due to the $-Q$ charge points toward that charge, as shown in Fig. 23.3. As in the solution of Example 6 of Chapter 22, the x components of the electric field at the point P in Fig. 23.3 cancel, and the y components are equal, giving

FIGURE 23.3 Two point charges $-Q$ and $+Q$ and a point P on the x axis. The net electric field $\mathbf{E}$ at the point P is the vector sum of the contributions $\mathbf{E}_1$ due to $-Q$ and $\mathbf{E}_2$ due to $+Q$.

$$E = E_y = -2E_1 \cos\theta = -2 \frac{1}{4\pi\epsilon_0} \frac{Q}{x^2 + d^2/4} \frac{d/2}{\sqrt{x^2 + d^2/4}} \quad (23.6)$$

$$= -\frac{1}{4\pi\epsilon_0} \frac{Qd}{(x^2 + d^2/4)^{3/2}}$$

For $x \gg d$ we can neglect d^2 compared with x^2 and obtain

$$E = -\frac{1}{4\pi\epsilon_0} \frac{Qd}{x^3}$$

COMMENTS: Note that since we already calculated the force due to the charges $\pm Q$ on a charge q in Example 6 of Chapter 22, we could equally well solve this problem using the definition $\mathbf{E} = \mathbf{F}/q$ and obtain the same result. Also note that for large distances, the dependence of the dipole field on distance is $E \propto 1/x^3$. It turns out that this $1/x^3$ behavior is characteristic of the electric dipole field at large distances *in any direction*, not just along the perpendicular bisector. Finally, note that here we have calculated the field *generated* by an electric dipole, that is, the field that would act on a third charge when placed near the dipole. Later, in Section 23.5, we will examine the *response* of an electric dipole to an *external* field. Be careful not to confuse the two situations.

EXAMPLE 2   The distribution of electric charge in a thundercloud can be roughly approximated by several pointlike charges placed at different heights. Figure 23.4a shows such an approximate charge distribution consisting of a charge of $+40\,\text{C}$ at a height of $10.0\,\text{km}$ and a charge of $-30\,\text{C}$ at a height of $4.0\,\text{km}$ in the thundercloud. What are the horizontal and the vertical components of the electric field that these two charges produce at a point P at a height of $10.0\,\text{km}$ and at a horizontal distance of $6.0\,\text{km}$ to the right? What is the magnitude of this field at that point?

Concepts — in — Context

SOLUTION: At the point P, the charge $Q_1 = +40\,\text{C}$ produces an electric field that is directed away from this charge, and the charge $Q_2 = -30\,\text{C}$ produces an electric field that is directed toward that charge. The net electric field is the vector sum of these two individual electric fields.

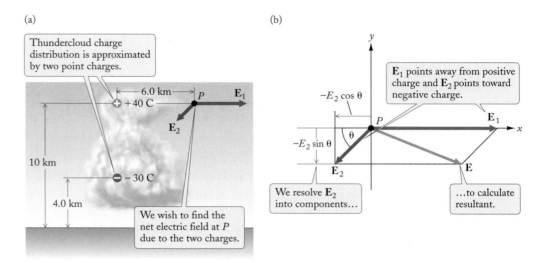

FIGURE 23.4 (a) Distribution of positive and negative electric charges in a thundercloud. (b) The net electric field is the vector sum of the individual electric fields $\mathbf{E}_1$ and $\mathbf{E}_2$.

If h represents the height difference between the charges and d the horizontal distance of the point P, then the radial distances from the charges Q_1 and Q_2 to P are, respectively, d and $\sqrt{d^2 + h^2}$. According to Eq. (23.2), the *magnitudes* of the two individual electric fields are

$$E_1 = \frac{1}{4\pi\epsilon_0} \frac{Q_1}{d^2} \quad \text{and} \quad E_2 = \frac{1}{4\pi\epsilon_0} \frac{|Q_2|}{d^2 + h^2}$$

From Fig. 23.4b we see that the horizontal components (x components) of the two individual electric fields are E_1 and $-E_2 \cos\theta$, and the vertical components (y components) are 0 and $-E_2 \sin\theta$, respectively. We therefore obtain the following components of the net electric field:

$$E_x = \frac{1}{4\pi\epsilon_0} \left(\frac{Q_1}{d^2} - \frac{|Q_2|}{d^2 + h^2} \cos\theta \right)$$

and

$$E_y = -\frac{1}{4\pi\epsilon_0} \frac{|Q_2|}{d^2 + h^2} \sin\theta$$

For this example, we have $d = h = 6.0$ km, and thus $\theta = 45°$, so we find

$$E_x = (9.0 \times 10^9 \text{ N·m}^2/\text{C}^2) \left[\frac{40 \text{ C}}{(6.0 \times 10^3 \text{ m})^2} - \frac{30 \text{ C}}{2 \times (6.0 \times 10^3 \text{ m})^2} \cos 45° \right]$$

$$= 7.3 \times 10^3 \text{ N/C}$$

and

$$E_y = -(9.0 \times 10^9 \text{ N·m}^2/\text{C}^2) \frac{30 \text{ C}}{2 \times (6.0 \times 10^3 \text{ m})^2} \sin 45°$$

$$= -2.7 \times 10^3 \text{ N/C}$$

The magnitude of the electric field is then

$$E = \sqrt{E_x^2 + E_y^2} = \sqrt{(7.3 \times 10^3 \text{ N/C})^2 + (-2.7 \times 10^3 \text{ N/C})^2}$$

$$= 7.8 \times 10^3 \text{ N/C}$$

EXAMPLE 3   Consider the thundercloud charge distribution of Example 2. To find the total electric field, we must take into account that the ground is a conductor and that the charges in the thundercloud induce charges in the ground. It can be shown that the effect of the charge induced by an above-ground point charge can be simulated by an opposite point charge the same distance *below ground*; these fictitious charges are called **image charges**. The resulting charge arrangement is shown in Fig. 23.5. By summing the electric fields due to the real charges in the thundercloud and due to the image charges, calculate the electric field at a point on the ground directly below the thundercloud charges.

SOLUTION: At the ground, the electric field due to the -30 C thundercloud charge is directed upward, toward this negative charge, and the electric field due to the $+30$ C below ground image charge is also upward, directed away from that positive charge. The charges are each 4.0 km from the ground, so these two electric field contributions have equal magnitudes and are both directed upward, which we choose as the positive y direction. Hence the sum is twice an individual contribution:

$$E_{y,\pm30\text{ C}} = 2 \times \frac{1}{4\pi\epsilon_0} \frac{|Q|}{r^2} = 2 \times (9.0 \times 10^9 \text{ N·m}^2/\text{C}^2) \frac{30 \text{ C}}{(4.0 \times 10^3 \text{ m})^2}$$

$$= +3.4 \times 10^4 \text{ N/C}$$

The electric field at the ground due to the $+40$ C thundercloud charge is directed downward, away from this positive charge, and the electric field due to the -40 C below ground image charge is also downward, directed toward that negative charge. These charges are 10 km from the ground, so the two electric field contributions again have equal magnitudes, but now are directed in the negative y direction. Hence

$$E_{y,\pm40\text{ C}} = -2 \times \frac{1}{4\pi\epsilon_0} \frac{|Q|}{r^2} = -2 \times (9.0 \times 10^9 \text{ N·m}^2/\text{C}^2) \frac{40 \text{ C}}{(1.0 \times 10^4 \text{ m})^2}$$

$$= -7.2 \times 10^3 \text{ N/C}$$

Since each of the ±30 C and ±40 C pairs of charges is an electric dipole, we could alternatively obtain these results by evaluating Eq. (23.6) in Example 1 directly between each pair of charges, at $x = 0$. In any case, the net electric field on the ground directly below the thundercloud charges is

$$E_y = E_{y,\pm30\text{ C}} + E_{y,\pm40\text{ C}} = +3.4 \times 10^4 \text{ N/C} - 7.2 \times 10^3 \text{ N/C}$$

$$= +2.7 \times 10^4 \text{ N/C}$$

The positive result indicates that the net electric field is directed vertically upward.

Concepts
— *in* —
Context

image charge

(a)

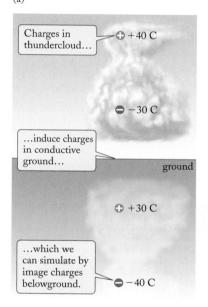

Charges in thundercloud...   ⊕ +40 C

⊖ −30 C

...induce charges in conductive ground...   ground

⊕ +30 C

...which we can simulate by image charges belowground.   ⊖ −40 C

(b)

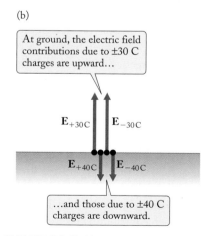

At ground, the electric field contributions due to ±30 C charges are upward...

$\mathbf{E}_{+30\text{C}}$ $\mathbf{E}_{-30\text{C}}$

$\mathbf{E}_{+40\text{C}}$ $\mathbf{E}_{-40\text{C}}$

...and those due to ±40 C charges are downward.

FIGURE 23.5 (a) Charges in a thundercloud and image charges below the ground. (b) Electric field contributions at ground level.

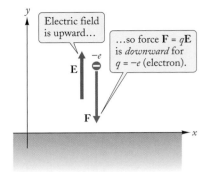

FIGURE 23.6 The upward electric field **E** exerts a downward force **F** on the electron.

| EXAMPLE 4 | An electron is near the ground, directly below the thundercloud charges of Example 3, in the net electric field of 2.7×10^4 N/C |

pointing vertically upward (see Fig. 23.6). What is the force on the electron, and what is its acceleration?

SOLUTION: By Eq. (23.4), the force on the electron equals the charge multiplied by the electric field. If the y axis is vertically upward, $E_y = 2.7 \times 10^4$ N/C, and

$$F_y = qE_y = -eE_y = (-1.60 \times 10^{-19}\ \text{C})(2.7 \times 10^4\ \text{N/C}) = -4.3 \times 10^{-15}\ \text{N}$$

This force gives the electron an acceleration

$$a_y = \frac{F_y}{m_e} = \frac{-4.3 \times 10^{-15}\ \text{N}}{9.11 \times 10^{-31}\ \text{kg}} = -4.7 \times 10^{15}\ \text{m/s}^2$$

This is an enormous acceleration.

COMMENTS: Note that, since the charge of the electron is negative, the direction of the force and of the acceleration is *opposite* to the direction of the electric field. Note also that a charge in the thundercloud is attracted to its own (opposite) image charge (the same image-charge attraction is responsible for the binding of free electrons within the volume of a metal). Also, note that in this calculation we have neglected the acceleration due to gravity; since this acceleration is much smaller than that due to the electric field, the neglect is justified.

PROBLEM-SOLVING TECHNIQUES   ELECTRIC FIELDS

- To find the electric field of an arrangement of several point charges (such as the point charges of the thundercloud in Example 2), you need to calculate the vector sum of the electric fields of the individual point charges. The magnitude of the electric field of an individual point charge q' is given by Eq. (23.2), $E = (1/4\pi\epsilon_0)q'/r^2$; and the direction of this electric field is radially outward if q' is positive, and radially inward if q' is negative.

- The techniques for evaluating the vector sum of the electric field vectors of several charges are the same as for the sum of several position vectors or any other kind of vectors. First, draw a careful diagram, showing each electric field vector along the line through the charge producing the field contribution and through a point P where you want to know the field. It may be easiest to put the tail of each contributing vector at P (as in Example 1) and then

use geometry to break each vector into its x and y (and, if necessary, z) components. You obtain the total electric field vector by separately summing the x, y, and z components of the contributing vectors. In many examples of calculations of electric fields, some components of the field will cancel. If you select the x, y, and z axes judiciously, you can often achieve cancellation of all components except one (see Example 1).

- When calculating the electric force $\mathbf{F} = q\mathbf{E}$ on charge q from the electric field $\mathbf{E}$ generated by a charge distribution, be sure to keep in mind the distinction between the charges that generate the electric field and the charge q that is placed in the resultant electric field (and experiences the force). The electric field of the latter charge is not to be included in the calculation of the electric field $\mathbf{E}$; only the *other* charges contribute to the electric field that acts on the charge q.

Checkup 23.1

QUESTION 1: In the discussion leading up to Eq. (23.3), we began with the Coulomb force between two point charges q and q', and we found the electric field generated by the charge q'. If, instead, we had wanted to find the electric field generated by the charge q, what would have been the result?

QUESTION 2: The magnitude of the electric field at a distance of 1 m from a point charge is 1×10^5 N/C. What is the magnitude of the electric field at a distance of 2 m? 3 m?

QUESTION 3: Given that an electric field of 10 N/C has a northward direction and another electric field of 10 N/C has an eastward direction, what are the magnitude and direction of the superposition of these two electric fields?

QUESTION 4: Four equal positive charges are located at the corners of a square. What is the magnitude of the electric field at the center of the square?

QUESTION 5: Three equal positive charges are located at three corners of a square, and a negative charge is located at the fourth corner. What is the direction (if any) of the electric field at the center of the square?

(A) Toward the negative charge   (B) No direction, since field is zero
(C) Away from the negative charge
(D) Perpendicular to the diagonal through the negative charge

23.2 THE ELECTRIC FIELD OF CONTINUOUS CHARGE DISTRIBUTIONS

We learned in Chapter 22 that charge is discrete, or quantized; that is, it always occurs in integer multiples of the fundamental charge e. However, in a description of the charge distribution on macroscopic bodies, the discrete nature of charge can often be ignored, and it is usually sufficient to treat the charge as a continuous "fluid" with a charge density (C/m^3) that varies more or less smoothly over the volume of the charged body. This is analogous to describing the mass distribution of a solid, a liquid, or a gas by a smooth mass density (kg/m^3), which ignores the fact that, on a microscopic scale, the mass is concentrated in atoms. A solid, a liquid, or a gas seems smooth because there are very many atoms in each cubic millimeter, and the distances between atoms are very small. Likewise, charge distributions placed on wires or other conductors seem smooth because they consist of very many electrons (or protons) in each cubic millimeter.

The Superposition Principle still holds for continuous charge distributions; we need merely sum the individual small contributions to the electric field from each small piece of charge. Suppose that a small region contains a charge dq and contributes a field of magnitude dE at some point where we want to know the total field. If this region is sufficiently small, we can treat the contribution as that of a point charge, with the usual Coulomb field:

$$dE = \frac{1}{4\pi\epsilon_0} \frac{dq}{r^2} \qquad (23.7)$$

magnitude of electric field contribution

Next, we must sum these contributions to the electric field from the various small regions of charge. Mathematically, a sum over a continuous region means we integrate the contributions to get the total field. Since the electric field is a vector, we have to perform the integration for each component of **E**. For example, if θ is the angle that an electric field contribution makes with the x axis, then the x component of the contribution is $dE_x = dE \cos \theta$, that is,

x component of electric field contribution

$$dE_x = \frac{1}{4\pi\epsilon_0} \frac{\cos\theta}{r^2} dq \qquad (23.8)$$

When integrating this over the charge distribution to obtain the total E_x, we must take into account that $\cos\theta$ and r^2 vary over the distribution. We can perform similar calculations for E_y and E_z and obtain the total field $\mathbf{E} = E_x\mathbf{i} + E_y\mathbf{j} + E_z\mathbf{k}$. In the general case, such a calculation can be very difficult, and numerical methods (and computers) often must be used to obtain accurate results. However, in some important examples, the integration is simple enough to be performed directly. In the examples that follow, such exact solutions provide significant insight into the behavior of the electric field due to continuous charge distributions.

In the examples below, we will encounter two of the three types of continuous charge distributions: a linear charge distribution (charge distributed on a string, a wire, or a thin rod) and a surface charge distribution (charge distributed over a flat or curved surface). In Chapter 24, we will also encounter the third type, a volume charge distribution (charge distributed throughout a three-dimensional volume). In all the cases of immediate interest, we will assume the charge distribution is *uniform*, that is, constant, over a specified region. For a uniform distribution, the small charge dq in a small linear region of length dL, or on a small piece of surface area dA, is equal to the length dL or the area dA multiplied by a constant:

$$\text{Linear:} \quad dq = \lambda \, dL$$

$$\text{Surface:} \quad dq = \sigma \, dA. \qquad (23.9)$$

where λ (lambda) is the charge per unit length in coulombs per meter (C/m) and σ (sigma) is the charge per unit area in coulombs per square meter (C/m^2). For example, a rod of length L with a total charge Q uniformly distributed along its length has $\lambda = Q/L$, and a sheet of area A with a total charge Q uniformly distributed over its area has $\sigma = Q/A$. The more general possibility of a *nonuniform* distribution, for example, where the value of λ may vary along a line of charge, is considered in Problem 43. In the examples, we restrict ourselves to uniform distributions.

EXAMPLE 5   Charge is distributed uniformly along an infinitely long, straight thin rod. If the charge per unit length of the rod is λ, what is the electric field at some distance from the rod?

SOLUTION: Figure 23.7 shows the charged rod lying along the y axis. The point P at which the electric field is to be evaluated is on the x axis, at a perpendicular distance x from the rod. To find the electric field, we must perform an integration, regarding the rod as made up of many infinitesimal line elements, each of which can be treated as a point charge. Before proceeding with such an integration, it is helpful to determine the direction of the field by a qualitative symmetry argument. The electric field generated by the line element dy shown below the origin in Fig. 23.7 has both x and y components. Upon integration, the y component will cancel against the y component contributed by a line element lying at the same distance above the origin, and the total y component of the electric field will be zero. Therefore the net electric field will have only an x component.

The line element dy carries a charge $dQ = \lambda\, dy$. Since this line element can be treated as a point charge, it generates an electric field of magnitude

$$dE = \frac{1}{4\pi\epsilon_0}\frac{dQ}{r^2} = \frac{1}{4\pi\epsilon_0}\frac{\lambda\, dy}{r^2}$$

This electric field has an x component

$$dE_x = dE\cos\theta = \frac{1}{4\pi\epsilon_0}\frac{\lambda\, dy\cos\theta}{r^2}$$

The net electric field is then

$$E_x = \int_{-\infty}^{+\infty} \frac{1}{4\pi\epsilon_0}\frac{\lambda\, dy\cos\theta}{r^2}$$

As we vary the position of the contribution along y in this integral, the quantities r and θ also change. To evaluate this integral, we must express all variables in terms of a single variable. It turns out that the most convenient choice in this particular case is the angle θ. From Fig. 23.7,

$$y = x\tan\theta \quad \text{and} \quad dy = x\,d(\tan\theta) = x\sec^2\theta\, d\theta$$

Furthermore,

$$r = \frac{x}{\cos\theta} = x\sec\theta$$

With these substitutions, we obtain an integral over the angle θ. In terms of this angle, the limits of integration are $\theta = -90°$ and $\theta = 90°$, or, in radians, $\theta = -\pi/2$ and $\theta = \pi/2$, and the integral becomes

$$E_x = \frac{1}{4\pi\epsilon_0}\int_{-\pi/2}^{\pi/2}\frac{\lambda\cos\theta}{(x\sec\theta)^2}x\sec^2\theta\, d\theta = \frac{1}{4\pi\epsilon_0}\frac{\lambda}{x}\int_{-\pi/2}^{\pi/2}\cos\theta\, d\theta$$

Since

$$\int_{-\pi/2}^{\pi/2}\cos\theta\, d\theta = \sin\theta\Big|_{-\pi/2}^{\pi/2} = 2$$

we find

$$E_x = \frac{1}{2\pi\epsilon_0}\frac{\lambda}{x} \tag{23.10}$$

The direction of the electric field is radially away from the rod (if λ is positive).

COMMENTS: The magnitude of this electric field decreases in inverse proportion to the distance from the rod (not the square of the distance). The result (23.10) is exact for an infinitely long rod, but it is also a reasonable approximation for a rod of finite length, provided the point P is near the rod and away from its ends (seen from such a point, the finite rod looks almost like an infinite rod).

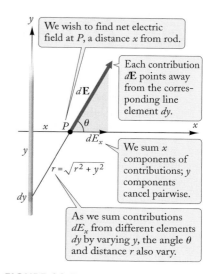

We wish to find net electric field at P, a distance x from rod.

Each contribution $d\mathbf{E}$ points away from the corresponding line element dy.

We sum x components of contributions; y components cancel pairwise.

As we sum contributions dE_x from different elements dy by varying y, the angle θ and distance r also vary.

$r = \sqrt{r^2 + y^2}$

FIGURE 23.7 A long, thin, charged rod lies along the y axis. A small line element dy of this rod contributes an electric field $d\mathbf{E}$ at the point P.

PROBLEM-SOLVING TECHNIQUES   ELECTRIC FIELD OF CHARGE DISTRIBUTION

- To find the electric field of a continuous distribution of charge (such as the charge distributed uniformly along an infinite straight rod in Example 5), you need to regard the charge distribution as a collection of many small (infinitesimal) charge elements, each of which can be treated as a point charge. The vector sum of the electric fields of all the small charge elements then involves integration over all these charge elements.

- To begin, make some convenient choice of coordinate axes. Choose a charge element dQ and look at a typical contribution $dE = k\, dQ/r^2$; then consider the x, y, and z components of the electric field contribution one by one, and try to decide if any component is zero because of cancellations brought about by the symmetry of the charge distribution. Any component that is not zero must be evaluated by integration over the small charge elements.

- The choice of variable of integration is often a delicate matter, since the integrand might become very simple for some choices of variable of integration, but very complicated and intractable for other choices. Thus, in Example 5, the integral is elementary if θ is used as variable, but it is much more complicated if y is used as variable. When faced with an apparently intractable integral, try substitutions of variables (or try finding the integral in a table, such as the one in Appendix 4).

- For some problems, the simplest approach is best; the variable of integration can remain the length element obtained when substituting $dQ = \lambda\, dx$. Be sure to specify the limits of integration and the quantity r in a manner consistent with the chosen variable of integration.

- Sometimes the electric field of a complicated charge distribution can be found by superposition of the individual electric fields of the separate pieces of the charge distribution. For example, the electric field of the two large charged sheets in Fig. 23.10 is the superposition of their individual electric fields, and the net electric field of two long lines of charge arranged parallel or arranged at some angle is the superposition of their individual electric fields.

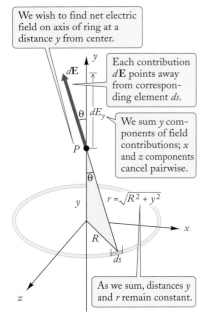

We wish to find net electric field on axis of ring at a distance y from center.

Each contribution $d\mathbf{E}$ points away from corresponding element ds.

We sum y components of field contributions; x and z components cancel pairwise.

$r = \sqrt{R^2 + y^2}$

As we sum, distances y and r remain constant.

FIGURE 23.8 A small segment ds of a ring of charge contributes an electric field $d\mathbf{E}$.

EXAMPLE 6   An amount Q of positive charge is distributed uniformly along the circumference of a thin ring of radius R. What is the electric field on the axis of the ring at some distance from the center?

SOLUTION: Figure 23.8 shows the ring of charge and the point P at which the electric field is to be evaluated. This point P is on the y axis, at a distance y above the ring. We can regard the ring as made up of many small elements, each of which can be treated as a point charge. Figure 23.8 shows one of these elements, of length ds. The electric field dE contributed by this element points away from the element. To find the net electric field, we must evaluate the vector sum of all the small contributions to the electric field arising from all the small elements of the ring. Before evaluating this vector sum, let us determine the direction of the net electric field. The field contributed by the small element ds shown in Fig. 23.8 has both a horizontal and a vertical component. For any given small element ds of the ring, there is an equal small element on the opposite side of the ring's center that contributes an electric field of opposite horizontal component; consequently, the horizontal components cancel pairwise. The net electric field is therefore vertical.

The small element ds has a small charge dQ. The distance from this charge dQ to the point P is $r = \sqrt{R^2 + y^2}$. By Coulomb's Law, the magnitude of the electric field contributed by dQ is

$$dE = \frac{1}{4\pi\epsilon_0}\frac{dQ}{r^2} = \frac{1}{4\pi\epsilon_0}\frac{dQ}{R^2 + y^2}$$

We are interested only in the vertical component of this electric field, since we already know that the horizontal component makes no net contribution to the net electric field. The vertical component of $d\mathbf{E}$ is

$$dE_y = dE \cos \theta$$

where $\cos \theta = y/\sqrt{R^2 + y^2}$ (see Fig. 23.8). Consequently,

$$dE_y = \frac{1}{4\pi\epsilon_0} \frac{dQ}{R^2 + y^2} \frac{y}{\sqrt{R^2 + y^2}}$$

$$= \frac{1}{4\pi\epsilon_0} \frac{dQ\, y}{(R^2 + y^2)^{3/2}}$$

To find the total electric field, we must integrate around the ring:

$$E_y = \int dE_y = \frac{1}{4\pi\epsilon_0} \int \frac{dQ\, y}{(R^2 + y^2)^{3/2}}$$

Notice that the quantities y and R do not vary around the ring, so

$$E_y = \frac{1}{4\pi\epsilon_0} \frac{y}{(R^2 + y^2)^{3/2}} \int dQ$$

The integral of dQ around the ring is simply the net electric charge Q of the ring, so we obtain the result

$$E_y = \frac{1}{4\pi\epsilon_0} \frac{Q\, y}{(R^2 + y^2)^{3/2}} \tag{23.11}$$

COMMENTS: Note that if $y = 0$, then $E_y = 0$; that is, at the center of the ring of charge, the electric field is zero. That we could have deduced from symmetry, since contributions from opposite sides of the ring exactly cancel at the center of the ring. Note also the behavior far from the ring: if y is very much larger than R, then Eq. (23.11) becomes approximately

$$E_y \approx \frac{1}{4\pi\epsilon_0} \frac{Q\, y}{(y^2)^{3/2}} = \frac{1}{4\pi\epsilon_0} \frac{Q}{y^2}$$

which is the usual inverse-square electric field of a point charge. Thus, the ring behaves like a point charge whenever the distance y is very large.

EXAMPLE 7   Positive charge is uniformly distributed over an infinite flat horizontal sheet, such as a very large sheet of paper. Suppose that the amount of charge per unit area on this sheet is σ. Find the electric field in the space above and below the sheet.

SOLUTION: The sheet can be regarded as made up of many concentric rings. Figure 23.9 shows one of these rings, with radius R and width dR; this ring has

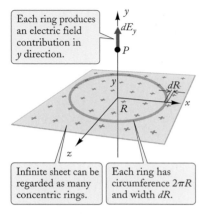

Each ring produces an electric field contribution in y direction.

Infinite sheet can be regarded as many concentric rings.

Each ring has circumference $2\pi R$ and width dR.

FIGURE 23.9 A very large sheet of charge lies in the x–z plane. A thin ring of charge within this sheet produces an electric field $d\mathbf{E}$.

electric field of flat sheet

an area $dA = (\text{length} \times \text{width}) = (2\pi R)\, dR$ and, from Eq. (23.9), a charge $dQ = \sigma\, dA = \sigma \times 2\pi R\, dR$. According to Eq. (23.11), the ring produces an electric field

$$dE_y = \frac{1}{4\pi\epsilon_0} \frac{(2\pi\sigma R\, dR)y}{(R^2 + y^2)^{3/2}}$$

The net electric field of the entire infinite sheet is then obtained by integrating over R, from $R = 0$ (the radius of the smallest ring) to $R = \infty$ (the radius of the largest ring):

$$E_y = \frac{\sigma y}{2\epsilon_0} \int_0^\infty \frac{R\, dR}{(R^2 + y^2)^{3/2}}$$

With the substitutions $u = R^2$ and $du = 2R\, dR$, this becomes

$$E_y = \frac{\sigma y}{2\epsilon_0} \int_0^\infty \frac{\frac{1}{2}du}{(u + y^2)^{3/2}} = \frac{\sigma y}{2\epsilon_0} \left[-\frac{1}{(u + y^2)^{1/2}} \right]\Bigg|_0^\infty = \frac{\sigma y}{2\epsilon_0} \frac{1}{y}$$

or simply

$$E_y = \frac{\sigma}{2\epsilon_0} \tag{23.12}$$

COMMENTS: This electric field is proportional to the charge density, and it is constant, that is, it is independent of the distance from the sheet. Although the result (23.12) is strictly valid only for the case of an infinitely large sheet, it is also a good approximation for a sheet of finite size, provided we stay near the sheet and we stay away from the vicinity of the edges.

Large charged flat sheets or plates are often used in laboratories to generate uniform electric fields. In practice, two parallel charged sheets with charges of opposite signs are preferred over a single sheet. To find the field due to two sheets, consider the electric fields due to each sheet separately, and then apply the Superposition Principle to sum them. In the space between the sheets, the individual electric fields of the two

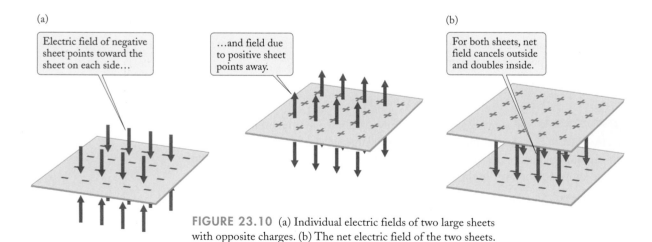

(a)

Electric field of negative sheet points toward the sheet on each side…

…and field due to positive sheet points away.

(b)

For both sheets, net field cancels outside and doubles inside.

FIGURE 23.10 (a) Individual electric fields of two large sheets with opposite charges. (b) The net electric field of the two sheets.

PHYSICS IN PRACTICE   ELECTROSTATIC PRECIPITATORS

Electrostatic precipitators are widely used to remove pollutants from the smoke produced by power plants, smelters, and other industrial installations. Before the installation of such cleaning devices, industrial smoke ejected huge amounts of pollutants into the atmosphere. For instance, in just one day, a large copper smelter would eject more than 3000 tons of sulfur compounds, 30 tons of arsenic compounds, and several tons of zinc, copper, lead, and antimony. The electrostatic precipitator consists of a metallic cylindrical chamber in which hangs a wire connected to a high-voltage generator (see Fig. 1). The high concentration of electric charge on the wire produces a strong electric field—as much as 10^6 N/C—in the space between the wire and the wall of the chamber. When dirty smoke, loaded with particles of pollutants, enters this strong electric field, it triggers a discharge, called a corona

discharge, from the wire into the smoke. This discharge deposits negative electric charges on the particles of pollutants. The electric field then pushes the charged particles sideways and drives them against the wall, where they adhere. Every so often, a motorized mechanical hammer strikes and shakes the chamber, dislodging the accumulated layers of pollutants, which fall down into hoppers for collection and disposal.

Electrostatic precipitators used in industrial installations contain large arrays of many parallel plates and charged wires (see Fig. 2). Automatic control systems connected to the wires and the generators continually adjust the charges on the wires and the resulting electric fields to optimize the collection of pollutants. Precipitators can collect solid particles, such as fly ash, soot, or dust; they can also collect liquid droplets, such as sulfuric acid mist or tar fumes.

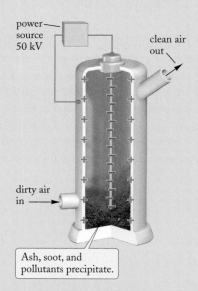

FIGURE 1 A schematic diagram of an electrostatic precipitator.

FIGURE 2 An electrostatic precipitator array.

sheets are in the same direction (see Fig. 23.10a); they therefore reinforce each other, giving a net field twice as large as the field of one sheet, that is,

$$E_y = \frac{\sigma}{\epsilon_0} \qquad (23.13)$$

electric field between two oppositely charged flat sheets

In the space above or below the two sheets, the individual electric fields of the two sheets are in opposite directions; thus, these electric fields cancel, and the space above and below has zero electric field. Figure 23.10b shows the net electric field of the two sheets.

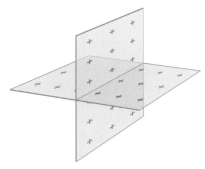

FIGURE 23.11 Two intersecting sheets of charge.

✔ Checkup 23.2

QUESTION 1: Two infinitely large flat sheets with equal uniform positive charge distributions intersect at right angles (see Fig. 23.11). What is the direction of the electric field in each quadrant?

QUESTION 2: A small disk (approximate diameter 1 cm) has a charge Q uniformly distributed over its area A. What is the electric field far from this disk (e.g., meters away)? What is the field very close to its surface (e.g., 10^{-3} cm) and far from the edges?

QUESTION 3: The electric field at a distance of 1 m from a uniformly charged infinite sheet has the value $E = E_0$. What is the value of E at a distance of 2 m? At 4 m?

QUESTION 4: Suppose that two large parallel sheets have equal surface charge densities σ with the same sign, say, positive. What is the magnitude of the electric field in the space between the sheets? Outside the sheets?

(A) $\sigma/\epsilon_0, 0$ (B) $0, \sigma/\epsilon_0$ (C) $\sigma/\epsilon_0, \sigma/\epsilon_0$ (D) $\sigma/2\epsilon_0, \sigma/2\epsilon_0$ (E) $0, 0$

23.3 LINES OF ELECTRIC FIELD

The electric field can be represented graphically by drawing, at any given point of space, a vector whose magnitude and direction are those of the electric field at that point. Figure 23.12 shows the electric field vectors in the space surrounding a positive point charge, and Fig. 23.13 shows the electric field vectors in the space surrounding a negative point charge. Note that, as demanded by Eq. (23.8), the magnitude of these vectors decreases in inverse proportion to the square of the distance from the point charge. Also note that, to avoid overcrowding the drawing, the field vectors at only a few representative points can be shown in such a diagram, although the electric field has a value at every point in space.

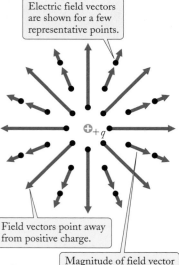

Electric field vectors are shown for a few representative points.

Field vectors point away from positive charge.

Magnitude of field vector decreases with inverse square of distance.

FIGURE 23.12 Electric field vectors surrounding a positive point charge. The field vectors are directed radially outward. The positions to which the field vectors belong are marked in black.

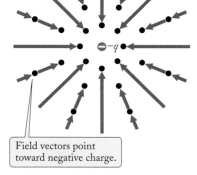

Field vectors point toward negative charge.

FIGURE 23.13 Electric field vectors surrounding a negative point charge. The field vectors are directed radially inward.

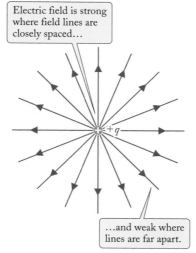

Electric field is strong where field lines are closely spaced…

…and weak where lines are far apart.

FIGURE 23.14 Electric field lines of a positive point charge. Note that in three dimensions the lines spread out in all three directions of space, whereas the diagram shows the lines spreading out in only the two directions within the page. This gives a misleading impression of the density of field lines as a function of distance. This limitation of two-dimensional representations should be kept in mind when looking at diagrams of field lines.

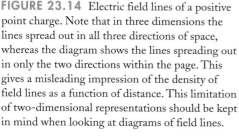

Direction of electric field is tangent to field line.

FIGURE 23.15 Electric field lines of a negative point charge.

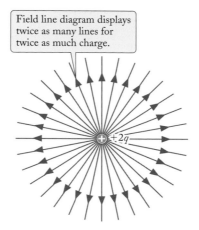

Field line diagram displays twice as many lines for twice as much charge.

FIGURE 23.16 Electric field lines of a positive point charge twice as large as in Fig. 23.14.

Alternatively, the electric field can be represented graphically by **field lines**. These lines are drawn in such a way that, at any given point, the tangent to the line has the direction of the electric field, that is, the direction of the field vector. Furthermore, the density of lines is directly proportional to the magnitude of the electric field; that is, where the lines are closely spaced the electric field is strong, and where the lines are far apart the electric field is weak. Figure 23.14 shows the electric field lines of a positive point charge and Fig. 23.15 shows those of a negative point charge. The arrows on these lines indicate the direction of the electric field along each line.

When we draw a pattern of field lines, we must begin each line on a positive point charge and end on a negative point charge (or the line may extend to infinity). Since the magnitude of the electric field is directly proportional to the amount of electric charge, the number of field lines that we draw emerging from a positive point charge must be proportional to the amount of charge. Figure 23.16 shows the electric field lines of a positive point charge twice as large as that of Fig. 23.14. Figure 23.17 shows the field lines generated jointly by a positive and a negative charge of equal magnitudes (an electric dipole), and Fig 23.18 shows those of a pair of equal positive charges (note the equal number of field lines in each case). Figure 23.19 shows the field lines of a pair of unequal positive and negative charges. Note that there are 2 times as many field lines originating on the $+2q$ charge (32 lines) as there are ending on the $-q$ charge (16 lines). If desired, we can vary the number of field lines actually drawn, but we must

field lines

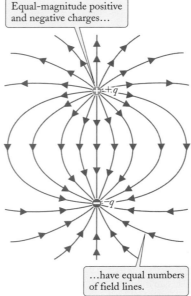

FIGURE 23.17 Field lines generated by positive and negative charges of equal magnitudes.

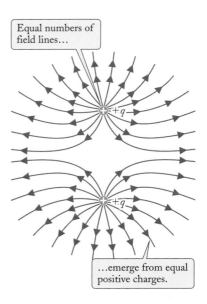

FIGURE 23.18 Field lines generated by two positive charges of equal magnitudes.

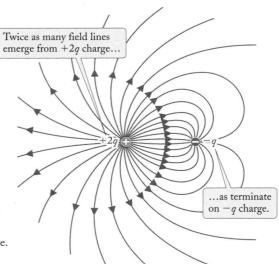

FIGURE 23.19 Field lines generated by positive and negative charges of unequal magnitudes. The positive charge has 2 times the magnitude of the negative charge.

be careful to maintain a fixed ratio of field lines, proportional to charge. It is necessary to keep such a normalization fixed through any given computation or series of drawings.

Note that the field lines start on positive charges and end on negative charges—the positive charges are sources of field lines and the negative charges are sinks. Also, note that the field lines never intersect (except where they start or end on point charges). If the lines ever were to intersect, the electric field would have *two* directions at the point of intersection; this is impossible.

The above pictures of field lines help us to develop some intuitive feeling for the spatial dependence of the electric fields surrounding diverse arrangements of electric charges. But we must refrain from thinking of the field lines as physical objects. The

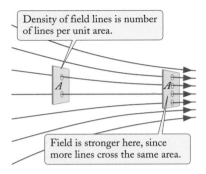

Density of field lines is number of lines per unit area.

Field is stronger here, since more lines cross the same area.

FIGURE 23.20 Two identical areas A intercept different numbers of field lines.

electric field is a physical entity, it is a form of matter; but the field lines are merely mathematical crutches to aid our imagination.

We said the electric field is proportional to the density of field lines. The precise definition is that the density of lines is the number of lines per unit area, that is, the number of lines intercepted by a small area A erected perpendicularly to the lines (see Fig. 23.20) divided by the magnitude of this area. So if a greater number of field lines cross a given amount of area A at one location than at a second location, the electric field is proportionally stronger at the first location.

The inverse-square law for the electric field of a point charge can be "derived" from the picture of field lines. By a simple argument, we can show that the density of lines necessarily obeys an inverse-square law. Since a point charge q is spherically symmetric and makes no distinction between one radial direction and another, symmetry tells us that the arrangement of field lines must also be spherically symmetric, and they must be uniformly distributed over all radial directions. At a distance r from the point charge, these lines are uniformly distributed over the area $A = 4\pi r^2$. For larger concentric spheres, the same number of field lines is distributed over increasingly larger areas; hence the density of lines decreases in proportion to the inverse square of the distance. This "derivation" of the inverse-square law from the picture of field lines is really no more than a consistency check—the electric field can be represented by field lines only because it is an inverse-square field; any other dependence on distance would make it impossible to draw continuous field lines that start and end only on charges.

✔ Checkup 23.3

QUESTION 1: Figure 23.21 shows diagrams of hypothetical (but incorrect) field lines corresponding to some static charge distributions, which are beyond the edge of the diagram. What is wrong with each of these diagrams of field lines?

QUESTION 2: A field line diagram shows a point charge with eight field lines emerging from it. Nearby in the same diagram is a metal sphere with four lines emerging and two lines entering the sphere. What can we say about the arrangement of charge on the sphere? The net charge on the sphere?

QUESTION 3: During a day of fair weather, the Earth has an atmospheric electric field that points vertically down. This electric field is due to charges distributed over the surface of the Earth. What must be the sign of these charges?

QUESTION 4: A field line diagram shows a +6.0 C point charge with eight field lines emerging from it. Nearby in the diagram is a short, charged rod with 12 field lines entering it. The charge on the rod is

   (A) +4.0 C   (B) −4.0 C   (C) +9.0 C   (D) −9.0 C   (E) −12.0 C

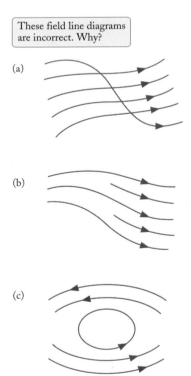

These field line diagrams are incorrect. Why?

(a)

(b)

(c)

FIGURE 23.21 Hypothetical electric field lines; the patterns are incorrect if due to static charges beyond the edges of the diagram.

23.4 MOTION IN A UNIFORM ELECTRIC FIELD

We have now discussed the electric field and the electric force at some length. A charged particle in a known electric field **E** experiences an electric force **F** = *q***E**, and from this we can calculate the acceleration of the charged particle, as in Example 4. To find the actual motion of a charged particle (position **r** as a function of time *t*), we apply the definitions of acceleration **a** = *d***v**/*dt* and velocity **v** = *d***r**/*dt*. As we learned in mechanics, solving these equations of motion can in general be quite difficult; however, the simplest case of a constant acceleration is easy. Recall that for a constant acceleration in two dimensions, the solution for the individual components of the velocity, Eqs. (4.20)–(4.21), gives

$$v_x = v_{0x} + a_x t \quad \text{and} \quad v_y = v_{0y} + a_y t \tag{23.14}$$

where v_{0x} and v_{0y} are the initial velocity components, $a_x = F_x/m = qE_x/m$, and $a_y = F_y/m = qE_y/m$. The solution for the components of the position, Eqs. (4.23)–(4.24), yields

$$x = x_0 + v_{0x}t + \tfrac{1}{2}a_x t^2 \quad \text{and} \quad y = y_0 + v_{0y}t + \tfrac{1}{2}a_y t^2 \tag{23.15}$$

where x_0 and y_0 specify the initial position. You might remember from Chapter 4 that in the case of a uniform gravitational field, the result was projectile motion, where the position of the particle traces out a parabola in two dimensions.

The same is true for uniform electric fields. As we saw above, two parallel sheets of charge with equal and opposite surface charge densities produce a region of constant electric field. This arrangement, typically attained with metal plates, is exploited in many electric devices. One difference with projectile motion is that electric charges can be positive or negative (and electric fields can point in any direction), so the parabolic path is not always concave down, as it is with gravity. Another feature is that, in practice, the charged particle's motion can be manipulated only in the limited region of space between the parallel sheets of charge.

To calculate the motion of a charged particle in a uniform **E** field, we use Eqs. (23.14) and (23.15). If the only motion is along the applied field, the problem is one-dimensional and simple. But if there is motion both along the field and perpendicular to it, two equations will probably need to be solved simultaneously. We can best simplify such problems if we choose our coordinate system with one axis parallel to the field; then, the motion along the other direction (perpendicular to the field) has zero acceleration.

FIGURE 23.22 Ion gun from an ion-milling machine.

EXAMPLE 8   An ion milling machine (see Fig. 23.22) uses a beam of gallium ions (*m* = 70 u) to carve microstructures from a target. A region of uniform electric field between parallel sheets of charge is used for precise control of the beam direction. Singly ionized gallium atoms with an initially horizontal velocity of 1.8×10^4 m/s enter a 2.0-cm-long region of uniform electric field which points vertically upward as shown in Fig. 23.23. The ions are redirected by the field and exit the field region at the angle θ shown. If the field is set to a value of $E = 90$ N/C, what is the exit angle θ?

SOLUTION: A singly ionized atom has charge $q = +e$. It is convenient to choose the origin of our coordinate system at the initial ion position at the left end of the field region ($x_0 = 0$, $y_0 = 0$) and the x axis horizontal. The desired exit angle θ is

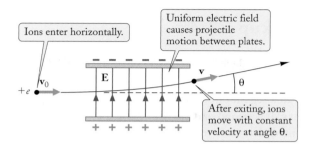

FIGURE 23.23 Motion of a charged particle; between the parallel plates is a region of uniform electric field.

given by the final direction of motion of the beam as it exits the electric field region:

$$\tan \theta = \frac{v_y}{v_x}$$

There is no horizontal electric field, so the x component of the acceleration is zero; Eq. (23.14) confirms that the x component of the velocity does not change, so

$$v_x = v_{0x}$$

The initial y component of the velocity is zero, so Eq. (23.14) for the y component of the velocity yields

$$v_y = a_y t$$

To obtain the time t that the ions spend in the region of uniform electric field, we rearrange Eq. (23.15) for the x component of the position, with $a_x = 0$:

$$t = \frac{x}{v_{0x}}$$

where x is the width of the field region. Inserting this expression for t in the previous equation, we have

$$v_y = a_y \frac{x}{v_{0x}}$$

Using $a_y = F_y/m = eE/m$, we find

$$v_y = \frac{eEx}{mv_{0x}}$$

We can now determine the exit angle:

$$\tan \theta = \frac{v_y}{v_x} = \frac{1}{v_{0x}} \frac{eEx}{mv_{0x}}$$
$$= \frac{eEx}{mv_{0x}^2}$$

Inserting the given numerical values and the value of the atomic mass unit, $1 \text{ u} = 1.66 \times 10^{-27}$ kg, we obtain

$$\tan \theta = \frac{eEx}{mv_{0x}^2} = \frac{1.60 \times 10^{-19} \text{ C} \times 90 \text{ N/C} \times 2.0 \times 10^{-2} \text{ m}}{70 \times 1.66 \times 10^{-27} \text{ kg} \times (1.8 \times 10^4 \text{ m/s})^2}$$
$$= 7.6 \times 10^{-3}$$

Using a calculator, we take the inverse tangent of 7.6×10^{-3} and obtain

$$\theta = 0.44°$$

Checkup 23.4

QUESTION 1: A charged particle moves along a straight line. Does this mean that no electric field is present?

QUESTION 2: In a region of uniform electric field **E**, a charged particle experiences an acceleration **a**. If a second particle with twice the charge and twice the mass of the first particle enters that same region, it will experience an acceleration

(A) $\frac{1}{4}$**a**　　(B) $\frac{1}{2}$**a**　　(C) **a**　　(D) 2**a**　　(E) 4**a**

23.5 ELECTRIC DIPOLE IN AN ELECTRIC FIELD

If an electrically neutral body is placed in a given electric field, we might expect that the body experiences no force. However, this expectation is not always realized. A neutral body may contain within it separate positive and negative charges (of equal magnitudes), and it is possible that the electric force on one of these charges is larger than that on the other; the body then experiences a net force. Such an imbalance of the forces on the positive and negative charges will happen if the electric field is stronger at the location of one kind of charge than at the location of the other. For example, in a nonuniform electric field, the neutral body shown in Fig. 23.24 with positive charges on one end and negative charges on the other end will be pushed to the left because the electric field that acts on the body is stronger at the location of the negative charges. Note that this electric field—indicated by the field lines in Fig. 23.24—is not the field generated by the body; rather it is an electric field generated by some other charges (not shown in Fig. 23.24). The electric field that acts on a body is usually called the **external field**; in contrast, the field generated by the body itself is called the **self-field**. The latter field exerts only internal forces within the body and does not contribute to the net force acting on the body from the outside.

If the external electric field is uniform, such as the electric field generated by the large flat charged sheets discussed in the preceding sections, then the forces on the positive and negative charges in a neutral body cancel, and there is no net force. However, there may still be a torque. Figure 23.25 shows a neutral body, with positive charge on one end and negative charge on the other end, in a uniform electric field. The body carries equal positive and negative charges $\pm Q$, with the average positions of these charges separated by a distance l. As we already mentioned in Example 1, such a body is called an **electric dipole**. Since the force on the positive charge in Fig. 23.25 is toward the right and the force on the negative charge is toward the left, there is a torque on the body. The torque τ of each force about the center of the body is $\tau = rF \sin \theta$, that is, the product of the distance from the center ($r = \frac{1}{2}l$), the magnitude of the force ($F = QE$), and the sine of the angle between the line from the center and the force ($\sin \theta$). Thus, the magnitude of the torque due to each force is $\tau = \frac{1}{2}lQE \sin \theta$, and the net torque of both forces together is

$$\tau = -lQE \sin \theta \qquad (23.16)$$

The minus sign in this equation is a reminder that here the torque is clockwise, in the sense of negative angles. The torque tends to align the body with the electric field.

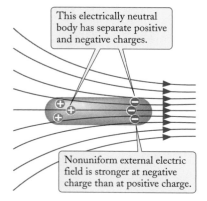

This electrically neutral body has separate positive and negative charges.

Nonuniform external electric field is stronger at negative charge than at positive charge.

FIGURE 23.24 An electric dipole placed in a nonuniform electric field.

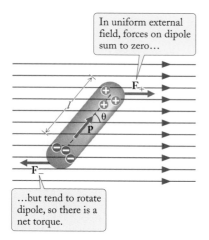

In uniform external field, forces on dipole sum to zero…

…but tend to rotate dipole, so there is a net torque.

FIGURE 23.25 Electric dipole placed in a uniform electric field. The dipole moment vector **p** is directed from the negative charge to the positive charge.

We can write Eq. (23.16) as

$$\tau = -pE \sin \theta \qquad (23.17)$$

where

torque on electric dipole

$$p = lQ \qquad (23.18)$$

dipole moment

The quantity p is called the **dipole moment** of the body; it is simply the magnitude of the charge at either end multiplied by the separation between the charges. The units of dipole moment are coulomb-meters (C·m). We can also introduce a **dipole moment vector p** whose magnitude is equal to the dipole moment p and whose direction is from the negative charge toward the positive charge. Then the torque vector can be simply expressed in terms of the vector **p** as

dipole moment vector p

$$\tau = \mathbf{p} \times \mathbf{E} \qquad (23.19)$$

where the direction of τ is, as usual, the axis of rotation.

Corresponding to the torque (23.17), there exists a potential energy that equals the amount of work that *you* must do against the electric forces to twist the dipole through some angle. If we choose $\theta = 90°$ as a reference position (when the dipole is perpendicular to the field), then, from Eq. (13.4), the work done to rotate the dipole to some other angle θ is

$$W = \int_{90°}^{\theta} \tau \, d\theta = -pE \int_{90°}^{\theta} \sin \theta \, d\theta = pE \cos \theta \Big|_{90°}^{\theta} = pE \, (\cos \theta - 0)$$

$$= pE \cos \theta$$

The potential energy is the negative of the work done, $U = -W$, so

$$U = -pE \cos \theta \qquad (23.20)$$

potential energy of dipole

In terms of the dipole moment vector **p**, the quantity $pE \cos \theta$ is $\mathbf{p} \cdot \mathbf{E}$, and the potential energy U becomes

$$U = -\mathbf{p} \cdot \mathbf{E} \qquad (23.21)$$

The potential energy (23.20) or (23.21) has a minimum $(-pE)$ when the dipole is oriented parallel to the electric field ($\theta = 0$), and it has a maximum $(+pE)$ when the dipole is oriented antiparallel to the electric field ($\theta = 180°$). Figure 23.26 is a plot of the potential energy vs. the angle θ. The equilibrium orientation of the dipole is the parallel orientation, corresponding to the minimum of the potential energy.

Many asymmetric molecules have **permanent dipole moments** due to an excess of electrons on one end of the molecule and a corresponding deficit on the other. This means that the molecule has an accumulation of negative charge on one end and of positive charge on the other. For example, Fig. 23.27 shows the structure of a water molecule. In this molecule, the electrons tend to concentrate on the oxygen atom; in Fig. 23.27, the left side of the molecule is negatively charged, and the right side positively charged. Because of this separation of the average positions of the negative and positive charges, the molecule has a dipole moment. The direction of the dipole vector is from the negative side of the molecule toward the positive side. For a water molecule in water vapor, the dipole moment is $p = 6.1 \times 10^{-30}$ C·m.

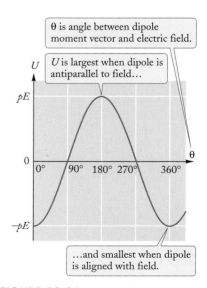

FIGURE 23.26 Potential energy of an electric dipole as a function of angle.

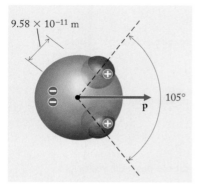

FIGURE 23.27 Arrangement of the atoms of hydrogen (red) and oxygen (blue) in a water molecule. The direction of the dipole moment vector **p** is from the negative side of the molecule toward the positive side.

induced dipole moment

EXAMPLE 9 A molecule of water is placed in an electric field of 2.0×10^5 N/C. What is the difference between the potential energies for the parallel and the antiparallel orientations of the molecule in this electric field?

SOLUTION: As given just above, the dipole moment of the water molecule is $p = 6.1 \times 10^{-30}$ C·m. When the dipole is parallel to the electric field, the angle θ in Eq. (23.20) is zero, and

$$U = -pE \cos 0° = (-6.1 \times 10^{-30} \text{ C·m})(2.0 \times 10^5 \text{ N/C})(1)$$

$$= -1.2 \times 10^{-24} \text{ J}$$

When the dipole is antiparallel to the electric field, the angle is 180°, and

$$U = -pE \cos 180° = (-6.1 \times 10^{-30} \text{ C·m})(2.0 \times 10^5 \text{ N/C})(-1)$$

$$= 1.2 \times 10^{-24} \text{ J}$$

Hence the energy difference between the parallel and antiparallel orientations is 2.4×10^{-24} J.

Molecules of atoms that do not have a permanent dipole moment may acquire a temporary dipole moment when placed in an electric field. The opposite electric forces on the positive and negative charges can distort the molecule and produce a charge separation. Such a dipole moment, which lasts only as long as the molecule is immersed in the electric field, is called an **induced dipole moment**. The magnitude of the induced dipole moment is usually directly proportional to the magnitude of the electric field causing it. This *linear* behavior of the induced dipole moment is useful in devices that store electric charge, discussed in Chapter 26.

✔ Checkup 23.5

QUESTION 1: The torque on a dipole in an electric field depends on the angle between the dipole and the electric field. For what angle is the torque maximum?

QUESTION 2: For what angle between the dipole and the electric field is the potential energy zero?

QUESTION 3: Consider a body with an induced dipole moment. If the body is in a nonuniform electric field (stronger in one region than in another), would you expect the body to be pulled into the region of stronger electric field or pushed into the region of weaker electric field?

QUESTION 4: The permanent electric dipole moment vector of a body is directed along the $+x$ axis. An electric field is applied in the $+y$ direction. The torque on the dipole will cause the dipole moment vector to rotate toward which direction?

(A) $-x$ (B) $+y$ (C) $-y$ (D) $+z$ (E) $-z$

SUMMARY

SUPERPOSITION PRINCIPLE   Electric forces and electric fields produced by different charges or by different charge distributions combine by vector addition.

DEFINITION OF ELECTRIC FIELD
$$\mathbf{E} = \frac{\mathbf{F}}{q}$$
(23.5)

SI UNIT OF ELECTRIC FIELD
$$1 \text{ newton/colomb} = 1 \text{ N/C}$$

ELECTRIC FIELD OF POINT CHARGE q'
Direction is radially outward for positive charge, radially inward for negative charge.
$$E = \frac{1}{4\pi\epsilon_0}\frac{q}{r^2}$$
(23.2)

ELECTRIC FIELD OF A CONTINUOUS CHARGE DISTRIBUTION   Magnitude of contribution:
$$dE = \frac{1}{4\pi\epsilon_0}\frac{dq}{r^2}$$
(23.7)

The total field $\mathbf{E} = E_x\mathbf{i} + E_y\mathbf{j} + E_z\mathbf{k}$ is obtained by summing contributions to each component; for example $E_x = \int dE_x$, with:

$$dE_x = \cos\theta\, dE = \frac{1}{4\pi\epsilon_0}\cos\theta\frac{dq}{r^2}$$
(23.8)

where θ is the angle between the electric field contribution and the x axis.

LINEAR AND SURFACE CHARGE DISTRIBUTIONS

Linear:
$$dq = \lambda\, dL \quad \text{where } \lambda \text{ is in coulombs per meter.}$$

Surface:
$$dq = \sigma\, dA \quad \text{where } \sigma \text{ is in coulombs per square meter.}$$
(23.9)

ELECTRIC FIELD OF INFINITE, UNIFORMLY CHARGED THIN ROD
Direction is perpendicular to rod (outward for $+\lambda$, inward for $-\lambda$).
$$E = \frac{1}{2\pi\epsilon_0}\frac{\lambda}{r}$$
(23.10)

ELECTRIC FIELD OF INFINITE, UNIFORMLY CHARGED FLAT SHEET
Direction is perpendicular to sheet (outward for $+\sigma$, inward for $-\sigma$).
$$E = \frac{\sigma}{2\epsilon_0}$$
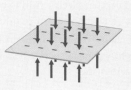
(23.12)

ELECTRIC FIELD OF A PAIR OF OPPOSITELY CHARGED, PARALLEL FLAT SHEETS

$$E = \frac{\sigma}{\epsilon_0} \quad \text{(between sheets)}$$

$$E = 0 \quad \text{(outside sheets)}$$

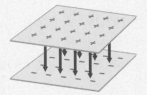

PROPERTIES OF ELECTRIC FIELD LINES

Lines are tangent to the electric field vector at any point.

Density of lines is proportional to the magnitude of the field.

Field lines do not cross.

Field lines start on positive charges and end on negative charges. The number of field lines emerging from (terminating on) a positive (negative) charge is proportional to the charge.

MOTION IN UNIFORM E

$$\mathbf{a} = \frac{\mathbf{F}}{m} = \frac{q\mathbf{E}}{m} \tag{23.14}$$

$$\mathbf{v} = \mathbf{v}_0 + \mathbf{a}t$$

$$\mathbf{r} = \mathbf{r}_0 + \mathbf{v}_0 t + \tfrac{1}{2}\mathbf{a}t^2$$

ELECTRIC DIPOLE MOMENT

Direction of dipole moment vector **p** is from − to +.

$$p = lQ \tag{23.18}$$

TORQUE ON DIPOLE

$$\tau = -pE \sin\theta \quad \text{or} \quad \boldsymbol{\tau} = \mathbf{p} \times \mathbf{E} \tag{23.17, 23.19}$$

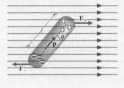

POTENTIAL ENERGY OF DIPOLE

$$U = -pE\cos\theta \quad \text{or} \quad U = -\mathbf{p} \cdot \mathbf{E} \tag{23.20, 23.21}$$

QUESTIONS FOR DISCUSSION

1. Does it make any difference whether the value of the charge q in the equation defining the electric field [Eq. (23.5)] is positive or negative?

2. How would you formally define a gravitational field vector? Is the unit of gravitational field the same as the unit of electric field? According to your definition, what are the magnitude and the direction of the gravitational field at the surface of the Earth?

3. A large, flat sheet measures $L \times L$; the sheet carries a uniform distribution of charge. Roughly how far from the center of the sheet would you expect the electric field to be markedly different from the uniform electric field of an infinitely large sheet?

4. If a positive point charge is released from rest in an electric field, will its trajectory coincide with a field line? What if the point charge has zero mass?

5. A **tube of force** is the volume enclosed by a bundle of adjacent field lines (Fig. 23.28). Such a tube of force is analogous to a stream tube in hydrodynamics, and the field lines are analogous to streamlines. Along such a tube of force, the magnitude of the electric field varies in inverse proportion to the cross-sectional area of the tube. Explain.

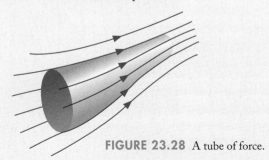

FIGURE 23.28 A tube of force.

6. A negative point charge $-q$ sits in front of a very large, flat sheet with a uniform distribution of positive charge. Make a rough sketch of the pattern of field lines. Is there any point where the electric field is zero?

7. A very long straight line of positive charge lies along the z axis. A very large flat sheet of positive charge lies in the x–y plane. Sketch a few of the field lines of the net electric field produced by both of these charge distributions acting together.

8. A large, flat, thick slab of insulator has positive charge uniformly distributed over its volume. Sketch the field lines on both sides and inside the slab; pay careful attention to the starting points of the field lines.

9. If our Universe is topologically closed, so that a straight line drawn in any direction returns on itself from the opposite direction, can the net charge in the Universe be different from zero?

10. How could you build a "compass" that indicates the direction of the electric field?

11. When a neutral metallic body, insulated from the ground, is placed in an electric field, it develops a charge separation, acquiring positive charge on one end and negative charge on the other. This means the body acquires an induced dipole moment. How is the direction of this dipole moment related to the direction of the electric field?

12. One electric dipole is at the origin, oriented parallel to the z axis. Another electric dipole is at some distance on the x axis. The electric field of the first dipole then exerts a torque on the second dipole. For what orientation of the second dipole is the potential energy minimum?

PROBLEMS

†23.1 The Electric Field of Point Charges

1. Electric fields as large as 3.4×10^5 N/C have been measured by airplanes flying through thunderclouds. What is the force on an electron exposed to such a field? What is the acceleration?

2. An electron moving through an electric field is observed to have an acceleration of 1.0×10^{16} m/s^2 in the x direction. What must be the magnitude and the direction of the electric field that produces this acceleration?

3. Example 2 gives the x and y components and the magnitude of the electric field near a thundercloud. What is the direction of this electric field?

4. The Earth has an atmospheric electric field. During days of fair weather (no thunderclouds), this atmospheric electric field has a strength of about 100 N/C and points down. Taking into account this electric field and also gravity, what will be the acceleration (magnitude and direction) of a grain of dust of mass 1.0×10^{-18} kg carrying a single *electron* charge?

5. **Millikan's experiment** measures the elementary charge e by the observation of the motion of small oil droplets in an electric field. The oil droplets are charged with one or several elementary charges, and if the (vertical) electric field has the right magnitude, the electric force on the droplet will balance its weight, holding the drop suspended in midair. Suppose that an oil droplet of radius 1.0×10^{-4} cm carries a single elementary charge. What electric field is required to balance the weight? The density of oil is 0.80 g/cm^3.

6. In an X-ray tube, electrons are exposed to an electric field of 8.0×10^5 N/C. What is the force on an electron? What is its acceleration?

7. According to a theoretical estimate, at the surface of a neutron star of mass 1.4×10^{30} kg and radius 1.0×10^4 m there is an electric field of magnitude 6.0×10^3 N/C pointing vertically up. Compare the electric force on a proton with the gravitational force on the proton.

8. A long hair, taken from a girl's braid, has a mass of 1.2×10^{-3} g. The hair carries a charge of 1.3×10^{-9} C distributed along its length. If we want to suspend this hair in midair, what (uniform) electric field do we need?

9. Electric breakdown (sparks) occurs in air if the electric field reaches 3.0×10^6 N/C. At this field strength, free electrons present in the atmosphere are quickly accelerated to such large speeds that upon impact on atoms they knock electrons off the atom and thereby generate an avalanche of electrons. How far must a free electron move under the influence of the above electric field if it is to attain a kinetic energy of 3.0×10^{-19} J (which is sufficient to produce ionization)?

10. The nuclei of the atoms in a chunk of metal lying on the surface of the Earth would fall to the bottom of the metal if their weight were the only force acting on them. Actually, within the interior of any metal exposed to gravity there exists a very small electric field that points vertically up. The corresponding electric force on a nucleus just balances the weight of the nucleus. Show that for a nucleus of atomic number Z and mass m, the required field has a magnitude $mg/(Ze)$. What is the numerical value of this electric field in a chunk of iron?

11. The hydrogen atom has a radius of 5.3×10^{-11} m. What is the magnitude of the electric field that the nucleus of the atom (a proton) produces at this radius?

†For help, see Online Concept Tutorial 27 at www.wwnorton.com/physics

12. What is the strength of the electric field at the surface of a uranium nucleus? The radius of the nucleus is 7.4×10^{-15} m and the electric charge is $92e$. For the purposes of this problem the electric charge may be regarded as concentrated at the center.

13. Figure 23.29 shows the arrangement of nuclear charges (positive charges) of a KBr molecule. Find the electric field that these charges produce at the center of mass at a distance of 9.3×10^{-11} m from the Br atom.

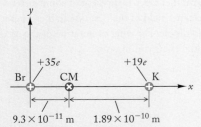

FIGURE 23.29 The positive (nuclear) charges in a KBr molecule.

14. Suppose that in a hydrogen atom the electron is (instantaneously) at a distance of 2.1×10^{-10} m from the proton. What is the net electric field that the electron and the proton produce jointly at a point midway between them?

*15. If a 1.0×10^{-10} C charge is placed on the x axis 0.15 m from the origin of a coordinate system (see Fig. 23.30), what is the magnitude of the electric field at a point 0.10 m up the y axis?

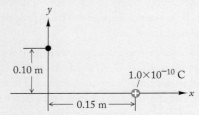

FIGURE 23.30 A point charge.

*16. Three point charges $-Q$, $2Q$, and $-Q$ are arranged on a straight line, as illustrated in Fig. 23.31. What is the electric field that the charges produce at a distance x to the right of the central charge?

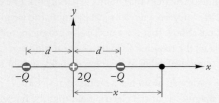

FIGURE 23.31 Three point charges.

*17. Four point charges $\pm Q$ are arranged at the corners of a square of side L as shown in Fig. 23.32. Consider the midpoint of each of the four sides of the square. At each midpoint, find the magnitude and the direction of the electric field.

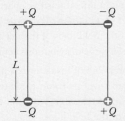

FIGURE 23.32 Four point charges.

*18. The distance between the oxygen nucleus and each of the hydrogen nuclei in an H_2O molecule is 9.58×10^{-11} m; the angle between the hydrogen atoms is 105° (Fig. 23.33). Find the electric field produced by the nuclear charges (positive charges) at the point P at a distance of 1.20×10^{-10} m to the right of the oxygen nucleus.

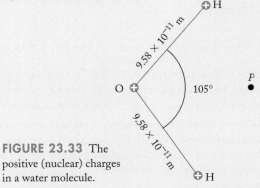

FIGURE 23.33 The positive (nuclear) charges in a water molecule.

*19. Figure 23.34 shows the charge distribution within a thundercloud. There is a charge of 40 C at a height of 10 km, -40 C at 5.0 km, and 10 C at 2.0 km. Treating these charges as pointlike, find the electric field (magnitude and direction) that they produce at a height of 8.0 km and a horizontal distance of 3.0 km.

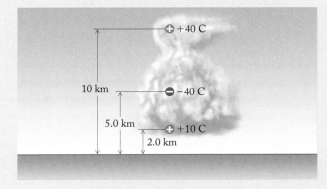

FIGURE 23.34 Charges in a thundercloud.

*20. Suppose that an airplane flies through the thundercloud described in Problem 19 at the 8.0-km level. Plot the magnitude of the electric field as a function of position along the path of the airplane; start with the airplane 10 km away from the thundercloud.

*21. For the four thundercloud and image charges of Fig. 23.5, calculate the magnitude of the electric field at horizontal distances of 2.0, 4.0, 6.0, 8.0, and 10 km from the point on the ground directly below the thundercloud charges. Plot the magnitude of the field vs. the horizontal distance.

*22. Consider eight of the ions of Cl^- and Na^+ on a crystal lattice of common salt. The ions are located at the vertices of a cube measuring 2.82×10^{-10} m on an edge (Fig. 23.35). Calculate the magnitude of the electric force that seven of these ions exert on the eighth.

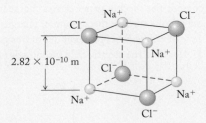

FIGURE 23.35 Ions of Cl^- and Na^+ in a salt crystal.

*23. For the three charges described in Problem 16, again consider the electric field at a point on the positive x axis. What is the approximate dependence of the electric field on the distance x for $x \gg d$? This is the **electric quadrupole** field. [Hint: For small $\delta = (d/x)$, use the approximation $(1 + \delta)^n \approx 1 + n\delta$.]

*24. Identical charges $+Q$ are at seven of the corners of a cube of side a. The eighth corner is empty. From symmetry considerations, what is the direction of the net electric field at the empty corner? Calculate the magnitude of the net electric field at the empty corner.

23.2 The Electric Field of Continuous Charge Distributions

25. (a) Equation (23.11) gives the electric field on the axis of a charged ring. Where is the strength of this electric field maximum?

(b) Roughly sketch the electric field in the space surrounding the ring.

26. You wish to generate a uniform electric field of 2.0×10^5 N/C in the space between two flat, parallel plates of metal placed face to face. The plates measure 0.30 cm $\times$ 0.30 cm. How much electric charge must you put on each plate? Assume that the gap between the plates is small so that the charge distribution and the electric field are approximately uniform, as for infinite plates.

27. A long, straight rod carries a uniform distribution of electric charge of 2.0×10^{-14} C per meter. What is the electric field at a perpendicular distance of 0.50 m from this rod? At 1.0 m? At 1.5 m?

*28. Each of two very long, straight rods carries a positive charge of 1.0×10^{-12} C per meter. One rod lies along the x axis, the other along the y axis. Find the electric field (magnitude and direction) at the point $x = 0.50$ m, $y = 0.20$ m.

*29. Two thin rods of length L, each with uniform linear charge distribution λ, form a cross. Find the electric field at a point a distance $L/2$ from each rod, in the plane of the cross.

*30. Consider an infinitely long line of charge of zero thickness, with a uniform linear charge distribution λ. Show that the tension in such a line is infinite (hence such a line is a physical impossibility). Show that the tension is also infinite in a line of charge of finite length.

*31. Each of two very long, straight, parallel rods carries a positive charge of λ coulombs per meter. The distance between the rods is d (Fig. 23.36). Find the electric field at a point equidistant from the rods, with a distance $2d$ from each rod. Draw a diagram showing the direction of the electric field.

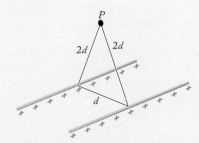

FIGURE 23.36 Two parallel charged rods.

*32. Two infinite lines of silk with uniform linear charge distribution λ lie along the x and the y axes, respectively. Find the electric field at a point with coordinates x, y, z; assume that $x > 0$, $y > 0$, $z > 0$.

*33. A semi-infinite line carrying a uniform charge distribution of λ coulombs per meter lies along the positive x axis from $x = 0$ to $x = \infty$. Find the components of the electric field at the point with coordinates x, y, with $z = 0$; assume $x > 0$, $y > 0$.

*34. A semi-infinite line with a uniform charge distribution of $+\lambda$ coulombs per meter lies along the positive x axis from $x = 0$ to $x = \infty$. Another semi-infinite line with a charge distribution of $-\lambda$ coulombs per meter lies along the negative x axis from $x = 0$ to $x = \infty$. Find the electric field at a point on the y axis.

*35. Electric charge is uniformly distributed over each of three large parallel sheets of paper (Fig. 23.37). The charges per unit area on the sheets are 2.0×10^{-6} C/m^2, 2.0×10^{-6} C/m^2, and -2.0×10^{-6} C/m^2, respectively. The distance between one sheet and the next is 1.0 cm. Find the strength of the electric field **E** above the sheets, below the sheets, and in the spaces between the sheets. Find the direction of **E** at each place.

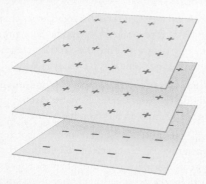

FIGURE 23.37 Three parallel charged sheets of paper.

*36. Each of two very large flat sheets of paper carries a uniform positive charge distribution of 3.0×10^{-4} C/m². The two sheets of paper intersect at an angle of 45° (Fig. 23.38). What are the magnitude and the direction of the electric field at a point between the two sheets?

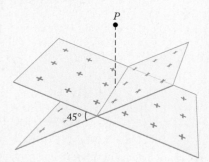

FIGURE 23.38 Two intersecting charged sheets of paper.

*37. Two large sheets of paper intersect at right angles. Each sheet carries a uniform charge distribution of positive charge (Fig. 23.39). The charge per unit area on the sheets is 3.0×10^{-6} C/m². Find the magnitude of the electric field in each of the four quadrants. Sketch the field lines in each quadrant.

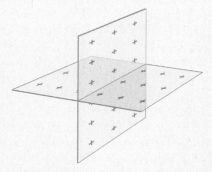

FIGURE 23.39 Another two intersecting charged sheets of paper.

*38. The electric field within a chunk of metal exposed to the Earth's gravity (see Problem 10) is due to a distribution of surface charge. Suppose that we have a slab of iron oriented horizontally (Fig. 23.40). What must be the surface charge densities of the upper and lower surfaces?

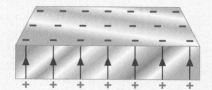

FIGURE 23.40 A horizontal slab of iron.

*39. A very large, flat sheet of paper carries charge uniformly distributed over its surface; the amount of charge per unit area is σ. A hole of radius R has been cut out of this paper (see Fig. 23.41). Find the electric field on the axis of the hole.

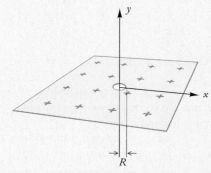

FIGURE 23.41 A charged sheet of paper with a hole.

*40. A paper annulus has an inner radius R_1 and an outer radius R_2. An amount of charge Q is uniformly distributed over the surface of the annulus. What is the electric field on the axis of the annulus at a distance z from the center?

*41. A total amount of charge Q is uniformly distributed along a thin, straight plastic rod of length l.

   (a) Find the electric field at the point P, at a distance x from one end of the rod (Fig. 23.42).

   (b) Find the electric field at point P', at a distance y from the midpoint of the rod (see Fig. 23.42).

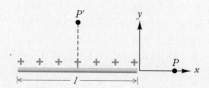

FIGURE 23.42 A charged straight rod.

*42. Two infinite parallel rods are separated a distance $2d$. One carries a uniformly distributed positive charge λ, the other an opposite charge $-\lambda$. The rods are parallel to the z axis and intersect the $x-y$ plane at $x = 0, y = \pm d$. Find the electric field for a point on the positive x axis. How does E behave for $x \gg d$?

*43. Consider the possibility of a **nonuniform charge distribution**. Specifically, consider the line segment of charge of Fig. 23.42, but now assume the total charge Q is distributed unevenly along the length l, with charge density λ that decreases linearly from left to right according to $\lambda = -\lambda_0 x / l$, where λ_0 is a constant, $x = -l$ is at the left end of the rod, and $x = 0$ is at the right end of the rod, as in Fig. 23.42. Note that a small charge dQ in a length element dx is still given by $dQ = \lambda \, dx$.

(a) By summing the charge along the rod, $Q = \int dQ$, determine the value of λ_0 (in terms of Q and l).

(b) Calculate the electric field E at the point P a distance x from the end of the rod (on the positive x axis).

(c) Check the behavior of your result for $x \gg l$. [Hint: You will need to expand the expressions from (b)].

*44. A thin plastic rod is bent so that it has the shape of a semicircle of radius R (Fig. 23.43). An amount of charge Q is uniformly distributed along the rod. What is the electric field at the center of the circle?

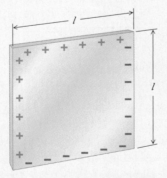

FIGURE 23.43 A charged semicircular rod.

*45. A Plexiglas square of dimension $l \times l$ has a uniform charge density of magnitude λ coulombs per meter along its edges. Two of the edges are positive and two are negative (Fig. 23.44). Find the electric field at the center of the square.

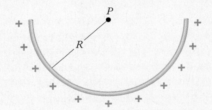

FIGURE 23.44 A square, with charge along its edges.

*46. A cylindrical Plexiglas tube of length l, radius R carries a charge Q uniformly distributed over its surface. Find the

electric field on the axis of the tube at one of its ends. [Hint: Sum field contributions from ringlike elements of the tube; see Eq. (23.11) from Example 6.]

*47. A thin, semi-infinite rod with a uniform charge distribution of λ coulombs per meter lies along the positive x axis from $x = 0$ to $x = \infty$; a similar rod lies along the positive y axis from $y = 0$ to $y = \infty$ (see Fig. 23.45). Calculate the electric field at a point in the $x-y$ plane in the first quadrant.

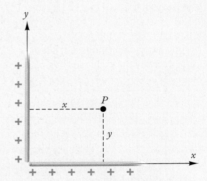

FIGURE 23.45 Two charged straight rods.

*48. (a) Consider an infinitely long rod of radius R with charge uniformly distributed over its volume at the rate of λ coulombs per meter of length. Show that the tension in such a rod is infinite.

(b) Consider a rod of finite length L; assume $R \ll L$. Estimate the tension in such a rod, at some point near its middle. Do not bother with an exact calculation, but try to discover the correct dependence on λ and R. What is the estimated tension for a long rod of radius 1.0 mm with a charge of 1.0×10^{-7} C/m?

**49. Two thin, semi-infinite rods lie in the same plane. They make an angle of 45° with each other, and they are joined by another thin rod bent along an arc of circle of radius R, with center P (see Fig. 23.46). All the rods carry a uniform charge distribution of λ coulombs per meter. Find the electric field at the point P.

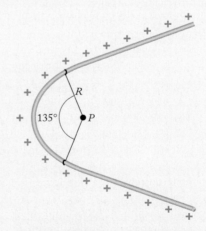

FIGURE 23.46 Two charged straight rods joined by a curved rod.

**50. A square of paper measuring $l \times l$ carries an amount of charge Q uniformly distributed over its surface. The square lies in the x–y plane with its center at the origin and its sides parallel to the x and y axes. Find the electric field at a point on the y axis; assume that the point is outside the square. (Hint: Sum the contributions from each line segment of charge along the square.)

†23.3 Lines of Electric Field

51. A positive charge $+2q$ and a negative charge $-3q$ are separated by a distance d. Sketch the field lines of the electric field produced jointly by these charges in a plane containing the two charges.

52. A very long, thin rod carries $+Q$ coulombs of charge uniformly distributed in each meter of its length, and a point charge $-Q$ is located one meter from the rod. Sketch the field lines of the electric field produced jointly by these charges in the plane containing the rod and the point charge.

53. Sketch the electric field lines due to a finite rod of length L which has a charge Q uniformly distributed over its length. (Hint: Consider the field both very close to the rod, where it looks like an infinite rod, and very far from the rod, where it looks like a point charge.)

54. Three positive charges are at the vertices of an equilateral triangle. Sketch the field line pattern in the plane of the triangle.

55. Two positive charges $+Q$ are at two of the vertices of an equilateral triangle; at the third vertex is a negative charge $-Q$. Sketch the field line pattern in the plane of the triangle.

56. Sketch the electric field line pattern for an infinite row of point charges $+q$, each separated by a distance d from its two nearest neighbors.

23.4 Motion in a Uniform Electric Field

57. The electric field in the electron gun of a TV tube is supposed to accelerate electrons uniformly from 0 to 3.3×10^7 m/s within a distance of 1.0 cm. What electric field is required?

58. An electron is initially at rest, 0.10 cm from a very large disk which carries a uniform surface charge density $\sigma = +3.0 \times 10^{-8}$ C/m². How long does it take the electron to strike the plate? What is its speed just before striking the plate?

59. A proton, initially 4.0 cm from a uniformly charged sheet with $\sigma = 5.0 \times 10^{-8}$ C/m², is fired with initial speed v_0 directly toward the sheet. The proton slows down and momentarily stops a distance 0.15 cm from the sheet before recoiling away from the sheet. What was the value of v_0?

60. In a **cathode-ray tube**, a beam of electrons (the cathode ray) is deflected in a region of electric field on its way to a fluorescent screen, as shown in Fig. 23.47. Consider the parallel-plate arrangement in the figure, and assume that the electric field $E = 400$ N/C is uniform between the plates and that $E = 0$ outside the plates. The beam of electrons is injected horizon-

†For help, see Online Concept Tutorial 27 at www.wwnorton.com/physics

tally with velocity $v_0 = 5.0 \times 10^6$ m/s. If the width of the plates is $L = 4.0$ cm, by what vertical distance y_1 is the beam deflected upon just exiting the plates? If the distance from the end of the plates to the screen is $D = 12.0$ cm, what is the total vertical deflection $y = y_1 + y_2$ upon reaching the screen?

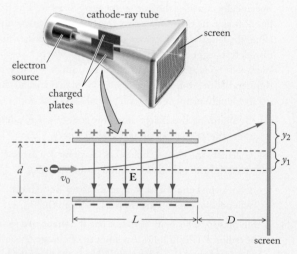

FIGURE 23.47 Schematic diagram of a cathode-ray tube.

61. In the cathode-ray tube of Fig. 23.47, a beam of electrons is injected horizontally into the exact center of the parallel-plate region. If $L = 3.0$ cm, $d = 0.20$ cm, and each plate carries a uniform charge density of magnitude $\sigma = 1.5 \times 10^{-7}$ C/m², what minimum initial speed v_0 must the electrons have to ensure that they do not strike the upper plate?

62. Consider the parallel plates of Fig. 23.48. Assume the injected particles are electrons with initial speed $v_0 = 4.0 \times 10^6$ m/s and the angle $\theta = 35°$. If the uniform vertical electric field has the value $E = 3000$ N/C, at what horizontal distance will the electrons strike the lower plate?

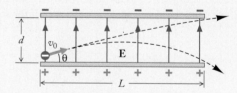

FIGURE 23.48 Charge injected at an angle between plates.

*63. Consider the parallel plates of Fig. 23.48. Assume that the injected particles are polystyrene beads, each of mass 2.0×10^{-11} g; the beads carry negative charges of hundreds to thousands of elementary charges. The uniform electric field between the plates is $E = 900$ N/C, directed vertically upward. The uniform field region is $L = 20$ cm long and $d = 1.0$ cm high. The balls have initial velocity $v_0 = 3.0$ m/s and enter the corner of the field region at $\theta = 5.0°$. For what numbers of elementary charges will the balls exit the field region without striking one of the plates?

23.5 Electric Dipole in an Electric Field

64. A small, straight bit of thread has a charge of $+1.0 \times 10^{-14}$ C at one end and a charge of -1.0×10^{-14} C at the other. The length of the thread is 2.0 mm.

    (a) What is the dipole moment?

    (b) What is the torque on this thread if it is placed in an electric field of 6.0×10^5 N/C at right angles to the field?

65. The two charges of ± 40 C in the thundercloud of Fig. 23.34 form a dipole. What is the dipole moment?

66. In a hydrogen atom, the electron is at a distance of 5.3×10^{-11} m from the proton.

    (a) What is the instantaneous dipole moment of this system?

    (b) Taking into account that the electron moves around the proton on a circular orbit, what is the time-average dipole moment of this system?

67. (a) Pretend that the HCl molecule consists of (pointlike) ions of H^+ and Cl^- separated by a distance of 1.0×10^{-10} m. If so, what would be the dipole moment of this system?

    (b) The observed dipole moment is 3.4×10^{-30} C·m. Can you suggest a reason for this discrepancy?

68. The dipole moment of an HCl molecule is 3.4×10^{-30} C·m. Calculate the magnitude of the torque that an electric field of 2.0×10^6 N/C exerts on this molecule when the angle between the electric field and the longitudinal axis of the molecule is 45°.

REVIEW PROBLEMS

69. To measure the magnitude of a horizontal electric field, an experimenter attaches a small charged cork ball to a string and suspends this device in the electric field. The electric force pushes the ball to one side, and the ball attains equilibrium when the string makes an angle of 35° with the vertical (see Fig. 23.49). The mass of the ball is 3.0×10^{-5} kg, and the charge on the ball is 4.0×10^{-7} C. What is the magnitude of the electric field?

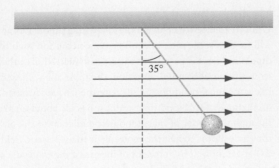

FIGURE 23.49 Cork ball suspended in an electric field.

70. When placed in a given uniform electric field, a proton develops an acceleration of 2.0×10^{10} m/s^2 in the eastward direction. What acceleration will be developed by an electron when placed in the same electric field?

71. A large sheet of paper with a uniform charge distribution of 2.0×10^{-4} C/m^2 is pierced perpendicularly by a very long straight rod with a uniform linear charge distribution of 3.0×10^{-7} C/m (see Fig. 23.50). Find the magnitude and direction of the electric field at a point above the sheet distance of 1.0 cm from the rod.

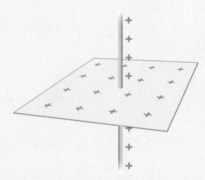

FIGURE 23.50 A charged sheet of paper and a charged rod.

72. Consider two parallel sheets with uniform, opposite charge distributions, as in Fig. 23.10. Suppose that the magnitude of the charge density on each sheet is 2.0×10^{-5} C/m^2. The upper sheet is positive and the lower is negative.

    (a) What is the magnitude of the electric field in the region between the sheets?

    (b) You now insert another large parallel sheet with positive charge density 1.0×10^{-5} C/m^2 in the space between the two charged sheets. What is the magnitude of the electric field in the space immediately above the middle sheet? Below the middle sheet?

73. Consider the charge distribution within the thundercloud illustrated in Fig. 23.4a. Find the electric field (magnitude and direction) that this charge distribution produces at a height of 10 km and a horizontal distance of 5.0 km to the right of the charges.

*74. Three charges $\pm Q$ are arranged at the vertices of an equilateral triangle of side L as shown in Fig. 23.51. Calculate the electric field that these charges produce at the center of the triangle.

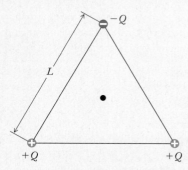

FIGURE 23.51 Three point charges.

*75. A semi-infinite thin rod with uniform linear charge density λ lies along the negative x axis, from $x = 0$ to $x = -\infty$ (see Fig. 23.52).

(a) Find the electric field at a point P a distance d on the positive x axis.

(b) What is the force that this field exerts on a charge q placed at the point P?

(c) What is the electric force that this point charge q exerts on the rod?

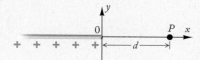

FIGURE 23.52 A semi-infinite charged rod.

*76. Two infinite thin rods intersect at an angle α (see Fig. 23.53). Each charge carries a uniform linear charge distribution λ. Find the electric field on the bisector of the angle α.

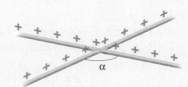

FIGURE 23.53 Two intersecting charged rods.

*77. Three point charges $-Q$, $2Q$, and $-Q$ are arranged on a straight line, as illustrated in Fig. 23.54. What is the electric field that these charges produce at a distance y above the central charge?

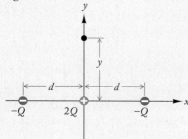

FIGURE 23.54 Three point charges.

*78. Three thin glass rods carry charges Q, Q, and $-Q$, respectively. The length of each rod is l, and the charge is uniformly distributed along each rod. The rods form an equilateral triangle. Calculate the electric field at the center of the triangle.

*79. A paper disk of radius R carries an amount of charge Q uniformly distributed over its surface. Find the electric field at a point on the axis of the disk at a distance z from the center. Show that in the limiting case $z \gg R$, the result reduces to that for a point charge. Hint: To estimate the repulsive force between the right and the left halves of the rod, divide each half into a short segment of whose length is a few times R (say nR) and a long segment of length L/2 − nR. The repulsive force between the long segments can be calculated as in Problem 22.50. Estimate the repulsive force contributed by the short segments by treating them as lumps of charge located at the centers of the segments, and show that this contribution is small if L ≫ R.)

*80. The dipole moment of the water molecule is 6.1×10^{-30} C·m. In an electric field, a molecule with a dipole moment will tend to settle into an equilibrium orientation such that the dipole moment is parallel to the electric field. If disturbed from this equilibrium orientation, the molecule will oscillate like a torsional pendulum. Calculate the frequency of small oscillations of this kind for a water molecule about an axis through the center of mass (and perpendicular to the plane of the three atoms) when the molecule is in an electric field of 2.0×10^6 N/C. The moment of inertia of the molecule about this axis is 1.93×10^{-47} kg·m².

Answers to Checkups

Checkup 23.1

1. In this case, the test charge is q', and the charge producing the field is q. So the magnitude of the field due to q is $E = F/q'$ (the force divided by the test charge), or $E = (1/4\pi\epsilon_0)q/r^2$. The direction is radially away from q for positive q, or radially toward q for negative q.

2. Since the field due to a point charge decreases with the inverse square of the distance, the field at 2 m is one-fourth as large,

or $(1/4) \times 10^5$ N/C. Similarly, at 3 m, the field is $(1/9) \times 10^5$ N/C.

3. Since these field contributions are in perpendicular directions, the resultant magnitude is $E = \sqrt{10^2 + 10^2}$ N/C $= \sqrt{2} \times 10$ N/C $= 14$ N/C; since the two perpendicular components are equal, the direction of the resultant is at 45°, i.e., precisely in the northeast direction.

4. At the center of the square, the contribution to the electric field from each point charge points away from that point charge. Since the charges are equal and are equidistant from the center, the electric fields due to charges diagonally across from each other exactly cancel, and the net field is zero.

5. (A) Toward the negative charge. The fields due to the two equal positive charges that are diagonally across from one another cancel at the center of the square. However, the electric field at the center of the square due to the remaining positive charge points away from it, and the electric field due to the negative charge points toward it; these two contributions are in the same direction, directly toward the negative charge.

Checkup 23.2

1. The contribution to the electric field from each positive sheet of charge points perpendicularly out from the corresponding sheet and is independent of distance from the sheet. For two perpendicular equivalent sheets, these two vectors are equal in magnitude and at right angles; they add to produce a field at 45° to the sheets (with magnitude $\sqrt{2}$ times as large as each individual sheet field).

2. As we saw in the examples, far away from a charge distribution, the field is essentially that of a point charge, in this case $E = (1/4\pi\epsilon_0)Q/r^2$. Very close to the surface, the disk looks like an infinite plane of charge, with field $E = \sigma/2\epsilon_0 = (Q/A)/2\epsilon_0$.

3. For an infinite sheet (Example 7), the electric field is independent of distance. So $E = E_0$ at both 2 m and 4 m from the sheet.

4. (B) 0; σ/ϵ_0. Since each sheet by itself produces a field that points perpendicularly away from the sheet on each side, the total field is zero in between the two sheets, where the two equal-magnitude contributions oppose. Outside the sheets, the two contributions are parallel, so the field is twice as strong as that due to either sheet alone. Outside, the sheets behave like a single sheet with twice the charge density.

Checkup 23.3

1. In Fig. 23.21a, there are electric field lines that cross; since the direction of the electric field is tangent to the field lines, crossing field lines (two directions at one point) are not allowed in field line diagrams. Figure 23.21b shows field lines beginning in empty space; a field line must begin and end on charges (or at infinity). Last, Fig. 23.21c shows a field line looping around and closing on itself; in electrostatics, field lines must have a beginning and an end.

2. The sphere has some positive charge (in the region where field lines are emerging) and some negative charge (where field lines are entering). Since the net number of field lines is two lines emerging (four emerging minus two entering), the net charge on the sphere must be one-quarter the value of the point charge (which had eight lines emerging).

3. Since the field is pointing downward, the field lines terminate at the Earth's surface, and the Earth's surface charge must be negative.

4. (D) −9.0 C. Since field lines only enter the rod, its charge is negative. Since there are 12 lines entering the rod, compared with 8 lines emerging from the point charge, the magnitude of the charge on the rod is 12/8 = 3/2 times that of the point charge. Thus the charge on the rod is $-(3/2) \times 6.0$ C = −9.0 C.

Checkup 23.4

1. No. If the motion is at constant velocity, then there is no net force and thus no electric field; but if the straight-line motion is accelerated, then there is an electric field directed along the line of motion.

2. (C) **a.** The acceleration of a charged particle in an electric field **E** is $\mathbf{a} = \mathbf{F}/m = q\mathbf{E}/m$. So if **E** is the same and the charge q and mass m of the particle both double, then the charge-to-mass ratio q/m remains the same, and the acceleration is unchanged.

Checkup 23.5

1. Since the torque is proportional to the sine of the angle between the field and the dipole moment ($\sin \theta$), it is maximum for $\theta = 90°$. That is, there is a maximum torque when the dipole moment is perpendicular to the field.

2. Since the potential energy of a dipole is defined by $U = -pE \cos \theta$, it is zero for $\theta = 90°$ (or $\theta = 270°$), when the dipole is perpendicular to the field.

3. For an induced dipole moment, the dipole moment is parallel to the applied field. For the equal-magnitude charges of a dipole, the greater force occurs where the field is stronger: if this is at the positive charge, the net force will be parallel to the field; if this is at the negative charge, the net force will be antiparallel to the field. In either case, the net force is toward the region of stronger field.

4. (B) +y. The torque tends to align the dipole moment of a body with the electric field, here the +y direction.

Gauss' Law

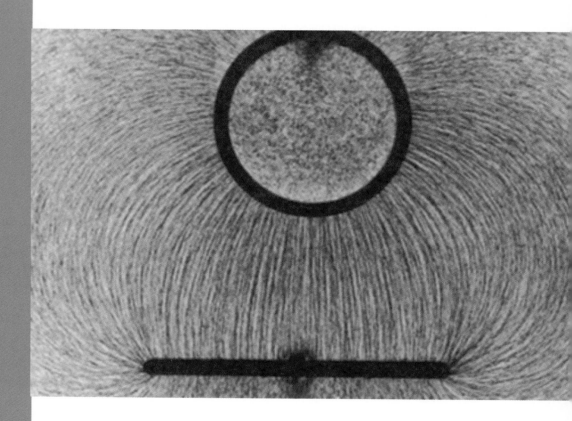

CONCEPTS IN CONTEXT

Concepts
— in —
Context

This photograph illustrates bits of thread dispersed in oil to reveal the pattern of the electric field generated by the electric charges residing on two parallel conductors. The bits of thread act as electric dipoles, which tend to align with the local electric field. The charge distribution in this photograph is highly symmetric, and correspondingly, the electric field pattern is symmetric.

We can exploit symmetry to facilitate the calculation of electric fields, which will permit us to answer the following questions:

? What is the electric field due to a spherically symmetric distribution of charge? (Examples 4 and 5, pages 764–766)

? What is the electric field near a flat sheet of charge, or near a flat conducting slab? (Examples 9 and 11, pages 770 and 775)

? What is the electric field inside a conductor in electrostatic equilibrium, and what does this tell us about the distribution of charge on the conductor? (Section 24.5, page 774, and Example 12, page 776)

The electric field of any charge distribution can be calculated by means of Coulomb's Law, as in the examples of the preceding chapter. The calculation is fairly simple if the charge distribution consists of an arrangement of just a few point charges. But as we saw in Section 23.2, this method often becomes complicated and tedious if the charge is distributed over a surface or volume, when we have to perform a difficult summation or integration of all the individual electric field vectors of the small pieces of the charge distribution. Fortunately, there is another method for calculating the electric field; this method relies on a theorem called Gauss' Law. This law is a consequence of Coulomb's Law, and therefore it contains no new physics. It does, however, contain some new mathematics, which supplies an elegant shortcut for calculating the electric field, provided that the charge distribution has a high degree of symmetry. This means that whereas Gauss' Law is always valid, it does not ease every calculation, but when it does, it works wonders.

In this chapter we will first introduce the concept of electric flux; we will then present Gauss' Law and show how it makes finding the electric field easy in many situations. We will also discuss how charge can be distributed on metals and insulators, and how such knowledge of allowed charge distributions helps in applying Gauss' Law.

24.1 ELECTRIC FLUX

Online
Concept
Tutorial

Consider a mathematical (that is, imagined) surface in the shape of a rectangle of area A. Suppose that this surface is immersed in a constant electric field $\mathbf{E}$ (see Fig. 24.1a). This electric field makes some angle with the surface; the electric field vector has a component tangential, or parallel, to the surface and a component normal, or perpendicular, to the surface. The **electric flux** Φ_E *through the surface is defined as the product of the area A and the normal component of the electric field*. If we designate this perpendicular component of the electric field by $E_\perp$, the electric flux is

$$\Phi_E = E_\perp A \qquad (24.1)$$

The SI unit for electric flux is the unit of electric field times the unit of area, that is, $(N/C)\cdot m^2$, or $N \cdot m^2/C$.

The normal component $E_\perp$ can also be expressed as $E \cos\theta$, where θ is the angle between $\mathbf{E}$ and a perpendicular to the surface (see Fig. 24.1a). With this expression for $E_\perp$, Eq. (24.1) becomes

$$\Phi_E = EA \cos\theta \qquad (24.2)$$

electric flux Φ_E

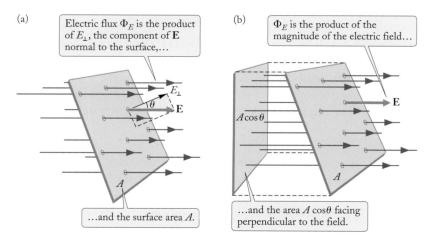

(a) Electric flux Φ_E is the product of $E_\perp$, the component of $\mathbf{E}$ normal to the surface,...

...and the surface area A.

(b) Φ_E is the product of the magnitude of the electric field...

...and the area $A \cos\theta$ facing perpendicular to the field.

FIGURE 24.1 (a) Flat rectangular surface immersed in a uniform electric field. The perpendicular to the surface makes an angle θ with the field lines. The normal component of the electric field is $E_\perp = E \cos\theta$. The electric flux is $\Phi = E_\perp A$. In terms of field lines, the electric flux is proportional to the number of lines intercepted by the area. (b) The area $A \cos\theta$ is the projection of the area A onto a plane perpendicular to the electric field.

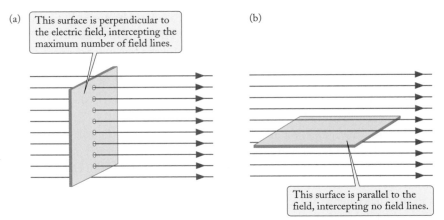

FIGURE 24.2 (a) A surface of area A is perpendicular to the electric field lines. It intercepts an electric flux $\Phi = EA$. (b) The surface is parallel to the field lines. It intercepts no electric flux.

The product $A \cos \theta$ can be interpreted as the projection of the area A onto a plane perpendicular to the electric field; that is, $A \cos \theta$ can be regarded as that portion of the area that effectively faces the electric field, as shown in Fig. 24.1b. Thus, Eq. (24.2) also states that the flux is the magnitude E of the electric field multiplied by the area facing perpendicular to the electric field. If we think in terms of the field lines introduced in Section 23.3, then the electric flux through an area A is proportional to the number of field lines intercepted by that area.

Note that if $\theta = 0$, the flux is simply $\Phi_E = EA$; in this case, the area A is exactly face on to the electric field, and it intercepts the maximum number of field lines (see Fig. 24.2a). On the other hand, if $\theta = 90°$, then the flux given by Eq. (24.2) is zero; in that case, the area A is parallel to the electric field, and all the field lines skim by the area without crossing it (see Fig. 24.2b). This dependence of flux, or the intercepted number of field lines, on the angle of orientation of the area can be understood by a simple analogy: Think of the electric field lines as analogous to the trajectories of falling raindrops, and think of the area as a sheet of paper. The flux is the flow of raindrops intercepted by the paper (*flux* comes from the Latin word for a flow). If you hold the paper face on to the rain, it will intercept a maximum number of drops; if you hold it parallel to the rain, it will intercept no drops. At intermediate angles, the number of drops will be proportional to $\cos \theta$, as in Eq. (24.2).

Equation (24.1) is mathematically equivalent to the dot product, or scalar product, of two vectors: the electric field vector **E** and the "vector area" **A**. The vector **A** is defined with a magnitude equal to the ordinary surface area A and a direction perpendicular to the surface; that is, $\mathbf{A} = A\,\hat{\mathbf{n}}$, where $\hat{\mathbf{n}}$ is a normal (perpendicular) unit vector. With this definition, the electric flux is

$$\Phi_E = EA \cos \theta = \mathbf{E} \cdot \mathbf{A} \qquad (24.3)$$

It is sometimes helpful to recall that in terms of the vector components the dot product is $\mathbf{E} \cdot \mathbf{A} = E_x A_x + E_y A_y + E_z A_z$. Whether the angular or the vector-component formula is easier to use depends on the particular situation.

If the surface of interest consists of several flat areas, each intercepting a uniform field **E**, then the total flux is simply given by the sum of the fluxes through each flat area:

$$\Phi_E = \sum E_\perp A = \sum \mathbf{E} \cdot \mathbf{A} \qquad (24.4)$$

Next, we need a general definition of the electric flux for a surface of arbitrary, curved shape immersed in an arbitrary, nonuniform electric field, that is, an electric field that may have different magnitudes and directions at different points on the surface (see Fig. 24.3). The surface can be regarded as consisting of many small (infinitesimal) flat

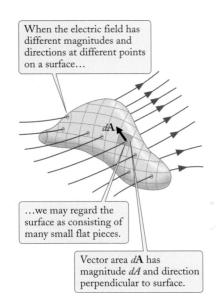

FIGURE 24.3 An arbitrary surface immersed in an arbitrary, nonuniform electric field.

pieces. For a small piece with a vector area $d\mathbf{A}$ in an electric field $\mathbf{E}$, the contribution to the electric flux is $\mathbf{E} \cdot d\mathbf{A}$. If we take the sum, or integral, of all these small amounts of electric flux for all the small pieces of the surface, we obtain the electric flux through the entire surface,

$$\Phi_E = \int \mathbf{E} \cdot d\mathbf{A} = \int E_\perp \, dA \qquad (24.5)$$

where again $E_\perp$ is the component of $\mathbf{E}$ perpendicular to the surface.

Keep in mind that the electric flux is proportional to the number of field lines intercepted by the surface. Note that whereas lines going through the surface from one side make a positive contribution to the flux, lines going through from the other side make a negative contribution (see Fig. 24.4 for an example). However, for the *open* surfaces thus far considered, the reference direction (the side for which outward flux is positive) can be chosen either one way or the other.

The definition of flux is also valid for a *closed* surface, such as the surface of a sphere or the surface of a cube. The electric flux through such a surface is

$$\Phi_E = \oint \mathbf{E} \cdot d\mathbf{A} = \oint E_\perp \, dA \qquad (24.6)$$

where the circle on the integral sign merely indicates a closed surface. For the case of closed surfaces, there is a sign convention: *the normal component $E_\perp$ is reckoned as positive if the direction of the electric field $\mathbf{E}$ is outward from the surface, and negative if $\mathbf{E}$ is inward, into the surface.* Thus, the electric flux through a closed surface is proportional to the *net* number of field lines emerging from the surface. The electric flux will be positive or negative depending on whether more field lines emerge from the surface or enter the surface.

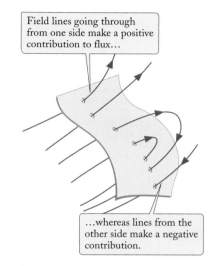

Field lines going through from one side make a positive contribution to flux…

…whereas lines from the other side make a negative contribution.

FIGURE 24.4 Arbitrary surface immersed in an arbitrary, nonuniform electric field. Although seven field lines cross the surface, four are in one direction and three are in the other; thus the net flux through the surface is one-seventh the flux one would have if all the field lines crossed in the same direction.

EXAMPLE 1   A flat sheet of paper measuring 22 cm × 28 cm is placed in a uniform electric field of 100 N/C. What is the flux through the paper if the paper makes an angle of 90° with the electric field? If the paper makes an angle of 30°? (See Fig. 24.5.)

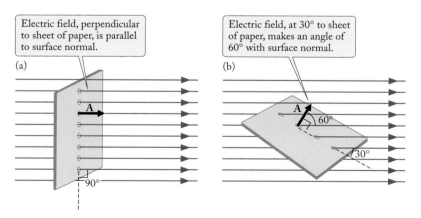

Electric field, perpendicular to sheet of paper, is parallel to surface normal.

Electric field, at 30° to sheet of paper, makes an angle of 60° with surface normal.

FIGURE 24.5 A sheet of paper placed in a uniform electric field (a) at an angle of 90° with the electric field and (b) at an angle of 30°. The black arrow $\mathbf{A}$ indicates the perpendicular to the surface.

SOLUTION: For a uniform field and a flat area, the flux is merely the product $\Phi_E = EA \cos\theta$, and since an angle of 90° between the sheet of paper and the electric field means $\theta = 0$ in Eq. (24.2) See Fig. 24.5a,

$$\Phi_E = EA \cos\theta = (100 \text{ N/C}) \times (0.22 \text{ m} \times 0.28 \text{ m}) \times 1 = 6.2 \text{ N·m}^2/\text{C}$$

An angle of 30° between the paper and the electric field means $\theta = 60°$ (see Fig. 24.5b), so

$$\Phi_E = EA \cos\theta = (100 \text{ N/C}) \times (0.22 \text{ m} \times 0.28 \text{ m}) \times \cos 60° = 3.1 \text{ N·m}^2/\text{C}$$

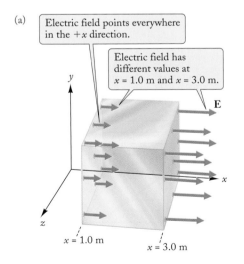

Consider an imaginary spherical surface with radius r.

Electric field is everywhere perpendicular to surface.

Magnitude of electric field is constant over area $4\pi r^2$.

FIGURE 24.6 A spherical surface with a point charge q at its center.

EXAMPLE 2   Consider an imaginary sphere of radius r which has a positive point charge q at its center (see Fig. 24.6). What electric flux does the electric field of this point charge produce through the surface of the sphere?

SOLUTION: The positive point charge q produces a radial, outward electric field of magnitude $E = q/(4\pi\epsilon_0 r^2)$ that crosses the spherical surface everywhere perpendicularly. The area of such a sphere is $4\pi r^2$, so

$$\Phi_E = EA \cos\theta = EA = \frac{q}{4\pi\epsilon_0 r^2} \times 4\pi r^2 = \frac{q}{\epsilon_0} \qquad (24.7)$$

COMMENT: We will see in the next section that this simple result is generally true; indeed, the same result, $\Phi_E = q/\epsilon_0$, would have been obtained for *any* surface enclosing the charge q, and for *any* way that the enclosed charge q was distributed. But in this example we see explicitly and simply how the result comes from the inverse-square nature of the electric field, which exactly cancels the surface area $4\pi r^2$.

EXAMPLE 3   Consider a cube of side $a = 2.0$ m with one corner at $(x, y, z) = (1.0 \text{ m}, 0, 0)$ and its sides parallel to the axes as shown in Fig. 24.7a. The cube is in a region where the electric field points everywhere in

(a)

Electric field points everywhere in the $+x$ direction.

Electric field has different values at $x = 1.0$ m and $x = 3.0$ m.

$x = 1.0$ m

$x = 3.0$ m

(b)

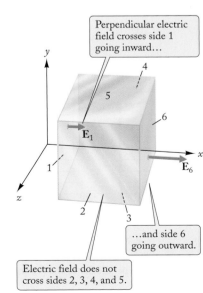

Perpendicular electric field crosses side 1 going inward...

...and side 6 going outward.

Electric field does not cross sides 2, 3, 4, and 5.

FIGURE 24.7 (a) A cubical surface immersed in a nonuniform electric field. (b) The numbering of the cube faces used when evaluating the net flux through the cube.

the $+x$ direction. The magnitude of this electric field varies as a function of x only, and has the values $E_x = 5.0$ N/C at $x = 1.0$ m and $E_x = 15$ N/C at $x = 3.0$ m. What is the electric flux through the cube?

SOLUTION: The total flux is the sum of the fluxes through the individual faces of the cube, labeled 1 through 6 in Fig. 24.7b. The area of each face is $A = a^2 = (2.0 \text{ m})^2 = 4.0 \text{ m}^2$. Note that for sides 2, 3, 4, and 5, the field skims along the surfaces without crossing them. Thus

$$\Phi_E = \oint E_\perp \, dA = \Phi_{E1} + \Phi_{E2} + \Phi_{E3} + \Phi_{E4} + \Phi_{E5} + \Phi_{E6}$$

$$= \Phi_{E1} + 0 + 0 + 0 + 0 + \Phi_{E6} \qquad (24.8)$$

$$= (E_\perp A)_1 + (E_\perp A)_6$$

Since the direction of **E** at side 1 of the cube is inward, the flux for side 1 is negative, whereas the flux for side 6 is positive. Also, the electric field is perpendicular to both areas, so the net flux through the cube is

$$\Phi_E = -(5.0 \text{ N/C}) \times (4.0 \text{ m}^2) + (15 \text{ N/C}) \times (4.0 \text{ m}^2) = 40 \text{ N·m}^2\text{/C}$$

$$(24.9)$$

 Checkup 24.1

QUESTION 1: Consider Fig. 24.1a. For the surface with the orientation shown in that figure, the flux has some particular value. Would the flux be larger or smaller if the surface were horizontal? Vertical?

QUESTION 2: Would the answer to Example 2 change if instead of the spherical surface we had the same charge inside a cube, as in Fig. 24.8? Inside a cylinder?

QUESTION 3: Consider the point charge and surface of Fig. 24.9. The net flux through this surface is

(A) Positive   (B) Negative   (C) Zero

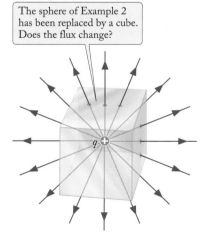

FIGURE 24.8 A cubical surface encloses a point charge q.

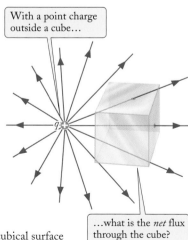

FIGURE 24.9 A cubical surface adjacent to a point charge q.

Online
29
Concept
Tutorial

24.2 GAUSS' LAW

We saw in Example 2 that a point charge q enclosed within a surface produces an electric flux $\Phi_E = q/\epsilon_0$ through the surface. This result can be generalized for an arbitrary charge distribution enclosed within an arbitrary surface. The general result is known as **Gauss' Law**:

> *If an arbitrary closed surface has a net electric charge Q_{inside} within it, then the electric flux through the surface is $Q_{\mathrm{inside}}/\epsilon_0$, that is,*

Gauss' Law

$$\Phi_E = \oint \mathbf{E} \cdot d\mathbf{A} = \frac{Q_{\mathrm{inside}}}{\epsilon_0} \quad \text{or} \quad \Phi_E = \oint E_\perp \, dA = \frac{Q_{\mathrm{inside}}}{\epsilon_0} \quad (24.10)$$

Gaussian surface

Gauss' Law says that the *total* electric flux through *any* closed surface is determined solely by the amount of charge *inside* the surface. The surface, called a **Gaussian surface**, can be a real one, but it is often an imaginary surface in space, chosen to pass through some point of interest. In general, as above in Figs. 24.8 and 24.9, a charge inside the surface generates a net flux through the surface, whereas any charge outside the surface makes zero contribution to the net flux through the surface.

The proof of Gauss' Law is easy, if we exploit the concept of field lines. The electric field appearing in Eq. (24.10) is the sum of individual electric fields of point charges, some *outside* the closed surface and some inside it. The individual electric field of a charge outside the closed surface generates no net flux through the closed surface—any such field line either does not touch the surface or else enters it at one point and leaves at another; neither case makes any net contribution to the flux. However, the individual electric field of a point charge within the surface does contribute to the flux. For example, a positive point charge q has a number of outward field lines proportional to its charge q, all of which will pierce the closed surface—such a charge thus contributes a flux proportional to q. We saw in Example 2 that the constant of proportionality between a flux and the charge producing it is $1/\epsilon_0$. Taking into account that positive charges generate positive flux and negative charges negative flux, we see that the net flux, proportional to the net number of field lines piercing the surface, is equal to the net charge divided by ϵ_0. This completes the proof of Gauss' Law.

Gauss' Law can be used to calculate the electric field, provided that the charge distribution has a high degree of symmetry. Essentially, Gauss' Law gives us some information about the electric field. Symmetry conditions give us some more information. By combining these two kinds of information, we can often evaluate the electric field easily, without the laborious process of summing small contributions of small pieces of charge using Coulomb's Law. In the next section, we will examine the three geometries where such an easy application of Gauss' Law is possible: spherical, cylindrical, and planar.

 Checkup 24.2

QUESTION 1: Given that a closed surface has zero electric flux through it, what can you conclude about the electric charge inside the surface? What if the surface is not quite closed (has a hole)?

QUESTION 2: Consider a closed surface that surrounds both of the charges in Fig. 24.10. What is the flux through this surface?

QUESTION 3: What is the charge enclosed by the cube of Example 3?

QUESTION 4: A sphere of radius R is in a region of uniform electric field E. The net flux through the surface of the sphere is

(A) $E \times \pi R^2$ (B) $E \times 2\pi R^2$ (C) $E \times 4\pi R^2$ (D) $E \times \frac{4}{3}\pi R^3$ (E) Zero

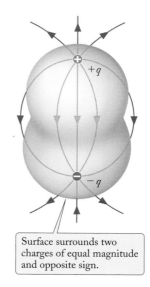

Surface surrounds two charges of equal magnitude and opposite sign.

FIGURE 24.10 Field lines generated by positive and negative charges of equal magnitudes.

24.3 APPLICATIONS OF GAUSS' LAW

Gauss' Law relates the normal component $E_\perp$ of the electric field on a closed surface to the total charge inside the surface. To apply Gauss' Law in each of the three geometries with high symmetry (spherical, cylindrical, and planar), we will first choose a spherical or cylindrical surface for which the flux on the left side of Gauss' Law can be rewritten as the simple product EA, where E is the electric field at a desired point and A is a known area containing that point. We will then need to calculate Q_{inside}, the charge inside our surface. Finally, we can equate the flux EA with Q_{inside}/ϵ_0 and solve for the desired quantity (usually E). To see how to calculate Q_{inside} for our surface, let us consider useful ways to quantify how charge may be distributed.

Macroscopically, the smallness of each elementary charge means that charge distributions can be viewed as *continuous* (even 1 μC contains several thousand billion elementary charges). We already saw two kinds of charge distributions in the previous chapter: a linear charge distribution, and a surface charge distribution. In the examples below, we will also encounter the third kind, a volume charge distribution. In the cases of immediate interest, we will again assume the charge distribution is *uniform*, that is, constant, over a specified region. For a uniform distribution, the charge q on a line of length L, or on a surface area A, or in a volume V, is proportional to the length, or the area, or the volume:

$$\text{Line:} \qquad q = \lambda L$$

$$\text{Surface:} \qquad q = \sigma A \qquad\qquad (24.11)$$

$$\text{Volume:} \qquad q = \rho V$$

where λ (lambda) is the charge per unit length in coulombs per meter, or C/m; σ (sigma) is the charge per unit area in coulombs per square meter, or C/m^2; and ρ (rho) is the charge per unit volume in coulombs per cubic meter, or C/m^3. For instance, a volume of 1 cm^3 with a charge of 1 μC uniformly distributed throughout it has $\rho = 1$ μC/cm$^3 = 1$ C/m^3. The more general possibility of a *nonuniform* distribution, where the value of ρ may vary as a function of position, is discussed in Problems 39–44. In the examples, we restrict ourselves to uniform distributions.

We will now examine some calculations of the electric field $\mathbf{E}$ in each of the three high-symmetry geometries: spherical, cylindrical, and planar; with these as background, Gauss' Law can then be applied to an enormous number of problems. We begin with two examples in spherical symmetry.

Spherical Symmetry

Charge distributions with spherical symmetry include points, spheres, thin or thick spherical shells, etc., and concentric layers of such objects.

KARL FRIEDRICH GAUSS (1777–1855) *German mathematician, physicist, and astronomer. Gauss was an indefatigable calculator, and he loved to perform enormously complicated computations, which today would be regarded as impossible without an electronic computer. He developed many new methods for calculations in celestial mechanics; later he became interested in electric and magnetic phenomena, which he researched in collaboration with Wilhelm Eduard Weber (1804–1891).*

Concepts
— in —
Context

EXAMPLE 4   Use Gauss' Law to deduce the electric field of a point charge.

SOLUTION: Although we already know the solution (Coulomb's Law, $E = kq/r^2$), it is instructive to obtain it from Gauss' Law. We begin with a symmetry argument: Since the point charge is spherically symmetric, so must be the electric field. This spherical symmetry requires that the magnitude of the electric field be the same at all points at equal distances from the center, that is, at all points on a spherical surface centered on the charge, since otherwise the electric field would make an unacceptable distinction between such points. Furthermore, the spherical symmetry requires that the direction of the electric field be rotationally symmetric around any radial line extending outward from the charge. The only direction of the electric field consistent with this requirement is *radial*, since any nonradial component would make an unacceptable distinction between different directions around this line.

Now consider a spherical Gaussian surface of radius r centered on the charge q, as shown in Fig. 24.6. According to the symmetry arguments, at all points on this surface, the electric field must be radial (radially outward for a positive charge and radially inward for a negative charge) and constant (for a given value of r). On this Gaussian surface, the normal component of the electric field is the same as the total field, $E_\perp = E$; and the flux is therefore

$$\oint E_\perp \, dA = EA = E \times 4\pi r^2 \tag{24.12}$$

The charge inside is merely the point charge q, so Gauss' Law tells us

$$E \times 4\pi r^2 = \frac{Q_{\text{inside}}}{\epsilon_0} = \frac{q}{\epsilon_0} \tag{24.13}$$

from which we immediately obtain Coulomb's Law:

$$E = \frac{1}{4\pi\epsilon_0} \frac{q}{r^2} \tag{24.14}$$

COMMENT: Notice that the same argument holds for *any* spherically symmetric charge distribution, provided $q = Q_{\text{inside}}$. Hence the general result: the electric field E at a radius r due to *any* spherically symmetric charge distribution is the same as if all the charge *inside the shell of radius r* were concentrated at a point charge in the center.

Concepts
— in —
Context

EXAMPLE 5   A spherical region of radius R has a total charge Q distributed uniformly thougout the volume of this region. (a) What is the electric field at points inside the sphere? (b) What is the electric field at points outside the sphere?

SOLUTION: Since there is charge throughout the volume of the sphere, the electric field lines of this charge distribution start within the volume of the spherical region, at varying distances from the center (see Fig. 24.11). Outside the sphere of charge, the number of electric field lines remains constant.

(a) Once again, since the charge distribution is spherically symmetric, the electric field must be radial and of constant magnitude over any concentric spherical surface of radius r. To find the magnitude of the electric field E at a radius

$r < R$ inside the charge distribution, consider the spherical Gaussian surface of Fig. 24.12a. As always in spherical symmetry, $E_\perp = E$ over the surface, so the flux through such a sphere is again the product $E \times 4\pi r^2$, and Gauss' Law again becomes

$$E \times 4\pi r^2 = \frac{Q_{inside}}{\epsilon_0} \qquad (24.15)$$

To express Q_{inside} in terms of the given quantities, we use the fact that the charge is uniformly distributed throughout the specified volume. Thus for $r < R$, the charge inside is

$$Q_{inside} = \rho V_{inside} \qquad (24.16)$$

where the volume charge density ρ (charge per unit volume) is determined by the total charge Q and total volume V of the *entire* charged sphere:

$$\rho = \frac{Q}{V} = \frac{Q}{\frac{4}{3}\pi R^3} \qquad (24.17)$$

In Eq. (24.16), the quantity V_{inside} is the volume inside our Gaussian surface of radius r:

$$V_{inside} = \frac{4}{3}\pi r^3$$

Substituting this in Eq. (24.16), we find

$$Q_{inside} = \rho V_{inside} = \frac{Q}{\frac{4}{3}\pi R^3} \cdot \frac{4}{3}\pi r^3 = Q \frac{r^3}{R^3}$$

and Gauss' Law then allows us to solve for E:

$$E \times 4\pi r^2 = \frac{Q_{inside}}{\epsilon_0} = \frac{Q}{\epsilon_0} \frac{r^3}{R^3}$$

$$E = \frac{Q}{4\pi\epsilon_0} \frac{r}{R^3} \qquad (r \le R) \qquad (24.18)$$

Note that this electric field inside a uniformly charged sphere increases in direct proportion to the radius r. It is zero at the center, and reaches its maximum value at the edge of the charge distribution ($r = R$), where $E = (1/4\pi\epsilon_0)Q/R^2$.

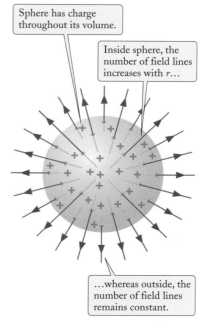

Sphere has charge throughout its volume.

Inside sphere, the number of field lines increases with r...

...whereas outside, the number of field lines remains constant.

FIGURE 24.11 A sphere of radius R with a uniform volume distribution of positive charge. The electric field lines start on positive charges, at different radii. The electric field is everywhere radially outward.

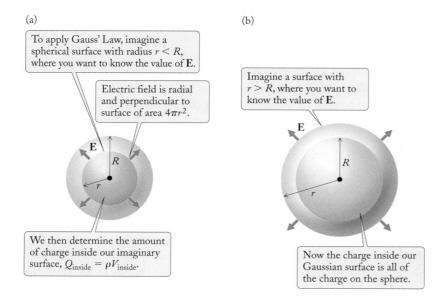

(a)

To apply Gauss' Law, imagine a spherical surface with radius $r < R$, where you want to know the value of **E**.

Electric field is radial and perpendicular to surface of area $4\pi r^2$.

We then determine the amount of charge inside our imaginary surface, $Q_{inside} = \rho V_{inside}$.

(b)

Imagine a surface with $r > R$, where you want to know the value of **E**.

Now the charge inside our Gaussian surface is all of the charge on the sphere.

FIGURE 24.12 (a) An imaginary Gaussian spherical surface of radius $r < R$ (green) inside a uniformly charged sphere of radius R (tan). (b) Same charged sphere as (a), but now the imaginary spherical surface has $r > R$.

(b) To find the electric field outside the sphere, we again take a spherical Gaussian surface of radius r, but now r is larger than R (see Fig. 24.12b). With the usual spherical symmetry argument, Gauss' Law again has the form

$$E \times 4\pi r^2 = \frac{Q_{\text{inside}}}{\epsilon_0} \qquad (24.19)$$

For any $r \geq R$, the charge inside the Gaussian surface is now equal to Q, the total charge. Thus,

$$E \times 4\pi r^2 = \frac{Q}{\epsilon_0} \qquad (24.20)$$

and

$$E = \frac{1}{4\pi\epsilon_0} \frac{Q}{r^2} \quad (r \geq R) \qquad (24.21)$$

Figure 24.13 is a plot of the electric field as a function of distance r from the center. Notice that it increases linearly inside the sphere, but decreases in proportion to $1/r^2$ outside the distribution.

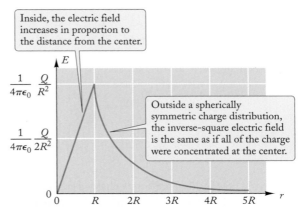

Inside, the electric field increases in proportion to the distance from the center.

Outside a spherically symmetric charge distribution, the inverse-square electric field is the same as if all of the charge were concentrated at the center.

FIGURE 24.13 Magnitude of the electric field as a function of radial distance r for a uniformly charged sphere.

The result (24.21) again implies that *outside a spherically symmetric region containing charge, the electric field is exactly the same as it would be if all of the charge were concentrated at a point in the center* (as already commented in Example 4). Note that the argument leading up to this result does not depend on the uniformity of the charge distribution—it depends only on the spherical symmetry. Hence the argument remains valid for any charge distribution that consists of several concentric shells of different charge densities (arranged like the layers of an onion), where the charge density varies with radius but is the same in all directions. The electric field outside such a charge distribution precisely mimics that of a point charge, Eq. (24.21). This result provides a proof of Newton's famous theorem, mentioned in Section 9.1, concerning the gravitational force exerted by a spherically symmetric planet: the force is as though all of the mass were concentrated at a point at the planetary center. The similarity between electricity and gravitation reflects the inverse-square nature of both force laws.

EXAMPLE 6   A uranium nucleus (charge $+92e$) is (approximately) a uniformly charged sphere of radius 7.4×10^{-15} m. What is the electric field inside a uranium nucleus at a point halfway from the center to the surface? What is the radial electric force if such a field acts on a proton at the same point?

SOLUTION: For the field halfway inside a uniformly charged sphere, we can use Eq. (24.18) with $r = R/2$:

$$E = \frac{Q}{4\pi\epsilon_0} \frac{r}{R^3} = \frac{92e}{4\pi\epsilon_0} \frac{1}{2R^2}$$

$$= \frac{(92)(1.60 \times 10^{-19}\ \text{C})(9.0 \times 10^9\ \text{N·m}^2/\text{C}^2)}{(2)(7.4 \times 10^{-15}\ \text{m})^2}$$

$$= 1.2 \times 10^{21}\ \text{N/C}$$

The force exerted on a proton by such a field is

$$F = eE = (1.6 \times 10^{-19}\ \text{C})(1.2 \times 10^{21}\ \text{N/C}) = 192\ \text{N}$$

In the nucleus of an atom, this large repulsive force is more than compensated by an even larger binding force—the "strong" force—that holds the nucleus together.

Cylindrical Symmetry

Charge distributions with cylindrical symmetry include infinitely long lines, rods (solid cylinders), tubes (cylindrical shells), etc., and concentric layers of such objects.

EXAMPLE 7   Using Gauss' Law, find the electric field of an infinitely long, thin straight rod of charge, with a uniform linear charge density λ.

SOLUTION: We calculated this electric field in Chapter 23 by direct integration of the electric fields of all the infinitesimal charge contributions, a tedious process. Now we will see that the calculation is much simpler by Gauss' Law. First we need to determine the direction of the electric field. The only direction consistent with a cylindrically symmetric charge distribution is *radial* (see Fig. 24.14); electric field lines must originate on charges in the rod, and if the electric field had any nonradial direction, it would make an unacceptable distinction between the upper and lower portions of the rod. Furthermore, the rotational symmetry around the rod dictates that the electric field has the same magnitude at any point on the surface of a cylinder concentric with the rod: otherwise, it would make an unacceptable distinction between different positions along or around the rod.

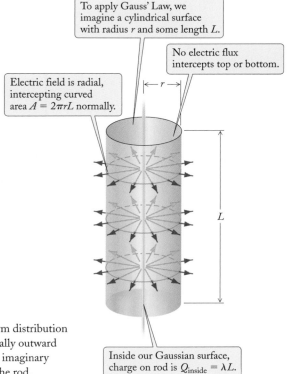

To apply Gauss' Law, we imagine a cylindrical surface with radius r and some length L.

No electric flux intercepts top or bottom.

Electric field is radial, intercepting curved area $A = 2\pi r L$ normally.

$\leftarrow r \rightarrow$

L

Inside our Gaussian surface, charge on rod is $Q_{\text{inside}} = \lambda L$.

FIGURE 24.14 A very long rod with a uniform distribution of positive charge (blue). The field lines are radially outward and are uniformly distributed along the rod. An imaginary cylindrical Gaussian surface (green) surrounds the rod.

For cylindrical problems, we take for our Gaussian surface an imaginary cylindrical surface of radius r (where we want to know E) and of some arbitrary length L. The closed cylindrical surface consists of two circles (top and bottom) and a curved side. Since the electric field is radial, it is parallel to the top and bottom circular surfaces, so no electric flux is intercepted by these surfaces. Thus the top and bottom ends of the cylinder make zero contribution to the flux in cylindrical symmetry. Also, the radial electric field is constant and normal over the curved side of the cylindrical surface, which has area $A = 2\pi rL$ (the circumference multiplied by the height). The total flux is

$$\oint E_\perp \, dA = \int_{\text{top}} E_\perp \, dA + \int_{\text{bottom}} E_\perp \, dA + \int_{\text{side}} E_\perp \, dA$$

$$= 0 + 0 + \int_{\text{side}} E_\perp \, dA$$

$$= E_\perp \int_{\text{side}} dA = E_\perp A_{\text{side}} = E_\perp \times 2\pi rL$$

So for all problems with cylindrical symmetry, Gauss' Law may always be written

$$E \times 2\pi rL = \frac{Q_{\text{inside}}}{\epsilon_0} \tag{24.22}$$

For the simple uniform linear distribution, the charge enclosed by our surface is the charge per unit length multiplied by the length:

$$Q_{\text{inside}} = \lambda L$$

Hence Gauss' Law implies

$$E \times 2\pi rL = \frac{\lambda L}{\epsilon_0}$$

and, solving for E, we find

$$E = \frac{\lambda}{2\pi\epsilon_0 r} \tag{24.23}$$

which agrees with the result from Chapter 23, Eq. (23.10).

COMMENT: Notice that the same relation (24.23) holds for *any* cylindrically symmetric charge distribution, provided λ represents the total charge per unit length inside a cylinder of radius r where we want to know the value of E. Thus we see another general result: the E field at a radius r due to *any* cylindrically symmetric charge distribution is the same as if all the charge *inside the cylinder of radius r* were concentrated on a line of charge along the axis.

EXAMPLE 8   A thick, insulating cylindrical shell with inner radius a and outer radius b has charge distributed thoughout its volume, with a uniform volume charge density ρ, as shown in Fig. 24.15. Find the electric field in the three regions (a) $r \le a$, (b) $a \le r \le b$, and (c) $r \ge b$.

SOLUTION: As just discussed, in cylindrical symmetry the electric field must be radial and of constant magnitude for any particular radius r. Thus Gauss' Law in cylindrical symmetry always becomes

$$E \times 2\pi r L = \frac{Q_{\text{inside}}}{\epsilon_0} \tag{24.24}$$

(a) For $r \leq a$, consider the cylindrical Gaussian surface shown in green in Fig. 24.15a. Since it is contained completely inside the cylindrical "hole" within the shell, it encloses no charge. Thus $Q_{\text{inside}} = 0$ and $E \times 2\pi r L = 0$, or

$$E = 0 \qquad (r \leq a) \tag{24.25}$$

(b) For $a \leq r \leq b$, consider a cylindrical Gaussian surface with its curved part inside the shell, as in Fig. 24.15b. In terms of the given volume charge density ρ,

$$Q_{\text{inside}} = \rho V_{\text{inside}}$$

but in calculating V_{inside}, we must include only that volume that *actually contains charge* (and is inside the radius r). The volume of our Gaussian cylinder is the area πr^2 of its end times its length L; if we subtract the volume that is empty, we have

$$V_{\text{inside}} = \pi r^2 L - \pi a^2 L = \pi L (r^2 - a^2)$$

so Gauss' Law becomes

$$E \times 2\pi r L = \frac{\rho V_{\text{inside}}}{\epsilon_0} = \frac{\rho \pi L (r^2 - a^2)}{\epsilon_0}$$

and from this we find

$$E = \frac{\rho}{2\epsilon_0} \left(r - \frac{a^2}{r} \right) \qquad (a \leq r \leq b) \tag{24.26}$$

(c) For $r \geq b$, the cylindrical Gaussian surface is outside the shell (see Fig. 24.15c). The charge inside is obtained using *only the volume that contains charge* (this is the volume of the charged shell, since we exclude the empty volume inside and outside the shell):

$$Q_{\text{inside}} = \rho V_{\text{inside}} = \rho (\pi b^2 L - \pi a^2 L)$$

so Gauss' Law gives us

$$E \times 2\pi r L = \frac{\rho \pi L (b^2 - a^2)}{\epsilon_0}$$

and

$$E = \frac{\rho (b^2 - a^2)}{2\epsilon_0 r} \qquad (r \geq b) \tag{24.27}$$

Note the $1/r$ dependence for this electric field outside the cylindrically symmetric charge distribution. Indeed, Eq. (24.27) is identical to the result (24.23) for a line of charge, $E = \lambda / 2\pi\epsilon_0 r$, where now the charge per unit length is the charge per unit volume times the cross-sectional area of the shell, that is, $\lambda = \rho(\pi b^2 - \pi a^2)$.

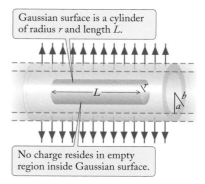

(a) $r \leq a$

Gaussian surface is a cylinder of radius r and length L.

No charge resides in empty region inside Gaussian surface.

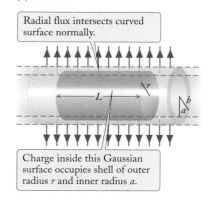

(b) $a \leq r \leq b$

Radial flux intersects curved surface normally.

Charge inside this Gaussian surface occupies shell of outer radius r and inner radius a.

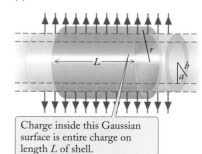

(c) $r \geq b$

Charge inside this Gaussian surface is entire charge on length L of shell.

FIGURE 24.15 (a) An infinite cylindrical shell (tan), with inner radius a and outer radius b, with a uniform volume distribution of positive charge. The field lines are radially outward. To find the electric field at different radii, we consider imaginary Gaussian cylindrical surfaces (green) of different radii, (a) Gaussian surface of radius $r \leq a$. (b) Gaussian surface of radius $a \leq r \leq b$. (c) Gaussian surface of radius $r \geq b$.

Planar Symmetry

Charge distributions with planar symmetry include infinite sheets and slabs of charge and symmetric layers of such sheets or slabs.

(a)

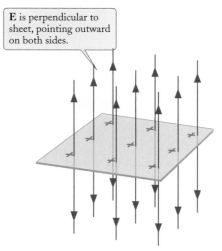

E is perpendicular to sheet, pointing outward on both sides.

(b)

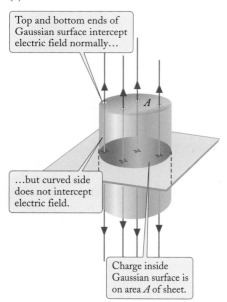

Top and bottom ends of Gaussian surface intercept electric field normally...

A

...but curved side does not intercept electric field.

Charge inside Gaussian surface is on area A of sheet.

FIGURE 24.16 (a) A very large sheet of uniformly distributed positive charge. The field lines are uniformly distributed. (b) A cylindrical Gaussian surface; the area of each end is A.

Concepts in Context

EXAMPLE 9   Using Gauss' Law, find the electric field of a very large uniform sheet of charge with σ coulombs per square meter.

SOLUTION: This is another instance of an electric field we have already calculated by direct integration in Chapter 23. Again, the calculation is much simpler by Gauss' Law. Symmetry tells us that the electric field is everywhere perpendicular to the sheet of charge and has a constant magnitude over any particular surface parallel to the sheet (see Fig. 24.16a). We take a Gaussian surface in the shape of a cylinder of base area A that extends symmetrically on either side of the sheet of charge (see Fig. 24.16b). In this geometry, the electric field is parallel to the curved surface of the cylinder, which therefore intercepts no flux. On each of the two circular ends, the electric field is perpendicular to the surface. Thus the flux through each end is EA. The total flux is

$$\oint E_\perp \, dA = \int_{\text{bottom}} E_\perp \, dA + \int_{\text{curved part}} E_\perp \, dA + \int_{\text{top}} E_\perp \, dA$$

$$= EA + 0 + EA = 2EA$$

so Gauss' Law for this planar geometry becomes

$$2EA = \frac{Q_{\text{inside}}}{\epsilon_0} \tag{24.28}$$

The amount of charge inside the Gaussian surface is the charge per unit area multiplied by the area of the sheet enclosed by the surface, or $Q_{\text{inside}} = \sigma A$. So Gauss' Law gives

$$2EA = \frac{\sigma A}{\epsilon_0}$$

Canceling the common factor of area A, we find

$$E = \frac{\sigma}{2\epsilon_0} \tag{24.29}$$

This result, of course, agrees with the result obtained in Section 23.2.

COMMENTS: In this geometry, we could have used a cubical or rectangular box as our Gaussian surface, instead of a cylinder. The shape of the ends is irrelevant; the crucial features of the Gaussian surface are that its ends must be parallel to the sheet, the surface around it must be perpendicular to the sheet, and it should be symmetric on either side of the sheet. Note that outside the planar charge distribution, the magnitude of the electric field is independent of the distance from the sheet. For each type of symmetry, it is useful to remember the behavior of the electric field outside the charge distribution: spherical, $E \propto 1/r^2$; cylindrical, $E \propto 1/r$; and planar, $E \propto$ constant.

EXAMPLE 10   An infinite, insulating slab of thickness d has a uniform volume charge density ρ, as shown in Fig. 24.17. Find the electric field inside and outside the slab.

SOLUTION: With $x = 0$ at the center of the slab, "inside" the slab means $|x| < d/2$ and "outside" means $|x| > d/2$. Consider the Gaussian surfaces shown in Fig. 24.17. To obtain E inside the slab, we use the smaller cylinder to evaluate $Q_{\text{inside}} = \rho V_{\text{inside}}$. Since its volume is $V_{\text{inside}} = [\text{length}] \times [\text{base area}] = 2x\,A$, Gauss' Law becomes

$$2EA = \frac{Q_{\text{inside}}}{\epsilon_0} = \frac{\rho \times 2x\,A}{\epsilon_0}$$

and solving for E we obtain

$$E = \frac{\rho x}{\epsilon_0} \quad (|x| \leq d/2) \tag{24.30}$$

Thus the E field is zero at the center of the slab and increases linearly with distance inside the slab.

Outside the slab, we use the larger Gaussian cylinder. The amount of volume inside it *which contains charge* is $V_{\text{inside}} = d \times A$, so Gauss' Law implies

$$2EA = \frac{Q_{\text{inside}}}{\epsilon_0} = \frac{\rho \times d \times A}{\epsilon_0}$$

and

$$E = \frac{\rho d}{2\epsilon_0} \quad (|x| \geq d/2) \tag{24.31}$$

Note that this constant electric field is the same as for a sheet of charge ($E = \sigma/2\epsilon_0$) with surface charge density $\sigma = \rho d$, as it must be.

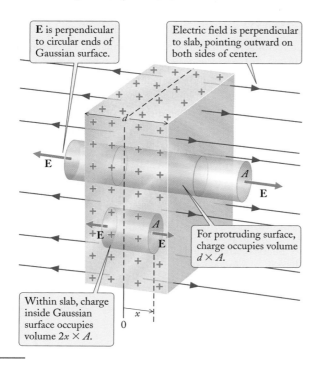

E is perpendicular to circular ends of Gaussian surface.

Electric field is perpendicular to slab, pointing outward on both sides of center.

For protruding surface, charge occupies volume $d \times A$.

Within slab, charge inside Gaussian surface occupies volume $2x \times A$.

FIGURE 24.17 An infinite slab of thickness d with a uniform volume distribution of positive charge. A small, imaginary Gaussian cylinder with end area A is used to find the field inside the slab, a distance x from the center; a longer Gaussian cylinder with end area A is used to find the field outside the slab.

✔ Checkup 24.3

QUESTION 1: A charge distribution consists of four equal charges at the corners of a square. Does this arrangement have enough symmetry so you can use Gauss' Law to calculate the electric field at some distance from the square?

QUESTION 2: A long wire of radius 10^{-5} m carries a uniform linear charge distribution. Another long wire of radius 10^{-6} m carries a linear charge distribution of the same magnitude. Which wire produces a larger electric field at its surface?

QUESTION 3: A charge Q is distributed uniformly over the volume of a sphere of radius R. What is the electric field at the center of the sphere? Where is the electric field strongest? What is the magnitude of this strongest electric field?

QUESTION 4: A charge Q is distributed over the the volume of a sphere of radius R. The charge distribution consists of several concentric shells of different charge densities; although the charge density varies with radius, it is the same in all directions. What is the electric field at the center of this sphere? At the surface?

QUESTION 5: The electric field at some distance from a long wire carrying a uniform charge distribution is 1.0×10^5 N/C. What is the electric field at twice the distance?

    (A) 2.5×10^4 N/C   (B) 5.0×10^4 N/C   (C) 1.0×10^5 N/C
    (D) 2.0×10^5 N/C   (E) 4.0×10^5 N/C

24.4 SUPERPOSITION OF ELECTRIC FIELDS

It is worthwhile to emphasize that the Superposition Principle holds also for extended charge distributions, the same way it did for point charges. That is, the electric field at any point due to several charge distributions is simply the vector sum of the electric fields of the individual charge distributions. With point charges, this is a straightforward sum of Coulomb fields. For a collection of extended objects (spheres, cylinders, shells, planes, etc.), the vector sum can also be readily calculated. The same can be true for more complicated objects, as long as we can view them as sums of simpler objects.

We already considered an important example of superposition in Fig. 23.10, where we saw that two parallel, infinite sheets with equal but opposite charge densities produce an electric field $E = \sigma/\epsilon_0$ between the two sheets and zero field outside them. For another example, consider two parallel, infinite sheets with equal positive charge densities, as shown in Fig. 24.18a. To find the total electric field in each region of space, we need merely consider the electric field due to either sheet separately. Figure 24.18b shows the field $E = \sigma/2\epsilon_0$ due to the left sheet, which points away from the sheet on each side; similarly, Fig. 24.18c shows the field due to the right sheet, which

(a)

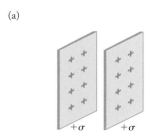

(b)

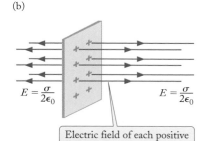

Electric field of each positive sheet points away from sheet on each side.

(c)

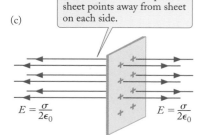

With both sheets, net electric field cancels inside and doubles outside.

(d)

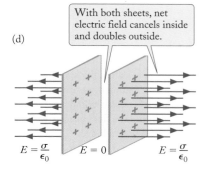

FIGURE 24.18 Superposition example. (a) Two positively charged infinite sheets. The total field can be found by considering the individual fields due to the left sheet (b) and the right sheet (c) separately. In each region of space, one can sum the fields due to the individual sheets to get the total field (d).

FIGURE 24.19 Superposition example. (a) A solid, infinite cylinder with a uniform volume distribution of positive charge, with a spherical hole. The field due to such an object with missing volume can be found by summing the field due to a solid cylinder (b) plus a field due to an oppositely charged sphere (c). At each point, the net field (d) is the vector sum of these two fields.

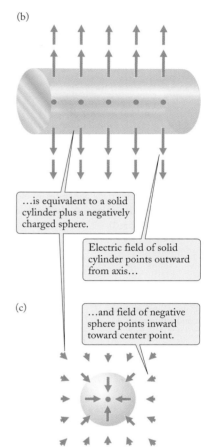

(a)

Uniformly charged cylinder with spherical hole…

(b)

…is equivalent to a solid cylinder plus a negatively charged sphere.

Electric field of solid cylinder points outward from axis…

(c)

…and field of negative sphere points inward toward center point.

points away from it on each side. When we sum the fields due to the two sheets, as in Fig. 24.18d, we see immediately that $E = 0$ between the two sheets, whereas $E = \sigma/2\epsilon_0 + \sigma/2\epsilon_0 = \sigma/\epsilon_0$ to the left and right of both sheets. These examples of sheets with charge densities of equal magnitudes lead to simple doublings or cancellations of the field. However, the Superposition Principle can just as easily be applied to any number of sheets with any values of uniform surface charge densities.

Another example is shown in Fig. 24.19a. Here we have a solid cylinder with uniform volume charge distribution ρ that has a spherical cavity inside of it. Such a hole may be viewed as consisting of equal amounts of positive and negative charge (zero net charge in the hole), so that the electric field at any point is the sum of the electric field due to a solid cylinder (no hole) with charge density ρ and that due to a solid sphere of charge density $-\rho$. Of course, at any position in space, we must take the vector sum of the cylindrical (radially outward from the axis, Fig. 24.19b) and spherical (radially toward the center point, Fig. 24.19c) fields. Such a vector sum is indicated schematically at various points in Fig. 24.19d. Note that for such an object, the highly symmetric field due to each "piece" is a simple application of Gauss' Law, whereas the vector sum of the two contributions is less symmetric and must be calculated for each point.

✔ Checkup 24.4

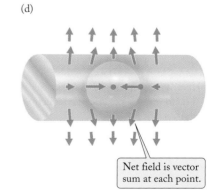

(d)

Net field is vector sum at each point.

QUESTION 1: Suppose that the surface of the spherical cavity in Fig. 24.19 has a uniform surface charge density σ. Can the electric field still be calculated from a superposition of simple high-symmetry applications of Gauss' Law?

QUESTION 2: Assume that the cavity in Fig. 24.19 is cubic instead of spherical. Can the electric field still be calculated from a superposition of simple high-symmetry applications of Gauss' Law?

QUESTION 3: Two parallel infinite sheets carry positive uniform surface charge densities σ and 2σ. What is the magnitude of the electric field between the sheets? Outside both sheets?

    (A) σ/ϵ_0; $3\sigma/\epsilon_0$ (B) $\sigma/2\epsilon_0$; $3\sigma/2\epsilon_0$ (C) $3\sigma/2\epsilon_0$; $\sigma/2\epsilon_0$
    (D) σ/ϵ_0; 0   (E) 0; σ/ϵ_0

Concepts
— in —
Context

24.5 CONDUCTORS AND ELECTRIC FIELDS

In insulators (or *dielectrics*), electrons and ions in atoms are firmly bound to each other, and there are no free, mobile charges. Extra charges can be placed on an insulator in an arbitrary arrangement; these charges also stay put and do not move. However, as we pointed out in Section 22.5, in conductors the charges can move. For instance, in metals—such as copper, silver, or aluminum—some of the electrons are free, that is, they can move about without restraint within the volume of the metal. If such a conductor is immersed in an electric field, the free electrons move in response to the electric force. As an excess of electrons accumulates on one part of the conductor, a deficit of electrons will appear on another part of the conductor; thus, negative and positive charges are induced on the conductor. Within the volume of the conductor, the electric field of the induced charges tends to cancel the external electric field in which the conductor was originally immersed (see Fig. 24.20). The accumulation of negative and positive charges on the surfaces of the conductor continues until the electric field generated by these charges exactly cancels the original electric field that initiated the motion of the electrons. Consequently, *when the charge distribution on a conductor reaches electrostatic equilibrium, the net electric field within the material of the conductor is exactly zero.* The proof of this statement is by contradiction: if the electric field were different from zero, the free electrons would continue to move, and the charge distribution would not (yet) be in equilibrium. For a good conductor (copper, aluminum, etc.), the equilibrium is reached quickly, in a very small fraction of a second.

Furthermore, *for a conductor in electrostatic equilibrium, any (extra) electric charge deposited on the conductor resides on the surfaces of the conductor.* We can prove this using Gauss' Law: Consider a small Gaussian surface *inside* the conducting material (see Fig. 24.21). Since at equilibrium **E** = 0 everywhere in this material, the left side of Gauss' Law vanishes and therefore the right side must also vanish—which means the charge enclosed by *any* arbitrary surface inside the conducting material is zero. Obviously, if the charges are not inside the conductor, they must be on the surfaces.

Finally, we can say something about the electric field just outside a conductor: *the electric field at the surface of a conductor in electrostatic equilibrium is perpendicular to the surface.* The proof is again by contradiction: if the electric field had a component tangential to the surface, the free electrons would move along the surface, and the charge distribution would not be in equilibrium.

Note that this argument does not exclude an electric field perpendicular to the surface of the conductor; such an electric field merely pushes the free electrons against the surface, where they are held in equilibrium by the combination of the force exerted by the electric field and the restraining force of the surface of the conductor. Figure 24.22

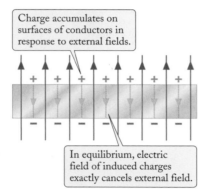

Charge accumulates on surfaces of conductors in response to external fields.

In equilibrium, electric field of induced charges exactly cancels external field.

FIGURE 24.20 A very large conducting slab immersed in a uniform electric field. The uniform distribution of charge that has accumulated on its surface generates an electric field (light red) opposite to the original electric field (dark red).

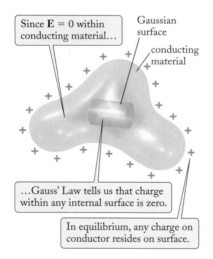

Since **E** = 0 within conducting material…

Gaussian surface

conducting material

…Gauss' Law tells us that charge within any internal surface is zero.

In equilibrium, any charge on conductor resides on surface.

FIGURE 24.21 A closed Gaussian surface (green) inside a volume of conducting material.

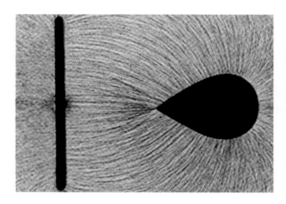

FIGURE 24.22 Field lines in the space surrounding charged conductors made visible by small bits of thread suspended in oil.

displays an experimental demonstration of electric fields perpendicular to the surfaces of conductors. The flat plate and the cylinder in this figure are conductors, and we see that the the field lines meet the surfaces at right angles.

EXAMPLE 11	Find the electric field above a very large, flat conducting slab (see Fig. 24.23) on which there is a surface charge density of σ coulombs per square meter.

SOLUTION: In view of the symmetry of the charge configuration, the electric field will be perpendicular to the conducting slab and will have a constant magnitude over any plane parallel to the conducting slab. As a Gaussian surface, take the cylinder shown in Fig. 24.23, with a base area A. The upper base contributes an amount EA to the flux. The lower base, inside the conductor, does not contribute to the flux, since the electric field is zero in this region. Finally, the curved surface does not contribute to the flux, since any electric field is parallel to this surface. The charge within the Gaussian surface is $Q_{\text{inside}} = \sigma A$. Hence, from Gauss' Law,

$$\oint E_\perp \, dA = EA = \frac{Q_{\text{inside}}}{\epsilon_0} = \frac{\sigma A}{\epsilon_0} \tag{24.32}$$

we obtain

$$E = \frac{\sigma}{\epsilon_0} \tag{24.33}$$

This is a constant electric field, independent of the distance from the conducting slab. This result is strictly valid for a slab of infinite extent, but it is approximately valid for the electric field near a slab of finite extent.

COMMENTS: Note that, for a given value of the surface charge density σ, the field generated by a conducting slab is twice as strong as the field generated by a sheet of charge [see Eq. (24.29)]. The reason is clear from a comparison of Figs. 24.16 and 24.23. In the latter case, the charges produce flux only on one side of the surface; in the former case, the flux is distributed over both sides. How is this possible? In the conducting case, there must always be another surface charge distribution that produces a field which cancels that due to the surface shown in Fig. 24.23. For example, a charged, conducting infinite slab can be considered as two charged sheets, such as the two we considered in Fig. 24.18.

Over a small region *near the surface,* any smooth, curved conducting surface can be approximated by a flat surface. Hence we can use the formula (24.33) to find the electric field in a region very near *any* smooth curved conducting surface (but $E = \sigma/\epsilon_0$ is *not* a good approximation near sharp edges). The formula is valid even if σ is not constant—if σ depends on position along the surface, then the electric field also depends on position.

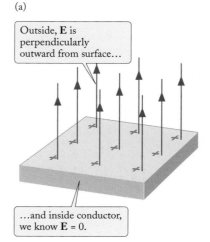

(a)

Outside, **E** is perpendicularly outward from surface…

…and inside conductor, we know **E** = 0.

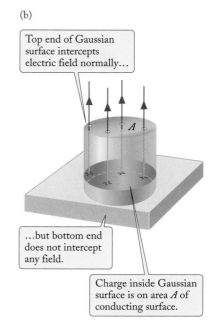

(b)

Top end of Gaussian surface intercepts electric field normally…

A

…but bottom end does not intercept any field.

Charge inside Gaussian surface is on area A of conducting surface.

FIGURE 24.23 (a) A large, flat conducting slab with a uniform distribution of charge on its surface. (b) The cylindrical Gaussian surface has one base outside the conductor and the other base in the conductor.

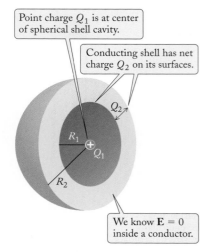

Point charge Q_1 is at center of spherical shell cavity.

Conducting shell has net charge Q_2 on its surfaces.

We know **E** = 0 inside a conductor.

FIGURE 24.24 A point charge Q_1 at the center of a thick spherical conducting shell with inner radius R_1 and outer radius R_2.

EXAMPLE 12

A point charge Q_1 is at the center of a spherical conducting shell of inner radius R_1 and outer radius R_2, as shown in Fig. 24.24. The shell has a net charge Q_2 on its surfaces. Find the electric field in the three regions $r < R_1$, $R_1 < r < R_2$, and $r > R_2$. How much charge is on the inner surface of the shell? The outer surface?

SOLUTION: For any spherical Gaussian surface of radius $r < R_1$, only the point charge is inside the surface, so $Q_{\text{inside}} = Q_1$. Gauss' Law in spherical symmetry gives us $E \times 4\pi r^2 = Q_1/\epsilon_0$ and

$$E = \frac{Q_1}{4\pi\epsilon_0 r^2} \quad (r < R_1) \tag{24.34}$$

The region $R_1 < r < R_2$ is inside the conducting material, so we know immediately

$$E = 0 \quad (R_1 < r < R_2) \tag{24.35}$$

For a spherical Gaussian surface of radius $r > R_2$, the enclosed charge is $Q_1 + Q_2$, so Gauss' Law gives us

$$E = \frac{Q_1 + Q_2}{4\pi\epsilon_0 r^2} \quad (r > R_2) \tag{24.36}$$

To determine the charge on the inner surface, we use Gauss' Law in reverse: since $E = 0$ in the conducting material, there must be zero net charge inside any Gaussian sphere in the conducting material ($R_1 < r < R_2$). Since a point charge Q_1 is at the center of the spherical shell, the charge on the inner surface of the shell must be the opposite of Q_1:

$$Q_{\text{inner surface}} = -Q_1 \tag{24.37}$$

We say the point charge *induces* an equal but opposite charge on the inner surface of the conductor. To find the charge on the outer surface, we merely use conservation of charge:

$$Q_{\text{outer surface}} = Q_2 - Q_{\text{inner surface}} = Q_2 - (-Q_1) = Q_2 + Q_1 \tag{24.38}$$

The electrostatic requirement that $E = 0$ in a conducting material can lead to induced charges on any number of concentric conducting shells in either spherical or cylindrical geometry. We can always consider a Gaussian surface within the conducting material of a particular shell, and the requirement of zero net charge inside the Gaussian surface implies that the charge on the inner surface of that shell is exactly opposite to any net charge interior to that shell. Such interior charges can reside on any number of other conducting or insulating shells, or on a central sphere or point charge (spherical geometry) or a central rod or line of charge (cylindrical geometry).

For a planar geometry, there is a difference: an infinite sheet with surface charge density σ will induce a charge density of only $-\sigma/2$ on the near face of a parallel, conducting slab; an opposite charge density $+\sigma/2$ must also be induced on the far face of the slab. In the spherical and cylindrical shells, this "outer" charge does not contribute to the field inside the conductor, but in the planar slab the contribution is equal to that of the other face. Finally, we saw in Example 12 that the value of the charge on the outer surface of any conducting spherical or cylindrical shell is equal to the net charge on both surfaces of the shell minus the value of any induced charge on its inner surface. For a planar conducting slab, however, any net charge on the surfaces divides equally between the two faces of the slab.

PROBLEM-SOLVING TECHNIQUES	GAUSS' LAW FOR CHARGE DISTRIBUTIONS WITH SYMMETRY

As in Examples 4 through 12, the calculation of the electric field is readily possible by means of Gauss' Law only if the charge distribution has high symmetry (spherical, cylindrical, or planar). If so, you use Gauss' Law,

$$\oint E_\perp \, dA = \frac{Q_{inside}}{\epsilon_0}$$

in the following way:

- Determine the *direction* of the electric field from symmetry: *radial* for spherical and cylindrical geometries; *normal* for planar geometries.

- Choose an (imaginary) Gaussian surface in the region where you want to know the electric field: a sphere of radius r, a cylinder of radius r and length L, or a cylinder of base area A, as in the examples.

- Rewrite the left side of Gauss' Law (the electric flux) as the product [electric field] × [area] in one of the following ways:

$$\oint E_\perp \, dA = \begin{cases} E \times 4\pi r^2 & \text{(spherical)} \\ E \times 2\pi r L & \text{(cylindrical)} \\ 2EA & \text{(planar, both ends in field)} \\ EA & \text{(planar, one end in field)} \end{cases}$$

- For the right side of Gauss' Law, find Q_{inside} for your surface. Be sure to include only the charge *inside* the surface. You may need to express the enclosed charge in terms of the appropriate charge density (see Examples 4 through 12). For uniform densities, use

Line: $q = \lambda L$

Surface: $q = \sigma A$

Volume: $q = \rho V$

- Apply Gauss' Law: equate the appropriate expression (for $\oint E_\perp \, dA$) with Q_{inside}/ϵ_0.

- Solve the resulting equation for the electric field E.

- If the problem involves conductors (metal) in electrostatic equilibrium, you have additional information (Section 24.5): $E = 0$ inside the conducting material, and any net charge resides on the surfaces of the conductors.

✔ Checkup 24.5

QUESTION 1: Can there be an electric field within the material of a conductor in electrostatic equilibrium?

QUESTION 2: A uniform horizontal electric field has horizontal field lines. Suppose we shove an uncharged copper ball into this electric field. Will this alter the direction of the field lines?

QUESTION 3: A large, horizontal sheet of paper carries a uniform charge distribution of 10^{-6} C/m^2. A large, horizontal aluminum plate carries a uniform charge distribution of 10^{-6} C/m^2 equally divided between its upper side and its lower side (0.5×10^{-6} C/m^2 on each side). Compare the electric fields produced above and below the paper sheet and the aluminum plate.

QUESTION 4: A point charge Q is at the center of an uncharged spherical conducting shell. How much charge is on the inner surface of the shell? The outer surface?

(A) $0; 0$ (B) $Q; 0$ (C) $-Q; 0$ (D) $-Q; Q$ (E) $Q; -Q$

SUMMARY

PROBLEM-SOLVING TECHNIQUES   Gauss' Law for Charge Distributions with Symmetry   **(page 777)**

ELECTRIC FLUX THROUGH AN OPEN SURFACE

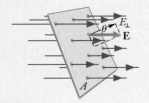

$$\Phi_E = \mathbf{E} \cdot \mathbf{A} = E_\perp A = EA \cos \theta \quad \text{(flat surface, uniform } \mathbf{E}\text{)} \quad \textbf{(24.1)}$$

$$\Phi_E = \int \mathbf{E} \cdot d\mathbf{A} = \int E_\perp \, dA \quad \text{(arbitrary surface, varying } \mathbf{E}\text{)} \quad \textbf{(24.5)}$$

ELECTRIC FLUX THROUGH A CLOSED SURFACE
(positive for **E** outward)

$$\oint \mathbf{E} \cdot d\mathbf{A} = \oint E_\perp \, dA \qquad \textbf{(24.6)}$$

GAUSS' LAW

$$\oint E_\perp \, dA = \frac{Q_{\text{inside}}}{\epsilon_0} \qquad \textbf{(24.10)}$$

where

$$\oint E_\perp \, dA = \begin{cases} E \times 4\pi r^2 & \text{(spherical)} \\ E \times 2\pi r L & \text{(cylindrical)} \\ 2EA & \text{(planar, both ends in field)} \\ EA & \text{(planar, one end in field)} \end{cases}$$

UNIFORM CHARGE DISTRIBUTIONS

Line: $q = \lambda L$

Surface: $q = \sigma A$ **(24.11)**

Volume: $q = \rho V$

CONDUCTORS IN ELECTROSTATIC EQUILIBRIUM

The electric field within the conducting material is zero. Any charge resides on the surface(s). The electric field at the surface is perpendicular and of magnitude $E = \sigma/\epsilon_0$

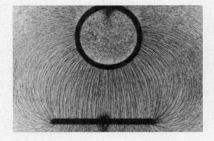

QUESTIONS FOR DISCUSSION

1. A point charge Q is inside a spherical Gaussian surface, which is enclosed in a larger, cubical Gaussian surface. Compare the fluxes through these two surfaces.

2. Figure 24.25 shows a Gaussian surface and lines of electric field entering and leaving the surface. What can you say about the sign of the electric charge within the surface?

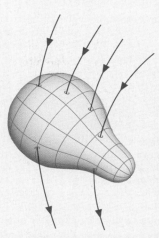

FIGURE 24.25 Gaussian surface and some field lines.

3. Suppose we had adopted some normalization for the number of flux lines per unit charge, say, $1/\epsilon_0$ lines per unit charge. Would this change Gauss' Law?

4. A Gaussian surface contains an electric dipole, and no other charge. What is the electric flux through this surface?

5. A point charge is at the center of a spherical shell. If it is moved to a position just inside the shell surface, does the electric flux through the shell change?

6. Suppose that the electric field of a point charge were not exactly proportional to $1/r^2$, but rather to $1/r^{2+a}$ where a is a small number, $a \ll 1$. Would Gauss' Law still be valid? (Hint: Consider Gauss' Law for the case of a point charge.)

7. Problems soluble by Gauss' Law fall into three categories, according to their symmetry: spherical, cylindrical, and planar. Give some examples in each category.

8. A hemisphere of radius R has a charge Q uniformly distributed over its volume. Can we use Gauss' Law to find the electric field?

9. A spherical rubber balloon of radius R has a charge Q uniformly distributed over its surface. The balloon is placed in a uniform electric field of 120 N/C. What is the net electric field inside the rubber balloon?

10. The electric field at the surface of a conductor in static equilibrium is perpendicular to the surface. Is the gravitational field at the surface of a mass in static equilibrium necessarily perpendicular to the surface? What if the surface is that of a fluid, such as water?

11. Suppose we drop a charged Ping-Pong ball into a cookie tin and quickly close the lid. What happens to the portions of the electric field lines that are outside the cookie tin when we close the lid?

12. When an electric current is flowing through a wire connected between a source and a sink of electric charge—such as the terminals of a battery—there is an electric field inside the wire, even though the wire is a conductor. Why does our conclusion about zero electric field inside a conductor not apply to this case?

13. The free electrons belonging to a metal are uniformly distributed over the entire volume of the metal. Does this contradict the result we derived in Section 24.5, according to which the charges are supposed to reside on the surface of a conductor?

PROBLEMS

24.1 Electric Flux[†]

1. Consider the thundercloud described in Problem 19 of Chapter 23 and Figure 23.34. What is the total electric flux coming out of the surface of the cloud?

2. A small square surface, 1.0 cm × 1.0 cm, is placed at a distance of 1.0 m from a point charge of 3.0×10^{-9} C. What is the approximate electric flux through this square if it is face on to the electric field (see Fig. 24.26)? If it is tilted by 30°? If it is tilted by 60°?

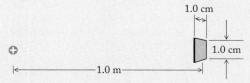

FIGURE 24.26 Point charge and small square surface.

[†]For help, see Online Concept Tutorial 28 at www.wwnorton.com/physics

3. The magnitude and the direction of a constant electric field are given by the vector $\mathbf{E} = 2.0\mathbf{i} - 1.0\mathbf{j} + 3.0\mathbf{k}$, where the field is measured in N/C. What is the flux that this electric field produces through the surface shown in Fig. 24.27, which consists of three squares $L \times L = 0.20$ m × 0.20 m each joined along their edges?

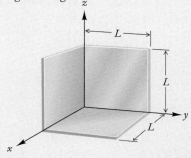

FIGURE 24.27 A surface consisting of three squares joined along their edges.

4. A thundercloud produces a vertical electric field of magnitude 2.8×10^4 N/C near ground level. You hold a 22 cm × 28 cm sheet of paper horizontally below the cloud. What is the electric flux through the sheet? What is the flux if you hold the sheet vertically?

5. The four faces of a tetrahedral pyramid are equilateral triangles of side a. Such a pyramid sits with one face flat on an infinite sheet of charge with surface charge density σ. What is the flux through the face of the pyramid that sits on the sheet? What is the flux through each of the other three faces?

6. A uniform electric field is given by $\mathbf{E} = (3.0$ N/C$)\mathbf{i} + (2.0$ N/C$)\mathbf{j} - (1.0$ N/C$)\mathbf{k}$. What is the flux through a flat, 4.0-m^2 area that lies in the y–z plane? What if that same area instead has its surface normal along an octant diagonal, so that the normal unit vector is $\hat{\mathbf{n}} = (1/\sqrt{3})\mathbf{i} + (1/\sqrt{3})\mathbf{j} + (1/\sqrt{3})\mathbf{k}$?

7. The direction of a uniform electric field is in the y–z plane at an angle of 30° from the $+y$ axis, 60° from the $+z$ axis. This uniform field extends throughout the region of a 2.0-m cube similar to the one in Fig. 24.7. What is the electric flux through each of the faces of the cube, labeled 1 through 6 in Fig. 24.7b? What is the net flux?

*8. A point charge of 6.0×10^{-8} C sits at some distance above the x–y plane. What is the electric flux that this charge generates through the (infinite) x–y plane? Does the electric flux depend on the distance?

*9. Consider the (infinite) midplane between two charges of equal magnitudes and opposite signs (see Fig. 23.17). If the charges are $\pm q$, what is the electric flux through this midplane?

*10. A point charge of 6.0×10^{-8} C sits on the wood floor of a room. You hold a small cardboard square measuring 1.0 cm × 1.0 cm face down, and carry it across the room at a height of 2.0 m, passing over the position of the point charge. Roughly plot the flux through the square as a function of position.

*11. A nonuniform electric field points everywhere in the $+x$ direction, so $\mathbf{E} = E_x\mathbf{i}$. The magnitude of the field is independent of the z coordinate and varies with x and y according to $E_x = C(x - 2y)$, where x and y are in meters and C is a constant with value $C = 3.6 \times 10^2$ N/(C·m). For a cube of side 2.0 m at the same position as in Fig. 24.7, what is the flux through the y–z face at $x = 1.0$ m (side 1)?

*12. A very small electric dipole is a distance z from the center of a loop of radius a, with its dipole moment vector $\mathbf{p}$ directed along the axis of the loop. The component of the electric field of the dipole perpendicular to the loop at any point P a distance r from the dipole is given by

$$E_\perp = \frac{p}{4\pi\epsilon_0 r^3}\,(3\cos^2\theta - 1)$$

where θ is the angle between the axis of the loop and the line

from the dipole to P. Show that the electric flux through the loop is given by

$$\Phi_E = \frac{p}{2\epsilon_0}\frac{a^2}{(z^2 + a^2)^{3/2}}$$

24.2 Gauss' Law†

13. A point charge of 2.0×10^{-12} C is located at the center of a cubical Gaussian surface. What is the electric flux through *each* of the faces of the cube?

14. A point charge of 1.0×10^{-8} C is placed inside an uncharged metallic can (say, a closed beer can) insulated from the ground. How much flux will emerge from the surface of the can when the point charge is inside?

15. A point charge q is placed at some distance from an infinitely large, flat sheet carrying a uniform distribution of charge σ coulombs per m^2.

   (a) What is the electric flux that this point charge produces through the sheet?

   (b) The electric flux Φ is the integral $\int E_\perp\, dA$; hence the product $\sigma\Phi$ is $\int \sigma E_\perp\, dA$, which equals the electric force that the point charge exerts on the sheet. Using the flux calculated in (a), evaluate this force.

   (c) From the force exerted by the point charge on the sheet, find the force that the sheet exerts on the point charge. From this, find the electric field that the sheet produces at the position of the point charge.

16. A point charge q is placed at a perpendicular distance $d/2$ from the center of a square of size $d \times d$ carrying a uniform charge distribution of σ coulombs per m^2 (see Fig. 24.28).

   (a) What is the electric flux that this point charge produces through the square? (Hint: The square is one of the six faces of a cube surrounding the point charge.)

   (b) By following the steps given in Problem 15, find the electric field that the square produces at the position of the point charge.

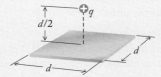

FIGURE 24.28 A point charge q and a charged square.

17. A point charge is placed at a distance $d/2$ from an infinitely long flat strip of width d carrying a uniform charge distribution of σ coulombs per m^2 (see Fig. 24.29).

†For help, see Online Concept Tutorial 29 at www.wwnorton.com/physics

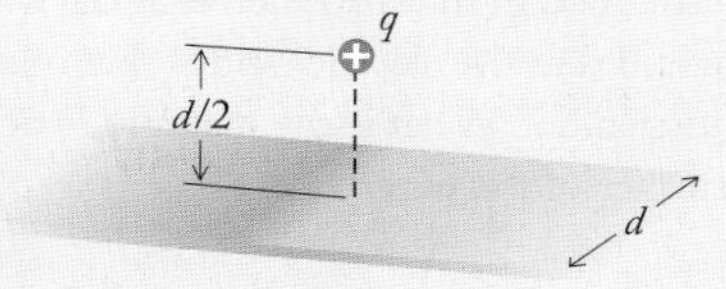

FIGURE 24.29 A point charge q and a charged strip.

(a) What is the electric flux that this point charge produces through the strip? (Hint: The strip is one of the four faces of a rectangular tube, or pipe, surrounding the point charge.)

(b) By following the steps in Problem 15, find the electric field that the strip produces at the position of the point charge.

18. A cube of side 2.0 m, similar to the one shown in Fig. 24.7, is in a region where the electric field is directed outward from both of two opposite cube faces, with uniform magnitude E_0 over each of those two faces. No flux crosses the other four faces. How much charge is inside the cube?

19. A short, thin rod of length 2.5 cm has charge distributed uniformly along its length. The rod is coaxial with and centered inside a much larger, uncharged cylindrical can. The flux through the curved part of the can is $+65\ \text{N}\cdot\text{m}^2/\text{C}$, and the flux through the bottom circular face is $+45\ \text{N}\cdot\text{m}^2/\text{C}$. What is the flux through the top circular face of the can? What is the value of the linear charge density λ of the rod?

20. Consider a cylindrical Gaussian surface of radius a and length $2z$ with a small electric dipole at its center, with its dipole moment **p** along the axis of the cylinder. According to the result of Problem 12, the flux though the upper circular face of the cylinder is given by

$$\Phi_E = \frac{p}{2\epsilon_0}\frac{a^2}{(z^2 + a^2)^{3/2}}$$

(a) What is the flux through the lower circular face?

(b) What is the flux through the curved surface of the cylinder?

*21. A spherical Gaussian surface of radius 1.00 m has a small hole of radius 10 cm. A point charge of 2.00×10^{-9} C is placed at the center of this spherical surface (see Fig. 24.30). What is the flux through the surface?

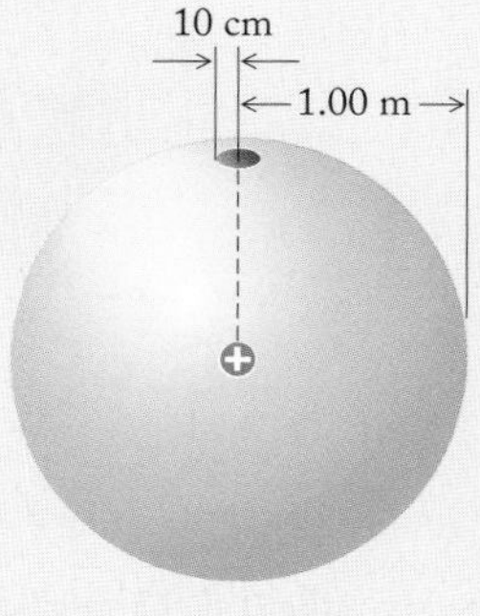

FIGURE 24.30 Spherical Gaussian surface with a small hole.

†For help, see Online Concept Tutorial 29 at www.wwnorton.com/physics

*22. Consider a point charge q located at one vertex of a cubical Gaussian surface, just inside the cube (see Fig. 24.31). What is the electric flux through each face of the cube?

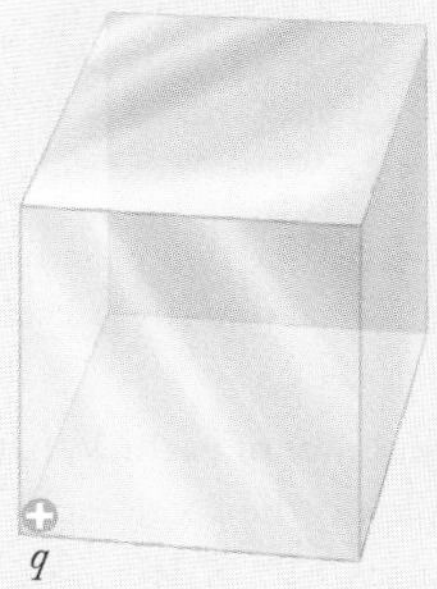

FIGURE 24.31 Point charge located just inside the vertex of a cubical Gaussian surface.

*23. Defining the gravitational field as the gravitational force per unit mass, formulate a Gauss' Law for gravity. Check that your law implies Newton's Law of Universal Gravitation.

*24. The electric field in a region has the following form as a function of x, y, z:

$$E_x = 5.0x \quad E_y = 0 \quad E_x = 0$$

where E is in newtons per coulomb and x is in meters. This represents an electric field in the x direction with a magnitude that increases in direct proportion to x. Show that such an electric field can exist only if the region is filled with some electric charge density. Find the value of the required charge density as a function of x, y, z.

**25. Prove that if the electric field is uniform in some region, then the charge density must be zero in that region. (Hint: Use Gauss' Law.)

†24.3 Applications of Gauss' Law

26. A uranium nucleus is a spherical ball of radius 7.4×10^{-15} m with a charge of $92e$ uniformly distributed over its volume. Find the electric field produced by this charge distribution at $r = 3.0 \times 10^{-15}$ m, 6.0×10^{-15} m, and 9.0×10^{-15} m.

27. Electrons can penetrate the nucleus because the nuclear material exerts no forces on them other than electric forces. Suppose that an electron penetrated a nucleus of lead. What is the electric force on the electron when it is at a distance of 3.0×10^{-15} m from the center of the nucleus? The nucleus of lead is a uniformly charged ball of radius 7.1×10^{-15} m with a total charge of $82e$.

28. In symmetric fission, a uranium nucleus splits into two equal pieces each of which is a palladium nucleus. The palladium nucleus is spherical with a radius of 5.9×10^{-15} m and a charge of $46e$ uniformly distributed over its volume. Suppose that, immediately after fission, the two palladium nuclei are barely touching (Fig. 24.32). What is the value of the total

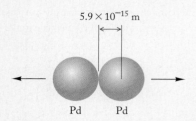

5.9 × 10⁻¹⁵ m

Pd   Pd

FIGURE 24.32 Two palladium nuclei in contact. Each nucleus is a sphere.

electric field at the center of each? What is the repulsive force between them? What is the acceleration of each? The mass of a palladium nucleus is 1.99×10^{-25} kg.

29. A spherical rubber balloon has a uniform distribution of charge over its surface. Show that the electric field that this charge produces in the (empty) interior of the balloon is exactly zero.

30. The diameter of a proton is about 1.0×10^{-15} m. How many diameters apart are two protons if the Coulomb force between them is 1.0 N?

31. Charge is uniformly distributed over the volume of a very long cylindrical plastic rod of radius R. The amount of charge per meter of length of the rod is λ. Find a formula for the electric field at a distance r from the axis of the rod. Assume $r < R$.

32. What is the maximum amount of electric charge per unit length that can be placed on a long and straight human hair of diameter 8.0×10^{-3} cm if the surrounding air is not to suffer electrical breakdown? The air will suffer breakdown if the electric field exceeds 3.0×10^6 N/C.

*33. A plastic spherical shell has inner radius a and outer radius b (Fig. 24.33). Electric charge is uniformly distributed over the region $a < r < b$. The amount of charge is ρ coulombs per cubic meter. Find the electric field in the regions $r \leq a$, $a \leq r \leq b$, and $r \geq b$.

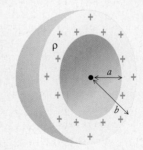

FIGURE 24.33 A spherical shell with uniform volume charge density ρ.

*34. The tube of a Geiger counter consists of a thin conducting wire of radius 1.3×10^{-3} cm stretched along the axis of a conducting shell of radius 1.3 cm (Fig. 24.34). The wire and the cylinder have equal and opposite charges of 7.2×10^{-10} C distributed along their length of 9.0 cm. Find a formula for

the electric field in the space between the wire and the cylinder; pretend that the electric field is that of an infinitely long wire and cylinder. What is the magnitude of the electric field at the surface of the wire?

1.3 cm

FIGURE 24.34 A thin wire stretched along the axis of a cylindrical conducting shell.

*35. A thick spherical shell of inner radius a and outer radius b has a charge Q uniformly distributed over its volume (Fig. 24.35). Find the electric field in the regions $r \leq a$, $a \leq r \leq b$, and $r \geq b$.

FIGURE 24.35 A thick spherical shell with total charge Q uniformly distributed over its volume.

*36. According to a (crude) model, the neutron consists of an inner core of positive charge surrounded by a shell of negative charge. Suppose that the positive charge has a magnitude $+e$ and is uniformly distributed over a sphere of radius 0.50×10^{-15} m; suppose that the negative charge has a magnitude $-e$ and is uniformly distributed over a concentric shell of inner radius 0.50×10^{-15} m and outer radius 1.0×10^{-15} m (Fig. 24.36). Find the magnitude and direction of the electric field at 1.0×10^{-15}, 0.75×10^{-15}, 0.50×10^{-15}, and 0.25×10^{-15} m from the center.

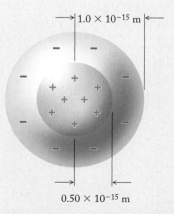

1.0 × 10⁻¹⁵ m

0.50 × 10⁻¹⁵ m

FIGURE 24.36 Hypothetical charge distribution inside a neutron.

*37. According to an old (and erroneous) model due to J. J. Thomson, an atom consists of a cloud of positive charge within which electrons sit like plums in a pudding. The electrons are supposed to emit light when they vibrate about their equilibrium positions in this cloud. Assume that in the case of the hydrogen atom the positive cloud is a sphere of radius $R = 0.050$ nm with a charge of e uniformly distributed over the volume of this sphere. The (pointlike) electron is held at the center of this charge distribution by the electrostatic attraction.

(a) Show that the restoring force on the electron is $e^2r/(4\pi\epsilon_0 R^3)$ when the electron is at a distance r from the center ($r \leq R$).

(b) What is the frequency of small oscillations of the electron moving back and forth along a diameter? Give a *numerical* answer.

*38. According to the Thomson model (see also Problem 37), the atom of helium consists of a uniform spherical cloud of positive charge within which sit two electrons. Assume that the positive cloud is a sphere of radius 5.0×10^{-11} m with a charge of $2e$ uniformly distributed over the volume. The two electrons are symmetrically placed with respect to the center (Fig. 24.37). What is the equilibrium separation of the electrons?

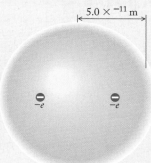

FIGURE 24.37 Thomson model of the helium atom.

*39. If a volume charge distribution varies continuously as a function of position the simple relation $Q = \rho V$ must be replaced by the integral $Q = \int \rho\, dV$. For spherical symmetry, $dV = 4\pi r^2\, dr$ (volume of a spherical shell with thickness dr and area $4\pi r^2$). Consider a sphere of radius R and total charge Q with a **nonuniform volume charge density** $\rho(r) = C/r$, where C is a constant to be determined.

(a) From $Q = \int \rho\, dV$, find the value of the constant C.

(b) Using Gauss' Law, find the electric field for $r \leq R$.

(c) What is the electric field for $r \geq R$?

*40. A thick, insulating spherical shell of inner radius a and outer radius b has a nonuniform volume charge density given by $\rho(r) = C/r^3$, where C is a known constant.

(a) What is the total charge Q in the shell? (Hint: See Problem 39.)

(b) Determine the value of the electric field for $r \leq a$, $a \leq r \leq b$, and $r \geq b$.

*41. A solid insulating sphere has a charge distribution with spherical symmetry, with a nonuniform charge density ρ coulombs per cubic meter described by the formula $\rho = kr^n$, where k is a constant and $n > -3$.

(a) What is the amount of charge $Q(r)$ inside a sphere of radius r? (Hint: See Problem 39.)

(b) What is the magnitude of the electric field as a function of r?

(c) For what value of n is the magnitude of the electric field constant?

(d) Why is it necessary to assume that $n > -3$?

*42. If a charge distribution varies continuously as a function of position, then the simple relation $Q = \rho V$ must be replaced by the integral $Q = \int \rho\, dV$. For cylindrical symmetry, $dV = 2\pi rL\, dr$ (volume of a cylindrical shell with radius r, length L, and thickness dr). Consider a solid cylinder of radius R with a nonuniform volume charge density $\rho(r) = B/r$, where B is a known constant

(a) What is the amount of charge $Q(r)$ inside a cylinder of radius $r \leq R$ and length L?

(b) What is the electric field for $r \leq R$?

*43. Consider a large, insulating slab of thickness d (similar to Example 10), but now with a nonuniform volume charge density $\rho = \rho(x)$, where $x = 0$ is at the center of the slab. Since ρ varies with position, the simple relation $Q = \rho V$ must be replaced by the integral $Q = \int \rho\, dV$. For the volume within the smaller Gaussian surface in Fig. 24.17, $dV = A\, dx$ (volume of a circular disk with area A and thickness dx). If the volume charge distribution is given by $\rho = Cx^2$, where C is a known constant, find the magnitude of the electric field for $|x| \leq d/2$ and for $|x| \geq d/2$.

**44. The electric charge of the proton is not concentrated in a point but, rather, distributed over a volume. According to experimental investigations at the Stanford Linear Accelerator, the charge distribution of the proton can be approximately described by a nonuniform volume charge density that is an exponential function of the radial distance:

$$\rho = \frac{e}{8\pi b^3} \exp\left(-\frac{r}{b}\right)$$

where e is the elementary charge and b is a constant, $b = 0.20 \times 10^{-15}$ m. Find the electric field as a function of the radial distance. What is the magnitude of the electric field at $r = 1.0 \times 10^{-15}$ m? [Hint: See Problem 39. Also, the following integral will be useful: $\int x^2e^{-x}\, dx = -x^2e^{-x} - 2e^{-x}(x + 1)$, where e here refers to the base of the natural logarithms.]

**45. When a point charge q moving at high speed passes by another stationary point charge q', the main effect of the electric forces is to give each charge a transverse impulse. Figure 24.38 shows the charge q moving at (almost) constant velocity v along the x axis in an almost straight line and shows the charge q' sitting at a distance R below the origin. The transverse impulse on q is

$$\int_{-\infty}^{\infty} F_y \, dt = \frac{q}{v} \int_{-\infty}^{\infty} E_y \, dx$$

Evaluate the integral $\int E_y \, dx$ by means of Gauss' Law, and prove that

$$\int_{-\infty}^{\infty} F_y \, dt = \frac{q}{v} \frac{q'}{2\pi\epsilon_0 R}$$

(Hint: Consider $2\pi R \int E_y \, dx$; show that this is the flux that q' produces through the infinite cylindrical surface indicated in Fig. 24.38.)

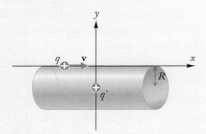

FIGURE 24.38 The charge q moves along the x axis. The charge q' sits at a distance R below the x axis. The cylinder, which extends to infinity, is the Gaussian surface.

*46. The formula derived in Problem 45 gives the transverse momentum that a high-speed charged particle acquires as it passes by a stationary charged particle.

(a) Calculate the transverse momentum that an electron of speed 4.0×10^7 m/s acquires as it passes by a stationary electron at a distance of 0.60×10^{-10} m.

(b) What transverse velocity corresponds to this transverse momentum?

(c) What will be the recoil velocity of the stationary electron (if it is free to move)?

24.4 Superposition of Electric Fields

*47. Positive charge Q is uniformly distributed over the volume of a solid sphere of radius R. Suppose that a spherical cavity of radius $R/2$ is cut out of the solid sphere, the center of the cavity being at a distance of $R/2$ from the center of the original solid sphere (Fig. 24.39); the cut-out material and its charge are discarded. What new electric field does the sphere with the cavity produce at the point P at a distance r from the original center? Assume $r > R$.

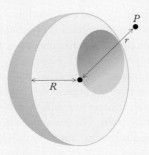

FIGURE 24.39 Sphere with a spherical cavity.

*48. Figure 24.19a shows an infinite cylinder with a positive uniform volume charge density ρ and radius R. The cylinder has a spherical empty cavity of radius $a < R$ centered on its axis. For points a radial distance r from the center of the spherical cavity along a line perpendicular to the cylindrical axis, find the magnitude and direction of the electric field for (a) $r \leq a$; (b) $a \leq r \leq R$; (c) $r \geq R$.

*49. Consider an infinite slab of thickness d parallel to the y–z plane and centered at $x = 0$, similar to the one in Fig. 24.17. This slab, however, has a spherical empty cavity of radius $d/2$ centered at the origin, just touching the slab faces; everywhere other than this cavity, the slab has a positive uniform volume charge density ρ. What are the magnitude and direction of the electric field at $\mathbf{r} = \frac{1}{4}d\mathbf{i}$? At $\mathbf{r} = 2d\mathbf{i}$? At $\mathbf{r} = \frac{1}{2}d\mathbf{i} + d\mathbf{j}$?

*50. A long cylinder of radius R has a uniform charge density ρ distributed over its volume. A cylindrical hole of radius R' has been drilled parallel to the axis of the cylinder along its full length, and the charge on this hole has been discarded. The axis of the hole is at a distance $h > R'$ from the axis of the cylinder (see Fig. 24.40). Find the electric field as a function of the radial distance r on the solid radial line shown in Fig. 24.40. Consider the cases $r \leq h - R'$, $h - R' \leq r \leq h + R'$, $h + R' \leq r \leq R$, and $r \geq R$ separately.

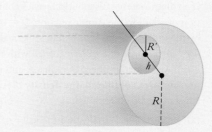

FIGURE 24.40 Cylinder with a cylindrical cavity.

**51. A charge q is uniformly distributed over the volume of a solid sphere of radius R. A spherical cavity is cut out of this solid sphere (Fig. 24.41), and the material and its charge are discarded. Show that the electric field in the cavity will then be uniform, of magnitude $(1/4\pi\epsilon_0)Qd/R^3$, where d is the distance between the centers of the spheres (Fig. 24.41). Make a drawing of the lines of electric field in the cavity.

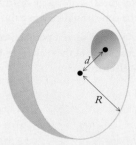

FIGURE 24.41 Sphere with a spherical cavity.

24.5 Conductors and Electric Fields

52. On a clear day, the Earth's atmospheric electric field near the ground has a magnitude of 100 N/C and points vertically down. Inside the ground, the electric field is zero, since the ground is a conductor. Consider a mathematical box of 1.0 m × 1.0 m × 1.0 m, half below the ground and half above. What is the electric flux through the sides of this box? What is the charge enclosed by the box?

53. Suppose we suspend a small ball carrying a charge of 1.0×10^{-6} C in the center of a safe and lock the door. The safe is uncharged and made of solid steel; it has inside dimensions of 0.30 m × 0.30 m × 0.30 m and outside dimensions 0.40 m × 0.40 m × 0.40 m. What is the electric flux through a cubical surface measuring 0.20 m × 0.20 m × 0.20 m centered on the ball? A cubical surface measuring 0.35 m × 0.35 × 0.35 m? A cubical surface measuring 0.50 m × 0.50 m × 0.50 m?

54. Charge is placed on a small metallic sphere which is surrounded by air. If the radius of the sphere is 0.50 cm, how much charge can be placed on the sphere before the air near the sphere suffers electric breakdown? The critical electric field strength that leads to breakdown in air is 3.0×10^6 N/C.

55. The surface of a long cylindrical copper pipe has a charge of λ coulombs per meter (Fig. 24.42). What is the electric field outside the pipe? Inside the pipe?

FIGURE 24.42 A long copper pipe with positive charge on its surface.

56. A solid copper sphere of radius 3.0 cm carries a charge of 1.0×10^{-6} C. This sphere is placed concentrically within a spherical, thin copper shell of radius 15 cm carrying a charge of 3.0×10^{-6} C. Find a formula for the electric field in the space between the sphere and the shell. Find a formula for the electric field outside the shell. Plot these electric fields as a function of radius.

57. A thick spherical shell made of solid metal has an inner radius a and an outer radius b and is initially uncharged. A point charge q is placed at the center of the shell. Find the electric field in the regions $r < a$, $a < r < b$, and $r > b$. Find the induced surface charge densities at $r = a$ and $r = b$.

58. Consider a completely isolated, thick, conducting infinite slab. One face of the slab has a uniform surface charge density $+\sigma$.

   (a) What is the electric field inside the slab?

   (b) What is the surface charge density on the other face of the slab?

   (c) What is the electric field outside the slab?

59. On days of fair weather, the atmospheric electric field of the Earth is about 100 N/C; this field points vertically downward (compare Problem 4 of Chapter 23). What is the surface charge density on the ground? Treat the ground as a flat conductor.

60. You wish to generate a uniform electric field of 2.0×10^5 N/C in the space between two flat parallel plates of metal placed face to face. The plates measure 0.30 cm × 0.30 cm. How much electric charge must you put on each plate? Assume that the gap between the plates is small so that the charge distribution and the electric field are approximately uniform, as for infinite plates. Assume the plates carry opposite charges of equal magnitudes.

61. A point charge Q_1 is at the center of two conducting spherical shells. The larger shell has a total charge Q_2 on it; the smaller shell has zero net charge.

   (a) What is the the charge on the inner surface of the smaller shell? On its outer surface?

   (b) What is the electric field at a distance r from the point charge, where r is between the two shells?

   (c) What is the charge on the inner surface of the larger shell? On its outer surface?

   (d) What is the electric field outside both shells, a distance r from the point charge?

*62. A solid dielectric sphere has radius R_1 and uniform volume charge density ρ. It is placed concentrically within a metal shell of inner radius R_2 and outer radius R_3. The shell carries a total charge Q_0. Find the magnitude of the electric field for (a) $r \le R_1$; (b) $R_1 \le r < R_2$; (c) $R_2 < r < R_3$; and (d) $r > R_3$.

REVIEW PROBLEMS

63. A point charge q is placed very close to (almost in contact with) the base of an infinitely long cylindrical Gaussian surface (see Fig. 24.43). What is the flux that the point charge produces through the base? What is the flux that this point charge produces though the curved lateral surface of the cylinder?

FIGURE 24.43 A point charge q inside a cylindrical Gaussian surface.

64. A point charge of 2.0×10^{-8} C is located at the center of a Gaussian surface in the shape of a cube of edge 8.0 cm. What is the average value of $E_{\perp}$ over one face of the cube?

65. Two very large plane surfaces intersect at 90°, forming a floor and a wall (see Fig. 24.44).

   (a) If a point charge q is placed just outside the edge formed by the intersecting surfaces, what is the electric flux through the combined surfaces?

   (b) If the point charge q is placed just inside the edge formed by the intersecting surfaces, what is the electric flux through the combined surfaces?

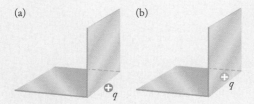

FIGURE 24.44 (a) Point charge located just outside the edge formed by the intersecting surfaces. (b) Point charge located just inside the edge.

66. Consider a cubical Gaussian surface of edge 5.0 cm. You do not know the electric charge or the electric field inside the cube, but you do know the electric field at the surface: at the top of the cube the electric field has a magnitude of 5.0×10^{5} N/C and points perpendicularly out of the cube; at the bottom of the cube, the electric field has a magnitude of 2.0×10^{5} N/C and points perpendicularly into the cube; on all other faces, the electric field is tangential to the surface of the cube.

(a) How much electric charge is inside the cube?

(b) Can you guess what charge distribution inside (and outside) the cube would generate this kind of electric field?

67. Figure 24.45 shows several point charges and several surfaces which enclose some of these charges. For each of the surfaces, what is the electric flux?

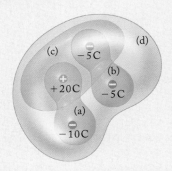

FIGURE 24.45 Several Gaussian surfaces (a–d).

68. In a xerographic copier, during the copying process, electric charge is deposited on the sheet of paper by means of a corona discharge from a thin wire carrying a concentration of charge (in some copiers you can see this corona wire if you open the top of the copier; see Fig. 24.46). The wire has a radius of 2.0×10^{-5} m. To produce the corona discharge, an electric field of 3.0×10^{6} N/C is required near the surface of the wire. What amount λ of charge per unit length must be placed on the wire to attain this electric field?

FIGURE 24.46 Corona wire in a xerographic copier.

69. A very long cylinder of radius R has positive charge uniformly distributed over its volume. The amount of charge is λ coulombs per meter of length of the cylinder. A spherical cavity of radius $R' \leq R$, centered on the axis of the cylinder, has been cut out of this cylinder, and the charge in this cavity has been discarded. Find the electric field as a function of distance from the center of the sphere along the axis of the cylinder.

70. A long, thin rod carries a charge of λ coulombs per meter. Consider a square surface of dimensions $d \times d$; the square surface and the rod are in the same plane. The nearest side of the square surface is a distance r from the rod (see Fig. 24.47). What is the electric flux through this square?

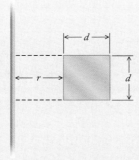

FIGURE 24.47 Charged rod and square surface.

71. A sphere of radius R and uniform volume charge density ρ remains stationary (levitates) when placed above an infinite sheet of paper with a uniform surface charge density σ. What is the mass of the sphere?

72. Two parallel infinite sheets have opposite uniform surface charge densities $\pm\sigma$. What is the electric force that one sheet exerts on a portion of area A of the other sheet?

73. Two parallel rods have the same radius R and opposite linear charge densities $\pm\lambda$. The centers of the rods are separated by a distance $4R$. What is the electric field halfway between the rods? What is the magnitude of the electric field at a point equidistant from the two rods, a distance R from the line parallel to and halfway between the rods?

74. The strength of the atmospheric electric field decreases with altitude. This decrease is due to the presence of positive ions in the atmosphere. Suppose that at ground level the atmospheric electric field has a magnitude of 130 N/C and a downward direction. If the positive ions contribute a volume charge density of 1.0×10^{-14} C/m³ throughout the air, what is the rate of decrease of the atmospheric electric field with height? What is the magnitude of the atmospheric electric field at a height of 1000 m?

75. A thick cylindrical shell of inner radius a and outer radius b has a charge of λ coulombs per unit length uniformly distributed over its volume. A thin line with a charge of $-\lambda$ coulombs per unit length is stretched along the axis of the shell (see Fig. 24.48). Find the electric field in the regions $r \leq a$, $a \leq r \leq b$, and $r \geq b$.

FIGURE 24.48 A thick cylindrical shell with charge distributed over its volume and a charged line along its axis.

76. Consider two large parallel metallic plates with uniform, opposite charge distributions, as in Fig. 24.49a. Suppose that the magnitude of the charge density on each plate is 2.0×10^{-5} C/m². The upper plate is positive and the lower negative.

(a) What is the magnitude of the electric field in the region between the plates?

(b) You now insert a neutral large parallel metallic plate in the space between the two charged plates, as in Fig. 24.49b. Suppose that this plate is 1.0 cm thick. What is the magnitude of the electric field inside this thick plate? What is the magnitude of the electric field in the remaining space above and below this thick plate?

(c) What is the charge density on the upper surface of the thick plate? The lower surface?

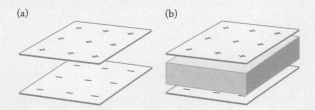

FIGURE 24.49 (a) Parallel plates with opposite charges. (b) Another plate inserted between the first two.

*77. The tau particle carries a negative fundamental charge similar to the electron, but is of much larger mass—its mass is 3.18×10^{-27} kg, about 3490 times the mass of an electron. Nuclear material is transparent to the tau: thus the tau can orbit around inside a nucleus, under the influence of the electric attraction of the nuclear charge. Suppose that a tau is in a circular orbit of radius 2.9×10^{-15} m inside a uranium nucleus. Treat the nucleus as a sphere of radius 7.4×10^{-15} m with a charge $92e$ uniformly distributed over its volume. Find the speed, the kinetic energy, the angular momentum, and the frequency of the orbital motion of the tau.

Answers to Checkups

Checkup 24.1

1. If the surface were horizontal, no field lines would cross it, so the flux would be zero (smaller); if the surface were vertical, more field lines would cross it, so the flux would be larger.

2. No change in either case; the same number of field lines would intersect any closed surface around the charge. As stated in the Comment in Example 2, the net flux depends only on the charge enclosed by the surface. In the next section, we formally explore this relation between the flux through a surface and the enclosed charge, known as *Gauss' Law*.

3. (C) Zero. Since all the field lines that enter the surface also leave it, the net flux through the surface is zero.

Checkup 24.2

1. Gauss' Law states that $\Phi_E = Q_{inside}/\epsilon_0$, so zero flux means zero total enclosed charge. If the surface has a hole, we can draw no conclusions: we don't know the flux through a closed surface, and "the charge inside" is ill defined for such a surface.

2. Since the two charges are of equal magnitude but opposite sign, the net charge inside the surface is zero, so the net flux through the surface is zero.

3. In Example 3, the net flux through the cube was calculated to be $\Phi_E = 40$ N·m^2/C, so Gauss' Law tells us that the charge inside that cube is

$$Q_{inside} = \epsilon_0 \Phi_E = (8.85 \times 10^{-12}\ \text{C}^2/\text{N·m}^2)(40\ \text{N·m}^2/\text{C})$$
$$= 0.35\ \text{n}C$$

4. (E) Zero. Since the field is uniform, field lines enter the sphere on one side and leave it on the other; the net flux through the sphere is zero.

Checkup 24.3

1. No. Although Gauss' Law is still valid, there is no surface through which the electric flux may be calculated easily. To calculate the electric field for four point charges, you would have to take the vector sum of the four Coulomb (point-charge) fields.

2. As in Example 7, $E = \lambda/(2\pi\epsilon_0 r) \propto \lambda/r$. So for equal values of λ, the thinner wire will produce a field 10 times larger at its surface.

3. As in Example 5 and Fig. 24.11, the field at the center is zero, and the field is largest at the surface of the sphere [where it takes the value $E = Q/(4\pi\epsilon_0 R^2)$].

4. As in the previous question, the field at the center is zero (zero enclosed charge for $r = 0$). At the surface, only the total charge enclosed matters (not how it is radially distributed), so the field there again takes the value $E = Q/(4\pi\epsilon_0 R^2)$. In this case, however, we do not know where the maximum value of E occurs; depending on the radial distribution, it could be anywhere in the region $0 < r \leq R$.

5. (B) 5.0×10^4 N/C. We saw in Example 7, as in Chapter 23, that the electric field E due to a linear charge distribution varies as $1/r$. Thus at twice the distance, the field is half as large.

Checkup 24.4

1. Yes. Inside the cavity, the field is the same as in Fig. 24.19; outside, there is an additional spherically symmetric contribution from the surface charge density.

2. No. The field due to a uniform cube of charge cannot be readily calculated from one of the high-symmetry (spherical, cylindrical, or planar) applications of Gauss' Law.

3. (B) $\sigma/2\epsilon_0$; $3\sigma/2\epsilon_0$. The field contribution due to the sheet with charge density σ is $\sigma/2\epsilon_0$; that due to the 2σ sheet is σ/ϵ_0. Since both charge densities are positive, each field contribution is directed away from its respective sheet. Thus between the two sheets these contributions oppose to give a net field of magnitude $\sigma/\epsilon_0 - \sigma/2\epsilon_0 = \sigma/2\epsilon_0$, and outside the sheets the two contributions are parallel and give a net field $\sigma/2\epsilon_0 + \sigma/\epsilon_0 = 3\sigma/2\epsilon_0$.

Checkup 24.5

1. No. It is possible in an insulator, but not in a conductor when in electrostatic equilibrium.

2. Yes. Any field lines that intersect the copper surface have to bend to meet the surface at right angles.

3. As we saw above, for a given charge density, the metal plate produces twice the electric field. So for half the charge density on each side of the plate, the electric fields due to the paper sheet and aluminum plate are the same. From far away, the division of the charge density is unimportant; the plate behaves like a single sheet of charge. The division of the charge density into two parts ensures that the electric field inside the metal will be zero.

4. (D) $-Q$; Q. The point charge Q induces an opposite charge $-Q$ on the inner surface of the conductor; this ensures $\mathbf{E} = 0$ inside the conducting shell. Since the shell is overall uncharged, the charge $-Q$ on the inner surface implies there is a charge Q on its outer surface.

Electrostatic Potential and Energy

CONCEPTS IN CONTEXT

Concepts *in* Context

This electrostatic generator at the Boston Science Museum has accumulated a large charge, distributed on the surface of its spherical conductor. When a charged particle is placed somewhere near this charge distribution, it will have a potential energy.

With the concept of electrostatic potential energy introduced in this chapter, we can consider such questions as:

? What is the potential energy of an electron placed near a proton? (Example 2, page 795)

? What is the electrostatic potential energy per unit charge at points outside and inside a charged conducting sphere? (Example 7, page 803)

? How does knowledge of the electrostatic potential energy enable us to calculate the electric field? (Example 9, page 807)

? What is the total work required to build up the charge on a conducting sphere? (Example 12, page 813)

From our study of mechanics we know that to formulate a law of conservation of energy for a particle moving under the influence of some force, we have to construct a potential energy corresponding to this force. In this chapter we will construct the electrostatic potential energy for a charged particle moving in the electric field generated by a static charge distribution. Like the potential energies we examined in Chapter 8, the electrostatic potential energy is helpful in the calculation of the motion of the particle. Furthermore, we will see that knowledge of the electrostatic potential energy at different distances from the charge distribution is equivalent to knowledge of the electric field—the potential energy can be calculated from the electric field and, conversely, the electric field can be calculated from the potential energy. It is therefore possible to calculate the electric field of a charge distribution by first evaluating the electrostatic potential energy that this charge distribution produces when acting on a charged particle placed at different distances. This means that we have available three alternative methods for the calculation of the electric field: via Coulomb's Law, via Gauss' Law, and via the potential energy. If a problem does not yield to the simple and elegant method based on Gauss' Law, it is usually best to use the method based on the potential energy, because the mathematics is likely to be less laborious than with the method based on Coulomb's Law.

In the final parts of this chapter, we will learn how to calculate the electric potential energy of systems of several point charges and of systems of conductors with charges placed on them. Since the charges exert electric forces on each other, it requires a certain amount of work to bring the charges together into their final configuration, starting with an initial configuration of very large (infinite) separation. This work is the electric potential energy of the configuration. It represents energy stored in the configuration during its assembly. We will see that this energy is actually stored in the electric field. The energy is concentrated in those regions of space where the electric field is strong.

Online
Concept
Tutorial

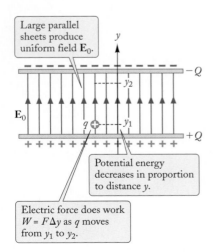

Large parallel sheets produce uniform field $\mathbf{E}_0$.

$\mathbf{E}_0$

Potential energy decreases in proportion to distance y.

Electric force does work $W = F\Delta y$ as q moves from y_1 to y_2.

FIGURE 25.1 Uniform electric field in the space between two charged parallel plates.

25.1 THE ELECTROSTATIC POTENTIAL

In Chapter 8 we learned that if a force is conservative, then the work performed by this force on a particle during a displacement can be expressed as a difference between two potential energies, one for the starting point, and one for the end point of the displacement. If we know the potential energy that corresponds to the force, we can immediately construct the conserved mechanical energy, which is simply the sum of the kinetic energy and the potential energy.

The electric force that a static distribution of charges exerts on a point charge is a conservative force. We can easily verify this for the electric force exerted by two uniform distributions of positive and of negative charges placed on two large parallel sheets or plates (see Fig. 25.1). We know from Section 23.2 that such parallel charged plates generate a uniform electric field E_0 in the space between them; the magnitude of this electric field is directly proportional to the amount of charge per unit area on each plate [see Eq. (23.13)]. The force that this electric field exerts on a point charge q is $F = qE_0$, and the work done by this constant force during a displacement from the point y_1 to the point y_2 illustrated in Fig. 25.1 is given by Eq. (7.1),

$$W = F\Delta y = qE_0(y_2 - y_1) \tag{25.1}$$

or

$$W = -qE_0 y_1 + qE_0 y_2 \tag{25.2}$$

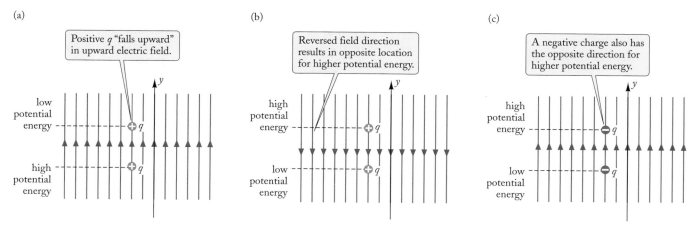

FIGURE 25.2 (a) For a positive charge in an electric field directed vertically upward, the potential energy decreases with height. (b) For a positive charge in an electric field directed vertically downward, the potential energy increases with height. (c) For a negative charge in an electric field directed vertically upward, the potential energy also increases with height.

This shows that if we identify the potential energy U as

$$U = -qE_0 y \qquad (25.3)$$

then the work is the difference between the potential energies corresponding to the points y_1 and y_2:

$$W = -qE_0 y_1 + qE_0 y_2 = U_1 - U_2 \qquad (25.4)$$

The conserved mechanical energy is the sum of the kinetic energy $K = \frac{1}{2}mv^2$ and the potential energy $U = -qE_0 y$:

$$[\text{energy}] = K + U = \tfrac{1}{2}mv^2 - qE_0 y = [\text{constant}] \qquad (25.5)$$

Note that the potential energy (25.3) is directly proportional to the distance y from the lower plate in Fig. 25.1. This direct proportionality of potential energy and height is reminiscent of the gravitational potential energy mgy. Mathematically, this electric potential energy and the gravitational potential energy are similar because both involve a constant force. But note that the signs of these two potential energies are opposite, because the electric force that the plates in Fig. 25.1 exert on a (positive) charge q is upward, whereas the gravitational force is downward; the signs of these two potential energies would be the same if the electric field in Fig. 25.1 were downward or if the charge q were negative (see Fig. 25.2). Note also the physical implications of changes in potential energy: for instance, in order for a positive charge to move against the electric field (that is, to a position of higher potential energy), the charge either must be pushed by an external agent (which does work) or must lose some kinetic energy.

We recall from Section 8.1 that a general criterion for a conservative force is that the work the force performs on any round trip must be zero. For the electric force exerted by the parallel charged plates in Fig. 25.1, this criterion is satisfied provided the round trip is confined to the region within the plates. For instance, consider the round trip consisting of the four straight segments shown in Fig. 25.3a. The work done by the electric field along the horizontal segments is zero, since the electric field is perpendicular to these segments; the work done along the upward vertical segment (on the right) is positive, and the work done along the downward vertical segment (on the left) is negative, resulting in zero work for the entire round trip.

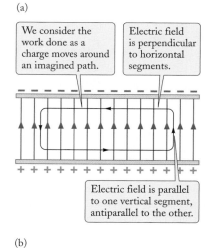

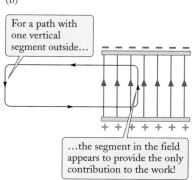

FIGURE 25.3 (a) A closed path between the plates. (b) A closed path with an upward segment between the plates and a downward segment outside the plates.

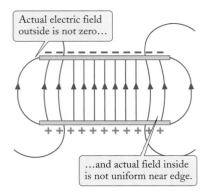

Actual electric field outside is not zero…

…and actual field inside is not uniform near edge.

FIGURE 25.4 The fringing field that extends beyond the space between the plates.

electrostatic potential

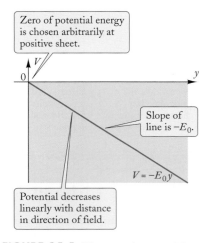

Zero of potential energy is chosen arbitrarily at positive sheet.

Slope of line is $-E_0$.

$V = -E_0 y$

Potential decreases linearly with distance in direction of field.

FIGURE 25.5 Electrostatic potential between a pair of oppositely charged sheets as a function of distance from the positive sheet.

If we attempt to apply the criterion of zero work to a different round trip that extends beyond the plates, such as the round trip shown in Fig. 25.3b, we see that now the work done along the downward segment on the far left is zero, since there is no electric field there. In this case it would seem we cannot achieve zero work for the round trip! However, the electric field distribution of Fig. 25.3 is only an approximation; an exact calculation of the electric field shows that near the edges of the plates the electric field is not uniform, and the electric field spills out into the space beyond the plates. The actual distribution of field lines is as shown in Fig. 25.4. The field that spills out beyond the edges of the plates is called the **fringing field**. The portions of the path that pass through this fringing field contribute to the work, and the net work for the complete round trip turns out to be zero, as required by our criterion for a conservative force.

We found it useful to define the electric field as the electric force divided by the charge q on which this force acts. Likewise, we find it useful to define the **electrostatic potential** V as the electric potential energy U divided by the charge q:

$$V = \frac{U}{q} \tag{25.6}$$

Thus, the electrostatic potential is the potential energy per unit charge. For example, in the case of a uniform electric field E_0, with a potential energy $U = -qE_0 y$, the electrostatic potential is

$$V = -E_0 y \tag{25.7}$$

Figure 25.5 is a plot of this linear potential as a function of the displacement y in the direction of the uniform field. Recall that it is the difference in potential energy between two points that is meaningful, as in Eq. (25.4). The reference position (the zero of potential energy) can be arbitrarily chosen. The same is true for the electrostatic potential V of Eq. (25.7).

The SI unit of electrostatic potential is the **volt** (V),[1] where

$$1 \text{ volt} = 1 \text{ V} = 1 \text{ joule/coulomb} = 1 \text{ J/C} \tag{25.8}$$

The unit of electric field we have employed in the preceding chapters is the N/C. This unit can be expressed in terms of volts as follows:

$$1 \frac{\text{N}}{\text{C}} = 1 \frac{\text{N·m}}{\text{C·m}} = 1 \frac{\text{J}}{\text{C}} \frac{1}{\text{m}} = 1 \frac{\text{V}}{\text{m}} \tag{25.9}$$

Thus, N/C and V/m are equal units; in practice, volt per meter is the preferred unit for the electric field. Table 25.1 gives some values of electrostatic potentials.

EXAMPLE 1   Suppose that near the ground directly below a thundercloud, the electric field is of a constant magnitude 2.0×10^4 V/m and points upward. What is the potential difference between the ground and a point in the air, 50 m above ground?

SOLUTION: For a constant (uniform) electric field, we use the linear potential of Eq. (25.7),

$$V = -E_0 y = -2.0 \times 10^4 \text{ V/m} \times 50 \text{ m} = -1.0 \times 10^6 \text{ volts}$$

[1] Note that the same letter V is used in physics both as a symbol for potential and as the abbreviation for *volt*. This leads to confusing equations such as $V = 3.0$ V (which is intended to mean that the potential $V = 3.0$ volts). If there is a possibility of confusion, it is best not to abbreviate *volt*.

COMMENTS: This calculation assumes that the ground is flat, without protuberances, such as trees or buildings. It would be wrong to conclude that the potential at the top of a 50-m-high building is -1.0×10^6 volts. Buildings are usually made of conducting materials, and, as we will see, the potential at all points of a conductor is the same. Thus, the building effectively acts as part of the ground, and the potential difference between the top of the building and the ground is zero. The presence of the building or of some other conducting protrusion modifies the electric field, so the assumption of a constant upward electric field is not valid near the building.

We can also verify that the electric force that one point charge exerts on another point charge is a conservative force. The electric force that a point charge q' exerts on another point charge q is $F = (1/4\pi\epsilon_0)qq'/r^2$. To find the potential energy corresponding to this force, we begin by calculating the work done as the charge q moves from, say, position P_1 to position P_2, and we then seek to express this work as a difference of two terms. In Fig. 25.6 the positions P_1 and P_2 are at radial distances r_1 and r_2, respectively, from the fixed charge q'. The work is the integral of the force $F = (1/4\pi\epsilon_0)qq'/r^2$ along the radial path from P_1 to P_2:

$$W = \int F\, dr = \int_{r_1}^{r_2} \frac{1}{4\pi\epsilon_0} \frac{qq'}{r^2}\, dr = \frac{qq'}{4\pi\epsilon_0}\left(-\frac{1}{r}\right)\Big|_{r_1}^{r_2} = \frac{qq'}{4\pi\epsilon_0}\left(\frac{1}{r_1} - \frac{1}{r_2}\right) \quad (25.10)$$

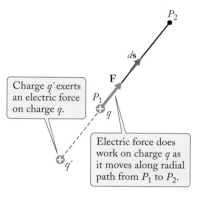

Charge q' exerts an electric force on charge q.

Electric force does work on charge q as it moves along radial path from P_1 to P_2.

FIGURE 25.6 Two points P_1 and P_2 in the electric field of a charge q'. The path connecting the points P_1 and P_2 is a straight radial line.

TABLE 25.1	SOME POTENTIALS AND POTENTIAL DIFFERENCES
Thundercloud to ground	5×10^7 V
Van de Graaff generator	10^7
High-voltage power line	5×10^5
At nucleus of uranium atom	2×10^5
Power supply for X-ray tube	10^5
Power supply for TV tube	2×10^4
Automobile ignition	10^4
Power supply for neon tube	2×10^3
Household outlet (Europe)	220
Household outlet (USA)	115
At electron orbit in hydrogen atom	27
Automobile battery	12
Dry cell	1.5
Single solar cell	0.6
Resting potential across nerve membrane	0.09
Potential changes on skin (measured by EKG or EEG)	5×10^{-5}
Potential changes due to thermal noise (typical)	10^{-8}

ALESSANDRO, CONTE VOLTA (1745–1827) *Italian physicist. Volta established that the "animal electricity" observed by Luigi Galvani, 1737–1798, in experiments with frog muscle tissue placed in contact with dissimilar metals, was not due to any exceptional property of animal tissues, but was also generated whenever any wet body was sandwiched between dissimilar metals. This led him to develop the first "voltaic pile," or battery, consisting of a large stack of moist disks of cardboard (electrolyte) sandwiched between disks of metal (electrodes).*

As expected, this result shows that the work is the difference between two potential energies. Accordingly, we can identify the electric potential energy as

$$U = \frac{1}{4\pi\epsilon_0} \frac{qq'}{r} \qquad (25.11)$$

In this calculation of the electric potential energy we assumed that the positions P_1 and P_2 lie on the same radial line (see Fig. 25.6). However, this assumption is not really necessary, because, if the points lie on different radial lines, then we can construct a path from one to the other with alternating radial segments and circular arcs (see Fig. 25.7). Recall from Chapter 7 that in general the work done by a force **F** is given by

$$W = \int \mathbf{F} \cdot d\mathbf{s} = \int F \cos\theta \, ds$$

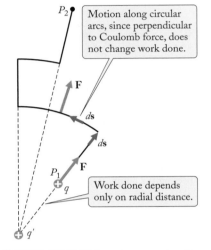

that is, a contribution to the work is the product of the component of the force parallel to the path times the displacement. The total work done along the radial segments is the same as for a single radial line, so this work is as given by Eq. (25.10). The work done along the circular arcs is zero, since the direction of the Coulomb force is perpendicular to these arcs. Hence, the net work, and the change of potential energy, is given by the same expression we found above, when we considered a single radial segment. Note that any arbitrary path between P_1 and P_2 can be approximated by small radial segments and circular arcs, and hence the work done by the electric force is independent of the path connecting P_1 and P_2; it depends only on the initial position P_1 and the final position P_2. This establishes that the electric force exerted by one point charge on another is conservative. More generally, the electric force exerted by any static distribution of electric charges on a point charge is conservative, since any distribution of electric charges consists of point charges, each of which exerts a conservative electric force.

For two charges of equal signs, the electric potential energy (25.11) is positive, and it decreases in inverse proportion to the distance. A decrease of potential energy with distance is characteristic of a repulsive force. *For two charges of opposite sign, the electric potential energy is negative,* and the magnitude of this negative potential energy decreases also in inverse proportion to distance (the potential energy *increases* from a large negative value to zero). Such an increase of potential energy with distance is characteristic of an attractive force.

The dependence of the electric potential energy of two point charges of opposite signs on distance is the same as the dependence of the gravitational potential energy of two point masses on distance [see Eq. (9.20)]—both are inversely proportional to the distance. This was to be expected, since the electric force is mathematically similar to the gravitational force—both are inversely proportional to the square of the distance.

The electrostatic potential produced by the charge q' is $V = U/q$; that is,

$$V = \frac{1}{4\pi\epsilon_0} \frac{q'}{r} \qquad (25.12)$$

FIGURE 25.7 This path connecting P_1 and P_2 consists of straight radial segments and circular arcs.

This electrostatic potential of a point charge is called the **Coulomb potential**.

Figure 25.8a is a plot of the Coulomb potential of a positive point charge q' vs. distance. Figure 25.8b is a plot of the Coulomb potential of a negative point charge q'

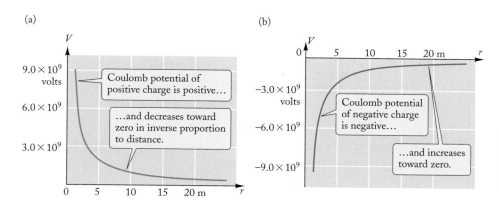

(a)

(b)

FIGURE 25.8 (a) Electrostatic potential of a positive charge q' vs. distance. For this plot, the magnitude of the charge is taken as $q' = 1$ C. (b) Electrostatic potential for a negative charge of magnitude $q' = -1$ C.

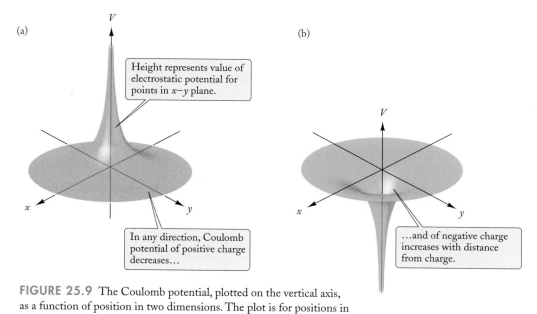

(a)

(b)

FIGURE 25.9 The Coulomb potential, plotted on the vertical axis, as a function of position in two dimensions. The plot is for positions in a plane containing (a) a positive and (b) a negative point charge.

vs. distance. Figure 25.9 shows plots of the Coulomb potentials of each of these point charges as a function of position in two dimensions.

Equation (25.12) also gives the electrostatic potential outside a solid charged sphere, or a hollow charged spherical shell, or any charge distribution with spherical symmetry. The electric field *outside* such spherical charge distributions is the same as that of a point charge, and therefore the electrostatic potential outside is also the same as for a point charge.

EXAMPLE 2   The electron in a hydrogen atom is at a distance of 5.3×10^{-11} m from the proton (see Fig. 25.10). The proton is a small ball of charge with $q' = e = 1.60 \times 10^{-19}$ C. What is the electrostatic potential generated by the proton at this distance? What is the potential energy of the electron?

Concepts
— *in* —
Context

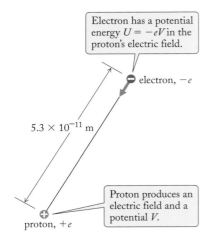

FIGURE 25.10 Proton ($+e$) and electron ($-e$) separated by a distance of 5.3×10^{-11} m.

SOLUTION: The electrostatic potential outside the proton is the same as for a point charge, so according to Eq. (25.12), the electrostatic potential produced by the proton is

$$V = \frac{1}{4\pi\epsilon_0} \frac{q'}{r} = 9.0 \times 10^9 \text{ N·m}^2/\text{C}^2 \times \frac{(1.60 \times 10^{-19} \text{ C})}{(5.3 \times 10^{-11} \text{ m})} \qquad (25.13)$$

$$= 27 \text{ volts}$$

The charge of the electron is $q = -e = -1.60 \times 10^{-19}$ C. From Eq. (25.6), the potential energy of the electron is then

$$U = qV = -e \times 27 \text{ volts} \qquad (25.14)$$

$$= -1.60 \times 10^{-19} \text{ C} \times 27 \text{ volts} = -4.3 \times 10^{-18} \text{ J} \qquad (25.15)$$

For the purposes of atomic physics, the joule is a rather large unit of energy and it is more convenient to leave the answer as in Eq. (27.14), $U = -27e \times$ volt.

The product of the elementary charge and the unit of potential, $e \times$ volt, or eV, is a unit of energy. This unit of energy is called an **electron-volt**. It can be converted to joules by substituting the numerical value for e:

electron-volt (eV)

$$1 \text{ eV} = 1.60 \times 10^{-19} \text{ C} \times 1 \text{ V} = 1.60 \times 10^{-19} \text{ J} \qquad (25.16)$$

In chemical reactions among atoms or molecules, the energy released or absorbed by each atom or molecule is typically around 1 or 2 eV. Such reactions involve a change in the arrangement of the exterior electrons of the atoms, and the energy of 1 or 2 eV represents the typical amount of energy needed for this rearrangement.

The mechanical energy of a point charge moving in the electric field of another point charge is the sum of the electric potential energy and the kinetic energy. The Law of Conservation of Energy for the motion of a point charge q in the electric field of a fixed point charge q' takes the form

$$[\text{energy}] = K + U = \tfrac{1}{2}mv^2 + \frac{1}{4\pi\epsilon_0} \frac{qq'}{r} = [\text{constant}] \qquad (25.17)$$

This total energy remains constant during the motion. As we saw in Chapter 8, examination of the energy reveals some general features of the motion. Obviously, if qq' is negative (opposite charges, attractive Coulomb force), then Eq. (25.17) implies that whenever r increases, v must decrease, and whenever r decreases, v must increase. For qq' positive (like charges, repulsive force), r and v increase together.

EXAMPLE 3   An electron is initially at rest at a very large distance from a proton. Under the influence of the electric attraction, the electron falls toward the proton, which remains (approximately) at rest. What is the speed of the electron when it has fallen to within 5.3×10^{-11} m of the proton? (See Fig. 25.11.)

SOLUTION: The word "initially" suggests using the before-and-after equality of the total energy. Since the electron is initially at rest, the initial kinetic energy is zero (see the discussion of the use of energy conservation in the Problem-Solving Techniques box). Furthermore, "a very large distance" implies the limit $r_0 \rightarrow \infty$, where the initial potential energy is also zero. The potential energy of the electron is $U = qV = -eV$. The total energy is the sum of the potential and kinetic energies:[2]

$$K + U = \tfrac{1}{2}m_e v^2 - eV \qquad (25.18)$$

This total energy is conserved. The initial value of the energy is zero; hence the final value of the energy must also be zero:

$$\tfrac{1}{2}m_e v^2 - eV = 0 \qquad (25.19)$$

Solving this for the speed v, we obtain

$$v = \sqrt{\frac{2eV}{m_e}} \qquad (25.20)$$

According to Example 2, $V = 27$ volts for $r = 5.3 \times 10^{-11}$ m, so

$$v = \sqrt{\frac{2 \times 1.60 \times 10^{-19}\ \text{C} \times 27\ \text{volts}}{9.11 \times 10^{-31}\ \text{kg}}}$$

$$= 3.1 \times 10^6\ \text{m/s}$$

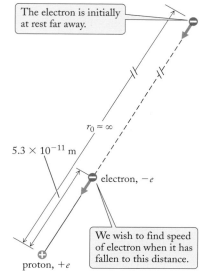

The electron is initially at rest far away.

$r_0 \approx \infty$

5.3×10^{-11} m

electron, $-e$

We wish to find speed of electron when it has fallen to this distance.

proton, $+e$

FIGURE 25.11 The approaching electron is instantaneously at a distance of 5.3×10^{-11} m from the proton.

For any nonuniform electric field, we can calculate the electric potential energy of a point charge q by the same method as for the electric field of a point charge: we express the work done by the electric field during a displacement from one position to another as a difference of two terms, one of which depends on the position of the starting point, and the other on the position of the end point. These two terms are the potential energies U_1 and U_2 corresponding to these two points. The work is then

$$W = U_1 - U_2 \qquad (25.21)$$

and the conserved mechanical energy is again [see Eq. (25.5)]

$$K + U = \tfrac{1}{2}mv^2 + U = [\text{constant}]$$

or, with $U = qV$,

$$K + U = \tfrac{1}{2}mv^2 + qV = [\text{constant}] \qquad (25.22) \qquad \text{conservation of energy}$$

PROBLEM-SOLVING TECHNIQUES

ENERGY CONSERVATION AND MOTION OF A POINT CHARGE

Note that the use of energy conservation with electric potential energy involves the familiar three steps we encountered with other forms of potential energy:

1. First write an expression for the energy of a charge (kinetic + potential) at one point of the motion [Eq. (25.18)].

2. Then write a similar expression for the energy at another point.

3. Then rely on energy conservation to equate the two expressions [Eq. (25.19)].

This yields one equation, which can be solved for the unknown final speed of the charge (if the final position is known) or the unknown final position (if the final speed is known).

---

[2] Do not confuse the symbol v (for speed) with the symbol V (for potential).

Checkup 25.1

QUESTION 1: Suppose that the potential of a point charge is 100 V at a distance of 1 m from this charge. What is the potential at a distance of 10 m? 100 m?

QUESTION 2: Given that the electric potential energy for two interacting point charges is positive, what can you conclude about the signs of their charges?

QUESTION 3: According to Fig. 25.8, where is the electric potential of a positive charge q' largest? Smallest? Where is the electric potential of a negative charge q' largest? Smallest?

QUESTION 4: Consider a point charge q moving in the electric field of a stationary point charge q'. If the charges are of the same sign and r increases, the speed of q will

   (A) Increase   (B) Decrease   (C) Remain the same

Online
Concept
Tutorial

25.2 CALCULATION OF THE POTENTIAL FROM THE FIELD

Any arbitrary static distribution of charge can be regarded as consisting of many point charges. Since the electric force generated by a stationary point charge is a conservative force, the electric force generated by such a distribution of charge must also be a conservative force. Thus, when a point charge q moves from a position P_0 to a position P in the electric field of a charge distribution, the work done by the electric force can always be expressed as a difference of two potential-energy terms,

$$W = U_0 - U \tag{25.23}$$

The work is related to the force by the general formula $W = \int \mathbf{F} \cdot d\mathbf{s}$ [Eq. (7.16)], so Eq. (25.23) becomes

$$\int_{P_0}^{P} \mathbf{F} \cdot d\mathbf{s} = U_0 - U$$

Next, we can express the force in terms of the electric field ($\mathbf{F} = q\mathbf{E}$) and the potential energy in terms of the electrostatic potential ($U = qV$):

$$\int_{P_0}^{P} q\mathbf{E} \cdot d\mathbf{s} = qV_0 - qV \tag{25.24}$$

We can then cancel the factor q on both sides of this equation, and solve the equation for the potential V:

$$V = -\int_{P_0}^{P} \mathbf{E} \cdot d\mathbf{s} + V_0 \tag{25.25}$$

For many calculations, we will find it convenient to write Eq. (25.25) in terms of the angle θ between the electric field and the displacement ds:

electrostatic potential from field

$$V = -\int_{P_0}^{P} E \cos\theta \, ds + V_0 \tag{25.26}$$

This formula permits us to calculate the electrostatic potential of any arbitrary charge distribution, if the electric field is known. In such a calculation, we must begin at some position P_0 at which the potential has a given value V_0 and evaluate the integral for some path connecting this position P_0 to the position P. Although the choice of the value of the reference potential V_0 is arbitrary, as discussed above, *it is conventional, whenever possible, to adopt the value $V_0 = 0$ at infinity, as in the Coulomb potential.* The following examples illustrate how such calculations are done.

EXAMPLE 4   A sphere of radius R carries a total positive charge Q distributed uniformly throughout its volume (Fig. 25.12). Find the electrostatic potential inside and outside the sphere.

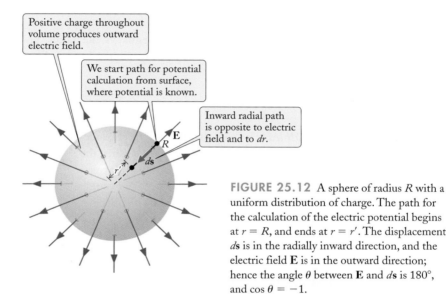

FIGURE 25.12 A sphere of radius R with a uniform distribution of charge. The path for the calculation of the electric potential begins at $r = R$, and ends at $r = r'$. The displacement $d\mathbf{s}$ is in the radially inward direction, and the electric field $\mathbf{E}$ is in the outward direction; hence the angle θ between $\mathbf{E}$ and $d\mathbf{s}$ is 180°, and $\cos \theta = -1$.

SOLUTION: Outside the sphere, the potential is the same as for a point charge,

$$V = \frac{1}{4\pi\epsilon_0} \frac{Q}{r} \tag{25.27}$$

To find the potential at some point r' inside the sphere, we want to evaluate Eq. (25.26) along a radial path that begins at some point where the potential is known. We choose a point on the surface of the sphere, and integrate the electric field from that point (at $r = R$) to a point inside where we want to find the potential (at $r = r'$). We know the potential at $r = R$ from Eq. (25.27),

$$V_0 = \frac{1}{4\pi\epsilon_0} \frac{Q}{R} \tag{25.28}$$

and we know the electric field inside the sphere from Eq. (24.18),

$$E = \frac{Q}{4\pi\epsilon_0} \frac{r}{R^3} \tag{25.29}$$

For our inward radial path, $ds = -dr$, and $\cos \theta = -1$ (see Fig. 25.12). Substituting all of this into Eq. (25.26), we obtain

$$V = -\int_{P_0}^{P} E \cos \theta \, ds + V_0 = -\int_{R}^{r'} \frac{Q}{4\pi\epsilon_0} \frac{r}{R^3} \, dr + \frac{1}{4\pi\epsilon_0} \frac{Q}{R} \tag{25.30}$$

$$= -\frac{1}{4\pi\epsilon_0} \frac{Q}{R^3} \int_{R}^{r'} r \, dr + \frac{1}{4\pi\epsilon_0} \frac{Q}{R} = -\frac{1}{4\pi\epsilon_0} \frac{Q}{R^3} \left(\frac{r^2}{2}\right)\Big|_{R}^{r'} + \frac{1}{4\pi\epsilon_0} \frac{Q}{R}$$

$$= -\frac{1}{4\pi\epsilon_0} \frac{Q}{R^3} \left(\frac{r'^2}{2} - \frac{R^2}{2}\right) + \frac{1}{4\pi\epsilon_0} \frac{Q}{R} \tag{25.31}$$

$$= -\frac{1}{4\pi\epsilon_0} \frac{Q}{2R^3} r'^2 + \frac{1}{4\pi\epsilon_0} \frac{3Q}{2R} \tag{25.32}$$

Figure 25.13 is a plot of the potential vs. the radius. The potential has a maximum at the center. The value of this maximum is, from Eq. (25.32) with $r' = 0$,

$$V = \frac{1}{4\pi\epsilon_0} \frac{3Q}{2R} \tag{25.33}$$

COMMENT: Note that the potential inside the sphere continues to increase as r' decreases, since we must do work against the electric field to move a positive point charge toward the center. However, it increases more slowly than the $\propto Q/r'$ behavior of the Coulomb potential, which would approach infinity as $r' \to 0$. Here, the potential attains the finite value of Eq. (25.33) at $r' = 0$, which is only a factor $3/2$ greater than the value at the surface of the sphere.

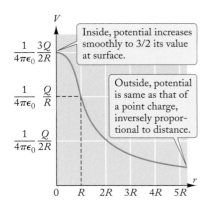

FIGURE 25.13 Electrostatic potential of a uniform spherical distribution of charge.

EXAMPLE 5   A coaxial cable consists of a long, cylindrical conductor of radius a concentric with a thin cylindrical shell of larger radius b (see Fig. 25.14). If the central conductor has a charge per unit length $\lambda = Q/L$ uniformly distributed on its surface, what is the potential difference between the inner and outer conductors? (Assume the space between them is empty.)

SOLUTION: We again calculate the potential by integrating the electric field along a path, now from the outer to the inner conductor. Since we are interested only in the potential *difference* between the inner and the outer conductor, we set the reference potential $V_0 = V_b$ in Eq. (25.26). We saw in Chapters 23 and 24 that the electric field outside a cylindrically symmetric charge distribution is given by Eqs. (23.10) and (24.23):

$$E = \frac{1}{2\pi\epsilon_0} \frac{\lambda}{r} \tag{25.34}$$

If we integrate radially inward from the diameter of the outer conductor ($r = b$) to that of the inner conductor ($r = a$), we again have $ds = -dr$, and $\cos \theta = -1$. So we obtain

$$V_a - V_b = -\int_{P_0}^{P} E \cos \theta \, ds = -\int_{b}^{a} \frac{1}{2\pi\epsilon_0} \frac{\lambda}{r} \, dr = -\frac{\lambda}{2\pi\epsilon_0} \int_{b}^{a} \frac{1}{r} \, dr$$

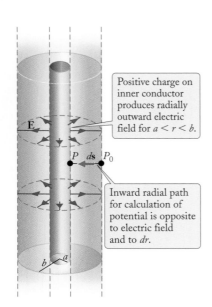

FIGURE 25.14 A coaxial cable.

$$= -\frac{\lambda}{2\pi\epsilon_0}(\ln r)\Big|_b^a = -\frac{\lambda}{2\pi\epsilon_0}(\ln a - \ln b)$$

$$= \frac{\lambda}{2\pi\epsilon_0}\ln\left(\frac{b}{a}\right) \tag{25.35}$$

where in the last equality we have used $\ln a - \ln b = \ln(a/b) = -\ln(b/a)$.

COMMENT: The outer cylinder played no role in the calculation, since for this symmetric distribution, it affects the electric field only beyond $r = b$. Note that because of the logarithmic dependence of the potential on distance, we could not have calculated the potential of the central conductor alone with respect to infinite distance—this would lead to an infinite result. This is a reflection of the fact that we have assembled an infinite charge distribution. For a coaxial cable in real use, the outer conductor has an opposite charge per unit length $\lambda = -Q/L$, so $E = 0$ outside, and we can set the reference potential $V_0 = 0$ everywhere outside the cable.

In the above examples of the calculation of the electrostatic potential of a charge distribution, we started with the known electric field. But it is also possible to start with the charge distribution and calculate the potential directly, without bothering with the electric field. Sometimes this is a much easier task. For such a direct calculation, we employ Eq. (25.12), which tells us how much a point charge contributes to the potential:

$$V = \frac{1}{4\pi\epsilon_0}\frac{Q}{r}$$

If the charge distribution consists of several point charges, then the electrostatic potential at some point is the sum of the individual Coulomb potentials evaluated at that point,

$$V = \frac{1}{4\pi\epsilon_0}\sum_i \frac{Q_i}{r_i} \tag{25.36}$$

Any other charge distribution can be regarded as consisting of small charge elements dQ, each of which can be treated as a point charge. Each such charge element will contribute an amount dV to the potential at a point,

$$dV = \frac{1}{4\pi\epsilon_0}\frac{dQ}{r} \tag{25.37}$$

The net potential of the charge distribution is then the sum or the integral of all the contributions from all these pointlike charge elements:

$$V = \frac{1}{4\pi\epsilon_0}\int \frac{dQ}{r} \tag{25.38}$$

potential of a continuous charge distribution

Note that since we are considering pointlike charge contributions using the Coulomb potential, the potential calculated will be with respect to $V_0 = 0$ at infinity, the conventional zero of potential.

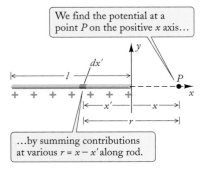

We find the potential at a point P on the positive x axis...

...by summing contributions at various $r = x - x'$ along rod.

FIGURE 25.15 A rod with a uniform distribution of charge.

EXAMPLE 6 A rod of length l has a charge Q uniformly distributed along its length (see Fig. 25.15). Find the electrostatic potential at a distance x from one end of the rod.

SOLUTION: Consider a small segment dx' of the rod. Since the charge per unit length is Q/l, the charge in this segment is $dQ = (Q/l)\,dx'$. The distance of this segment from the point P is $r = x - x'$ (here, x' is a negative quantity; see Fig. 25.15). According to Eq. (25.37), the segment then makes the following contribution to the potential at the point P:

$$dV = \frac{1}{4\pi\epsilon_0} \frac{(Q/l)\,dx'}{x - x'}$$

The integral of this over the length of the rod, from $x' = -l$ to $x' = 0$, gives us the total potential at the position P:

$$V(P) = V(x) = \int_{-l}^{0} \frac{1}{4\pi\epsilon_0} \frac{Q/l}{x - x'}\,dx'$$

$$= \frac{1}{4\pi\epsilon_0} \frac{Q}{l} \left[-\ln(x - x') \right]\Big|_{-l}^{0} = \frac{1}{4\pi\epsilon_0} \frac{Q}{l} \ln\left(\frac{x + l}{x} \right)$$

COMMENT: The distribution of charge along the rod produces a net potential proportional to $\ln[(x + l)/x]$, which, like the Coulomb potential, diverges at $x = 0$. Also, $V(x)$ approaches the Coulomb potential at large distances, where it is difficult to distinguish the rod from a point charge. This can be seen by expanding the logarithm for small values of l/x, since for small z, $\ln(1 + z) \approx z$. Thus

$$V(x) = \frac{1}{4\pi\epsilon_0} \frac{Q}{l} \ln\left(\frac{x + l}{x} \right) = \frac{1}{4\pi\epsilon_0} \frac{Q}{l} \ln\left(1 + \frac{l}{x} \right) \approx \frac{1}{4\pi\epsilon_0} \frac{Q}{l} \frac{l}{x} = \frac{1}{4\pi\epsilon_0} \frac{Q}{x}$$

which is identical to the Coulomb potential.

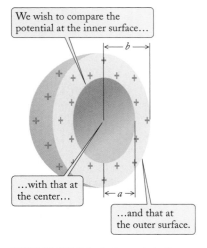

We wish to compare the potential at the inner surface...

...with that at the center...

...and that at the outer surface.

FIGURE 25.16 A charged shell.

✔ Checkup 25.2

QUESTION 1: For a sphere of radius R with a uniform volume distribution of positive charge, where is the electric potential maximum and where is it minimum? For a sphere of radius R with a uniform distribution of negative charge, where is the electric potential maximum and where is it minimum?

QUESTION 2: A hollow spherical charge distribution produces zero electric field in its interior. Does this mean that the electric potential is also zero in its interior?

QUESTION 3: A uniform electric field is directed along the x axis. Does the corresponding electric potential increase or decrease with x?

QUESTION 4: A thin ring of radius R has a charge Q distributed around its circumference. What is the electrostatic potential at the center of the ring? Does the charge need to be distributed uniformly around the ring?

QUESTION 5: A thick, insulating spherical shell with inner radius a and outer radius b has a positive charge Q distributed uniformly throughout its volume; the interior of the shell is empty (see Fig. 25.16). Is the electrostatic potential at $r = a$ larger, smaller, or the same as at $r = b$? Is the electrostatic potential at $r = 0$ larger, smaller, or the same as at $r = a$?

(A) Larger; larger   (B) Larger; smaller   (C) Larger; the same
(D) Smaller; larger   (E) Smaller; the same

25.3 POTENTIAL IN CONDUCTORS

Since the electric field in a conducting body in electrostatic equilibrium is zero, Eq. (25.26) implies that the potential difference between any two points within a conducting body is zero. Thus, *all points within a conducting body are at the same electrostatic potential.* For instance, since the ground is a conductor, all points on the surface of the Earth or within the Earth are at the same electrostatic potential; that is, the surface of the Earth is an equipotential. In experiments with electric circuits, it is usually convenient to adopt the convention that the potential of the surface of the Earth is zero, $V = 0$. The surface of the Earth is said to be the **electric ground**, and any conductors connected to it are said to be grounded. For example, the third terminal (round hole) in an ordinary household electric outlet (see Fig. 25.17) is grounded—it is connected to a plate or a rod buried in the ground outside the house. This grounded terminal is intended as a safety feature—if the insulation in your electric drill or some other appliance fails, the electric current from the "live" (flat) terminals can leak away harmlessly into the ground instead of leaking into your hands.

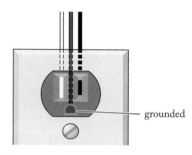

grounded

FIGURE 25.17 Grounded terminal of a three-hole outlet.

EXAMPLE 7   A conducting sphere of radius R carries a total positive charge Q uniformly distributed over its surface. What is the electrostatic potential inside and outside the sphere?

Concepts
— *in* —
Context

SOLUTION: As we found in Section 25.1, outside the sphere, the potential is the same as for a point charge,

$$V = \frac{1}{4\pi\epsilon_0} \frac{Q}{r} \quad (r \geq R) \tag{25.39}$$

Inside the conductor, the potential is the same everywhere, and so takes the same value as at the surface of the sphere, where $r = R$,

$$V = \frac{1}{4\pi\epsilon_0} \frac{Q}{R} \quad (r \leq R)$$

A plot of this potential is shown in Fig. 25.18. Note that for such a sphere, the electric field drops discontinuously to zero inside the sphere, but the electrostatic potential is continuous, remaining constant inside the sphere.

COMMENT: We would have obtained the same result if we had considered a hollow conducting shell; in that case, there would be no charge inside the shell, and thus no field and a constant potential.

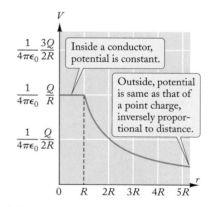

FIGURE 25.18 Electrostatic potential of a solid metal (conducting) sphere.

PHYSICS IN PRACTICE   ELECTRIC SHIELDING

The absence of electric fields in closed cavities surrounded by conductors has important practical applications. Delicate electric instruments can be shielded from atmospheric electric fields, and other stray electric fields, by placing them in a box made of sheet metal. Such a box is called a **Faraday cage**. Figure 1 shows a room-sized Faraday cage, also known as a *screen room*. Often, the box is made of fine copper wire mesh rather than sheet metal; although such wire mesh does not have the perfect shielding properties of solid sheet metal, it provides good enough shielding for most purposes. A mesh is convenient, since it is flexible and one can see through it.

The shielding of buildings against lightning relies on the same principle. Figure 2 shows a protective canopy, consisting of an array of wires, for the shielding of a building used for the storage of inflammable or explosive materials. The gaps between the wires permit the penetration of some field lines, and the shielding provided by such a canopy is even less perfect than that of a wire mesh, but the strength of the external atmospheric fields is attenuated to such an extent that lightning flashes are unlikely to reach the building.

FIGURE 1 A screen room, or room-sized Faraday cage.

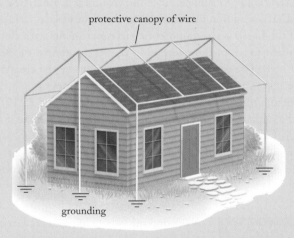

FIGURE 2 A building covering that provides protective electrostatic shielding.

For empty cavity, any field line would have to begin and end at surface.

V_0

V

Path parallel to field line would imply a potential difference: impossible!

Electric field is zero everywhere in cavity.

FIGURE 25.19 Empty cavity in a volume of conducting material. The red line is a hypothetical field line in the cavity.

With Eq. (25.26) we can prove an interesting theorem about the electric field in a conducting body with an empty, completely enclosed cavity (see Fig. 25.19): *within a closed, empty cavity inside a homogeneous conductor, the electric field is exactly zero.* The proof of the theorem is by contradiction. If there were an electric field inside this cavity, then there would have to be field lines in the cavity. Consider one of these field lines. Since the cavity is empty (contains no charge), the field line cannot end or begin within the space of the cavity—it must therefore begin and end on the surface of the cavity, as shown in Fig. 25.19 (note that the field line cannot penetrate the conducting material, since the electric field is zero in this material). Now, suppose that at the point where the field line begins, the potential has the value V_0, and evaluate Eq. (25.26) for a path that follows the field line from its beginning to its end. The electric field is everywhere tangent to the field line, so $\theta = 0$ and $\cos \theta = 1$, and therefore $-E \cos \theta \, ds$ is negative. According to Eq. (25.26), this implies that V is smaller than V_0. But such a difference between V and V_0 is impossible, since these potentials are evaluated at the surface of

In the same way, the sheet metal of an automobile provides adequate shielding against lightning. The windows of the automobile leave large gaps, and permit the penetration of some field lines. But the strength of the external atmospheric field is attenuated to such an extent that lightning flashes are unlikely to reach the occupants of the automobile. Figure 3 shows a direct lightning hit on an automobile.

Conductive suits are used to provide shielding for utility workers who are in direct contact with live power lines (see Fig. 4). Many high-voltage power lines are operated at 400 kV, and they are surrounded by fairly strong electric fields. For the routine repair and cleaning of insulators, the worker wears a protective suit of a heavy fabric that incorporates a woven mesh of stainless steel wires, which make it conducting. This suit shields the body from the external electric fields.

FIGURE 3 Lightning striking an automobile.

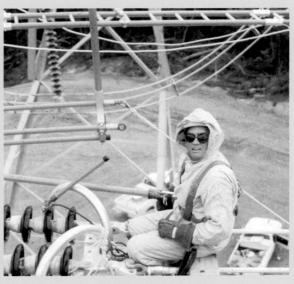

FIGURE 4 A conductive suit worn by a utility worker performing "bare-handed" repairs on a live power line.

the conductor, and all points on a conductor are necessarily at the same potential. This contradiction establishes the theorem. The fact that the electric field is zero in a conductor's cavity is quite useful, as discussed in Physics in Practice: Electric Shielding.

 Checkup 25.3

QUESTION 1: Two conducting spheres are connected by a conducting wire (Fig. 25.20). If one sphere has a radius R_1 and a positive charge Q_1, is this charge uniformly distributed over its surface? Is its potential $V = (1/4\pi\epsilon_0)Q_1/R_1$? Is its potential the same as that of the other sphere?

QUESTION 2: An airplane is flying through a thundercloud, where there are strong electric fields. Do these electric fields affect the occupants of the airplane?

QUESTION 3: An isolated 1.0-cm-diameter steel bearing ball and an isolated 1.0-m-diameter steel demolition ball each carry the same amount of charge. Compared with the small ball, the electrostatic potential of the large steel ball is

  (A) Larger   (B) Smaller   (C) The same

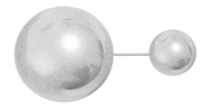

FIGURE 25.20 Two conducting spheres connected by a conducting wire.

25.4 CALCULATION OF THE FIELD FROM THE POTENTIAL

In Section 25.2 we saw how to calculate the electrostatic potential from the electric field. Now we will see how to calculate the electric field from the potential. For this purpose, we begin with Eq. (25.26), and we assume for the moment that ds is a small displacement in the direction of the electric field. With $\theta = 0$, Eq. (25.26) can be written as

$$\int E \, ds = -(V - V_0) \tag{25.40}$$

If the integration extends over only a small interval, the electric field is approximately constant, and the change in potential is small, $V - V_0 = dV$. Then Eq. (25.40) reduces to

$$E \, ds = -dV$$

which we can solve for E:

electric field from potential

$$E = -\frac{dV}{ds} \tag{25.41}$$

This says that the electric field equals the negative of the derivative of the potential with respect to the displacement.

EXAMPLE 8   The electrostatic potential generated by a pair of oppositely charged, parallel conducting plates is [Eq. (25.7)]

$$V = -E_0 y \tag{25.42}$$

For this potential, verify that Eq. (25.41) yields the correct value for the electric field.

SOLUTION: According to Eq. (25.41),

$$E = -\frac{dV}{ds} = -\frac{dV}{dy} = -\frac{d}{dy}(-E_0 y) = E_0$$

which is the expected result.

If the small displacement ds is not in the direction of the electric field, then we must retain the factor $\cos \theta$ in Eq. (25.26), and instead of Eq. (25.41) we obtain

$$E \cos \theta = -\frac{dV}{ds}$$

Since $E \cos \theta$ is the component of the electric field in the direction of the displacement, this formula tells us that the component of the electric field in any direction equals the negative of the derivative of the potential with respect to the displacement in that direction. Accordingly, we obtain E_x if we perform a small displacement in the x direction, E_y if we perform a small displacement in the y direction, and E_z if we perform a small displacement in the z direction. In each case, it is assumed that we are holding the other variables constant (e.g., for a displacement in the x direction, y and z are treated as constant). Mathematicians use the *partial derivative* symbol for this operation, for instance, $\partial/\partial x$ instead of d/dx. The components of the electric field are thus given by

$$E_x = -\frac{\partial V}{\partial x} \quad E_y = -\frac{\partial V}{\partial y} \quad E_z = -\frac{\partial V}{\partial z} \qquad (25.43)$$

To obtain each component of **E** for a given V, one merely treats the other variables like constants when one takes the derivative with respect to the variable of interest.

EXAMPLE 9   We found in Section 25.1 that the potential outside a charged sphere is the Coulomb potential, $V = (1/4\pi\epsilon_0)Q/r$. The Coulomb potential can be written in rectangular coordinates by using the identity $r = \sqrt{x^2 + y^2 + z^2}$. From that form of the potential, obtain the x, y, and z components of the electric field.

SOLUTION: The Coulomb potential in rectangular coordinates is simply

$$V = \frac{Q}{4\pi\epsilon_0} \frac{1}{\sqrt{x^2 + y^2 + z^2}}$$

Using Eq. (25.43), we evaluate E_x:

$$E_x = -\frac{\partial}{\partial x} \frac{Q}{4\pi\epsilon_0} \frac{1}{\sqrt{x^2 + y^2 + z^2}} = -\frac{Q}{4\pi\epsilon_0} \frac{\partial}{\partial x} (x^2 + y^2 + z^2)^{-1/2}$$

$$= -\frac{Q}{4\pi\epsilon_0} \left(-\frac{1}{2}\right)(x^2 + y^2 + z^2)^{-3/2} \times (2x) = \frac{Q}{4\pi\epsilon_0} \frac{x}{(x^2 + y^2 + z^2)^{3/2}}$$

Similarly, for E_y and E_z we have

$$E_y = \frac{Q}{4\pi\epsilon_0} \frac{y}{(x^2 + y^2 + z^2)^{3/2}} \quad \text{and} \quad E_z = \frac{Q}{4\pi\epsilon_0} \frac{z}{(x^2 + y^2 + z^2)^{3/2}} \qquad (25.44)$$

COMMENT: These can indeed be recognized as the x, y, and z components of the usual radial Coulomb field. For example, the second relation in Eq. (25.44) is

$$E_z = \frac{Q}{4\pi\epsilon_0} \frac{z}{r^3} = \frac{1}{4\pi\epsilon_0} \frac{Q}{r^2} \frac{z}{r} = \frac{1}{4\pi\epsilon_0} \frac{Q}{r^2} \cos\theta$$

where here θ is the angle between the electric field vector and the z axis, as shown in Fig. 25.21.

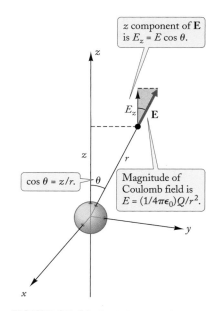

FIGURE 25.21 Radial electric field vector and its z component.

The results obtained in the preceding two examples do not tell us anything new—since we already know the electric field in these examples, the calculation merely verifies the consistency of our formulas. However, if we do not know the electric field, then Eq. (25.41) can be useful in a calculation of the electric field, provided we have some means of finding the potential. As discussed above in Section 25.2, if the charge distribution is specified, we can find the potential by treating the charge distribution as a collection of point charges, each of which contributes a potential of the form (25.37). The net potential is then the sum of all these point-charge-like contributions, given by Eq. (25.38).

The process of calculating the potential from the charge distribution, and then calculating the electric field from the potential, can be be significantly easier than calculating the electric field directly from the charge distribution. This is because the potential is a scalar quantity, so one need not consider vector components when one sums its contributions.

EXAMPLE 10   An amount of charge Q is uniformly distributed along the circumference of a thin ring of radius R (see Fig. 25.22). Find the potential on the axis of the ring, and find the electric field.

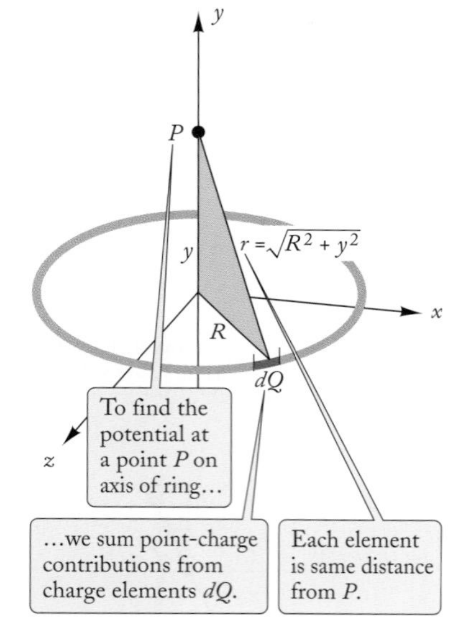

SOLUTION: To find the potential, we regard the ring as made up of many small charge elements, each of which can be treated as a point charge (see Fig. 25.22). If one of these elements has a charge dQ, its contribution to the potential at a distance r is

$$dV = \frac{1}{4\pi\epsilon_0}\frac{dQ}{r} = \frac{1}{4\pi\epsilon_0}\frac{dQ}{\sqrt{R^2 + y^2}} \qquad (25.45)$$

Since the distance $r = \sqrt{R^2 + y^2}$ does not depend on where along the circumference the charge dQ is located, all the small elements of charge on the ring contribute in the same way, and the net contribution to the potential therefore has the form of Eq. (25.45), with the charge contribution dQ replaced by the total charge $\int dQ = Q$:

$$V = \frac{1}{4\pi\epsilon_0}\frac{Q}{\sqrt{R^2 + y^2}} \qquad (25.46)$$

The electric field along the axis is the negative of the derivative of the potential with respect to y:

$$E_y = -\frac{\partial V}{\partial y} = -\frac{1}{4\pi\epsilon_0}\frac{\partial}{\partial y}\frac{Q}{\sqrt{R^2 + y^2}} = \frac{1}{4\pi\epsilon_0}\frac{Qy}{(R^2 + y^2)^{3/2}}$$

We already found this same result in Chapter 23 by the much more difficult technique of integration of the vector electric field, according to Coulomb's Law. Here, the calculation is much easier.

FIGURE 25.22 A uniformly charged ring.

equipotential surface

A mathematical surface on which the electrostatic potential has a fixed, constant value is called an **equipotential surface**. Figure 25.23 shows the equipotential surfaces belonging to the potential of a uniformly charged flat sheet—the equipotential surfaces are parallel planes. Figure 25.24 shows the equipotential surfaces belonging to the potential of a point charge—the equipotential surfaces are concentric spheres. Figure 25.25 shows the equipotential surfaces of a pair of positive and negative charges of equal magnitudes. The equipotential surfaces associated with a given potential provide a graphical representation of the potential.

Note that the electric field is everywhere perpendicular to the equipotentials. This is an immediate consequence of Eq. (25.41): along any equipotential surface, the potential is constant, that is, $dV = 0$; consequently, for a displacement ds along the surface, $E = -dV/ds = 0$, which says that the electric field in the direction tangential to the surface is zero. Thus, the electric field must be entirely perpendicular to the surface (see Fig. 25.26).

Conversely, if the electric field is everywhere perpendicular to a given surface, then this surface must be an equipotential surface. This, also, is a consequence of Eq. (25.41); if the electric field is zero in the direction parallel to the surface, then for a displacement ds parallel to the surface, $dV = 0$, and the potential is constant. Since we already know (from Section 24.5) that along the surface of any conductor in electrostatic equilibrium the electric field is perpendicular to the surface, we can conclude that any conducting surface is an equipotential surface. (This conclusion agrees with the general statement made in Section 25.3—the potential is constant throughout any conductor.)

Keep in mind that equipotential surfaces are surfaces of constant energy for a given charge. Thus it takes no energy to move a charge along a surface of constant potential, since such motion does not act against any electric field (see Fig. 25.26). However, to move perpendicular to such surfaces does require work, since the electric field is perpendicular to the equipotential surface.

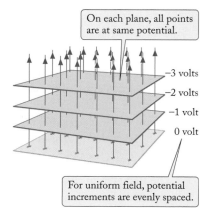

FIGURE 25.23 Equipotential surfaces for a very large flat sheet with a uniform distribution of charge. The equipotentials are flat planes. This indicates that all points at the same height above the sheet are at the same potential.

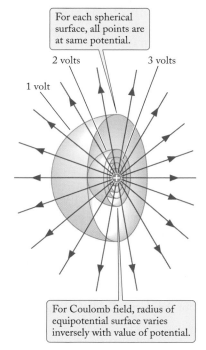

FIGURE 25.24 Equipotential surfaces for a positive point charge. The equipotentials are concentric spheres. This indicates that points at the same distance from the central charge are at the same potential.

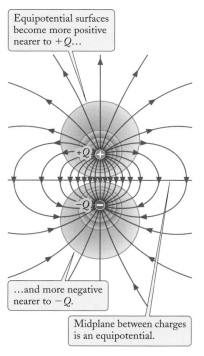

FIGURE 25.25 Equipotential surfaces for a positive and a negative point charge of equal magnitudes. Note that the midplane between the charges is an equipotential.

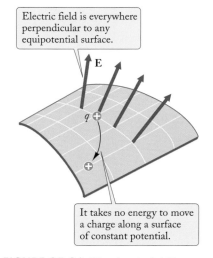

FIGURE 25.26 The electric field is everywhere perpendicular to the equipotential surface. If a charge q moves along this surface, the electric field does no work on it.

PROBLEM-SOLVING TECHNIQUES   ELECTROSTATIC POTENTIAL AND FIELD

In previous chapters we became acquainted with two methods for the calculation of the electric field of a charge distribution: from Coulomb's Law, and from Gauss' Law. The calculation of the electric field from the electrostatic potential, as in Example 10, gives us a third method for the calculation of the electric field.

- The first step in this method is to find the electrostatic potential of the charge distribution. For this purpose it is often convenient to introduce x, y, and z coordinates with the origin somewhere near the charge distribution (compare Fig. 25.22). If the charge distribution consists of an arrangement of several point charges, the net electrostatic potential is simply the sum of the individual electrostatic potentials of the point charges. Note that since the potential is a scalar, the sum of the individual potentials is an ordinary sum of ordinary numbers, which is much easier to evaluate than the vector sum required for a calculation of the net electric field via Coulomb's Law. If the charge distribution is continuous (such as the charge distribution along the circumference of the ring in Example 10), you need to regard it as a collection of many small (infinitesimal) charge elements, each of which can be treated as a point charge. The net potential is then the integral over the electrostatic potentials of all these small elements.

- The second step is the calculation of the electric field by differentiation of the potential, according to Eq. (25.41). If the symmetry of the problem permits you to determine the direction of the electric field, then the magnitude of the electric field is the negative derivative of the potential V with respect to the displacement s in that direction,

$$E = -\frac{dV}{ds}$$

- If the direction of the electric field is not obvious, then you can calculate the x, y, and z components of the electric field by differentiating the potential V in the x, y, and z directions:

$$E_x = -\frac{\partial V}{\partial x} \quad E_y = -\frac{\partial V}{\partial y} \quad E_z = -\frac{\partial V}{\partial z}$$

The calculation of all three components requires knowledge of the complete dependence of V on x, y, and z. For instance, Eq. (25.46) tells us the dependence of V on y along the axis of a charged ring, but does not tell us the dependence on x or z at points off the axis; hence Eq. (25.46) cannot be used to find the x or z components of the electric field for points off the axis.

 ## Checkup 25.4

QUESTION 1: Figure 25.23 shows the equipotential surfaces for 0 V, −1 V, −2 V, etc., for a uniformly charged flat sheet. How would the positions of these equipotential surfaces change if the charge density on the sheet were twice as large?

QUESTION 2: Figure 25.24 shows the spherical equipotential surfaces for 1 V, 2 V, 3 V, etc., for a point charge. How would the radii of these equipotential surfaces change if the point charge had twice the amount of charge?

QUESTION 3: Describe the equipotential surfaces for the electric field of a long, straight line of charge.

QUESTION 4: The midplane between a pair of positive and negative charges of equal magnitudes is an equipotential surface (see Fig. 25.25). Is this also true if both charges are of equal signs?

QUESTION 5: Can there be a nonzero electric field at a location where the potential is zero? Can there be a nonzero potential at a location where the electric field is zero?
      (A) No; no   (B) No; yes   (C) Yes; no   (D) Yes; yes

25.5 ENERGY OF SYSTEMS OF CHARGES

Arrangements of electric charges are used in physics and engineering to produce electric fields, and also to store electric energy. For instance, the energy for the operation of the flashlamp of a photographic camera is stored in a charge distribution accumulated on a system of parallel plates (called a *capacitor*) in the flash unit. It is therefore often important to calculate the electric energy of an arrangement of charges.

The simplest such arrangement consists of two point charges q and q' separated by a distance r. The electric potential energy of this system is

$$U = \frac{1}{4\pi\epsilon_0} \frac{qq'}{r} \tag{25.47}$$

This potential energy can be regarded as the work required to move q from an infinite distance to within a distance r of q'. It is a *mutual* potential energy that belongs to both q and q'; that is, it is an energy associated with the relative configuration of the pair (q, q').

For arrangements consisting of more than two charges, the net potential energy can be calculated by writing down a term similar to that in Eq. (25.47) for *each pair* of charges. For instance, if we are dealing with three charges q_1, q_2, and q_3 (see Fig. 25.27), we have three possible pairs (q_1, q_2), (q_2, q_3), and (q_1, q_3), so the net potential energy of the system is

$$U = \frac{1}{4\pi\epsilon_0} \frac{q_1 q_2}{r_{12}} + \frac{1}{4\pi\epsilon_0} \frac{q_2 q_3}{r_{23}} + \frac{1}{4\pi\epsilon_0} \frac{q_1 q_3}{r_{13}} \tag{25.48}$$

potential energy of a system of point charges

where r_{12}, r_{23}, and r_{13} are the distances indicated in Fig. 25.27. This potential energy is the work required to assemble the charges in the final configuration shown in Fig. 25.27, starting from an initial condition of infinite separation of all charges.

That this sum of the potential energies of each pair gives the correct total energy can be seen by assembling the charges sequentially. For example, for three charges, it takes no energy to bring a first charge into position (since there are no electric fields to act against). To bring the second charge into position involves moving it in the Coulomb potential of the first charge (thus the $q_1 q_2$ term). Finally, to bring the third charge into position involves working against the Coulomb fields of the first two charges (thus the $q_2 q_3$ and $q_1 q_3$ terms).

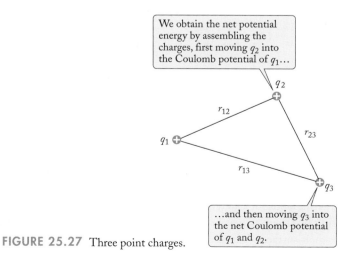

We obtain the net potential energy by assembling the charges, first moving q_2 into the Coulomb potential of q_1...

q_2

r_{12}

q_1

r_{23}

r_{13}

q_3

...and then moving q_3 into the net Coulomb potential of q_1 and q_2.

FIGURE 25.27 Three point charges.

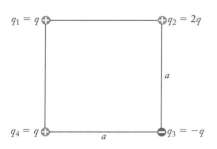

$q_1 = q$ $q_2 = 2q$

a

$q_4 = q$ a $q_3 = -q$

FIGURE 25.28 Four point charges.

EXAMPLE 11

Four charges, $q_1 = q$, $q_2 = 2q$, $q_3 = -q$, and $q_4 = q$, are at the corners of a square of side a, as shown in Fig. 25.28. If $q = 2.0 \ \mu\text{C}$ and $a = 7.5 \text{ cm}$, what was the total energy required to assemble this system of charges?

SOLUTION: For the total energy, we sum all the potential energies of the individual pairs:

$$U = \frac{1}{4\pi\epsilon_0}\left(\frac{q_1 q_2}{r_{12}} + \frac{q_1 q_3}{r_{13}} + \frac{q_2 q_3}{r_{23}} + \frac{q_1 q_4}{r_{14}} + \frac{q_2 q_4}{r_{24}} + \frac{q_3 q_4}{r_{34}}\right)$$

$$= \frac{1}{4\pi\epsilon_0}\left[\frac{(q)(2q)}{a} + \frac{(q)(-q)}{\sqrt{2}a} + \frac{(2q)(-q)}{a} + \frac{(q)(q)}{a} + \frac{(2q)(q)}{\sqrt{2}a} + \frac{(-q)(q)}{a}\right]$$

$$= \frac{1}{4\pi\epsilon_0}\frac{q^2}{a}\left(2 + \frac{-1}{\sqrt{2}} - 2 + 1 + \frac{2}{\sqrt{2}} - 1\right) = \frac{1}{4\pi\epsilon_0}\frac{q^2}{a}\frac{1}{\sqrt{2}}$$

$$= (9.0 \times 10^9 \text{ N·m}^2/\text{C}^2)\frac{(2.0 \times 10^{-6} \text{ C})^2}{(0.075 \text{ m})}\frac{1}{\sqrt{2}} = 0.34 \text{ J}$$

In practical applications, point charges are usually deposited on conductors, each of which carries many point charges. By summing the contributions as we build up such a system, we can calculate the potential energy of a system of charges placed on conductors.

First consider a single conducting sphere of radius R. When it has some charge q, its electrostatic potential $V = (1/4\pi\epsilon_0)(q/R)$. If we imagine changing this amount of charge by a small amount dq, then the potential energy will change by an amount dU given by

$$dU = V \, dq$$

If we build up a total charge Q in small steps from $q = 0$ to $q = Q$, we can sum the contributions to the potential energy during such a process:

$$U = \int dU = \int V \, dq = \int_0^Q \frac{1}{4\pi\epsilon_0}\frac{q}{R}\, dq$$

The radius R of the sphere is not changing while we add charge to it, so this is simply

$$U = \frac{1}{4\pi\epsilon_0}\frac{1}{R}\int_0^Q q \, dq = \frac{1}{4\pi\epsilon_0}\frac{1}{R}\frac{1}{2}Q^2$$

or, writing this in terms of the final potential $V = (1/4\pi\epsilon_0)(Q/R)$,

potential energy of a conductor

$$U = \tfrac{1}{2}QV \tag{25.49}$$

The factor of one-half arises because we are not considering the potential energy of a charge Q in some *external* potential, but instead the energy of interaction of the pieces of Q. The result (25.49) holds for any fixed geometry of conductor.

A similar result holds for a system of conductors. For instance, Fig. 25.29 shows several conducting bodies—such as bodies of metal—with charges $Q_1, Q_2, Q_3, \ldots$ distributed over their surfaces. We recall that on each conducting body, the potential has

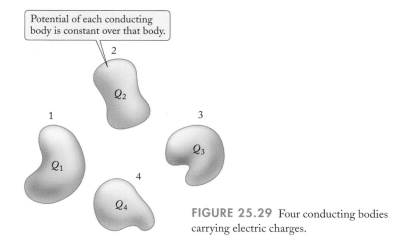

FIGURE 25.29 Four conducting bodies carrying electric charges.

a constant value over the entire volume of the body. Suppose that these electrostatic potentials of the bodies are $V_1, V_2, V_3, \ldots$ It can be shown, although we will not attempt to prove it here, that the net electric potential energy of such a system of conductors is

$$U = \tfrac{1}{2}Q_1 V_1 + \tfrac{1}{2}Q_2 V_2 + \tfrac{1}{2}Q_3 V_3 + \cdots \qquad (25.50)$$

potential energy of a system of conductors

EXAMPLE 12   The metallic sphere on the top of a large Van de Graaff generator has a radius of 3.0 m. Suppose that the sphere carries a charge of 5.0×10^{-5} C uniformly distributed over its surface. How much electric energy is stored in this charge distribution?

Concepts in Context

SOLUTION: In this problem, there is only one conductor, the metallic sphere. Accordingly, we need only the first term in Eq. (25.50):

$$U = \tfrac{1}{2}Q_1 V_1$$

or, with $Q_1 = Q$ and $V_1 = V$, we recover Eq. (25.49):

$$U = \tfrac{1}{2}QV$$

The potential outside a spherically symmetric charge distribution is given by Eq. (25.12). At $r = R$, this potential is

$$V = \frac{1}{4\pi\epsilon_0}\frac{Q}{R}$$

According to Eq. (25.49), the electric energy is then

$$U = \frac{1}{2}QV = \frac{1}{2}Q\frac{1}{4\pi\epsilon_0}\frac{Q}{R} = \frac{1}{8\pi\epsilon_0}\frac{Q^2}{R} \qquad (25.51)$$

With $Q = 5.0 \times 10^{-5}$ C and $R = 3.0$ m, this gives

$$U = \frac{1}{8\pi \times 8.85 \times 10^{-12}\ \text{C}^2/\text{N·m}^2}\frac{(5.0 \times 10^{-5}\ \text{C})^2}{3.0\ \text{m}} = 3.7\ \text{J}$$

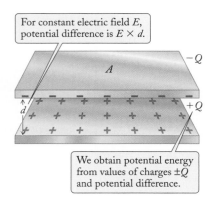

For constant electric field E, potential difference is $E \times d$.

We obtain potential energy from values of charges $\pm Q$ and potential difference.

FIGURE 25.30 Two parallel plates, with charges $+Q$ and $-Q$.

EXAMPLE 13

Two large, parallel metallic plates of area A are separated by a distance d. Charges $+Q$ and $-Q$ are placed on the plates, respectively (see Fig. 25.30). What is the electric potential energy?

SOLUTION: Here, we are dealing with two conductors, with $Q_1 = Q$ and $Q_2 = -Q$. Thus, Eq. (25.49) becomes

$$U = \tfrac{1}{2}Q_1 V_1 + \tfrac{1}{2}Q_2 V_2 = \tfrac{1}{2}QV_1 - \tfrac{1}{2}QV_2 = \tfrac{1}{2}Q(V_1 - V_2) \qquad (25.52)$$

To proceed, we need the potential difference $V_1 - V_2$ between the plates. The electric field in the region between the pair of plates is given by Eq. (24.29),

$$E = \frac{\sigma}{\epsilon_0}$$

where σ is the charge per unit area on each plate. Since the total charge on a plate is Q and the area is A, the charge per unit area is $\sigma = Q/A$, and

$$E = \frac{Q}{\epsilon_0 A} \qquad (25.53)$$

This expression fails near the edges of the plates, where there is an electric fringing field that is not constant. But if the plates are very large, then the edge region is only a very small fraction of the total region between the plates and we can ignore this region without introducing excessive errors in our calculation.

With the constant electric field (25.53) we can calculate the electrostatic potential difference between two plates, using our standard formula for the potential, Eq. (25.26):

$$V_2 = -\int_0^d E \cos \theta \, ds + V_1$$

Since we move parallel to the field going from the positive to the negative plate, $\cos \theta = 1$. For a constant electric field, we then have, similar to Eq. (25.7),

$$V_2 - V_1 = -Ed \qquad (25.54)$$

With $E = Q/\epsilon_0 A$, this is

$$V_1 - V_2 = \frac{Q}{\epsilon_0 A}d$$

Substituting this into Eq. (25.52), we obtain

$$U = \frac{1}{2}Q\frac{Q}{\epsilon_0 A}d = \frac{1}{2}\frac{Q^2 d}{\epsilon_0 A} \qquad (25.55)$$

Equation (25.55) can be rewritten in the following interesting way, in terms of the electric field:

$$U = \frac{1}{2}\frac{Q^2 d}{\epsilon_0 A} = \frac{1}{2}\epsilon_0 \left(\frac{Q}{\epsilon_0 A}\right)^2 Ad \qquad (25.56)$$

$$= \tfrac{1}{2}\epsilon_0 E^2 \times Ad$$

Here the product $Ad = $ [area] $\times$ [separation] is the volume of the space between the plates, that is, the volume filled with electric field. Thus, the electric potential energy has the form

$$U = \tfrac{1}{2}\epsilon_0 E^2 \times \text{[volume]} \qquad (25.57)$$

This expression suggests that the energy is distributed over the volume in which there is electric field, with an energy density, or amount of energy per unit volume, $\epsilon_0 E^2/2$. According to this formula, the energy is concentrated where the electric field is strong.

Note that although Eq. (25.57) expresses the energy in terms of the electric field and suggests that the energy is located in the electric field, Eq. (25.49) expresses the energy in terms of the electric charges and suggests that the energy is located on these charges. Thus, these two equations, which are mathematically equivalent, suggest conflicting physical interpretations. To decide which alternative is correct, we need some extra information. The clue is the existence of electric fields that are independent of electric charges. As we will see in Chapter 33, radio waves and light waves consist of electric and magnetic fields traveling through space. Such fields are originally created by electric charges, but they persist even when the charges disappear. For instance, a radio wave or a light beam continues to travel through space long after the radio transmitter has been shut down or the flashlight has been switched off—this indicates that the energy of a radio wave resides in the radio wave itself, in its electric and magnetic fields, and not in the electric charges in the antenna of the radio transmitter. We can then argue that if energy is associated with the traveling electric fields of a radio wave, energy should also be associated with the electric fields of a static charge distribution. A more detailed calculation confirms that the expression we found above for the energy density in the special case of a uniform electric field is valid in general, for uniform or nonuniform electric fields. In any electric field (in vacuum) the energy density u is

$$u = \tfrac{1}{2}\epsilon_0 E^2 \qquad (25.58)$$

energy density in electric field

EXAMPLE 14   What is the energy density in the electric field of a thundercloud where the electric field $E = 2.0 \times 10^6$ V/m?

SOLUTION: According to Eq. (25.58), the energy density u is

$$u = \tfrac{1}{2}\epsilon_0 E^2 = \tfrac{1}{2} \times 8.85 \times 10^{-12} \text{ C}^2/\text{N·m}^2 \times (2.0 \times 10^6 \text{ V/m})^2$$

$$= 18 \text{ J/m}^3$$

 Checkup 25.5

QUESTION 1: What happens to expression (25.48) if there are only two charges, q_1 and q_2?

QUESTION 2: Is the electric energy U of a system of charges necessarily positive if all the charges q_1, q_2, q_3, etc., are positive? If they are all negative? If some are positive and some are negative?

QUESTION 3: Suppose we have a system of point charges with the electric potential energy given by Eq. (25.48). By what factor will this energy change if we increase the values of all the charges by a factor of 2?

QUESTION 4: Suppose that an energy of 0.50 J is stored in the electric field of a charged spherical balloon coated with a conducting layer. What will the energy in the electric field of this balloon be if we inflate it so its radius becomes twice as large?

QUESTION 5: An electric field has an energy density of 10 J/m^3. What will be the energy density if we increaese the electric field by a factor of 2? A factor of 3?

QUESTION 6: An electric energy of 0.010 J is stored in the electric field between two parallel charged plates. How does this energy change if we increase the plate separation by a factor of 2, while holding the charge constant? How does the energy density change?

QUESTION 7: Suppose that an energy of 1.0 J is stored in the electric field of a charged metallic sphere. What will the energy in the electric field of this sphere be if we allow one-half of the charge to leak away?

(A) 4.0 J   (B) 2.0 J   (C) 1.0 J   (D) 0.50 J   (E) 0.25 J

SUMMARY

PROBLEM-SOLVING TECHNIQUES Energy Conservation and Motion of a Point Charge	**(page 797)**
PHYSICS IN PRACTICE Electric Shielding	**(page 804)**
PROBLEM-SOLVING TECHNIQUES Electrostatic Potential and Field	**(page 810)**

ELECTROSTATIC POTENTIAL (POTENTIAL ENERGY PER UNIT CHARGE)	$V = \dfrac{U}{q}$	**(25.6)**
SI UNIT OF POTENTIAL	1 volt $= 1$ V $= 1$ J/C	**(25.8)**
ELECTROSTATIC POTENTIAL IN A UNIFORM ELECTRIC FIELD E_0 where the distance y is along the field direction.	$V = -E_0 y$	**(25.7)**
ALTERNATE UNIT OF ENERGY Electron-volt:	1 eV $= 1.60 \times 10^{-19}$ J	**(25.16)**
CONSERVATION OF ENERGY	$K + U = \frac{1}{2}mv^2 + qV =$ [constant]	**(25.22)**
POTENTIAL ENERGY OF TWO POINT CHARGES	$U = \dfrac{1}{4\pi\epsilon_0}\dfrac{qq'}{r}$	**(25.11)**

ELECTROSTATIC POTENTIAL OF POINT CHARGE

$$V = \frac{1}{4\pi\epsilon_0} \frac{q'}{r}$$

(25.12)

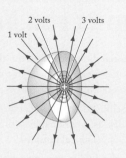

CALCULATION OF POTENTIAL FROM ELECTRIC FIELD   If V_0 is the potential at a point P_0, then at a point P,

$$V = -\int_{P_0}^{P} \mathbf{E} \cdot d\mathbf{s} + V_0 \quad \text{or} \quad V = -\int_{P_0}^{P} E \cos\theta \, ds + V_0$$

(25.25, 25.26)

POTENTIAL OF A CONTINUOUS CHARGE DISTRIBUTION

$$V = \frac{1}{4\pi\epsilon_0} \int \frac{dQ}{r}$$

(25.38)

POTENTIAL AND CONDUCTORS   For static charge distributions, the potential throughout a conductor is constant. Within an empty cavity in a conductor, the electric field is exactly zero.

CALCULATION OF THE ELECTRIC FIELD FROM THE POTENTIAL

Along the direction of the field:

$$E = -\frac{dV}{ds}$$

(25.41)

Along x, y, and z directions:

$$E_x = -\frac{\partial V}{\partial x} \quad E_y = -\frac{\partial V}{\partial y} \quad E_z = -\frac{\partial V}{\partial z}$$

(25.43)

EQUIPOTENTIAL SURFACE   An imaginary surface on which the electrostatic potential is constant. The electric field is everywhere perpendicular to an equipotential surface.

POTENTIAL ENERGY OF A SYSTEM OF POINT CHARGES

$$U = \frac{1}{4\pi\epsilon_0} \frac{q_1 q_2}{r_{12}} + \frac{1}{4\pi\epsilon_0} \frac{q_2 q_3}{r_{23}} + \frac{1}{4\pi\epsilon_0} \frac{q_1 q_3}{r_{13}} + \cdots \text{[all pairs]}$$

(25.48)

POTENTIAL ENERGY OF A CONDUCTOR

$$U = \tfrac{1}{2}QV$$

(25.49)

POTENTIAL ENERGY OF A SYSTEM OF CONDUCTORS

$$U = \tfrac{1}{2}Q_1 V_1 + \tfrac{1}{2}Q_2 V_2 + \tfrac{1}{2}Q_3 V_3 + \cdots$$

(25.50)

ENERGY DENSITY IN ELECTRIC FIELD

$$u = \tfrac{1}{2}\epsilon_0 E^2$$

(25.58)

QUESTIONS FOR DISCUSSION

1. The potential difference between the poles of an automobile battery is 12 volts. Explain what this means in terms of the definition of potential as work per unit charge.

2. An old-fashioned word for electrostatic potential is electrostatic *tension*. Is it reasonable to think of the potential as analogous to a mechanical tension?

3. If the electric field is zero in some region, must the potential also be zero? Give an example.

4. A bird sits on a high-voltage power line which is at a potential of 345 000 volts. Does this harm the bird?

5. How would you define the gravitational potential? Are the units of gravitational potential the same as the units of electric potential? According to your definition, what is the gravitational potential difference between the ground and a point 50 m above the ground?

6. Consider an electron moving in the vicinity of a proton. Where is the electrostatic potential produced by the proton highest? Where is the potential energy of the electron highest?

7. Suppose that the electrostatic potential has a minimum at some point. Is this an equilibrium point for a positive charge? For a negative charge?

8. Consider a sphere of radius R with a charge Q uniformly distributed over its volume. Where does the potential have a maximum? Where does the magnitude of the electric field have a maximum?

9. Give an example of a conductor that is not an equipotential. Is this conductor in electrostatic equilibrium?

10. If the potential in a three-dimensional region of space is known to be constant, what can you conclude about the electric field in this region? If the potential on a two-dimensional surface is known to be constant, what can you conclude about the electric field on this surface?

11. In many calculations it is convenient to assign a potential of 0 volt to the ground. If so, what is the potential at the top of the Eiffel Tower? What is the potential at the top of your head? (Hint: Your body is a conductor.)

12. Is it true that the surface of a mass in static equilibrium is a gravitational equipotential surface? What if the surface is that of a fluid, such as water?

13. If a high-voltage power cable falls on top of your automobile, you will probably be safest if you remain inside the automobile? Why?

14. Suppose that several separate metallic bodies have been placed near a charge distribution. Is it necessarily true that all of these bodies will have the same potential?

15. Consider the patterns of field lines displayed in Figs. 23.17 and 23.18. Roughly sketch the equipotentials for these field lines.

16. If we surround some region with a conducting surface, we shield it from external electric fields. Why can we not shield a region from gravitational fields in a similar manner?

17. A cavity is completely surrounded by conducting material. Can you create an electric field in this cavity?

18. Show that different equipotential surfaces cannot intersect.

19. Consider the pattern of field lines shown in Fig. 24.22. Roughly, sketch some of the equipotential surfaces for this case.

20. Consider a metallic sphere carrying a given amount of charge. Explain why the electric energy is large if the radius of the sphere is small. Would you expect a similar inverse proportion between the electric energy and the size of a conductor of arbitrary shape?

21. Equation (25.50) suggests that the electric energy is located at the charges, whereas Eq. (25.57) suggests it is located in the field. How could we perform an experiment to test where the energy is located? (Hint: Energy gravitates.)

22. Figure 25.31 shows a sequence of deformations of a nucleus as it undergos fission. The volume of the nucleus and the electric charge remain constant during these deformations. Which configuration has the highest electric energy? The lowest?

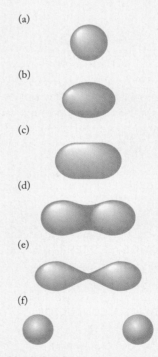

FIGURE 25.31 Fission of a nucleus.

23. Consider a sphere with uniform distribution of charge over its volume. Where is the energy density within the sphere highest? Lowest?

24. Since the electric energy density is never negative, how can the mutual electric potential energy of a pair of opposite charges be negative?

PROBLEMS

†25.1 The Electrostatic Potential

1. The electric potential difference between the positive and negative poles of an automobile battery is 12 volts. In order to charge the battery fully, the charging device must force $+2.0 \times 10^5$ coulombs from the negative terminal of the battery to the positive terminal. How much work must the charging device do during this process?

2. An ordinary flashlight battery has a potential difference of 1.5 V between its positive and negative terminals. How much work must you do to transport an electron from the positive terminal to the negative terminal?

3. On days of fair weather, the atmospheric electric field of the Earth is about 100 V/m; this field points vertically downward (compare Problem 4 of Chapter 23). What is the electric potential difference between the ground and an airplane flying at 600 m? What is the potential difference between the ground and the tip of the Eiffel Tower? Treat the ground as a flat conductor.

4. Consider the arrangement of parallel sheets of charge described in Problem 35 of Chapter 23 (see Fig. 23.37). Find the potential difference between the upper sheet and the lower sheet.

5. A proton is accelerated from rest through a potential of 2.50×10^5 V. What is its final speed?

6. At the Stanford Linear Accelerator (SLAC), electrons are accelerated from an energy of 0 eV to 20×10^9 eV as they travel in a straight evacuated pipe 1600 m in length (Fig. 25.32). The acceleration is due to a strong electric field pushing the electrons along. Assume that the electric field is uniform. What must be its strength?

FIGURE 25.32 Beam pipe at the Stanford Linear Accelerator (SLAC).

7. The potential difference between the two poles of an automobile battery is 12.0 V. Suppose that you place such a battery in empty space and that you release an electron at a point next to the negative pole of the battery. The electron will then be pushed away by the electric force and move off in some direction.

   (a) If the electron strikes the positive pole of the battery, what will be its impact speed?

   (b) If, instead, the electron moves away toward infinity, what will be its ultimate speed?

8. The gap between the electrodes of a spark plug in an automobile is 0.64 mm. In order to produce an electric field of 3.0×10^6 V/m (required to initiate an electric spark), what minimum potential difference must you apply to the spark plug?

9. An electron in a neon tube is accelerated from rest through a 2000-V potential. What velocity does the electron attain?

10. A charge of 2.0×10^{-12} C is placed on a small (pointlike) cork ball. What is the electrostatic potential at a distance of 30 cm from the ball? At a distance of 60 cm?

11. In an electron-beam heater, electrons at rest near a tungsten filament are accelerated toward a metal target at a high electrostatic potential. If the electrons strike the target to be heated with a speed of 1.8×10^7 m/s, what is the potential difference between the target and the filament?

12. An electron in a region of uniform electric field E_0 has an initial velocity $\mathbf{v}_0$ in the direction of the field. How far does the electron travel before stopping?

13. An electron is initially a distance of $r_0 = 4.3 \times 10^{-9}$ m from a proton, traveling directly away from the proton at speed 4.0×10^5 m/s. What is its speed when it is very far from the proton?

14. A conducting spherical shell of radius 12.0 cm is charged to a potential of 50 000 V. What is the value of the electrostatic potential 5.0 cm outside the surface of the sphere?

*15. In a demonstration, a small Styrofoam "worm" of mass 0.20 g sits on a Van de Graaff generator, a conducting spherical shell of radius 15 cm. The shell is at a potential of 75 000 V. When the worm acquires a charge Q, it is repelled by the sphere and travels vertically under the influence of gravity and the electric force. The worm moves upward and achieves equilibrium 0.50 m above the surface of the sphere. What is the charge Q?

†25.2 Calculation of the Potential from the Field

16. The nucleus of lead is a uniformly charged sphere with a charge of $82e$ and a radius of 7.1×10^{-15} m. What is the electrostatic potential at the nuclear surface? At the nuclear center?

17. The nucleus of platinum is a uniformly charged sphere with a charge of $78e$ and a radius of 7.0×10^{-15} m. What is the electric potential energy of an incident proton arriving at the nuclear surface? At the nuclear center?

18. An alpha particle of kinetic energy 1.7×10^{-12} J is shot directly toward a platinum nucleus from a very large distance. What will be the distance of closest approach? The electric charge of the alpha particle is $2e$ and that of the platinum nucleus is $78e$. Treat the alpha particle as a point particle, and the nucleus as a spherical charge distribution of radius 5.1×10^{-15} m, and disregard the motion of the nucleus.

19. An alpha particle is initially at a very large distance from a plutonium nucleus. What is the minimum kinetic energy with which the alpha particle must be launched toward the nucleus if it is to make contact with the nuclear surface? The plutonium nucleus is a sphere of radius 7.5×10^{-15} m with a charge of $94e$ uniformly distributed over the volume. For the purpose of this problem, the alpha particle may be regarded as a particle (of negligible radius) with a charge of $2e$.

20. Two positive point charges Q are on the y axis at $y = \pm d/2$. Find the potential for points on the positive x axis.

*21. Eight point charges $+Q$ sit at the corners of a cube of side a. What is the potential at the center of the cube? At the center of a cube face? At the center of a cube edge?

*22. A positive point charge Q is on the y axis at $y = D$; a negative point charge $-2Q$ is at the point $x = D, y = D$. Find the potential for points on the x axis.

*23. An arc of a circle of radius R subtends an angle θ. The arc is a thin rod with a uniform linear charge density λ. What is the electrostatic potential at the center of curvature?

*24. Three large charged sheets are parallel to the x–z plane. The sheets are at $y = 0$, $y = d$, and $y = 2d$; these sheets have uniform surface charge densities $+\sigma$, -2σ, and $+\sigma$, respectively. Assuming the reference potential is zero at $y = 0$, determine the potential as a function of y.

*25. A thorium nucleus emits an alpha particle according to the reaction

$$\text{thorium} \rightarrow \text{radium} + \text{alpha}$$

Assume that the alpha particle is pointlike and that the residual radium nucleus is spherical with a radius of 7.4×10^{-15} m. The charge on the alpha particle is $2e$, and that on the radium nucleus is $88e$.

(a) At the instant the alpha particle emerges from the nuclear surface, what is its electrostatic potential energy?

(b) If the alpha particle has no initial kinetic energy, what will be its final kinetic energy and speed when far away from the nucleus? Assume that the radium nucleus does not move. The mass of the alpha particle is 6.7×10^{-27} kg.

*26. Consider again the arrangement of charges within the thundercloud of Fig. 23.34. Find the electric potential due to these charges at a point which is at a height of 8.0 km and on the vertical line passing through the charges. Find the electric potential at a second point which is at the same height and has a horizontal distance of 5.0 km from the first point.

*27. In a helium atom, at some instant one of the electrons is at a distance of 3.0×10^{-11} m from the nucleus and the other electron is at a distance of 2.0×10^{-11} m, 90° away from the first (Fig. 25.33). Find the electric potential produced jointly by the two electrons and the nucleus at a point P beyond the first electron and at a distance of 6.0×10^{-11} m from the nucleus.

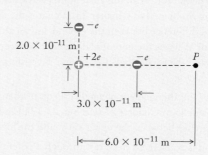

FIGURE 25.33 Instantaneous configuration of the nucleus $(+2e)$ and the electrons $(-e)$ of an atom of helium.

*28. A total charge Q is distributed uniformly along a straight rod of length l. Find the potential at a point P a distance h from the midpoint of the rod (see Fig. 25.34).

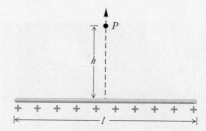

FIGURE 25.34 A uniformly charged rod of length l.

*29. Three thin rods of glass of length l carry charges uniformly distributed along their lengths. The charges on the three rods are $+Q$, $+Q$, and $-Q$, respectively. The rods are arranged along the sides of an equilateral triangle. What is the electrostatic potential at the midpoint of this triangle?

*30. A uniformly charged sphere of radius a is surrounded by a uniformly charged concentric spherical shell of inner radius b and outer radius c (Fig. 25.35). The total charge in the sphere is Q, and that on the outer shell $-Q$. Find the potential at $r = b$, at $r = a$, and at $r = 0$.

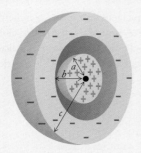

FIGURE 25.35 Charged sphere and concentric spherical shell.

*31. Four rods of length *l* are arranged along the edges of a square. The rods carry charges $+Q$ uniformly distributed along their lengths (Fig. 35.36). Find the potential at the point P at a distance x from one corner of the square.

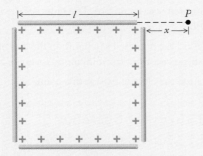

FIGURE 25.36 Four charged rods.

*32. Two semicircular rods and two short, straight rods are joined in the configuration shown in Fig. 25.37. The rods carry a charge of λ coulombs per meter. Calculate the potential at the center of this configuration.

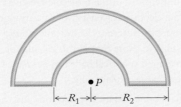

FIGURE 25.37 Two semicircular rods connected by two straight rods.

*33. A long plastic pipe has an inner radius a and an outer radius b. Charge is uniformly distributed over the volume $a \leq r \leq b$. The amount of charge is λ coulombs per meter of length of the tube. Find the potential difference between $r = b$ and $r = 0$. Assume that the plastic has no effect on the electric field.

*34. A flat disk of radius R has a charge Q uniformly distributed over its surface. Find a formula for the potential along the axis of the disk.

*35. The tube of a Geiger counter consists of a thin, straight wire surrounded by a coaxial conducting shell. The diameter of the wire is 0.0025 cm, and that of the shell is 2.5 cm. The length of the tube is 10 cm; however, in your calculation, use the formula for the electric field of an infinitely long line of charge. If the potential difference between the wire and the shell is 1.0×10^3 volts, what is the electric field at the surface of the wire? At the cylinder?

**36. An infinite charge distribution with spherical symmetry has a charge density ρ coulombs per cubic meter given by the formula $\rho = kr^{-5/2}$, where k is a constant. Find the potential as a function of the radius. Assume $V = 0$ at $r = \infty$.

**37. A point charge Q is on the positive z axis at the point $z = h$. A point charge $-Q \times R/h$ (where R is a positive length, $0 < R < h$) is on the z axis at the point $z = R^2/h$. Show that the surface of the sphere of radius R about the origin is an equipotential surface.

25.3 Potential in Conductors

*38. Two large flat parallel sheets have opposite uniform surface charge densities $\pm\sigma$ and are separated a distance d. A large, uncharged conducting slab of thickness $d/3$ is parallel to the charged sheets, centered between them. Find the electrostatic potential as a function of distance y perpendicular to the sheets. Take the reference potential $V_0 = 0$ and the origin $y = 0$ to be at the negative sheet.

*39. A solid conducting sphere of radius R has a charge Q on its surface; the sphere is concentric with a larger, thick conducting spherical shell with inner radius R_1, outer radius R_2, and net charge $-Q$. Find the electrostatic potential for (a) $r \geq R_2$, (b) $R_1 \leq r \leq R_2$, (c) $R \leq r \leq R_1$, and (d) $r \leq R$.

*40. A point charge $-Q$ is at the center of a thick conducting spherical shell of inner radius a and outer radius b. The shell has a net charge of $+3Q$ on it. What is the potential for $r \geq b$? For $a \leq r \leq b$? For $r \leq a$?

*41. A dielectric sphere of radius a has a charge $+Q$ uniformly distributed throughout its volume. The sphere is concentrically placed within a thick conducting shell of inner radius b and outer radius c; the shell has a net charge of $+2Q$. What is the potential outside the shell, for $r \geq c$? In the shell material, for $b \leq r \leq c$? Between the shell and the sphere, for $a \leq r \leq b$? Inside the sphere, for $r \leq a$?

†25.4 Calculation of the Field from the Potential

42. In some region of space, the electrostatic potential is the following function of x and y, but not of z:

$$V = x^2 + 2xy$$

where the potential is measured in volts and the distance in meters. Find the electric field at the point $x = 2, y = 2$.

43. Using the components E_x, E_y, and E_z obtained in Example 9, show that the magnitude $\sqrt{E_x^2 + E_y^2 + E_z^2}$ agrees with the usual expression for the electric field of a point charge.

44. A rod of length l has a charge Q uniformly distributed along its length. In Example 6, we found the potential $V(P)$ at the point P at a distance x from one end of the rod. Find the electric field at this point.

45. Equation (25.32) in Example 4, gives the potential inside a uniformly charged sphere of charge Q and radius R as a function of the distance r from the center:

$$V = -\frac{1}{4\pi\epsilon_0}\frac{Q}{2R^3}r^2 + \frac{1}{4\pi\epsilon_0}\frac{3Q}{2R} \quad (r \leq R)$$

Verify that this potential gives the radial electric field of Eq. (24.18).

46. In a region of space, the electrostatic potential is described by

$$V = x^2 y + 3xyz + zy^2$$

Find the electric field in this region.

47. The potential in a region of space is given by $V = \cos(2\pi x/a)$ $\times \cos(2\pi y/b) \times \cos(2\pi z/c)$, where a, b, and c are constants. What are the x, y, and z components of the electric field in this region?

*48. A nucleus of carbon (charge $6e$) and one of helium (charge $2e$) are separated by a distance of 1.2×10^{-13} m and instantaneously at rest. The center of mass of this system is at a distance of 4.0×10^{-14} m from the carbon nucleus. Take this point as origin and take the x axis along the line joining the nuclei, with the carbon nucleus on the negative x axis.

(a) Find the potential V as a function of x, y, and z.

(b) Find E_x and E_y as a function of x, y, and z.

*49. It can be shown that the electrostatic potential of a dipole p placed at the origin and oriented parallel to the z axis is

$$V = \frac{p}{4\pi\epsilon_0}\frac{z}{(x^2 + y^2 + z^2)^{3/2}}$$

By differentiating this potential, find the components E_x, E_y, and E_z of the electric field produced by the dipole. What are the magnitude and the direction of the electric field at points on the z axis? At points on the x axis?

†For help, see Online Concept Tutorial 30 at www.wwnorton.com/physics

*50. Two rods of equal lengths l form a symmetric cross. The rods carry charges $\pm Q$ uniformly distributed along their lengths. Calculate the potential at the point P at a distance x from one end of the cross (Fig. 25.38). Calculate the electric field at this point.

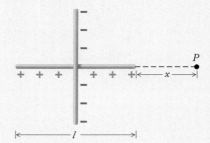

FIGURE 25.38 Two charged rods forming a cross.

**51. A thin cylindrical cardboard tube has a charge Q uniformly distributed over its surface. The radius of the tube is R and the length is l.

(a) Find the potential at a point on the axis of the tube at a distance x from the midpoint. Assume $x > l$.

(b) Find the electric field at this point.

25.5 Energy of Systems of Point Charges

52. Consider once more the distribution of charges within the thundercloud shown in Fig. 23.34. What is the electric potential energy of this charge distribution?

53. In the water molecule, the hydrogen atoms tend to give up their electrons to the oxygen atom. Crudely, the molecule may be regarded as consisting of a uniformly charged ball of charge $-2e$ and two smaller uniformly charged balls of charge $+e$ each. The dimensions of the molecule are given in Fig. 23.27. Calculate the electrostatic energy of this arrangement of three charges; ignore the internal electrostatic energy of the individual balls of charge. Note that in this calculation each of the uniformly charged spherical balls can be treated as though it were a point charge; explain why.

54. Suppose that at one instant the electrons and the nucleus of a helium atom occupy the positions shown in Fig. 25.39; at this instant, the electrons are at a distance of 2.0×10^{-11} m from the nucleus. What is the electric potential energy of this arrangement? Treat the electrons and the nucleus as point charges.

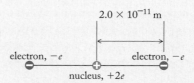

FIGURE 25.39 Instantaneous configuration of the nucleus ($+2e$) and the electrons ($-e$) of an atom of helium.

55. A pair of parallel conducting plates, each measuring 30 cm × 30 cm, are separated by a gap of 1.0 mm. How much work must you do against the electric forces to charge these plates with $+1.0 \times 10^{-6}$ C and -1.0×10^{-6} C, respectively?

56. Two large parallel conducting plates of area 0.20 m^2 are separated by a distance of 0.50 mm. The plates carry opposite charges, and the electric field in the space between them is 5.0×10^5 V/m. What is the electric energy?

57. A penny coin is hung from a silk thread inside a closed tin can placed on the ground. Given that the penny coin carries a charge of 2.0×10^{-6} C and that the potential difference between the tin can and the penny coin is 3.0×10^4 V, find the electric potential energy of this system of two conductors.

58. Near the surface of the nucleus of a lead atom, the electric field has a strength of 3.4×10^{21} V/m. What is the energy density in this field?

59. The atmospheric electric field near the surface of the Earth has a strength of 100 V/m.

    (a) What is its energy density?

    (b) Assuming that the field has the same magnitude everywhere in the atmosphere up to a height of 10 km, what is the corresponding total energy?

60. Six point charges, alternating in sign $\pm Q$, are at the vertices of a hexagon of side a. What is the electrostatic potential energy of this arrangement?

61. Short, extremely intense laser pulses can be produced with electric fields as high as 2.0×10^{10} V/m. What is the energy density in the region of such an electric field?

*62. Consider an infinite line of ions of alternating charge $\pm e$ separated by a distance a.

    (a) Write down the electrostatic potential energy U of one typical ion in the line by summing its potential-energy contributions from increasingly distant ions.

    (b) By comparing your expression with the expansion of the logarithmic function $\ln(1 + x) = x - \frac{1}{2}x^2 + \frac{1}{3}x^3 - \frac{1}{4}x^4 + \cdots$, show that the potential energy of (a) can be written

$$ U = -\frac{e^2 \ln 2}{2\pi\epsilon_0 a} $$

    (c) Evaluate the result in (b) for a typical atomic separation, $a = 3.0 \times 10^{-10}$ m. Express this value in eV.

*63. Four equal positive charges of magnitude Q are placed on the four corners of a square of side d. What is the electric energy of this system of charges?

*64. Four positive and four negative point charges of equal magnitudes $\pm Q$ are arranged alternately on the corners of a cube of edge d (see Fig. 25.40). What is the electric energy of this arrangement?

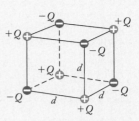

FIGURE 25.40 Point charges at the corners of a cube.

*65. According to the alpha-particle model of the nucleus, some nuclei consist of a regular geometric arrangement of alpha particles. For instance, the nucleus of ^{12}C consists of three alpha particles arranged in an equilateral triangle (Fig. 25.41). Assuming that the distance between pairs of alpha particles is 3.0×10^{-15} m, what is the electric energy (in eV) of this arrangement of alpha particles? Treat the alpha particles as pointlike particles with charge $+2e$.

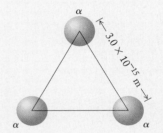

FIGURE 25.41 Three alpha particles.

*66. According to the alpha-particle model (see also the preceding problem), the nucleus of ^{16}O consists of four alpha particles arranged on the vertices of a tetrahedron (Fig. 25.42). If the distance between pairs of alpha particles is 3.0×10^{-15} m, what is the electric energy (in eV) of this configuration of alpha particles? Treat the alpha particles as pointlike particles with charge $+2e$.

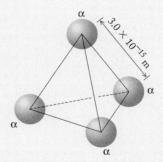

FIGURE 25.42 Four alpha particles.

*67. Problem 37 of Chapter 24 describes the Thomson model of the helium atom. The equilibrium separation of the electrons is 5.0×10^{-11} m. Calculate the electric energy of this configuration. Take into account both the electric energy between each electron and the positive charge, and the electric energy between the two electrons; ignore the energy of the cloud and of the electrons by themselves.

*68. A charge of 7.5×10^{-6} C can be placed on a metallic sphere of radius 15 cm before the surrounding air suffers electrical breakdown. What is the electric energy of the sphere with this charge?

*69. A sphere of radius R has a charge Q uniformly distributed over its volume. A thin conducting shell of radius $2R$ surrounds the sphere concentrically. The shell carries a charge $-Q$ on its interior surface and has no charge on its exterior surface. What is the electric energy of this system?

*70. Pretend that an electron is a conducting sphere of radius R with a charge e distributed uniformly over its surface. In terms of e and the mass m_e of the electron, what must be the radius R if the electric energy is to equal the rest-mass energy $m_e c^2$ of the electron? Numerically, what is the value of R?

*71. A sphere of radius R has a charge Q distributed uniformly over its volume. Show that the electric potential energy of this configuration is $U = (1/4\pi\epsilon_0)\frac{3}{5}Q^2/R$. [Hint: Sum energies of the form (25.49): $U = \int \frac{1}{2} V \, dQ$.]

*72. The nuclei of ^{235}Pu, ^{235}Np, ^{235}U, and ^{235}Pa all have the same radii, about 7.4×10^{-15} m, but their electric charges are $94e$, $93e$, $92e$, and $91e$, respectively. Treating these nuclei as uniformly charged spheres, calculate their electric energies; express your answer in electron-volts. Use the result of Problem 71.

*73. According to a crude model, a proton can be regarded as a uniformly charged sphere of charge e and radius 1.0×10^{-15} m. Find the electric self-energy of the proton. Express your answer in eV. Use the result of Problem 71.

*74. A solid sphere of copper of radius 10 cm with a charge of 1.0×10^{-6} C is placed at the center of a thin, spherical copper shell of radius 20 cm with a charge of -1.0×10^{-6} C. Find a formula for the energy density in the space between the solid sphere and the shell. Find the total electric energy.

*75. In analogy to the electric field **E** (electric force per unit charge), we can define a gravitational feld **g** (gravitational force per unit mass).

(a) The energy density in the electric field is $\epsilon_0 E^2/2$. By analogy, show that the energy density in the gravitational field is $g^2/8\pi G$.

(b) Calculate the gravitational field energy of the Moon due to its own gravity; treat this body as a sphere of uniform density. What is the ratio of the gravitational field energy to the rest-mass energy of the Moon?[3]

[3] Warning: This problem does not take the gravitatioinal *interaction* energy into account. The gravitational interaction energy density is [mass density] × [gravitational potential]. The total gravitational energy is the sum of the field energy and the interaction energy; this total gravitational energy is always negative.

*76. In symmetric fission, the nucleus of uranium (^{238}U) splits into two nuclei of palladium (^{119}Pd). The uranium nucleus is spherical with a radius of 7.4×10^{-15} m. Assume that the two palladium nuclei adopt a spherical shape immediately after fission; at this instant, the configuration is as shown in Fig. 25.43. The size of the nuclei in Fig. 25.43 can be calculated from the size of the uranium nucleus because the nucleus material maintains a constant density (the initial nuclear volume equals the final nuclear volume).

(a) Calculate the electric energy of the uranium nucleus before fission.

(b) Calculate the total electric energy of the palladium nuclei in the configuration shown in Fig. 25.43, immediately after fission. Take into account the mutual electric potential energy of the two nuclei and also the individual electric energies of the two palladium nuclei by themselves.

(c) Calculate the total electric energy a long time after fission when the two palladium nuclei have moved apart by a very large distance.

(d) Ultimately, how much electric energy is released into other forms of energy in the complete fission process (a) through (c)?

(e) If 1.0 kg of uranium undergoes fission, how much electric energy is released?

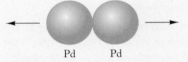

Pd　　Pd

FIGURE 25.43 Two palladium nuclei in contact. The nuclei are spheres.

**77. Consider the Geiger-counter tube described in Problem 35. If the tube is initially uncharged, how much work must be done to bring the tube to its operating voltage of 1.0×10^3 V?

**78. Using the model described in Problem 36 of Chapter 24 for the charge distribution of a neutron, calculate the electric self-energy of a neutron. Express your answer in eV.

**79. A long, thin rod of metal of radius a has a charge of λ coulombs per unit length distributed uniformly over its surface. The rod is surrounded by a concentric cylinder of sheet metal of radius b with a charge of $-\lambda$ coulombs per unit length on its interior surface.

(a) What is the energy density (as a function of radius) in the space between the rod and the cylinder?

(b) What is the total electric energy per unit length?

REVIEW PROBLEMS

80. A positive point charge q, with mass m, is released at a distance d from a fixed positive point charge Q. How fast is the charge q moving when the distance has grown to 3 times the initial value?

81. A proton is decelerated to rest from an initial speed of 6.9×10^6 m/s by a uniform electric field of 2.5×10^6 V/m. How far did the proton travel during this motion? What is the difference in electrostatic potential between the initial and final positions of the proton?

82. A proton sits at the origin of coordinates. How much work must you do against the electric force of the proton to push an electron from the point $x = 1.0 \times 10^{-10}$ m, $y = 0$ in the x–y plane to the point $x = 2.5 \times 10^{-10}$ m, $y = 2.5 \times 10^{-10}$ m?

83. Suppose that, as a function of x, an electric field has an x component

$$E_x = 6x^2y$$

where the electric field is measured in volts per meter and the distances are measured in meters. Find the potential difference between the origin and the point $x = 3$ on the x axis.

*84. The tau particle is similar to an electron and has the same electric charge, but its mass is 3490 times as large as the electron mass. Like the electron, the tau can penetrate nuclear material, experiencing no forces except the electric force. Suppose that a tau is initially at rest a large distance from a lead nucleus. Under the influence of the electric attraction, the tau accelerates toward the nucleus. What is the speed when it crosses the nuclear surface? When it reaches the center of the nucleus? The nucleus of lead is a uniformly charged sphere of radius 7.1×10^{-15} m and charge $82e$.

*85. A long, straight wire of radius 0.80 mm is surrounded by an evacuated concentric conducting shell of radius 1.2 cm. The wire carries a charge of -5.5×10^{-8} coulomb per meter of length. Suppose that you release an electron at the surface of the wire. With what speed will this electron hit the conducting shell?

86. A total charge Q is uniformly distributed along a straight rod of length l.

    (a) Find the electrostatic potential at a point P at a distance y from one end of the rod (see Fig. 25.44).

    (b) Find the y component of the electric field at this point.

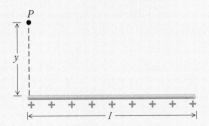

FIGURE 25.44 A charged rod.

*87. An annulus (a disk with a hole) made of paper has an outer radius R and an inner radius $R/2$ (Fig. 25.45). An amount Q of electric charge is uniformly distributed over the paper.

    (a) Find the potential as a function of distance on the axis of the annulus.

    (b) Find the electric field on the axis of the annulus.

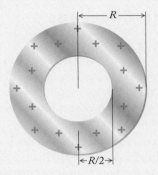

FIGURE 25.45 A charged annulus.

*88. Four equal particles of positive charges q and masses m are initially held at the four corners of a square of side L. If these particles are released simultaneously, what will be their speeds when they have separated by a very large distance?

*89. Two thin rods of length l carry equal charges Q uniformly distributed over their lengths. The rods are aligned, and their nearest ends are separated by a distance x (Fig. 25.46). Calculate the mutual electric potential energy. Ignore the self-energy of each rod.

FIGURE 25.46 Two aligned charged rods.

*90. One method for the determination of the radii of nuclei makes use of the known difference of electric energy between two nuclei of the same size but different electric charges. For instance, the nuclei ^{15}O and ^{15}N have the same size, but their charges are $8e$ and $7e$, respectively. Given that the difference in electric energy is 3.7×10^6 eV, what is the nuclear radius?

*91. A spherical shell of inner radius a, outer radius b carries a charge Q uniformly distributed over its volume. What is the electric energy of this charge distribution?

*92. Suppose that a nucleus of charge q, radius R, and electric energy $(1/4\pi\epsilon_0)(3q^2/5R)$ fissions into two equal parts of charge $q/2$ each. The nuclear material in the original nucleus and the final two nuclei has the same density. What is the radius of each of the final two nuclei? How does the sum of the individual electric energies of the two final, separated nuclei compare with the initial electric energy?

*93. In Problem 71 we integrated the potential energy of the charges in a uniformly charged sphere and found that the electric potential is $(1/4\pi\epsilon_0)\frac{3}{5}Q^2/r$. Obtain this result by integrating the energy density $\frac{1}{2}\epsilon_0 E^2$ of the electric field of the sphere. (Hint: The integration needs to be extended over the interior and the exterior of the sphere, since the electric field exists in both the interior and the exterior.)

Answers to Checkups

Checkup 25.1

1. Since the potential of a point charge varies inversely with distance (*not* as the inverse square), at 10 m the potential is reduced a factor of 10 to 10 V; at 100 m it is reduced a factor of 100 to 1 V.

2. Since the electric potential energy for two point charges q and q' involves their product, $U = kqq'/r$, a positive potential energy requires that both charges have the same sign: q and q' are either both positive or both negative.

3. For a positive charge, the potential $V = kq'/r$ is largest for $r = 0$ (where $V = +\infty$) and smallest for $r = \infty$ (where $V = 0$). For a negative charge, the potential is largest, $V = 0$, at $r = \infty$; and the potential is smallest, $V = -\infty$, at $r = 0$.

4. (A) Increase. Since the charges are of the same sign, the potential energy is positive ($U = kqq'/r$). When r increases, such a positive potential energy decreases. For the total energy $K + U$ to remain constant, the kinetic energy $K = \frac{1}{2}mv^2$ must increase. Thus the speed v increases.

Checkup 25.2

1. As calculated in Example 4 and displayed in Fig. 25.13, the potential increases from a minimum of zero at infinite distance to a maximum of $V = (3/2)kQ/R$ at the center of the positively charged sphere, $r = 0$. For the negatively charged sphere, this is inverted, so the minimum is $V = -(3/2)kQ/R$ at $r = 0$, and the maximum is $V = 0$ at $r = \infty$.

2. Zero electric field throughout a region implies that the potential is not *changing* throughout the region. However, the constant value of the potential can be nonzero; indeed, it can take any value, depending on the distribution of charge in the shell.

3. The electric potential decreases in the direction of the field, since a positive charge would lose potential energy as it moved along the field (as the electric force acts on it); it would gain kinetic energy.

4. The potential is obtained by summing, using Eq. (25.38), $V = (1/4\pi\epsilon_0) \int dQ/r$. Since at the center of the ring $r = R$ is constant, and since $\int dQ = Q$, (the total charge), the potential at the center is $V = (1/4\pi\epsilon_0)Q/R$. The result is independent of how the charge is distributed around the ring.

5. (C) Larger; the same. For $b \geq r \geq a$, we have positive charge in a spherically symmetric shell, producing an outward electric field; since we must do work against this electric field to move a positive point charge from $r = b$ to $r = a$, the potential at $r = a$ must be greater than the potential at $r = b$. There is no charge (and thus no field) for $r \leq a$, so no work is required to move in this region, and thus the potential is the same at $r = 0$ as at $r = a$.

Checkup 25.3

1. The charge on the other sphere and the charge on the wire connecting the spheres destroy spherical symmetry, so the charge on the sphere of interest will not be uniformly distributed. Due to contributions from the wire and the other sphere, the potential will not be $V = (1/4\pi\epsilon_0)Q_1/R_1$, but will be larger. Since the spheres are conducting and are connected by a conductor, they are indeed at the same potential.

2. No, not noticeably. The airplane is nearly a closed box of metal, and as such acts as an electric shield; although airplane windows permit some field penetration, external electric fields are severely attenuated.

3. (B) Smaller. The electrostatic potential of a solid metal sphere is simply $V = kQ/R$, so with the same charge, the larger sphere is at a smaller potential.

Checkup 25.4

1. If the charge density doubled, so would the electric field. If the electric field were twice as large, it would take half the distance for the same change in potential (since $V = -E_0 y$).

2. If the charge Q were doubled, the radii for given potentials $V = kQ/r$ would also double.

3. Since the field is radial with respect to the cylindrical axis and the equipotential surfaces are perpendicular to the field, such surfaces are cylinders with the line of charge as their common axis.

4. For charges of equal sign, the electric field points parallel to their midplane (except at their midpoint, where $E = 0$; see Fig. 23.18). Since the electric field is not perpendicular to this surface, it is not an equipotential surface.

5. (D) Yes; yes. The value of the field is given by the negative of the *derivative* of the potential with respect to distance; the particular value of the potential at a point is unimportant. Second, where the electric field is zero, the potential is constant, but not necessarily zero. For example, a charged conducting sphere has zero electric field inside, but the potential is constant there, $V = kQ/R$ (see Fig. 25.18).

Checkup 25.5

1. The expression reduces to a single term, like Eq. (25.47), namely, $U = (1/4\pi\epsilon_0)q_1q_2/r$.

2. If either all are positive or all are negative, then each term in the potential energy (25.48), which contains the product of two charges, is positive, and the total must be positive. However, if some charges are positive and some are negative, then some terms are negative, and the sign of the total depends on the relative sizes of the different terms.

3. Since each term involves the product of two charges, increasing the values of all the charges a factor of 2 will increase the value of each term, and thus the total, by a factor of 4.

4. According to Eq. (25.51), the stored energy varies as $1/R$, so doubling the radius will result in half as much stored electrical energy, or 0.25 J.

5. Since the energy density $u = \frac{1}{2}\epsilon_0 E^2$, doubling the field will quadruple the energy to 40 J/m^3; tripling the electric field will increase the energy density a factor of 9 to 90 J/m^3.

6. Since the charge density on the plates is constant, the electric field does not change. Thus, Eq. (25.56) tells us that when doubling the plate separation d, the total stored energy increases a factor of 2 to 0.020 J. The energy density depends only on the electric field, and so does not change.

7. (E) 0.25 J. The energy of a charged sphere, Eq. (25.51), is proportional to the charge squared. Thus half the charge will have one-fourth of the stored energy, or 0.25 J.

CONCEPTS IN CONTEXT

Concepts
— *in* —
Context

At the National Ignition Facility (NIF) in Livermore, California, intense light pulses from 192 lasers are focused and combined to attain the extremely high energy density required for nuclear fusion. The main amplification for each laser comes from flashlamps powered by the sudden discharge of capacitors, arrangements of conductors that can store charge and energy. The 192 banks of capacitors at the NIF (left photo), each with 20 advanced capacitors (the horizontal cylinders in the right photo), are the highest-energy array ever built.

As we learn about capacitance in this chapter, we can consider such questions as:

? What is the combined capacitance of each 20-capacitor bank? Of all 192 banks? (Example 4, page 835)

? How much charge can be stored on each NIF capacitor? The conductors in each capacitor are separated by an insulator, known as a dielectric. How does the presence of a dielectric modify the properties of a capacitor? (Example 7, page 841)

? How much energy can be stored in each NIF capacitor? In the entire array? (Example 10, page 845)

Any arrangement of conductors that is used to store electric charge is called a **capacitor**. Since work must be done while depositing charge on the conductors, the capacitor also stores electric potential energy whenever it stores electric charge. In our electronic technology, capacitors find widespread application—they are part of the circuitry of radios, CD players, computers, automobile ignition systems, and so on.

The first part of this chapter deals with the properties of capacitors; we will see how the ability of the capacitor to store charge depends on the arrangement and the sizes of the conductors. The second part of the chapter deals with the properties of electric fields in regions of space filled with an electrically insulating material, or **dielectric**. Since capacitors are usually filled with such a dielectric material, the study of the mutual effects between the electric field and the dielectric material is closely linked to the study of capacitors. But the effects of the electric fields and dielectric materials upon one another are also interesting in their own right. For instance, air is a dielectric material, and since many of the electric fields that we encounter in our study of electricity are in air, we ought to inquire how the electric fields in air differ from those in vacuum.

26.1 CAPACITANCE

As a first example of a capacitor, let us consider an isolated metallic sphere of radius R (see Fig. 26.1). Obviously, charge can be stored on this sphere. If the amount of charge placed on the sphere is Q, then the potential of the sphere will be, according to Eq. (25.28),

$$V = \frac{1}{4\pi\epsilon_0} \frac{Q}{R} \qquad (26.1)$$

This says that the amount of charge Q stored on the sphere is directly proportional to the potential V.

This proportionality of Q and V holds in general for any conductor of arbitrary shape. The charge on the conductor produces an electric field whose strength is directly proportional to the amount of charge (twice the charge produces twice the field), and the electric field yields a potential which is directly proportional to the field strength (twice the field strength yields twice the potential); hence, charge and potential are proportional. We write this relationship as

$$Q = CV \qquad \text{or} \qquad C = \frac{Q}{V} \qquad (26.2)$$

where C is the constant of proportionality. This constant is called the **capacitance** of the conductor*. *The capacitance is large if the conductor is capable of storing a large amount of charge at a low potential.* For instance, the capacitance of a spherical conductor is

$$C = \frac{Q}{V} = \frac{Q}{(1/4\pi\epsilon_0)Q/R} = 4\pi\epsilon_0 R \qquad (26.3)$$

*Do not confuse the capacitance C with the abbreviation C for the coulomb, the unit of charge.

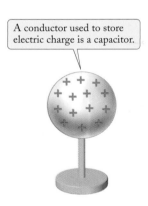

A conductor used to store electric charge is a capacitor.

FIGURE 26.1 An isolated metallic sphere.

capacitance of a single conductor

Thus, the capacitance of the sphere increases with its radius—a sphere of large radius can store a large amount of charge at a low potential. Note that the value of the capacitance depends only on geometrical properties of the conductor, and not on any particular value of Q or V.

The SI unit of capacitance is the **farad** (F):

farad (F)

$$1 \text{ farad} = 1 \text{ F} = 1 \frac{\text{coulomb}}{\text{volt}} = 1 \frac{\text{C}}{\text{V}} \qquad (26.4)$$

This unit of capacitance is rather large; in practice, electrical engineers prefer the **microfarad** and the **picofarad.** A microfarad equals 10^{-6} farad and a picofarad equals 10^{-12} farad:

$$1 \, \mu\text{F} = 10^{-6} \text{ F} \quad \text{and} \quad 1 \text{ pF} = 10^{-12} \text{ F}$$

Note that since $1 \text{ F} = 1 \text{ C/V} = 1 \text{ C}^2/\text{J} = 1 \text{ C}^2/(\text{N·m})$, the constant ϵ_0 can be written

$$\epsilon_0 = 8.85 \times 10^{-12} \frac{\text{C}^2}{\text{N·m}^2} = 8.85 \times 10^{-12} \text{ F/m} \qquad (26.5)$$

The latter expression is the one usually listed in tables of physical constants.

EXAMPLE 1   What is the capacitance of an isolated metallic sphere of radius 20 cm?

SOLUTION: According to Eq. (26.3),

$$C = 4\pi\epsilon_0 R = 4\pi \times 8.85 \times 10^{-12} \text{ F/m} \times 0.20 \text{ m}$$

$$= 2.2 \times 10^{-11} \text{ F} = 22 \text{ pF}$$

EXAMPLE 2   The ground and the oceans are conductors, and the Earth can therefore be regarded as a conducting sphere. What is its capacitance?

SOLUTION: The radius of the Earth is 6.4×10^6 m; therefore

$$C = 4\pi\epsilon_0 R = 4\pi \times 8.85 \times 10^{-12} \text{ F/m} \times 6.4 \times 10^6 \text{ m}$$

$$= 7.1 \times 10^{-4} \text{ F}$$

COMMENTS: As capacitances go, this is a rather large capacitance. However, note that to alter the potential of the Earth by 1 volt requires a charge of only $Q = CV = 7.1 \times 10^{-4} \text{ F} \times 1 \text{ volt} = 7.1 \times 10^{-4}$ coulomb.

The most common variety of capacitor consists of *two* metallic conductors, insulated from one another and carrying opposite amounts of charge of equal magnitudes, that is, a charge $+Q$ on one conductor and $-Q$ on the other. The capacitance of such a pair of conductors is defined in terms of the *difference* of potential ΔV between the two conductors:

capacitance of pair of conductors

$$Q = C \, \Delta V \qquad \text{or} \qquad C = \frac{Q}{\Delta V} \qquad (26.6)$$

In this expression, both Q and ΔV are taken as positive quantities. Note that the quantity Q is *not* the net charge in the capacitor, but the magnitude of the charge on each plate. The net charge that establishes the potential difference in any two-conductor capacitor is zero.

Figure 26.2 shows such a two-conductor capacitor consisting of two large, parallel metallic plates, each of area A, separated by a distance d. The plates carry charges $+Q$ and $-Q$, respectively, on their facing surfaces. The electric field in the region between the plates is [see Eq. (25.53)]

$$E = \frac{Q}{\epsilon_0 A} \qquad (26.7)$$

and the potential difference between the plates is [from Eq. (25.54)]

$$\Delta V = Ed = \frac{Qd}{\epsilon_0 A} \qquad (26.8)$$

Hence the capacitance of this configuration is

$$C = \frac{Q}{\Delta V} = \frac{Q}{Qd/\epsilon_0 A} = \frac{\epsilon_0 A}{d}$$

This is the capacitance of a parallel-plate capacitor,

$$C = \frac{\epsilon_0 A}{d} \qquad (26.9)$$

capacitance of parallel-plate capacitor

We again see that capacitance depends only on the geometry of the conductors.

From Eq. (26.9) we recognize that to store a large amount of charge at a low potential, we want a large plate area A, but a small plate separation d. Parallel-plate capacitors are often manufactured out of two parallel sheets of aluminum foil, a few centimeters wide, but several meters long. The sheets are placed very close together, but kept from contact by a thin sheet of plastic sandwiched between them (see Fig. 26.3a). For convenience, the entire sandwich is covered with another sheet of plastic and rolled up like a roll of two-ply toilet paper (see Fig. 26.3b). The two sheets of

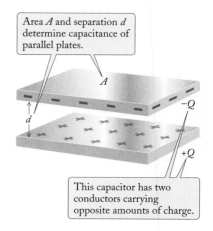

Area A and separation d determine capacitance of parallel plates.

This capacitor has two conductors carrying opposite amounts of charge.

FIGURE 26.2 Two very large parallel plates with opposite electric charges.

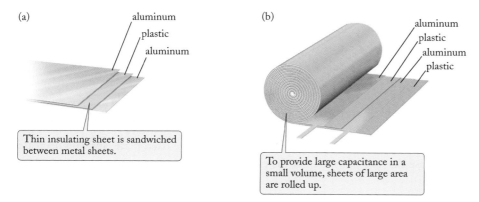

(a)
aluminum
plastic
aluminum

Thin insulating sheet is sandwiched between metal sheets.

(b)
aluminum
plastic
aluminum
plastic

To provide large capacitance in a small volume, sheets of large area are rolled up.

FIGURE 26.3 (a) Sheets of aluminum foil separated by a sheet of plastic. (b) Rolled capacitor.

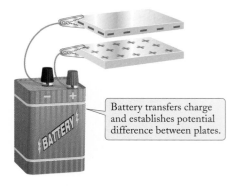

FIGURE 26.4 A two-conductor capacitor connected to the terminals of a battery.

aluminum are connected to the external terminals of the capacitor. To charge such a capacitor, we connect its terminals to the terminals of a battery, which transfers charge from one plate to the other, establishing equal amounts of charge of opposite sign on the capacitor plates, and produces a potential difference between the plates equal to the potential difference (the voltage) of the battery (Fig. 26.4).

Once a capacitor is charged, we can disconnect the battery, and the capacitor will then hold the charge (and the electric potential energy) in storage for a long time. How long depends on how good the insulation is. In some capacitors the insulation between the plates permits some leakage of charge from one plate to another. When the opposite charges of the two plates meet, they neutralize each other, and a capacitor can thereby discharge itself in a few minutes. But some capacitors hold their charge for hours or days. Electronic devices with large capacitors, such as television receivers or computers, often have labels on their casings warning users not to open the casing even when the equipment is disconnected from the power supply, because the capacitors hold their electric charge for a long time and they can give painful and dangerous electric shocks if their terminals accidentally come into contact with the skin of the user.

EXAMPLE 3   A parallel-plate capacitor consists of two strips of aluminum foil, each with an area of 0.20 m^2, separated by a distance of 0.10 mm. The space between the foils is empty. The two strips are connected to the terminals of a battery, which produces a potential difference of 200 volts between them. What is the capacitance of this capacitor? What is the electric charge on each plate? What is the strength of the electric field between the plates?

SOLUTION: According to Eq. (26.9), the capacitance of a parallel-plate capacitor is

$$C = \frac{\epsilon_0 A}{d} = \frac{8.85 \times 10^{-12} \text{ F/m} \times 0.20 \text{ m}^2}{1.0 \times 10^{-4} \text{ m}} = 1.8 \times 10^{-8} \text{ F}$$

$$= 0.018 \ \mu\text{F}$$

The charge on each plate is

$$Q = C \, \Delta V = 1.8 \times 10^{-8} \text{ F} \times 200 \text{ volts} = 3.6 \times 10^{-6} \text{ coulomb}$$

and the electric field between the plates is, from Eq. (26.8),

$$E = \frac{\Delta V}{d} = \frac{200 \text{ volts}}{1.0 \times 10^{-4} \text{ m}} = 2.0 \times 10^6 \text{ volts/m}$$

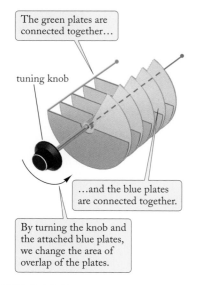

The green plates are connected together...

tuning knob

...and the blue plates are connected together.

By turning the knob and the attached blue plates, we change the area of overlap of the plates.

FIGURE 26.5 A variable capacitor.

Variable capacitors are used in the tuning circuits of radios. These capacitors consist of a fixed plate (or plates) and a movable plate (or plates). When you turn the tuning knob, you shift the moving plate parallel to the fixed plate, and you decrease or increase the area of overlap of the plates, and thereby change the capacitance (see Fig. 26.5).

PHYSICS IN PRACTICE   CAPACITOR MICROPHONE

A special kind of capacitor is used in the **capacitor micro-phone**, illustrated in the figure. Another name for a capacitor is a **condenser**; thus, this microphone is also often called a condenser microphone. The flexible diaphragm of this microphone forms one plate of the capacitor, and a rigid disk forms the other plate. When a sound wave impinges on the diaphragm, the periodic fluctuations of the air pressure alternately push and pull the diaphragm toward and away from the rigid plate. The change of distance between the plates produces a change in capacitance, according to Eq. (26.9). Since the plates are connected to a battery, which maintains a steady potential difference between the plates, the change in capacitance results in a change in the amount of electric charge on the plates. The charge that leaves the capacitor plates flows along the wires, forming an electric current. Thus, the capacitor microphone transforms a sound signal into an electric signal, which can be fed into an amplifier and from there into a loudspeaker, a tape recorder, or a digitizer. This kind of microphone has a good sensitivity to a wide range of frequencies, and it is often used in recording studios and telephones.

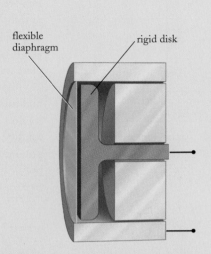

flexible diaphragm   rigid disk

FIGURE 1 A capacitor microphone. When a sound wave pushes on the flexible diaphragm, it alters the distance between the diaphragm and the rigid plate.

 ## Checkup 26.1

QUESTION 1: Is Eq. (26.9) valid for a parallel-plate capacitor with square plates? Rectangular plates? Circular plates?

QUESTION 2: One capacitor has a capacitance of 1 pF; another has a capacitance of 3 pF. If both are charged with 6×10^{-12} C, which has the larger potential difference?

QUESTION 3: Do we increase or decrease the capacitance of a parallel-plate capacitor when we increase the spacing between the plates? When we increase the area of the plates?

QUESTION 4: Suppose that instead of storing the usual equal amounts of charge of opposite signs on the plates of a parallel-plate capacitor, we attempt to store equal charges of the same sign, say, positive, on both plates. Is the capacitance then still given by Eq. (26.9)?

QUESTION 5: If we decrease the separation between the plates of a parallel-plate capacitor by a factor of 2 while holding the electric charge constant, by what factors will we change the electric field, the potential difference, and the capacitance?

QUESTION 6: A parallel-plate capacitor has plates measuring 10 cm × 10 cm and a plate separation of 2.0 mm. If we want to construct a parallel-plate capacitor of the same capacitance but with plates measuring 5.0 cm × 5.0 cm, what plate separation do we need?

   (A) 8.0 mm   (B) 4.0 mm   (C) 2.0 mm   (D) 1.0 mm   (E) 0.50 mm

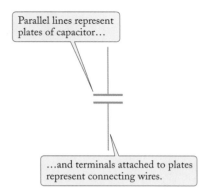

FIGURE 26.6 Symbol for a capacitor in a circuit diagram.

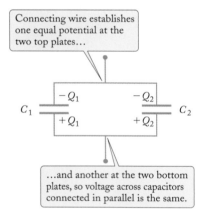

FIGURE 26.7 Two capacitors connected in parallel.

capacitors in parallel

26.2 CAPACITORS IN COMBINATION

Capacitors used in practical applications in electric circuitry in radios, televisions, computers, etc., commonly are of the two-conductor variety. Schematically, such capacitors are represented in a circuit diagram as two parallel lines with terminals attached to their middles (see Fig. 26.6). The terminals represent wires, each of which we will assume is an ideal conductor. In a circuit, several such capacitors are often wired together and it is then necessary to calculate the net capacitance of the combination. The simplest ways of wiring capacitors together are in **parallel** and in **series**.

Figure 26.7 shows two capacitors connected in *parallel*. If charge is fed into this combination via the two terminals, some of the charge will be stored on the first capacitor and some on the second. The net capacitance of the combination can be found as follows: Since the corresponding plates of the capacitors are joined by conductors, the potentials of corresponding plates must be equal and the potential differences across both capacitors must also be equal. This is true in general: *circuit components connected in parallel have the same voltage across each component.* Thus,

$$\Delta V = \frac{Q_1}{C_1} \quad \text{and} \quad \Delta V = \frac{Q_2}{C_2} \tag{26.10}$$

Therefore the net charge for the capacitor combination can be expressed as

$$Q = Q_1 + Q_2 = C_1 \Delta V + C_2 \Delta V$$

that is,

$$Q = (C_1 + C_2)\Delta V \tag{26.11}$$

Comparing this with the definition for capacitance given in Eq. (26.6), we recognize that the combination is equivalent to a single capacitor of capacitance

$$C = C_1 + C_2 \tag{26.12}$$

Thus, *the net capacitance, or equivalent capacitance, of the parallel combination is simply the sum of the individual capacitances.*

It is easy to obtain a similar result for any number of capacitors connected in parallel (see Fig. 26.8). The net capacitance for such a parallel combination is

$$C = C_1 + C_2 + C_3 + \cdots \tag{26.13}$$

Next, we consider the alternative way of connecting capacitors. Figure 26.9 shows two capacitors connected in *series*. Since charge cannot flow across the gap between the plates of the capacitors, any charge fed into the capacitor combination via the two

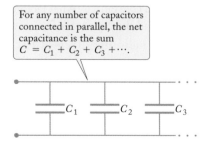

FIGURE 26.8 Several capacitors connected in parallel.

outside terminals will have to remain on the *outside* plates [the lower plate of the bottom capacitor (C_1) and the upper plate of the top one (C_2) in Fig. 26.9]. Thus, the bottom plate will have a charge $+Q$ and the top plate a charge $-Q$. But these charges on the outside plates will induce charges on the inside plates (the upper plate of C_1 and the lower plate of C_2). The charge $+Q$ on the bottom plate of C_1 will attract electrons to the facing plate and a charge $-Q$ will accumulate on this plate. Corresponding to the excess electrons on the upper plate of C_1, there will be a deficit of electrons on the lower plate of C_2, and a charge $+Q$ will accumulate there. In general, *capacitors in series have the same magnitude of charge on each plate.*

The capacitance of the combination can then be found as follows. The individual potential differences across the two capacitors are

$$\Delta V_1 = \frac{Q}{C_1} \quad \text{and} \quad \Delta V_2 = \frac{Q}{C_2} \tag{26.14}$$

Since potential is the energy per unit charge to move from one point to another, the net potential difference between the external terminals of two capacitors in series is the sum of the potential differences across the two individual capacitors:

$$\Delta V = \Delta V_1 + \Delta V_2 = \frac{Q}{C_1} + \frac{Q}{C_2} \tag{26.15}$$

from which

$$\Delta V = Q\left(\frac{1}{C_1} + \frac{1}{C_2}\right) \tag{26.16}$$

Comparing this, again, with the definition of capacitance given by Eq. (26.6), we see that the combination is equivalent to a single capacitor with

$$\frac{1}{C} = \frac{1}{C_1} + \frac{1}{C_2} \tag{26.17}$$

or

$$C = \frac{C_1 C_2}{C_1 + C_2}$$

According to Eq. (26.17), the inverse of the net capacitance of the series combination is obtained by taking the sum of the inverses of the individual capacitances. Note that in series, the net capacitance is always *less* than the individual capacitances; for example, if $C_1 = C_2$, then $C = \frac{1}{2}C_1 = \frac{1}{2}C_2$.

A similar result applies to any number of capacitors connected in series (see Fig. 26.10). The net capacitance C of such a series combination is given by

$$\frac{1}{C} = \frac{1}{C_1} + \frac{1}{C_2} + \frac{1}{C_3} + \cdots \tag{26.18}$$

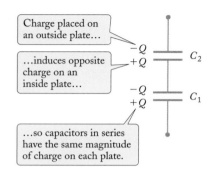

Charge placed on an outside plate…

…induces opposite charge on an inside plate…

…so capacitors in series have the same magnitude of charge on each plate.

FIGURE 26.9 Two capacitors connected in series.

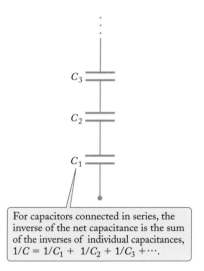

For capacitors connected in series, the inverse of the net capacitance is the sum of the inverses of individual capacitances, $1/C = 1/C_1 + 1/C_2 + 1/C_3 + \cdots$.

FIGURE 26.10 Several capacitors connected in series.

capacitors in series

EXAMPLE 4 Each of the advanced capacitors at the National Ignition Facility has a capacitance of 300 μF. Each of the 192 laser amplifiers is powered by a bank of 20 such capacitors connected in parallel. What is the net capacitance of each bank? What is the sum of the capacitances of the 192 banks of the entire laser power system?

Concepts
— *in* —
Context

SOLUTION: Since the capacitors in each bank are connected in a simple parallel combination, the net capacitance of a bank is simply the sum of the 20 individual capacitances:

$$C_{\text{bank}} = C_1 + C_2 + C_3 + \cdots = 20 C_1$$

$$= 20 \times 300 \times 10^{-6} \, \text{F} = 6.0 \times 10^{-3} \, \text{F}$$

The capacitance of the entire laser power system is

$$C_{\text{total}} = 192 \times C_{\text{bank}} = 192 \times 6.0 \times 10^{-3} \, \text{F}$$

$$= 1.2 \, \text{F}$$

This is an immense capacitance, more than a thousand times the capacitance of the Earth, calculated in Example 2. A capacitance value of the order of a farad is hard to achieve.

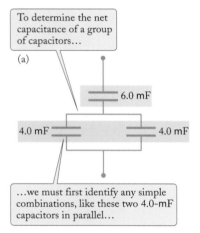

To determine the net capacitance of a group of capacitors...

(a)

6.0 mF

4.0 mF   4.0 mF

...we must first identify any simple combinations, like these two 4.0-mF capacitors in parallel...

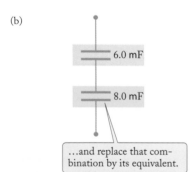

(b)

6.0 mF

8.0 mF

...and replace that combination by its equivalent.

FIGURE 26.11 (a) The two 4.0-μF capacitors (green) are connected in parallel, and the 6.0-μF capacitor (orange) is connected in series with these two. (b) The two 4.0-μF capacitors are equivalent to a single 8.0-μF capacitor, which is connected in series with the 6.0-μF capacitor.

EXAMPLE 5   Two capacitors of 4.0 μF each are connected in parallel and then a third capacitor of 6.0 μF is connected in series with the first two (see Fig. 26.11a). What is the net capacitance of this combination?

SOLUTION: We first look for a group of capacitors that is either a simple series or simple parallel combination. Accordingly, we first calculate the combined capacitance of the two 4.0-μF capacitors. Since they are connected in parallel, Eq. (26.13) tells us that their combined capacitance is the sum 4.0 μF + 4.0 μF = 8.0 μF.

Next, we consider the combined capacitance of this effective 8.0-μF capacitor connected in series with the 6.0-μF capacitor (see Fig. 26.11b). For this series combination, Eq. (26.18) tells us that the net capacitance is given by

$$\frac{1}{C} = \frac{1}{8.0 \, \mu\text{F}} + \frac{1}{6.0 \, \mu\text{F}} = \frac{6.0 \, \mu\text{F} + 8.0 \, \mu\text{F}}{8.0 \, \mu\text{F} \times 6.0 \, \mu\text{F}} = \frac{14}{48 \, \mu\text{F}}$$

so

$$C = \frac{48}{14} \, \mu\text{F} = 3.4 \, \mu\text{F}$$

This is the net capacitance of the entire combination.

EXAMPLE 6   One way to generate a high voltage is to take a large number of capacitors, charge them while they are connected in parallel, and then reconnect them in series (see Fig. 26.12). Suppose you take 140 capacitors of 0.50 μF and connect them in parallel to a 9.0-volt battery. Once they are fully charged, you reconnect the capacitors in series (without the battery). What is the potential difference across the series combination? How much charge did the capacitors absorb from the battery during the charging? How much charge will they release if you discharge the series combination by connecting the external terminals of the first and last capacitors in the series?

SOLUTION: The potential difference across the series combination is 140 times the potential difference across each capacitor; that is,

$$\Delta V' = 140 \times 9.0 \, \text{volts} = 1260 \, \text{volts}$$

The capacitance of the original parallel combination is the sum of the capacitances of all the capacitors:

$$C = C_1 + C_2 + C_3 + \cdots = 140 \times 0.50 \, \mu F = 70 \, \mu F$$

Hence the charge absorbed from the battery is

$$Q = C \, \Delta V = 70 \, \mu F \times 9.0 \text{ volts} = 6.3 \times 10^{-4} \text{ coulomb}$$

The capacitance of the series combination is given by

$$\frac{1}{C} = \frac{1}{C_1} + \frac{1}{C_2} + \frac{1}{C_3} + \cdots = 140 \times \frac{1}{0.50 \, \mu F}$$

so

$$C = \frac{0.50 \, \mu F}{140} = 3.57 \times 10^{-3} \, \mu F$$

Hence the charge of the series combination is

$$Q = C \, \Delta V' = 3.57 \times 10^{-3} \, \mu F \times 1260 \text{ volts} = 4.5 \times 10^{-6} \text{ C}$$

Thus, when you discharge the series combination, the charge that flows out of the positive external terminal is $+4.5 \times 10^{-6}$ C, and the charge that flows out of the negative external terminal is -4.5×10^{-6} C.

COMMENTS: In this example, the charge of the series combination is simply the charge of a single capacitor. The charges on all plates except the two connected to the external terminals simply cancel each other and are not counted as charge for the series combination. The initial charge of the parallel combination is 6.3×10^{-4} C, but the charge of the series combination is only 4.5×10^{-6} C, or 140 times smaller.

(a)

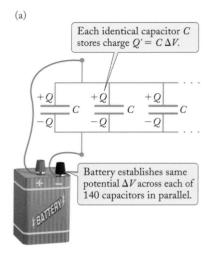

Each identical capacitor C stores charge $Q' = C \, \Delta V$.

Battery establishes same potential ΔV across each of 140 capacitors in parallel.

(b)

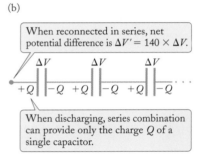

When reconnected in series, net potential difference is $\Delta V' = 140 \times \Delta V$.

When discharging, series combination can provide only the charge Q of a single capacitor.

FIGURE 26.12 (a) First the capacitors are connected in parallel and are charged by a battery. (b) Then the capacitors are reconnected in series.

PROBLEM-SOLVING TECHNIQUES   COMBINATIONS OF CAPACITORS

Keep in mind that when we speak of the charge Q on a parallel-plate capacitor, we always mean the magnitude of the charge on each plate.

If several such capacitors are connected in *series*:

- They all have the same Q (that is, each capacitor has a charge $+Q$ on the positive plate and $-Q$ on the negative plate), and

- The net potential difference ΔV is the sum $\Delta V_1 + \Delta V_2 + \cdots$ of the potential differences of the individual capacitors.

If several capacitors are connected in *parallel*:

- They all have the same ΔV, and

- The net charge Q is the sum $Q_1 + Q_2 + \cdots$ of the charges of the individual capacitors.

When dealing with a circuit containing several capacitors connected in some complicated manner, proceed in two steps:

1. In the first step, look for groups of capacitors that form simple parallel or series combinations (such as the group of the two 4.0-μF capacitors in Fig. 26.11a). Evaluate the capacitance of each such group.

2. In the second step, see how these effective group capacitances are connected to each other (as in Fig. 26.11b), and evaluate the net capacitance of the combination of the group(s).

In some cases, it may be necessary to repeat these steps until a single equivalent capacitance is obtained.

✔ Checkup 26.2

QUESTION 1: If several capacitors are connected in series, do they all have the same electric field between their plates? What if several capacitors are connected in parallel?

QUESTION 2: Qualitatively, why is the net capacitance of a parallel combination of equal capacitors larger than the individual capacitances? Why is the net capacitance of a series combination of equal capacitors smaller than the individual capacitances?

QUESTION 3: If you connect 10 capacitors of 1.0 μF each in parallel, what is the net capacitance? What if you connect them in series?

    (A) 10 μF, 10 μF   (B) 10 μF, 0.10 μF

    (C) 0.10 μF, 10 μF   (D) 0.10 μF, 0.10 μF

26.3 DIELECTRICS

So far, in dealing with problems in electrostatics we have assumed that the medium surrounding the electric charges consisted of a vacuum or of air. Vacuum has no effect on the electric field; and air, as we will see, has only a small and often negligible effect on the electric field. However, in dealing with the capacitors used in practical applications, we cannot neglect the effects of the medium that surrounds the electric charges. The space between the plates of such capacitors is usually filled with an electric insulator, or **dielectric**. Such a dielectric drastically changes the electric field from what it would be in a vacuum: *the dielectric reduces the strength of the electric field.*

    To understand this, consider a parallel-plate capacitor whose plates carry some amount of charge per unit area. Suppose that a slab of dielectric, such as glass or nylon, fills most of the space between the plates (see Fig. 26.13). This dielectric contains a large number of atomic nuclei and electrons, but, of course, these positive and negative charges balance each other, so the material is electrically neutral. *In an insulator, all the charges are bound*—the electrons are confined within their atoms or molecules and they cannot wander about as in a conductor. Nevertheless, in reponse to the force exerted by the electric field, the charges will move very slightly without leaving their atoms. The electrons move in a direction opposite to that of the electric field; the nuclei move in the direction of the electric field. These opposite displacements separate the positive and negative charges and thereby create electric dipoles within the dielectric. In most dielectrics, the magnitudes of the charge separations and the magnitudes of the induced dipole moments are directly proportional to the strength of the electric field; such dielectrics are said to be **linear**.

    The details of the mechanism of displacement and separation of charge depend on the dielectric. In some dielectrics—such as glass, nylon, and other solids—the creation of dipole moments involves a distortion of the molecules or atoms. By tugging on the electrons and nuclei in opposite directions, the electric field stretches the molecule while producing a charge separation within it (see Fig. 26.14). In other dielectrics—such as distilled water[1] or carbon monoxide—the creation of dipole moments results mainly from a realignment of existing dipoles. In such dielectrics, the molecules have permanent dipole moments which are randomly oriented when the dielectric is left to itself. The randomness of the orientation of the dipoles means that, on the average,

dielectric

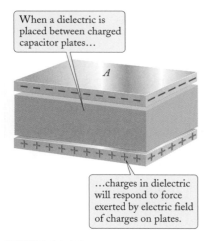

When a dielectric is placed between charged capacitor plates…

A

…charges in dielectric will respond to force exerted by electric field of charges on plates.

FIGURE 26.13 A slab of dielectric between the plates of a capacitor.

---

[1] Remember that distilled water is an insulator.

(a) Atoms and many molecules have no electric dipole moment in zero field.

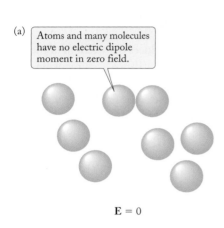

$$\mathbf{E} = 0$$

(a) Some molecules have a permanent electric dipole moment.

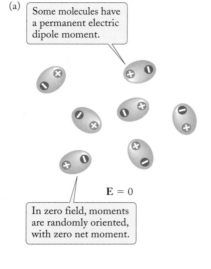

$$\mathbf{E} = 0$$

In zero field, moments are randomly oriented, with zero net moment.

(b) In an electric field, electrons and nuclei remain bound together…

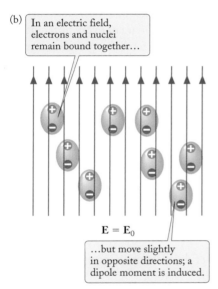

$$\mathbf{E} = \mathbf{E}_0$$

…but move slightly in opposite directions; a dipole moment is induced.

(b) In an electric field, torque on dipoles tends to align them with the field…

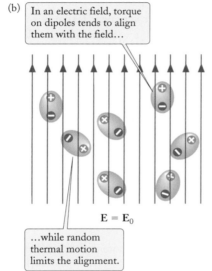

$$\mathbf{E} = \mathbf{E}_0$$

…while random thermal motion limits the alignment.

FIGURE 26.14 (a) Undistord molecules. (b) The electric field produces a distortion of the molecules.

FIGURE 26.15 (a) Unaligned molecules. (b) The electric field produces a (partial) alignment of already distorted molecules.

Displacement of dielectric charges results in a positive layer of charge near negative plate…

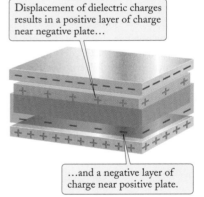

…and a negative layer of charge near positive plate.

FIGURE 26.16 The distributions of positive charge (brown) and of negative charge (green) of the slab of dielectric fail to overlap precisely. Thus, the electric field has produced a separation of the charges.

there is no charge separation in the dielectric. But as we saw in Section 23.5, when placed in an electric field, permanent dipoles experience a torque that tends to align them with the electric field (see Fig. 26.15). Random thermal motions oppose this alignment, and the molecules achieve an average equilibrium state in which the average amount of alignment is (approximately) proportional to the strength of the electric field. This average alignment is equivalent to an average charge separation.

The displacement of the negative and the positive charges in opposite directions implies that the positive and negative charge distributions of the dielectric cease to overlap precisely (see Fig. 26.16). Consequently, there will be an excess of positive bound charge on one surface of the slab of dielectric and an excess of negative bound charge on the opposite surface. The slab of dielectric is then said to be **polarized**. These surface charges act just like a pair of parallel sheets of positive and negative charge; between the sheets, the charges generate an electric field that is *opposite* to the

polarized dielectric

original applied electric field. The total electric field, consisting of the sum of the field of the free charges on the conducting plates plus the opposing field of the induced bound charges on the dielectric surfaces, is therefore *smaller* than the field of the free charges alone (see Fig. 26.17).

In a linear dielectric, the amount by which the dielectric reduces the strength of the electric field can be characterized by the **dielectric constant** κ (kappa), a dimensionless number. This constant is merely the factor by which the electric field in the dielectric between the parallel plates is reduced; that is, if E_{free} is the electric field that the free charges (on the plates) produced by themselves and E the electric field that these free charges and the bound charges (in the dielectric) produce together, then

dielectric constant κ

electric field in dielectric

$$E = \frac{1}{\kappa} E_{\text{free}} \tag{26.19}$$

where κ is larger than 1.

Table 26.1 lists the values of the dielectric constants of some materials. Note that air has a value very near 1; that is, the dielectric properties of air are not very different from those of a vacuum, and the electric fields produced by (free) charges placed in air are almost the same as those produced in vacuum. This justifies our neglect of the presence of air in many of the problems in preceding chapters.

If the slab of dielectric fills the space between the plates entirely, then the formula (26.19) for the reduction of the strength of the electric field applies throughout all of this space. Since the potential difference between the capacitor plates is directly proportional to the strength of the electric field, it follows that, for a given amount of free charge on the plates, the presence of the dielectric also reduces the potential difference by the factor κ:

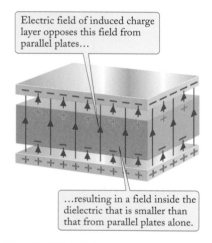

Electric field of induced charge layer opposes this field from parallel plates…

…resulting in a field inside the dielectric that is smaller than that from parallel plates alone.

FIGURE 26.17 Some electric field lines stop on the negative charges at the bottom of the slab of dielectric. The density of field lines is smaller in the dielectric than in the empty gaps adjacent to the plates; that is, the electric field is smaller.

$$\Delta V = \frac{1}{\kappa} \Delta V_0 \tag{26.20}$$

where ΔV_0 is the potential difference in the absence of the dielectric. Consequently, the presence of the dielectric *increases* the capacitance by a factor κ:

$$C = \frac{Q}{\Delta V} = \kappa \frac{Q}{\Delta V_0}$$

or simply

capacitance of capacitor filled with dielectric

$$C = \kappa C_0 \tag{26.21}$$

where $C_0 = Q/\Delta V_0$ is the capacitance in the absence of dielectric. For example, the capacitance of a parallel-plate capacitor filled with Plexiglas (see Table 26.1) is

$$C = \kappa C_0 = \kappa \frac{\epsilon_0 A}{d}$$

$$= 3.4 \times \frac{\epsilon_0 A}{d} \tag{26.22}$$

By filling the space between the capacitor plates with dielectric, we can therefore obtain a substantial gain of capacitance. Furthermore, the dielectric can prevent electric breakdown in the space between the plates. If this space contains air, sparking will

FIGURE 26.18 Electric breakdown in a Plexiglas block in a very strong electric field caused minute perforations in the block and created this beautiful arboreal pattern.

occur between the plates when the electric field reaches a value of about 3×10^6 V/m, and the capacitor will discharge spontaneously. Most dielectrics are better insulators than air and they can tolerate larger electric fields. For instance, Plexiglas can tolerate an electric field of up to 40×10^6 V/m before it suffers electric breakdown (see Fig. 26.18). The maximum electric field that a dielectric can tolerate is called its **dielectric strength**; thus, air has a dielectric strength of 3×10^6 V/m, and Plexiglas a dielectric strength of 40×10^6 V/m.

dielectric strength

EXAMPLE 7 The 300-μF capacitors at the National Ignition Facility are made of two strips of metal foil with an effective parallel-plate area of 123 m². The plates are separated by a layer of polypropylene dielectric material 8.0×10^{-6} m thick. What is the dielectric constant? If a potential difference of 24 kV is applied to each of these capacitors what is the magnitude of the free charge on each plate? What is the electric field in the dielectric?

SOLUTION: From Eq. (26.21), the dielectric constant is given by the ratio of the actual capacitance to the capacitance in the absence of dielectric:

$$\kappa = \frac{C}{C_0} = \frac{C}{\epsilon_0 A/d} = \frac{300 \times 10^{-6} \text{ F}}{(8.85 \times 10^{-12} \text{ F/m} \times 123 \text{ m}^2)/(8.0 \times 10^{-6} \text{ m})}$$

$$= 2.2$$

The magnitude of the free charge on each plate is

$$Q_{\text{free}} = C \, \Delta V = 300 \times 10^{-6} \text{ F} \times 24 \times 10^3 \text{ V} = 7.2 \text{ coulombs}$$

The electric field produced by the free charges on the plates in the absence of dielectric would be [see Eq. (26.18)]

$$E_{\text{free}} = \frac{Q}{\epsilon_0 A}$$

and the field in the dielectric is therefore, according to Eq. (26.19),

$$E = \frac{1}{\kappa} E_{\text{free}} = \frac{1}{\kappa} \frac{Q}{\epsilon_0 A} = \frac{1}{2.2} \times \frac{7.2 \text{ coulombs}}{8.85 \times 10^{-12} \text{ F/m} \times 123 \text{ m}^2}$$

$$= 3.0 \times 10^9 \text{ volts/m}$$

This is a larger electric field, by a factor of about 100, than most common dielectrics can withstand without electric breakdown.

TABLE 26.1

DIELECTRIC CONSTANTS OF SOME MATERIALS[a]

MATERIAL	κ
Vacuum	1
Air	1.000 54
Carbon dioxide	1.000 98
Polyethylene	2.3
Polystyrene	2.5
Rubber, hard	2.8
Transformer oil	≈ 3
Plexiglas	3.4
Nylon	3.5
Epoxy resin	3.6
Paper	≈ 4
Glass	≈ 6
Porcelain	≈ 7
Water, distilled	80
Strontiun titanate	320

[a] At room temperature (20°C) and 1 atm.

COMMENTS: Alternatively, we can calculate the electric field in the dielectric from the usual relationship between a uniform electric field and its potential,

$$E = \frac{\Delta V}{d} = \frac{24 \times 10^3 \text{ V}}{8.0 \times 10^{-6} \text{ m}} = 3.0 \times 10^9 \text{ volts/m}$$

This relationship between electric field and potential is not affected by the presence of the dielectric.

The simple formula (26.19) for the reduction of the electric field by a dielectric applies to any configuration in which the dielectric and the distribution of free charges have the same symmetry—for instance, a pair of flat charged plates with a flat slab of dielectric (as in the parallel-plate capacitor) or a spherical distribution of charge surrounded by a concentric spherical shell of dielectric. However, if the symmetries of the charge distribution and the dielectric differ—for instance, a parallel-plate capacitor with a spherical ball of dielectric placed between its plates—then the simple formula (26.19) is not applicable, and the reduction of the electric field in the dielectric becomes rather more difficult to calculate.

EXAMPLE 8   What is the electric field generated by a point charge q surrounded by a large volume of dielectric, for instance, a point charge placed in a large volume of gas?

SOLUTION: If the volume of the dielectric surrounding the point charge is large, then the electric field in the neighborhood of this charge is not significantly influenced by the shape of the (remote) surfaces of the dielectric. Thus, the dielectric can be regarded as providing a spherically symmetric environment for the spherically symmetric point charge. Consequently, Eq. (26.19) is applicable to this problem and

$$E = \frac{1}{\kappa} E_{\text{free}} = \frac{1}{\kappa} \frac{q}{4\pi\epsilon_0 r^2} \tag{26.23}$$

where, as usual, r is the distance from the point charge.

Note that since this inverse-square electric field produced by the point charge in the large volume of dielectric differs from the electric field of a point charge in vacuum only by the factor $1/\kappa$, the arguments that led us to Gauss' Law (in vacuum) in Section 24.2 will now lead us to a modified version of Gauss' Law for a distribution of point charges placed in a large volume of dielectric:

Gauss' Law in dielectrics

$$\oint \kappa E_\perp \, dA = \frac{Q_{\text{free, inside}}}{\epsilon_0} \tag{26.24}$$

This result, which we have reached here by a rather specialized argument involving a large volume of dielectric, is actually valid in general. It holds true for any configuration of free charges and dielectrics, regardless of their symmetry. Equation (26.24) is **Gauss' Law in dielectrics**. We could opt to use this Law, but for the simple problems we will be dealing with, we can always find the electric field from the formula (26.19), by first calculating the electric field of the free charges by themselves.

EXAMPLE 9   A common type of cable, the **coaxial cable**, consists of a solid cylindrical conductor on the axis of a hollow cylindrical conducting shell; the two conductors are separated by a dielectric (see Fig. 26.19). Calculate the capacitance per centimeter of a coaxial cable with a polystyrene dielectric. The inner conductor of the cable has a diameter of 1.5 mm, and the thin conducting shell has a radius of 3.0 mm.

SOLUTION: The capacitance of a cylindrical capacitor *without* dielectric is defined in the usual way [Eq. (26.6)]:

$$C_0 = \frac{Q}{\Delta V_0}$$

As more length is added, the additional capacitance is in parallel, and so simply adds in proportion to length. So the capacitance per unit length *l* is

$$\frac{C_0}{l} = \frac{\lambda}{\Delta V_0}$$

where $\lambda = Q/l$ is the linear charge density. In Example 5 of Chapter 25, we calculated the potential difference without dielectric from the field obtained from Gauss' Law; this gave Eq. (25.35),

$$\Delta V_0 = \frac{\lambda}{2\pi\epsilon_0} \ln\left(\frac{b}{a}\right)$$

where *b* and *a* are the radii of the outer and inner conductors. Thus we have

$$\frac{C_0}{l} = 2\pi\epsilon_0 \frac{1}{\ln(b/a)} \qquad (26.25)$$

For the capacitor *with* dielectric, we then obtain, with $\kappa = 2.5$ for polystyrene,

$$\frac{C}{l} = \kappa \frac{C_0}{l} = \kappa 2\pi\epsilon_0 \frac{1}{\ln(b/a)}$$

$$= 2.5 \times 2\pi \times 8.85 \times 10^{-12} \text{ F/m} \times \frac{1}{\ln(3.0 \text{ mm}/0.75 \text{ mm})}$$

$$= 1.0 \times 10^{-10} \text{ F/m} = 1.0 \text{ pF/cm}$$

A capacitance close to one picofarad per centimeter of length is typical of many coaxial cables. This kind of capacitance plays a role when coaxial cables are used to connect electronic devices, and it is known as **cable capacitance**.

coaxial cable

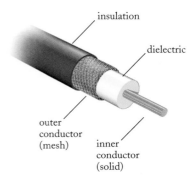

insulation

dielectric

outer conductor (mesh)

inner conductor (solid)

FIGURE 26.19 Coaxial cable.

capacitance per unit length of cylindrical capacitor

✔ Checkup 26.3

QUESTION 1: Explain why the mechanism of alignment of dipoles described above leads to a dielectric constant $\kappa > 1$.

QUESTION 2: Suppose that the capacitance of a metallic sphere is 3.0×10^{-12} F when this sphere is surrounded by vacuum. What will be the capacitance of this sphere if it is immersed in a large volume of oil with a dielectric constant $\kappa = 3.0$?

(A) 1.0×10^{-12} F   (B) 3.0×10^{-12} F   (C) 9.0×10^{-12} F

26.4 ENERGY IN CAPACITORS

Capacitors store not only electric charge, but also electric energy. We saw in Section 25.5 that any arrangement of conductors with electric charges has an electric potential energy, which represents the amount of work we must do to bring the electric charges to their positions on the conductors.

Consider a two-conductor capacitor, with charges $\pm Q$ on its plates. If the capacitor contains no dielectric, then we can calculate the electric potential energy directly from Eq. (25.52):

$$U = \tfrac{1}{2} Q(V_1 - V_2)$$

where V_1 and V_2 are the potentials of the plates. Thus, the potential energy can be expressed in terms of the charge and the potential difference:

$$U = \tfrac{1}{2} Q \, \Delta V \tag{26.26}$$

By means of the definition of capacitance, $Q = C\Delta V$, this can be put in the alternative forms

potential energy in capacitor

$$U = \frac{1}{2} C(\Delta V)^2 \quad \text{or} \quad U = \frac{1}{2} \frac{Q^2}{C} \tag{26.27}$$

If the capacitor contains a dielectric, then the calculation of the energy is a bit more involved. The trouble is that the dielectric, with its bound charges, contributes to the electric potential energy. However, in practice we are usually not interested in the total potential energy, but only in that part of the potential energy that changes as we charge (or discharge) the capacitor; that is, we are interested only in the amount of work required to charge (or discharge) the capacitor. It turns out that this amount of work is correctly given by Eqs. (26.26) and (26.27), regardless of whether the capacitor contains a dielectric or not. The quantity Q in these equations is the charge on the plates; i.e., it is the *free* charge. To see this, let us derive Eq. (26.27) from a different starting point, imitating the method we used in section 5 of Chapter 25 for a charged conducting sphere [Eq. (25.49)].

Imagine that we charge the capacitor gradually, starting with an initial charge $q = 0$, adding small amounts of charge step by step, and ending with a final charge $q = Q$. When the plates carry charges $\pm q$, the potential difference between them is q/C and the work that we must perform to increase the charges on the plates by a small amount $\pm dq$ is the product of the transported charge dq and the potential q/C:

$$dU = \frac{q}{C} \, dq \tag{26.28}$$

To find the final potential energy, we add up these small changes in energy by integrating from the initial value of the charge ($q = 0$) to the final value of the charge ($q = Q$):

$$U = \int dU = \frac{1}{C} \int_0^Q q \, dq = \frac{1}{C} \left(\frac{q^2}{2} \right) \Bigg|_0^Q = \frac{1}{C} \left(\frac{Q^2}{2} - 0 \right)$$

that is,

$$U = \frac{1}{2} \frac{Q^2}{C} \tag{26.29}$$

This formula is the same as the second relation in Eq. (26.27). This confirms that the relations in Eq. (26.27) remain valid for a capacitor containing a dielectric, provided we use the appropriate value of the capacitance (with the dielectric constant included).

EXAMPLE 10 Consider one of the National Ignition Facility capacitors filled with dielectric, described in Example 7. What is the stored potential energy in this capacitor? What is the total energy stored in all $192 \times 20 = 3840$ capacitors in the array? What is the energy density in the capacitors?

Concepts — in — Context

SOLUTION: In Example 7, we were given $\Delta V = 24$ kV and we found the free charge on each plate, $Q = 7.2$ C. By Eq. (26.26), the stored energy in one capacitor is

$$U = \tfrac{1}{2} Q \, \Delta V = \tfrac{1}{2} \times 7.2 \text{ coulombs} \times 24 \times 10^3 \text{ volts} = 8.6 \times 10^4 \text{ J}$$

The total energy stored in the array is the energy stored in each capacitor times the number of capacitors:

$$U_{\text{array}} = 8.6 \times 10^4 \text{ J} \times 3840 = 3.3 \times 10^8 \text{ J}$$

The capacitors can deliver this stored energy, one-third of a billion joules, in a time much less than a nanosecond, corresponding to a peak power greater than billions of gigawatts.

The energy density u is the energy divided by the volume between the plates [Eq. (25.58)]. Using the values from Example 7,

$$u = \frac{U}{Ad} = \frac{8.6 \times 10^4 \text{ J}}{123 \text{ m}^2 \times 8.0 \times 10^{-6} \text{ m}} = 8.7 \times 10^7 \text{ J/m}^3$$

Note that we here calculated the energy density directly from the energy and the volume, instead of using the formula $u = \epsilon_0 E^2 / 2$ for the energy density from Chapter 25. If we had used this formula, we would have obtained a wrong answer, because the formula from Chapter 25 is not applicable in a dielectric. It is easy to check that in a dielectric material, the correct formula for the energy density is $u = \kappa \epsilon_0 E^2 / 2$. For example, for a capacitor filled with dielectric, the potential energy is

$$U = \frac{1}{2} C (\Delta V)^2 = \frac{1}{2} \kappa C_0 (\Delta V)^2 = \frac{1}{2} \kappa \frac{\epsilon_0 A}{d} (Ed)^2 = \frac{1}{2} \kappa \epsilon_0 E^2 \cdot Ad$$

and so the energy density is

$$u = \frac{U}{Ad} = \frac{1}{2} \kappa \epsilon_0 E^2 \tag{26.30}$$

energy density in dielectric

EXAMPLE 11 An isolated, charged parallel-plate capacitor with plate separation $d = 0.10$ mm and area $A = 1.0$ m^2 is initially empty (Fig. 26.20a). You slide a Plexiglas dielectric slab with thickness equal to the plate separation but with only half the area of the plates into the region between the plates, as shown. What is the final capacitance? Does the stored energy increase or decrease? Does the capacitor pull the dielectric in or do you have to force the dielectric into the capacitor?

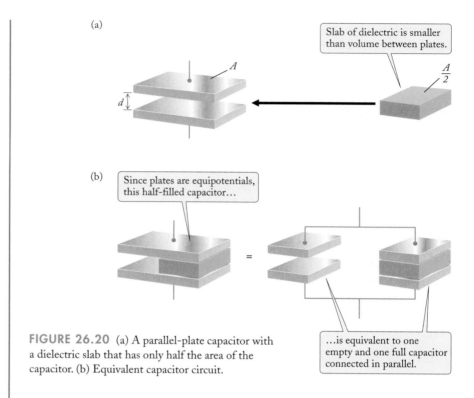

FIGURE 26.20 (a) A parallel-plate capacitor with a dielectric slab that has only half the area of the capacitor. (b) Equivalent capacitor circuit.

SOLUTION: Because the dielectric fills the capacitor only partially, we must consider the right and the left halves of the capacitor separately to determine the equivalent capacitance. Each plate is an equipotential, so we may separate the capacitor into two pieces connected in parallel by wires (which are also equipotentials), as shown in Fig. 26.20b. Each of these two capacitors has half the area of the original, and capacitances in parallel add, so the empty capacitance is

$$C_0 = \frac{\epsilon_0 A}{d} = \frac{\epsilon_0 A/2}{d} + \frac{\epsilon_0 A/2}{d} = \frac{1}{2}\, C_0 + \frac{1}{2}\, C_0$$

When one of the two halves is filled, its capacitance increases by a factor of the dielectric constant, so the final capacitance is

$$C = \frac{1}{2}\, C_0 + \kappa\, \frac{1}{2}\, C_0 = \frac{1 + \kappa}{2}\, C_0 = \frac{1 + \kappa}{2}\, \frac{\epsilon_0 A}{d}$$

$$= \frac{1 + 3.4}{2}\, \frac{8.85 \times 10^{-12}\ \text{F/m} \times 1.0\ \text{m}^2}{10^{-4}\ \text{m}} = 1.9 \times 10^{-7}\ \text{F}$$

Since the charge is held constant, the energy is most conveniently written

$$U = \frac{1}{2}\, \frac{Q^2}{C}$$

Because the capacitance increases by a factor of $(1 + \kappa)/2$, the stored energy decreases. Since the dielectric attains a position of lower energy, the capacitor is doing work on the dielectric, that is, it produces a force which pulls the dielectric in.

COMMENTS: Note that this example involved an isolated capacitor, that is, the charge on the plates was constant. If the plates had instead been held at a constant potential by keeping them connected to a battery, then the charge would have increased as the dielectric moved into place. Also, for a constant potential difference ΔV between the plates, the stored energy $U = C(\Delta V)^2/2$ would increase by $\Delta U = \Delta C(\Delta V)^2/2$ with increasing C. The increase ΔU in stored energy comes from the battery. The battery actually delivers a larger energy $\Delta Q\,\Delta V = (\Delta C\Delta V)\Delta V = 2\Delta U$; the extra energy is delivered as work on the slab. This work is positive, indicating that the force exerted by the capacitor plates on the slab pulls the slab in. Thus this force has the same direction in the case of constant potential as in the case of constant charge.

 ## Checkup 26.4

QUESTION 1: Two parallel-plate capacitors are identical, except that one has dielectric between its plates and the other does not. If charged to the same voltage, which will have the larger charge on its plates? The larger electric field? The larger energy density? The larger energy?

QUESTION 2: Two parallel-plate capacitors are identical, except that one has dielectric between its plates and the other does not. If equal amounts of charge are placed on their plates, which will have the larger voltage? The larger electric field? The larger energy density? The larger energy?

QUESTION 3: Two identical empty capacitors, each with capacitance C_0, are connected in series. Both are then filled with a material with dielectric constant $\kappa = 2.0$. What is the final net capacitance of the series combination?

   (A) $4.0C_0$ (B) $2.0C_0$ (C) C_0 (D) $0.50C_0$

SUMMARY

SI UNIT OF CAPACITANCE	$1\ \mathrm{F} = 1\ \mathrm{farad} = 1\ \mathrm{coulomb/volt}$	(26.4)
CAPACITANCE OF A SINGLE CONDUCTOR	$C = \dfrac{Q}{V}$	(26.2)
CAPACITANCE OF A PAIR OF CONDUCTORS	$C = \dfrac{Q}{\Delta V}$	(26.6)
CAPACITANCE OF AN ISOLATED SPHERE	$C = 4\pi\epsilon_0 R$	(26.3)
CAPACITANCE OF PARALLEL PLATES	$C = \dfrac{\epsilon_0 A}{d}$	(26.9)

PARALLEL COMBINATION OF CAPACITORS
Capacitors in parallel have the *same potential difference* across each.

$$C = C_1 + C_2 + C_3 + \cdots$$

(26.13)

SERIES COMBINATION OF CAPACITORS
Capacitors in series have the *same charge* on each.

$$\frac{1}{C} = \frac{1}{C_1} + \frac{1}{C_2} + \frac{1}{C_3} + \cdots$$

(26.16)

CAPACITANCE PER UNIT LENGTH OF CYLINDRICAL CAPACITOR

$$\frac{C}{l} = \frac{2\pi\epsilon_0}{\ln(b/a)}$$

(26.25)

ELECTRIC FIELD IN A DIELECTRIC BETWEEN PARALLEL PLATES

where κ, the dielectric constant, is larger than 1. The relation $E = E_{\text{free}}/\kappa$ also applies in any dielectric with the same symmetry as the distribution of free charges.

$$E = \frac{1}{\kappa} E_{\text{free}}$$

(26.19)

CAPACITANCE WITH DIELECTRIC

$$C = \kappa C_0$$

(26.21)

ENERGY STORED IN CAPACITOR

$$U = \frac{1}{2} Q \, \Delta V = \frac{1}{2} \frac{Q^2}{C} = \frac{1}{2} C(\Delta V)^2$$

(26.26–27)

ENERGY DENSITY IN DIELECTRIC

$$u = \frac{1}{2} \kappa\epsilon_0 E^2$$

(26.30)

QUESTIONS FOR DISCUSSION

1. The large capacitors at the National Ignition Facility have a total capacitance of more than a farad. How is it possible that the capacitance of such a device is larger than the capacitance of the Earth?

2. Suppose we enclose the entire Earth in a conducting shell of radius slightly larger than the Earth's radius. Explain why this would make the capacitance of the Earth much larger than the value calculated in Example 2.

3. Equation (26.9) shows that $C \to \infty$ as $d \to 0$. In practice, why can we not construct a capacitor of arbitrarily large C by making d sufficiently small? (Hint: What happens to E as $d \to 0$ while ΔV is held constant?)

4. If you put more charge on one plate of a parallel-plate capacitor than on the other, what happens to the extra charge?

5. Taking the fringing field into account, would you expect the capacitance of a parallel-plate capacitor to be larger or smaller than the value given by Eq. (26.9)? (Hint: How does the fringing field affect the density of field lines between the plates?)

6. Figure 26.21 shows a capacitor with **guard rings**. These rings fit snugly around the edges of the capacitor plates. In use, the

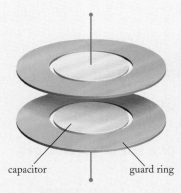

FIGURE 26.21 Capacitor with guard rings.

potential on the rings is adjusted to the same value as the potential on the plates. Explain how these rings keep the field of the plates from fringing.

7. Figure 26.5 shows the design of an adjustable capacitor used in the tuning circuit of a radio. This capacitor can be regarded as several connected capacitors. Are these several capacitors connected in series or in parallel? If we turn the tuning knob

(and the attached blue plates) counterclockwise, does the capacitance increase or decrease?

8. Consider a parallel-plate capacitor. Does the capacitance change if we insert a thin conducting sheet between the two plates, parallel to them?

9. Suppose we insert a thick slab of metal between the plates of a parallel-plate capacitor, parallel to the plates, but not in contact with them. Does the capacitance increase or decrease?

10. Consider a fluid dielectric that consists of molecules with permanent dipole moments. Will the dielectric constant increase or decrease if the temperature increases?

11. Figure 26.22 shows a dielectric slab partially inserted between the plates of a charged, isolated capacitor. Will the electric

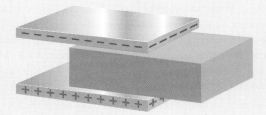

FIGURE 26.22 Dielectric slab partially inserted between the plates of a capacitor.

forces between the slab and the plates pull the slab into the region between the plates or push it out? Explain this in terms of the fringing field.

PROBLEMS

26.1 Capacitance

1. Consider an isolated metallic sphere of radius R and another isolated metallic sphere of radius $3R$. If both spheres are at the same potential, what is the ratio of their charges? If both spheres carry the same charge, what is the ratio of their potentials?

2. The collector of an electrostatic machine is a metal sphere of radius 18 cm.

   (a) What is the capacitance of this sphere?

   (b) How many coulombs of charge must you place on this sphere to raise its potential to 2.0×10^5 V?

3. Your head is (approximately) a conducting sphere of radius 10 cm. What is the capacitance of your head? What will be the charge on your head if, by means of an electrostatic machine, you raise your head (and your body) to a potential of 100 000 V (see Fig. 26.23)?

FIGURE 26.23 A charged head.

4. A capacitor consists of a metal sphere of radius 5.0 cm placed at the center of a thin metal spherical shell of radius 12 cm. The space between is empty. What is the capacitance?

5. A capacitor consists of two parallel conducting disks of radius 20 cm separated by a distance of 1.0 mm. What is the capacitance? How much charge will this capacitor store if connected to a battery of 12 V?

6. What is the electric field in a 3.0-μF capacitor with parallel plates of area 15 m^2 charged to 4.4 volts?

7. A 4.00-μF capacitor has been charged by a 9.00-volt battery. How many electrons must be moved from the negative plate to the positive plate of the capacitor to reverse the electric field inside the capacitor?

8. A capacitor is made from two concentric conducting spherical shells; the inner shell has radius a and the outer shell has radius b. What is the capacitance of this arrangement?

9. A parallel-plate variable capacitor has a fixed plate separation of 0.50 mm, but the plate area can be changed by moving one plate. If the capacitance can be varied from 10.0 pF to 120 pF, what are the corresponding minimum and maximum overlapping plate areas?

10. In digital circuits, "de-spiking" capacitors are often placed near each semiconductor chip to provide a local reservoir of charge. Typically, a 0.10-μF capacitor is placed between the chip's 5.0-volt power-supply connection and ground. How much charge is stored on such a capacitor?

11. Some modern wireless and mobile electronic devices use "supercapacitors" with extremely large capacitance values. How much charge is stored in a supercapacitor with capacitance 50.0 F and potential difference 2.50 V?

12. Spot welders use the sudden discharge of a large capacitor to melt and bond metals. A spot welder uses a 51-mF capacitor at a potential difference of 250 V. How much charge is stored?

13. The local electronic properties of surfaces are sometimes studied using a **scanning capacitance microscope**, where a metal

probe is moved over a surface and acts as one capacitor plate; the other plate is the portion of the surface below the probe. If the effective plate area is 200 nm $\times$ 200 nm and the probe is 20 nm from the surface, what is the capacitance? How much charge is on the capacitor when a voltage of 0.10 V is applied between the probe and the surface?

14. In many computer keyboards the switches under the keys consist of small parallel-plate capacitors (see Fig. 26.24). The key is attached to the upper plate, which is movable. When you push the key down, you push the upper plate toward the lower plate, and you alter the plate separation d and the capacitance. The capacitor is connected to an external circuit that maintains a constant potential difference ΔV across the plates. The change of capacitance therefore sends a pulse of charge from the capacitor into the computer circuit. Suppose that the initial plate separation is 5.0 mm and the initial capacitance is 6.0×10^{-13} F. The final plate separation (with the key fully depressed) is 0.20 mm. The constant potential difference is 8.0 V. What is the change in capacitance when you depress the key? What is the amount of electric charge that flows out of the capacitor into the computer circuit?

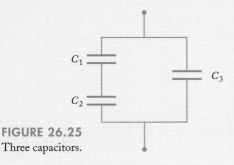

FIGURE 26.24 Capacitive keyboard switch.

*15. What is the capacitance of the Geiger-counter tube described in Problem 35 of Chapter 25? Pretend that the space between the conductors is empty.

26.2 Capacitors in Combination

16. What is the combined capacitance if three capacitors of 3.0, 5.0, and 7.5 μF are connected in parallel? What is the combined capacitance if they are connected in series?

17. Three capacitors with capacitances $C_1 = 5.0$ μF, $C_2 = 3.0$ μF, and $C_3 = 8.0$ μF are connected as shown in Fig. 26.25. Find their combined capacitance.

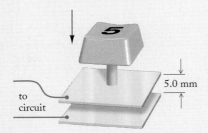

FIGURE 26.25
Three capacitors.

18. Two capacitors of 5.0 μF and 8.0 μF are connected in series to a 24-V battery. What is the potential difference across each capacitor?

19. What is the total charge stored on the three capacitors connected to a 30-V battery as shown in Fig. 26.26?

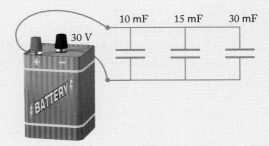

FIGURE 26.26 Three capacitors connected to a battery.

20. Six identical capacitors of capacitance C are connected as shown in Fig. 26.27. What is the net capacitance of the combination?

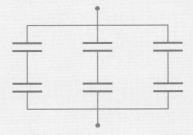

FIGURE 26.27 Six identical capacitors.

21. Six identical capacitors of capacitance C are connected as shown in Fig. 26.28. What is the net capacitance of the combination?

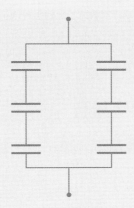

FIGURE 26.28 Six identical capacitors.

22. Seven capacitors are connected as shown in Fig. 26.29. What is the net capacitance of this combination?

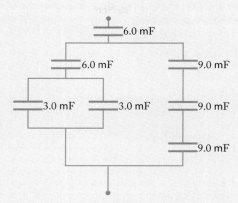

FIGURE 26.29 Seven connected capacitors.

23. When the arrangement of capacitors shown in Fig. 26.11 is connected to a voltage source, each of the 4.0-μF capacitors stores a charge of 6.0 μC. What charge is stored on the 6.0-μF capacitor? What is the voltage of the source?

*24. Two capacitors of 2.0 μF and 5.0 μF are connected in series; the combination is connected to a 1.5-V battery. What is the charge stored on each capacitor? What is the potential difference across each capacitor?

*25. A 2.5-μF capacitor is connected to a 9.0-V battery. The capacitor is then disconnected from the battery and connected to an initially uncharged 5.0-μF capacitor. What is the final charge on each capacitor? What is the final potential difference across the capacitors?

*26. A multiplate capacitor, such as used in radios, consists of four parallel plates arranged one above the other as shown in Fig. 26.30. The area of each plate is A, and the distance between adjacent plates is d. What is the capacitance of this arrangement?

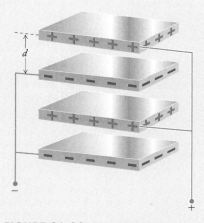

FIGURE 26.30 A multiplate capacitor.

*27. How can you connect four 1.0-μF capacitors so the net capacitance is 1.0 μF?

*28. Two capacitors, of 2.0 and 6.0 μF, respectively, are initially charged to 24 V by connecting each, for a few instants, to a 24-V battery. The battery is then removed and the charged capacitors are connected in a closed series circuit, the positive terminal of each capacitor being connected to the negative terminal of the other (Fig. 26.31). What is the final charge on each capacitor? What is the final potential difference across each?

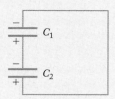

FIGURE 26.31 Two capacitors connected in series after they have been charged.

**29. Three capacitors, of capacitances $C_1 = 2.0$ μF, $C_2 = 5.0$ μF, and $C_3 = 7.0$ μF, are initially charged to 36 V by connecting each, for a few instants, to a 36-V battery. The battery is then removed and the charged capacitors are connected in a closed series circuit, with the positive and negative terminals joined as shown in Fig. 26.32. What is the final charge on each capacitor? What is the voltage across the points PP' in Fig. 26.32?

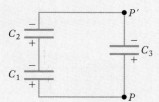

FIGURE 26.32 Three capacitors connected after they have been charged.

26.3 Dielectrics

30. You wish to construct a capacitor out of a sheet of polyethylene of thickness 5.0×10^{-2} mm and $\kappa = 2.3$ sandwiched between two aluminum sheets. If the capacitance is to be 3.0 μF, what must be the area of the sheets?

31. What is the capacitance of a sphere of radius R immersed in a large volume of gas of dielectric constant κ?

32. A parallel-plate capacitor has plates of area 0.050 m^2 separated by a distance of 0.20 mm. The space between the plates is filled with Plexiglas.

(a) What is the capacitance?

(b) What is the maximum potential difference that this capacitor can withstand? The maximum electric field that

Plexiglas can tolerate without electric breakdown is 40×10^6 V/m.

(c) What is the corresponding maximum amount of charge we can place on the plates?

33. High-dielectric-constant thin-film capacitors are attractive for digital memory applications; for example, when barium strontium titanate ($BaSrTi_2O_6$) is used as a 50-nm-thick dielectric material, a capacitance per unit area of 90 $\mu F/cm^2$ can be achieved. What is the dielectric constant?

*34. Polystyrene has a dielectric strength of 24×10^6 V/m and nylon has a dielectric strength of 14×10^6 V/m. We wish to make a 1.0-μF parallel-plate capacitor which can withstand 25 V. What is the smallest area for which this can be achieved for each material?

*35. In order to measure the dielectric constant of a dielectric material, a slab of this material 1.5 cm thick is slowly inserted between a pair of parallel conducting plates separated by a distance of 2.0 cm. Before insertion of the dielectric, the potential difference across these capacitor plates is 3.0×10^5 V. During insertion, the charge on the plates remains constant. After insertion, the potential difference is 1.8×10^5 V. What is the value of the dielectric constant?

*36. A parallel-plate capacitor of plate area A and spacing d is filled with two parallel slabs of dielectric of equal thickness with dielectric constants κ_1 and κ_2, respectively (Fig. 26.33). What is the capacitance? (Hint: Check that the configuration of Fig. 26.33 is equivalent to two capacitors in series.)

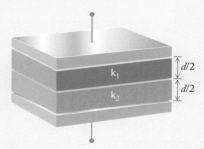

FIGURE 26.33 Parallel-plate capacitor with two slabs of dielectric.

*37. A capacitor with two large parallel plates of area A separated by a distance d is filled with two equal-size slabs of dielectric side by side (Fig. 26.34). The dielectric constants are κ_1 and κ_2. What is the capacitance?

FIGURE 26.34 Parallel-plate capacitor with two slabs of dielectric side by side.

*38. A parallel-plate capacitor of plate area A and separation d contains a slab of dielectric of thickness $d/2$ and constant κ (see Fig. 26.35). What is the capacitance of this capacitor? (Hint: Regard this capacitor as two capacitors in series, one with dielectric, one without.)

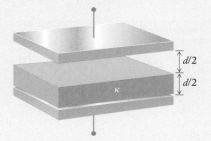

FIGURE 26.35 Parallel-plate capacitor, partially filled with dielectric.

*39. A parallel-plate capacitor of plate area A and separation d contains a slab of dielectric of thickness $d/2$ (Fig. 26.35) and dielectric constant κ. The potential difference between the plates is ΔV.

(a) In terms of the given quantities, find the electric field in the empty region of space between the plates.

(b) Find the electric field inside the dielectric.

(c) Find the density of bound charges on the surface of the dielectric.

*40. Within some limits, the difference between the dielectric constants of air and of vacuum is proportional to the pressure of the air, i.e., $\kappa - 1 \propto p$. Suppose that a parallel-plate capacitor is held at a constant potential difference by means of a battery. What will be the percentage change in the amount of charge on the plates as we increase the air pressure between the plates from 1.0 atm to 3.0 atm?

*41. A sensor for measuring liquid levels is made from a cylindrical capacitor (Fig. 26.36) of length $L = 50$ cm. The inner conductor has radius $a = 1.0$ mm and the outer conducting shell has radius $b = 4.0$ mm. If the sensor is used to detect the level of liquid nitrogen ($\kappa = 1.433$), what is its capacitance when it is (a) empty and (b) full?

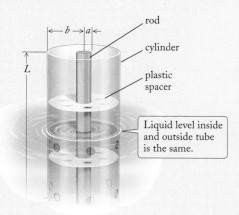

FIGURE 26.36 A capacitance liquid-level sensor.

*42. Two identical capacitors with $C_0 = 2.0 \ \mu F$ are in series, initially empty, and connected to a potential difference of $\Delta V = 9.0$ volts (Fig. 26.37).

   (a) What is the charge on each capacitor? What is the potential difference across each capacitor?

   (b) One capacitor is then filled with a dielectric slab with $\kappa = 2.5$. What are the charge on each capacitor and the potential difference across each capacitor now?

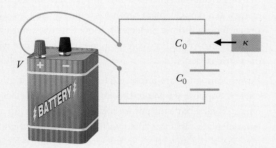

FIGURE 26.37 Two capacitors, connected to a battery.

*43. Two identical capacitors with $C_0 = 2.0 \ \mu F$ are in series, initially empty, and are briefly connected to a potential difference of $\Delta V = 9.0$ volts. After charging, the capacitors are disconnected from the potential difference and electrically isolated. One capacitor is then filled with a dielectric slab with $\kappa = 2.5$. What are the charge on each capacitor and the potential difference across each capacitor now?

*44. A spherical capacitor consists of a metallic sphere of radius R_1 surrounded by a concentric thin metallic spherical shell of radius R_2. The space between R_1 and R_2 is filled with a dielectric having a constant κ. Suppose that the free surface-charge density on the metallic sphere at R_1 is $\sigma_{free(1)}$.

   (a) What is the free surface-charge density on the metallic shell at R_2?

   (b) What is the bound surface-charge density on the dielectric at R_1?

   (c) What is the bound surface-charge density on the dielectric at R_2?

*45. A long cylindrical copper wire of radius 0.20 cm is surrounded by a cylindrical sheath of rubber of inner radius 0.20 cm and outer radius 0.30 cm. The rubber has $\kappa = 2.8$. Suppose that the surface of the copper has a free charge density of $4.0 \times 10^{-6} \ C/m^2$.

   (a) What will be the bound charge density on the inside surface of the rubber sheath? On the outside surface?

   (b) What will be the electric field in the rubber near its inner surface? Near its outer surface?

   (c) What will be the electric field just outside the rubber sheath?

*46. A metallic sphere of radius R is surrounded by a concentric dielectric shell of inner radius R, outer radius $3R/2$. This is surrounded by a concentric thin metallic shell of radius $2R$ (Fig. 26.38). The dielectric constant of the dielectric shell is κ. What is the capacitance of this contraption?

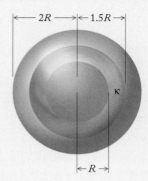

FIGURE 26.38 Spherical capacitor, partially filled with dielectric.

*47. Two small metallic spheres are submerged in a large volume of transformer oil of dielectric constant $\kappa = 3.0$. The spheres carry electric charges of 2.0×10^{-6} C and 3.0×10^{-6} C, respectively, and the distance between them is 0.60 m. What is the force on each?

**48. A hollow sphere of brass floats in a large lake of oil of dielectric constant $\kappa = 3.0$. The sphere is exactly halfway immersed in the oil (Fig. 26.39). The sphere has a net charge of 2.0×10^{-6} C. What fraction of this electric charge will be on the upper hemisphere? On the lower? (Hint: The electric fields in the oil and in the air above the oil are exactly the same.)

FIGURE 26.39 Brass sphere afloat in a lake of oil.

**49. A spherical capacitor consists of two concentric spherical shells of metal of radii R_1 and R_2. The space between these shells is filled with two kinds of dielectric (Fig. 26.40); the dielectric in the upper hemisphere has a constant κ_1, and the dielectric in the lower hemisphere has a constant κ_2. What is the capacitance of this device? (Hint: The electric fields in both dielectrics are exactly the same.)

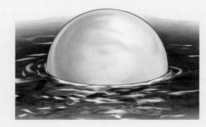

FIGURE 26.40 Spherical capacitor with two kinds of dielectric.

26.4 Energy in Capacitors

50. How much energy is stored in a $3.0 \times 10^3 \ \mu F$ capacitor charged to 100 volts?

51. A parallel-plate capacitor has a plate area of 900 cm^2 and a plate separation of 0.50 cm.

    The space between the plates is empty.

    (a) What is the capacitance?

    (b) What is the potential difference if the charges on the plates are $\pm 6.0 \times 10^{-8}$ C?

    (c) What is the electric field between the plates?

    (d) The energy density?

    (e) The total energy?

52. Repeat Problem 51, under the assumption that the space between the plates is filled with Plexiglas.

53. A TV receiver contains a capacitor of 10 μF charged to a potential difference of 2.0×10^4 V. What is the amount of charge stored in this capacitor? The amount of energy?

54. Two parallel conducting plates of area 0.50 m^2 placed in a vacuum have a potential difference of 2.0×10^5 V when charges of $\pm 4.0 \times 10^{-3}$ C are placed on them, respectively.

    (a) What is the capacitance of the pair of plates?

    (b) What is the distance between them?

    (c) What is the electric field between them?

    (d) What is the electric energy?

55. A large capacitor has a capacitance of 20 μF. If you want to store an electric energy of 40 J in this capacitor, what potential difference do you need?

56. An isolated, charged parallel-plate capacitor has plate separation d. If the plates are pulled apart to a separation $3d$, does the stored energy increase or decrease? By what factor?

57. A "supercapacitor" has a huge capacitance of 6.8 F, but can withstand a potential difference of only 2.5 V. A power-supply capacitor has a capacitance of 820 μF and can operate up to 400 V. Which can store more charge? Which can store more energy?

*58. Two identical, empty capacitors are connected in series; the combination is permanently connected to a potential source. When one of the capacitors is filled with a material of dielectric constant κ, does the total stored energy increase or decrease? By what factor?

*59. A parallel-plate capacitor has plates of area 0.040 m^2 separated by a distance of 0.50 mm. Initially, the space between the plates is empty and the capacitor is uncharged. A 12-V battery is available for connection to the capacitor. For each of the following cases, find the resulting capacitance C, potential difference ΔV, and charge Q on the plates; for cases (a) and (b), also find the energy delivered by the battery:

    (a) The battery is connected to the capacitor.

    (b) While the battery remains connected, a slab of dielectric with $\kappa = 3.0$ is inserted between the plates, completely filling the space between them.

    (c) The battery is disconnected, and the dielectric is then pulled out from the capacitor.

    (d) The capacitor is discharged, and the battery is again connected to the empty capacitor for some moments. The battery is then disconnected, and the dielectric is then inserted between the plates of the capacitor.

*60. Two capacitors of 5.0 μF and 8.0 μF are connected in series to a 24-V battery. What is the energy stored in the capacitors?

61. Starting from the general expression $\frac{1}{2}Q \, \Delta V$ for the electric energy in a parallel-plate capacitor (equally valid for a capacitor with or without a dielectric), show that the energy density in the electric field between the plates of a capacitor with a dielectric is $\frac{1}{2}\kappa\epsilon_0 E^2$.

*62. Power companies are interested in the storage of surplus energy. Suppose we wanted to store 10^6 kW·h of electric energy (half a day's output for a large power plant) in a large parallel-plate capacitor filled with a plastic dielectric with $\kappa = 3.0$. If the dielectric can tolerate a maximum electric field of 5.0×10^7 V/m, what is the minimum total volume of dielectric needed to store this energy?

*63. Three capacitors are connected as shown in Fig. 26.41. Their capacitances are $C_1 = 2.0 \ \mu F$, $C_2 = 6.0 \ \mu F$, and $C_3 = 8.0 \ \mu F$. If a voltage of 200 V is applied to the two free terminals, what will be the charge on each capacitor? What will be the energy in each?

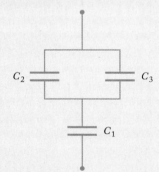

FIGURE 26.41 Three capacitors.

*64. Ten identical 5.0-μF capacitors are connected in parallel to a 240-V battery. The charged capacitors are then disconnected from the battery and reconnected in series, the positive terminal of each capacitor being connected to the negative terminal of the next. What is the potential difference between the negative terminal of the first capacitor and the positive terminal of the last capacitor? If these two terminals are connected via an external circuit, how much charge will flow around this circuit as the series arrangement discharges? How much energy will be released in the discharge? Compare this charge and this energy with the charge and energy stored in the original, parallel arrangement, and explain any discrepancies.

**65. A large dielectric slab partially fills an isolated parallel-plate capacitor with charges $\pm 2.0 \ \mu C$. The slab thickness is equal to the plate separation. The parallel plates are squares with 10-cm sides and 1.0 mm separation, and the dielectric constant is 2.5. Find the value of the force on the dielectric when the capacitor is half full.

REVIEW PROBLEMS

66. A parallel-plate capacitor consists of two square conducting plates of area 0.040 m² separated by a distance of 0.20 mm. The plates are connected to the terminals of a 12-V battery.

   (a) What is the charge on each plate? What is the electric field between the plates?

   (b) If you move the plates apart, so their separation becomes 0.30 mm, how much charge will flow from each plate to the terminals of the battery? What will be the new electric field?

67. Three capacitors are connected as shown in Fig. 26.42. Their capacitances are $C_1 = 4.0\ \mu F$, $C_2 = 6.0\ \mu F$, and $C_3 = 3.0\ \mu F$. If a voltage of 400 V is applied to the two free terminals, what will be the charge on each capacitor? What will be the potential energy of each?

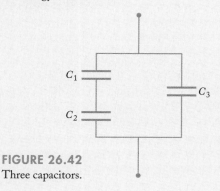

FIGURE 26.42
Three capacitors.

68. You have three capacitors of 1.0 μF, 2.0 μF, and 3.0 μF. If you connect these three capacitors in series, in parallel, or in some combination of series and parallel, what different net capacitances can you construct?

69. Figure 26.43 shows five capacitors of 4.0 μF each connected together.

   (a) Find the net capacitance of this combination between the terminals A and A'.

   (b) Find the net capacitance of this combination between the terminals B and B'.

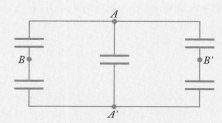

FIGURE 26.43 Five capacitors.

70. A capacitor of 5.0 μF is initially charged by connecting it briefly to a 40-V battery, and a capacitor of 8.0 μF is initially charged by connecting it briefly to a 60-V battery. The batteries are then removed, and the two capacitors are connected in parallel (see Fig. 26.44). What is the final charge on each capacitor? What is the final potential difference across each?

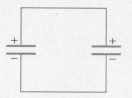

FIGURE 26.44 Two capacitors connected in parallel after they have been charged.

71. The plates of a parallel-plate capacitor are movable and are initially separated by an air gap of width d. A piece of dielectric of dielectric constant κ and a thickness of $3d$ is inserted between the plates. If the ratio of the capacitance before inserting the dielectric to the capacitance after inserting the dielectric is 1.5, what is the value of κ?

72. A parallel-plate capacitor has a capacitance of 25 μF when filled with air and it can withstand a potential difference of 50 V before it suffers electrical breakdown.

   (a) What is the maximum amount of charge we can place on this air-filled capacitor? The dielectric strength of air is 3.0×10^6 V/m.

   (b) If we fill this capacitor with polyethylene, what will be its new capacitance?

   (c) What will be the maximum potential difference that this new capacitor can withstand? What will be the corresponding maximum amount of charge we can place on this capacitor? The dielectric strength of polyethylene is 18×10^6 V/m.

73. A capacitor used to power a strobe light has a capacitance of 200 μF and it is charged to a potential difference of 360 V.

   (a) What is the energy stored in this capacitor?

   (b) Suppose that the dielectric in the capacitor has a dielectric constant of 2.2 and can withstand a maximum electric field of 70×10^6 V/m (the dielectric strength). What is the maximum permitted energy density in the dielectric? What is the minimum volume that the dielectric must have if it is to store the energy calculated in part (a)?

74. A primitive capacitor, originally called a **Leyden jar**, consists of a glass bottle filled with water and wrapped on the outside with metal foil (see Fig. 26.45). The foil plays the role of one plate of the capacitor, and the surface of the water facing the foil plays the role of the other plate; these "plates" are separated by the layer of glass (dielectric). Suppose that, as shown in Fig. 26.45, the wrapped portion of the bottle is

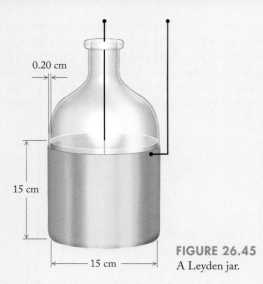

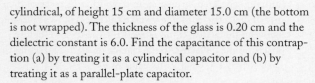

FIGURE 26.45
A Leyden jar.

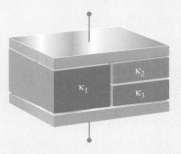

FIGURE 26.46 Parallel-plate capacitor filled with three dielectrics.

cylindrical, of height 15 cm and diameter 15.0 cm (the bottom is not wrapped). The thickness of the glass is 0.20 cm and the dielectric constant is 6.0. Find the capacitance of this contraption (a) by treating it as a cylindrical capacitor and (b) by treating it as a parallel-plate capacitor.

75. Two capacitors are connected in series, and the combination is connected to a 9.0-V battery. One capacitor has a capacitance of 4.0 μF, and the potential across the other is 3.0 V. What is the capacitance of the other capacitor? What is the energy stored in each capacitor?

76. Two identical, empty, isolated capacitors are connected in series; each carries the same charge. If one of the capacitors is then filled with a material of dielectric constant κ, by what factor does the total stored energy decrease?

77. A cylindrical capacitor has an inner conductor of radius a and a coaxial outer conducting shell of radius b. The region between the conductors is filled with two cylindrical shells of dielectric, one of constant κ_1 for $a < r < c$ and another of constant κ_2 for $c < r < b$. What is the capacitance per unit length of this layered cylindrical capacitor?

78. A parallel-plate capacitor has capacitance C_0 when empty. Three slabs, each with area equal to half the plate area, are inserted between the plates as shown in Fig. 26.46: one slab with dielectric constant κ_1 has thickness equal to the plate separation, and the other two slabs with dielectric constants κ_2 and κ_3 have half that thickness. What is the capacitance of the filled capacitor?

*79. A parallel-plate capacitor is filled with carbon dioxide at 1.0 atm pressure. Under these conditions the capacitance is 0.50 μF. We charge the capacitor by means of a 48-V battery and then disconnect the battery so that the electric charge remains constant thereafter. What will be the change in the potential difference if we now pump the carbon dioxide out of the capacitor, leaving it empty?

*80. A spherical capacitor consists of a metallic sphere of radius R_1 surrounded by a concentric metallic shell of radius R_2. The space between R_1 and R_2 is filled with dielectric having a constant κ.

(a) If the free charge on the surface of the inner sphere is Q and that on the outer spherical shell is $-Q$, what is the potential difference between the two conductors?

(b) What is the capacitance of this spherical capacitor?

*81. A parallel-plate capacitor without dielectric has an area A and a charge $\pm Q$ on each plate.

(a) What is the electric force F of attraction between the plates? (Hint: One-half of the electric field between the plates is due to one plate, and one-half is due to the other. Consequently, in the calculation of the electric force on a plate from the product of field times charge, only one-half of the field must be used; the other half gives the electric force of the plate on *itself* and is of no interest.)

(b) How much work must you do against this force in order to increase the plate separation by an amount Δd?

(c) By means of Eq. (26.29), calculate the change ΔU in potential energy during this change.

(d) By comparing (a) and (c), check that $F = -\Delta U/\Delta d$.

Answers to Checkups

Checkup 26.1

1. Yes for all three—as long as the plate separation is small compared with the lateral dimensions, fringing fields will be small, and the particular shape determining the area A in Eq. (26.9) does not matter.

2. A smaller capacitance means it is more difficult to store the same amount of charge, so it is the smaller, 1-pF capacitor that requires the larger potential difference.

3. Increasing the plate separation decreases the capacitance [see Eq. (26.9)]. Increasing the area enables proportionally more charge to be stored, increasing the capacitance [Eq. (26.9)].

4. No. With a single sign of charge, the pair of plates more closely resembles a single isolated conductor, like a charged sphere. We would expect the capacitance to be similar to Eq. (26.3), where the sphere radius R would be replaced by a length on the order of the plate dimension.

5. Holding the charge constant means holding the electric field constant ($E = \sigma/\epsilon_0 = Q/\epsilon_0 A$). The potential decreases a factor of 2 ($\Delta V = Ed$), and the capacitance increases a factor of 2 ($C = Q/\Delta V = \epsilon_0 A/d$).

6. (E) 0.50 mm. Since the area is one-fourth of the original, for a constant capacitance ($C = \epsilon_0 A/d$), the separation needs to decrease by the same factor to 0.50 mm.

Checkup 26.2

1. Not necessarily in either case. For capacitors with the same charge (in series), the electric fields will be the same only if they have the same area ($E = Q/\epsilon_0 A$). For capacitors with the same potential (in parallel), the electric field will be the same only if they have the same plate separation ($E = \Delta V/d$).

2. For a given potential across them, capacitors in parallel have a greater area and so can store a greater amount of charge—thus a larger capacitance. For capacitors in series, a larger total potential difference is required to have a voltage drop across each capacitor sufficient to store a given charge, so the capacitance is smaller.

3. (B) 10 μF; 0.10 μF. Capacitors in parallel add, so ten 1.0-μF capacitors give a net capacitance of 10 μF. Capacitors in series add inverses, so the equivalent capacitance is $1/[10 \times (1/1.0\ \mu\text{F})] = 0.10\ \mu$F.

Checkup 26.3

1. The electric dipoles in the dielectric tend to align in the electric field generated by the free charges on the conductors, producing a field that partially cancels the original field. A smaller field is what led to our definition (26.19), which requires $\kappa > 1$.

2. (C) 9.0×10^{-12} F. Since $C = \kappa C_0$, where C_0 is the capacitance in the absence of dielectric, we simply have $C = 3.0 \times 3.0 \times 10^{-12}$ F $= 9.0 \times 10^{-12}$ F.

Checkup 26.4

1. Since a dielectric always increases the capacitance according to $C = \kappa C_0$, and since $Q = C \Delta V$, at constant potential the capacitor with the dielectric will have the larger stored charge. The electric fields will be the same, since the potential difference and spacing are the same ($\Delta V = Ed$). Since the electric fields are the same, the energy density, $u = \kappa \epsilon_0 E^2/2$, will be greater for the capacitor with dielectric. Also, $U = C(\Delta V)^2/2$, so the one with dielectric has the larger stored energy.

2. Again, the dielectric always increases the capacitance, because $C = \kappa C_0$. So $\Delta V = Q/C$ implies that the empty capacitor will require the larger voltage to store the same charge (larger by a factor of κ). For an identical geometry, $E = \Delta V/d$ requires that the empty capacitor also have the larger electric field (larger by the same factor of κ). For the energy density, $u = \kappa \epsilon_0 E^2/2$; thus the larger electric field in the capacitor without dielectric results in a larger energy density. Finally, $U = Q^2/2C$ tells us that for the same charge, more energy is stored in the capacitor without dielectric.

3. (C) C_0. For capacitors in series, we add inverses to obtain the inverse of the net capacitance [see Eq. (26.17)]. For two identical empty capacitors, the net capacitance is thus $\frac{1}{2}C_0$. When filled, the net capacitance increases by a factor of κ, so $C = 2.0 \times \frac{1}{2}C_0 = C_0$.

27

Currents and Ohm's Law

CONCEPTS IN CONTEXT

Concepts
— *in* —
Context

The electricity we use to power our appliances and electronic devices involves the motion of charge, or current, which flows to our homes along a power-line cable, through the wires and devices in our homes, and back out again. Our appliances, and even ordinary wires and cables, offer an opposition to the flow of charge known as resistance. Special wires and cables called superconductors offer no resistance to the flow of charge.

As we learn about current and resistance in this chapter, we can consider such questions as:

? For a known current, how much total charge flows in a given amount of time? (Examples 1 and 2, pages 860–861)

? How is the resistance of a piece of cable measured? (Example 3, page 867)

? How does the resistance of a piece of wire change with temperature? (Section 27.3, page 877 and Example 6, page 870)

? How do we decide if it is safe to plug several appliances into the same outlet? (Example 10, page 877)

Under static conditions, no electric field can exist inside a conductor. But suppose that we suddenly deposit opposite amounts of electric charge on the opposite ends of a long metallic conductor, such as a copper wire connected to the opposite plates of a capacitor (see Fig. 27.1). The conductor will then not be in electrostatic equilibrium, and the charges will generate an electric field along and inside the conductor (see Figs. 27.2 and 27.3). This electric field propels the charges toward each other. When the charges meet, they cancel, and the electric field then disappears—the conductor reaches equilibrium.

For a good conductor, such as copper, the approach to equilibrium is fairly rapid: typically, the time required to achieve equilibrium is a very small fraction of a second. However, we can permanently keep a conductor from achieving equilibrium if we continually supply more electric charge to its ends. For example, we can connect the two ends of a copper wire to the terminals of a battery or an electric generator. The terminals of such a device act as a source and sink of electric charge, just as the outlet and the intake of a pump act as source and sink of water. Under these conditions electric charge will continually flow from one terminal to the other. Such a flow of electric charge is called an electric current. In this chapter, we will examine some general properties of electric currents, and in the next chapter we will discuss the inner workings of batteries, generators, and other "pumps" that serve to maintain electric currents.

(a) Separate conductors are each in electrostatic equilibrium.

(b) Joined conductor is momentarily not in electrostatic equilibrium.

FIGURE 27.1 (a) A charged capacitor and a separate, uncharged piece of copper wire. (b) We suddenly connect the ends of the wire to the capacitor plates. The wire and the plates now form a single conductor, with opposite amounts of charge on its ends.

27.1 ELECTRIC CURRENT

When a wire is connected between the two terminals of a battery or generator, the electric charges are propelled from one end of the wire to the other by the electric field that exists along and within the wire. Most of the field lines originate at the terminals of the battery or generator, but some field lines originate at charges that have accumulated on the wire itself. As Figs. 27.2 and 27.3 show, the field lines tend to concentrate within the conductor, and they tend to follow the conductor. If the conductor has no sharp kinks, the field lines are uniformly distributed over the cross-sectional

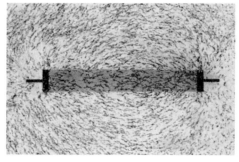

FIGURE 27.2 Electric field lines in and near a straight conductor not in equilibrium. The conductor consists of a strip of metallic paint on a paper surface. The field lines have been made visible by sprinkling grass seeds on the paper.

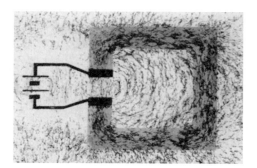

FIGURE 27.3 Electric field lines in and near a rectangular conductor carrying an electric current.

area of the conductor. For instance, if the conductor is a more or less straight wire of constant thickness, then the electric field inside the wire will be of constant magnitude and have a direction parallel to the wire. If the length of the wire is l and if the battery or generator maintains a difference of potential ΔV across its ends, then this constant electric field in the wire has a magnitude [compare Eq. (25.7)]

electric field in the wire

$$E = \frac{\Delta V}{l} \tag{27.1}$$

This electric field causes the flow of charge, or **electric current**, from one end of the wire to the other. Before we can explore the dependence of the current on the electric field, we need a precise definition of the current. Suppose that in a time Δt an amount of charge ΔQ flows past some given place on the wire (for instance, the end of the wire). *The electric current is defined as charge divided by time*:

$$I = \frac{\Delta Q}{\Delta t} \tag{27.2}$$

If the flow is not steady, the current is defined in terms of the small amount of charge dq that flows past a place in a small interval of time dt, that is, it is the instantaneous rate

electric current

$$I = \frac{dq}{dt} \tag{27.3}$$

Note that if the sides of the wire do not leak (good insulation), then the conservation of electric charge requires that *the current is the same everywhere* along the wire; that is, the current is simply the rate at which charge enters the wire at one end or the rate at which charge leaves at the other end.

The SI unit of current is the **ampere** (A); this is a flow of charge of one coulomb per second:

ampere (A)

$$1 \text{ ampere} = 1 \text{ A} = 1 \text{ C/s} \tag{27.4}$$

Table 27.1 (page 862) gives the values of some selected currents.

In metallic conductors, the charge carriers are electrons—a current in a metal is nothing but a flow of electrons. In electrolytes, such as salt water, the charge carriers are positive ions, negative ions, or both—a current in such a conductor is a flow of ions. For the sake of mathematical uniformity, whenever there is need to indicate the direction of the current along a conductor, physicists and engineers *adopt the convention that the current has the direction of a hypothetical flow of positive charge*. This means that we pretend that the moving charges are always positive charges. Of course, in metals the moving charges are actually negative charges (electrons), and hence the above convention assigns to the current a direction opposite to that of the true motion of the charges. However, as regards the transfer of charge, the transport of negative charge in one direction is equivalent to the transfer of positive charge in the opposite direction (see Fig. 27.4). The convention for labeling the direction of the current as the direction of the flow of positive charge takes advantage of this equivalence.

Concepts — in — Context

EXAMPLE 1   In the headlamp of an automobile, a current of 8.0 A flows through the filament of the lightbulb. How much electric charge flows through the filament in a minute? How many electrons?

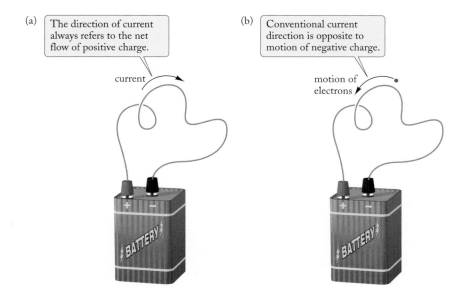

(a) The direction of current always refers to the net flow of positive charge.

current

(b) Conventional current direction is opposite to motion of negative charge.

motion of electrons

FIGURE 27.4 Flow of charge in a wire connected to a battery. (a) The motion of hypothetical positive charges from the positive terminal to the negative terminal. (b) The actual motion of the electrons in the wire from the negative terminal to the positive. The motions illustrated in (a) and (b) are equivalent in that both transfer positive charge to the negative terminal and negative charge to the positive terminal. According to our convention, the direction of the current is assumed to be from the positive terminal to the negative, even though the actual motion of the electrons is opposite to this.

SOLUTION: The amount of charge that flows through the filament in 1.0 min is, from Eq. (27.2),

$$\Delta Q = I \, \Delta t = 8.0 \text{ A} \times 1.0 \text{ min} = 8.0 \text{ A} \times 60 \text{ s} = 480 \text{ C}$$

A charge of 480 C moving through the light bulb in one direction is equivalent to an electron charge of -480 C moving in the opposite direction. The number of electrons equals the total charge divided by the charge per electron:

$$[\text{number of electrons}] = \frac{-480 \text{ C}}{-e} = \frac{-480 \text{ C}}{-1.6 \times 10^{-19} \text{ C}} = 3.0 \times 10^{21} \text{ electrons}$$

EXAMPLE 2   Superconductors can carry large currents without opposition, but to prevent damage, these currents must be changed gradually. For example, a power supply for a superconducting magnet provides a current which starts from zero at time $t = 0$ and steadily increases with time at a gradual rate of 0.080 A/s. How long does it take the current to reach a value of 60 A? At the instant when the current reaches 60 A, how much total charge has flowed through the magnet?

Concepts — in — Context

SOLUTION: Since the constant rate of increase of current is known, the time to reach 60 A can be obtained by dividing the final value by the rate:

$$\Delta t = \frac{\Delta I}{\Delta I / \Delta t} = \frac{60 \text{ A}}{0.080 \text{ A/s}} = 750 \text{ s}$$

The charge may be obtained by rearranging Eq. (27.3) to solve for dq,

$$dq = I \, dt$$

and then integrating the changing contributions dq to get the total charge q that has flowed:

$$q = \int dq = \int_0^{\Delta t} I \, dt \tag{27.5}$$

TABLE 27.1	SOME CURRENTS

Lightning stroke (a)	10^4 A
High-tension power line (b)	10^3
Large transformer (c)	10^3
Large electromagnet	200
Starter motor of automobile (d)	100
Alternator of automobile	30
Fuse blows	30
Defibrillation treatment for heart	20
Air conditioner	12
Hair dryer	10
Ordinary lightbulb	1
Flashlight bulb	0.5
Lethal fibrillation of heart	0.1
Barely perceptible by skin	1×10^{-3}
Electronic calculator (e)	1×10^{-4}
Scanning tunneling microscope	1×10^{-12}

(a)
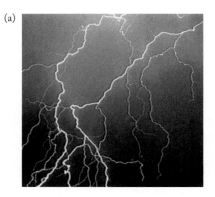

(b)

(c)

(d)

(e)

For a constant rate of change, the current I is simply the rate times the time t, $I = (0.080 \text{ A/s}) \times t$, so

$$q = \int_0^{\Delta t} (0.080 \text{ A/s}) \times t \, dt = (0.080 \text{ A/s}) \left(\frac{t^2}{2} \right) \Big|_0^{\Delta t}$$

$$= (0.080 \text{ A/s}) \left[\frac{(\Delta t)^2}{2} - 0 \right] = (0.040 \text{ A/s})(\Delta t)^2$$

Evaluating this expression at $\Delta t = 750$ s, we obtain

$$q = 0.040 \text{ A/s} \times (750 \text{ s})^2 = 22\,500 \text{ C} = 2.3 \times 10^4 \text{ C}$$

 ## Checkup 27.1

QUESTION 1: The flow of an electric current in a wire is analogous to the flow of water in a pipe. However, if we cut the pipe off, the water continues to flow, and it spills out of the pipe. Do electrons continue to flow when we cut a wire carrying a current? Do they spill out? Why or why not?

QUESTION 2: Suppose that a wire has a nonuniform cross section (thicker in some parts than in others). If an electric current of 5 A enters the wire at one end, is the current 5 A everywhere along this wire, even where the wire is thick?

QUESTION 3: In a lightning bolt (see Fig. 22.13) the electric current flows in the upward direction, from the ground to the thundercloud. What is the direction of the electric field in this lightning bolt? What is the direction of the motion of the electrons making up this current?

QUESTION 4: Suppose some currents consist of both positive and negative moving elementary charges, as shown in Fig. 27.5. Which of the currents has the largest magnitude? The smallest magnitude? (Assume all speeds are equal.)

(A) (a), (b)   (B) (a), (c)   (C) (b), (c)   (D) (c), (a)   (E) (c), (b)

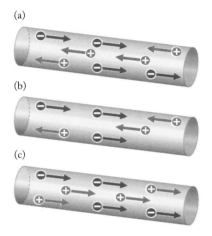

(a)

(b)

(c)

FIGURE 27.5 Several motions of positive and negative charges. All charges have the same magnitude and the same speed.

27.2   RESISTANCE AND OHM'S LAW

We will now examine in detail the behavior of a current in a metallic conductor. Such conductors contain a vast number of free electrons, which are not bound to any atom but are free to roam inside the entire volume of the metal; for example, copper has about 8×10^{22} free electrons per cubic centimeter. These electrons behave like particles of a gas, which are free to roam inside a container. By analogy, we say that the electrons form an **electron gas**, which fills the entire volume of the metal. Of course, in an electrically neutral conductor, the negative charge of the free electrons is exactly balanced by the positive charge of the ions that make up the crystal lattice of the metal. A current in this metallic conductor is simply a flow of the gas of electrons, while the ions remain at rest.

The flow of the gas of electrons along a metallic wire is analogous to the flow of water along a canal leading down a gentle slope. In such a canal, the force of gravity

electron gas

acting on the water has a component along the canal; this component pushes the water along. But the water does not accelerate—the friction between the water and the walls of the canal opposes the motion, and the water moves at a constant speed because the friction exactly matches the push of gravity.

Likewise, the electric field in the wire pushes the gas of electrons along. But the gas of electrons does not accelerate—friction between the gas and the crystal lattice of the wire opposes the motion, and the gas moves at a constant speed because such friction exactly matches the push of the electric field.

The analogy between the motion of water and the motion of the electron gas extends to the motion of the individual water molecules and individual electrons. Although the water in a canal usually has a fairly low speed, perhaps a few meters per second, the individual molecules within the water have a rather high speed—the typical speed of the random thermal motion of water molecules is about 600 m/s at ordinary temperature. But since this thermal motion consists of rapid random zigzags, which are just as likely to move the molecule backward as forward, this high speed does not contribute to the net downhill motion of water. Figure 27.6 shows the motion of a water molecule in the canal; on a microscopic scale, this motion consists of rapid zigzags on which is superimposed a much slower "drift" along the canal.

Likewise, the electron gas moves along the wire at a rather low speed, perhaps 10^{-4} m/s, but the individual electrons have a much higher speed—the typical speed of the random motion of electrons in a metal is about 10^6 m/s (this very high speed is due to quantum-mechanical effects, which we discuss in Chapter 39). Thus, the net motion of an electron in a metal also consists of rapid zigzags on which is superimposed a much slower drift motion along the wire. Qualitatively, the motion resembles the path of a water molecule shown in Fig. 27.6, but the amount of drift per zigzag is even less than shown in this figure.

The friction between the electron gas and the wire is caused by collisions between the electrons and the ions of the crystal lattice of the wire. An electron moving through a wire of copper will suffer about 10^{14} collisions with ions per second. These collisions change the direction of motion of the electron and produce the zigzags. Because of the disturbing effects of these collisions, the electron never gains much velocity from the electric field that is attempting to accelerate it, and the average drift velocity of the electron along the wire remains low. The collisions dissipate the kinetic energy that the electron acquires from the electric field. This dissipated kinetic energy of the electrons remains in the crystal lattice in the form of random kinetic and potential energy of the ions; that is, it remains in the form of thermal energy or heat. In some instances, the amount of heat produced in a wire is so large as to make the wire glow. The bright glow of a lightbulb is produced in this way, and so is the dull red glow of the coils of an electric range.

We can find the average motion of an electron by examining the losses and gains of momentum of this electron. If the average velocity, or **drift velocity**, of an electron is v_d, then the average momentum is $p = m_e v_d$. We expect that, on the average, a collision will absorb all of this momentum, that is, a collision will destroy the forward drift velocity and leave the electron with only random thermal motion. This means that, in the average time interval τ (tau) between collisions, the electron loses a momentum $m_e v_d$. The averate rate at which the electron loses momentum in collisions is therefore

$$\left(\frac{\Delta p}{\Delta t}\right)_{\text{loss}} = \frac{m_e v_d}{\tau} \tag{27.6}$$

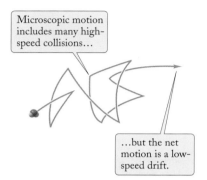

Microscopic motion includes many high-speed collisions...

...but the net motion is a low-speed drift.

FIGURE 27.6 Path of a water molecule in a canal. The molecule gradually drifts from left to right.

On the other hand, the rate at which the electron gains momentum by the action of the electric force is

$$\left(\frac{\Delta p}{\Delta t}\right)_{\text{gain}} = -eE \qquad (27.7)$$

Under steady-state conditions, the rate of loss of momentum must match the rate of gain. By setting the right sides of Eqs. (27.6) and (27.7) equal, we immediately obtain

$$v_d = \frac{-eE\tau}{m_e} \qquad (27.8)$$

drift velocity (average velocity)

This is the average velocity with which the electron gas flows along the wire. This relation merely reflects the fact that if the electric field is stronger, the electron gains more velocity between one collision and the next, and therefore attains a proportionally larger average velocity. Similarly, if the time between collisions is longer, a larger average velocity is attained.

The electric current carried by the wire is proportional to the average velocity of the electrons, and it is therefore proportional to the electric field. This is evident if we examine Fig. 27.7, where the charge ΔQ that passes a given area A can be obtained from

$$\Delta Q = [\text{charge per electron}] \times [\text{electrons per unit volume}] \times [\text{volume}] \quad (27.9)$$

Let us assume the material is a metal with n electrons per unit volume. From Fig. 27.7, we can see that the volume of moving charge that sweeps past a cross-sectional area of the wire in a time Δt is equal to the area A times the length l, that is, [volume] $= Al = Av_d \, \Delta t$. Thus Eq. (27.9) becomes

$$\Delta Q = (-e) \times (n) \times (A v_d \, \Delta t)$$

The current is then given by

$$I = \frac{\Delta Q}{\Delta t} = -en A v_d \qquad (27.10)$$

current of free electrons

or, with Eq. (27.8) for v_d,

$$I = \frac{ne^2\tau}{m_e} AE = \frac{1}{\rho} AE \qquad (27.11)$$

where ρ is a constant of proportionality that depends on the characteristics of the material of the wire. This constant ρ is called the **resistivity** of the material. It is an intrinsic property of the *type of material*, but does not depend on the shape or size of the particular piece of wire. It is a measure of how strongly a material opposes the flow of charge. Equation (27.11) indicates that the resistivity is determined by the electron's mass, density, charge, and average collision time τ:

resistivity ρ

$$\rho = \frac{m_e}{ne^2\tau} \qquad (27.12)$$

resistivity in terms of average collision time

Table 27.2 lists the resistivities of some conducting materials.

With $E = \Delta V/l$ from Eq. (27.1), the current of Eq. (27.11) becomes

$$I = \frac{1}{\rho} A \frac{\Delta V}{l} \qquad (27.13)$$

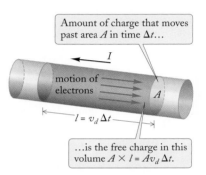

Amount of charge that moves past area A in time Δt...

I

motion of electrons

A

$l = v_d \Delta t$

...is the free charge in this volume $A \times l = A v_d \Delta t$.

FIGURE 27.7 The current carried by the wire is uniformly distributed over its entire cross section. A wire of twice the cross-sectional area would carry twice the current (with the same electric field).

For our wire of area A and length l, it is customary to define the **resistance** R of the wire as

resistance R in terms of resistivity

$$R = \rho \frac{l}{A} \qquad (27.14)$$

Thus, the resistance is directly proportional to the length and inversely proportional to the cross-sectional area. Equation (27.13) can then be expressed in the convenient form

Ohm's Law

$$I = \frac{\Delta V}{R} \qquad (27.15)$$

This equation is called **Ohm's Law**. It asserts that *the current is proportional to the potential difference between the ends of the conductor.* Ohm's law can be written in three useful forms:

$$I = \frac{\Delta V}{R} \quad \Delta V = IR \quad \text{and} \quad R = \frac{\Delta V}{I} \qquad (27.16)$$

Note that in Eq. (27.15), the resistance R plays the role of a constant of proportionality; the resistance is a quantitative measure of the opposition that the piece of metal presents to the flow of current. For a wire of uniform cross section, the resistance can be calculated from the simple formula (27.14). But Ohm's Law is also valid for conductors of arbitrary shape—such as wires of nonuniform cross section—for which the resistance must be calculated from a more complicated formula tailored to the shape and the size of the conductor (see, for example, Problem 43).

Ohm's Law is valid for metallic conductors and also for many nonmetallic conductors (for example, carbon) in which the current is carried by a flow of electrons. It is also valid for plasmas and for electrolytes, in which the current is carried by a flow

GEORG SIMON OHM (1787–1854)
German physicist. Ohm was led to his law by an analogy between the conduction of electricity and the conduction of heat.

TABLE 27.2	RESISTIVITIES AND TEMPERATURE COEFFICIENTS OF RESISTANCE OF METALS[a]	
MATERIAL	ρ	α
Silver	1.6×10^{-8} $\Omega \cdot$m	$3.8 \times 10^{-3}/°$C
Copper	1.7×10^{-8}	3.9×10^{-3}
Aluminum	2.8×10^{-8}	3.9×10^{-3}
Brass	$\approx 7 \times 10^{-8}$	2×10^{-3}
Nickel	7.8×10^{-8}	6×10^{-3}
Iron	10×10^{-8}	5×10^{-3}
Steel	$\approx 11 \times 10^{-8}$	4×10^{-3}
Constantan	49×10^{-8}	1×10^{-5}
Nichrome	100×10^{-8}	4×10^{-4}

[a] At a temperature of 20°C.

of both electrons and ions. However, we ought to keep in mind that in spite of its wide range of applicability, Ohm's Law is not a general law of nature, such as Gauss' Law. In many materials, Ohm's Law fails when the current is large; and in inhomogeneous materials, such as the layered materials used in semiconductor devices, Ohm's Law fails even when the current is small (in fact, the operation of semiconductor devices usually hinges on the exploitation of this non-ohmic behavior). Thus, Ohm's Law is not of universal applicability; it is merely an assertion about the electrical properties of many conducting materials.

EXAMPLE 3

To measure the resistance of a long piece of cable, a physicist connects this wire between the terminals of a 6.0-volt battery. She finds that this produces a current of 30 A in the wire. What is the resistance of the wire?

Concepts
— *in* —
Context

SOLUTION: According to Ohm's Law,

$$R = \frac{\Delta V}{I}$$

Since the potential difference across the wire is $\Delta V = 6.0$ volts and the current is 30 A, we obtain

$$R = \frac{6.0 \text{ V}}{30 \text{ A}} = 0.20 \text{ V/A}$$

The unit V/A is called the *ohm* (see Section 27.3); thus, the resistance is 0.20 ohm.

EXAMPLE 4

In silver, there is one free electron per atom. Calculate the value of the average collision time τ. The mass density of silver is 10.5×10^3 kg/m^3.

SOLUTION: We can solve for the collision time τ from Eq. (27.12), in terms of the quantities ρ, m_e, e, and n in that relation. From Table 27.2, we see that the resistivity of silver is $\rho = 1.6 \times 10^{-8}$ ohm·m. We know the charge e and the electron mass m_e, so we need only find the electron number density n. The periodic table (Appendix 8) indicates that the atomic mass of silver is 108 g/mole = 0.108 kg/mole. Thus there are $10.5 \times 10^3/0.108$ moles/m^3, and since each mole has 6.02×10^{23} atoms (Avogadro's number), there are $(10.5 \times 10^3/0.108) \times 6.02 \times 10^{23}$ atoms/m^3, each with one free electron. The number density of free electrons is then

$$n = \frac{1 \text{ electron/atom} \times 6.02 \times 10^{23} \text{ atoms/mole} \times 10.5 \times 10^3 \text{ kg/m}^3}{0.108 \text{ kg/mole}}$$

$$= 5.85 \times 10^{28} \text{ electrons/m}^3$$

Solving Eq. (27.12) for the average collision time, we obtain

$$\tau = \frac{m_e}{ne^2 \rho}$$

$$= \frac{9.11 \times 10^{-31} \text{ kg}}{5.85 \times 10^{28} \text{ electrons/m}^3 \times (1.60 \times 10^{-19} \text{ C})^2 \times 1.6 \times 10^{-8} \text{ ohm·m}}$$

$$= 3.8 \times 10^{-14} \text{ second}$$

Checkup 27.2

QUESTION 1: Suppose that a wire has a nonuniform cross section (thicker in some parts than in others). Is the drift velocity of the electrons the same everywhere along this wire?

QUESTION 2: Two copper wires have the same diameter, but one is 3 times as long as the other. How much larger is its resistance? Two copper wires have the same length, but one has twice the diameter of the other. How much smaller is its resistance?

QUESTION 3: When a thin wire is connected across the terminals of a 6-V battery, the current is 0.1 A. What will be the current if we connect the same wire across the terminals of a 12-V battery? An 18-V battery?

QUESTION 4: When a wire is connected across the terminals of an automobile battery, the current is 0.2 A. What will be the current if we cut this wire to half of its original length and connect such a half wire across the terminals of the battery?

QUESTION 5: If the average collision time gets longer, what happens to the resistivity? What happens to the resistivity when the density of electrons increases?

(A) Increases, increases　　(B) Increases, decreases
(C) Decreases, increases　　(D) Decreases, decreases

Online
Concept
Tutorial

27.3 RESISTIVITY OF MATERIALS

As we saw in the preceding section, the resistance R of a uniform wire is directly proportional to the resistivity ρ and the length l, and inversely proportional to the cross-sectional area A:

$$R = \rho \frac{l}{A} \qquad (27.17)$$

We can use this formula to calculate the resistance if the resistivity of the material is known, and we can also use it to calculate the resistivity if the resistance has been measured experimentally. The latter calculation is important in the experimental determination of the resistivity of a material, which is done by measuring the potential difference and current in a wire of given length and cross section made of a sample of the material.

As is obvious from Ohm's Law, Eq. (27.15), *the SI unit of resistance is 1 volt per ampere; this unit is called the* **ohm,** *abbreviated* Ω:

ohm (Ω)

$$1 \text{ ohm} = 1\ \Omega = 1\ \text{V/A} \qquad (27.18)$$

The unit of resistivity is the ohm-meter ($\Omega \cdot$m). Table 27.2 lists the resistivities of diverse conducting materials.

EXAMPLE 5　A wire commonly used for electrical installations in homes is No. 10 copper wire, which has a radius of 0.129 cm. What is the resistance of a piece of this wire 30 m long? What is the potential difference along this wire if it carries a current of 10 A?

SOLUTION: The cross-sectional area of the wire is

$$A = \pi r^2 = \pi \times (0.129 \times 10^{-2} \text{ m})^2 = 5.2 \times 10^{-6} \text{ m}^2$$

The resistivity of copper listed in Table 27.2 is $\rho = 1.7 \times 10^{-8}$ Ω·m. By Eq. (27.17), the resistance is

$$R = \rho \frac{l}{A} = 1.7 \times 10^{-8} \text{ Ω·m} \times \frac{30 \text{ m}}{5.2 \times 10^{-6} \text{ m}^2} = 0.098 \text{ Ω}$$

For a current of 10 A, Ohm's Law then gives a potential difference

$$\Delta V = IR = 10 \text{ A} \times 0.098 \text{ Ω} = 0.98 \text{ volt}$$

By rearranging Eq. (27.13) we readily find

$$\frac{I}{A} = \frac{1}{\rho} \frac{\Delta V}{l} \qquad (27.19)$$

The ratio of the current I to the cross-sectional area A is the **current density** j,

$$j = \frac{I}{A} \qquad (27.20)$$

current density j

This is the current per unit area at a point in the conductor. From Eq. (27.1), $\Delta V/l$ is the electric field E in the conductor, so Eq. (27.19) can be written

$$j = \frac{1}{\rho} E \qquad (27.21)$$

Thus, the current density is directly proportional to the electric field. This is an alternative expression for Ohm's Law. This equation relates two local quantities: the current density at one point within the conductor, and the electric field at that point.

The resistivity of materials depends on the temperature. *In ordinary metals, the resistivity increases with the temperature.* This is due to an increase of the rate of collisions between the moving electrons and the atoms of the lattice—at higher temperatures, the atoms jump more violently about their positions in the lattice, and they are then more likely to disturb the motion of the electrons, decreasing the average collision time. The numbers in Table 27.2 give the resistivities at room temperature (20°C). At very low temperatures, the collision rate and resistivity approach constant values, due to the remaining collisions with impurities in the material. Figure 27.8 shows the typical increase of resistivity with increasing temperature in a metal.

A drastic increase of the resistance with temperature is observed in ordinary incandescent lightbulbs. The resistance of the filament of a hot lightbulb (at its operating temperature) is about 10 times as large as the resistance of the filament of a cold lightbulb (at room temperature). When you switch on a cold lightbulb, there is a large initial surge of current through the filament, and this current then quickly decreases and levels off while the filament attains its operating temperature. The inital surge, due to the low "cold resistance" of the filament, is the reason filaments often burn out at the instant a light is turned on.

Concepts — *in* — Context

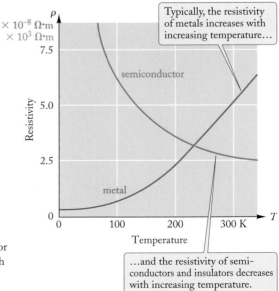

FIGURE 27.8 Resistivity as a function of temperature for a metal and for a semiconductor (silicon). The resistivity of metals increases with increasing temperature, whereas that of semiconductors decreases.

Typically, the resistivity of metals increases with increasing temperature…

…and the resistivity of semiconductors and insulators decreases with increasing temperature.

For a reasonably small increment in temperature, the increase in the resistivity and in the resistance of a metal is directly proportional to the increment in temperature. Mathematically, we can express this increase of resistance as

change of resistance with temperature

$$\Delta R = \alpha R_0 \, \Delta T \qquad (27.22)$$

where R_0 is the resistance at the initial temperature. [Note that this equation is reminiscent of the equation for the thermal expansion of the length of a solid; see Eq. (20.6).]

temperature coefficient of resistance α

The constant of proportionality α is called the **temperature coefficient of resistance**. Table 27.2 (page 866) lists values of this coefficient for some metallic conductors.

Concepts
— in —
Context

EXAMPLE 6   Suppose that because of a current overload, the temperature of the copper wire in Example 5 increases from 20°C to 50°C. How much does the resistance increase?

SOLUTION: The initial resistance is $R_0 = 0.098 \; \Omega$, from Example 5. The temperature increment is $\Delta T = 30°C$. According to Table 27.2, the temperature coefficient for copper is $\alpha = 3.9 \times 10^{-3}/°C$. The change of resistance is therefore

$$\Delta R = \alpha R_0 \, \Delta T = 3.9 \times 10^{-3}/°C \times 0.098 \; \Omega \times 30°C$$

$$= 0.012 \; \Omega$$

The new resistance of the wire will then be

$$R = R_0 + \Delta R = 0.098 \; \Omega + 0.012 \; \Omega = 0.110 \; \Omega$$

resistance thermometer

The change of electrical resistance with temperature is exploited in the operation of the **resistance thermometer**. Figure 27.9 shows such a thermometer, made of a coil of fine platinum wire. We can calibrate this thermometer in the same way as, say, a mercury thermometer by first immersing it into an ice–water mixture (0°C) and then into boiling water (100°C). Measurements of the resistance at these two temperatures tell us how many ohms correspond to 100°C, and, by extrapolation, how many ohms correspond to any other temperature.

superconductivity

At very low temperatures, the resistivity of a metal will be substantially less than at room temperature. Some metallic elements, such as lead, tin, zinc, and niobium, as well as many alloys and compounds, exhibit the phenomenon of **superconductivity**: *their resistance vanishes completely at some critical temperature above absolute zero.* For example, Fig. 27.10 displays a plot of resistivity vs. temperature for tin; at an absolute temperature of 3.72 K, the resistivity abruptly vanishes. In one experiment, a current of several hundred amperes was started in a superconducting ring; the current continued on its own with undiminished strength for over a year, without any battery or generator to maintain it.

Concepts
— in —
Context

In some copper-oxide–based superconductors, the resistance vanishes at temperatures of up to 138 K (under normal pressure) or even up to 162 K (under high pressure). Such **high-temperature superconductors** were first discovered in 1986, and are now in use in some devices, magnets, and high-current power lines. Although superconducting power cables require the added expense and complications of cryogenic cooling to operate, their benefits include a much higher current capacity than conventional cables, which is especially important in city utility tunnels, where space is at a premium.

high-temperature superconductors

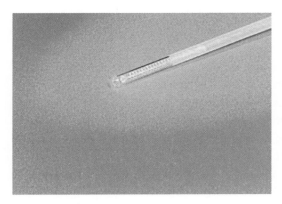

FIGURE 27.9 A coil of fine platinum wire serves as sensor in a resistance thermometer.

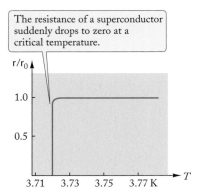

The resistance of a superconductor suddenly drops to zero at a critical temperature.

FIGURE 27.10 Resistivity of tin as a function of temperature. Below 3.72 K, the resistivity is zero. For the pupose of this plot, the resistivity has been expressed as a fraction of the resistivity at 4.2 K, the boiling point of helium.

According to the definition we gave in Section 22.5, an ideal insulator is a material that does not permit any motion of electric charge. Real **insulators**, such as porcelain or glass, do permit some very slight motion of charge. What distinguishes them from conductors is their enormously large resistivity. Typically, the resistivities of insulators are more than 10^{20} times as large as those of conductors (compare the values in Table 27.3 with those in Table 27.2). This means that even when we apply a high voltage to a piece of glass, the flow of current will be insignificant (provided, of course, that the material does not suffer electrical breakdown). In fact, on a humid day it is likely that more current will flow along the microscopic film of water that tends to form on the surface of the insulator than through the insulator itself. The resistivities of **semiconductors**, such as carbon and silicon, are between those of conductors and insulators (see Table 27.4).

In both insulators and semiconductors, the temperature coefficient of resistivity α is usually negative, opposite to that of metals. This is because heating the material tends to shake free some of the bound electrons, increasing the density of free electrons [n in Eq. (27.12)], and thus decreasing the resistivity. Values of α for semiconductors at temperatures near room temperature are given in Table 27.4. The temperature dependence of the resistivity of the semiconductor silicon is included in Fig. 27.8. Because of the rapid increase of resistivity with decreasing temperature, semiconductor resistors are often used as thermometers, particularly at low temperatures.

TABLE 27.3	RESISTIVITIES OF INSULATORS
MATERIAL	ρ
Polyethylene	2×10^{11} $\Omega\cdot$m
Glass	$\approx 10^{12}$
Porcelain, unglazed	$\approx 10^{12}$
Rubber, hard	$\approx 10^{13}$
Epoxy	$\approx 10^{15}$

TABLE 27.4	RESISTIVITIES AND TEMPERATURE COEFFICIENTS OF RESISTANCE OF SEMICONDUCTORS[a]	
MATERIAL	ρ	α
Carbon (graphite)	3.5×10^{-5} $\Omega\cdot$m	$-5 \times 10^{-4}/°C$
Silicon	2.6×10^{3}	-8×10^{-2}
Germanium	4.2×10^{-1}	-5×10^{-2}

[a]At a temperature of 20°C.

✔ Checkup 27.3

QUESTION 1: Two wires of iron and of nichrome have the same lengths and same diameters. Which has the larger resistance, and by what factor?

QUESTION 2: Two wires of iron and of silver have equal lengths but different diameters. If these wires are to be of equal resistances, what must be the ratio of their diameters?

QUESTION 3: For a temperature increase of 100°C, what is the percentage increase of resistance of a constantan wire? Of a nickel wire?

QUESTION 4: Equation (27.12), $\rho = m_e/(ne^2\tau)$, relates the resistivity to the density n and average collision time τ of free electrons. Which of these two quantities dominates the temperature dependence of the resistivity for a metal? For a semiconductor?

    (A) n, n (B) n, τ (C) τ, τ (D) τ, n

27.4 RESISTANCES IN COMBINATION

The metallic wires of any electric circuit have some resistance. But in electronic devices—CD players, televisions, computers, etc.—the main contribution to the resistance is usually due to devices that are specifically designed to have a high resistance. These devices are called **resistors**, and they are used to control and modify the currents. For instance, the manual volume controls on radios are adjustable resistors. Such a volume control is made of a long, coiled piece of high-resistance wire on which rests a sliding contact (see Fig. 27.11); by moving the sliding contact, you increase or decrease the length of wire that lies between the two external terminals, and you increase or decrease the resistance, and you thereby control the current reaching the loudspeaker. Adjustable, or variable, resistors of this kind are called **rheostats** or **potentiometers** ("pots") when used to limit current or select a voltage, respectively. The name potentiometer is also used in a more technical sense for a device used to measure potential, which incorporates a sliding contact similar to that shown in Fig. 27.11.

Resistors used in the circuits of electronic devices are often made of a piece of pure carbon (graphite) connected between two terminals (Fig. 27.12a). Carbon has a high resistivity, and hence a small piece of carbon can have a higher resistance than a long piece of metallic wire. Such resistors obey Ohm's Law (current proportional to potential difference) over a wide range of values of the current; but if the resistor is overloaded with current, it will heat up, possibly even burn, and Ohm's Law will fail.

resistor

rheostat

potentiometer

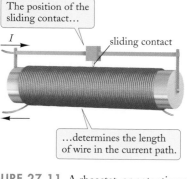

The position of the sliding contact...

I sliding contact

...determines the length of wire in the current path.

FIGURE 27.11 A rheostat, or potentiometer ("pot"), consisting of a long coiled wire with a sliding contact.

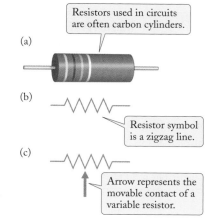

Resistors used in circuits are often carbon cylinders.

(a)

(b)

Resistor symbol is a zigzag line.

(c)

Arrow represents the movable contact of a variable resistor.

FIGURE 27.12 (a) A resistor consisting of a cylinder of carbon with two terminals attached. (b) Symbol for a resistor in a circuit diagram. (c) Symbol for a variable resistor, also known as a potentiometer.

FIGURE 27.13 Resistor color code. The first two stripes represent the first digits of the resitance value; these are multiplied by the power of 10 corresponding to the third stripe. The stripes on this resistor are yellow–violet–orange (4–7–3), corresponding to a resistance value of 47×10^3 Ω, or 47 kΩ. A fourth stripe represents the tolerance, or accuracy, of the specified value (see Table 27.5).

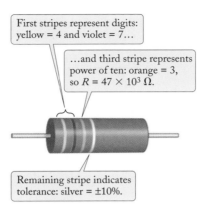

First stripes represent digits: yellow = 4 and violet = 7…

…and third stripe represents power of ten: orange = 3, so $R = 47 \times 10^3$ Ω.

Remaining stripe indicates tolerance: silver = ±10%.

TABLE 27.5		RESISTOR COLOR CODE[a]	
COLOR	**NUMBER**	**POWER**	**TOLERANCE**
Black	0	10^0	
Brown	1	10^1	
Red	2	10^2	
Orange	3	10^3	
Yellow	4	10^4	
Green	5	10^5	
Blue	6	10^6	
Violet	7	10^7	
Gray	8	10^8	
White	9	10^9	
Gold		10^{-1}	±5%
Silver		10^{-2}	±10%
None			±20%

[a]Figure 27.13 describes the use of this color code.

In circuit diagrams the symbol for a resistor is a zigzag line, reminiscent of the path of an electron inside a conducting material (see Fig. 27.12b). The potentiometer symbol is the symbol for a resistor with an arrow to represent the sliding contact (Fig. 27.12c). The value of the resistance is sometimes printed on the side of a resistor; more often, the value is encoded in a series of color stripes around the resistor. Table 27.5 gives the numerical values corresponding to the various colors, and Fig. 27.13 shows an example of a color-coded carbon resistor.

Even circuits without resistors, such as the wiring of a house, often have the main contribution to their resistance concentrated, or lumped, in one place. For instance, when an electric toaster or a lightbulb is plugged into an electric outlet, the current flowing through the circuit encounters the most resistance in the relatively short piece of wire within the toaster or the lightbulb. The short, high-resistance piece of wire in the appliance has the same effect on the flow of current as a carbon resistor, and in a circuit diagram it can be treated as though it were a resistor. The external wires of the appliance can often be treated as if they had no resistance; their resistance is negligible compared with that of the device or resistor.

For resistors, as for capacitors, the two simplest ways of connecting several resistors are in series or in parallel. Figure 27.14 shows two resistors connected in **series**. Since each of these resistors offers a resistance to the current entering the wire, our intuition suggests that the net resistance of this combination is the sum of the individual resistances,

$$R = R_1 + R_2 \tag{27.23}$$

The formal derivation of this result begins with the observation that if the potential differences across the individual resistors are ΔV_1 and ΔV_2, then the net potential difference across the combination is

$$\Delta V = \Delta V_1 + \Delta V_2 \tag{27.24}$$

This additivity of the potentials is a direct consequence of the definition of the potential as work per unit charge—the net work done by the electric field on a unit charge that moves through the first resistor and then through the second is simply the sum of the work done in the first and the work done in the second. Furthermore, *the currents in both resistors are exactly the same*, since any charge that flows through the first resistor continues to flow through the second. Hence, using Ohm's Law, $\Delta V = IR$, we find

$$\Delta V = \Delta V_1 + \Delta V_2 = IR_1 + IR_2 = I(R_1 + R_2)$$

From this it is clear that the resistance of the combination is equivalent to a single resistance $R = R_1 + R_2$, as given in Eq. (27.23).

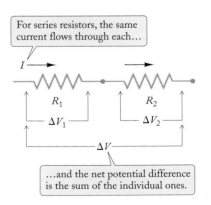

For series resistors, the same current flows through each…

…and the net potential difference is the sum of the individual ones.

FIGURE 27.14 Two resistors connected in series.

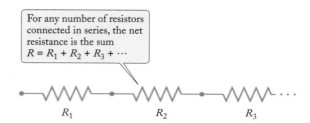

FIGURE 27.15 Several resistors connected in series.

We can easily generalize this result to any number of resistors in series (see Fig. 27.15). The net resistance, or equivalent resistance, of the series combination is

resistors in series

$$R = R_1 + R_2 + R_3 + \cdots \qquad (27.25)$$

Figure 27.16 shows two resistors in **parallel**. As with any circuit elements connected *in parallel, the potential difference across each resistor is the same* as the potential difference across the combination. Hence, from Ohm's Law, the currents are

$$I_1 = \frac{\Delta V}{R_1} \quad \text{and} \quad I_2 = \frac{\Delta V}{R_2} \qquad (27.26)$$

Resistors in parallel are analogous to the several parallel lanes of a highway. The total flow of automobiles is the sum of the flows in the individual lanes; likewise, the total flow of charge through a combination of parallel resistors is the sum of the flows in the individual resistors. Thus, the total current through the combination is the sum of the individual parallel currents,

$$I = I_1 + I_2 = \frac{\Delta V}{R_1} + \frac{\Delta V}{R_2} = \left(\frac{1}{R_1} + \frac{1}{R_2}\right)\Delta V \qquad (27.27)$$

Comparison of Eq. (27.27) with Ohm's law, $I = \Delta V/R$, shows that the resistance of the combination is therefore equivalent to a single resistance R given by

$$\frac{1}{R} = \frac{1}{R_1} + \frac{1}{R_2} \qquad (27.28)$$

Note that the resistance of the parallel combination is less than each of the individual resistances. For example, if $R_1 = R_2 = 1.0\ \Omega$, then $R = 0.5\ \Omega$.

We can also generalize this to any number of resistors in parallel (see Fig. 27.17). The net resistance R of the parallel combination is given by

resistors in parallel

$$\frac{1}{R} = \frac{1}{R_1} + \frac{1}{R_2} + \frac{1}{R_3} + \cdots \qquad (27.29)$$

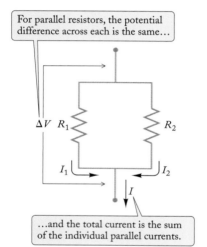

FIGURE 27.16 Two resistors connected in parallel.

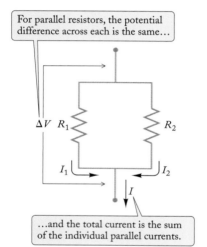

FIGURE 27.17 Several resistors connected in parallel.

EXAMPLE 7   Two resistors, with $R_1 = 10\ \Omega$ and $R_2 = 20\ \Omega$, are connected in series (see Fig. 27.18), and a current of 1.8 A flows through this combination. What is the potential difference across the combination? What is the potential difference across each resistor?

SOLUTION: According to Eq. (27.25), the net resistance is given by

$$R = R_1 + R_2 = 10\ \Omega + 20\ \Omega = 30\ \Omega$$

Ohm's Law applied to this combined system of two resistors tells us that the potential difference is

$$\Delta V = IR = 1.8\ \text{A} \times 30\ \Omega = 54\ \text{volts}$$

Furthermore, Ohm's Law applied to each individual resistor tells us that the individual potential differences are

$$\Delta V_1 = IR_1 = 1.8\ \text{A} \times 10\ \Omega = 18\ \text{volts}$$

$$\Delta V_2 = IR_2 = 1.8\ \text{A} \times 20\ \Omega = 36\ \text{volts}$$

As expected, the sum of these two potential differences is 54 volts, which is the potential difference across the combination. The arrangement of Fig. 27.18 is known as a **voltage divider**, since it serves to divide the total potential difference between the two series-connected resistors in proportion to their resistance.

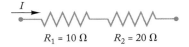

FIGURE 27.18 A current flows through two resistors connected in series.

voltage divider

EXAMPLE 8   Suppose that the same two resistors, with $R_1 = 10\ \Omega$ and $R_2 = 20\ \Omega$, are connected in parallel (see Fig. 27.19) and a net current of 1.8 A flows through this combination. What is the potential difference across the combination? What is the current in each resistor?

SOLUTION: According to Eq. (27.29), the net resistance is given by

$$\frac{1}{R} = \frac{1}{R_1} + \frac{1}{R_2} = \frac{1}{10\ \Omega} + \frac{1}{20\ \Omega} = \frac{30}{200\ \Omega}$$

which yields $R = 200\ \Omega/30 = 6.7\ \Omega$. Hence, Ohm's Law applied to the combination as a whole tells us that the potential difference is

$$\Delta V = IR = 1.8\ \text{A} \times 6.7\ \Omega = 12\ \text{volts}$$

and Ohm's Law applied to the two branches tells us that the individual currents are

$$I_1 = \frac{\Delta V}{R_1} = \frac{12\ \text{volts}}{10\ \Omega} = 1.2\ \text{A}$$

$$I_2 = \frac{\Delta V}{R_2} = \frac{12\ \text{volts}}{20\ \Omega} = 0.60\ \text{A}$$

As expected, the sum of the two currents is 1.8 A, which is the net current through the combination.

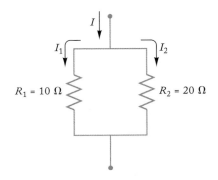

FIGURE 27.19 A current flows through the same two resistors connected in parallel.

In the calculations of the preceding example, we neglected the resistance of the wires connecting the resistors. This is a good approximation if the resistance of these wires is small compared with the resistance of the resistors. If this is not so, then we must take the resistance of the wires into account in our calculation. In circuit diagrams, it

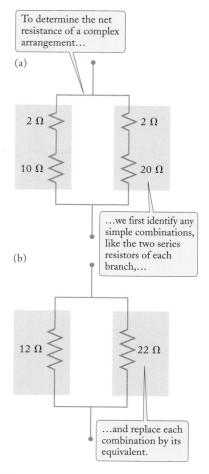

To determine the net resistance of a complex arrangement…

(a)

2 Ω   2 Ω

10 Ω   20 Ω

…we first identify any simple combinations, like the two series resistors of each branch,…

(b)

12 Ω   22 Ω

…and replace each combination by its equivalent.

FIGURE 27.20 (a) In this circuit diagram, the resistances of the wires are represented by the two small resistors. (b) The right (green) and the left (orange) branches of the circuit are equivalent to two resistors of 12 Ω and 22 Ω connected in parallel.

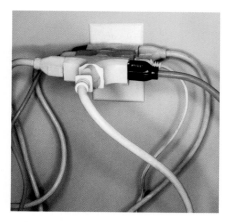

FIGURE 27.21 Several appliances plugged into the power outlets in a house.

is customary to represent the resistance of a wire schematically by an equivalent resistor of suitable magnitude; this means we pretend that all of the resistance of a wire is concentrated, or lumped, in one place. The line segments connecting these equivalent resistors to the rest of the circuit can then be assumed to have zero resistance exactly. For instance, if each of the two wires connecting the resistors in Fig. 27.19 is long and each has a resistance of 2 Ω, then the schematic circuit diagram that includes the resistance of the wires to the connections is as shown in Fig. 27.20. Note that in this diagram, we placed the resistors of 2 Ω that represent the resistances of the wires immediately above each of the resistors R_1 and R_2. We could equally well place these equivalent resistors immediately below each of R_1 and R_2, or one half of each above and one half below—all these arrangements are effectively equivalent, and lead to the same distribution of current among the two branches of the circuit. Also note that we neglected the resistance of the short leads sticking out of the circuit above and below; if these have a significant resistance, then we would also need to take these into account.

EXAMPLE 9   If, as in the preceding example, a current of 1.8 A flows through the combination of resistors and resistive wires shown schematically in Fig. 27.20a, what is the potential difference across the entire combination? What is the current in each resistor?

SOLUTION: To solve this problem we proceed in two steps. First, we find the resistances of the left and the right branches of the circuit. The left branch consists of a resistor R_1 of 10 Ω in series with 2 Ω; the right branch consists of a resistor R_2 of 20 Ω in series with 2 Ω. Hence, the resistance of the left branch of the circuit is $R_1' = R_1 + 2\ \Omega = 12\ \Omega$ and that of the right branch is $R_2' = R_2 + 2\ \Omega = 22\ \Omega$.

Next, we find the net resistance of the circuit by combining R_1' and R_2' in parallel (Fig. 27.20b). This parallel combination of the resistances of the left and the right branches gives

$$\frac{1}{R} = \frac{1}{R_1'} + \frac{1}{R_2'} = \frac{1}{12\ \Omega} + \frac{1}{22\ \Omega} = \frac{34}{264\ \Omega}$$

From this, we find $R = 264\ \Omega/34 = 7.8\ \Omega$ for the net resistance of the circuit. Hence, by Ohm's Law, the potential difference across the combination is

$$\Delta V = IR = 1.8\ \text{A} \times 7.8\ \Omega = 14\ \text{volts}$$

and the individual currents in the branches are

$$I_1 = \frac{\Delta V}{R_1'} = \frac{14\ \text{volts}}{12\ \Omega} = 1.2\ \text{A}$$

$$I_2 = \frac{\Delta V}{R_2'} = \frac{14\ \text{volts}}{22\ \Omega} = 0.64\ \text{A}$$

Ordinary appliances plugged into the power outlets in your home are connected in parallel (see Fig. 27.21). The potential difference across each of these appliances is 115 V, which is the potential difference supplied across the wires that feed electric power into the house.[1] Some of these appliances—such as lightbulbs, electric toasters,

---

[1] The potential difference supplied across the power wires is actually an oscillating potential difference, which periodically reverses sign. We will deal with the details of such oscillating, or alternating, potentials and currents in Chapter 32.

irons, hot plates, heaters, and blankets—simply consist of a piece of resistive wire in which the dissipation of the electric energy produces heat. More complicated appliances—such as refrigerators, fans, washers, radios, TVs, camcorders, CD players—may contain electric motors and a variety of electronic devices. Such complicated devices usually do not obey Ohm's Law; for instance, if you plug your TV into a 120-V outlet instead of a 115-V outlet, the current will not increase in proportion to the potential. However, for purposes of comparison, it is often useful to assign to a complicated device an effective resistance $R = \Delta V/I$, as though Ohm's Law were valid. Since this effective resistance of a complicated device is *not* constant, it can be used only in calculations at the given, fixed potential difference of, say, 115 V.

EXAMPLE 10   A vacuum cleaner, a hair dryer, and an electric iron are simultaneously plugged into a single outlet (see Fig. 27.22a). Their resistances are 9.0 Ω, 10 Ω, and 12 Ω, respectively. The outlet supplies a potential difference of 115 V. What is the net current flowing through the outlet? If the maximum safe current that the outlet can carry is 30 A, is it advisable to plug all of these appliances into the outlet?

Concepts
— in —
Context

SOLUTION: All the appliances are connected in parallel (see Fig. 27.22b); thus the potential difference across each is 115 V. By Ohm's Law, $I = \Delta V/R$, the individual currents are then, respectively,

$$I_1 = 115 \text{ V}/9.0 \text{ Ω} = 12.8 \text{ A}$$

$$I_2 = 115 \text{ V}/10 \text{ Ω} = 11.5 \text{ A}$$

$$I_3 = 115 \text{ V}/12 \text{ Ω} = 9.6 \text{ A}$$

The net current is the sum of the individual parallel currents, 12.8 A + 11.5 A + 9.6 A = 34 A. This is more than the outlet can carry safely. These devices should not be plugged into the same outlet, nor into several outlets supplied by the same wire.

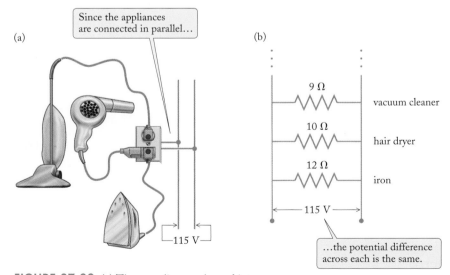

FIGURE 27.22 (a) Three appliances plugged into the same outlet. (b) Schematic diagram showing how the resistances are connected.

PROBLEM-SOLVING TECHNIQUES | COMBINATIONS OF RESISTORS

When seeking the net resistance of a circuit containing several resistors connected in some complicated manner, proceed in steps, as in the case of a circuit with capacitors.

1 In the first step, look for groups of resistors that form simple parallel or series combinations (such as the group of two resistors in the left branch and the group of two resistors in the right branch in Fig. 27.20a). Evaluate the resistance of each such group.

2 In the second step, see how these effective group resistances are connected to each other (as in Fig. 27.20b), and evaluate the net resistance of the combination of the groups.

3 In some cases, you may have to repeat these steps until you obtain a single equivalent resistance.

 Checkup 27.4

QUESTION 1: A wire has a resistance of 3 Ω. If we cut this wire in half and connect the two pieces in parallel, what will be the new resistance?

QUESTION 2: Qualitatively, why is the net resistance of a series combination of equal resistors larger than the individual resistances? Why is the net resistance of a parallel combination of equal resistors smaller than the individual resistances?

QUESTION 3: A wire of copper and a wire of silver are connected in parallel to a battery. Both wires have the same lengths and the same diameters. Which carries more current?

QUESTION 4: If you connect 10 resistors of 1 Ω in series, what is the net resistance? What if you connect them in parallel?

(A) 10 Ω, 10 Ω (B) 0.1 Ω, 10 Ω (C) 10 Ω, 0.1 Ω (D) 0.1 Ω, 0.1 Ω

SUMMARY

PROBLEM-SOLVING TECHNIQUES   Combinations of Resistors   **(page 878)**

ELECTRIC FIELD IN UNIFORM WIRE OF LENGTH l	$E = \dfrac{\Delta V}{l}$	**(27.1)**
ELECTRIC CURRENT (FLOW OF CHARGE PER UNIT TIME)	$I = \dfrac{\Delta Q}{\Delta t}$ or $I = \dfrac{dq}{dt}$	**(27.2, 3)**
SI UNIT OF ELECTRIC CURRENT	1 ampere = 1 A = 1 C/s	**(27.4)**
DRIFT VELOCITY (AVERAGE VELOCITY) IN TERMS OF THE AVERAGE COLLISION TIME τ	$v_d = \dfrac{-eE\tau}{m_e}$	**(27.8)**
CURRENT OF FREE ELECTRONS IN A CONDUCTOR where n is the number density of free electrons, v_d is the drift velocity, and A is the cross-sectional area.	$I = -nev_dA$	**(27.10)**
SI UNIT OF RESISTANCE	1 ohm = 1 Ω = 1 V/A	**(27.18)**

RESISTIVITY IN TERMS OF THE AVERAGE COLLISION TIME τ	$\rho = \dfrac{m_e}{ne^2\tau}$	(27.12)

RESISTANCE IN TERMS OF RESISTIVITY	$R = \rho\dfrac{l}{A}$	(27.14)

OHM'S LAW	$I = \dfrac{\Delta V}{R} \quad \Delta V = IR \quad R = \dfrac{\Delta V}{I}$	(27.16)

CURRENT DENSITY	$j = \dfrac{I}{A}$	(27.20)

OHM'S LAW AND CURRENT DENSITY	$j = \dfrac{1}{\rho}E$	(27.21)

CHANGE OF RESISTANCE WITH TEMPERATURE
where α is the temperature coefficient of resistivity (Table 27.2).

$$\Delta R = \alpha R_0\,\Delta T \qquad (27.22)$$

SERIES COMBINATION OF RESISTORS
For resistors connected in series, the same current flows through each, and the net potential difference is the sum of the individual potential differences.

$$R = R_1 + R_2 + R_3 + \cdots \qquad (27.25)$$

$R_1 \qquad R_2 \qquad R_3$

PARALLEL COMBINATION OF RESISTORS
For resistors connected in parallel, the total current is the sum of the individual parallel currents, and the potential difference across each is the same.

$$\frac{1}{R} = \frac{1}{R_1} + \frac{1}{R_2} + \frac{1}{R_3} + \cdots \qquad (27.29)$$

$R_1 \quad R_2 \quad R_3$

QUESTIONS FOR DISCUSSION

1. A wire is carrying a current of 15 A. Is this wire in electrostatic equilibrium?

2. Can a current flow in a conductor when there is no electric field? Can an electric field exist in a conductor when there is no current?

3. Figure 27.23 shows a putative mechanical analog of an electric conductor with resistance. A marble rolling down the inclined board is stopped every so often by a collision with a pin and therefore maintains a constant average velocity v_d. Is this a good analog, that is, is the average velocity v_d proportional to g? (Hint: Assume that the marble rolls an average distance d between collisions.)

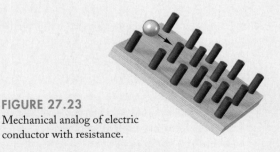

FIGURE 27.23
Mechanical analog of electric conductor with resistance.

4. By what factor must we increase the diameter of a wire to decrease its resistance by a factor of 2?

5. Show that, for a wire of given length made of a given material, the resistance is inversely proportional to the mass of the wire.

6. What deviations from Ohm's Law do you expect if the current is very large?

7. Ohm's Law is an approximate statement about the electrical properties of a conducting body, just as Hooke's Law is an approximate statement about the mechanical properties of an elastic body. Is there an electric analog of the elastic limit?

8. An automobile battery has a potential difference of 12 V between its terminals even when there is no current flowing through the battery. Does this violate Ohm's Law?

9. Figure 27.24 shows two alternative experimental arrangements for determining the resistance of a carbon cylinder. A known potential difference is applied via the copper terminals, the current is measured, and the resistance is calculated from $R = \Delta V / I$. The arrangement in Fig. 27.24a yields a higher resistance than that in Fig. 27.24b. Why?

(a)

(b)

FIGURE 27.24 (a) The wires touch the carbon cylinder directly. (b) The wires end in contact plates which touch the carbon cylinder.

10. Why is it bad practice to operate a high-current appliance off an extension cord?

11. The installation instructions for connecting an outlet to the wiring of a house recommend that the wire be wrapped at least three-quarters of the way around the terminal post (Fig. 27.25). Explain.

FIGURE 27.25 Terminal post of an outlet.

12. The temperature coefficient of resistivity α is defined as the fractional increase of resistivity per degree Celsius, $\alpha = (1/\rho)d\rho/dT$. According to Table 27.2, what is the value of α for copper?

13. Figure 27.9 shows a platinum resistance thermometer consisting of a coil of fine platinum wire inside a glass tube that can be put in thermal contact with a body. How would you measure the resistance of this wire to determine the temperature of the body?

14. Aluminum wire should never be connected to terminals designed for copper wire. Why not? (Hint: Aluminum has a considerably higher coefficient of thermal expansion than copper.)

15. Two copper wires are connected in parallel. The wires have the same length, but one has twice the diameter of the other. What fraction of the total current flows in each wire?

PROBLEMS

†27.1 Electric Current
27.2 Resistance and Ohm's Law

1. The electric current in the lightbulb of a flashlight is 0.50 A. How much electric charge flows through the lightbulb in one hour? How many electrons pass through the lightbulb?

2. In a typical lightning stroke, the electric current is about 20 000 A and it lasts about 1.0×10^{-4} s. The direction of the current is upward, from the ground to the cloud. What is the charge (magnitude and sign) that this stroke deposits on the ground?

3. A 40-μF capacitor is initially charged with a 9.0-V battery. To reverse the voltage on the capacitor, how long must a constant 3.0-A current flow from the positive to the negative plate of the capacitor?

4. What is the capacitance of a capacitor that charges to 1.4 V in 0.50 μs by a constant current of 25 mA?

5. A conducting wire of length 2.0 m is connected between the terminals of a 12-V battery. The resistance of the wire is 3.0 Ω. What is the electric current in the wire? What is the electric field in the wire?

6. When a thin copper wire is connected between the poles of a 1.5-V battery, the current in the wire is 0.50 A. What is the resistance of this wire? What will be the current in the wire if it is connected between the terminals of a 7.5-V battery?

7. A current begins to flow at $t = 0$ and increases with time according to $I(t) = At + Bt^2$, where $A = 0.50$ C/s^2 and $B = 0.20$ C/s^3. What is the current at $t = 5.0$ s? What total charge has flowed by $t = 5.0$ s?

†For help, see Online Concept Tutorial 31 at www.wwnorton.com/physics

8. At $t = 0$, a current begins to flow around a simple circuit. The total charge that has flowed clockwise around the circuit is given by $Q(t) = 5.0t - 2.0t^2$, where Q is in coulombs and t is in seconds. What total charge has flowed clockwise around the circuit at $t = 1.0$ s? At $t = 2.0$ s? At $t = 3.0$ s? What current is flowing at each of these times? Comment on the meaning of the algebraic sign of your results at each time.

9. A current begins flowing into an initially uncharged capacitor at $t = 0$ and decreases to zero during $0 \leq t \leq 2.0$ s with a time dependence given by $I = A \times (t - 2.0 \text{ s})^2$, where $A = 0.25 \text{ C/s}^2$. What is the initial current? The current at 1.0 s? How much charge is on the capacitor when the current stops at $t = 2.0$ s?

10. The collision time for electrons in the metal sodium (atomic symbol Na) is 8.8×10^{-15} s at room temperature. Assuming one free electron per atom, calculate the electrical resistivity of sodium.

11. In silver, the number density of free electrons is $n = 5.8 \times 10^{28}$ per cubic meter.

    (a) Using the value of the resistivity given in Table 27.2, calculate the average time between collisions of one of these electrons.

    (b) Assuming that the electric field in a current-carrying silver wire is 8.0 V/m, calculate the average drift velocity of an electron.

12. The **mean free path** l of an electron is the average distance traveled between collisions, $l = v\tau$. Due to quantum effects, the speed v of an electron in a metal is much greater than the thermal speed of an ideal-gas molecule given by Eq. (19.23); for silver, the actual speed is 12 times greater than the thermal speed. Assuming one free electron per atom, find the mean free path in silver at 20°C.

13. Constantan wire has a high resistivity and is often used as a heater. What is the resistance of a 30-cm length of constantan wire with a diameter of 0.25 mm?

14. Photodiodes supply a current proportional to the power of the light incident on them. The minimum current provided by a low-noise photodiode is 4.0×10^{-14} A. If this current passes though a 1.0-MΩ resistor, what is the voltage difference across that resistor?

15. The resistance of a 150-W, 115-V lightbulb is 88 Ω when the lightbulb is at its operating temperature. What current passes through this lightbulb when in operation? How many electrons per second does this amount to?

16. The resistance of the wire in the windings of an electric starter motor for an automobile is 3.0×10^{-2} Ω. The motor is connected to a 12-V battery. What current will flow through the motor when it is stalled (does not turn)?

17. A long, thin wire of resistance R is cut into eight pieces. Four of these pieces are then placed side by side to form a new wire $\frac{1}{8}$ of the original length. What is the resistance of the new wire?

*18. A circular loop of superconducting material has a radius of 2.0 cm. It carries a current of 4.0 A. What is the orbital angular momentum of the moving electrons in the wire? Take the center of the loop as origin.

*19. An aluminum wire has a resistance of 0.10 Ω. If you draw this wire through a die, making it thinner and twice as long, what will be its new resistance?

*20. A table lamp is connected to an electric outlet by a copper wire of diameter 0.20 cm and length 2.0 m. Assume that the current through the lamp is 1.5 A, and that this current is steady. How long does it take an electron to travel from the outlet to the lamp?

*21. When the starter motor of an automobile is in operation, the cable connecting it to the battery carries a current of 80 A. This cable is made of copper and is 0.50 cm in diameter. What is the electric field in the cable?

*22. A copper cable in a high-voltage transmission line has a diameter of 3.0 cm and carries a current of 750 A. What is the electric field in the wire?

27.3 Resistivity of Materials

23. The electromagnet of a bell is constructed by winding copper wire around a cylindrical core, like thread on a spool. The diameter of the copper wire is 0.45 mm, the number of turns in the winding is 260, and the average radius of a turn is 5.0 mm. What is the resistance of the wire?

24. The following is a list of some types of copper wire manufactured in the United States (Fig. 27.26):

GAUGE NO.	DIAMETER
8	0.3264 cm
9	0.2906
10	0.2588
11	0.2305
12	0.2053

For each type, calculate the resistance for a 100-m segment.

FIGURE 27.26 Copper wire.

25. To measure the resistivity of a metal, an experimenter takes a wire of this metal of diameter 0.500 mm and length 1.10 m and applies a potential difference of 12.0 V to the ends. He finds that the resulting current is 3.75 A. What is the resistivity?

26. A high-voltage transmission line has an aluminum cable of diameter 3.0 cm, 200 km long. What is the resistance of this cable?

27. Consider the aluminum cable described in Problem 26. If the temperature of this cable increases from 20°C to 50°C, how much will its resistance increase?

28. What increase of temperature will increase the resistance of a nickel wire from 0.50 Ω to 0.60 Ω?

29. What is the current density in a 0.259-cm-diameter copper wire carrying a current of 30 A? What is the electric field in the wire?

30. In a scanning tunneling microscope, a small, picoampere current, 1.0×10^{-12} A, flows across an atomic-sized region, roughly 0.30 nm in diameter. What is the current density?

31. Equation (27.21), the alternative form of Ohm's Law, can also be written $j = \sigma E$, where σ is called the **electrical conductivity**. What is the electrical conductivity of copper?

32. Commercial platinum resistance thermometers are often manufactured with a resistance of 100.0 Ω at 0.0°C. The temperature coefficient of resistivity is 3.9×10^{-3}/°C. If the resistance of such a thermometer reads 109.8 Ω, what is the temperature?

33. A piece of silicon has a resistance of 12.0 Ω at a temperature of 20°C. What is the temperature when its resistance is 10.0 Ω?

*34. You want to make a resistor of 1.0 Ω out of carbon rod of diameter 1.0 mm. How long a piece of carbon do you need?

*35. A lightning rod of iron has a diameter of 0.80 cm and a length of 0.50 m. During a lightning stroke, it carries a current of 1.0×10^4 A. What is the potential drop along the rod?

*36. The air conditioner in a home draws current of 12 A.

(a) Suppose that the pair of wires connecting the air conditioner to the fuse box are No. 10 copper wire with a diameter of 0.259 cm and a length of 25 m each. What is the potential drop along each wire? Suppose that the voltage delivered to the home is exactly 115 V at the fuse box. What is the voltage delivered to the air conditioner?

(b) Some older homes are wired with No. 12 copper wire with a diameter of 0.205 cm. Repeat the calculation of part (a) for this wire.

*37. Although aluminum has a somewhat higher resistivity than copper, it has the advantage of having a considerably lower density. Find the mass of a 100-m segment of aluminum cable 3.0 cm in diameter. Compare this with that of a copper cable of the same length and the same resistance. The densities of aluminum and of copper are 2.7×10^3 kg/m^3 and 8.9×10^3 kg/m^3, respectively.

*38. According to the National Electrical Code, the maximum permissible current in a No. 12 copper wire (diameter 0.205 cm) with rubber insulation is 25 A.

(a) What is the potential drop along a 1.0-meter segment of the wire carrying this current?

(b) What is the electric field in the wire?

*39. An aluminum wire of length 15 m is to carry a current of 25 A with a potential drop of no more than 5.0 V along its length. What is the minimum acceptable diameter of this cable?

*40. According to safety standards set by the American Boat and Yacht Council, the potential drop along a copper wire connecting a 12-V battery to an item of electrical equipment should not exceed 10%, that is, it should not exceed 1.2 V. Suppose that a 9.0-m wire (length measured around the circuit) carries a current of 25 A. What gauge of wire is required for compliance with the above standard? Use the table of wire gauges given in Problem 24. Repeat the calculation for currents of 35 A and 45 A.

*41. A parallel-plate capacitor with a plate area of 8.0×10^{-2} m^2 and a plate separation of 1.0×10^{-4} m is filled with polyethylene. If the potential difference between the plates is 2.0×10^4 V, what will be the current flowing through the polyethylene from one plate to the other?

*42. The windings of high-current electromagnets are often made of copper pipe. The current flows in the walls of the pipe, and cooling water flows in the interior of the pipe. Suppose the copper pipe has an outside diameter of 1.20 cm and an inside diameter of 0.80 cm. What is the resistance of 30 m of this copper pipe? What voltage must be applied to it if the current is to be 600 A?

*43. A 10.0-m length of coaxial cable with a solid inner conductor of diameter 1.00 mm and an outer conducting shell of inner diameter 4.00 mm has a dielectric of polyethylene between the conductors (see Fig. 26.19). What is the resistance of the dielectric cylindrical shell for current flow from one conductor to the other? What current flows when a potential difference of 300 V is applied between the conductors? (Hint: Treat the dielectric as a series combination of thin cylindrical shells connected radially.)

27.4 Resistances in Combination

44. Three resistors of 4.0 Ω, 6.0 Ω, and 8.0 Ω, respectively, are connected in series. What is the resistance of this combination? If the combination is connected to a 12-V battery, what is the current? What is the potential difference across each resistor?

45. Three resistors of 5.0 Ω, 7.0 Ω, and 9.0 Ω, respectively, are connected in parallel. What is the resistance of this combination? If this combination is connected to a 12-V battery, what is the net current? What is the current in each resistor?

46. A brass wire and an iron wire of equal diameters and of equal lengths are connected in parallel. Together they carry a current of 6.0 A. What is the current in each?

47. Consider the brass and iron wires described in Problem 46. What is the current in each if they are at a temperature of 90°C instead of the temperature of 20°C assumed in Table 27.2?

48. Two resistors are connected in parallel; their color codes are brown–red–orange and yellow–violet–red (see Fig. 27.13). What would be the color code of a single, equivalent resistor?

49. You have three resistors: 2.0 Ω, 3.0 Ω, and 4.0 Ω. What values of resistance can be obtained from combinations of some or all of these resistors?

50. When two resistors are connected in series, the net resistance is 80 Ω. When they are connected in parallel, the net resistance is 15 Ω. What are the values of the two resistors?

51. During operation, the effective resistances of a toaster, a microwave oven, and an electric frying pan are 11 Ω, 16 Ω, and 12 Ω, respectively. If the three are all plugged into a 115-volt outlet, what is the total current?

*52. An ordinary lightbulb draws 0.87 A of current when connected to a 115-V outlet. What is the resistance of the bulb? When a series arrangement of two such bulbs is connected to the 115-V outlet, the current through the combination is 0.69 A. What is the resistance of each bulb now? Why did it change?

*53. An electric cable of length 12.0 m consists of a copper wire of diameter 0.30 cm surrounded by a cylindrical layer of rubber insulation of thickness 0.10 cm. A potential difference of 6.0 V is applied to the ends of the cable.

   (a) What will be the current in the copper?

   (b) Taking into account the finite resistivity of the rubber (see Table 27.3), what will be the current in the rubber?

*54. A copper wire, of length 0.50 m and diameter 0.259 cm, has been accidentally cut by a saw. The region of the cut is 0.40 cm long, and in this region the remaining wire has a cross-sectional area of only one-quarter of the original area. What is the percentage increase of the resistance of the wire caused by this cut?

*55. An underground telephone cable, consisting of a pair of wires, has suffered a short somewhere along its length (Fig. 27.27). The telephone cable is 5.0 km long, and in order to discover where the short is, a technician first measures the resistance across terminals AB; then he measures the resistance across terminals CD. The first measurement yields 30 Ω; the second, 70 Ω. Where is the short?

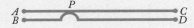

FIGURE 27.27 A pair of wires with a short at the point P (the wires are touching at P).

*56. The air of the atmosphere has a slight conductivity due to the presence of a few free electrons and positive ions.

   (a) Near the surface of the Earth, the vertical atmospheric electric field has a strength of about 100 V/m and the atmospheric current density is 4.0×10^{-12} A/m². What is the resistivity?

   (b) The potential difference between the ionosphere (upper layer of atmosphere) and the surface of the Earth is 4.0×10^5 V. What is the total resistance of the atmosphere? (Hint: For the purposes of this problem assume that the vertical electric field has a constant strength throughout the atmsphere.)

*57. A flexible wire for an extension cord for electric appliances is made of 24 strands of fine copper wire, each of diameter 0.053 cm, tightly twisted together. What is the resistance of a length of 1.0 m of this kind of wire?

*58. Two copper wires of diameters 0.26 cm and 0.21 cm, respectively, are connected in parallel. What is the current in each if the combined current is 18 A?

*59. Commercially manufactured superconducting cables consist of filaments of superconducting wire embedded in a matrix of copper (see Fig. 27.28). As long as the filaments are superconducting, all the current flows in them, and no current flows in the copper. But if superconductivity suddenly fails because of a temperature increase, the current can spill into the copper; this prevents damage to the filaments of the superconductor. Calculate the resistance per meter of length of the copper matrix shown in Fig. 27.28. The diameter of the copper matrix is 0.70 mm, and each of the 896 filaments has a diameter of 0.010 mm.

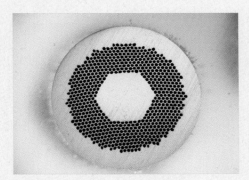

FIGURE 27.28 A cross-section of a strand of a superconducting cable, consisting of 896 filaments of a niobium–titanium superconductor embedded in a copper matrix.

*60. What is the net resistance of the combination of four resistors shown in Fig. 27.29? Each of the resistors has a resistance of 3.0 Ω.

FIGURE 27.29 Four resistors.

*61. Three resistors with $R_1 = 2.0$ Ω, $R_2 = 4.0$ Ω, and $R_3 = 6.0$ Ω are connected as shown in Fig. 27.30.

   (a) Find the net resistance of the combination.

   (b) Find the current that passes through the combination if a potential difference of 8.0 V is applied to the terminals.

   (c) Find the potential drop and the current for each individual resistor.

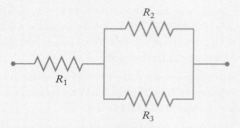

FIGURE 27.30 Three resistors.

*62. Three resistors with $R_1 = 4.0\ \Omega$, $R_2 = 6.0\ \Omega$, and $R_3 = 8.0\ \Omega$ are connected as shown in Fig. 27.31.

   (a) Find the net resistance of the combination.

   (b) Find the current that passes through the combination if a potential difference of 12.0 V is applied to the terminals.

   (c) Find the potential drop and the current for each individual resistor.

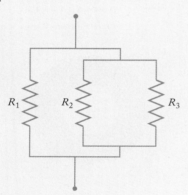

FIGURE 27.31 Three resistors.

*63. Consider the combination of three resistors described in the preceding problem. If we want a current of 6.0 A to flow through resistor R_2, what potential difference must we apply to the external terminals?

**64. Twelve resistors, each of resistance R, are connected along the edges of a cube (Fig. 27.32). What is the resistance between diagonally opposite corners of this cube? (Hint: Calculate $R_{\text{net}} = \Delta V/I_{\text{total}}$, where ΔV sums the three voltage drops along any path between opposite corners. Use symmetry to determine each current in terms of I_{total}.)

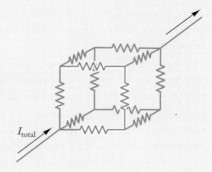

FIGURE 27.32 A cube with a resistor on each edge.

**65. What is the resistance of an infinite ladder of 1.0-Ω resistors connected as shown in Fig. 27.33a? (Hint: The ladder can be regarded as made of two pieces connected in parallel; see Fig. 27.33b.)

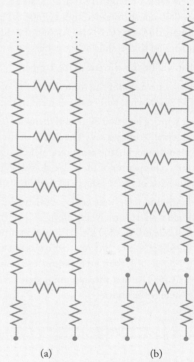

(a)   (b)

FIGURE 27.33 (a) An infinite "ladder" of resistors. The terminals are marked by the pair of dots. (b) One rung has been cut off from the ladder.

REVIEW PROBLEMS

66. Three resistors, with resistances of 3.0 Ω, 5.0 Ω, and 8.0 Ω, are connected in parallel. If this combination is connected to a 12.0-V battery, what is the current through each resistor? What is the current through the combination?

67. A fully charged automobile battery delivers 40 A for 1.0 h before it runs down. How much electric charge flows through the battery during this time? How many electrons pass through the battery?

68. A high-voltage transmission line consists of a copper cable of diameter 3.0 cm and length 250 km. Assume the cable carries a steady current of 1500 A.

    (a) What is the resistance of the cable?
    (b) What is the electric field inside the cable?

69. The filament of a lightbulb consists of a piece of tungsten wire, 5.0 cm long. If the potential difference across this filament is 115 V, what is the electric field in the wire?

70. Two wires of silver and of copper have the same lengths and the same diameters. If they are to have the same resistance, by how many degrees must the silver wire be warmer than the copper wire?

71. The resistivity of a certain material is found to increase by 4.45% when the material is heated from 20.0°C to 70.0°C. By what percentage will the resistivity of the material increase at 120.0°C compared with 20.0°C?

72. The electromagnet of a bell is wound with 8.2 m of copper wire of diameter 0.45 mm. What is the resistance of the wire? What is the current through the wire if the electromagnet is connected to a 12-V source?

73. The copper cable connecting the positive pole of a 12-V automobile battery to the starter motor is 0.60 m long and 0.50 cm in diameter.

    (a) What is the resistance of this cable?

    (b) When the starter motor is stalled, the current in the cable may be as much as 600 A. What is the potential drop along the cable under these conditions?

74. Ten equal strands of thin wire are held together side by side to form a thick wire of resistance R. Six of these strands are then separated and tied together end to end to form a new wire six times as long as the original wire. What is the resistance of the long, thin wire?

75. How can you connect four 1.0-Ω resistors so the net resistance of the combination is 1.0 Ω?

76. The resistance of the filament of the lightbulb of a flashlight is 8.0 Ω (at its operating temperature). The two batteries of the flashlight, of 1.5 V each, are connected in series, and they therefore supply 3.0 V. What is the current through the filament? What will be the current if you rewire the flashlight so it operates on a single battery? Assume that the resistance of the filament remains the same (actually, with less current, the filament will be cooler, and its resistance will be lower).

77. When connected in parallel across a 12-V battery, two resistors carry currents of 0.50 A and 0.75 A, respectively. What currents will these resistors carry if they are connected in series with the same battery?

78. A water pipe is made of iron with an outside diameter of 2.5 cm and an inside diameter of 2.0 cm. The pipe is used to ground an electric appliance. If a current of 20 A flows from the appliance into the water pipe, what fraction of this current will flow in the iron? What fraction in the water? Assume that water has a resistivity of 0.010 $\Omega \cdot$m.

79. An arc-welder uses No. 2 gauge copper wire (diameter 0.654 cm); during use, the cable carries 110 A. What is the current density? What is the drift velocity of the electrons?

80. Two resistors, 3.0 Ω and 5.0 Ω, are connected in series. The current though the 5.0 Ω resistor is 0.75 A. What is the potential difference across the series combination?

*81. The resistance of a square centimeter of dry human epidermis is about 1.0×10^5 Ω. Suppose that a (foolish) man firmly grasps two wires in his fists. The wires have a radius of 0.13 cm, and the skin of each hand is in full contact with the surface of the wire over a length of 8.0 cm (see Fig. 27.34).

    (a) Calculate the resistance the man offers to a current flowing through his body from one wire to the other. In this calculation you can neglect the resistance of the internal tissues of the human body, because the body fluids are reasonably good conductors and their resistance is small compared with the skin resistance.

    (b) What current will flow through the body of the man if the potential difference between the wires is 12 V? If it is 115 V? If it is 240 kV? What is your prognosis in each case? (A current above 0.001 A can cause injury, and a current above 0.02 A can be fatal; see Section 28.8.)

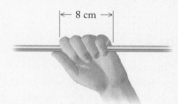

FIGURE 27.34 Contact between hand and wire.

*82. Three resistors, with $R_1 = 4.0$ Ω, $R_2 = 6.0$ Ω, and $R_3 = 2.0$ Ω, are connected as shown in Fig. 27.35. This combination is connected to a 1.5-V battery.

    (a) What is the net resistance of this combination?

    (b) What is the current through the combination?

    (c) What are the potential drop and the current for each resistor?

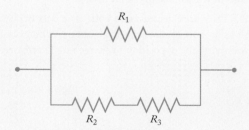

FIGURE 27.35 Three resistors.

Answers to Checkups

Checkup 27.1

1. When we cut a wire, the electrons do not continue to flow, nor do they spill out. Cutting a wire is analogous to closing off a water pipe; since air is not a conductor, the charges remain confined to the wire. Since the current is the same everywhere along a wire, cutting the wire stops the flow everywhere.

2. Yes—current is the flow past a given place; whether the cross-sectional area at that point is small or large does not matter. Of course, where the wire is thicker, that current is spread over a larger cross-sectional area.

3. Since current is defined as the direction of the flow of positive charge, an upwardly flowing current must have been pushed by the force of an upward electric field. Since the current in a lightning bolt is actually a flow of electrons, the motion of the (negative) electrons is downward, which is equivalent to an upward flow of positive charge.

4. (B) (a); (c). In (a), the positive charges move to the left and the negative to the right, so both contribute to a conventional current (the flow of positive charge) to the left. The same is true in (b), but there are fewer charges, so the current in (a) is larger. In (c) equal numbers of positive and negative charges move to the right, so there is zero net current.

Checkup 27.2

1. No. Where the wire is thinner, fewer electrons must move faster to maintain the same total current. One can also see from Eq. (27.11) that where the area is smaller, the electric field must be larger; thus the drift velocity, proportional to the electric field [Eq. (27.8)], is also larger.

2. From $R = \rho l/A$, we see that tripling the length triples the resistance. Also, doubling the diameter increases the area fourfold, and so decreases the resistance by a factor of 4.

3. From Ohm's Law, for a given resistance (for a given piece of wire), the current and voltage are proportional. So the 12-V battery results in 0.2 A of current, and the 18-V battery, 0.3 A.

4. The resistance of the half-length will be half as much as originally, so Ohm's Law tells us that for the same voltage, the current will double to 0.4 A.

5. (D) Decreases; decreases. In both cases (less frequent collisions or more electrons), charge can move more easily, so the resistivity (the opposition to charge flow) decreases.

Checkup 27.3

1. Since the geometrical factors are the same, the resistance will be proportional to the resistivity, by Eq. (27.17). From Table 27.2, the resistivity of nichrome ($100 \times 10^{-8}\ \Omega\cdot m$) exceeds that of iron ($10 \times 10^{-8}\ \Omega\cdot m$) by a factor of 10.

2. For equal lengths and equal resistances, the ratio of areas must be the same as the ratio of resistivities, according to Eq. (27.17). The ratio of the diameter of the iron wire to that of the silver one varies as the square root of the ratio of areas, and so is equal to $(\rho_{iron}/\rho_{silver})^{1/2} = (10/1.6)^{1/2} = (100/16)^{1/2} = 10/4 = 2.5$.

3. The percentage increase is 100 times the fractional increase of $\Delta R/R_0 = \alpha\,\Delta T$. For $\Delta T = 100°C$, the values of α in Table 27.2 give 0.1% for constantan and 60% for nickel.

4. (D) τ; n. In metals, the free-electron density n is essentially constant; it is the collision time τ that decreases with increasing temperature due to the increasing thermal motion of lattice atoms. When increasing the temperature of a semiconductor, a similar decrease in τ is overwhelmed by a much larger increase in n, because heating shakes many bound electrons free.

Checkup 27.4

1. Cutting the wire into two equal lengths results in two resistances, each with half the original resistance. Connecting two equal resistors in parallel again halves the resistance, resulting in a new resistance one-quarter of the original, or $\frac{3}{4}\ \Omega$.

2. One easily pictured reason is that resistances in series are analogous to a long wire—the longer the wire, the higher the resistance. Resistances in parallel are analogous to a thicker wire—the thicker the wire, the lower the resistance.

3. Since the potential difference (from the battery), the lengths, and the areas are all the same, the wire with the lower resistivity will have more current (using $I = \Delta V/R$ and $R = \rho l/A$). From Table 27.2, silver has a (slightly) lower resistivity than copper, and so will carry more current.

4. (C) 10 Ω; 0.1 Ω. In series, resistances add, so ten 1-Ω resistors provide a net resistance of 10 Ω. Resistors in parallel add such that the inverse of the net resistance R is the sum of the inverse resistances, so $R = 1/[10 \times 1/(1\ \Omega)] = 0.1\ \Omega$.

Direct Current Circuits

CONCEPTS IN CONTEXT

Concepts
in
Context

Batteries such as these power our portable tools, appliances, and electronics. We can often represent a device connected to a battery as a resistor.

In this chapter, we will consider circuits with one or more batteries and one or more resistors. We can ask:

? How much energy can a typical battery supply? (Example 1, page 889)

? How does a battery operate? (Section 28.2, page 890)

? In a circuit with several batteries and resistors, how do we determine the current through each resistor? (Example 2, page 894; and Example 5, page 898)

? For such a circuit, what is the power supplied by each battery? What is the power dissipated in each resistor? (Example 6, page 902; and Example 7, page 904)

The electric circuits installed in automobiles, homes, and factories carry one or the other of two different kinds of currents: **direct currents (DC)** and **alternating currents (AC)**. The direct current flows steadily along the wires of the circuit; it remains constant, except when it is switched on or off. The alternating current periodically reverses its direction of flow along the wire; like a pendulum swinging back and forth, the alternating current oscillates sinusoidally from one direction (positive) to the opposite direction (negative).

To keep any current flowing in the wires of the circuit, we must connect the ends of the wire to a "pump of electricity," a device that continuously supplies electric charges to one end of the wire and removes them from the other. What kind of current flows in the wire depends on what kind of pump we use. A steady pump—such as an automobile battery—produces a steady current. An alternating pump—such as the generator of the power station to which the circuits in your home are connected—produces an alternating current. In this chapter we will deal with direct currents produced by batteries or by other pumps of electricity that behave like batteries.

Online
Concept
Tutorial

28.1 ELECTROMOTIVE FORCE

Figure 28.1 shows a simple circuit consisting of a single (resistive) wire connected to the terminals of a battery. A steady, time-independent current will then flow around this circuit; in Fig. 28.1, we have indicated the direction of the current according to our convention that it is the direction of flow of (hypothetical) positive charges. The current flowing in this circuit is an example of a direct current—as long as the "strength" of the battery and the resistance of the wire remain constant, the current will also remain constant.

The battery must do work on the charges in order to keep them flowing around the circuit. Suppose that a (hypothetical) positive charge is originally at the point P, at one terminal of the battery. Pushed along by the electric field, the charge moves along the wire. On the average, the kinetic energy that the charge gains from the electric field is dissipated by friction due to collisions within the wire, and the charge reaches the point P', at the other terminal of the battery, with no more kinetic energy than it had originally. Thus, on the average, the kinetic energy does not change. But the potential energy of the charge does change. The electric field is directed along the wire, and it does work on the charge; hence *the electric potential steadily decreases with distance along the wire*, and the charge reaches the point P' with a potential energy lower than its original potential energy. In order to keep the current flowing, the battery must "pump" the charge from the low-potential terminal to the high-potential terminal; that is, the battery must supply electric potential energy to the charge.

Our simple electric circuit is analogous to the hydraulic circuit illustrated in Fig. 28.2, consisting of a channel in which water runs down a hill and a pump that lifts the water from the bottom to the top of the hill. The water then flows around a closed hydraulic circuit, just as charge flows around the closed electric circuit. The wire is analogous to the channel in which the water runs down the hill, losing gravitational potential energy. The battery is analogous to the water pump that lifts water from the bottom to the top of the hill, thereby increasing the gravitational potential energy of the water. The water pump of Fig. 28.2 can be regarded as a source of gravitational potential energy—it produces this energy from an external supply of chemical or mechanical energy. Likewise, the battery, or "pump of electricity," in Fig. 28.1 can be regarded as a source of electric potential energy—it produces this energy from a supply of chemical energy.

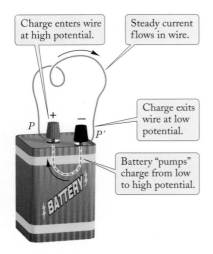

FIGURE 28.1 A resistive wire connected between the terminals of a battery. The potential energy of a (positive) charge is high when it is at the positive terminal P of the battery. The potential energy gradually decreases as the charge moves along the wire toward the negative terminal P'. The potential energy then again increases as the charge passes through the battery (dashed line), from P' to P.

To characterize the "strength" of a source of electric potential energy, we introduce the concept of **electromotive force**, or **emf.** The emf of a source of electric potential energy is defined as *the amount of electric energy delivered by the source per coulomb of positive charge as this charge passes through the source from the low-potential terminal to the high-potential terminal.* Since the emf is energy per unit charge, its units are volts. Note that the electromotive "force" is not a force, but an energy per coulomb—the confusing name became attached to it a long time ago, when physicists were not yet making a sharp distinction between force and energy. Because the units of emf are volts, the emf is often simply called the **voltage** of the source.

If a steady, time-independent current carries one coulomb of charge around the circuit of Fig. 28.1 from P to P' along the wire and from P' to P through the source of emf, then the energy that this charge receives from the source of emf must exactly match the energy it loses within the wire. If so, the charge returns to its starting point with exactly the same energy it had originally, and it can repeat this round trip again and again, in exactly the same manner. We can write this energy balance as

$$\mathcal{E} + \Delta V = 0 \tag{28.1}$$

where $\mathcal{E}$ represents the emf, or the increase of potential energy per coulomb of charge, due to the source and ΔV represents the change of potential energy along the wire (here, $\mathcal{E}$ is positive and ΔV is negative; see Fig. 28.3).

According to Eq. (28.1), the emf $\mathcal{E}$ has the same magnitude as the potential drop in the external circuit connected between the terminals of the source of emf. For example, a battery with an emf of 1.5 V (or 1.5 J/C) connected to an external circuit will do 1.5 J of work on a coulomb of positive charge that passes through the battery in the forward direction (from the − terminal to the + terminal), and the resistors and other devices in the external circuit will do −1.5 J of work on the charge as it flows around this circuit (from the + terminal to the − terminal).

electromotive force (emf)

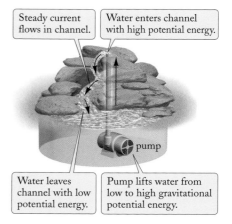

Steady current flows in channel.

Water enters channel with high potential energy.

pump

Water leaves channel with low potential energy.

Pump lifts water from low to high gravitational potential energy.

FIGURE 28.2 Mechanical analog of the battery–wire circuit of Fig. 28.1. The potential energy of a parcel of water is high at the top of the hill; it gradually decreases as the water flows down the hill; and it again increases as the pump lifts the water to the top.

EXAMPLE 1   A fresh flashlight battery with a voltage of 1.5 V will deliver a current of 1.0 A for about 1.0 h before running down completely. How much work does the battery do in this time interval?

SOLUTION: A 1.5-V battery does 1.5 J of work on each coulomb that passes through. If the current is 1.0 A, the charge that passes through in 1.0 h is 1.0 A × 3600 s = 3600 C. The total work is then the work per coulomb multiplied by the number of coulombs, that is, 1.5 J/C × 3600 C = 5400 J.

Concepts
— *in* —
Context

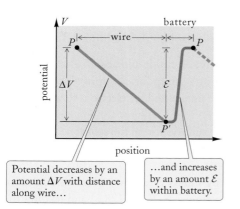

Potential decreases by an amount ΔV with distance along wire…

…and increases by an amount $\mathcal{E}$ within battery.

FIGURE 28.3 Plot of the electric potential, or the potential energy per coulomb of charge, as a function of position along the wire for the circuit illustrated in Fig. 28.1. The potential is highest at the point P (the positive pole of the battery), lowest at the point P' (the negative pole).

Note that if one coulomb of positive charge is forced through the battery in the *reverse* direction (from the + terminal to the − terminal), then the charge will deliver electric potential energy to the battery. The charge will then emerge from the battery at a potential that is 1.5 volts lower than the potential with which it entered. The energy delivered by the charge either will be stored within the battery (if it is a reversible, or rechargeable, battery) or else will merely be wasted as heat within the battery (if it is not a rechargeable battery).

 Checkup 28.1

QUESTION 1: When a charge moves around a circuit such as shown in Fig. 28.1, the battery delivers energy to the charge. What happens to this energy?

QUESTION 2: The battery delivers positive energy to the current that flows around the circuit as illustrated in Fig. 28.1. Since this current actually consists of a flow of negative electrons, will the battery deliver negative energy to the electrons?

QUESTION 3: Suppose that we change the current through the battery in Example 1 to 2.0 A. As it runs down completely, the battery will then deliver:

   (A) The same energy   (B) Less energy   (C) More energy

Online
Concept
Tutorial

28.2 SOURCES OF ELECTROMOTIVE FORCE

The most important kinds of sources of emf are batteries, electric generators, fuel cells, and solar cells. We will now briefly discuss the physical principles underlying the operation of some of these.

Concepts
— *in* —
Context

Batteries

lead–acid battery

Batteries convert chemical energy into electric energy. A very common type of battery is the **lead–acid battery**, which finds widespread use in automobiles. In its simplest form, this battery consists of two plates of lead—the positive electrode and the negative electrode—immersed in a solution of sulfuric acid (see Fig. 28.4). The positive electrode is covered with a layer of lead dioxide, PbO_2. When the external circuit is closed, the sulfuric acid reacts with the immersed surfaces of the electrodes. As already mentioned in Section 22.4, the reactions that occur at the negative and the positive electrodes are, respectively,

$$Pb + SO_4^{2-} \rightarrow PbSO_4 + 2e^- \qquad (28.2)$$

$$PbO_2 + 4H^+ + SO_4^{2-} + 2e^- \rightarrow PbSO_4 + 2H_2O \qquad (28.3)$$

These reactions deposit electrons on the negative electrode and absorb electrons from the positive electrode. Thus, the battery acts as a pump for electrons—the negative electrode is the outlet, the positive electrode the intake, and the electrons flow from one to the other via the external circuit.

The reactions (28.2) and (28.3) deplete the sulfuric acid in the solution and deposit lead sulfate on the electrodes. The depletion of sulfuric acid finally halts the reaction—the battery is then "discharged."

The lead–acid battery can be "charged" by simply forcing a current through it in the backward direction. This reverses the reactions (28.2) and (28.3) and restores the

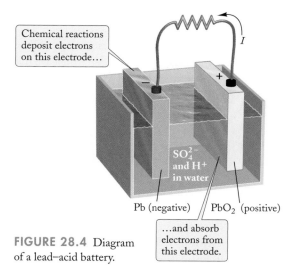

Chemical reactions deposit electrons on this electrode…

SO_4^{2-} and H^+ in water

Pb (negative) PbO₂ (positive)

…and absorb electrons from this electrode.

FIGURE 28.4 Diagram of a lead–acid battery.

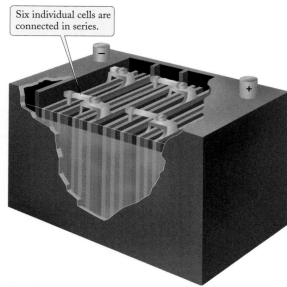

Six individual cells are connected in series.

FIGURE 28.5 An automobile battery.

sulfuric acid solution. Note that what is stored in the battery during the charging process is not electric charge, but chemical energy. The number of positive and negative electric charges (protons and electrons) in the battery remains constant; what changes is the concentration of chemical compounds. A charged battery contains chemical compounds (lead, lead dioxide, sulfuric acid) of relatively high internal energy; a discharged battery contains chemical compounds (lead sulfate, water) of lower internal energy. Charging a battery is analogous to winding up the spring of a clock or pumping water into a reservoir on a high hill—in all these cases we are storing energy, which we can subsequently release upon demand.

In the single-cell battery shown in Fig. 28.4, the reactions (28.2) and (28.3) generate an emf of 2.0 V. In an automobile battery (see Fig. 28.5) six such cells are stacked together and connected in series to give an emf of 12.0 V (historically, the word *battery* originated from such stacking of single cells). The energy stored in such a battery is typically about 2×10^6 J.

Another familiar type of battery is the **dry cell**, or flashlight battery. In an alkaline dry cell, the positive electrode consists of a carbon rod and the negative electrode of a cylinder of powdered zinc. The electrolyte in which these electrodes are "immersed" is a moist paste of potassium hydroxide and manganese dioxide (see Fig. 28.6). When this battery is connected to an external circuit, the chemical reactions at the electrodes convert chemical energy into electric energy and pump electrons from one electrode to the other via the external circuit. The emf of such a dry cell is 1.5 V. Since there is no liquid to slosh around, these batteries are particularly suitable for portable devices. The energy stored in a flashlight battery is typically of the order of 5000 J.

dry cell

(a)

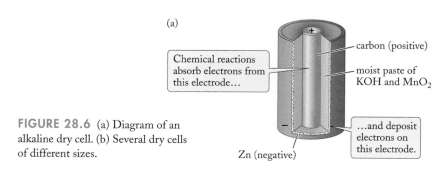

Chemical reactions absorb electrons from this electrode…

carbon (positive)

moist paste of KOH and MnO₂

…and deposit electrons on this electrode.

Zn (negative)

(b)

FIGURE 28.6 (a) Diagram of an alkaline dry cell. (b) Several dry cells of different sizes.

Electric Generators

Generators convert mechanical energy (kinetic energy) into electric energy. Their operation involves magnetic fields and the phenomenon of induction. We will leave the description of electric generators for Sections 31.2 and 31.3.

Fuel Cells

Fuel Cells resemble batteries in that they convert chemical energy into electric energy. However, in contrast to a battery, neither the high-energy chemicals nor the low-energy reaction products are stored inside the fuel cell. The former are supplied to the fuel cell from external tanks, and the latter are ejected. Essentially, the fuel cell acts as a combustion chamber in which a controlled chemical reaction takes place. The fuel cell "burns" a high-energy fuel, but produces electric energy rather than heat energy.

Figure 28.7a shows a fuel cell that burns a hydrogen–oxygen fuel. The electrodes of the fuel cell are hollow cylinders of porous carbon; oxygen at high pressure is pumped into the positive electrode and hydrogen into the negative electrode. The electrodes are immersed in a potassium hydroxide electrolyte. The reactions at the negative and positive electrode are, respectively,

$$2H_2 + 4OH^- \rightarrow 4H_2O + 4e^- \qquad (28.4)$$

$$O_2 + 2H_2O + 4e^- \rightarrow 4OH^- \qquad (28.5)$$

These reactions deposit electrons on the negative electrode and remove electrons from the positive electrode. This pumps electrons from one electrode to the other via the external circuit.

Note that the net result of the sequence of reactions (28.4) and (28.5) is the conversion of oxygen and hydrogen into water. This reaction is the reverse of the electrolysis of water (decomposition of water by an electric current). The excess water is removed from the cell in the form of water vapor.

All fuel cells produce a certain amount of waste heat. The best available fuel cells convert about 45% of the chemical energy of the fuel into electric energy and waste the remainder. Fuel cells have been put to practical use as power sources aboard the Apollo spacecraft and on Skylab (see Fig. 28.7b), and they have also been installed as power sources in experimental models of automobiles (Fig. 28.7c). They are compact and clean; on Skylab, the waste water eliminated from the fuel cell was used both for drinking and for washing.

(a)

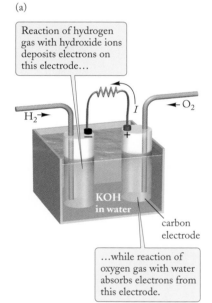

Reaction of hydrogen gas with hydroxide ions deposits electrons on this electrode…

…while reaction of oxygen gas with water absorbs electrons from this electrode.

(b)

(c)

FIGURE 28.7 (a) Diagram of a fuel cell. (b) Fuel cell used on Skylab. (c) Automobile using fuel cells.

Solar Cells

Solar cells convert the energy of sunlight directly into electric energy. They are made of thin wafers of a semiconductor, such as silicon. We will discuss semiconductors and their applications in Chapter 39.

 ## Checkup 28.2

QUESTION 1: Does a charged battery have more electric charge than a discharged battery?

QUESTION 2: A battery stores energy. In what form is this energy stored?

QUESTION 3: The fuel of a fuel cell stores energy. In what form is this energy stored?

QUESTION 4: A battery is being used to charge a capacitor, and during this charging the battery delivers energy to the capacitor. What is the form of this energy while it is still in the battery? What is the form of this energy when it is stored in the capacitor?

QUESTION 5: Two identical single-cell batteries are stacked in series; another two are connected in parallel. The two arrangements have a different:

   (A) Amount of stored energy   (B) Electromotive force   (C) Form of stored energy

28.3 SINGLE-LOOP CIRCUITS

Online
Concept
Tutorial

The electric circuits in automobiles and in battery-operated tools or appliances, such as electric drills, flashlights, laptop computers, and portable telephones, contain one or several batteries or other sources of emf connected by wires to lightbulbs, electric motors, display screens, etc. In schematic diagrams of electric circuits, the latter devices can be represented by their resistances. Hence the schematic circuit diagram consists of one or several sources of emf connected to one or several resistors (see Fig. 28.8). In such a diagram, any source with a time-independent emf is represented by a stack of parallel short and long lines suggesting the plates of a lead–acid battery. The high-potential terminal is represented by a long thin line (marked with a plus sign), and the low-potential terminal by a short thick line (marked with a minus sign). If the terminals of such a source are connected to a network with resistances, a steady, direct current,

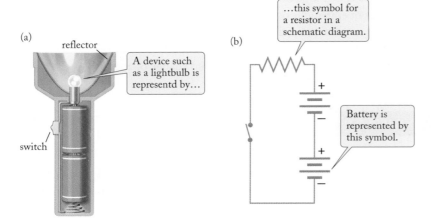

FIGURE 28.8 (a) A flashlight containing a lightbulb connected in series to two batteries. (b) Schematic circuit diagram for the flashlight. Each of the two batteries is represented by a stack of parallel short and long lines with plus and minus signs.

or DC, will flow through the network. In this section and the next we will learn how to calculate the currents that flow through the different parts of a circuit. We begin with simple circuits, consisting of a single closed loop. In a single-loop circuit, there is only one pathway, so *the same current flows past any point on a single loop.*

Figure 28.9 shows a schematic circuit diagram for such a single-loop circuit, consisting of a source of emf, such as a battery, connected to a resistor. The emf of the battery is $\mathcal{E}$ and the resistance of the resistor is R. The wires from the resistor to the battery are assumed to have negligible resistance (if higher accuracy is required, the resistance of the wires must be included in the circuit diagram). The resistance R in Fig. 28.9 could equally well represent a carbon resistor or some other device, such as a light-bulb, endowed with an electrical resistance; the resistance R could even represent the resistance of a wire by itself connected between the terminals of the battery. To find the current that flows through the circuit, we note that, according to Ohm's Law, the potential change across the resistor must be

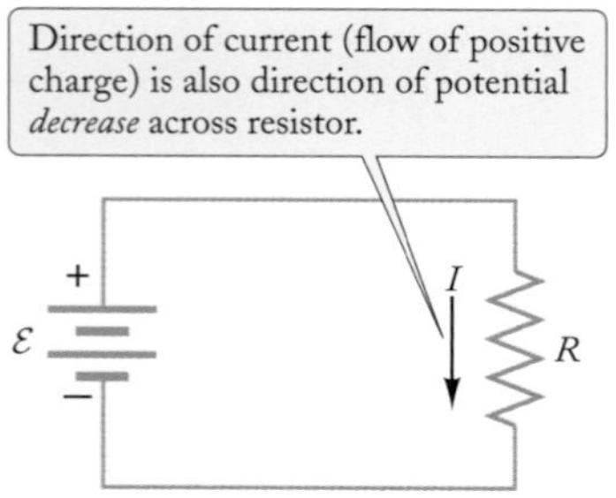

$$\Delta V = -IR \tag{28.6}$$

Here the potential change has been reckoned in the direction of the arrow (see Fig. 28.9), from the upper end to the lower end of the resistor. The negative sign on the right side of Eq. (28.6) means that for a positive charge that moves around the circuit in the direction of the arrow, the potential decreases across the resistor. According to Eq. (28.1), the emf plus the potential change must equal zero; hence

$$\mathcal{E} - IR = 0 \tag{28.7}$$

from which

$$I = \frac{\mathcal{E}}{R} \tag{28.8}$$

Equation (28.7) is an instance of **Kirchhoff's voltage rule**, which states that when we go *around any closed loop in a circuit, the sum of all the emfs and all the potential changes across resistors and other circuit elements must equal zero.* In this sum, the emf must be considered positive (gain of potential energy) whenever we go through a source of emf in the forward direction, from the − terminal to the + terminal; and negative (loss of potential energy) whenever we go through a source in the backward direction, from the + terminal to the − terminal. Similarly, the potential change across a resistor is reckoned as negative (−IR, a voltage drop) when we go through the resistor in the same direction as the current, and as positive (+IR, a voltage rise) when we go through the resistor in the direction opposite to the current.

The proof of Kirchhoff's voltage rule is similar to the proof of Eq. (28.1). If one coulomb of positive charge flows once around a closed loop in a circuit with one or several sources of emf and resistors, it will gain or lose potential energy while passing through each source of emf and lose potential energy while passing through each resistor. Under steady conditons, the sum of gains and losses must equal zero, since the charge must return to its starting point with no change of energy.

Kirchhoff's voltage rule

Direction of current (flow of positive charge) is also direction of potential *decrease* across resistor.

FIGURE 28.9 A simple circuit with a source of emf and a resistor.

Concepts — in — Context

EXAMPLE 2

Figure 28.10a shows a circuit with two batteries and two resistors. The emfs of the batteries are $\mathcal{E}_1 = 12.0$ V and $\mathcal{E}_2 = 15.0$ V; the resistances are $R_1 = 4.0\ \Omega$ and $R_2 = 2.0\ \Omega$. What is the current in the circuit?

SOLUTION: To apply Kirchhoff's voltage rule, we must decide in which direction the current flows around the loop. We will arbitrarily assume that the current flows in the clockwise direction (Fig. 28.10a). If this hypothesis is wrong, our calculation of the current, while still correct, will yield a negative value, and this will indicate that the direction of the actual current is opposite to the hypothetical direction.

According to Kirchhoff's voltage rule, the sum of all emfs and all potential changes across resistors is zero. To apply Kirchhoff's voltage rule, we can start at any point on a loop and go around the loop in either direction. If we choose to go around the loop in a clockwise direction, starting and ending at the lower left corner (Fig. 28.10b), we have

$$\mathcal{E}_1 - IR_1 - \mathcal{E}_2 - IR_2 = 0 \tag{28.9}$$

Note that $\mathcal{E}_2$ enters with a negative sign into this equation, since our clockwise path passes through this source of emf in the backward direction. Also, both resistive potential changes are here reckoned as negative, since our clockwise path is in both cases parallel to the assumed current direction. Solving Eq. (28.9) for the current I provides the solution

$$I = \frac{\mathcal{E}_1 - \mathcal{E}_2}{R_1 + R_2} \tag{28.10}$$

or

$$I = \frac{12.0 \text{ V} - 15.0 \text{ V}}{4.0 \ \Omega + 2.0 \ \Omega} = -0.50 \text{ A} \tag{28.11}$$

Here, the negative sign indicates that the current is *not* clockwise as originally chosen, but counterclockwise instead.

COMMENTS: We could have guessed the direction of the current, since it is obvious that the stronger battery on the right will force the current backward through the weaker battery on the left. But in more complicated circuits the direction of the current will not be so obvious, and we will have to discover the direction by carefully keeping track of the signs in our calculations.

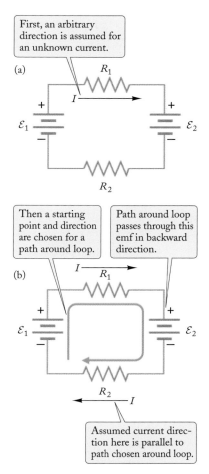

First, an arbitrary direction is assumed for an unknown current.

Then a starting point and direction are chosen for a path around loop.

Path around loop passes through this emf in backward direction.

Assumed current direction here is parallel to path chosen around loop.

FIGURE 28.10 (a) Two sources of emf and two resistors. (b) Same sources of emf and resistors, with a path chosen for the application of Kirchhoff's voltage rule.

In Example 2 we neglected the **internal resistance** of the batteries. The electrolyte in a battery always has some resistance, and this causes the current to suffer a voltage drop even before it leaves the external terminals of the battery. The nominal emf $\mathcal{E}$ quoted on the labels of batteries refers to the potential difference between the terminals when no current is flowing; this is often called the "open-circuit" voltage. The internal resistance R_i of the battery may be regarded as connected in series with the emf $\mathcal{E}$ (Fig. 28.11). When a current I is flowing, the voltage drops by $\Delta V = -IR_i$ across the internal resistance, and hence the remaining voltage at the external terminals of the battery will be $\mathcal{E} - IR_i$. The internal resistance of a good battery is small, and can often be neglected. But if we need to take it into account, we can do so by simply placing the appropriate internal resistance in series with each battery in the circuit diagram; then we can proceed as usual with the calculation of the currents.

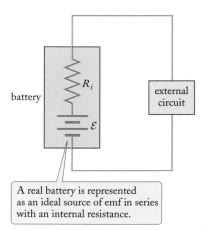

A real battery is represented as an ideal source of emf in series with an internal resistance.

FIGURE 28.11 The internal resistance R_i is in series with the nominal emf $\mathcal{E}$.

(a)

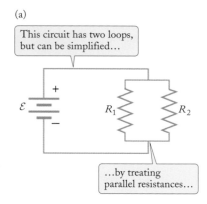

(b)

FIGURE 28.12 (a) A circuit containing two resistances in parallel. (b) Equivalent circuit with one resistance.

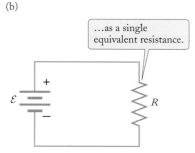

(a)

(b)

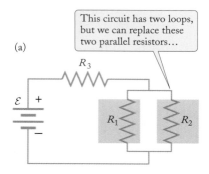

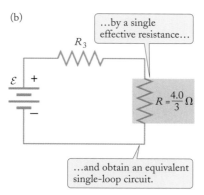

FIGURE 28.13 (a) A circuit containing two resistances in parallel (blue). (b) The two parallel resistances are equivalent to a single resistance of $R = 4.0/3\ \Omega$.

EXAMPLE 3 An alkaline flashlight battery of nominal emf 1.5 V has an internal resistance of 0.12 Ω. What will be the potential difference across its terminals if the battery is delivering a current of 1 A? What if the battery is delivering a current of 5 A?

SOLUTION: As discussed above, the potential difference available at the terminals will be the emf of the source minus the internal resistive voltage drop. For a current of 1 A, the voltage across the terminals will be

$$\mathcal{E} - IR_i = 1.5\ \text{V} - 1\ \text{A} \times 0.12\ \Omega = 1.5\ \text{V} - 0.1\ \text{V} = 1.4\ \text{V}$$

For a current of 5 A, the voltage across the terminals will be

$$\mathcal{E} - IR_i = 1.5\ \text{V} - 5\ \text{A} \times 0.12\ \Omega = 1.5\ \text{V} - 0.6\ \text{V} = 0.9\ \text{V}$$

A circuit that contains two or more resistors in parallel, such as the circuit shown in Fig. 28.12, is not a genuine single-loop circuit, since the parallel resistors and their connecting wires form additional loops. The general method for dealing with multi-loop circuits will be discussed in the next section. However, circuits with parallel resistors, such as the circuit in Fig. 28.12, can be handled by the same method as single-loop circuits because, as we know from the preceding chapter, the parallel resistances are effectively equivalent to a single resistance [see Eq. (27.29)].

EXAMPLE 4 Suppose that, in the circuit shown in Fig. 28.13a, the emf of the battery is 12 V and the resistances are $R_1 = 4.0\ \Omega$, $R_2 = 2.0\ \Omega$, and $R_3 = 3.0\ \Omega$. What is the current through the battery in this circuit?

SOLUTION: The current through the source flows through R_3 and branches between R_1 and R_2. The net resistance of the parallel combination is given by Eq. (27.29),

$$\frac{1}{R} = \frac{1}{R_1} + \frac{1}{R_2} = \frac{1}{2.0\ \Omega} + \frac{1}{4.0\ \Omega} = \frac{3}{4.0\ \Omega}$$

or

$$R = \frac{4.0}{3}\ \Omega = 1.3\ \Omega$$

The circuit shown in Fig. 28.13a is therefore effectively equivalent to the single-loop circuit shown in Fig. 28.13b. Kirchhoff's rule for this single-loop circuit with resistances of $R = 1.3\ \Omega$ and $R_3 = 3.0\ \Omega$ in series with an emf of 12 V tells us

$$\mathcal{E} - IR - IR_3 = 0$$

We can now solve for the unknown current I:

$$I = \frac{\mathcal{E}}{R + R_3} = \frac{12\ \text{V}}{1.3\ \Omega + 3.0\ \Omega} = 2.8\ \text{A}$$

COMMENTS: The currents in the individual parallel resistors can be found from this total current in the same way as in Example 8 of Chapter 27. Namely, the potential across each parallel resistor is $\Delta V = IR = 2.8\ \text{A} \times 4/3\ \Omega = 3.7$ volts, so the individual currents are $I_1 = \Delta V/R_1 = 3.7$ volts$/4.0\ \Omega = 0.9$ A and $I_2 = \Delta V/R_2 = 3.7$ volts$/2.0\ \Omega = 1.9$ A.

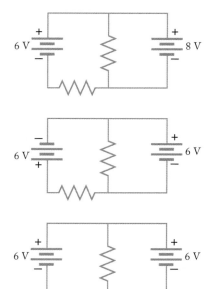

 Checkup 28.3

QUESTION 1: A circuit contains several batteries and several resistors. If we replace all the batteries by new batteries of twice the emf, how will the current change? If we next replace all the resistors by new resistors of twice the resistance?

QUESTION 2: Suppose that in a circuit we reverse all the batteries, that is, we turn them end for end. What happens to the current? What if we reverse all the resistors, that is, turn them end for end?

QUESTION 3: Figure 28.14 shows several circuits containing batteries. In each case, what are the directions of the electric currents?

QUESTION 4: Suppose that a battery with emf $\mathcal{E}$ has a somewhat large internal resistance, $R_i \approx 2\ \Omega$. If we connect a thick piece of wire with a resistance $R \approx 0.1\ \Omega$ between the terminals of the battery, the voltage at the terminals of the battery is

(A) Exactly equal to $\mathcal{E}$ (B) Slightly less than $\mathcal{E}$ (C) Much less than $\mathcal{E}$

FIGURE 28.14 Several circuits with batteries.

28.4 MULTI-LOOP CIRCUITS

Online *Concept* Tutorial

If several sources of emf and several resistors are connected in some complicated circuit with several branches, then the currents will flow along several alternative paths. Kirchhoff's voltage rule can be applied to any loop in the complicated circuit. In doing so, however, we must be able to identify the separate currents in each separate branch. If there is only one branch in some part of a circuit, the current is the same past every point. But when the current encounters a junction, it will split along two or more branches, as illustrated in Fig. 28.15. We can relate the separate currents using **Kirchhoff's current rule**: *the total current flowing into a junction is equal to the total current flowing out of the junction*, or

$$\sum I_{\text{in}} = \sum I_{\text{out}} \qquad (28.12)$$

Kirchhoff's current rule

This rule asserts that charge is not accumulating at a junction, nor is it being depleted from the junction; the current is merely flowing through it.

Thus, for the currents flowing into and out of the junction in Fig. 28.15, with the directions as indicated, Kirchhoff's current rule provides the relation

$$I_1 = I_2 + I_3$$

As noted in Example 2, in many applications we might not initially know the directions of the currents. In that case, we choose an arbitrary direction for each separate current. We can then write down the simultaneous set of equations for the unknown currents according to Kirchhoff's rules, assured that when we solve these equations, the algebraic sign of each result will tell us whether that current is in the chosen direction (positive) or opposite to the chosen direction (negative).

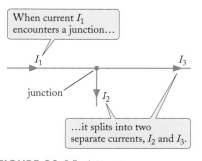

FIGURE 28.15 A junction.

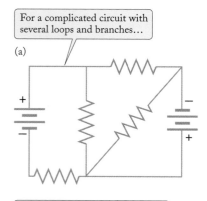

For a complicated circuit with several loops and branches...

(a)

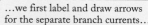

...we first label and draw arrows for the separate branch currents...

(b)

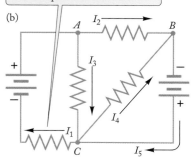

...and then identify several loops, choosing a starting point and direction to go around each loop.

(c)

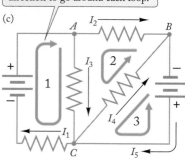

FIGURE 28.16 (a) A multiloop circuit. (b) The same circuit with its five separate currents. Three junctions (A, B, and C) are also labeled. (c) The same circuit with three loops (1, 2, and 3) labeled.

For instance, Fig. 28.16a shows a complicated circuit with several branches. For each branch, we can identify the current and choose a direction for that current, as shown in Fig. 28.16b. Kirchhoff's current rule can be applied at several junctions to relate these currents. Also, the circuit has several loops, and we can apply Kirchhoff's voltage rule to any of these, for example, the rectangular loop (number 1) and two triangular loops (numbered 2 and 3) in Fig. 28.16c. After we choose an arbitrary starting point and direction to go around a particular loop, we sum the potential changes for that path around the loop and set the sum equal to zero. Recall from the previous section that if our path around a loop takes us through a resistor in the same direction as the current, we must write the voltage change as $-IR$ (a voltage drop); if our path takes us through a resistor in the opposite direction to the current, we must write the voltage change as $+IR$ (a voltage rise).

So the procedure for obtaining the necessary equations is as follows:

1. Identify the separate currents through each branch in the circuit, and assign each current a label (I_1, I_2, I_3, . . .) and an arbitrary direction (by drawing an arrow).
2. Write down Kirchhoff's current rule, $\Sigma\,I_{\text{in}} = \Sigma\,I_{\text{out}}$, at various junctions. Choose enough junctions so that every current is included in at least one equation, and choose only junctions with at least one current not in any other equation.
3. Regard the circuit as a collection of closed loops. Identify enough loops so that every part of the circuit is included in at least one of the loops. The loops may overlap, but each loop must have at least some circuit element that does not overlap with any other loop.
4. Apply Kirchhoff's voltage rule to each loop: the sum of all the emfs and all the potential changes across resistors must add to zero as we go around a loop. Note that when writing the potential change across a resistor, $\Delta V = \pm IR$, we must take the negative sign if we traverse the resistor in the direction of the current, and the positive sign if we traverse in a direction opposite to the current. Similarly, a source of emf $\mathcal{E}$ must be considered positive, $+\mathcal{E}$, whenever we go through the source in the forward direction, from the $-$ to the $+$ terminals; and negative, $-\mathcal{E}$, whenever we go though the source in the reverse direction, from the $+$ to the $-$ terminal.

This procedure will result in the right number of equations for the unknown currents $I_1, I_2, I_3, \ldots$ We can then solve the equations for these unknowns by the standard mathematical methods for the solution of simultaneous equations for several unknowns (see Appendix A2). If a current turns out to be negative, its direction is opposite to the direction assigned in step 1.

The following example illustrates this method for the two-loop circuit shown in Fig. 28.17.

Concepts
— in —
Context

EXAMPLE 5   Suppose that the emfs and the resistors in the circuit shown in Fig. 28.17a are $\mathcal{E}_1 = 12.0\text{ V}$, $\mathcal{E}_2 = 8.0\text{ V}$, $R_1 = 4.00\,\Omega$, $R_2 = 4.00\,\Omega$, and $R_3 = 2.00\,\Omega$. Find the current in each of the resistors.

SOLUTION: First, we can label and assign directions to the unknown currents at the junction at the top center of the figure, point A (see Fig. 28.17b). We apply Kirchhoff's current rule at point A:

$$I_1 = I_2 + I_3 \qquad (28.13)$$

Note that for this example, these are the only separate currents in the circuit; there are only three branches in this circuit, so applying Kirchhoff's current rule at a single junction is sufficient. [If we were to apply Kirchhoff's current rule at the junction at the bottom center of the circuit, we would obtain the same information as in Eq. (28.13), since the same three currents encounter that junction.]

Now let us apply Kirchhoff's voltage rule to the left and right loops of the circuit. In this example, we will choose to start at point A and traverse each loop in a clockwise direction, although any starting point and either direction will work for any given loop. For the left loop (number 1) we then obtain

$$-I_2 R_2 + \mathcal{E}_1 - I_1 R_1 = 0 \tag{28.14}$$

and for the right loop (number 2), we obtain

$$-\mathcal{E}_2 - I_3 R_3 + I_2 R_2 = 0 \tag{28.15}$$

Note that the signs in Eq. (28.14) and (28.15) have been chosen according to step (4) of the procedure outlined above. For example, the emf in the first equation is written $+\mathcal{E}_1$, since our path went from the $-$ to the $+$ terminal of $\mathcal{E}_1$; whereas the emf in the second equation is $-\mathcal{E}_2$, since that path went through $\mathcal{E}_2$ from the $+$ to the $-$ terminal. Similarly, in the second equation, the voltage change across resistor R_3 is written as $-I_3 R_3$, since our path was parallel to the direction chosen for I_3; whereas the voltage change across resistor R_2 is written as $+I_2 R_2$, since the path was opposite to the direction chosen for I_2.

Kirchhoff's rules have thus provided the three equations (28.13), (28.14), and (28.15) for the three unknown currents I_1, I_2, and I_3.

Before proceeding to the solution of these equations, it is convenient to substitute the numerical values for the known quantities $\mathcal{E}$ and R. With these numerical values, Eqs. (28.14) and (28.15) become

$$-4I_2 + 12 - 4I_1 = 0 \tag{28.16}$$

$$-8 - 2I_3 + 4I_2 = 0 \tag{28.17}$$

To solve these equations for the three unknowns I_1, I_2, and I_3, we first eliminate I_1 by substituting $I_1 = I_2 + I_3$ from Eq. (28.13) into Eq. (28.16):

$$-4I_2 + 12 - 4(I_2 + I_3) = 0$$

or

$$12 - 8I_2 - 4I_3 = 0 \tag{28.18}$$

Now Eqs. (28.17) and (28.18) contain only the unknowns I_2 and I_3. If we solve Eq. (28.17) for I_2, we find

$$I_2 = \frac{2I_3 + 8}{4} = \frac{1}{2}I_3 + 2 \tag{28.19}$$

We can now eliminate I_2 by substituting Eq. (28.19) into Eq. (28.18):

$$12 - 8(\tfrac{1}{2}I_3 + 2) - 4I_3 = 0$$

or

$$-4 - 8I_3 = 0$$

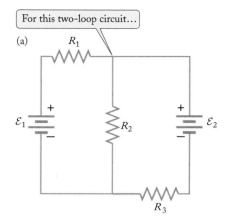

For this two-loop circuit...

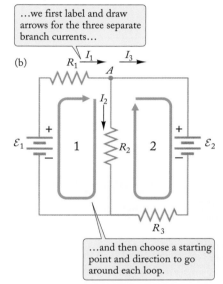

...we first label and draw arrows for the three separate branch currents...

...and then choose a starting point and direction to go around each loop.

FIGURE 28.17 (a) A two-loop circuit. Any two loops are sufficent to include every part of the circuit. (b) The separate currents I_1, I_2, and I_3 and two loops have been labeled.

This immediately gives

$$I_3 = -\tfrac{1}{2} = -0.50$$

Then Eq. (28.19) gives

$$I_2 = \tfrac{1}{2}I_3 + 2 = -\tfrac{1}{4} + 2 = 1.75$$

and Eq. (28.13) gives

$$I_1 = I_2 + I_3 = 1.75 + (-0.50) = 1.25$$

These currents are, of course, measured in amperes; that is, $I_1 = 1.25$ A, $I_2 = 1.75$ A, and $I_3 = -0.50$ A. The negative sign on I_3 indicates the current in that branch is actually opposite to the direction chosen in Fig. 28.17b.

COMMENT: We could have applied Kirchhoff's rules to a different junction or a different loop in the circuit, but we would have obtained the same results. For example, the larger loop around the outside of the entire circuit could have been substituted for one of the loops used above. However, once loops have been chosen which include every part of the circuit, no new information is obtained by considering other possible loops.

PROBLEM-SOLVING TECHNIQUES

KIRCHHOFF'S RULES AND MULTILOOP CIRCUITS

In setting up the equations for the several junctions and loops in a circuit, you must handle the directions of the currents and the sign conventions for the currents and potentials consistently. To maintain this consistency, follow these steps:

1   First, label the currents in the separate branches and assign a direction for each by drawing an arrow.

2   Choose a junction and apply Kirchhoff's current rule, $\Sigma I_{in} = \Sigma I_{out}$. Continue to do this for other junctions until every separate current is included in at least one equation. Choose only junctions that include at least one current not in any previously chosen junction.

3   Then identify a sufficient number of loops so that all circuit elements of the circuit are included in at least one loop, and each loop has at least some circuit element that is not included in any other loop.

4   Start from any point on each loop and proceed around the loop in either direction, and apply Kirchhoff's voltage rule: the sum of all the potential differences is zero. Note that each emf contributes a positive term if the

direction of the chosen path is the forward direction for that emf (from − to + terminal), and it contributes a negative term if the direction is the backward direction. Each resistor contributes a term of the form $\mp$ [current] × [resistance], where the − sign is used if the direction of the chosen path is the same as the current and the + sign is used if the direction is opposite to the current.

5   Once Kirchhoff's rules have been properly applied, you have enough algebraic equations to find the unknown branch currents. To solve such a system of equations with several unknowns, eliminate unknowns successively, until you obtain one equation with only one unknown [as illustrated by the solution of the system of three equations (28.13), (28.16), and (28.17) in Example 5].

6   If the potential difference between two given points in the circuit is needed, simply sum all the emfs and voltage changes in resistors along a path from one point to another, complying with the same sign conventions as for a closed loop.

Checkup 28.4

QUESTION 1: If we want to use Kirchhoff's rules for the circuit shown in Fig. 28.16, how many closed loops do we need? How many junctions must be evaluated?

QUESTION 2: If we want to use Kirchhoff's rules in Fig. 28.12 (instead of the simple treatment exploiting the parallel resistors used in Example 4), how many loops do we need?

QUESTION 3: You wish to evaluate Kirchhoff's voltage rule for loop number 1 in Fig. 28.16c, using the indicated current directions; the resistances are numbered according to their corresponding currents. You start at point A and proceed clockwise. The correct equation is:

(A) $-I_3R_3 - I_1R_1 + \mathcal{E}_1 = 0$ (B) $-I_3R_3 - I_1R_1 - \mathcal{E}_1 = 0$

(C) $-I_3R_3 + I_1R_1 + \mathcal{E}_1 = 0$ (D) $I_3R_3 - I_1R_1 + \mathcal{E}_1 = 0$

28.5 ENERGY IN CIRCUITS; JOULE HEAT

As we saw in Section 28.1, to keep a current flowing in a circuit, the batteries or other sources of emf must do work. If an amount of charge dq passes through a source of emf, the amount of work the source does will be [compare Eq. (25.6)]

$$dW = \mathcal{E}\, dq \qquad (28.20)$$

Hence the rate at which the source does work is

$$\frac{dW}{dt} = \mathcal{E}\frac{dq}{dt} \qquad (28.21)$$

The rate of work is the power P [see Eq. (8.34)]; the rate of flow of charge is the current. Equation (28.21) therefore asserts that the electric power P delivered by the source of emf to a current I is

$$P = \mathcal{E}I \qquad (28.22)$$

power delivered by emf

Thus, the power delivered by the "pump of electricity" is large if it pumps a large current through a large potential difference. In terms of our analogy with a hydraulic pump, this merely means that the power delivered by a hydraulic pump is large if it pumps water at a fast rate (large water current) and lifts the water to a large height (large gravitational potential difference).

Note that consistency of units in Eq. (28.22) demands that the product of the unit of current (1 ampere) and the unit of emf (1 volt) be equal to the unit of power (1 watt). Indeed,

$$1\,\text{A} \times 1\,\text{V} = 1\,\text{C/s} \times 1\,\text{J/C} = 1\,\text{J/s} = 1\,\text{W}$$

Also note that in Eq. (28.22) we have not yet taken into account the algebraic sign of the power. We will have to attach a positive sign to the power if the current passes through the source in the forward direction and a negative sign if the current passes through in the backward direction. In the former case, the source delivers energy to the current, and in the latter case the source receives energy from the current.

Concepts
— in —
Context

| EXAMPLE 6 | What power do the two batteries described in Example 5 deliver? |

SOLUTION: The current through the first battery is 1.25 A, and its emf is 12.0 V. This current passes through the battery in the forward direction; hence the power delivered by the battery is

$$P = 1.25 \text{ A} \times 12.0 \text{ V} = 15.0 \text{ W}$$

Likewise, the 0.50-A current passing through the second battery is also in the forward direction, so the power delivered by this battery is

$$P = 0.50 \text{ A} \times 8.0 \text{ V} = 4.0 \text{ W}$$

The net power delivered by both batteries is 19.0 W.

COMMENT: In this example the power delivered by each battery is positive, so the two simply add to 19.0 W. But in other examples one battery may be delivering power while another receives it. Thus, in the single-loop circuit of Fig. 28.10 (Example 2), with a counterclockwise current of 0.50 A, the 15.0-V battery delivers 7.5 W, while the 12.0-V battery receives 6.0 W. Thus for the circuit of Fig. 28.10, the *net* power delivered by both batteries is 7.5 W − 6.0 W = 1.5 W.

The electric potential energy acquired by the charges is carried along the circuit to the resistors, and it is continually dissipated in the resistors. If within a given resistor the charge dq suffers a potential change ΔV (regarded as a positive quantity), then the loss of potential energy is $dU = \Delta V \, dq$, and the rate at which energy is dissipated is

$$\frac{dU}{dt} = \Delta V \frac{dq}{dt}$$

Hence the power dissipated in the resistor is

power dissipated in resistor

$$P = \Delta V I \qquad (28.23)$$

By means of Ohm's Law, $\Delta V = IR$, we can also write the power two more ways,

Joule heating

$$P = I^2 R \qquad \text{and} \qquad P = \frac{(\Delta V)^2}{R} \qquad (28.24)$$

Figure 28.18 provides a summary of the equations relating power, current, potential difference, and resistance. When any two of these quantities are known, the other two can be calculated.

The energy lost by the charges during their passage through a resistor generates thermal energy; that is, it generates disordered, microscopic kinetic and potential energy of the atoms in the resistor. This *conversion of electric energy into thermal energy in a resistor is called* **Joule heating.**

Many simple electric appliances—such as electric toasters, heaters, hot plates, irons, blankets, lightbulbs—rely on Joule heating (see Fig. 28.19). All these devices merely convert electric energy into thermal energy by means of a heating element consisting of a coiled wire of fairly high resistance. For instance, in a typical electric toaster, the ribbon of wire you can see wound around the plates of insulating material inside the bread slots has a resistance of 10 or 15 Ω, which is much larger than the resistance of the wires connecting the toaster to the outlet, or the wires connecting the outlet to the circuit breaker panel of your house. The resistance of the internal wire in the toaster

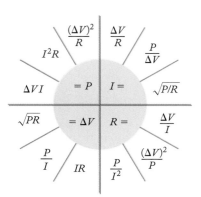

FIGURE 28.18 A summary of relations between current, resistance, potential difference, and power.

must be kept larger than the resistance of the outside wire because otherwise the outside wire would produce more heat than the internal wire, which would be wasteful and hazardous, since overheating of the outside wire could start a fire.

Ordinary incandescent lightbulbs also rely on Joule heating. Within the lightbulb, the current passes through a very fine coiled tungsten filament that acts like a miniature electric heater. The temperature of the tungsten filament reaches 3000°C, and the filament glows, white-hot. The bulb is either partially evacuated or else filled with a chemically inert gas, such as argon. Some of the heat escapes by conduction and convection through the gas in the bulb and through the wires connected to the filament. But a large fraction of the heat escapes by radiation, mostly in the form of infrared light and visible light.

Note that all such appliances consume electric power, but they do not consume electric current or electric charge—any electric current or electric charge that enters the appliance at the high-voltage terminal leaves at the other terminal. However, the electric charge leaves the appliance with less potential energy than it had when it entered, and this decrease represents the electric energy consumed by the appliance.

In any circuit consisting of several sources of emf and several resistors with steady currents, the total power delivered by the sources of emf must equal the total power dissipated in the resistors. This equality can be shown to be mathematically equivalent to Kirchhoff's rules. The net result of the flow of current in such a circuit is therefore a conversion of energy of the sources of emf into an equal amount of energy of heat.

(a)

(b)

(c)

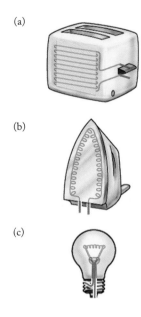

FIGURE 28.19 Some electric appliances that rely on Joule heating. (a) Toaster, (b) iron, (c) incandescent lightbulb.

PHYSICS IN PRACTICE   FUSES AND CIRCUIT BREAKERS

Joule heating can also be used for the control of currents by means of fuses and circuit breakers. These safety devices are designed to cut off the current if it exceeds a preset level. In an ordinary fuse (see Fig. 1), the current passes through a thin ribbon of metal of low melting point. The current heats this ribbon, and if the current is excessive, the ribbon melts away and thus cuts off the current. In a circuit breaker, the current sensor is a bimetallic strip (see Fig. 2), which bends when heated by the current. If the bimetallic strip bends far enough upward, a lever permits a contact arm to flip to one side. The contact arm acts as a switch, and when it flips, it opens the contact points and cuts off the current. After the bimetallic strip cools, the contact arm can be pushed back into place, so as to close the contact points and again permit the current to flow.

Maximum currents permitted for wires of different thicknesses are specified in the National Electrical Code. For instance, the maximum current permitted for the typical wire (No. 12 copper) installed in modern houses is 25 A, and the circuit breakers protecting such wire are therefore designed to cut off the current before it exceeds this level.

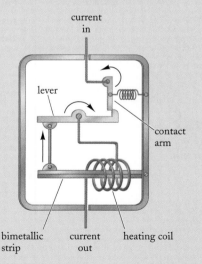

FIGURE 2   A circuit breaker. When heated, the bimetallic strip bends upward and pushes the lever, which releases the spring-loaded contact arm and breaks the circuit.

FIGURE 1   An ordinary fuse.

**Concepts
— in —
Context**

EXAMPLE 7 What is the rate at which Joule heat is produced in the resistors of Example 5?

SOLUTION: The resistances are $R_1 = 4.00\ \Omega$, $R_2 = 4.00\ \Omega$, and $R_3 = 2.00\ \Omega$, the corresponding currents are 1.25 A, 1.75 A, and 0.50 A. Equation (28.24) then gives the power in each resistor:

$$P_1 = I_1^2 R_1 = (1.25\ \text{A})^2 \times 4.00\ \Omega = 6.25\ \text{W}$$

$$P_2 = I_2^2 R_2 = (1.75\ \text{A})^2 \times 4.00\ \Omega = 12.25\ \text{W}$$

$$P_3 = I_3^2 R_3 = (0.50\ \text{A})^2 \times 2.00\ \Omega = 0.50\ \text{W}$$

COMMENTS: Note that the net power dissipated is $P_1 + P_2 + P_3 = 19.0\ \text{W}$, which agrees with the net power delivered by the batteries (compare with Example 6).

EXAMPLE 8 A high-voltage transmission line that connects a city to a power plant consists of a pair of aluminum cables, each with a resistance of 4.0 Ω. The current flows to the city along one wire and back along the other.

(a) The transmission line delivers to the city 1.7×10^5 kW of power at 2.3×10^5 V. What is the current in the transmission line? How much power is lost as Joule heat in the transmission line? (b) If the transmission line were to deliver the same 1.7×10^5 kW of power at 115 V, how much power would be lost as Joule heat? Is it more efficient to transmit power at high voltage or at low voltage?

SOLUTION: (a) Figure 28.20 shows the circuit consisting of power plant, transmission line, and city. In terms of the power and the voltage delivered to the city, the current through the city is, according to Eq. (28.23),

$$I = \frac{P_{\text{delivered}}}{\Delta V_{\text{delivered}}} = \frac{1.7 \times 10^8\ \text{W}}{2.3 \times 10^5\ \text{V}} = 7.4 \times 10^2\ \text{A}$$

The current in both portions of the transmission line must be the same, since any current that flows toward the city must return to the power plant. The combined resistance of both wires is 4.0 Ω + 4.0 Ω = 8.0 Ω, and hence the power lost in the transmission line is, according to Eq. (28.24),

$$P_{\text{lost}} = I^2 R = (7.4 \times 10^2\ \text{A})^2 \times 8.0\ \Omega = 4.4 \times 10^6\ \text{W}$$

Thus, the power lost is about 3% of the power delivered.

(b) For $\Delta V_{\text{delivered}} = 115$ V, the current is

$$I = \frac{P_{\text{delivered}}}{\Delta V_{\text{delivered}}} = \frac{1.7 \times 10^8\ \text{W}}{115\ \text{V}} = 1.5 \times 10^6\ \text{A}$$

and the power lost is

$$P_{\text{lost}} = I^2 R = (1.5 \times 10^6\ \text{A})^2 \times 8.0\ \Omega = 1.8 \times 10^{13}\ \text{W}$$

Thus, the power lost is much larger than the power delivered! Comparing the results of (a) and (b), we see that transmission at high voltage is much more efficient than transmission at low voltage.

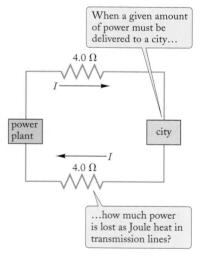

When a given amount of power must be delivered to a city...

4.0 Ω

$I \longrightarrow$

power plant

city

$\longleftarrow I$

4.0 Ω

...how much power is lost as Joule heat in transmission lines?

FIGURE 28.20 Circuit diagram for a high-voltage transmission line connecting a city to a power plant.

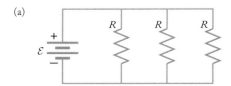

(a)

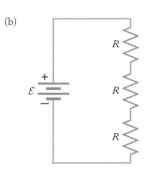

(b)

Checkup 28.5

QUESTION 1: A resistor of 10 Ω connected to a battery produces 0.4 W of Joule heat. If we replace this resistor by one of 20 Ω, what will be the rate of production of heat?

QUESTION 2: If we replace both batteries in the circuit described in Example 5 by batteries of twice the emf, by what factor do we increase the rate of production of Joule heat?

QUESTION 3: In Example 8(a), the city receives an electric current of 7.4×10^2 A. Does the city consume this current? Does the city consume electric charge? Electric energy?

QUESTION 4: Figure 28.21a shows three identical resistors connected in parallel to a battery with emf $\mathcal{E}$; Fig. 28.21b shows the same resistors and battery, but now connected in series. What is the ratio of the power dissipated by one of the resistors in Fig. 28.21b to that dissipated by one of the resistors in Fig. 28.21a?

(A) 9   (B) 3   (C) 1   (D) $\frac{1}{3}$ (E) $\frac{1}{9}$

FIGURE 28.21 An emf $\mathcal{E}$ and three identical resistors R connected (a) in parallel and (b) in series.

Online
Concept
Tutorial

28.6 ELECTRICAL MEASUREMENTS

Measurements of currents, potentials, and resistances in electrical circuits require various specialized instruments. Here we will briefly discuss some of the instruments used in DC circuits.

Ammeter and Voltmeter

Most electrical measurements are performed with ammeters and voltmeters (see Fig. 28.22). The ammeter measures the electric current flowing through its terminals, and the voltmeter measures the potential difference applied to its terminals.

The internal mechanisms of the ammeter and the voltmeter are similar. In the old moving-coil instruments, the sensitive element is a small coil of wire, delicately suspended between the poles of a magnet, which experiences a deflection when a current passes through it (see Fig. 30.18 for a view of the internal mechanism of these instruments). Thus, such ammeters and voltmeters both respond to an electric current passing through the instrument. In a modern digital voltmeter, a stable counter measures time while a calibrated current charges a capacitor; the capacitor voltage is compared with the input voltage being measured. When the two are equal, the counter stops; the count is proportional to the input voltage, providing digital output. In a modern ammeter, the current is passed through a known internal resistor, providing a voltage drop that is again measured digitally.

The resistance that the meter offers to a current is called the **internal resistance** of the meter. As in the case of a battery, the internal resistance can be regarded as placed in series with the meter. Hence, the current registered by the ammeter equals the current through its internal resistance, and the voltage registered by the voltmeter equals the potential difference across its internal resistance.

Ammeters and voltmeters differ in that *the ammeter has a low internal resistance and permits the passage of whatever current enters its terminals with little hindrance, whereas the voltmeter has a very large internal resistance and draws only an extremely small current,* even when the potential difference applied to its terminals is large.

FIGURE 28.22 Two multimeters, which can be used either as ammeters or as voltmeters. The switches are used to select the function and the measurement scale.

internal resistance of meter

(a)

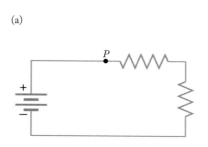

(b)

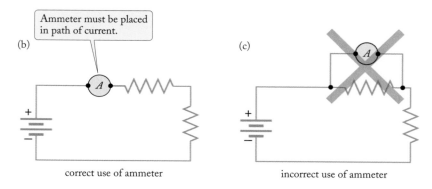

Ammeter must be placed in path of current.

correct use of ammeter

(c)

incorrect use of ammeter

FIGURE 28.23 (a) An electric circuit. (b) Correct connection of the ammeter. The symbol for the ammeter is a circle labeled *A*, with two terminals. The current to be measured enters through one terminal of the ammeter and leaves through the other. (c) Incorrect connection of the ammeter.

Figure 28.23a shows an example of an electric circuit. To measure the electric current flowing past, say, the point *P* in this circuit, the experimenter must cut the wire apart and insert the ammeter. Figure 28.23b shows the correct way of connecting the ammeter (for comparison, Fig. 28.23c shows a wrong way). Since the ammeter has a very low internal resistance, its insertion in the circuit usually has an insignificant effect on the current. However, if the resistances in the circuit are very small, then the insertion of an ammeter can have a significant inhibiting effect on the current. For the accurate measurement of the current, the experimenter must select an ammeter whose internal resistance is much lower than the resistances in the circuit.

To measure the potential difference between, say, the points *P* and *P'* in a circuit, the experimenter must connect the terminals of the voltmeter to these points. Figure 28.24a shows the correct way of connecting the voltmeter (for comparison, Fig. 28.24b shows a wrong way). Since the voltmeter has a very large internal resistance, it draws only a very small current, and the alteration in the flow of current through the resistance *PP'*, and the consequent alteration of the potential difference between *P* and *P'*, are usually insignificant. However, if the resistance *PP'* across which the voltmeter is connected is large, then the voltmeter can draw a significant fraction of the current, with a consequent decrease of potential across the resistance *PP'*. For an accurate measurement of the potential, the experimenter must select a voltmeter with an internal resistance much larger than the resistance *PP'*.

(a)

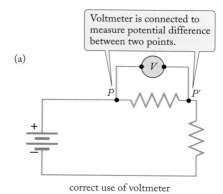

Voltmeter is connected to measure potential difference between two points.

correct use of voltmeter

(b)

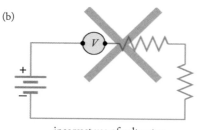

incorrect use of voltmeter

FIGURE 28.24 (a) Correct connection of the voltmeter. The symbol for the voltmeter is a circle labeled *V*, with two terminals. One terminal is connected to the point *P*, the other is connected to the point *P'*. (b) Incorrect connection of the voltmeter.

Wheatstone Bridge

The Wheatstone bridge is used for precise comparisons of an unknown resistance with known reference resistances. Figure 28.25 shows a schematic diagram of a Wheatstone bridge. The resistance R_x is the unknown resistance to be determined. Resistor R_1 is a high-precision variable reference resistor, whereas resistors R_2 and R_3 are fixed reference resistors. In practice, the bridge is "balanced" by varying the resistance R_1 until the current through the ammeter is zero. In that case, a common current flows through R_1 and R_2 in the left leg of the bridge, say, a current I_l. Similarly, a common current I_r flows through R_x and R_3 in the right leg. Also, the potential at points P and Q must be the same, since no current through the ammeter implies no voltage drop across it. Thus we can equate the left and right resistive potential changes for each of the upper and lower parts of the bridge:

$$I_l R_1 = I_r R_x \qquad \text{and} \qquad I_l R_2 = I_r R_3$$

If we divide the two equations to eliminate I_l and I_r, we see that the ratio of R_1 to R_2 must be the same as the ratio of R_x to R_3, and we can solve for the unknown resistance:

$$R_x = \frac{R_3}{R_2} R_1 \qquad (28.25)$$

This equation permits the calculation of the unknown resistance. One advantage of the bridge technique is that the ammeter need not be calibrated; it need only provide a measurement of the zero, or null, value of current, which can be achieved with very high accuracy.

In essence, the operation of the Wheatstone bridge hinges on an adjustment of the ratio of resistances in the circuit. Often, the Wheatstone bridge is set up with $R_3 = R_2$, so that the bridge will balance when $R_1 = R_x$; in this case, the dials on the adjustable, calibrated resistance directly tell us the value of the unknown resistance. Similar bridge circuits can be used to measure the values of other circuit components, such as capacitance.

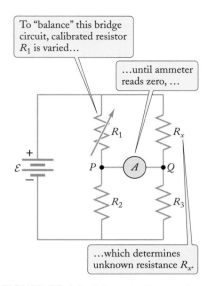

FIGURE 28.25 Schematic diagram for a Wheatstone bridge.

 Checkup 28.6

QUESTION 1: What is the main difference between an ammeter and a voltmeter?

QUESTION 2: Exactly what is wrong with the way the ammeter is connected in Fig. 28.23c and the voltmeter in Fig. 28.24b?

QUESTION 3: Consider the incorrect use of the ammeter in Fig. 28.23c. Compared with the original current flowing in the circuit without the ammeter (Fig. 28.23a), the ammeter in Fig. 28.23c will indicate a current that is

(A) Larger   (B) The same   (C) Smaller

28.7 THE *RC* CIRCUIT

Online
Concept
32
Tutorial

Throughout this chapter we have dealt only with steady, time-independent currents. However, *Kirchhoff's rules and the methods for solving circuits we have developed in this chapter apply also to time-dependent currents.* The only restriction is that the emfs and the currents in the circuit must not vary too quickly. (A rough criterion for the applicability of Kirchhoff's rules is that the currents and emfs must not change significantly in an interval equal to the travel time for a light signal around the circuit.)

In Chapter 32 we will deal with a variety of circuits with time-dependent currents. Here we will deal with the simple case of a time-dependent current in a circuit consisting of a resistor R and a capacitor C connected in series and being charged by a battery. Figure 28.26 shows a schematic diagram for such an *RC* circuit. We assume that the capacitor is initially uncharged, and that the battery is suddenly connected at time $t = 0$. Initially, the potential difference across the capacitor is then zero. When the battery is connected, charge flows from the terminals of the battery to the plates of the capacitor. As the charge accumulates on the plates, the potential difference across the plates gradually increases. The flow of charge will come to a halt when the potential difference across the plates matches the emf of the battery. This qualitative discussion of the charging process indicates that the current is initially large, but gradually tapers off, and ultimately approaches zero.

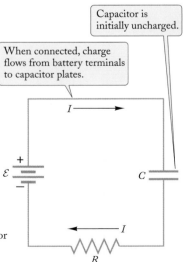

FIGURE 28.26 An *RC* circuit, consisting of a resistor and a capacitor connected in series to a battery.

For a mathematical treatment of the current in the circuit we turn to Kirchhoff's voltage rule: the sum of all the emfs and voltage drops around the circuit must be zero. The emf of the battery is $\mathcal{E}$. If the current at some instant is I, the potential drop across the resistor is $\Delta V_R = -IR$. And if the charge on the capacitor plates at some instant has a magnitude Q, the potential drop across the plates is $\Delta V_C = -Q/C$. Hence

$$\mathcal{E} - IR - \frac{Q}{C} = 0 \qquad (28.26)$$

At the initial time, $Q = 0$ and Eq. (28.26) then tells us that the initial current is $I = \mathcal{E}/R$. At the completion of the charging process, $I = 0$ and Eq. (28.26) then tells us that $Q = C\mathcal{E}$. Thus, at the initial time, the entire potential drop occurs in the resistor, and at the completion of the charging process, the entire potential drop occurs in the capacitor. At intermediate times, both the resistor and the capacitor contribute to the potential drop.

To calculate the current and the charge at intermediate times, let us rewrite Eq. (28.26) using $I = dQ/dt$:

$$\mathcal{E} - \frac{dQ}{dt} R - \frac{Q}{C} = 0 \qquad (28.27)$$

This equation can be solved by direct integration if we rearrange it to make one side proportional to dQ and the other proportional to dt as follows:

$$\frac{dQ}{Q - C\mathcal{E}} = -\frac{1}{RC} dt$$

We now integrate each side from its initial zero value to the value at some later time t:

$$\int_0^Q \frac{dQ}{Q - C\mathcal{E}} = -\frac{1}{RC} \int_0^t dt' \qquad (28.28)$$

The integral on the left is just the natural logarithm $\ln(Q' - C\mathcal{E})$, and the integral on the right is simply t. Evaluation of these at their limits and using $\ln A - \ln B = \ln(A/B)$ gives

$$\ln\left(\frac{Q - C\mathcal{E}}{-C\mathcal{E}} \right) = -\frac{t}{RC} \qquad (28.29)$$

If we take the exponential function of each side and recall that $e^{\ln x} = x$, where $e = 2.718 \ldots$ is the base of the natural logarithms[1] (see Math Help: The Exponential Function, page 910) we obtain

$$\left(\frac{Q - C\mathcal{E}}{-C\mathcal{E}} \right) = e^{-t/(RC)} \qquad (28.30)$$

or, solving for Q, we finally have

charging RC circuit

$$Q = C\mathcal{E}\left(1 - e^{-t/(RC)} \right) \qquad (28.31)$$

---

[1] The pure number $e \approx 2.718$ should not be confused with the elementary charge $e \approx 1.60 \times 10^{-19}$ C.

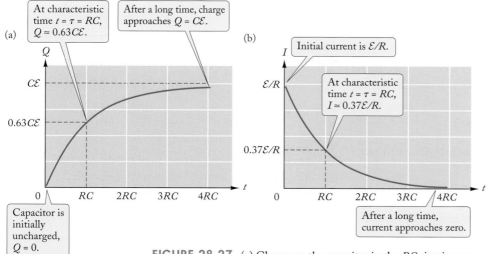

FIGURE 28.27 (a) Charge on the capacitor in the *RC* circuit as a function of time. (b) Current in the *RC* circuit as a function of time.

This behavior of the charge on the capacitor as a function of time is shown in Fig. 28.27a; as expected, it starts at $Q = 0$ (at $t = 0$, we have $e^0 = 1$), and it approaches $Q = C\mathcal{E}$ as the charging process nears completion (for $t = \infty$, we have $e^{-\infty} = 0$).

The current in the circuit is the time derivative of the charge:

$$I = \frac{dQ}{dt} = \frac{d}{dt}\left[C\mathcal{E}\left(1 - e^{-t/(RC)}\right)\right] = C\mathcal{E}\frac{d}{dt}\left(-e^{-t/(RC)}\right) = \frac{C\mathcal{E}}{RC}e^{-t/(RC)} \quad (28.32)$$

or, simply,

$$I = \frac{\mathcal{E}}{R}e^{-t/(RC)} \quad (28.33)$$

current in *RC* circuit

Thus, as shown in Fig. 28.27b, the current is an exponentially decaying function of time.

The product *RC* that occurs in the exponential has the units of time, and it is designated by τ:

$$\tau = RC \quad (28.34)$$

characteristic time (*RC* time constant)

This time is called the **characteristic time** for the *RC* circuit; it is also called the *RC* **time constant**. From Fig. 28.27, we see that most of the charging occurs within the characteristic time. More precisely, at the characteristic time $t = \tau = RC$ the charge reaches the value

$$Q = C\mathcal{E}\left(1 - e^{-1}\right) = C\mathcal{E}\left(1 - \frac{1}{2.718}\right) = C\mathcal{E} \times 0.6321$$

that is, *at the characteristic time, the charge reaches* ≈ 63% *of its final value*. In terms of the characteristic time, the charge (28.31) and the current (28.33) can be written

$$Q = C\mathcal{E}\left(1 - e^{-t/\tau}\right) \quad \text{and} \quad I = \frac{\mathcal{E}}{R}e^{-t/\tau} \quad (28.35)$$

***RC* circuit**

| **MATH HELP** | **THE EXPONENTIAL FUNCTION** |

The exponential function $\exp(x)$, or e^x, is defined as e to the power x, where the number $e = 2.7183\ldots$ is the base of natural logarithms,

$$\exp(x) = e^x = (2.7183\ldots)^x$$

For instance, if $x = 2$, then the value of the exponential function is

$$e^2 = (2.7183\ldots)^2 = 7.3891$$

The natural logarithm $\ln(x)$ is the inverse of the exponential function. Thus,

$$\ln(1) = \ln(e^0) = 0$$

$$\ln(e) = 1$$

$$\ln(e^2) = 2$$

$$\ln(e^3) = 3$$

and, in general,

$$\ln(e^x) = x$$

Conversely,

$$e^{\ln x} = x$$

On many calculators, the exponential function is performed by taking the inverse natural logarithm; for example, to obtain the number e, we take the inverse-ln of 1.

As in the case of common logarithms (with base 10), the natural logarithm obeys

$$\ln(y/x) = \ln(y) - \ln(x)$$

For example,

$$\ln(1/2) = \ln(1) - \ln(2) = 0 - \ln 2 = -\ln 2$$

and

$$1/2 = e^{\ln(1/2)} = e^{-\ln 2}$$

EXAMPLE 9 Suppose that in the circuit illustrated in Fig. 28.26, the resistance is $R = 8.0 \times 10^3\ \Omega$, the capacitance is $C = 2.0\ \mu\text{F}$, and the emf of the battery is $\mathcal{E} = 1.5$ V. What is the initial value of the current, at the instant after the battery is connected? What is the final value of the charge on the capacitor? What is the charge at $t = \tau$? At $t = 2\tau$? At $t = 5\tau$?

SOLUTION: The initial value of the current is [see the discussion following Eq. (28.26)]

$$I = \frac{\mathcal{E}}{R} = \frac{1.5\ \text{V}}{8.0 \times 10^3\ \Omega} = 1.9 \times 10^{-4}\ \text{A}$$

The final value of the charge is

$$Q = C\mathcal{E} = 2.0 \times 10^{-6}\ \text{F} \times 1.5\ \text{V} = 3.0 \times 10^{-6}\ \text{C}$$

At the time $t = \tau = RC$, the charge is [using Eq. (28.35)]

$$Q = C\mathcal{E}(1 - e^{-1}) = 3.0 \times 10^{-6}\ \text{C} \times \left(1 - \frac{1}{2.718}\right) = 1.9 \times 10^{-6}\ \text{C}$$

at the time $t = 2\tau = 2RC$, the charge is

$$Q = C\mathcal{E}(1 - e^{-2}) = 3.0 \times 10^{-6}\ \text{C} \times \left[1 - \frac{1}{(2.718)^2}\right] = 2.6 \times 10^{-6}\ \text{C}$$

and at the time $t = 5\tau = 5RC$, the charge is

$$Q = C\mathcal{E}(1 - e^{-5}) = 3.0 \times 10^{-6}\ \text{C} \times \left[1 - \frac{1}{(2.718)^5}\right] = 3.0 \times 10^{-6}\ \text{C}$$

After five time constants, the charge has reached within 1% of its final value.

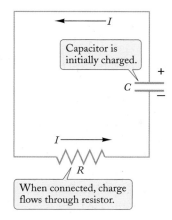

FIGURE 28.28 An *RC* circuit, with the capacitor discharging through the resistor. The arrow indicates the direction of the current.

Once the charging process has been completed and the current in the circuit has stopped, we can disconnect the battery. The charge will then remain on the capacitor plates (except for a slow leakage through the capacitor or into the air). However, suppose we now connect the free terminals of the capacitor and the resistor, as shown in Fig. 28.28. The capacitor will then discharge through the resistor. The current will be large at the initial time, and it will gradually taper off and approach zero as the potential difference across the capacitor decreases. This circuit is simpler than the one we considered above when charging the capacitor. A similar analysis using Kirchhoff's voltage rule and direct integration yields formulas that describe the discharge of the capacitor. The relations involve the same exponential function as in Eqs. (28.31) and (28.33), with the same characteristic time, except now the charge decays to zero with time:

$$Q = C\mathcal{E}e^{-t/(RC)} \qquad (28.36)$$

discharging *RC* circuit

and

$$I = -\frac{dQ}{dt} = \frac{\mathcal{E}}{R}e^{-t/(RC)} \qquad (28.37)$$

The current now is the negative of the rate of change of the charge because during the discharging process, a decrease of charge, that is, a negative dQ/dt, results in a positive current in the direction shown in Fig. 28.28. Figure 28.29 gives plots of the charge vs. time and the current vs. time for this discharging process.

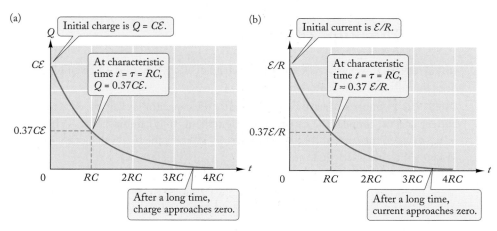

FIGURE 28.29 (a) Charge on a discharging capacitor as a function of time. (b) Current in the discharging *RC* circuit as a function of time.

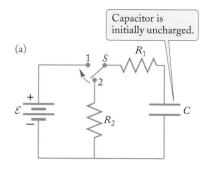

(a)

Capacitor is initially uncharged.

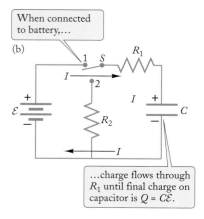

(b)

When connected to battery,...

...charge flows through R_1 until final charge on capacitor is $Q = C\mathcal{E}$.

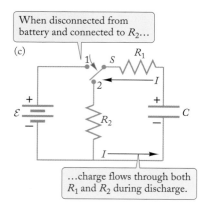

(c)

When disconnected from battery and connected to R_2...

...charge flows through both R_1 and R_2 during discharge.

FIGURE 28.30   A switchable *RC* circuit.

EXAMPLE 10   Consider the circuit of Fig. 28.30a. The switch S has been in position 2 for a long time. (a) At $t = 0$, it is switched to position 1. What is the charge on the capacitor as a function of time t? (b) Much later, at some time $t' = 0$, the switch is returned to position 2. What is the charge on the capacitor as a function of the time t'? What is the voltage across resistor R_2 as a function of the time t'?

SOLUTION: (a) Since the switch has been in position 2 for a long time, the capacitor is uncharged (otherwise it would discharge through the resistors). When the switch is moved to position 1, current flows only in the outer loop (Fig. 28.30b), through R_1, as the capacitor C charges to its final value $Q = C\mathcal{E}$ just as in our previous discussion of a charging *RC* circuit. Thus the relevant characteristic time is $\tau = R_1 C$, and the charge on the capacitor is given by Eq. (28.31) with $R = R_1$:

$$Q = C\mathcal{E}\left(1 - e^{-t/(R_1 C)}\right)$$

(b) Much later, at $t \gg R_1 C$, the switch is moved to position 2. Let this new initial time be $t' = 0$. The capacitor is initially charged with $Q = C\mathcal{E}$. Since the emf is now effectively disconnected, a discharge occurs only in the right loop (Fig. 28.30c), and the current flows counterclockwise through the series combination of resistors, so now the equivalent resistance is $R = R_1 + R_2$. Thus the charge on the capacitor will be given as a function of t' by the discharging function of Eq. (28.36), with $\tau = (R_1 + R_2)C$:

$$Q = C\mathcal{E}\,e^{-t'/[(R_1 + R_2)C]}$$

The current is counterclockwise in the right loop with magnitude [Eq. (28.37)]

$$I = \frac{\mathcal{E}}{R_1 + R_2}\,e^{-t'/[(R_1 + R_2)C]}$$

and the voltage across R_2 is given by Ohm's Law:

$$\Delta V = IR_2 = \frac{\mathcal{E}R_2}{R_1 + R_2}\,e^{-t'/[(R_1 + R_2)C]}$$

✔ Checkup 28.7

QUESTION 1: If we use a 3.0-V battery instead of a 1.5-V battery in Example 9, how does this affect the characteristic time?

QUESTION 2: Suppose that in Example 9, after the capacitor is charged, we remove the battery and we connect the capacitor to the resistor, so it can discharge. What is the initial value of the current through the resistor? What is the final value of the current? What is the characteristic time?

QUESTION 3: Consider the charging of the capacitor C in Fig. 28.26. Suppose we increase the original resistance R to $2R$ and we again begin to charge the capacitor at time $t = 0$. Compared with its final value $Q = C\mathcal{E}$, the charge on the capacitor at time $t = RC$ will be

(A) Less than 63% of $C\mathcal{E}$ (B) ≈ 63% of $C\mathcal{E}$ (C) More than 63% of $C\mathcal{E}$

28.8 THE HAZARDS OF ELECTRIC CURRENTS

As a side effect of the widespread use of electric machinery and devices in factories and homes, each year in the United States about 1000 people die by accidental electrocution. A much larger number suffer nonfatal electric shocks. Fortunately, the human skin is a fairly good insulator, providing a protective barrier against injurious electric currents. The resistance of a square centimeter of dry human epidermis in contact with a conductor can be as much as 10^5 Ω. However, the resistance varies in a rather sensitive way with the thickness, moisture, and temperature of the skin, and with the magnitude of the potential difference.[2]

The electric power supplied to factories and homes in the United States is usually in the form of alternating currents, or AC. These are oscillating currents, which periodically reverse direction (the standard period of the alternating current supplied by power companies is $\frac{1}{60}$ second). Since most accidental electric shocks involve alternating currents, the following discussion of the effects of currents on the human body will emphasize alternating currents.

In the typical accidental electric shock, the current enters the body through the hands (in contact with one terminal of the source of emf) and exits through the feet (in contact with the ground, which constitutes the other terminal of the source of emf for most AC circuits). Thus, the body plays the role of a resistor, closing an electric circuit (see Fig. 28.31).

The damage to the body depends on the magnitude of the current passing through it. An alternating current of about 0.001 A produces only a barely detectable tingling sensation. Higher currents produce pain and strong muscular contractions. If the victim has grasped an electric conductor—such as an exposed power cable—with the hand, the muscular contraction may prevent the victim from releasing the hold on the conductor. The magnitude of the "let-go" current, at which the victim can just barely release the hold on the conductor, is about 0.01 A. Higher currents lock the victim's hand to the conductor. Unless the circuit is broken within a few seconds, the skin in contact with the conductor will then suffer burns and blister. Such damage to the skin drastically reduces its resistance and so can lead to a fatal increase of the current.

An alternating current of about 0.02 A flowing through the body from the hands to the feet produces a contraction of the chest muscles that halts breathing; this leads to death by asphyxiation if it lasts for a few minutes. A current of about 0.1 A lasting just a few seconds induces fibrillation of the heart muscles, with cessation of the natural rhythm of the heartbeat and cessation of the pumping of blood. Fibrillation usually continues even when the victim is removed from the electric circuit; the consequences are fatal unless immediate medical assistance is available. The treatment for fibrillation involves the deliberate application of a severe electric shock to the heart by means of electrodes placed against the chest; this arrests the motion of the heart completely. When the shock ends, the heart usually resumes beating with its natural rhythm.

A current of a few amperes produces a seizure of the nervous system and paralysis of the respiratory muscles. Victims of such currents can sometimes be saved by prompt recourse to artificial respiration. At these high values of the current, the effects of AC and DC are not very different. But at lower values, a DC current poses less of a hazard than the comparable AC current, because the former does not trigger the strong muscular contractions triggered by the latter.

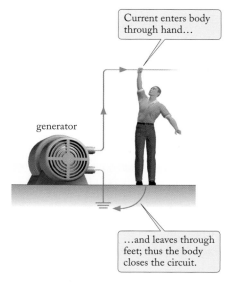

Current enters body through hand...

generator

...and leaves through feet; thus the body closes the circuit.

FIGURE 28.31 A human body closing an electric circuit. The current enters through the hand and exits through the feet.

---

[2] The variation of resistance with potential difference implies that skin does not obey Ohm's Law.

In the above we assumed that the path of the current through the body is from the hands to the feet. If the current enters through one leg and exits through the other, no vital organs lie in its path, and the threat to life is lessened. However, an intense current through a limb tends to kill the tissue through which it passes, and may ultimately require the surgical excision of large amounts of dead tissue, and even the amputation of the limb.

Other things being equal, a higher voltage will result in a higher current. The hazard posed by contact with high-voltage sources is therefore obvious. But under exceptional circumstances, even sources of low voltage can be hazardous. Several cases of electrocution by contact with sources of a voltage as low as 12 V have been reported. It seems that in these cases death resulted from an unusually sensitive response of the nervous system; it is also conceivable that an unusually small skin resistance was a contributing factor. Thus, it is advisable to treat even sources of low voltage with respect!

first aid for electric shock   *Aid to victims of electric shock should begin with switching off the current. When no switch, plug, or fuse for cutting off the current is accessible, the victim must be pushed or pulled away from the electric conductor by means of a piece of insulating material, such as a piece of dry wood or a rope. The rescuer must be careful to avoid electric contact. If the victim is not breathing, artificial respiration must be started at once. If there is no heartbeat, cardiac massage must be applied by trained personnel until the victim can be treated with a defibrillation apparatus.*

SUMMARY

PROBLEM-SOLVING TECHNIQUES Kirchhoff's Rules and Multiloop Circuits	**(page 900)**
PHYSICS IN PRACTICE Fuses and Circuit Breakers	**(page 903)**
MATH HELP The Exponential Function	**(page 910)**

ELECTROMOTIVE FORCE (emf)   Energy per unit charge, or voltage, provided by a source.

KIRCHHOFF'S VOLTAGE RULE   The sum of emfs and potential changes across resistors around any closed loop in a circuit must equal zero.

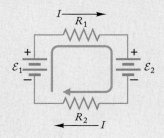

SIGN CONVENTIONS   When traversing a source of emf in the forward direction ($-$ to $+$ terminal), the voltage is positive, ($+\mathcal{E}$); in the reverse direction ($+$ to $-$ terminal), the voltage is negative ($-\mathcal{E}$). Traversing a resistor in the direction of the current provides a voltage drop (negative, $-IR$); doing so in the direction opposite to the current provides a voltage increase (positive, $+IR$).

KIRCHHOFF'S CURRENT RULE The total current flowing into any junction must equal the total current flowing out of the junction:

$$\sum I_{\text{in}} = \sum I_{\text{out}}$$

(28.12)

MULTILOOP CIRCUIT ANALYSIS After labeling all separate currents, Kirchhoff's voltage and current rules are applied until enough equations are obtained to solve for the desired unknown quantities.

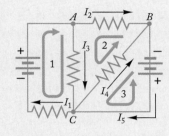

POWER DELIVERED BY A SOURCE OF emf

$$P = \mathcal{E}I$$

(28.22)

POWER DISSIPATED BY A RESISTOR (JOULE HEAT)

(From these relations between current, resistance, potential difference, and power, when any two are known, the other two can be calculated.)

$$P = \Delta V I = I^2 R = \frac{(\Delta V)^2}{R}$$

(28.23, 28.24)

ELECTRICAL MEASUREMENTS Ammeters measure current, have a low internal resistance, and are placed in line; voltmeters measure voltage, have a high internal resistance, and are connected across terminals.

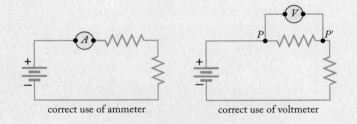

correct use of ammeter   correct use of voltmeter

CHARACTERISTIC TIME FOR *RC* CIRCUIT

$$\tau = RC$$

(28.34)

CHARGE ON CAPACITOR AND CURRENT IN *RC* CIRCUIT

Charging:

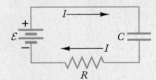

$$Q = C\mathcal{E}\left(1 - e^{-t/\tau}\right) \qquad I = \frac{dQ}{dt} = \frac{\mathcal{E}}{R}e^{-t/\tau}$$

(28.35)

Discharging:

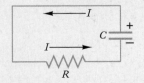

$$Q = C\mathcal{E}e^{-t/\tau} \qquad I = -\frac{dQ}{dt} = \frac{\mathcal{E}}{R}e^{-t/\tau}$$

(28.36)

QUESTIONS FOR DISCUSSION

1. Can we use a capacitor as a pump of electricity in a circuit? In what way would such a pump differ from a battery?

2. Does a fully charged battery have the same emf as a partially charged battery?

3. How would you measure the internal resistance of a battery?

4. Kirchhoff's voltage rule is equivalent to energy conservation in the electric circuit. Explain.

5. What would happen if we were to connect both an ammeter and a voltmeter in series with a source of emf?

6. What would happen if we were to connect the voltmeter incorrectly, as shown in Fig. 28.24b?

7. A mechanic determines the internal resistance of an automobile battery by connecting a rugged, high-current ammeter (of nearly zero resistance) directly across the poles of the battery. The internal resistance of the battery is inversely proportional to the ammeter reading, $R_1 \propto 1/I$. Explain.

8. An **ohmmeter** consists of a reference source of emf (a battery), connected in series with a reference resistance and an ammeter. When the terminals of the ohmmeter are connected to an unknown resistor (Fig. 28.32), the current registered by the ammeter permits the evaluation of the unknown resistance. Explain.

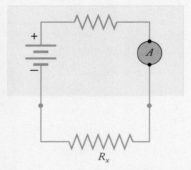

FIGURE 28.32 An ohmmeter, consisting of a battery, a reference resistance, and an ammeter enclosed in a box (color).

9. If a voltmeter and an ammeter are connected to a resistor simultaneously, their readings can be used to evaluate the resistance, according to $R = \Delta V/I$. Figure 28.33 shows two ways of connecting the ammeter and the voltmeter to the resistor. Explain why both of these methods yield a value of R slightly different from the actual value.

(a)

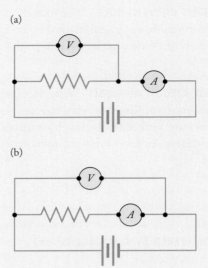

(b)

FIGURE 28.33 Simultaneous ammeter and voltmeter connections.

10. A homeowner argues that he should not pay his electric bill, since he is not keeping any of the electrons that the power company delivers to his home—any electron that enters the wiring of his home sooner or later leaves and returns to the power station. How would you answer?

11. The spiral heating elements commonly used in electric ranges *appear* to be made of metal (Fig. 28.34). Why do they not short-circuit when you place an iron pot on them?

FIGURE 28.34 Electric range heating element.

12. What are the advantages and what are the disadvantages of high-voltage power lines?

13. In many European countries, electric power is delivered to homes at 230 V, instead of the 115 V customary in the United States. What are the advantages and what are the disadvantages of 230 V?

PROBLEMS

28.1 Electromotive Force†
28.2 Sources of Electromotive Force†

1. The smallest batteries have a mass of 1.5 grams and store an electric energy of about 5.0×10^{-6} kW·h. The largest batteries (used aboard submarines) have a mass of 270 metric tons and store an electric energy of 5×10^3 kW·h. What is the amount of energy stored per kilogram of battery in each case?

2. A size D flashlight battery will deliver 1.2 A·h at 1.5 V (that is, it will deliver 1.2 A for 1.0 h, or a larger or smaller current for a correspondingly shorter or longer time). An automobile battery will deliver 55 A·h at 12 V. The flashlight battery is cylindrical with a diameter of 3.3 cm, a length of 5.6 cm, and a mass of 0.086 kg. The automobile battery is rectangular with dimensions 30 cm × 17 cm × 23 cm and a mass of 23 kg.

   (a) What is the available electric energy stored in each battery?

   (b) What is the amount of energy stored per cubic centimeter of each battery?

   (c) What is the energy stored per kilogram of each battery?

3. A heavy-duty 12-V battery for a truck is rated at 160 A·h; that is, this battery will deliver 1.0 A for 160 h (or a larger current for a correspondingly shorter time). What is the amount of electric energy that this battery will deliver?

4. Rechargeable 1.2-V nickel–cadmium batteries of sizes AAA, AA, C, and D are rated at 0.22, 0.70, 2.2, and 4.4 A·h, respectively. How much charge in coulombs passes through each battery when fully used as rated? How much energy does each deliver?

5. A 14.4-V metal hydride battery for a laptop computer can deliver a current of 0.65 A for 4.0 h. What is the total energy delivered?

*6. The electric starter motor in an automobile equipped with a 12-V battery draws a current of 80 A when in operation.

   (a) Suppose it takes the starter motor 3.0 s to start the engine. What amount of electric energy has been withdrawn from the battery?

   (b) The automobile is equipped with a generator that delivers 5.0 A to the battery when the engine is running. How long must the engine run so that the generator can restore the energy in the battery to its original level? Assume that all the power delivered to the battery is stored.

28.3 Single-Loop Circuits†

7. In a flashlight, two 1.5-V batteries are connected in series. An 8.0-Ω lightbulb closes the circuit. What is the current in the circuit? What is the electric energy delivered to the lightbulb in 1.0 h?

† For help, see Online Concept Tutorial 32 at www.wwnorton.com/physics

8. Find the currents in the circuit shown in Fig. 28.10 if the battery on the right is reversed.

9. Two small lightbulbs are connected in parallel to a 6.0-V battery. If the resistance of each lightbulb, at its operating temperature, is 2.0 Ω, what is the current in the circuit? What is the current in each lightbulb?

10. An array of solar cells is to be used to charge a 12-V lead–acid battery. If each solar cell generates 0.60 V, how many cells must we connect in series to charge the battery?

11. Consider the circuit shown in Fig. 28.35. When the resistance of R_2 is decreased, do the currents in R_1 and R_3 increase or decrease?

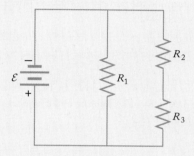

FIGURE 28.35 A source of emf and three resistors.

12. In order to reduce the effective internal resistance of an ammeter, a physicist connects a resistor in parallel across the terminals (see Fig. 28.36). If the resistance of this parallel resistor, or *shunt* resistor, is 1/10 of the resistance of the ammeter, by what factor does the shunt reduce the effective resistance of the modified ammeter? How must the physicist recalibrate the dial of the ammeter?

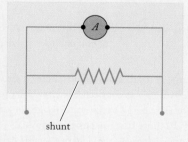

shunt

FIGURE 28.36 An ammeter with a parallel resistor, or shunt resistor.

13. A voltmeter of internal resistance 5.0×10^4 Ω is connected across the poles of a 12-V battery of internal resistance 0.020 Ω.

   (a) What is the current flowing through the battery?

   (b) What is the voltage drop across the internal resistance of the battery?

14. A battery has an emf of 9.00 V when no current flows. When the battery is connected to a 47.0-Ω resistor, a current of 0.187 A flows. What is the internal resistance of the battery?

15. In a set of holiday lights, 25 identical miniature lightbulbs are connected in series to a 115-V source; a 0.074-A current flows through the circuit. What are the voltage drop across and the resistance of each bulb? One bulb burns out. What is the potential across that bulb, which then has infinite resistance? For some lights, this high voltage soon causes an automatic short to develop across the leads to the burned filament, subsequently reducing the resistance of that bulb to zero; thus, "when one burns out, the rest stay lit." Now what is the potential across each of the other bulbs?

16. When a battery is connected to a 22.0-Ω resistor, the voltage drop across the resistor is 5.74 V; when connected to a 47.0-Ω resistor, the voltage drop is 6.19 V. What is the emf of the source? What is the internal resistance of the battery?

17. In Fig. 28.37, $\mathcal{E} = 12$ V, $R_1 = 3.0\,\Omega$, $R_2 = 8.0\,\Omega$, $R_3 = 3.0\,\Omega$, and $R_4 = 5.0\,\Omega$. Determine the currents in the circuit with single-loop analysis as follows:

   (a) Find the current through both the source of emf and R_1 by first finding the equivalent resistance of the entire resistor network.

   (b) Find the currents though R_2, R_3, and R_4 by first subtracting the voltage drop across R_1 from the emf $\mathcal{E}$ to obtain the voltage drop across R_2.

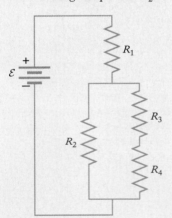

FIGURE 28.37 A source of emf and four resistors.

18. In an effort to recharge a "dead" automobile battery, a strong battery with emf $\mathcal{E}_1 = 12.6$ V is connected to a weak battery with $\mathcal{E}_2 = 11.1$ V, forcing charge backward through the weak battery, as in Fig. 28.10 (note that the positive terminals of the two batteries are connected, as are the negative terminals). Assume that the resistance of the wires and the internal resistance of the batteries total 0.12 Ω, and that the emfs remain constant. How much current flows? How much energy is delivered to the weak batttery in the first 30 seconds? Suppose that, by mistake, the positive terminal of each battery is connected to the negative terminal of the other, a dangerous arrangement. What current flows?

19. To measure the emf and the internal resistance of a battery, an experimenter connects the battery in series to a (resistanceless) ammeter and a variable resistor. She finds that when the variable resistor is set at 1.0 Ω, the current in the circuit is 0.40 A;

and when the variable resistor is set at 2.0 Ω, the current is 0.22 A. What emf and what internal resistance for the battery can she deduce from this? (Hint: Draw a circuit diagram including the internal resistance of the battery. What is the total resistance when the variable resistor is set at 1.0 Ω? When it is set at 2.0 Ω?)

20. Four resistors are connected in parallel to a voltage source as shown in Fig. 28.38. The 25-Ω resistor carries a current of 5.0 A. What is the total current supplied by the battery to the resistors?

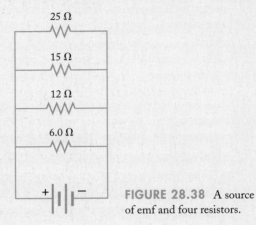

FIGURE 28.38 A source of emf and four resistors.

21. What is the current through the 200-Ω resistor in the circuit shown in Fig. 28.39?

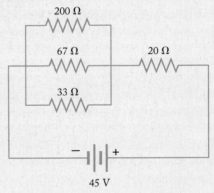

45 V

FIGURE 28.39 A source of emf and four resistors.

*22. Four resistors, with $R_1 = 25\,\Omega$, $R_2 = 15\,\Omega$, $R_3 = 40\,\Omega$, and $R_4 = 20\,\Omega$, are connected to a 12-V battery as shown in Fig. 28.40.

   (a) Find the combined resistance of the four resistors.

   (b) Find the current in each resistor.

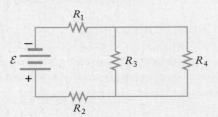

FIGURE 28.40 A source of emf and four resistors.

28.4 Multiloop Circuits†

23. Four wires are connected at a junction. Three of the wires carry currents of 10.0 A, 8.5 A, and 12.5 A into the junction. What are the magnitude and direction of the current in the fourth wire?

24. In the circuit shown in Fig. 28.41, $\mathcal{E}_1$ = 6.0 V, $\mathcal{E}_2$ = 4.0 V, R_1 = 3.0 Ω, and R_2 = 5.0 Ω. What is the current in R_1? In R_2? What is the current through the source $\mathcal{E}_1$?

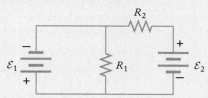

FIGURE 28.41 Two sources of emf and two resistors.

25. Consider the circuit shown in Fig. 28.41. Given that $\mathcal{E}_1$ = 6.0 V and $\mathcal{E}_2$ = 10 V, what must be the value of the resistance R_2 if the current through this resistance is to be 2.0 A?

26. Consider the circuit shown in Fig. 28.42. Given that $\mathcal{E}_1 = \mathcal{E}_2$ = 4.0 V, $R_1 = R_2$ = 2.0 Ω, and R_3 = 1.0 Ω, what is the current in each source? (Hint: Use the symmetry.)

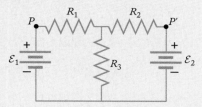

FIGURE 28.42 Two sources of emf and three resistors.

27. In the circuit shown in Fig. 28.43, $R_1 = R_2$ = 2.0 Ω, $R_3 = R_4$ = 4.0 Ω, and $\mathcal{E}_1 = \mathcal{E}_2$ = 10 V. The value of R_5 is unknown. What is the current through each resistor? (Hint: Use the symmetry.)

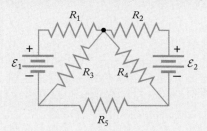

FIGURE 28.43 Two sources of emf and five resistors.

28. Find the currents in the circuit shown in Fig. 28.17a if the battery on the left is reversed.

*29. When using Kirchhoff's voltage rule to obtain the equations for a circuit, we can make several choices for the loops. In Fig. 28.17b, we made one possible choice and obtained Eqs.

†For help, see Online Concept Tutorial 32 at www.wwnorton.com/physics

(28.14) and (28.15). Suppose that instead we make the choice shown in Fig. 28.44. What are the two loop equations in this case? Show that the new loop equations lead to the same result as the old ones.

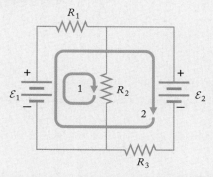

FIGURE 28.44 Alternative choice for voltage loops.

*30. Consider the circuit of Fig. 28.42. Given that $\mathcal{E}_1$ = 5.0 V, $\mathcal{E}_2$ = 3.0 V, R_1 = 4.0 Ω, R_2 = 3.0 Ω, and R_3 = 2.0 Ω, what is the current in each resistor? What is the potential difference between points P and P'?

*31. Consider the circuit shown in Fig. 28.17 with the resistances and emfs given in Example 5. Suppose we replace the emf $\mathcal{E}_1$ = 12.0 V by a larger emf. How large must we make $\mathcal{E}_1$ if the current I_3 is to charge the battery $\mathcal{E}_2$?

*32. Two batteries with internal resistances are connected in the circuit shown in Fig. 28.45. Given that R_1 = 0.50 Ω, R_2 = 0.20 Ω, $\mathcal{E}$ = 12.0 V, $\mathcal{E}'$ = 6.0 V, R_i = 0.025 Ω, and R'_i = 0.020 Ω, find the currents in the resistances R_1 and R_2.

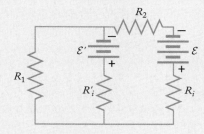

FIGURE 28.45 Batteries with internal resistances R_i and R'_i.

**33. Five resistors, of resistances R_1 = 2.0 Ω, R_2 = 4.0 Ω, R_3 = 6.0 Ω, R_4 = 2.0 Ω, and R_5 = 3.0 Ω, are connected to a 12-V battery as shown in Fig. 28.46

(a) What is the current in each resistor?

(b) What is the potential difference between the points P and P'?

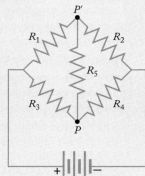

FIGURE 28.46 A source of emf and five resistors.

**34. Two batteries, with $\mathcal{E}_1 = 6.0$ V and $\mathcal{E}_2 = 3.0$ V, are connected to three resistors, with $R_1 = 6.0$ Ω, $R_2 = 4.0$ Ω, and $R_3 = 2.0$ Ω, as shown in Fig. 28.47. Find the current in each resistor and the current in each battery.

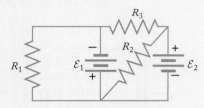

FIGURE 28.47 Two sources of emf and three resistors.

28.5 Energy in Circuits; Joule Heat

35. An electronic calculator operates from a 3.0-V battery. The calculator uses a current of 2.0×10^{-4} A. What power does it use?

36. An electric water heater of resistance 7.5 Ω draws 15 A of current when connected to the voltage supply. What is the cost of operating the water heater for 4.0 h if the electric company charges 15¢ per kilowatt-hour?

37. How much does it cost you to operate a 100-watt lightbulb for 24 hours? The price of electric energy is 15¢ per kilowatt-hour.

38. An air conditioner operating on 115 V uses 1500 W of electric energy. What is the electric current through the air conditioner? What is the resistance of the air conditioner?

39. A cryogenic refrigerator can maintain a low temperature provided that less than 350 microwatts of heat flows into the refrigerator. What maximum current can we permit through a 25-Ω resistor in this refrigerator?

40. Tabletop lasers can produce short pulses of power of 1.0 terawatt (1.0×10^{12} W). If we wish to deliver the same power with a 115-V source of emf, what current must flow? What resistance must be used? The laser pulses are short, approximately 15 femtoseconds (1.5×10^{-14} s). What is the energy per pulse? Suppose that 20 pulses can be produced each second. What is the average power? What current must flow through a 115-V source to produce this average power?

41. An electric heater has two settings, low and high, controlled by a switch. When switched to low, two identical resistors are connected in series and 275 watts are delivered. When switched to high, the two resistors are connected in parallel. What power is delivered then?

42. A photodiode produces a current of 0.50 A per watt of light incident on its surface. In sunlight, 0.10 W of light is incident. The two terminals of the photodiode are connected to a 5.0-Ω resistor. What is the voltage across the resistor? What is the ratio of the power dissipated in the resistor to the power of the incident light? Ignore the internal resistance of the photodiode.

43. A small electric motor operating on 115 V delivers 0.75 hp of mechanical power. Ignoring friction losses within the motor, what current does this motor require?

44. The rate of flow of water over Niagara Falls is 2800 m^3/s; this water falls a vertical distance of 51 m. At night, one-half of the water is diverted to a power plant. If the plant converts all of the gravitational potential energy of this diverted water into electric power, what is the electric power in kilowatts? If this power is fed into a power line at 240 kV, what is the current?

45. A cyclotron accelerator produces a beam of protons of an energy of 700 million eV. The average current of this beam is 1.0×10^{-6} A. What is the number of protons per second delivered by the accelerator? What is the corresponding power delivered by the accelerator?

46. An electric toaster uses 1200 W at 115 V. What is the current through the toaster? What is the resistance of its heating coils?

47. The banks of batteries in a submarine store an electric energy of 5.0×10^3 kW·h. If the submarine has an electric motor developing 1000 hp, how long can it run on these batteries?

48. While cranking the engine, the starter motor of an automobile draws 80 A at 12 V for a time interval of 2.5 s. What is the electric power used by the starter motor? How many horsepower does this amount to? What is the electric energy used up in the given time interval?

49. A 12-V automobile battery stores an energy of 2.5×10^6 J. How long will this battery last if it delivers 1.0 A to a resistor?

50. A house contains appliances with the following power consumptions:

40 lightbulbs, each 75 W	2 air conditioners, 1.2 kW each
1 cooking range, 9.0 kW	2 fans, 0.2 kW each
1 microwave oven, 1.4 kW	1 electric iron, 1.1 kW
1 dishwasher, 1.0 kW	1 hair dryer, 1.4 kW
1 clothes washer, 0.7 kW	1 stereo system, 250 W
1 clothes dryer, 5.0 kW	1 TV, 200 W
1 water pump, 0.7 kW	1 computer, 150 W
1 vacuum cleaner, 1.1 kW	

If all are operating simultaneously, what is the total power required? If the electric power is supplied at 115 V, what is the current?

51. The aluminum cable of a high-voltage transmission line carries a current of 600 A. The cable is 60 km long, and it has a diameter of 2.5 cm. What is the power lost to Joule heating in this cable?

*52. A resistor R, connected to a source of emf $\mathcal{E}$, is dissipating a power P. The voltage is increased to a value 25% larger. How should the resistance be changed to keep the power dissipated by the resistor the same?

*53. A resistor R, connected to a source of emf $\mathcal{E}$, is dissipating a power P. The resistor is then changed to a resistance value 20%

larger. How should the emf be changed to keep the power dissipated by the resistor the same?

*54. To heat 1.00 gram of water (density 1.00 g/cm^3) 1.00°C takes 4.18 joules. How long does it take to heat one liter of water by 50.0°C with an electric heating element of resistance 2.00 Ω carrying a current of 32.0 A?

*55. A hair dryer intended for travelers operates at 115 V and also at 230 V. A switch on the dryer adjusts the dryer for the voltage. At each voltage, the dryer delivers 1000 W of heat. What must be the resistance of the heating coils for each voltage? For such a dryer, design a circuit consisting of two identical heating coils connected to a switch and to the power outlet.

*56. An electric toothbrush draws 7.0 watts. If you use it 4.0 minutes per day and if electric energy costs you 15¢/kW·h, what do you have to pay to use your toothbrush for one year?

*57. In a small electrostatic generator, a rubber belt transports charge from the ground to a spherical collector at 2.0×10^5 V. The rate at which the belt transports charge is 2.5×10^{-6} C/s. What is the rate at which the belt does work against the electrostatic forces?

*58. A solar panel (an assemblage of solar cells) measures 58 cm × 53 cm. When facing the Sun, this panel generates 2.7 A at 14 V. Sunlight delivers an energy of 1.0×10^3 W/m^2 to an area facing it. What is the efficiency of this panel, that is, what fraction of the energy in sunlight is converted into electric energy?

*59. A battery of emf $\mathcal{E}$ and internal resistance R_i is connected to an external circuit of resistance R. In terms of $\mathcal{E}$, R_i, and R, what is the power delivered by the battery to the external circuit? Show that this power is maximum if $R = R_i$.

*60. A 3.0-V battery with an internal resistance of 2.5 Ω is connected to a lightbulb of a resistance of 6.0 Ω. What is the voltage delivered to the lightbulb? How much electric power is delivered to the lightbulb? How much electric power is wasted in the internal resistance?

*61. Suppose that a 12-V battery has an internal resistance of 0.40 Ω.

(a) If this battery delivers a steady current of 1.0 A into an external circuit until it is completely discharged, what fraction of the initial stored energy is wasted in the internal resistance?

(b) What if the battery delivers a steady current of 10.0 A? Is it more efficient to use the battery at low current or at high current?

*62. Two heating coils have resistances 12.0 Ω and 6.0 Ω, respectively.

(a) What is the Joule heat generated in each if they are connected in parallel to a source of emf of 115 V?

(b) What if they are connected in series?

*63. A 40-m cable connecting a lightning rod on a tower to the ground is made of copper and has a diameter of 7.0 mm. Suppose that during a stroke of lightning the cable carries a current of 1.0×10^4 A.

(a) What is the potential drop along the cable?

(b) What is the rate at which Joule heat is produced?

*64. The maximum current recommended for No. 10 copper wire, of diameter 0.259 cm, is 30 A. For such a wire with this current, what is the rate of production of Joule heat per meter of wire? What is the potential drop per meter of wire?

*65. A large electromagnet draws a current of 200 A at 400 V. The coils of the electromagnet are cooled by a flow of water passing over them. The water enters the electromagnet at a temperature of 20°C, absorbs the Joule heat, and leaves at a higher temperature. If the water is to leave with a temperature no higher than 80°C, what must be the minimum rate of flow of water (in liters per minute) through the electromagnet? See Problem 54.

28.6 Electrical Measurements†

66. Most ammeters have different input terminals for measuring large currents and small currents. These terminals have different values of the internal resistance. For one popular digital multimeter, the internal resistance is 11 Ω when measuring 40 mA, but only 0.030 Ω when measuring 3.0 A. What is the power dissipated internally in each case? What would be the power dissipated internally if the 40-mA connections were accidentally used to measure a 3.0-A current?

67. For the Wheatstone bridge of Fig. 28.25, $R_2 = 1.0$ kΩ and $R_3 = 2.0$ kΩ. If the ammeter reads zero when R_1 is adjusted to the value 350 Ω, what is the value of the unknown resistance R_x?

*68. A voltmeter of internal resistance 5000 Ω is connected across the poles of a battery of internal resistance 0.20 Ω. The voltmeter reads 1.4993 volts. What is the actual zero-current emf of the battery?

*69. A circuit consists of a resistor of 3.000 Ω connected to a (resistanceless) battery. To measure the current in this circuit, you insert an ammeter of internal resistance 2.0×10^{-3} Ω. This ammeter then reads 3.955 A. What was the current in the circuit before you inserted the ammeter?

*70. A voltmeter reads 11.9 V when connected across the poles of a battery. The internal resistance of the battery is 0.020 Ω. What must be the minimum value of the internal resistance of the voltmeter if the reading of the instrument is to coincide with the emf of the battery to within better than 1.0%?

*71. Experimentalists usually determine resistance with a **four-wire measurement**. Two wires carry a known current to and from the specimen to be measured, and two other wires connect the specimen to a voltmeter, so that the resistance of the wires does not contribute to the measured voltage drop. For a 5.0-Ω specimen, suppose we instead make a two-wire measurement, that is, we include the resistance of the connecting wires. What total resistance is measured if the wires are 0.25-mm-diameter copper with total length 3.0 m? What if the wire was made from constantan? (Constantan wires are often used for thermal isolation.)

---

†For help, see Online Concept Tutorial 32 at www.wwnorton.com/physics

28.7 The RC Circuit†

72. For a laboratory demonstration, you want to construct an *RC* circuit with a characteristic time of 15 s. You have available a capacitor of 20 μF. What resistance do you need?

73. Consider the circuit shown in Fig. 28.48. How long does it take after the switch is closed to charge the capacitor to within 63% of its final voltage?

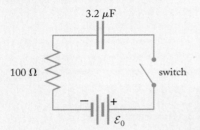

FIGURE 28.48
Charging a capacitor.

74. A capacitor charged through a 75-Ω resistor takes 2.7 ms to come within 63% of its final voltage. What is its capacitance?

75. How long does it take for a fully charged 4.0-μF capacitor to decay to 37% of its initial charge when discharged through a 100-Ω resistor?

76. An *RC* circuit consists of a resistance *R*, a capacitance *C*, and a battery connected in series. At what time is the current 1/10 of its initial (maximum) value? At what time is the current 1/100 of its initial value?

77. Two capacitors of 2.0 μF and 4.0 μF and a resistor of 8.0×10^3 Ω are connected to a battery as shown in Fig. 28.49. What is the characteristic time of this circuit?

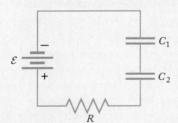

FIGURE 28.49
Charging two capacitors.

78. For the case of a charging capacitor [Eq. (28.31)], what percentage of the final value has the charge reached after two time constants? After three? After five?

79. For the circuit shown in Fig. 28.50, the switch *S* has been in position 2 for a long time. At *t* = 0, it is moved to position 1. (a) Write an expression for the charge *Q* on the capacitor as a function of the time *t*. (b) Determine the voltage across resistor R_1 as a function of the time *t*.

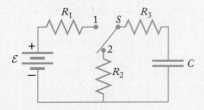

FIGURE 28.50
A switchable *RC* circuit.

†For help, see Online Concept Tutorial 32 at www.wwnorton.com/physics

*80. For the circuit of Fig. 28.50, assume that the switch has been in position 1 for a long time, and that at *t* = 0 it is moved to position 2. What is the current in resistor R_2 as a function of time? What is the power dissipated in resistor R_2 as a function of time? What is the total energy dissipated in resistor R_2 for $0 \le t \le \infty$?

*81. In many *RC* circuits, the charge on the capacitor can be written as $Q = A + Be^{-t/\tau}$, where *A*, *B*, and τ can be determined by inspecting the circuit. If we evaluate this expression at *t* = 0, we see that the initial charge equals *A* + *B*, and if we do so at *t* = ∞, we see that the final charge equals *A*. Consider the circuit shown in Fig. 28.51.

(a) Suppose the switch has been in position 1 for a long time. At *t* = 0, it is moved to position 2. What is *Q* as a function of time, that is, what are *A*, *B*, and τ for this case?

(b) Now suppose that the switch has been in position 2 for a long time. We reset our clock, and at *t* = 0 the switch is moved to position 1. What is *Q* as a function of time for this case?

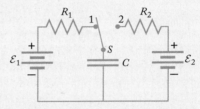

FIGURE 28.51   A switchable
RC circuit with two sources of emf.

*82. A capacitor with *C* = 0.25 μF is initially charged to a potential of 6.0 V. The capacitor is then connected across a resistor and allowed to discharge. After a time of 5.0×10^{-3} s, the potential across the capacitor has dropped to 1.2 V. What value of the resistance can you deduce from this?

*83. Consider the *RC* circuit described in Example 9. The final (asymptotic) value of the charge in the capacitor is $\mathcal{E}C$. At what time is the charge in the capacitor one-half of this final value? At what time is the electric energy in the capacitor one-half of its final value?

*84. Equation (28.27) is a differential equation for the charge. Substitute the expression (28.31) into this equation and verify explicitly that it is a solution.

**85. Equations (28.31) and (28.33) describe the charge and current in an *RC* circuit with a battery. From these equations, deduce the total energy dissipated as Joule heat in the resistor in the time interval *t* = 0 to *t* = ∞, and deduce the total energy that the battery has to deliver in this time interval.

**86. An *RC* circuit, with $R = 6.0 \times 10^5$ Ω and *C* = 9.0 μF, is connected to a 5.0-V battery until the capacitor is fully charged. Then, the battery is suddenly replaced with a new 3.0-V battery of opposite polarity. At what time after this replacement will the voltage across the capacitor be zero? What will be the current in the resistor at this time? (Hint: see Problem 81.)

REVIEW PROBLEMS

87. Figure 28.52 shows the path of an electric current through a human body when the right hand is in good contact with a high-voltage conductor, such as a power cable, and the right foot is in good contact with the ground. The numbers give the resistances of the body parts in ohms (assume that the current punctures the skin, so the skin resistance is zero).

(a) What is the net resistance of the body?

(b) What is the current if the potential at the hand is 600 kV?

(c) What would be the resistance and the current if instead of contact at the hand and the foot, the body made contact with the cable at the shoulder joint and with the ground at the middle of the thigh?

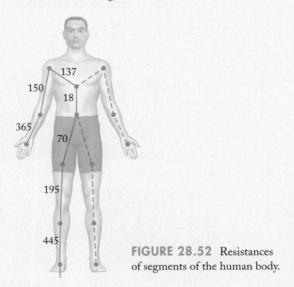

FIGURE 28.52 Resistances of segments of the human body.

88. Two wires of equal lengths and equal diameters are connected in parallel to a source of emf. One wire is made of copper, and the other of aluminum. What is the ratio of the Joule heats produced in these two wires?

89. When connected in parallel across a 115-V source, two lightbulbs deliver 75 W and 130 W, respectively. What powers do these lightbulbs deliver if instead they are connected in series across the same source? Which of the lightbulbs now delivers the larger power? Assume that the resistance of each lightbulb is a constant.

*90. Find the currents in the two resistors in the circuit shown in Fig. 28.53. The resistances are $R_1 = 10\ \Omega$, $R_2 = 8.0\ \Omega$; the emfs are $\mathcal{E}_1 = 10$ V and $\mathcal{E}_2 = 12$ V. Find the power delivered by the 12-V battery.

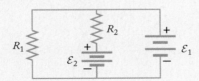

FIGURE 28.53 Two sources of emf and two resistors.

*91. Two batteries of emf $\mathcal{E}$ and of internal resistances R_i and R_i', respectively, are combined in parallel. The combination is connected to an external resistance R. Find the current through each battery.

*92. Two batteries and three resistors are connected as shown in Fig. 28.54. Given that $R_1 = 0.25\ \Omega$, $R_2 = 0.20\ \Omega$, $R_3 = 0.50\ \Omega$, $\mathcal{E}_1 = 6.0$ V, and $\mathcal{E}_2 = 12.0$ V, find the currents in the resistors R_1 and R_3.

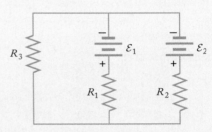

FIGURE 28.54 Two sources of emf and three resistors.

*93. Four resistors, with $R_1 = 10\ \Omega$, $R_2 = 8.0\ \Omega$, $R_3 = 20\ \Omega$, and $R_4 = 10\ \Omega$, are connected to two batteries with $\mathcal{E}_1 = 12$ V and $\mathcal{E}_2 = 16$ V as shown in Fig. 28.55. Find the currents in the batteries, and find the power they deliver.

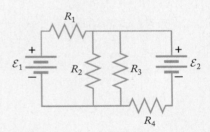

FIGURE 28.55 Two sources of emf and four resistors.

*94. A 12-V battery of internal resistance $0.20\ \Omega$ is being charged by an external source of emf delivering 6.0 A.

(a) What must be the minimum emf of the external source?

(b) What is the rate at which heat is developed in the internal resistance of the battery?

95. An engineer wants to design a high-voltage transmission line 100 km long. The aluminum cable used in this transmission line is to have a resistance of no more than $6.0\ \Omega$.

(a) What diameter of the cable is required?

(b) For a cable of this diameter carrying a current of 750 A, what is the power lost to Joule heating?

96. You want to design a hot-water heater for use in a house. The heater is to operate at 115 V, and it is to heat 150 liters of water from an initial temperature of 10°C to a final temperature of 60°C. If the heater is to take no more than 1.0 h to reach this final temperature, what electric power is required?

What must be the resistance of the electric heating coil in this heater? Assume that no water is taken out of the heater before it reaches its final temperature, and assume that the heater is well insulated and loses no heat to the environment. See Problem 54.

*97. An electric automobile (Fig. 28.56) is equipped with an electric motor supplied by a bank of sixteen 12-V batteries connected in series. When fully charged, each battery stores an energy of 2.2×10^6 J.

   (a) What current is required by the motor when it is delivering 12 hp? Ignore friction losses.

   (b) With the motor delivering 12 hp, the car has a speed of 65 km/h (on a level road). How far can the automobile travel before its batteries run down?

FIGURE 28.56 An electric automobile operated by batteries.

*98. An electric clothes dryer operates on a voltage of 230 V and draws a current of 20 A. How long does the dryer take to dry a full load of clothes? The clothes weigh 6.0 kg when wet and 3.7 kg when dry. Assume that all the electric energy going into the dryer is used to evaporate the water (the heat of evaporation is 2.26×10^6 J/kg).

*99. The cable connecting the electric starter motor of an automobile with the 12.0-V battery is made of copper and has a diameter of 0.50 cm and a length of 0.60 m. If the starter motor draws 500 A (while stalled), what is the rate at which Joule heat is produced in the cable? What fraction of the power delivered by the battery does this Joule heat represent?

100. Ideal capacitors have an infinite internal resistance between their plates (i.e., the material between the plates is a perfect insulator). However, real capacitors have a finite internal resistance, and consequently the charge will gradually leak from one plate to the opposite, and the capacitor will gradually discharge when left to itself. If a capacitor of 8.0 μF has an internal resistance of 5.0×10^8 Ω, how long does it take for one-half of its initial charge to leak away?

101. A capacitor with $C = 20$ μF and a resistor with $R = 100$ Ω are suddenly connected in series to a battery with $\mathcal{E} = 6.0$ V.

   (a) What is the charge on the capacitor at $t = 0$? At $t = 0.0010$ s? At $t = 0.0020$ s?

   (b) What is the final value of the charge?

   (c) What is the rate of increase of the charge at $t = 0$?

Answers to Checkups

Checkup 28.1

1. The energy is dissipated as heat in the wire.

2. No, the electrons pass through the battery in the direction opposite to the current (from the + terminal to the − terminal), and so they gain potential energy.

3. (A) The same energy. A battery has some given amount of stored chemical energy; in Example 1, the amount is 5400 J. Doubling the current means that the battery will deliver the same energy in half the time.

Checkup 28.2

1. Both charged and uncharged batteries have zero net charge. It is the arrangement of the charges, and thus the stored energy of the system, that is different.

2. A battery stores energy in the arrangement and type of chemical bonds, i.e., chemical energy.

3. Like a battery, the fuel of a fuel cell stores energy in the arrangement and type of chemical bonds, i.e., chemical energy.

4. The energy is chemical in the battery; the capacitor (two oppositely charged conductors) stores electric potential energy in the arrangement of charges.

5. (B) Electromotive force. Both arrangements store the same total energy of the two batteries in the form of chemical energy. For the series-stacked batteries, the potential energy of a charge increases for each battery it passes through (as in the case of the cells of an automobile battery), while for the parallel arrangement, a charge can pass through only one battery or the other. Thus the emf of the series arrangement is twice that of the parallel pair.

Checkup 28.3

1. Doubling the emf of every battery will double each current in the circuit, since current depends linearly on emf (Ohm's Law,

$I = \mathcal{E}/R$). Similarly, if we follow the emf doubling by doubling all of the resistances, the currents will return to their original values, since current is inversely proportional to resistance.

2. Reversing the potentials of the batteries will also reverse the directions of all of the currents in the circuit, but will leave the magnitudes of the currents unchanged. Reversing all or any of the resistors changes nothing, since they still offer the same opposition to current flow; the orientation of a resistor does not matter.

3. The current directions may be determined by considering Kirchhoff's voltage rule for the right and outer loops. From the three right loops, we see that current must flow downward through the center resistor in each circuit. From the outer-loop sources of emf, we see that in the first and second circuits, the currents around the outside must travel counterclockwise. In the third circuit, the balanced potentials result in zero current in the lower resistor; this implies that the downward current in the center resistor must travel back upward through the right source of emf.

4. (C) Much less than $\mathcal{E}$. Since the same current I flows through both the internal resistance and the wire, and since R_i is much greater than R, the terminal voltage IR will be much less than the voltage drop IR_i across the internal resistance.

Checkup 28.4

1. From Fig. 28.16, we see that three closed loops are enough so that every part of the circuit is included in at least one of the loops. However, any such three loops will suffice—it is not required to choose the three shown. We also see that five separate currents can be identified, that is, we have five unknowns. Since we have three voltage equations, we need two more equations; thus, two of the three junctions must be evaluated.

2. For this circuit, any two loops are enough to include every part of the circuit.

3. (A) $-I_3R_3 - I_1R_1 + \mathcal{E}_1 = 0$. Starting at point A and proceeding clockwise, we go through both R_3 and R_1 in the same direction as chosen for each current, so each is a voltage drop and thus includes a minus sign. As we continue around, we go through the source of emf in the forward direction ($-$ to $+$ terminal), so that potential change is an increase, and is written as positive.

Checkup 28.5

1. Since $P = (\Delta V)^2/R$, doubling the resistance will halve the power, so the rate of production of heat will be 0.2 W.

2. Doubling each emf will increase the rate of production of Joule heat by a factor of 4, since the power is $P = (\Delta V)^2/R$.

3. The city does not consume electric charge or current; the charge flows through the city and back to the power plant, and the current is the flow of that charge. The city does consume electric energy; it receives the charge at a high potential energy and returns it at a low potential energy.

4. (E) $\frac{1}{9}$. The power dissipated by one resistor, $P = (\Delta V)^2/R$, is proportional to the square of the potential across the resistor. The full emf $\mathcal{E}$ appears across each of the parallel resistors in Fig. 28.21a, but for the three series resistors of Fig. 28.21b, the potential difference across each is only $\frac{1}{3}\mathcal{E}$. Thus the power per resistor in Fig. 28.21b is $(\frac{1}{3})^2 = \frac{1}{9}$ of that in Fig. 28.21a.

Checkup 28.6

1. An ammeter has a low internal resistance, and measures the current passing through it with (ideally) no appreciable voltage drop across it; a voltmeter has a very high internal resistance, and measures the potential difference between its terminals without (ideally) drawing any appreciable current.

2. In Fig. 28.23c, the ammeter is not in series with the resistance, and so will not measure the current through the resistance (the current to be measured must flow through the ammeter). In Fig. 28.24b, the voltmeter is not connected across the two end terminals of the resistor, and so will not measure the potential difference across the resistor (the potential difference to be measured must be applied across the terminals of the voltmeter).

3. (A) Larger. Since an ammeter has a low internal resistance, placing it in parallel with one of the resistors as in Fig. 28.23c will decrease the overall resistance of the circuit; nearly all of the resulting larger current will flow through the low-resistance ammeter.

Checkup 28.7

1. The characteristic time depends only on the product of the values of R and C, and does not depend on the emf of the voltage source.

2. By Eq. (28.37), the initial current is $I = \mathcal{E}/R = (1.5 \text{ V})/(8.0 \times 10^3 \text{ }\Omega) = 1.9 \times 10^{-4} \text{ A}$ and the final current is zero, as in the charging case. The characteristic time is $\tau = RC = 8.0 \times 10^3 \text{ }\Omega \times 2.0 \times 10^{-6} \text{ F} = 1.6 \times 10^{-2} \text{ s}$.

3. (A) Less than 63% of $C\mathcal{E}$. Doubling the resistance to $2R$ doubles the time constant, so the charge now takes a time $t = 2RC$ to reach $\approx 63\%$ of its final value. Thus at half that time, $t = RC$, the charge on the capacitor will be appreciably less than 63% of its final value.

Magnetic Force and Field

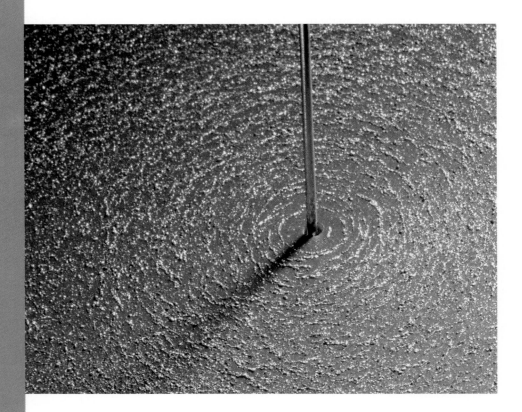

CONCEPTS IN CONTEXT

Concepts
— in —
Context

The straight wire carries a strong current, and this current forces the small iron particles in the photo into the circular patterns shown. We will see in this chapter that a current exerts a magnetic force on other moving charges or currents, such as the microscopic currents in the iron particles; and we will describe this force in terms of the magnetic field produced by the current.

For a current in a straight wire, we will consider such questions as

? What is the force on a moving charge near a long wire? (Section 29.1, page 928; and Example 1, page 930)

? What is the direction of the magnetic field near the wire? (Section 29.2, page 936; and Example 4, page 940)

? How does the magnetic field outside the wire vary with distance? (Section 29.2, page 939; and Example 4, page 940)? The magnetic field inside the wire? (Example 5, page 940)

? What is the total magnetic field due to two wires? (Example 3, page 938)

? What is the magnetic field near a short, straight wire? (Section 29.5, page 949)

The magnetic forces most familiar from everyday experience are the forces that permanent magnets exert on each other or on pieces of iron or other "magnetic" materials. If you bring two bar magnets together, they will attract or repel. Each magnet has two distinct ends, called its north pole and its south pole. If you bring like poles together, they repel; and if you bring opposite poles together, they attract (see Fig. 29.1). This attraction or repulsion between permanent magnets also accounts for the behavior of a compass needle, which is simply a small magnet free to rotate on a pivot. When we place a compass needle near the, say, south pole of a bar magnet, this south pole will attract the north pole of the compass needle, and the needle will settle into an equilibrium configuration, pointing roughly toward the south pole of the bar magnet (see Fig. 29.2). When we remove the compass needle from the disturbing influence of nearby bar magnets, it will respond to the magnetic force exerted by the Earth, and it will point northward. The core of the Earth acts like a large permanent magnet, whose poles roughly coincide with the geographic poles (see Fig. 29.3; the north pole of the compass needle is attracted by the geographic north pole of the Earth, because this geographic north pole is actually the magnetic south pole of the Earth).

The magnetic forces between permanent magnets were known for many centuries, but only during the nineteenth century did experimenters discover that electric currents also exert magnetic forces on permanent magnets (see Fig. 29.4) and that electric currents exert magnetic forces on one another. Finally, physicists came to understand that *the magnetic force is simply an extra electric force acting between charges in motion.* The ordinary electric force, given by Coulomb's Law, always acts between electric charges, when they are at rest and when they are in motion. The extra magnetic force acts between electric charges only when they are in motion. Whether the charges are moving along a wire, forming currents, or whether the charges are moving through empty space, on their own, makes no essential difference; in both cases *moving charges exert magnetic forces on one another.*

The magnetic forces between permanent magnets involve the same fundamental mechanism as the magnetic forces between moving charges or between currents. The magnetic forces between permanent magnets arise from microscopic currents in the atoms of the magnet. Within these atoms, currents flow around in closed loops.

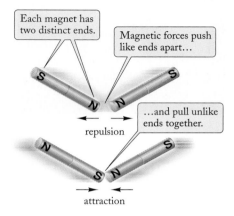

FIGURE 29.1 Equal ends of two bar magnets repel; opposite ends attract.

FIGURE 29.2 A compass needle placed near a bar magnet.

FIGURE 29.3 The Earth behaves like a large permanent magnet with magnetic poles nearly opposite to the geographic poles.

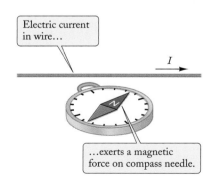

FIGURE 29.4 A compass needle placed near a wire carrying a current. The needle tends to orient at right angles to the wire.

FIGURE 29.5 A magnet placed near a TV tube distorts the image.

Although the individual microscopic rings of current are too weak to produce a noticeable force, the magnet contains many atoms and many rings of current, and the combined effect of all these microscopic rings gives a noticeable macroscopic force on another nearby magnet or on a nearby moving charge or current. You can observe the force exerted by a magnet on a moving charge if you place the magnet near the screen of your TV tube or computer monitor.[1] The magnetic force on the moving electrons in the tube will then deflect the electrons and produce strange distortions of the image on your screen (see Fig. 29.5).

In this chapter we will see that the magnetic force acting on a moving charge can be expressed in terms of a magnetic field, in much the same way that the electric force can be expressed in terms of an electric field. We will examine how magnetic fields are generated by currents, and we will examine several ways to calculate the magnetic fields generated by given distributions of currents.

29.1 THE MAGNETIC FORCE

In principle, it would be desirable to begin the study of magnetic forces with the law for the magnetic force between two moving point charges, just as we began our study of electric forces with the law for the electric force between two point charges at rest (Coulomb's Law). However, the magnitude and direction of the magnetic force that two moving charges exert on each other depends in a complicated manner on the magnitudes and directions of their velocities, and the mathematical formula for the force is rather messy.

Instead of dealing with this complicated case, it is easier to begin with the magnetic force exerted on one moving point charge q by a steady current I flowing on a long, straight wire. We will take this case as the basis for our study of magnetic forces, and we will regard the formula for the magnetic force exerted on a moving point charge by a current flowing on a long, straight wire as a fundamental law of physics, justified by experiments. The formula for this magnetic force is somewhat similar to the formula for the electric force exerted on a point charge by a long, straight charged rod. According to Example 5 of Chapter 23, the electric force that such a charged rod exerts on the point charge is inversely proportional to the distance r between the point charge and the rod. This inverse proportionality to the distance r also holds for the magnetic force **F** that a current in a long wire exerts on a moving point charge. However, in contrast to the electric force, the magnetic force also depends on the velocity **v** of the moving charge—the magnitude and the direction of the magnetic force **F** depend on the magnitude and direction of the velocity **v**. To spell out this dependence in detail, it is useful to consider three separate cases:

1. v *parallel to the current.* As illustrated in Fig. 29.6a, the current I is in the x direction and the velocity **v** of the point charge q is in the same direction. In this case, the magnetic force is in the radial direction, toward or away from the current. The *magnitude of the magnetic force is directly proportional to the product of the current I, the charge q, and the speed v, and it is inversely proportional to the distance r:*

$$F = -[\text{constant}] \times \frac{qvI}{r} \tag{29.1}$$

[1]You should try this experiment only with an old, unwanted screen or monitor, since you might cause a permanent discoloration of the screen. Some computer monitors possess a "degauss" capability that can reverse such discoloration.

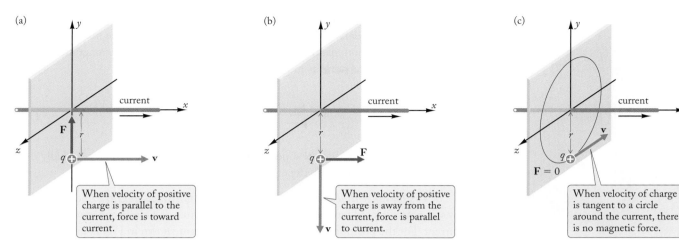

FIGURE 29.6 Direction of the magnetic force for three possible directions of the velocity of the charge q (in these diagrams, q is assumed positive). The current on the wire is in the x direction in each case. (a) The velocity of the charge q is in the x direction. The magnetic force is in the radially inward direction. (b) The velocity of the charge q is perpendicularly away from the current. The magnetic force is in the x direction. (c) The velocity of the charge q is in the tangential direction. The magnetic force is then zero.

The minus sign in this formula indicates that the force is attractive if the charge q is positive and the velocity **v** and the current I are in the same direction. Thus, in contrast to the repulsive electric force, the magnetic force between a positive charge and a current is attractive when the velocity and the current are in the same direction. However, that force is repulsive when the velocity and the current are in opposite directions. More generally, if we reckon v as positive when parallel to the current and negative when antiparallel, then the magnetic force is attractive if the product qvI is positive and repulsive if qvI is negative. Note that the magnetic force is zero unless both the speed v and the current I are different from zero—the charge q and the charges on the wire must *both* be in motion if there is to be a magnetic force between them.

In the SI system of units, the numerical value of the constant of proportionality in Eq. (29.1) is exactly

$$[\text{constant}] = 2 \times 10^{-7} \ \frac{\text{N·s}^2}{\text{C}^2} \tag{29.2}$$

This constant is conventionally written in the form

$$[\text{constant}] = \frac{\mu_0}{2\pi} \tag{29.3}$$

with

$$\mu_0 = 4\pi \times 10^{-7} \ \frac{\text{N·s}^2}{\text{C}^2} \approx 1.26 \times 10^{-6} \ \frac{\text{N·s}^2}{\text{C}^2} \tag{29.4}$$

permeability constant

The quantity μ_0 (pronounced mu-nought) is called the **magnetic constant** or the **permeability constant**. With this constant, our equation for the magnetic force on the point charge q becomes

$$F = -\frac{\mu_0}{2\pi} \frac{qvI}{r} \quad \text{for } \mathbf{v} \text{ parallel to current} \tag{29.5}$$

magnetic force exerted on moving charge by current in long wire

2. v *in the radial direction.* This case is illustrated in Fig. 29.6b. The velocity **v** is directed perpendicularly away from the direction of the current. The magnetic force is now in the *x* direction, parallel to the direction of the current. The magnitude of the magnetic force is the same as in the first case:

$$F = +\frac{\mu_0}{2\pi}\frac{qvI}{r} \quad \text{for } \mathbf{v} \text{ in radial direction} \tag{29.6}$$

The + sign indicates that for a positive charge with positive velocity (radially outward), the force is parallel to the current.

3. v *in the tangential direction.* This final case is illustrated in Fig. 29.6c. The velocity **v** is now directed tangentially to the circumference of a circle concentric with the current. For the position of the charge *q* at the instant shown in Fig. 29.6c, the direction of the velocity is the −*z* direction. For the tangential case, the magnetic force is *zero* (and has no direction):

$$F = 0 \quad \text{for } \mathbf{v} \text{ in tangential direction} \tag{29.7}$$

If the direction of the velocity **v** is *not* as described in one of the three basic cases 1, 2, or 3 above, then the velocity can be split into three components along the directions described in these three basic cases, and the magnetic force can be obtained by calculating the contribution for each direction separately and then taking the vector sum of the separate magnetic forces. Thus, in principle, we can calculate the magnetic force on a charge *q* moving in some arbitrary direction by taking a suitable combination of the three basic cases discussed above (in the next section we will learn about a more convenient way of calculating the force on a charge moving in an arbitrary direction, by means of the magnetic field). Note that in all cases, *the magnetic force is perpendicular to the velocity of the point charge*; this perpendicularity of force and velocity is a characteristic feature of the magnetic force.

The formulas (29.5) and (29.6) for the magnetic force exerted by a current on a moving charge are to be regarded as basic laws of physics, rooted in experiments with currents and with moving charges. In our study of magnetism, these laws for the magnetic force play a role analogous to that played in electricity by Coulomb's Law.

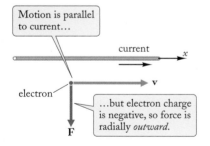

FIGURE 29.7 An electron moves parallel to a wire carrying a current. The magnetic force is directed perpendicularly away from the current.

 Concepts —in— Context

EXAMPLE 1   A long, straight wire carries a current of 50 A. An electron of speed 2.0×10^6 m/s is (instantaneously) moving parallel to this wire at a distance of 0.030 m. What magnetic force does the current in the wire exert on the electron?

SOLUTION: This is essentially case 1 above, but the charge *q* is now negative, $q = -e$. If the current *I* is in the *x* direction, as in Fig. 29.7, and the velocity of the electron is also in the *x* direction, then the magnetic force is repulsive, radially outward from the current, and is of magnitude

$$F = -\frac{\mu_0}{2\pi}\frac{qvI}{r} = +\frac{\mu_0}{2\pi}\frac{evI}{r}$$

$$= 2 \times 10^{-7}\,\frac{\text{N·s}^2}{\text{C}^2} \times \frac{1.6 \times 10^{-19}\,\text{C} \times 2.0 \times 10^6\,\text{m/s} \times 50\,\text{A}}{0.030\,\text{m}}$$

$$= 1.1 \times 10^{-16}\,\text{N}$$

Another instance of a magnetic force is illustrated in Fig. 29.8, which shows two long, straight, parallel wires carrying currents. The conditions here are similar to those described in case 1. The moving charges of one wire exert magnetic forces on the moving charges of the other wire. If the currents are in the same direction, as in Fig. 29.8a, the magnetic force is attractive. If the currents are in opposite directions (antiparallel), as in Fig. 29.8b, then the magnetic force is repulsive. In a later section we will see how to calculate the net magnitude of the magnetic force between the currents on such wires.

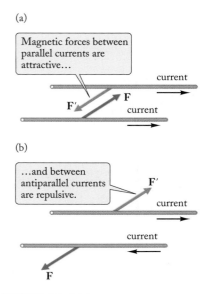

FIGURE 29.8 Two long, straight, parallel wires carrying currents (a) in the same direction and (b) in opposite directions.

 ## Checkup 29.1

QUESTION 1: For each of the following four cases, what is the direction of the magnetic force **F** on the charge q?

   (a) In Fig. 29.6a, the positive charge q is moving toward the left (in the $-x$ direction) instead of the right.

   (b) In Fig. 29.6b, the positive charge q is moving upward (in the $+y$ direction) instead of downward.

   (c) In Fig. 29.6b, a negative charge q is moving downward.

   (d) In Fig. 29.6c, the positive charge q is moving in the $+z$ direction, instead of the $-z$ direction.

QUESTION 2: Suppose we reverse the current in Fig. 29.6. How does this change the magnetic force exerted on the charge q?

QUESTION 3: Suppose we increase the distance between the wire and the charge q to twice the distance shown in Fig. 29.7. How does this change the magnetic force exerted on the charge q?

QUESTION 4: Consider a current flowing along the x axis in the $+x$ direction and a charge q a distance r below the origin on the y axis, as in Fig. 29.6. However, now the velocity **v** of the charge is not parallel to an axis. If the speed of the charge is v_0, the force on the charge is largest when

   (A) $\mathbf{v} = v_0(\mathbf{i} + \mathbf{j})/\sqrt{2}$ (B) $\mathbf{v} = v_0(\mathbf{i} + \mathbf{k})/\sqrt{2}$ (C) $\mathbf{v} = v_0(\mathbf{j} + \mathbf{k})/\sqrt{2}$

29.2 THE MAGNETIC FIELD

In Section 23.1 we saw that the electric force is communicated from one charge q' to another charge q by an electric field. Likewise, *the magnetic force is communicated from one moving charge q' to another moving charge q through a magnetic field*, which serves as the mediator of the magnetic force according to a scheme similar to that for the electric field:

$$\begin{array}{ccccc} \text{current } I & & \text{magnetic field} & & \text{force} \\ (\text{or moving charge } q') & \Rightarrow & \text{of current } I & \Rightarrow & \text{on moving charge } q \end{array}$$

Starting with our formulas for the magnetic force (29.5) or (29.6) exerted by a current, we define the magnetic field by separating the expression for the force into two factors: one factor comprising quantities associated only with the moving point charge (its charge q, its velocity **v**), and another factor comprising quantities associated with the current (its magnitude I and its distance r from the point charge). Thus, in the case of a point charge moving parallel to the current, we write

$$F = (qv)\left(\frac{\mu_0 I}{2\pi r}\right) \tag{29.8}$$

We identify the second factor (including the constant of proportionality μ_0) as the magnetic field of the current, and we designate this magnetic field by B:

magnetic field of current in long wire

$$B = \frac{\mu_0 I}{2\pi r} \tag{29.9}$$

The expression for the magnitude of the magnetic force (29.5) or (29.6) then becomes

$$F = qvB \tag{29.10}$$

Concepts in Context

For a long wire, the direction of the magnetic field of the current is tangent to a circle with the current at its center. Figure 29.9 shows the direction of the magnetic field at several positions around the current. Note that the direction of the magnetic field coincides with that direction of the velocity **v** for which the magnetic force on the moving charge is zero (compare case 3, above). Thus there is no magnetic force when the velocity is parallel to the magnetic field.

right-hand rule for the magnetic field

The direction of the magnetic field of a current on a wire can be determined by a simple **right-hand rule for the magnetic field**: *if the thumb of the right hand is placed along the direction of the current, then the fingers will curl around the wire in the direction of the magnetic field* (see Fig. 29.10). This rule is consistent with the direction of the magnetic field shown in Fig. 29.9.

According to Eq. (29.9), the magnetic field of the current on a long, straight wire is inversely proportional to the distance r. In this regard the magnetic field of the long, straight wire is similar to the electric field of a long straight line of charge [see Eq. (24.23)]. But the directions of these fields differ—the magnetic field is in the tangential direction, whereas the electric field is in the radial direction.

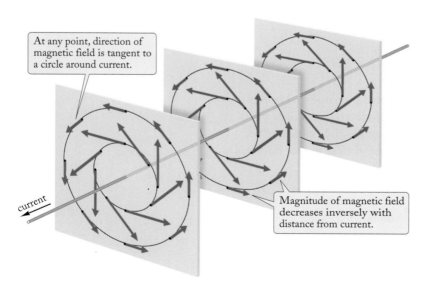

At any point, direction of magnetic field is tangent to a circle around current.

current

Magnitude of magnetic field decreases inversely with distance from current.

FIGURE 29.9 Magnetic field surrounding a current on a wire.

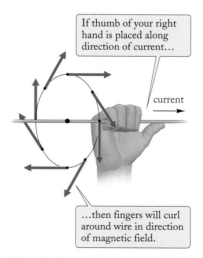

If thumb of your right hand is placed along direction of current...

current

...then fingers will curl around wire in direction of magnetic field.

FIGURE 29.10 The right-hand rule indicates the direction of the magnetic field of a current on a wire.

Equation (29.10) is valid for the case of a point charge moving parallel to the current. But we can easily generalize this equation for all the other cases of motion of the point charge by recognizing that in cases 1 and 2 the point charge is moving perpendicularly to the magnetic field, and in case 3 the point charge is moving parallel to the magnetic field (compare Figs. 29.6 and 29.9). The absence of a magnetic force in the case of motion parallel (or antiparallel) to the magnetic field means that if the point charge moves at some general angle α with respect to the magnetic field (see Fig. 29.11), only the component of its velocity perpendicular to the magnetic field generates a magnetic force, whereas the component parallel to the magnetic field does not. But the component of the velocity $\mathbf{v}$ perpendicular to the magnetic field is $v \sin \alpha$ (see Fig. 29.11). Hence, a general expression for the magnitude of the magnetic force is

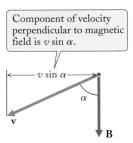

FIGURE 29.11 A point charge q moving at some angle α with respect to the direction of the magnetic field.

$$F = qvB \sin \alpha \qquad (29.11)$$

magnitude of magnetic force

This expression automatically gives zero if $\alpha = 0$ (motion parallel to $\mathbf{B}$), and it gives $F = qvB$ if $\alpha = 90°$ [motion perpendicular to $\mathbf{B}$; see Eq. (29.10)].

By means of the cross product for vectors introduced in Chapter 3, we can write a concise formula that gives both the magnitude and the direction of the force:

$$\mathbf{F} = q\mathbf{v} \times \mathbf{B} \qquad (29.12)$$

magnetic force vector

According to the general formula (3.29) for the magnitude of the cross product, the magnitude of $\mathbf{v} \times \mathbf{B}$ is $vB \sin \alpha$, in agreement with Eq. (29.11). Figure 29.12 illustrates how the cross product of the two vectors $\mathbf{v}$ and $\mathbf{B}$ yields the direction of the force vector $\mathbf{F}$. Note that you must orient your right hand so that the fingers can curl from the first vector $\mathbf{v}$ to the second vector $\mathbf{B}$; the thumb then gives the direction of the cross product $\mathbf{v} \times \mathbf{B}$, and thus, for a positive charge q, the direction of the force $\mathbf{F}$.

If both electric and magnetic fields exert forces on a charge q, the net force on the charge is the vector sum of the individual forces. The total force due to both an electric field $\mathbf{E}$ and a magnetic field $\mathbf{B}$ is called the **Lorentz force**:

$$\mathbf{F} = q\mathbf{E} + q\mathbf{v} \times \mathbf{B} \qquad (29.13)$$

Lorentz force

We will examine situations involving both electric and magnetic fields in the next chapter. For now, we consider magnetic fields only.

Using the definition (29.9) of the magnetic field produced by a current on a long, straight wire as guidance, we can now proceed to define the magnetic field produced by any general distribution of moving charges or currents. To discover the magnetic field at a given position, we use a test charge q and let it move repeatedly through that position,

FIGURE 29.12 This right-hand rule for the magnetic force on a moving point charge relates the directions of the magnetic field $\mathbf{B}$, the magnetic force $\mathbf{F}$, and the velocity $\mathbf{v}$. The vector $\mathbf{F}$ is always perpendicular to the plane defined by the vectors $\mathbf{B}$ and $\mathbf{v}$ (horizontal plane in this figure).

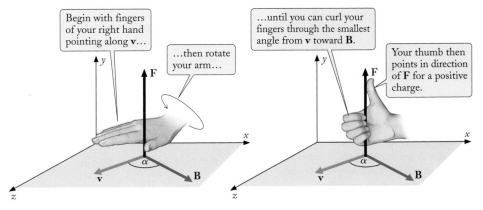

with different velocities, and we measure the force on the test charge. The direction and magnitude of the magnetic field **B** are then determined according to the following rules:

- The direction of the magnetic field is parallel or antiparallel to that direction of motion that results in zero force.
- The magnitude of the magnetic field is obtained by dividing the magnitude of the maximum force (the force that acts when the motion of the test charge is perpendicular to the direction of the magnetic field) by the charge and the speed:

$$B = \frac{F}{qv} \qquad \text{for } \mathbf{v} \text{ perpendicular to } \mathbf{B} \qquad (29.14)$$

right-hand rule for the magnetic force

- The remaining ambiguity in the direction of the magnetic field is resolved by the **right-hand rule for the magnetic force**, which is equivalent to the right-hand rule for the cross product of **v** and **B**: if you place the fingers of your right hand along the direction of the velocity **v** and curl the fingers toward the direction of the magnetic field **B** through the smallest angle between **v** and **B**, your thumb will lie along the direction of the force **F** experienced by a positive test charge (see Fig. 29.12). For a negative test charge, the magnetic force is opposite to that for a positive test charge. You can check that the direction of the magnetic force exerted by the current in a wire (see cases 1 and 2 in Section 29.1) agrees with this right-hand rule.

With these definitions of the magnitude and the direction of a general magnetic field, the force on a charged particle moving in some arbitrary direction will always be given by the cross-product formula (29.12).

According to Eq. (29.14), the magnetic field is the force per unit charge and unit velocity. The SI unit of magnetic field is N/(C·m/s), the unit of force divided by the unit of charge and of velocity; this unit is called the **tesla** (T):

tesla (T)

gauss (G)

$$1 \text{ tesla} = 1 \text{ T} = 1 \text{ N/(C·m/s)}$$

A non-SI unit of magnetic field in common use is the **gauss** (G):

$$1 \text{ gauss} = 1 \text{ G} = 10^{-4} \text{ T}$$

Table 29.1 lists the values of some typical magnetic fields.

NIKOLA TESLA (1856–1943) *American electrical engineer and inventor. He made many brilliant contributions to high-voltage technology, ranging from new motors and generators to transformers and a system for radio transmission. Tesla designed the power-generating station at Niagara Falls.*

| **EXAMPLE 2** | In Florida, the Earth's magnetic field is in the north–south vertical plane (and toward the north), but directed downward from |

the horizontal at an angle of 58° (see Fig. 29.13a). The magnitude of this magnetic field is 5.3×10^{-5} T. Suppose that an electron in a TV tube is moving with an (instantaneous) horizontal velocity of 2.0×10^{6} m/s in the south to north direction. What are the magnitude and the direction of the force that the magnetic field of the Earth exerts on this electron?

SOLUTION: According to the right-hand rule, the magnetic force would be horizontally toward west for a *positive* charge (see Fig. 29.13b). Hence, for the *negative* electron, the magnetic force is horizontally toward east. The magnitude of this force is given by Eq. (29.11):

$$F = evB \sin\alpha = 1.6 \times 10^{-19} \text{ C} \times 2.0 \times 10^{6} \text{ m/s} \times 5.3 \times 10^{-5} \text{ T} \times \sin 58°$$

$$= 1.4 \times 10^{-17} \text{ N}$$

(a)

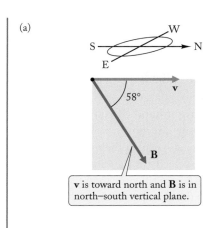

58°

v

B

v is toward north and **B** is in north–south vertical plane.

(b)

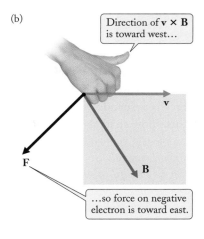

Direction of **v** × **B** is toward west…

v

B

F

…so force on negative electron is toward east.

FIGURE 29.13 (a) Direction of the magnetic field of the Earth in Florida. An electron is moving south to north in this magnetic field. (b) Right-hand rule for a *positive* charge. For the *negative* electron, the magnetic force (blue) is opposite to the direction given by the right-hand rule.

TABLE 29.1	**SOME MAGNETIC FIELDS**
At surface of pulsar	$\approx 10^8$ T
Maximum achieved in laboratory:	
Explosive compression of field lines	1×10^3
Steady	45
In particle accelerator magnet	8
In large bubble-chamber magnet	2
In MRI magnet (a)	1.5
In sunspot (b)	≈ 0.3
Near small ceramic magnet	$\approx 2 \times 10^{-2}$
At surface of Sun	$\approx 10^{-2}$
Near household wiring (c)	$\approx 10^{-4}$
At surface of Earth	$\approx 5 \times 10^{-5}$
In sunlight (rms)	3×10^{-6}
In Crab Nebula (d)	$\approx 10^{-8}$
In radio wave (rms)	$\approx 10^{-9}$
In interstellar galactic space	$\approx 10^{-10}$
Produced by human body	3×10^{-10}
In shielded antimagnetic chamber	2×10^{-14}

(a)

(b)

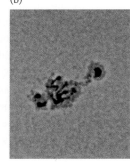

(c)

(d)

PROBLEM–SOLVING TECHNIQUES   DIRECTION OF MAGNETIC FORCE

To determine the direction of a magnetic force $\mathbf{F} = q\mathbf{v} \times \mathbf{B}$ on a charge q, draw the velocity vector $\mathbf{v}$ and the magnetic field vector $\mathbf{B}$ tail to tail. The magnetic force is perpendicular to the plane defined by these two vectors. If the charge q is positive, the direction of the magnetic force is given by the right-hand rule: *place the fingers of your right hand along the direction of the velocity* $\mathbf{v}$ *and curl the fingers toward the direction of the magnetic field* $\mathbf{B}$ *through the smallest angle between* $\mathbf{v}$ *and* $\mathbf{B}$; *your thumb will then point along the direction of the force* $\mathbf{F}$ (see Fig. 1 for several examples). If the

charge q is negative, the direction of the magnetic force is opposite to that for a positive charge.

Be careful to use your **right hand** (if you use your left hand, you will get the opposite direction).

By trial and error, you can also exploit the right-hand rule to find the direction of the velocity, if the directions of the force and the magnetic field are specified; or to find the direction of the magnetic field, if the directions of the force and the velocity are specified.

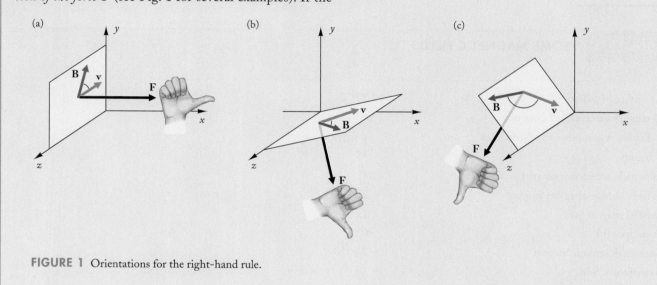

FIGURE 1 Orientations for the right-hand rule.

Concepts
— *in* —
Context

The magnetic field can be represented graphically by field lines. As in the case of the electric field, the tangent to the field lines indicates the direction of the field, and the density of field lines indicates the relative strength of the field. Figure 29.14 shows the pattern of magnetic field lines for the magnetic field produced by a current on a long, straight wire. The decrease of the strength of the magnetic field with distance is indicated by the decrease of density of the field lines. The magnetic field of the straight wire in the chapter photo is made visible by sprinkling small iron filings on a sheet of paper placed around the wire. The iron filings behave like small compass needles such as those shown in Fig. 29.15, and they align in the direction of the magnetic field. Figures 29.16 and 29.17 show the pattern of magnetic field lines for the magnetic field produced by a bar magnet. *The end of the magnet from which the field lines emerge is the north pole of the magnet, and the end into which the field lines enter is the south pole.*

Note that the magnetic field lines in Figs. 29.14 and 29.16 form closed loops, that is, the magnetic field lines do not begin or end anywhere in the way that electric field lines begin and end on positive and negative charges. Isolated magnetic poles, or magnetic monopoles, do not exist. Since there is no such "magnetic charge" that acts as source or sink of magnetic field lines, the magnetic field lines of any kind of magnetic

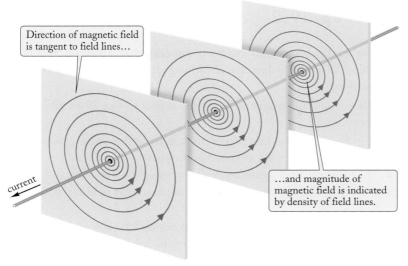

Direction of magnetic field is tangent to field lines…

…and magnitude of magnetic field is indicated by density of field lines.

current

FIGURE 29.14 Magnetic field lines around a current on a long, straight wire.

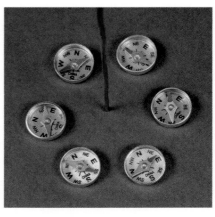

FIGURE 29.15 Circular magnetic field line around a current in a long, straight wire indicated by alignment of compass needles.

field must always form closed loops. Mathematically, we can express this feature of the magnetic field by stating *that the total magnetic flux* Φ_B *through any closed surface is zero*:

$$\Phi_B = 0 \quad \text{or} \quad \oint \mathbf{B} \cdot d\mathbf{A} = 0 \tag{29.15}$$

Gauss' Law for the magnetic field

Here the magnetic flux is defined in the same way as the electric flux: the magnetic flux through a surface is the integral of the perpendicular component of the magnetic field over the surface. Accordingly, the magnetic flux is proportional to the net number of magnetic field lines intercepted by the surface. Thus, Eq. (29.15) states that the number of magnetic field lines entering any closed surface matches the number leaving the surface (see Fig. 29.18). Equation (29.15) is **Gauss' Law for the magnetic field**.

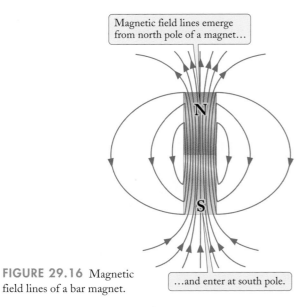

Magnetic field lines emerge from north pole of a magnet…

N

S

FIGURE 29.16 Magnetic field lines of a bar magnet.

…and enter at south pole.

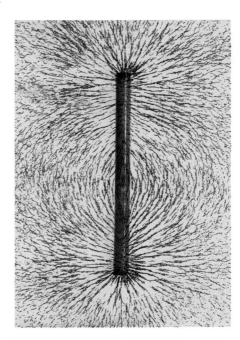

FIGURE 29.17 Magnetic field lines of a bar magnet made visible by iron filings sprinkled on a sheet of paper placed over the magnet.

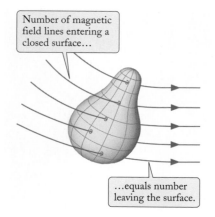

FIGURE 29.18 A closed surface inter-cepting magnetic field lines.

 Checkup 29.2

QUESTION 1: You are standing directly under a power cable that carries a current in the northward direction. What is the direction of the magnetic field that this current produces at your position?

QUESTION 2: How would the magnetic field lines shown in Fig. 29.14 differ if the current were reversed?

QUESTION 3: If in Fig. 29.15 the direction of the current on the wire is upward, what is the direction of the magnetic field around this current?

QUESTION 4: The magnetic field produced by a long, straight power cable is 2×10^{-5} T at a distance of 6 m. What is the magnitude of the magnetic field at a distance of 3 m? 2 m?

QUESTION 5: A magnetic field is directed vertically downward. A proton is moving eastward in this magnetic field. What is the direction of the magnetic force?

  (A) Up   (B) North   (C) South   (D) East   (E) West

29.3 AMPÈRE'S LAW

The magnetic fields produced by currents flowing in wires are of great practical interest, because many applications of magnetism, such as electromagnets and electric motors, rely on currents in wires to produce magnetic fields. However, the arrangements of wires used in practical applications are usually much more complicated than the single long, straight wire we dealt with in the preceding sections. We therefore need to develop a more general method for the calculation of magnetic fields.

Superposition Principle

The net magnetic field produced by several wires or other current distributions, each of which produces an individual magnetic field, obeys the **Superposition Principle**. This principle states that *the net magnetic field produced by several currents is the (vector) sum of the individual magnetic fields of the individual currents*. The following example illustrates this Superposition Principle.

Concepts — in — Context

EXAMPLE 3   A high-voltage transmission line consists of two long parallel wires separated by a distance of 2.0 m. The wires carry currents of 800 A in opposite directions (see Fig. 29.19a). What is the net magnetic field that these wires produce jointly at a point midway between them?

(a)

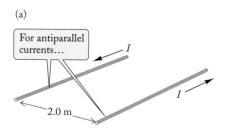

(b)

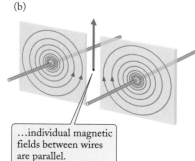

FIGURE 29.19 (a) Two long, parallel wires carrying opposite currents. (b) Magnetic field lines of the wires.

SOLUTION: Figure 29.19b shows the magnetic field lines of each wire. Since the currents are opposite, the field lines curl around the two wires in opposite directions. At the midpoint between the wires, the individual magnetic fields of the two wires are parallel. These individual magnetic fields therefore add, and the net magnetic field is twice as large as the individual magnetic field of each wire. By Eq. (29.9), the individual magnetic field of each wire is $\mu_0 I/2\pi r$, and hence the net magnetic field is

$$B = 2 \times \frac{\mu_0}{2\pi} \frac{I}{r} = 2 \times \frac{4\pi \times 10^{-7}\,\text{N·s}^2/\text{C}^2}{2\pi} \times \frac{800\,\text{A}}{1.0\,\text{m}}$$
$$= 3.2 \times 10^{-4}\,\text{T}$$

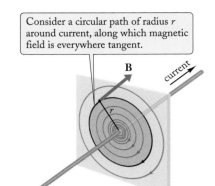

Consider a circular path of radius r around current, along which magnetic field is everywhere tangent.

FIGURE 29.20 A circular path of radius r around a long, straight wire carrying a current I. This circular path coincides with a magnetic field line, and the magnetic field is everywhere tangent to this path.

The expression (29.9) for the magnetic field produced by a current on a long, straight wire leads to an interesting relationship between field and current. We can rewrite Eq. (29.9) as

$$2\pi r B = \mu_0 I$$

or as

$$Bs = \mu_0 I \qquad\qquad (29.16)$$

where $s = 2\pi r$ is the circumference of the circle of radius r (see Fig. 29.20). In words, Eq. (29.16) states that the magnetic field along the circumference of a circle multiplied by the length of this circumference equals μ_0 times the current intercepted by the area within the circle.

In a slightly modified form, this statement is valid for a closed path of arbitrary shape in the magnetic field of an arbitrary distribution of currents. Consider some closed mathematical path (see Fig. 29.21), and designate by $B_\parallel$ the component of the magnetic field along the path, that is, the component parallel, or tangent, to the path. For a path element $d\mathbf{s}$ that makes an angle θ with the magnetic field, the product of the component along the path and the length of the path is $B_\parallel\, ds = B\cos\theta\, ds = \mathbf{B}\cdot d\mathbf{s}$. Then **Ampère's Law** states

The integral around a closed path of the component of the magnetic field tangent to the direction of the path equals μ_0 times the current intercepted by the area within the path:

$$\oint \mathbf{B}\cdot d\mathbf{s} = \mu_0 I \qquad\qquad (29.17)$$

or, more simply,

$$\oint B_\parallel\, ds = \mu_0 I \qquad\qquad (29.18)$$

where the circle on the integral symbol here indicates that we perform the integration of $B_\parallel$ around a *closed* path; that is, we sum [field] × [length] contributions around a closed path.

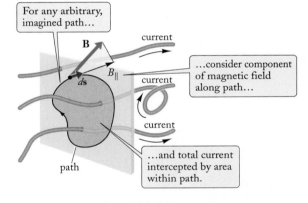

For any arbitrary, imagined path…

…consider component of magnetic field along path…

…and total current intercepted by area within path.

FIGURE 29.21 A closed path in a magnetic field. The area (orange) within the path intercepts some currents flowing on wires or on other conductors.

Ampère's Law

Ampère's Law is one of the general laws of magnetism. The only restriction on this law is that the currents must be steady; that is, they must be constant in time. Note that the current I in Ampère's Law refers to the current crossing the area *inside* the path; currents outside make no net contribution.

Ampère's Law can be used to calculate the magnetic field of a given distribution of currents, provided that the distribution has a high degree of symmetry. The technique for calculating magnetic fields by means of Ampère's Law is similar to the technique for calculating electric fields by means of Gauss' Law. Using Ampère's Law is somewhat easier, since it involves a length instead of an area. As in Section 24.3, the calculation involves two steps: first determine the *direction* of the magnetic field by appealing to symmetry arguments, and then determine the *magnitude* of the magnetic field by evaluating Ampère's Law along some suitable, cleverly chosen path.

The following example illustrates this procedure in the simple case of a current on a straight wire. For the purposes of this example, we will pretend that we do not know the magnetic field (29.9) of the straight wire, and we will see how this magnetic field can be deduced from Ampère's Law.

Concepts
— *in* —
Context

EXAMPLE 4   From Ampère's Law, deduce the magnetic field produced by a current on a very long, straight wire.

SOLUTION: The arrangement of magnetic field lines has to match the symmetry of the current. Since the current has cylindrical symmetry, the arrangement of magnetic field lines must also have cylindrical symmetry. Thus, the field lines must be either concentric circles around the wire, radial lines, or parallel lines in the same direction as the wire. Radial lines would require that the field lines start on the wire, which is impossible, since the field lines must form closed loops. Parallel lines in the direction of the wire are likewise inconsistent with closed loops. Thus, the field lines must necessarily be concentric circles. Furthermore, by symmetry, the magnetic field must have a constant magnitude along each circle.

Taking a path that follows one of these circles, of radius r (see Fig. 29.20), the magnetic field is everywhere parallel to the path, so $B_\parallel = B =$ constant, and the left side of Ampère's Law is

$$\oint B_\parallel \, ds = B \oint ds = B \times 2\pi r$$

The right side is $\mu_0 I$; hence

$$2\pi r B = \mu_0 I$$

As expected, this yields the magnetic field of Eq. (29.9):

$$B = \frac{\mu_0}{2\pi} \frac{I}{r}$$

ANDRÉ-MARIE AMPÈRE (1775–1836)
French physicist and mathematician. After Oersted's discovery of the generation of magnetic fields by electric currents, Ampère demonstrated experimentally that currents exert magnetic forces on each other. He carefully investigated the relationship between currents and magnetic fields, and he established that a magnet is equivalent to a distribution of currents.

Concepts
— *in* —
Context

EXAMPLE 5   A very long, straight conducting wire has a circular cross section of radius R. The wire carries a current I_0 uniformly distributed over this cross-sectional area. (a) What is the magnetic field inside the wire? (b) What is the magnetic field outside the wire?

SOLUTION: The symmetry of this thick wire and its current is the same as the symmetry of the thin wire treated in the preceding example. Hence the magnetic field lines must be concentric circles, both inside the wire and outside the wire.

(a) To calculate the magnetic field inside the wire, we take a path that follows one of the circular field lines, of radius r (see Fig. 29.22). As in the preceding example, $B_\parallel = B$ and the left side of Ampère's Law is again

$$\oint B_\parallel \, ds = B \times 2\pi r$$

Recall that the current I on the right side of Ampère's Law is only the current *inside* our path. Since the current is uniformly distributed over the cross section of the wire, the amount of current intercepted by the area πr^2 within the path is directly proportional to this area:

$$I = \frac{\pi r^2}{\pi R^2} \times I_0 = \frac{r^2}{R^2} I_0$$

Ampère's Law then becomes

$$2\pi r B = \mu_0 I = \mu_0 \frac{r^2}{R^2} I_0$$

which yields

$$B = \frac{\mu_0}{2\pi} \frac{Ir}{R^2}$$

(b) To calculate the magnetic field outside the wire, we proceed as in the preceding example. Since the magnetic field of a thick wire has the same symmetry as for a thin wire, the calculation presented in the preceding example is equally valid for a thick wire and for a thin wire, and we find the same result as before, Eq. (29.9), except that the net current on the entire wire is now designated by I_0:

$$B = \frac{\mu_0}{2\pi} \frac{I_0}{r}$$

COMMENT: Note that the magnetic field is zero at the center of the wire ($r = 0$); it increases in proportion to the radius r and reaches a maximum value of $\mu_0 I_0/(2\pi R)$ at the surface of the wire. Beyond this point, the magnetic field decreases in proportion to $1/r$.

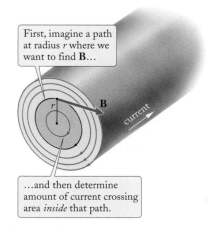

First, imagine a path at radius r where we want to find **B**...

...and then determine amount of current crossing area *inside* that path.

FIGURE 29.22 Circular field lines for the magnetic field inside a long, straight wire of cross-sectional radius R. The path used for Ampère's Law follows one of these field lines, of radius r.

The magnetic field outside a thick cylindrical wire with a uniform distribution of current over its interior is given by the same formula (29.9) as for a thin wire. More generally, this is true of any assembly of concentric cylindrical shells (like those defined by the rings of an ideal, cylindrical tree trunk) when each carries uniformly distributed current. For instance, the magnetic field produced by a power line consisting of a thick aluminum cable, typically a few centimeters across, can be calculated from the formula (29.9).

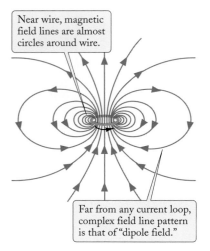

FIGURE 29.23 Magnetic field lines of a ring of current.

FIGURE 29.24 Magnetic field lines of a ring of current made visible with iron filings.

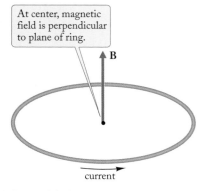

FIGURE 29.25 Direction of the magnetic field at the center of a ring of current.

Although circular concentric field lines and a magnetic field with the strength given by the formula (29.9) are characteristic of a current on a long, straight wire (thin or thick), this formula also provides an approximation for the magnetic field in the immediate vicinity of any segment of wire, straight or curved, except where the wire has a sharp kink. From points very near the wire, any wire looks almost straight, and the magnetic field at such points is then approximately that of a long straight wire. The magnetic field of a current flowing on a ring illustrates this behavior. The field lines for such a ring are shown in Figs. 29.23 and 29.24; very near the wire, the field lines are almost circles, concentric with the wire. However, far from the wire, the pattern of field lines is much more complex. Similar behavior for two rings is illustrated on the book cover.

The calculation of the complete magnetic field of a ring of current at points near and far from the ring is quite complicated, and we will not attempt to do this calculation. We will show in Section 29.5 that the magnitude of the magnetic field at the center of a ring of radius R carrying a current I is

$$B = \frac{\mu_0 I}{2R} \tag{29.19}$$

The direction of the magnetic field at the center is perpendicular to the plane of the ring (see Fig. 29.25), and is given by a **right-hand rule for a current loop**: if the fingers of the right hand curl around the loop in the direction of the current, the thumb gives the direction of the magnetic field at the center of the loop.

Note that the pattern of magnetic field lines at large distance from the ring of current resembles the pattern of electric field lines of an electric dipole (compare Fig. 23.17). This resemblance is also found in the pattern of magnetic field lines at large distance from a closed loop of current of any other shape. At large distance, square loops, rectangular loops, oval loops all produce the same pattern of field lines as a circular ring—the exact shape of the loop has little effect on the distant magnetic field. For example, in the central core of the Earth, currents flow in loops of some kind, and the magnetic field that these currents produce at the surface of the Earth and in the space above is pretty much that of a ring of current (see Fig. 29.26).

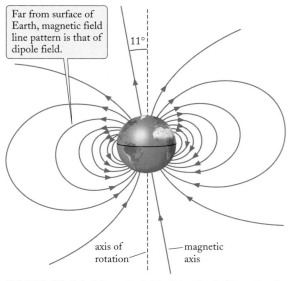

FIGURE 29.26 Magnetic field of the Earth. The axis of the magnetic field makes an angle of about 11° with the axis of rotation of the Earth.

Checkup 29.3

QUESTION 1: Two long parallel wires carry currents of equal magnitudes in the same direction. What is the net magnetic field that these wires produce jointly at a point midway between them?

QUESTION 2: If the direction of the current in the visible segment of the ring in Fig. 29.24 is right to left, what is the direction of the magnetic field along the axis of the ring?

QUESTION 3: If a charged particle is moving vertically upward along the axis of the ring of current shown in Fig. 29.25, what is the magnetic force on this particle?

QUESTION 4: A circular path is located as shown in Fig. 29.27 in the magnetic field of a long straight wire. What is the integral of the tangential magnetic field around this path?

QUESTION 5: A thick wire of radius R carries a current I uniformly distributed over its cross section. What is the magnetic field at the surface of the wire?

(A) $\mu_0 I/(2\pi R)$ (B) $\mu_0 I/(4\pi R)$ (C) $\mu_0 I/R$
(D) $\mu_0 I/(2R)$ (E) $\mu_0 I/(4R)$

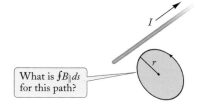

What is $\oint B_\parallel \, ds$ for this path?

FIGURE 29.27 A path near a current-carrying wire.

29.4 SOLENOIDS AND MAGNETS

A solenoid is a conducting wire wound in a tight helical coil of many turns (see Fig. 29.28). A current in this wire will produce a strong magnetic field within the coil (see Figs. 29.29 and 29.30). Because of the similarity of the current distributions, such a tight coil produces essentially the same magnetic field as a large number of rings stacked next to one another. The calculation of the magnetic field of such a solenoid of finite length is fairly difficult, and we will not attempt it here. Instead, we will calculate the magnetic field of an **ideal solenoid**, that is, a very long (infinitely long) solenoid with very tightly wound coils, so the current distribution on the surface of the solenoid is nearly uniform.

To find this magnetic field, we begin with an appeal to symmetry, as in Example 4. The ideal solenoid has translational symmetry (along the axis of the solenoid) and cylindrical symmetry (around the axis). For consistency with these symmetries, the magnetic field lines inside the solenoid will then have to be either concentric circles, or radial lines, or lines parallel to the axis. Concentric circles and radial lines are unacceptable; the former would require the presence of a current along the axis (compare Example 4), and the latter would require that the field lines begin on the

ideal solenoid

FIGURE 29.30 Iron filings sprinkled on a sheet of paper inserted in a solenoid. Note that inside the solenoid, the distribution of field lines is nearly uniform.

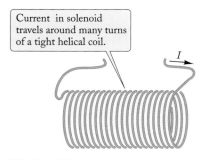

Current in solenoid travels around many turns of a tight helical coil.

I

FIGURE 29.28 A solenoid.

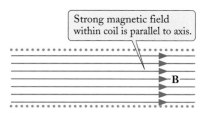

Strong magnetic field within coil is parallel to axis.

B

FIGURE 29.29 Magnetic field lines in a segment of a long solenoid.

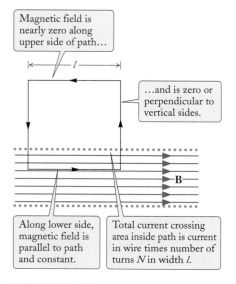

Magnetic field is nearly zero along upper side of path...

...and is zero or perpendicular to vertical sides.

Along lower side, magnetic field is parallel to path and constant.

Total current crossing area inside path is current in wire times number of turns N in width l.

FIGURE 29.31 Magnetic field lines of an ideal, very long solenoid. The path for the application of Ampère's Law is a square of side l.

axis, which is impossible, since magnetic field lines cannot begin or end anywhere. Thus, the field lines inside the solenoid must all be parallel to the axis (see Fig. 29.31). These lines emerge from the (distant) end of the solenoid, and they curve around the exterior of the solenoid and return to the other (distant) end. For an ideal, very long solenoid, these external field lines will then be spread out over a very large region of space; hence, the density of field lines and the magnetic field outside the solenoid are nearly zero.

We can now use Ampère's Law to determine the magnitude of the magnetic field. To evaluate the left and the right sides of Ampère's Law, we must choose a closed path. A convenient choice is the square path shown in Fig. 29.31, which is partly inside the solenoid and partly outside. The magnetic field has a component parallel to the path only along the lower side of the square, within the solenoid; along this side $B_\parallel = B$. Note that by symmetry the field cannnot vary along the length l of this lower part of the path, and is thus $B =$ constant. Along all other sides, either the magnetic field is zero (outside the solenoid) or the component of the magnetic field tangent to the path is zero (the field is perpendicular to the other parts of the path inside the solenoid). Hence the integral appearing in Ampère's Law involves only the lower side of the square:

$$\oint B_\parallel \, ds = 0 + 0 + 0 + \int_0^l B_\parallel \, ds = Bl$$

The net current intercepted by the area of the square is $I_0 \times N$, where I_0 is the current in one wire and N is the number of wires intercepted by the area of the square. Thus, Ampère's Law becomes

$$Bl = \mu_0 I_0 N$$

from which

$$B = \mu_0 I_0 \frac{N}{l} \tag{29.20}$$

The ratio N/l is the number of turns of wire per unit length of the solenoid, commonly designated by $n = N/l$. Thus,

magnetic field inside solenoid

$$B = \mu_0 n I_0 \tag{29.21}$$

This shows that to obtain a large magnetic field, we want a solenoid with a large current and a large number of turns of wire per unit length (a densely wound solenoid).

right-hand rule for solenoids

The direction of the magnetic field inside a solenoid is given by a **right-hand rule for solenoids**: if the fingers of the right hand curl around the solenoid in the direction of the current, the thumb gives the direction of the magnetic field inside the solenoid. This rule is consistent with the direction of the magnetic field in Figs. 29.29 and 29.31 for the direction of the current in Fig. 29.28. This rule is similar to the right-hand rule for a current loop given in the previous section, as it must be, since a solenoid is a coil of many current loops.

Note that the result (29.21) is independent of the "depth" to which the square is immersed in the solenoid—we can slide the square deeper into the solenoid (provided its upper side remains outside the solenoid and its lower side remains inside), and we still obtain the same result (29.21) for the magnetic field. Hence the magnetic field has the same magnitude everywhere within the solenoid. This means that the magnetic field within an ideal solenoid is perfectly uniform.

Furthermore, the result (29.21) is also independent of the shape of the cross section of the solenoid. In Fig. 29.28, the solenoid has a circular cross section (circular coils). But it could equally well have a square cross section, or a rectangular cross section, or some other cross section (see Fig. 29.32), since this does not affect the application of Ampère's Law.

A variation on the solenoid is the toroid, shown in Fig. 29.33. This doughnut- or torus-shaped coil resembles a solenoid that has been bent around in a circle so that its two ends meet. The magnetic field lines remain parallel to the axis of the original solenoid and thus now form closed loops in the interior of the toroid. This geometry is useful in electronic devices, since such "contained" field lines will not interfere with nearby circuit components. Although the toroid does not have as high a symmetry as the infinite solenoid, we can still apply Ampère's Law to a circular path of radius r inside the toroid, such as the one through the toroid's midplane shown in Fig. 29.33. For a given value of r, symmetry requires that the field must be constant and tangent to such a path, so the left side of Ampère's Law becomes

$$\oint B_{\parallel}\,ds = B \times 2\pi r$$

For a toroid with N loops of current, each carrying the same current I_0, the right side of Ampère's Law is $\mu_0 N I_0$, so equating the two sides gives

$$2\pi r B = \mu_0 N I_0$$

Solving for the magnetic field, we obtain

$$B = \mu_0 I_0 \frac{N}{2\pi r} \tag{29.22}$$

Note that Eq. (29.22) has the same form as Eq. (29.20), with the solenoid length l replaced by the path length around the torus interior, $2\pi r$. The field varies somewhat with position inside the torus; the field is larger for small r and smaller for large r. Outside, including the region of the doughnut "hole," there are no magnetic field lines, and the field there is zero.

EXAMPLE 6   A solenoid used for research consists of 180 turns of wire wound on a narrow cardboard tube 19 cm long. The current in the wire is 5.0 A. What is the strength of the magnetic field within the tube?

SOLUTION: The number of turns per unit length is

$$n = 180/0.19 \text{ m} = 9.5 \times 10^2/\text{m}$$

Although this solenoid is not infinitely long, its length is large compared with its width, and therefore the magnetic field at all points within it except those near the ends will be approximately given by Eq. (29.21):

$$B = \mu_0 n I_0 = 1.26 \times 10^{-6} \text{ N·m}^2/\text{C}^2 \times 9.5 \times 10^2/\text{m} \times 5.0 \text{ A}$$

$$= 6.0 \times 10^{-3} \text{ T}$$

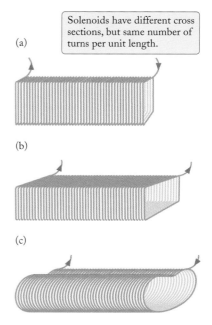

FIGURE 29.32 (a) Square, (b) rectangular, and (c) irregular solenoids. All long solenoids produce equal uniform magnetic fields (for equal currents per unit length).

Solenoids have different cross sections, but same number of turns per unit length.

(a)

(b)

(c)

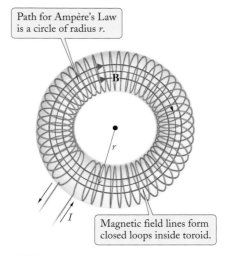

Path for Ampère's Law is a circle of radius r.

Magnetic field lines form closed loops inside toroid.

FIGURE 29.33 A toroid.

(a)

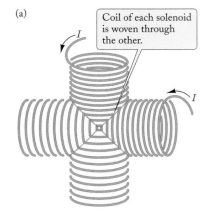

Coil of each solenoid is woven through the other.

(b)

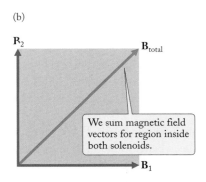

We sum magnetic field vectors for region inside both solenoids.

FIGURE 29.34 (a) Two long solenoids oriented at right angles. (b) The vector sum of their magnetic fields. (c) The field lines in the combined solenoids.

(c)

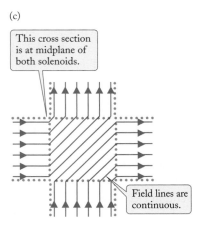

This cross section is at midplane of both solenoids.

Field lines are continuous.

EXAMPLE 7 Two long, identical solenoids such as described in Example 6 are oriented at right angles. The coils of one of these solenoids are woven through the coils of the other (see Fig. 29.34a). What are the magnitude and the direction of the magnetic field in the region of intersection?

SOLUTION: The magnetic field of each solenoid has a magnitude $B = \mu_0 n I_0$. The magnetic fields are at right angles to each other, and the magnitude of their vector sum is (see Fig. 29.34b)

$$B_{\text{total}} = \sqrt{(B_1)^2 + (B_2)^2} = \sqrt{(\mu_0 n I_0)^2 + (\mu_0 n I_0)^2} = \sqrt{2}\mu_0 n I_0$$

$$= \sqrt{2} \times 6.0 \times 10^{-3}\,\text{T} = 8.5 \times 10^{-3}\,\text{T}$$

This magnetic field makes an angle of 45° with the axis of each solenoid. The field lines are as shown in Fig. 29.34c.

For a more precise calculation of the magnetic field of a short solenoid, we must take the sum of the magnetic fields of the individual rings in this solenoid. But even a solenoid of just a few rings produces a fairly uniform magnetic field in its interior. For instance, Fig. 29.30 shows a photograph of the magnetic field lines, made visible with iron filings, of a short solenoid of just six widely-spaced turns. The magnetic field in the interior is not far from uniform.

electromagnet

An **electromagnet** is essentially a solenoid with a gap, or, what amounts to the same thing, a pair of solenoids with their ends placed close together (see Fig. 29.35a). Magnetic field lines come out of one solenoid and go into the other solenoid (of course, field lines will also have to come out of the solenoids at their other ends, curve around, and close on themselves). The first solenoid is called the north pole of the electromagnet, and the second the south pole, so magnetic field lines emerge from the north pole and enter the south pole. If the gap is small, then the magnetic field in this gap is almost the same as inside the solenoids. The gap makes the magnetic field more accessible; the gap makes it easier to immerse wires, loops, or other equipment in the field.

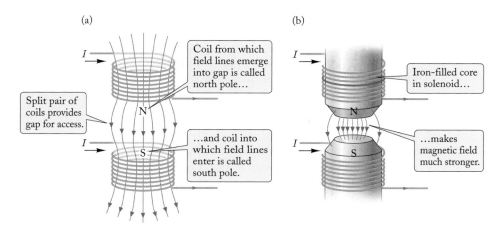

FIGURE 29.35 (a) An electromagnet with two coils. (b) An electromagnet with iron pole pieces.

In most electromagnets, the space inside the solenoids is filled with an iron core (Fig. 29.35b). Iron, and other **ferromagnetic materials**, enhance the magnetic field, making it much stronger than the value given by Eq. (29.21). It is not unusual for the magnetic field to be enhanced by a factor of several thousand. We will discuss this enhancement of magnetic fields in Section 30.4.

ferromagnetic material

PROBLEM–SOLVING TECHNIQUES   AMPÈRE'S LAW FOR CURRENT DISTRIBUTIONS WITH SYMMETRY

The calculation of the magnetic field of a given current distribution by means of Ampère's Law is possible only if the current distribution and the field are highly symmetric. If so, the calculation involves the following steps, which are analogous to those we used in the calculation of the electric field by means of Gauss' Law:

1 Determine the direction of the magnetic field from considerations of symmetry. For example, for current flow along a straight line, the field lines will be loops around the line; for current flow around a cylinder, the field lines will resemble those of an infinite solenoid.

2 Select a closed path such that for some portion of the path the magnetic field is parallel to the path and of constant magnitude B, and for the remaining portion (if any) the magnetic field is perpendicular to the path or is zero. Keep in mind that the path for Ampère's Law is a mathematical path—it does not have to coincide with a wire

or with the surface of some physical body. Instead, the path is chosen to coincide with points where the magnetic field B is to be evaluated.

3 Express the integral $\oint B_\parallel \, ds$ in terms of the magnitude B of the magnetic field, $\oint B_\parallel \, ds = Bl$, where l is the length of that portion of the path where the magnetic field is tangent to the path [the portion (if any) where the magnetic field is perpendicular does not contribute to the integral].

4 Use Ampère's Law to relate Bl to the current enclosed by the path:

$$Bl = \mu_0 I$$

When evaluating I, make sure to include only the portion of the total current that is *enclosed* by the path.

5 Solve the resulting equation for the magnitude of the magnetic field.

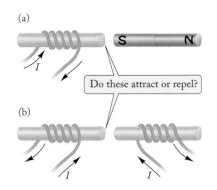

FIGURE 29.36 (a) Electromagnet and permanent magnet. (b) Two electromagnets.

Checkup 29.4

QUESTION 1: If the current in the visible segments of the coil in Fig. 29.30 is from left to right, what is the direction of the magnetic field inside the coil?

QUESTION 2: Suppose that we wind a second layer of 180 turns around the solenoid described in Example 6. If the current in this second layer is of the same magnitude and direction as in the first layer, how will the magnetic field change?

QUESTION 3: Suppose we grasp the coiled wire of the solenoid shown in Fig. 29.28 at its ends and stretch the coil, so it is twice as long. How does this affect the magnetic field, for the same current?

QUESTION 4: Suppose we squeeze the sides of the coiled wire of the solenoid shown in Fig. 29.28, so the coil is flattened uniformly along its full extent and the cross sectional area is reduced to one-half of its original value. How does this affect the magnetic field, for the same current? How does it affect the magnetic flux intercepted by the cross section of the solenoid?

QUESTION 5: In Fig. 29.36a, will the magnets attract or repel each other? In Fig. 29.36b?

(A) Attract, attract   (B) Attract, repel   (C) Repel, attract   (D) Repel, repel

29.5 THE BIOT–SAVART LAW

Although Ampère's Law provides us with a quick way of determining the magnetic field for a current distribution of high symmetry, it is sometimes necessary to calculate the field of an arbitrary current distribution. The prescription for doing this is contained in the **Biot–Savart Law**. Although the implementation of this law usually requires complicated computational methods, there are some simple applications of this technique that readily yield results. Figure 29.37 shows a segment of a current I of length ds and a point P at which we wish to evaluate the magnetic field **B**. The Biot–Savart Law states,

The contribution $d\mathbf{B}$ to the magnetic field $\mathbf{B}$ from a length $d\mathbf{s}$ of a current I is given by

Biot–Savart Law

$$d\mathbf{B} = \frac{\mu_0}{4\pi}\frac{I\,d\mathbf{s}\times\mathbf{r}}{r^3} \qquad (29.23)$$

where $\mathbf{r}$ is the displacement vector from the current element to the point P.

The contribution is thus in the direction of $d\mathbf{s}\times\mathbf{r}$; by the right-hand rule, this is perpendicularly into the plane of the paper for the point P shown in Fig. 29.37. If the point P and all contributing current elements are in the same plane, the contributions

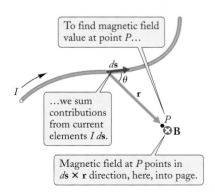

FIGURE 29.37 Evaluation of the magnetic field **B** at the point P according to the Biot–Savart Law. The current element $I\,d\mathbf{s}$ contributes to the field, the vector $\mathbf{r}$ points from the current element to the point P, and the angle between $d\mathbf{s}$ and $\mathbf{r}$ is θ.

$d\mathbf{B}$ will all be perpendicular to that plane. For such contributions, we need consider only the magnitude of Eq. (29.23):

$$dB = \frac{\mu_0 I}{4\pi} \frac{ds \, \sin\theta}{r^2} \tag{29.24}$$

Let us apply the Biot–Savart Law in the form (29.24) to find the magnetic field at the center point P of a circular loop of radius R. As shown in Fig. 29.38, the direction of $\mathbf{B}$ will again be perpendicular to the plane containing the loop and its center. The value of θ for each contribution around the loop is $\theta = 90°$, so $\sin\theta = 1$, and each contribution is equidistant from the center; that is, $r = R = $ constant. So the total field is

$$B = \int dB = \int \frac{\mu_0 I}{4\pi} \frac{ds \, \sin\theta}{r^2} = \frac{\mu_0 I}{4\pi R^2} \int ds \tag{29.25}$$

where we have brought all quantities that are not varying around the loop outside the integral. There remains only the integral of the length around the loop, $\int ds = 2\pi R$, which yields

$$B = \frac{\mu_0 I}{2R} \tag{29.26}$$

This result was already mentioned in Eq. (29.19).

The result (29.26) can also be extended to find the field at the center of curvature of any circular arc. If an arc subtends an angle $\Delta\theta$, the field at its center of curvature will be reduced by a factor of $\Delta\theta/2\pi$, compared with the result (29.26) for a full loop:

$$B = \frac{\mu_0 I \, \Delta\theta}{4\pi R} \tag{29.27}$$

A second configuration calculable with the Biot–Savart Law is the case of a finite length of straight wire, as shown in Fig. 29.39. We need to sum the contributions of the form (29.24); now, however, the quantities r, s, and θ all vary along the wire. If we want to sum the contributions, we must put them in terms of a single variable. It so happens that using the angle θ is easiest. From Fig. 29.39, we can relate

$$\tan\theta = R/(-s)$$

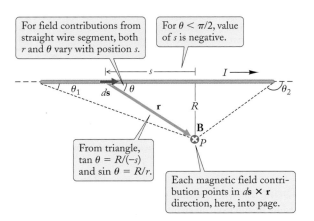

FIGURE 29.39 Geometry used to find the magnetic field of a finite straight wire using the Biot–Savart Law.

For field at center of circular current loop, $\mathbf{r}$ is always perpendicular to $d\mathbf{s}$, …

…distance $r = R$ is constant, …

…and each field contribution points in $d\mathbf{s} \times \mathbf{r}$ direction, here, into page.

FIGURE 29.38 Geometry used to find the magnetic field at the center of a circular ring of current (also used for the field at the center of curvature of a circular arc).

magnetic field at center of circular current loop

magnetic field at center of arc

Concepts — *in* — Context

and we can take the derivative of $s = -R/\tan\theta$ with respect to θ to obtain

$$ds = \frac{R}{\sin^2\theta}\, d\theta$$

Also from Fig. 29.39, $r = R/\sin\theta$. Substituting these relations into Eq. (29.24), we can write

$$B = \frac{\mu_0 I}{4\pi}\int_{\theta_1}^{\theta_2} \frac{R}{\sin^2\theta}\frac{\sin^2\theta}{R^2}\sin\theta\, d\theta = \frac{\mu_0 I}{4\pi R}\int_{\theta_1}^{\theta_2}\sin\theta\, d\theta$$

Using $\int \sin\theta\, d\theta = -\cos\theta$, we obtain

magnetic field of finite wire segment

$$B = \frac{\mu_0 I}{4\pi R}(\cos\theta_1 - \cos\theta_2) \tag{29.28}$$

The result (29.28) can be used repeatedly to sum contibutions from various straight segments of a conductor. Note also that in the case of an infinite wire, we have $\theta_1 = 0$ and $\theta_2 = \pi$, so $\cos\theta_1 - \cos\theta_2 = 1 - (-1) = 2$, and the result (29.28) reduces to the field of an infinite wire, $B = \mu_0 I/2\pi R$, as it must.

EXAMPLE 8   A magnetic probe tip consists of two long wires separated by a distance R and an arc of a circle with the radius also equal to R as shown in Fig. 29.40. If the wire carries a current I, what is the magnetic field at the center of curvature of the arc?

SOLUTION: There are three contributions to the field: one each from the straight segments, and one from the arc. In the top straight segment, the current I is flowing directly toward the point in question, so $\sin\theta = 0$ in Eq. (29.24) and there is no contribution. By the right-hand rule, the other straight segment and the arc each produce a field directed into the plane of the paper, so we may simply sum them. The arc is missing $90° = \pi/2$ with repect to a full loop, and so subtends an angle $3\pi/2$. By Eq. (29.27), the contribution from the arc is

$$B_{\text{arc}} = \frac{\mu_0 I}{4\pi R}\frac{3\pi}{2} = \frac{3}{4}\times\frac{\mu_0 I}{2R}$$

JEAN-BAPTISTE BIOT (BEOH)
(1774–1862) *French physicist, professor at the Collège de France. His most important work dealt with the refraction and polarization of light, but he was also interested in a broad range of problems in the physical sciences. With Félix Savart, 1791–1841, he confirmed Oersted's discovery of magnetic fields generated by electric currents, and formulated the equation (29.23) for the strength of the magnetic field.*

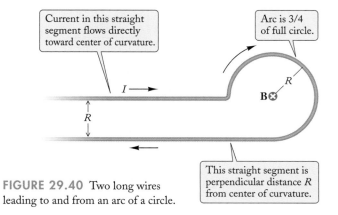

Current in this straight segment flows directly toward center of curvature.

Arc is 3/4 of full circle.

$I \rightarrow$

R

R

B ⊗

This straight segment is perpendicular distance R from center of curvature.

FIGURE 29.40 Two long wires leading to and from an arc of a circle.

The bottom straight segment produces a contribution of the form (29.28), with $\theta_1 = 0$ at the far left end and $\theta_2 = \pi/2$ at the right end, directly below the center of curvature. Thus

$$B_{\text{segment}} = \frac{\mu_0 I}{4\pi R}\left[\cos(0) - \cos\left(\frac{\pi}{2}\right)\right] = \frac{\mu_0 I}{4\pi R}$$

which is half the field of the infinite wire. The total field at the center of curvature of the arc is

$$B = B_{\text{arc}} + B_{\text{segment}} = \left(\frac{3}{4} + \frac{1}{2\pi}\right)\frac{\mu_0 I}{2R} \approx 0.91 \times \frac{\mu_0 I}{2R}$$

 ## Checkup 29.5

QUESTION 1: A circular loop of wire carries a current I and produces a magnetic field B_0 at its center. What is the field at the center when the same current flows in a loop of twice the radius?

QUESTION 2: Suppose that two concentric circular loops of wire are in the plane of this page; the larger one carries a clockwise current, the smaller one carries the same current, but counterclockwise. What is the direction of the magnetic field at the center of the loops?

QUESTION 3: Consider a square current loop. The contribution to the magnetic field at the center of the square from a single side of the square is B_1. What is the net magnetic field at the center of the square?

    (A) 0   (B) $2B_1$ (C) $2\sqrt{2}B_1$ (D) $4B_1$

SUMMARY

PROBLEM-SOLVING TECHNIQUES	Direction of Magnetic Force (right-hand rule)	**(page 936)**
PROBLEM-SOLVING TECHNIQUES	Ampère's Law for Current Distributions with Symmetry	**(page 947)**
MAGNETIC FORCE EXERTED ON MOVING CHARGE BY CURRENT IN LONG WIRE where the $-$ sign means **F** is attractive for **v** parallel to I and $+$ means **F** is parallel to I for **v** radially outward. For **v** tangent to circles around the wire, $F = 0$.	$F = \pm\dfrac{\mu_0}{2\pi}\dfrac{qvI}{r}$ or $F = 0$	**(29.5, 29.6)**
PERMEABILITY CONSTANT	$\mu_0 = 4\pi \times 10^{-7}\ \text{N·s}^2/\text{C}^2 \approx 1.26 \times 10^{-6}\ \text{N·s}^2/\text{C}^2$	**(29.4)**
FORCE EXERTED BY MAGNETIC FIELD ON MOVING CHARGE	$\mathbf{F} = q\mathbf{v} \times \mathbf{B}$	**(29.12)**

MAGNITUDE OF MAGNETIC FORCE
$$F = qvB \sin \alpha$$
(29.11)

MAGNETIC FIELD OF CURRENT IN LONG WIRE
Right-Hand Rule   With thumb along current, fingers curl in direction of field, tangent to circles around current.

$$B = \frac{\mu_0 I}{2\pi r}$$
(29.9)

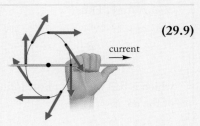

SI UNIT OF MAGNETIC FIELD
$$1 \text{ tesla} = 1 \text{ T} = 1 \text{ N/(C·m/s)}$$

LORENTZ FORCE
$$\mathbf{F} = q\mathbf{E} + q\mathbf{v} \times \mathbf{B}$$
(29.13)

GAUSS' LAW FOR MAGNETISM
(for closed surface)
$$\Phi_B = 0$$
(29.15)

SUPERPOSITION PRINCIPLE   Magnetic forces and magnetic fields produced by different currents combine by vector addition.

AMPÈRE'S LAW
$$\oint B_\parallel \, ds = \mu_0 I$$

where, for cylindrical symmetry,
$$\oint B_\parallel \, ds = \begin{cases} B \times 2\pi r & \text{(current flow along a line)} \\ B \times l & \text{(current flow around a cylinder)} \end{cases}$$
(29.18)

MAGNETIC FIELD INSIDE SOLENOID
Right-hand rule: With fingers curled around in direction of current, thumb gives direction of axial field.

$$B = \mu_0 n I \quad \text{where} \quad n = N/l$$
(29.21)

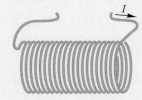

BIOT–SAVART LAW
Contribution to magnetic field: where **r** is the vector from the current element $I\,d\mathbf{s}$ to the point P.

$$d\mathbf{B} = \frac{\mu_0}{4\pi} \frac{I \, d\mathbf{s} \times \mathbf{r}}{r^3}$$

(29.23, 29.24)

If the current and P are in the same plane, then
$$B = \int dB = \int \frac{\mu_0 I}{4\pi} \frac{ds \sin\theta}{r^2}$$

MAGNETIC FIELD AT CENTER CIRCULAR CURRENT LOOP

Right-hand rule: With fingers curled around in direction of current, thumb gives direction of field at center of loop.

$$B = \frac{\mu_0 I}{2R}$$

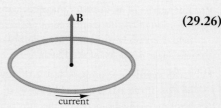

(29.26)

MAGNETIC FIELD AT CENTER OF ARC

$$B = \frac{\mu_0 I \Delta\theta}{4\pi R}$$

(29.27)

MAGNETIC FIELD OF FINITE WIRE SEGMENT

$$B = \frac{\mu_0 I}{4\pi R}(\cos\theta_1 - \cos\theta_2)$$

(29.28)

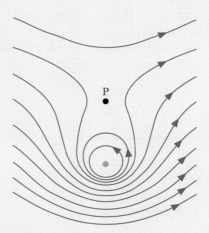

QUESTIONS FOR DISCUSSION

1. Theoretical physicists have proposed the existence of **magnetic monopoles**, which are sources and sinks of magnetic field lines, just as electric charges are sources and sinks of electric field lines. What would the pattern of magnetic field lines of a positive magnetic monopole look like?

2. The Earth's magnetic field at the equator is horizontal, in the northward direction. What is the direction of the magnetic force on an electron moving vertically up?

3. An electron with a vertical velocity passes through a magnetic field without suffering any deflection. What can you conclude about the magnetic field?

4. At an initial time, a charged particle is at some point P in a magnetic field and it has an initial velocity. Under the influence of the magnetic field, the particle moves to a point P'. If you now reverse the velocity of the particle, will it retrace its orbit and return to the point P?

5. An electron moving northward in a magnet is deflected toward the east by the magnetic field. What is the direction of the magnetic field?

6. A Faraday cage made of wire mesh shields electric fields. Does it also shield magnetic fields?

7. A long straight wire carrying a current I is placed in a uniform magnetic field B_0 at right angles to the direction of the field. The net magnetic field is the superposition of the field of the wire and the uniform field; the field lines of this net field are shown in Fig. 29.41. At the point P, the net magnetic field is zero. What is the distance of this point from the wire?

FIGURE 29.41 These magnetic field lines result from the superposition of the field of a long straight wire (directed out of the page and indicated by the colored dot) and a uniform horizontal magnetic field, directed from left to right.

8. Strong electric fields are hazardous—if you place some part of your body in a strong electric field, you are likely to receive an electric shock. Are strong magnetic fields hazardous? Do they produce any effect on your body?

9. Figure 29.26 shows the magnetic field of the Earth. What must be the direction of the currents flowing in the loops inside the Earth to give this magnetic field?

10. The needle of an ordinary magnetic compass indicates the direction of the horizontal component of the Earth's magnetic field. Explain why the magnetic compass is unreliable when used near the poles of the Earth.

11. A **dip needle** is a compass needle that swings about a horizontal axis. If the axis is oriented east–west, then the equilibrium position of the dip needle is the direction of the Earth's magnetic field. The **dip angle** of the dip needle is the angle that it makes with the horizontal. How does the dip angle vary as you transport a dip needle along the surface of the Earth from the South Pole to the North Pole?

12. Suppose we replace the single ring shown in Fig. 29.25 by a coil of N loops. How does this change the formula [Eq. (29.19)] for the magnetic field?

13. In order to eliminate or reduce the magnetic field generated by the pair of wires that connect a piece of electric equipment to an outlet, a physicist twists these wires tightly about each other. How does this help?

14. Consider a circular loop of wire carrying a current. Describe the direction of the magnetic field at different points in the plane of the loop, both inside and outside the loop.

15. An infinite flat conducting sheet lies in the x–y plane. The sheet carries a current in the y direction; this current is uniformly distributed over the entire sheet. What is the direction of the magnetic field above the sheet? Below the sheet?

16. Suppose you evaluate the integral $\int B_{\parallel} \, ds$ for the magnetic field of the Earth along a closed circular path along a meridian passing through the magnetic north and south poles. What is the value of the integral?

17. Consider a long solenoid and a long straight wire along the axis of the solenoid, both carrying some current. Describe the field lines within the solenoid.

18. The drawing of Fig. 29.35 shows the field lines near the gap of an electromagnet. Describe the pattern of field lines beyond the edges of the drawing.

19. A **tangent galvanometer** is an old form of ammeter, consisting of an ordinary magnetic compass mounted at the center of a coil whose axis is horizontal and oriented along the east–west line (Fig. 29.42). If there is no current in the coil, the compass needle points north. Explain how the compass will deviate from north when there is a current in the coil.

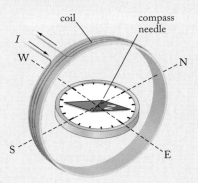

FIGURE 29.42 Tangent galvanometer.

20. A simple indicator of electric current, first used by H. C. Oersted in his early experiments, consists of a compass needle placed below a wire stretched in the northward direction. Explain how the angle of the compass needle indicates the electric current in the wire.

PROBLEMS

29.1 The Magnetic Force

1. Suppose that, instead of moving parallel to the wire, the electron in Example 1 is moving radially away from the wire. In this case, what is the magnitude of the magnetic force on this electron? What is its direction?

2. A proton is instantaneously at a distance of 0.10 m from a long straight wire carrying a current of 30 A. The speed of the proton is 5.0×10^6 m/s, and its direction of motion is radially toward the wire. What is the magnitude of the force on the proton? What is the magnitude of the instantaneous acceleration of the proton? Draw a diagram displaying the direction of the current, and the proton's velocity and acceleration.

3. In a TV tube, electrons of energy 3.0×10^{-15} J are moving on a straight path from the back of the tube to the front. This TV tube is placed near a (single) straight cable carrying a current of 12 A parallel to the path of the electrons, at a radial distance of 0.30 m from the path of the electrons. What is the magnetic force on each electron? What is the corresponding transverse acceleration?

4. A copper atom is 1.5 cm from a long, straight wire and moves parallel to a current of 25 A in the wire with speed 7.0×10^3 m/s. What magnetic force does the current in the wire exert on one electron in the copper atom? On the copper nucleus? On the entire copper atom?

5. A wire lying along the x axis carries a current of 30 A in the $+x$ direction. A proton at $\mathbf{r} = 2.5\mathbf{j}$ has instantaneous velocity $\mathbf{v} = 2.0\mathbf{i} - 3.0\mathbf{j} + 4.0\mathbf{k}$, where $\mathbf{r}$ is in meters and $\mathbf{v}$ is in meters per second. What is the instantaneous magnetic force on this proton?

6. In terms of m, s, and kg, what are the units of $1/\sqrt{\epsilon_0\mu_0}$?

29.2 The Magnetic Field

7. A charge of q is traveling with a velocity **v** at an angle θ with respect to the direction of a magnetic field **B** that points along the x axis (see Fig. 29.43). For what angle θ is the magnitude of the magnetic force one-third of the maximum magnetic force?

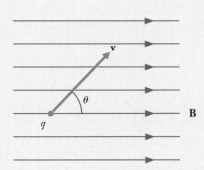

FIGURE 29.43 Velocity vector of a charge q in a uniform magnetic field.

8. An electron is traveling with a velocity of 2.0×10^5 m/s at an angle of 120° to the direction of a 0.33-T magnetic field that points along the x axis (see Fig. 29.44). What are the magnitude and direction of the force on the electron?

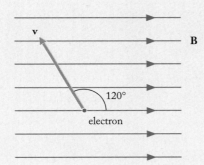

FIGURE 29.44 Velocity vector of an electron in a uniform magnetic field.

9. A proton is moving through a vertical magnetic field. The (instantaneous) velocity of the proton is 8.0×10^5 m/s horizontally in the north direction. The (instantaneous) acceleration produced by the magnetic force is 3.2×10^{14} m/s² in the west direction. What is the magnitude of the magnetic field? Is the direction of this field up or down?

10. Suppose you want to momentarily balance the downward gravitational force on a proton by a magnetic force. If the proton is moving horizontally in the east direction with a speed of 6.0×10^4 m/s, what magnetic field do you need (in magnitude and direction)?

11. The current in a lightning bolt may be as much as 2.0×10^4 A. What is the magnetic field at a distance of 1.0 m from a lightning bolt? The bolt can be regarded as a straight line of current.

12. The cable of a high-voltage power line is 25 m above the ground and carries a current of 1.8×10^3 A.

    (a) What magnetic field does this current produce at the ground?

    (b) The strength of the magnetic field of the Earth is 0.60×10^{-4} T at the location of the power line. By what factor do the fields of the power line and of the Earth differ?

13. The magnetic field surrounding the Earth typically has a strength of 5.0×10^{-5} T. Suppose that a cosmic-ray electron of kinetic energy 3.0×10^4 eV is instantaneously moving in a direction perpendicular to the lines of this magnetic field. What is the force on this electron?

14. An alpha particle (charge $+2e$) moves with velocity **v** = $5.0\mathbf{i} - 3.0\mathbf{j}$ in a magnetic field **B** = $-4.0\mathbf{i} + 2.5\mathbf{j}$, where **B** is in teslas and **v** is in meters per second. What is the magnetic force on the alpha particle?

15. In a region where the magnetic field is **B** = $2.5\mathbf{i} + 3.6\mathbf{j} + 1.5\mathbf{k}$, an electron moves with velocity **v** = $-3.0\mathbf{i} + 4.0\mathbf{j} - 3.5\mathbf{k}$, where **B** is in teslas and **v** is in meters per second. What is the magnetic force on this electron?

16. A proton is just above the surface of the Earth on the magnetic equator, where the magnetic field points north and has magnitude $B = 4.2 \times 10^{-5}$ T. In what direction and with what velocity should the proton move in order for the magnetic force to balance the gravitational force?

*17. In New York, the magnetic field of the Earth has a vertical (down) component of 6.0×10^{-5} T and a horizontal (north) component of 1.7×10^{-5} T. What are the magnitude and direction of the magnetic force on an electron of velocity 1.0×10^6 m/s moving (instantaneously) in an east-to-west direction in a television tube?

*18. At the surface of a pulsar, or neutron star, the magnetic field may be as strong as 1.0×10^8 T. Consider the electron in a hydrogen atom on the surface of such a neutron star. The electron is at a distance of 5.3×10^{-11} m from the proton and has a speed of 2.2×10^6 m/s. Compare the electric force that the proton exerts on the electron with the magnetic force that the magnetic field of the neutron star exerts on the electron. Is it reasonable to expect that the hydrogen atom will by strongly deformed by the magnetic field?

*19. The electric field of a long, straight line of charge with λ coulombs per meter is [see Eq. (24.23)]

$$E = \frac{1}{2\pi\epsilon_0} \frac{\lambda}{r}$$

Suppose that we move this line of charge parallel to itself at speed v.

    (a) The moving line of charge constitutes an electric current. What is the magnitude of the current?

(b) What is the magnitude of the magnetic field produced by this current? Show that the magnitude of the magnetic field is proportional to the magnitude of the electric field; that is,

$$B = \mu_0 \epsilon_0 v E$$

(However, the directions of the electric and magnetic fields differ. The electric field is radial, whereas the magnetic field is tangential.)

*20. According to the preceding problem, when a line of charge is made to move at speed v parallel to itself, it produces a magnetic field of magnitude proportional to its electric field, $B = \mu_0 \epsilon_0 v E$. Use this result to find the magnetic field of a large charged flat sheet of paper with a charge of σ coulombs per square meter moving at a speed v in a direction parallel to itself.

29.3 Ampère's Law

21. A wire of superconducting niobium, 0.20 cm in diameter, can carry a current of up to 1900 A. What is the strength of the magnetic field just outside the wire when it carries this current?

22. A long, straight wire of copper with a radius of 1.0 mm carries a current of 20 A. What are the instantaneous magnetic force and the corresponding acceleration of one of the conduction electrons moving at 1.0×10^6 m/s along the surface of the wire in a direction opposite to that of the current? What is the direction of the acceleration?

23. In a proton accelerator, protons of velocity 3.0×10^8 m/s form a beam of current of 2.0×10^{-3} A. Assume that the beam has a circular cross section of radius 1.0 cm and that the current is uniformly distributed over the cross section. What is the magnetic field that the beam produces at its edge? What is the magnetic force on a proton at the edge of the beam?

24. A ring of superconducting wire carries a current of 2.0 A. The radius of the ring is 1.5 cm. What is the magnitude of the magnetic field at the center of the ring?

25. A circular coil consists of 60 turns of wire wound around the rim of a plywood disk of radius 0.15 m (see Fig. 29.45). If a current of 2.0 A is sent through this coil, what is the magnetic field produced at the center of this disk?

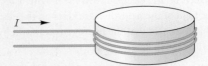

FIGURE 29.45 A circular coil of wire.

26. At the boiling point of liquid helium, a niobium–titanium alloy is superconducting only when the magnetic field is less than 9.5 T. If a wire made of this alloy is 3.0 mm in diameter, what maximum current can it carry before superconductivity at its surface is lost? Assume the current is uniformly distributed over the cross section of the wire.

27. A current I flows in a thin wire bent into a circle of radius R. The axis of the circle coincides with the z axis. What is the value of the integral $\int B_\parallel \, ds$ along the z axis from $z = -\infty$ to $z = +\infty$?

*28. Six parallel aluminum wires of small, but finite, radius lie in the same plane. The wires are separated by equal distances d, and they carry equal currents I in the same direction. Find the magnetic field at the center of the first wire. Assume that the current in each wire is uniformly distributed over its cross section.

*29. A circular ring of wire of diameter 0.60 m carries a current of 35 A. What acceleration will the magnetic force generated by this ring give to an electron that is passing through the center of the ring with a velocity of 1.2×10^6 m/s in the plane of the ring?

*30. Two very long straight wires carry currents I at right angles. One of the wires lies along the x axis; the other lies along the y axis (see Fig. 29.46). Find the magnetic field at a point in the x–y plane in the first quadrant.

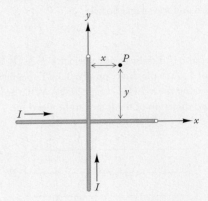

FIGURE 29.46 Two long, straight wires.

*31. In a motorboat, the compass is mounted at a distance of 0.80 m from a cable carrying a current of 20 A from an electric generator to a battery.

(a) What magnetic field does this current produce at the location of the compass? Treat the cable as a long, straight wire.

(b) The horizontal (north) component of the Earth's magnetic field is 1.8×10^{-5} T. Since the compass points in the direction of the net horizontal magnetic field, the current will cause a deviation of the compass. Assume that the magnetic field of the current is horizontal and at right angles to the horizontal component of the Earth's magnetic field. Under these circumstances, by how many degrees will the compass deviate from north?

*32. Two very long, straight, parallel wires separated by distance d carry currents of magnitude I in opposite directions. Find the magnetic field at a point equidistant from the lines, with a distance $2d$ from each line. Draw a diagram showing the direction of the magnetic field.

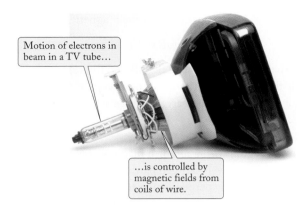

Motion of electrons in beam in a TV tube…

…is controlled by magnetic fields from coils of wire.

FIGURE 30.1 A TV tube. The beam of electrons from the cathode at the back of the tube strikes the rear side of the screen at the front.

In a TV tube or a cathode-ray tube (CRT) electrons move in a beam from the cathode at the rear end of the tube to the screen at the front end, where their impact causes the screen to glow, forming a visible display (Fig. 30.1). The motion of the electrons is controlled by magnetic fields. These magnetic fields, produced by coils of wire within the tube, deflect the electrons up or down and left or right and thereby determine where they will strike the screen, and what display you will see on the screen.

The magnetic fields in the TV tube are nonuniform, and the motion of the electrons is rather hard to calculate. However, the motion of electrons or other charged particles in uniform magnetic fields is fairly easy to calculate, and in this chapter we will deal with this motion in detail. Such uniform fields are not found in TV tubes, but they are found in magnets used for experiments and applications in science and engineering, such as the magnets in cyclotrons and other accelerators that produce high-energy particles for research in physics, the magnets for materials-science experiments to determine the electrical and magnetic properties of matter, and also the magnets for special kinds of radiation therapy in medicine.

In this chapter we will also examine the magnetic forces acting on wires and on current loops; the enhancement of magnetic fields by the presence of iron or other magnetic materials; and, finally, the potential difference that arises between the sides of a wire carrying a current while placed in a magnetic field (the Hall effect).

Online
Concept
Tutolial

30.1 CIRCULAR MOTION IN A UNIFORM MAGNETIC FIELD

The direction of the force exerted by a magnetic field on a moving charged particle is always perpendicular to both the magnetic field and the velocity, as specified by the right-hand rule. Since the magnetic force is always perpendicular to the velocity, the acceleration—and the small change of velocity in a small time interval—is perpendicular to the velocity. If the change of velocity is always perpendicular to the velocity, then the velocity can never change in magnitude, but only in direction (see Fig. 30.2). We therefore recognize that *the magnetic force acting on a particle will deflect the particle, keeping the speed constant.* We can also recognize this in another way: the force is always perpendicular to the displacement; therefore it does no work, and therefore the kinetic energy and the speed remain constant.

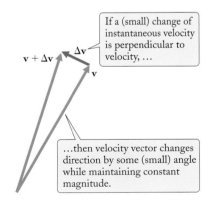

If a (small) change of instantaneous velocity is perpendicular to velocity, …

$\mathbf{v} + \Delta\mathbf{v}$ $\Delta\mathbf{v}$

$\mathbf{v}$

…then velocity vector changes direction by some (small) angle while maintaining constant magnitude.

FIGURE 30.2 Instantaneous velocity **v**, small change in velocity $\Delta\mathbf{v}$ perpendicular to **v**, and instantaneous velocity $\mathbf{v} + \Delta\mathbf{v}$. The speed is constant.

Magnetic field points perpendicularly into plane of page, …

B

…and velocity is in plane of page, …

v

r

F

q

…so force on charge is also in plane of page, perpendicular to velocity.

FIGURE 30.3 Positively charged particle moving in a uniform magnetic field. The crosses show the tails of the magnetic field vectors.

According to Eq. (29.11) the magnitude of the magnetic force is

$$F = qvB \sin \alpha \qquad (30.1)$$

where α is the angle between the magnetic field **B** and the velocity **v**. For motion parallel to the magnetic field, this force is zero. For motion perpendicular to the magnetic field, the force is not zero. In the latter case, the force is

$$F = qvB \qquad (30.2)$$

Figure 30.3 shows a region with a uniform magnetic field, directed into the plane of the page. Suppose that a positively charged particle has an initial velocity in the plane of the page; this initial velocity is perpendicular to the magnetic field. The magnetic force is then in the plane of the page, perpendicular to both the velocity and the magnetic field; its direction is shown in Fig. 30.3. According to Eq. (30.2), the acceleration caused by this force has a constant magnitude

$$a = \frac{F}{m} = \frac{qvB}{m} \qquad (30.3)$$

and its direction is also perpendicular to the velocity. Such a constant acceleration perpendicular to the velocity is characteristic of uniform circular motion. Thus, the particle will move in a circle of some radius r, and the acceleration given by Eq. (30.3) will play the role of centripetal acceleration; that is, $a = v^2/r$ [see Eq. (4.49)] and

$$\frac{qvB}{m} = \frac{v^2}{r} \qquad (30.4)$$

This leads to the following formula for the radius of the circular orbit:

radius of circular orbit

$$r = \frac{mv}{qB} \qquad (30.5)$$

Figure 30.4 is a photograph of a beam of electrons executing such uniform circular motion in a cathode-ray tube placed in a magnetic field.

In terms of the momentum $p = mv$, our equation for the radius of the circular orbit becomes

$$r = \frac{p}{qB} \qquad (30.6)$$

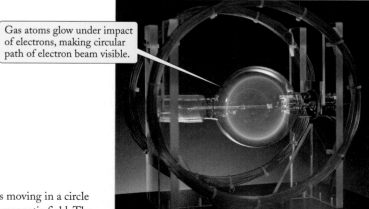

Gas atoms glow under impact of electrons, making circular path of electron beam visible.

FIGURE 30.4 Electrons moving in a circle in a cathode-ray tube in a magnetic field. The tube contains a gas at a very low pressure.

In this form, the equation is valid even for relativistic particles (that is, particles of a speed close to the speed of light). As we will see in Chapter 36, for such particles, the momentum is not equal to mv, and Eq. (30.5) is not valid; however, Eq. (30.6) remains valid.

The time per revolution is the distance (the circumference) divided by the speed. Thus the frequency f of the circular motion, or the number of revolutions per second, is

$$f = \frac{[\text{speed}]}{[\text{circumference}]} = \frac{v}{2\pi r} = \frac{v}{2\pi\,(mv/qB)}$$

or

$$f = \frac{qB}{2\pi m} \tag{30.7}$$

cyclotron frequency

This is called the **cyclotron frequency** because the operation of cyclotrons (described below) involves particles moving with this frequency in a magnetic field. The cyclotron frequency is independent of the speed of the circular motion—in a uniform magnetic field, slow particles and fast (but nonrelativistic) particles of a given charge and mass move around circles at the same frequency, but the slow particles move along smaller circles than the fast particles.

Note that this purely circular motion occurs when the velocity of a charged particle is perpendicular to a uniform magnetic field. If the velocity is parallel to the field, the magnetic force is zero, and the particle moves in a straight line at constant speed. If the particle velocity has components both parallel and perpendicular to the uniform magnetic field, then both motions occur: circular motion perpendicular to the field and linear motion parallel to the field. The path of a particle executing such motion traces out a helix. These three possible motions of a charged particle in a uniform magnetic field are illustrated in Fig. 30.5.

The **cyclotron** is a device for the acceleration of protons or other ions. It consists of an evacuated container placed between the poles of a large electromagnet; within the container there is a flat metallic can cut into two D-shaped pieces, or *dees* (Fig. 30.6). An oscillating high-voltage generator is connected to the dees; this creates an oscillating electric field in the gap between the dees. The frequency of the voltage generator is adjusted so that it coincides with the cyclotron frequency of Eq. (30.7). An ion source at the center of the cyclotron releases protons or other ions. The electric field in the gap

cyclotron

Concepts
— in —
Context

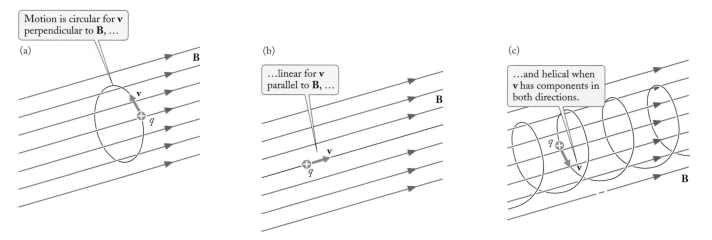

FIGURE 30.5 The three possible motions of a charge in a uniform magnetic field: (a) circular, for **v** perpendicular to **B**; (b) linear, for **v** parallel to **B**; and (c) helical, for **v** with components in both directions.

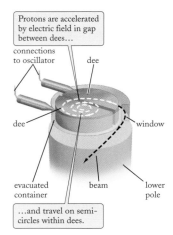

FIGURE 30.6 Trajectory of a particle within the dees of a cyclotron. In this diagram, the upper pole of the electromagnet has been omitted for the sake of clarity.

between the dees gives each of these protons a push, and the uniform magnetic field in the cyclotron then makes the proton travel on a semicircle inside the first dee. When the proton returns to the gap after one half period, the high-voltage generator will have reversed the electric field in the gap; the proton therefore receives an additional push, which sends it into the second dee. There, it travels on a semicircle of slightly larger radius corresponding to its slightly larger energy, and so on. Each time the proton crosses the gap between the dees, it receives an extra push and extra energy. The proton therefore travels along semicircles of stepwise increasing radius. When the protons reach the outer edge of the dees they leave the cyclotron as a high-energy beam.

Concepts
— *in* —
Context

EXAMPLE 1   One of the first cyclotrons, built by E. O. Lawrence at Berkeley in 1932, had dees with a diameter of 28 cm, and its magnet was capable of producing a magnetic field of 1.4 T. What was the maximum energy of the protons accelerated by this cyclotron? What was the frequency of the circular motion?

SOLUTION: The energy of the proton is its kinetic energy, $K = \frac{1}{2} m_p v^2$. When the proton reaches its maximum energy, its orbit has a radius of 14 cm. Since the magnetic field is 1.4 T, the speed of such a proton is, according to Eq. (30.5),

$$v = \frac{qBr}{m} = \frac{eBr}{m_p} = \frac{1.6 \times 10^{-19}\,\text{C} \times 1.4\,\text{T} \times 0.14\,\text{m}}{1.67 \times 10^{-27}\,\text{kg}}$$

$$= 1.9 \times 10^7\,\text{m/s}$$

and the energy is

$$K = \tfrac{1}{2} m_p v^2 = \tfrac{1}{2} \times 1.67 \times 10^{-27}\,\text{kg} \times (1.9 \times 10^7\,\text{m/s})^2 = 2.9 \times 10^{-13}\,\text{J}$$

or, in electron-volts,

$$K = 2.9 \times 10^{-13}\,\text{J} \times \frac{1\,\text{eV}}{1.6 \times 10^{-19}\,\text{J}} = 1.8 \times 10^6\,\text{eV} = 1.8\,\text{MeV}$$

The frequency is given by Eq. (30.7),

$$f = \frac{qB}{2\pi m} = \frac{eB}{2\pi m_p} = \frac{1.6 \times 10^{-19}\,\text{C} \times 1.4\,\text{T}}{2\pi \times 1.67 \times 10^{-27}\,\text{kg}} = 2.1 \times 10^7/\text{s} = 21\,\text{MHz}$$

Checkup 30.1

QUESTION 1: Two electrons are moving in a uniform magnetic field. One has an orbital radius of 15 cm. The other has twice the speed of the first. What is its orbital radius?

QUESTION 2: The magnetic field exerts forces on charged particles, but is incapable of doing work on them. Why?

QUESTION 3: A proton and an electron are moving in a uniform magnetic field. Which has the larger cyclotron frequency?

QUESTION 4: Can electrons be accelerated in a cyclotron? In what regard does a cyclotron used for electrons differ from one used for protons, if the magnetic field is the same?

QUESTION 5: Two protons are moving in the same uniform magnetic field. The orbital radius of the first proton is 10 cm, and the orbital radius of the second is 20 cm. Which proton has the larger speed? Which has the larger orbital frequency?

(A) First, first   (B) First, second   (C) Second, first
(D) Second, second   (E) Second, both same

30.2 FORCE ON A WIRE

If a wire carrying a current is placed in a magnetic field, the moving charges within the wire will experience a force. Since the motion of the charges in the wire is constrained by the wire (the charges must move along the wire, regardless of the magnetic force they experience), any force acting on these charges is merely transferred to the wire, and therefore the wire as a whole will experience a force equal to the total force acting on the charges.

Consider a segment of the wire of length dl. To find the amount of moving charge in this segment, suppose that the speed of these moving charges is v; then they take a time $dt = dl/v$ to move out of the segment (see Fig. 30.7). Since the amount of charge is the product of the time and the current, we find that

$$dq = I\, dt = I\, dl/v$$

where, as always, we pretend that the moving charge is positive. If the wire segment is oriented at right angles to the magnetic field, then the magnetic force is

$$dF = dq\, vB = I(dl/v)vB = IB\, dl \qquad (30.8)$$

The direction of this force is related to the direction of the current and the direction of the magnetic field by the right-hand rule. For instance, the force on the wire segment oriented perpendicularly to the magnetic field, as illustrated in Fig. 30.8, is perpendicular to the wire and to the magnetic field, out of the plane of the page.

If the wire segment is oriented parallel to the magnetic field, then the force is zero. Other orientations of the wire segment give forces of a magnitude somewhere between zero and the maximum magnitude $IB\,dl$. We can find the dependence of the force on the orientation of the wire segment by noting that, for a charged particle moving at an angle α relative to the magnetic field, the force is reduced by a factor $\sin\alpha$, as in Eq. (30.1). Hence this same factor must appear in the general expression for the force on a wire segment oriented at an angle α to the magnetic field (Fig. 30.9):

$$dF = IB\, dl \sin\alpha \qquad (30.9)$$

This equation yields Eq. (30.8) as a special case (with $\alpha = 90°$).

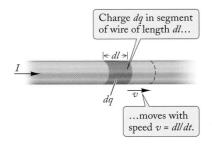

FIGURE 30.7 In a time $dt = dl/v$, the moving charge moves a distance dl toward the right. Hence, in this time, all the moving charge in the segment dl of the wire moves out of this segment toward the right (this charge is replaced by charge entering the segment from the left).

force on wire segment

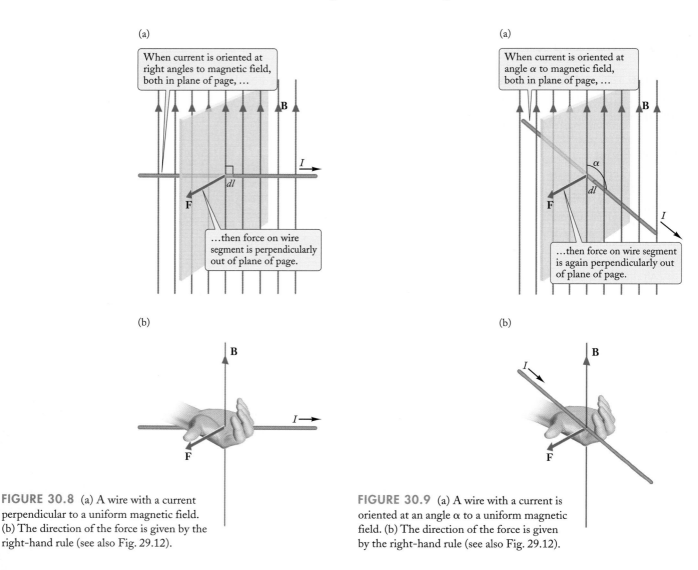

FIGURE 30.8 (a) A wire with a current perpendicular to a uniform magnetic field. (b) The direction of the force is given by the right-hand rule (see also Fig. 29.12).

FIGURE 30.9 (a) A wire with a current is oriented at an angle α to a uniform magnetic field. (b) The direction of the force is given by the right-hand rule (see also Fig. 29.12).

To find the force on an entire wire, we must add the contributions (30.9) from all the small segments of this wire vectorially. For a straight wire of length L in a uniform magnetic field, the contributions simply add to give a force of the same form as (30.9),

force on straight wire in uniform field

$$F = ILB \sin\alpha \tag{30.10}$$

If we define a vector $d\mathbf{l}$ of length dl (or $\mathbf{L}$ of length L) in the direction of the current I, then the magnetic force on a wire segment or straight wire can be expressed as, respectively, the cross product

$$d\mathbf{F} = Id\mathbf{l} \times \mathbf{B} \quad \text{or} \quad \mathbf{F} = I\mathbf{L} \times \mathbf{B} \tag{30.11}$$

EXAMPLE 2   Figure 30.10 shows a balance used for the measurement of a magnetic field. A loop of wire carrying a precisely known current is partially immersed in the magnetic field. The force that the magnetic field exerts on the loop can be measured with the balance, and this permits the calculation of the strength of the magnetic field. Suppose that the short side of the loop measures 10.0 cm, the current in the wire is 0.225 A, and the magnetic force is 5.35×10^{-2} N. What is the strength of the magnetic field?

SOLUTION: The upper part of the loop is not in the magnetic field and hence experiences no magnetic force. The vertical segments of the loop carry the same current in opposite directions, and so these force contributions cancel. Thus the net magnetic force is that experienced by the lower, short side of the loop. This segment and the magnetic field are perpendicular, so Eq. (30.10) with $\alpha = 90°$ gives

$$F = ILB$$

The strength of the magnetic field is then

$$B = \frac{F}{IL} = \frac{5.35 \times 10^{-2}\,\text{N}}{0.225\,\text{A} \times 0.100\,\text{m}} = 2.38\,\text{T}$$

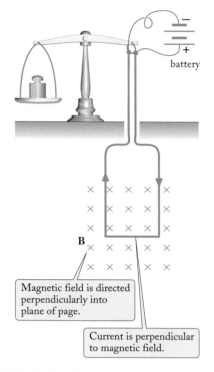

Magnetic field is directed perpendicularly into plane of page.

Current is perpendicular to magnetic field.

FIGURE 30.10 A current balance.

EXAMPLE 3   Two very long, parallel wires separated by a distance r carry currents I_1 and I_2, respectively. Find the magnetic force that each wire exerts on a segment dl of the other wire.

SOLUTION: Each wire generates a magnetic field, which exerts a force on the other wire. Figure 30.11 shows the magnetic field $\mathbf{B}_1$ that the current I_1 produces in the vicinity of the current I_2. By Eq. (29.9), this magnetic field has a magnitude

$$B_1 = \frac{\mu_0}{2\pi} \frac{I_1}{r}$$

and it is perpendicular to wire 2. By Eq. (30.9), the force on a segment dl of wire 2 has a magnitude

$$dF = I_2 B_1\, dl = \frac{\mu_0}{2\pi} \frac{I_1 I_2}{r}\, dl \tag{30.12}$$

The force on a segment of wire 1 is of the same magnitude, but opposite direction. The force between the wires is attractive if the currents in the wires are parallel and repulsive if they are antiparallel. Note that for parallel wires, the force dF is a constant, independent of the position of dl. Thus the total force F on a larger segment L is obtained by replacing dl with L in Eq. (30.12), as given in Eq. (30.11).

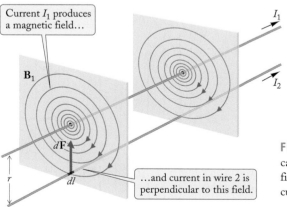

Current I_1 produces a magnetic field...

$\mathbf{B}_1$

$d\mathbf{F}$

dl

...and current in wire 2 is perpendicular to this field.

I_1

I_2

FIGURE 30.11 A long, straight wire carrying a current I_2 in the magnetic field of a long, straight wire carrying a current I_1.

The official definition of the SI **unit of current** is based on the force per meter of length between two long parallel wires. This force can be measured very precisely by holding one wire stationary and suspending the other from a balance; the wires are connected in series so that the currents are exactly equal ($I_1 = I_2$). The force per unit length and the distance r can be measured experimentally, and the value of I_1 (or I_2) calculated from Eq. (30.12) is then the current in amperes. The constant μ_0 appearing in Eq. (30.12) is assigned the value $\mu_0 = 4\pi \times 10^{-7}$ N·s^2/C^2 *by definition.*

The official definition of the SI unit of charge is based on the unit of current. *The* **coulomb** *is defined as the amount of charge that a current of one ampere delivers in one second,* 1 coulomb = (1 ampere $\times$ 1 second).

coulomb (C)

✔ Checkup 30.2

QUESTION 1: A power cord for an electric appliance consists of two parallel straight wires carrying currents in opposite directions. Is the magnetic force between them attractive or repulsive?

QUESTION 2: A straight wire carrying a current is placed in a uniform horizontal magnetic field. For what orientation of the wire is the magnetic force on the wire zero?

QUESTION 3: A wire hangs in a horizontal magnetic field, held taut by a weight (see Fig. 30.12). We connect this wire to a battery, so a current flows vertically upward. The deflection of the hanging wire will be

(A) To the right   (B) To the left   (C) Into the page
(D) Out of the page   (E) Zero

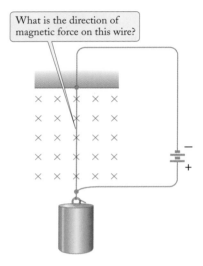

FIGURE 30.12 A wire hanging in a uniform magnetic field.

30.3 TORQUE ON A LOOP

If a loop of wire with a current is placed in a magnetic field, the action of the magnetic field can in general result not only in a net force on the loop, but also in a torque. The force and the torque depend on the shape and the orientation of the loop, and they can be zero for some orientations of the loop. For example, the net force on a current loop in a *uniform* magnetic field is always zero. To see this, let us consider the simple case of a rectangular loop oriented perpendicularly to the magnetic field, illustrated in Fig. 30.13. The forces on the four sides of the loop are given by Eq. (30.10). The forces

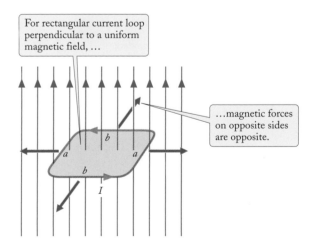

FIGURE 30.13 Forces on a loop of current at right angles to a magnetic field.

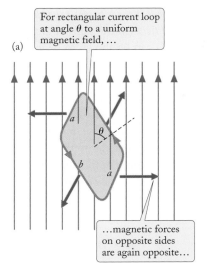

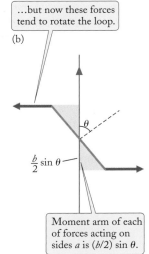

(a) For rectangular current loop at angle θ to a uniform magnetic field, …

…magnetic forces on opposite sides are again opposite…

(b) …but now these forces tend to rotate the loop.

Moment arm of each of forces acting on sides a is $(b/2)\sin\theta$.

FIGURE 30.14 (a) Forces on a rectangular loop of current oriented at an angle θ with respect to a magnetic field. (b) Side view of the loop showing forces acting on sides a.

on opposite sides are opposite (because the currents are opposite); hence the forces cancel in pairs, and the loop will be in equilibrium. The same is true for a loop of arbitrary shape if we consider all the small contributions to the force: the net force is zero.

Figure 30.14 shows a loop oriented at an angle θ to a uniform magnetic field. In this figure, the angle θ is measured between the magnetic field and the normal to the loop ($\theta = 0$ means the loop is perpendicular to the magnetic field; $\theta = 90°$ means the loop is parallel). Again the forces on opposite sides are opposite, but now the lines of action of the forces pulling on the left and on the right sides do not coincide—these forces exert a torque, which tends to rotate the loop. Since the magnetic field is constant, Eq. (30.10) tells us that the magnitude of the forces on the top and bottom sides is

$$F = IaB \tag{30.13}$$

where a is the length of each of these sides. We recall from Eq. (13.14) that the torque is the product of the force and the moment arm. If, as shown in Fig. 30.14b, the length of each of the other two sides is b, then the moment arm about the center of the loop is $(b/2)\sin\theta$, and the torque due to the pair of forces IaB is therefore

$$\tau = F(b/2)\sin\theta + F(b/2)\sin\theta$$

$$= IabB\sin\theta \tag{30.14}$$

The product of the area $A = ab$ of the loop and the current flowing around it is called the **magnetic dipole moment** of the loop:

$$\mu = [\text{current}] \times [\text{area}] = IA \tag{30.15}$$

magnetic dipole moment

Note that if the loop consists of several turns of wire, then Eq. (30.15) must be evaluated with the net current flow around the loop. This net current equals the number N of turns of wire times the current I_0 in one turn, so

$$\mu = NI_0A \tag{30.16}$$

In any case, the magnitude of the torque is

$$\tau = \mu B\sin\theta \tag{30.17}$$

torque on current loop

The direction of the torque is such that it tends to twist the plane of the loop into an orientation perpendicular to the magnetic field (so the normal to the loop is parallel to the field, or $\theta = 0$). The formula (30.17) can be shown to be valid not only for the rectangular loop, but also for loops of any other shape. The magnetic moment of any (flat) loop of arbitrary shape is the product of the area and the net current around the loop.

We know from Section 13.4 that the torque is a vector with direction along the axis around which the torque acts. To obtain a vectorial expression for the torque, we need to treat the magnetic moment as a vector. The magnetic moment vector $\boldsymbol{\mu}$ has a magnitude $\mu = IA$ and a direction perpendicular to the plane of the loop, according to the following **right-hand rule for the magnetic moment**: *if you wrap the fingers of your right hand around the current loop in the direction of the current, your thumb will point in the direction of $\boldsymbol{\mu}$* (see Fig. 30.15). The magnitude and the direction of the torque are then given by the vector formula

$$\boldsymbol{\tau} = \boldsymbol{\mu} \times \mathbf{B} \tag{30.18}$$

A loop of current suitably pivoted on an axis acts as a compass needle; the normal to the loop seeks to align itself with the magnetic field. This similarity is no accident. A compass needle is a small permanent magnet, and as we will see in Section 30.4, a permanent magnet contains a large number of electrons acting as small current loops. The mechanism underlying the alignment of a compass needle with a magnetic field is therefore the same as for a current loop.

The torque on a current loop pivoted on an axis and placed in a strong magnetic field is exploited in electric motors. This torque brings about the rotational motion of the current loop and provides the means of converting electric energy into mechanical energy of rotational motion. In small electric motors, the magnetic field is produced by permanent magnets; but in large motors, the magnetic field is produced by electromagnets. The following example illustrates the operation of a simple electric motor.

right-hand rule for magnetic moment

magnetic torque vector

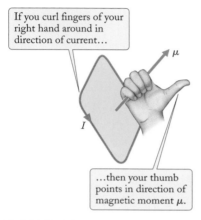

If you curl fingers of your right hand around in direction of current…

…then your thumb points in direction of magnetic moment μ.

FIGURE 30.15 Right-hand rule for the magnetic moment of a current loop.

EXAMPLE 4 A simple electric motor consists of a rectangular coil of wire that rotates on a longitudinal axle in a magnetic field of 0.50 T (see Fig. 30.16). The coil measures 10 cm × 20 cm; it has 40 turns of wire, and the current in the wire is 8.0 A. (a) In terms of the angle θ betweeen the magnetic field and the normal to the coil, what is the torque that the magnetic field exerts

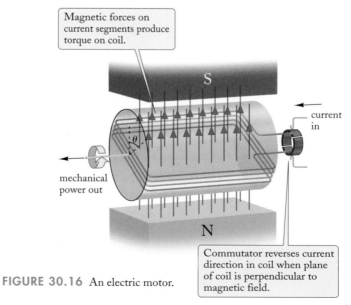

Magnetic forces on current segments produce torque on coil.

current in

mechanical power out

Commutator reverses current direction in coil when plane of coil is perpendicular to magnetic field.

FIGURE 30.16 An electric motor.

on the coil? (b) In order to keep the sign of the torque constant, a switch (commutator) mounted on the axle reverses the current in the coil whenever θ passes through 0° and 180°. Plot this torque vs. the angle θ.

SOLUTION: (a) According to Eq. (30.17), the torque on the coil is

$$\tau = N I_0 abB \sin\theta$$

$$= 40 \times 8.0 \text{ A} \times 0.10 \text{ m} \times 0.20 \text{ m} \times 0.50 \text{ T} \times \sin\theta$$

$$= 3.2 \text{ N·m} \sin\theta$$

(b) If the torque always has the same sign (say, positive), then the plot of the torque for the interval 180° to 360° merely repeats the plot for the interval 0° to 180° Figure 30.17 shows this plot.

COMMENT: Practical electric motors consist of many such coils, each with its commutator, arranged at regular intervals around the axle. The plot of the net torque of this arrangement is a sum of plots such as shown in Fig. 30.17, but with different starting angles. This averages out the ups and downs of Fig. 30.17 and yields a torque that is nearly constant at all angles.

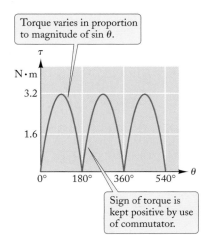

FIGURE 30.17 Torque as a function of angle, according to Example 4.

The torque on a current loop is also exploited in instruments for the measurement of current, generally called **galvanometers**. In such an instrument, the current to be measured is sent into a coil of several turns pivoted on an axis and placed in the magnetic field of a permanent magnet (see Fig. 30.18). A fine spiral spring attached to the moving coil provides a restoring torque that opposes the magnetic torque. Under the influence of these two opposing torques, the moving coil attains an angular deflection that increases with the magnitude of the current. A pointer attached to the moving coil indicates the magnitude of the current on a calibrated scale. Many types of meters operate on this principle.

The potential energy of a current loop in a magnetic field can be obtained from the torque as follows. If an external agent pushes against the magnetic torque, turning a loop through an angle $d\theta$, it does work on the loop [see Eq. (13.3)]. The potential energy of the loop will increase:

$$dU = \tau \, d\theta = \mu B \sin\theta \, d\theta \qquad (30.19)$$

If only the angle θ between $\boldsymbol{\mu}$ and $\mathbf{B}$ is changing, the potential energy stored when the loop is turned from some reference angle θ_0 to some final angle θ is

$$U = \int dU = \int \mu B \sin\theta \, d\theta = \mu B \int_{\theta_0}^{\theta} \sin\theta \, d\theta = -\mu B(\cos\theta - \cos\theta_0) \qquad (30.20)$$

If we choose $\theta_0 = 90°$, when $\boldsymbol{\mu}$ is perpendicular to $\mathbf{B}$, as our reference angle, we obtain a simple form for the potential energy:

$$U = -\mu B \cos\theta \qquad (30.21)$$

This can be written in terms of the scalar product (dot product) as

$$U = -\boldsymbol{\mu} \cdot \mathbf{B} \qquad (30.22)$$

This potential energy of a magnetic moment in a magnetic field is smallest (negative) when $\boldsymbol{\mu}$ is parallel to $\mathbf{B}$. The energy difference between parallel and antiparallel

galvanometer

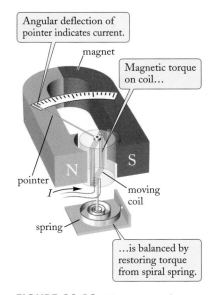

FIGURE 30.18 Mechanism of a galvanometer.

potential energy of current loop

alignment of $\boldsymbol{\mu}$ with respect to **B** is $\Delta U = 2\mu B$. The magnetic energy (30.21) is an important quantity in the study of atomic physics, magnetic resonance, and magnetic alignment, for example, in magnetic recording materials.

 Checkup 30.3

QUESTION 1: The torque given by Eq. (30.17) has a maximum magnitude at $\theta = 90°$ and at $\theta = -90°$. What is the direction of the torque on the loop shown in Fig. 30.14 for $\theta = 90°$? For $\theta = -90°$?

QUESTION 2: The torque given by Eq. (30.17) is zero at $\theta = 0°$ and at $\theta = 180°$, so the loop is in equilibrium. Is the equilibrium stable at $0°$? At $180°$?

QUESTION 3: What would happen if the motor in Fig. 30.16 had no commutator?

QUESTION 4: For the two loops in Figs. 30.13 and 30.14, what is the direction of the magnetic moment vector?

QUESTION 5: What is the maximum torque delivered by the motor described in Example 4? At what orientation is it maximum? If the motor had 120 turns of wire instead of 40, what would be the maximum torque?

QUESTION 6: A coil in the x-z plane carries a current in the direction shown in Fig. 30.19 and is in a region of uniform magnetic field with direction in the x-y plane as indicated. The direction of the torque is

(A) $-\mathbf{i}$ (B) $\mathbf{j}$ (C) $-\mathbf{j}$

(D) $\mathbf{k}$ (E) $-\mathbf{k}$

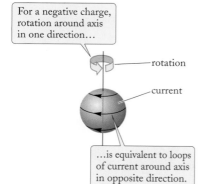

FIGURE 30.19 A coil in a uniform magnetic field.

FIGURE 30.20 A small ball of negative charge spinning about an axis.

30.4 MAGNETISM IN MATERIALS

In all the calculations of magnetic fields in Chapter 29 we assumed that the wires carrying the currents were placed in air or in a vacuum. However, in practical applications of magnetism, the current-carrying wires are often placed around cores of solid iron. The presence of iron or other ferromagnetic materials enhances the magnetic field, and often makes it several thousand times stronger than what it would be without the iron. This enhancement arises from an alignment of the axes of spin of the electrons.

Crudely, we can picture the electron as a small ball of negative electric charge spinning about an axis (see Fig. 30.20). The rotational motion of the electron charge is equivalent to currents flowing in loops around the axis of spin, and the magnetic field produced by the electron is therefore that of a small loop of current. The magnetic moment of this loop of current has a fixed value, called a **Bohr magneton**

Bohr magneton

$$\mu_B = \frac{eh}{4\pi m_e} = 9.27 \times 10^{-24}\,\text{J/T} \tag{30.23}$$

Planck's constant

where $h = 6.63 \times 10^{-34}$ J·s is a fundamental constant of nature known as **Planck's constant**, discussed in detail in Chapter 37. In some materials, there are also rings of current due to the orbital motions of electrons around the nuclei in atoms, so in addition to spin magnetic moments, there are also orbital magnetic moments.

The nucleus of an atom can also have a magnetic moment, always much smaller than that of the electron. The proton, the nucleus of the hydrogen atom, has the largest magnetic moment of all nuclei, but even this moment is about 650 times smaller than a Bohr magneton. Nonetheless, the small nuclear magnetic moment can be separately sensed using resonance techniques to provide vivid images of subsurface structures in materials and body tissues; such **magnetic resonance imaging (MRI)** is commonly used in hospitals and in research (see Fig. 30.21).

The net magnetic moment of an atom is obtained by adding the various spin and orbital magnetic moments, taking into account the directions of these moments. In ferromagnetic materials, the magnetic moment is completely dominated by the contributions from the electron spins.

When a piece of iron is left to itself, the electron current loops are oriented at random, and the magnetic fields of the electrons average to zero. But when the iron is placed in an external magnetic field, such as the field of a solenoid, the axes of the spins of the electrons align with the magnetic field. Although this alignment is triggered by the external magnetic field, which exerts a torque on the small electron current loops (see Section 30.3), the amount of alignment is strongly magnified by a natural tendency for spin alignment in iron and other ferromagnetic materials. This alignment of spins makes all the electron current loops parallel, and *the combined magnetic fields of all these parallel current loops add to the original magnetic field and enhance it drastically.*

The enhancement of the magnetic field by this effect is measured by the **magnetic susceptibility** χ (the Greek letter chi). If we denote the magnetic field of the external currents by B_{external} and the field due to the electrons in the material by B_{matter}, then the magnetic susceptibility is defined as the ratio

$$\chi = \frac{B_{\mathrm{matter}}}{B_{\mathrm{external}}} \tag{30.24}$$

Ferromagnets have very large values of the susceptibility, often on the order of 1000 or more, because of the cooperative alignment of many electron spins.

For most ordinary "nonmagnetic" materials, χ is very small, with a magnitude around 10^{-5} to 10^{-4}. Such materials with small positive susceptibilities are called **paramagnets**. The external field does not do much to overcome the random orientations of the electron current loops in a paramagnet, except at very low temperatures. Some materials have small negative susceptibilities and are called **diamagnets**.

An alternative measure of magnetism in a material is the **magnetic permeability** μ:[1]

$$\mu = \mu_0(1 + \chi) \tag{30.25}$$

Thus the magnetic permeabilities of most paramagnetic and diamagnetic materials do not differ appreciably from μ_0 (the permeability constant), because for most materials, $|\chi| \ll 1$. But the magnetic permeability of ferromagnets is much greater than μ_0.

A piece of iron with aligned electron spins is said to be magnetized. If the iron is removed from the solenoid, it will retain some of its magnetization and act as a **permanent magnet**. Some ferromagnetic materials, such as alnico (an alloy mostly of iron, nickel, and cobalt, with some aluminum and copper), retain more magnetization than pure iron, and they make better permanent magnets. Figure 30.22 shows several kinds of permanent magnets, including bar magnets and horseshoe magnets.

Electromagnets and permanent magnets are used in many kinds of electric machinery, such as electric motors (see Example 4 above), electric generators, tape recorders, videocassette recorders, computer disk drives, etc.

---

[1]Do not confuse the magnetic permeability with the magnetic dipole moment. Both are represented by the same letter μ. But they are different quantities.

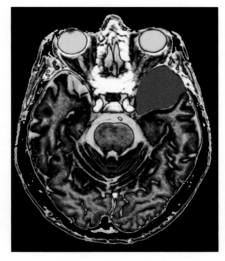

magnetic resonance imaging (MRI)

FIGURE 30.21 Nuclear magnetic resonance image (MRI) of a human head.

magnetic susceptibility

ferromagnet

paramagnet and diamagnet

magnetic permeability μ

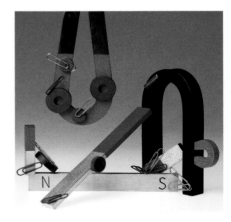

FIGURE 30.22 Different kinds of permanent magnets.

PHYSICS IN PRACTICE   MAGNETIC RECORDING MEDIA

In the recording head of a magnetic tape recorder, a small electromagnet produces a magnetic field whose strength varies in proportion to the amplitude of the sound signal arriving at the microphone (see Fig. 1). The tape passing over the recording head contains a magnetizable material. Initially, the tape is blank and unmagnetized. When the tape is subjected to the magnetic field of the recording head, it becomes magnetized. The strength of the induced magnetization in the tape is proportional to the strength of the magnetic field of the recording head, which is proportional to the intensity of the sound signal. Thus, as the tape passes over the recording head, it stores the information about the varying sound signal in the form of varying zones of high and low magnetization.

The gap between the poles of the electromagnet in the recording head must be extremely narrow—typically no more than 10^{-5} m—so an instant of the sound signal is recorded on a very narrow, almost pointlike spot or band on the tape. The quality of the recording improves if the speed of the tape over the recording head is high, since this leads to a wider separation between the magnetizations recorded at successive instants. The tape recorders in broadcasting studios typically use a tape speed of 15 in./s.

Digital information is stored on magnetic media by orienting the magnetization of each bit, or unit of information, in either of two directions; since the exact magnitude of the magnetization of each bit is unimportant, digital magnetic

(a)

(b)

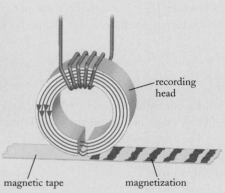

recording head

magnetic tape

magnetization

FIGURE 1 Recording head of magnetic tape recorder. (a) Close-up view of reel-to-reel transport, showing the head block. (b) Diagram of interface between recording head and tape.

EXAMPLE 5

Perminvar is an alloy of iron, nickel, and cobalt with a relative permeability of $\mu/\mu_0 = 2200$. Suppose that a long solenoid has 450 turns per meter and carries a current of 1.5 A. If the solenoid is filled with Perminvar, what is the total magnetic field in the solenoid?

SOLUTION: The solenoid provides the external field for the material, given by Eq. (29.21):

$$B_{\text{external}} = \mu_0 nI = 1.26 \times 10^{-6} \text{ N·m}^2/\text{C}^2 \times 450/\text{m} \times 1.5 \text{ A} = 8.5 \times 10^{-4} \text{ T}$$

This external field induces the magnetization in the ferromagnet, and according to Eq. (30.24) produces a field

$$B_{\text{matter}} = \chi B_{\text{external}}$$

The total field inside the solenoid is thus

$$B = B_{\text{external}} + B_{\text{matter}} = (1 + \chi)B_{\text{external}}$$

media are relatively free from random errors, or "noise." So-called floppy disks store more than 10 million bits of information, each in small rectangular regions of magnetization. The magnetic bits on computer hard disks (Fig. 2) can be 100 to 1000 times more dense, with individual bits as small as 10^{-6} m across (see Fig. 3). Such disks often use a cobalt–nickel alloy as their magnetizable material. One of the advantages (and disadvantages!) of magnetic tapes and disks over optical media, such as CDs, is that the magnetic media can be erased and rewritten easily, for an essentially unlimited number of cycles, whereas CDs are harder to rewrite and

can tolerate only a limited number of rewriting cycles. Because the data stored on magnetic tapes and disks can be easily erased, it is important to keep such media (including the magnetic stripe on ATM and credit cards) away from strong permanent magnets, which could destroy the recorded information.

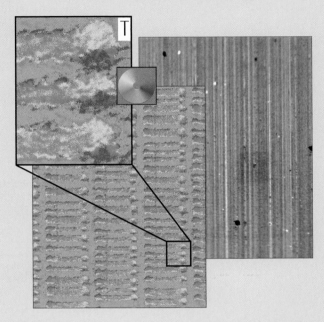

FIGURE 3 Individual bits of a digital hard disk. The magnetic image (color) reveals the bits which are not visible on the corresponding topographic image (gray). The scale bar (top) is 1 μm.

FIGURE 2 Computer hard disk.

But by Eq. (30.25), $(1 + \chi)$ is just the relative permeability μ/μ_0, so

$$B = \frac{\mu}{\mu_0} B_{\text{external}} = 2200 \times 8.5 \times 10^{-4}\,\text{T}$$

$$= 1.9\,\text{T}$$

EXAMPLE 6   The electromagnet of a cyclotron consists of two iron pole pieces, each surrounded by a large wire coil. Suppose such a coil has 200 turns, and that each turn is a concentric circular loop of radius $R = 30$ cm. To produce a total field of $B = 1.4$ T in the iron at the center of the current loops, what field should the coil alone provide? What should the current in the wire be? The relative permeability of iron is 35 when $B = 1.4$ T.

Concepts
— in —
Context

SOLUTION: As in Example 5, the total field B in the iron is related to the externally provided field B_{external} from the current loops by

$$B = \frac{\mu}{\mu_0} B_{\text{external}} \tag{30.26}$$

Solving for B_{external} and using the given permeability gives

$$B_{\text{external}} = \frac{\mu_0}{\mu} B = \frac{1}{35} \times 1.4\,\text{T} = 4.0 \times 10^{-2}\,\text{T}$$

This is the field that must be produced by the wire loops.

For a number N of current loops with radius R, the field produced is N times the field for an individual loop, given by Eq. (29.19):

$$B_{\text{external}} = N\frac{\mu_0 I}{2R}$$

Solving for the current I, and substituting the required value of B_{external}, gives

$$I = \frac{2RB_{\text{external}}}{N\mu_0} = \frac{2 \times 0.30\,\text{m} \times 4.0 \times 10^{-2}\,\text{T}}{200 \times 4\pi \times 10^{-7}\,\text{N·s}^2/\text{C}^2} = 95\,\text{A}$$

Electromagnets carrying such large currents require water cooling to prevent overheating.

 Checkup 30.4

QUESTION 1: Consider the magnetic field of the rotating ball of negative charge shown in Fig. 30.20. What is the direction of the magnetic field that this rotating ball produces along its axis?

QUESTION 2: At high values of the externally applied field, a ferromagnet *saturates*; that is, all the magnetic moments fully align with the field, and no further alignment is possible. In this regime, how does the permeability vary with increasing external field?

  (A) Increases   (B) Decreases   (C) Remains constant

30.5 THE HALL EFFECT

When charges flow along a wire or a conducting rod placed in a uniform magnetic field, a potential difference is produced along opposite sides of the wire. This phenomenon is called the **Hall effect**; it is important in the study of materials and in various practical applications.

Consider a rectangular rod carrying an electric current I. The rod is immersed in a uniform magnetic field **B** perpendicular to the current (see Fig. 30.23). The magnetic force qvB diverts the moving charges within the rod to one side. Since the current is confined to the rod, the diverted charges build up at one side and are depleted from the other side until such charges produce a transverse electric force that exactly balances the magnetic force. If we let $E_\perp$ denote the corresponding electric field, in the direction perpendicular to the current, equilibrium implies

$$qE_\perp = qvB \tag{30.27}$$

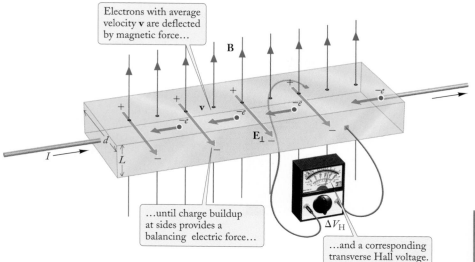

Electrons with average velocity **v** are deflected by magnetic force...

...until charge buildup at sides provides a balancing electric force...

...and a corresponding transverse Hall voltage.

FIGURE 30.23 Geometry used for the Hall effect. A longitudinal current I flows down the length of a conductor in a perpendicular magnetic field **B**; a transverse Hall voltage ΔV_H is generated.

If the width of the rod is d, the transverse electric field produces a potential difference between the sides of the rod known as the **Hall voltage** ΔV_H, where

$$\Delta V_H = E_\perp d$$

or, with Eq. (30.27),

$$\Delta V_H = vBd \qquad (30.28)$$

The average velocity v of the charges is related to the current I; for electrons, it has a magnitude given by Eq. (27.10),

$$v = \frac{I}{neA} \qquad (30.29)$$

where n is the density of electrons and A is the cross-sectional area of the rod. The Hall voltage is thus

$$\Delta V_H = \frac{IBd}{neA}$$

If the thickness of the rod is L, then the area is $A = Ld$, which gives

$$\Delta V_H = \frac{IB}{neL} \qquad (30.30)$$

Thus the Hall voltage is proportional to the applied current and magnetic field, and is inversely proportional to the density of charge carriers and the thickness of the conductor.

The Hall effect is used to determine the charge carrier density n in a material, as well as the sign of the charge carriers. The Hall effect provides empirical evidence that the charge carriers in most metals are negative charges, that is, electrons. In some

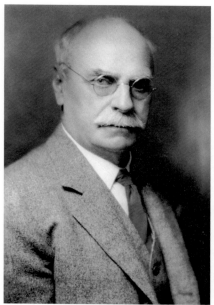

EDWIN HERBERT HALL (1855–1938)
American physicist. While working on his Ph.D. thesis, Hall investigated a question Maxwell had posed: Does the magnetic force act on the conductor or the current? He demonstrated that the force was indeed on the current, and realized that electric charge was pushed to one side of the conductor. His observation of the generation of a transverse electric field in materials carrying current in a magnetic field, or Hall effect, provides the basis for many modern instruments and for our understanding of conduction in solids.

Hall voltage

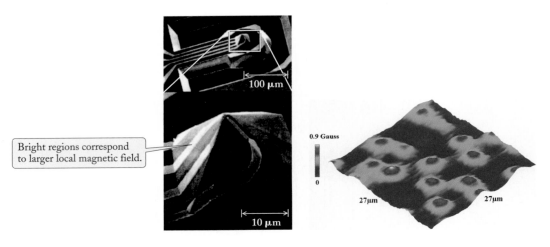

100 μm

Bright regions correspond
to larger local magnetic field.

10 μm

0.9 Gauss

0

27μm   27μm

FIGURE 30.24 (a) A microfabricated Hall sensor. (b) A scanning Hall microscope image of the magnetic field as a function of lateral position just above a high-temperature superconductor. The red features represent individual superconducting vortices.

materials with mobile ions, the sign of the Hall voltage identifies positive ions as the charge carriers. Even in some metals, such as aluminum at high magnetic fields, the Hall effect can reveal that the charge carriers are effectively positive. Such "anomalous" behavior is understood in terms of the collective behavior of electrons moving in the electrostatic potential of the crystal lattice of ions (see Section 39.4); the effectively positive charges are known as **holes**.

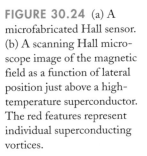

hole

For metals, with large values of n, Eq. (30.30) tells us that the Hall voltage will be small unless thin wires are used. But for semiconductors, with small values of n, the Hall voltage is large, and thus very sensitive to changes in B. Semiconductors are used as **Hall sensors**, both in simple instruments that measure B, and in devices such as the scanning Hall microscope, where a microfabricated Hall sensor (see Fig. 30.24a) detects small local changes in B. Such a mapping of the local field is shown in Fig. 30.24b, which depicts the magnetic field as a function of position above the surface of a high-temperature superconductor.

Hall sensor

A related phenomenon is observed in thin, nearly two-dimensional samples at very low temperatures. Modern quantum physics reveals that the cyclotron orbits discussed in Section 30.1, like many other quantities, can actually take on only certain discrete sizes. Quantum calculations show that the number of allowed orbits in which electrons may travel, and thus the allowed values of the transverse electric field in the material, depends on the magnetic field B. This gives rise to the **quantum Hall effect**, where precisely spaced steps are observed in the Hall resistance, $\Delta V_{\mathrm{H}}/I$, as a function of the field B, instead of the simple linear relation predicted by Eq. (30.30); both behaviors are shown in Fig. 30.25. These steps occur at exactly inverse-integer multiples of a fundamental Hall resistance given by $h/e^2 = 25\,816.802\ \Omega$, where h is Planck's constant (see Section 30.4 and Chapter 37). The quantum Hall effect permits the precise determination of fundamental constants and also provides an easily reproducible; fundamental standard of resistance.

quantum Hall effect

EXAMPLE 7   A flat metal ribbon of width 2.0 mm and thickness 150 μm is immersed in a perpendicular uniform magnetic field of strength 8.2 T. A current of 1.5 A is passed along the length of the ribbon, and a transverse voltage equal to 375 μV is measured across its width. What is the average velocity of the electrons in the metal? What is the number of conduction electrons per unit volume in this metal?

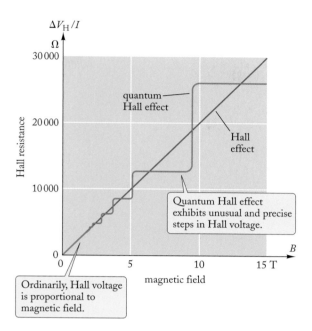

FIGURE 30.25 A plot of the Hall resistance $\Delta V_{\mathrm{H}}/I$ as a function of magnetic field. The straight line is for an ordinary conductor; the stepped curve is for a two-dimensional conductor at low temperatures (the quantum Hall effect).

SOLUTION: The given transverse voltage in a uniform magnetic field is just the Hall voltage, $\Delta V_{\mathrm{H}} = 375~\mu\mathrm{V}$. The drift velocity can be obtained immediately from Eq. (30.28):

$$v = \frac{\Delta V_{\mathrm{H}}}{Bd} = \frac{375 \times 10^{-6}~\mathrm{V}}{8.2~\mathrm{T} \times 2.0 \times 10^{-3}~\mathrm{m}}$$

$$= 0.023~\mathrm{m/s}$$

The number of electrons per unit volume, or the charge carrier density, is given by Eq. (30.30).

$$n = \frac{IB}{\Delta V_{\mathrm{H}e}L} = \frac{1.5\mathrm{A} \times 8.2~\mathrm{T}}{375 \times 10^{-6}~\mathrm{V} \times 1.6 \times 10^{-19}~\mathrm{C} \times 150 \times 10^{-6}~\mathrm{m}}$$

$$= 1.4 \times 10^{27}~\mathrm{m}^{-3}$$

 ## Checkup 30.5

QUESTION 1: Two wires of the same material are in the same uniform magnetic field, and each carries the same current. If the first wire is 10 times thicker than the second (in the direction of the magnetic field), how do their Hall voltages differ?

QUESTION 2: The Hall effect geometry of Fig. 30.23 shows the case for negative charge carriers. For the current and magnetic field directions shown, consider instead positive charge carriers. Which vectors would reverse direction?

    (A) $\mathbf{v}$ only   (B) $\mathbf{E}_{\perp}$ only   (C) Both $\mathbf{v}$ and $\mathbf{E}_{\perp}$ (D) Neither $\mathbf{v}$ nor $\mathbf{E}_{\perp}$

SUMMARY

PHYSICS IN PRACTICE   Magnetic Recording Media   **(page 978)**

CIRCULAR ORBIT IN MAGNETIC FIELD

$$r = \frac{mv}{qB}$$

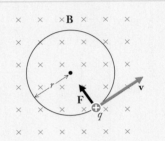

(30.5)

CYCLOTRON FREQUENCY

$$f = \frac{qB}{2\pi m}$$

(30.7)

FORCE ON WIRE SEGMENT

$$dF = I \, dl \, B \sin\alpha$$
or
$$d\mathbf{F} = I \, d\mathbf{l} \times \mathbf{B}$$

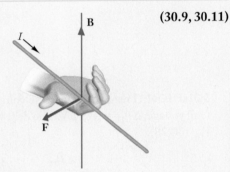

(30.9, 30.11)

FORCE ON A STRAIGHT WIRE

$$F = ILB \sin\alpha \quad \text{or} \quad \mathbf{F} = I\mathbf{L} \times \mathbf{B}$$

(30.10, 30.11)

FORCE PER UNIT LENGTH BETWEEN PARALLEL WIRES

$$\frac{F}{l} = \frac{\mu_0}{2\pi} \frac{I_1 I_2}{r}$$

(30.12)

MAGNETIC DIPOLE MOMENT OF CURRENT LOOP

$$\mu = [\text{current}] \times [\text{area}] = IA$$

(30.15)

TORQUE ON CURRENT LOOP

$$\boldsymbol{\tau} = \boldsymbol{\mu} \times \mathbf{B}$$

magnitude of torque

$$\tau = \mu B \sin\theta$$

(30.18)

(30.17)

POTENTIAL ENERGY OF CURRENT LOOP

$$U = -\boldsymbol{\mu} \cdot \mathbf{B} = -\mu B \cos\theta$$

(30.21, 30.22)

BOHR MAGNETON

$$\mu_B = \frac{eh}{4\pi m_e} = 9.27 \times 10^{-24}\,\text{J/T} \qquad \textbf{(30.23)}$$

MAGNETIC SUSCEPTIBILITY

$$\chi = \frac{B_{matter}}{B_{external}} \qquad \textbf{(30.24)}$$

χ positive and $\chi \ll 1$: paramagnet

χ positive and $\chi \gg 1$: ferromagnet

χ negative: diamagnet

MAGNETIC PERMEABILITY

$$\mu = \mu_0(1 + \chi) \qquad \textbf{(30.25)}$$

HALL VOLTAGE

where n is the number density of charges and d and L are the width and thickness of the conductor.

$$\Delta V_H = vBd = \frac{IB}{neL} \qquad \textbf{(30.28, 30.30)}$$

QUESTIONS FOR DISCUSSION

1. Cosmic rays are high-speed charged particles—mostly protons—that crisscross interstellar space and strike the Earth from all directions. Why is it easier for the cosmic rays to penetrate through the magnetic field of the Earth near the poles than anywhere else?

2. Is it possible to define a magnetic potential energy for a charged particle moving in a magnetic field?

3. If we want a proton to orbit all the way around the equator in the Earth's magnetic field, must we send it eastward or westward?

4. Consider the strip of metal placed in a magnetic field, as shown in Fig. 30.23. How does the Hall potential difference between the sides of the strip change if we reverse the current? If we reverse the magnetic field? If we reverse both?

5. If we replace the metallic strip in Fig. 30.23 by a semiconducting strip of p-type semiconductor, that is, a material with positive charge carriers, will the back side remain at the higher potential?

6. The Earth's magnetic field at the equator is horizontal and in the northward direction. Assume that the atmospheric electric field is downward. The electric and magnetic fields are then "crossed" fields. In what direction must we launch an electron if the magnetic and the electric forces are to balance?

7. If a strong current flows through a thick wire, it tends to cause a compression of the wire. Explain.

8. If a wire carrying a current is placed in a magnetic field, the field exerts forces on the moving electrons in the wire. How does this cause a push on the wire?

9. A horizontal wire carries a current in the eastward direction. The wire is located in a magnet producing a uniform magnetic field. What must be the direction of the field if it is to compensate for the weight of the wire?

10. In the SI system, we first define the unit of current (by means of the magnetic force between wires), and we then define the unit of charge in terms of the current. Could we proceed in the opposite manner: first define the unit of electric charge (by means of the electric force between charges), and then define the unit of current as a flow of charge? What would be the advantages and the disadvantages?

11. Suppose that an infinitely long straight wire lies along the axis of a circular loop. Both the wire and the loop carry currents. Explain why the force from the wire on the loop is zero, and explain why the force from the loop on the wire is also zero.

12. Positive charge is uniformly distributed over a sphere. If this sphere spins about a vertical axis in a counterclockwise direction as seen from above, what is the direction of the magnetic moment vector?

13. According to Eq. (30.15), the magnetic moment of a square loop is the same as the magnetic moment of a circular loop of the same area. Why is the shape unimportant? [Hint: The circular loop can be regarded as made of many small (infinitesimal) square loops.]

PROBLEMS

†30.1 Circular Motion in a Uniform Magnetic Field

1. A bubble chamber, used to make the tracks of protons and other charged particles visible, is placed between the poles of a large electromagnet that produces a uniform magnetic field of 2.0 T. A high-energy proton passing through the bubble chamber makes a track that is an arc of a circle of radius 3.5 m. According to Eq. (30.6), what is the momentum of the proton?

2. A proton of kinetic energy 1.0×10^7 eV moves in a circular orbit in the magnetic field near the Earth. The strength of the field is 5.0×10^{-5} T. What is the radius of the orbit?

3. In principle, a proton of the right energy can orbit the Earth in an equatorial orbit under the influence of the Earth's magnetic field. If the orbital radius is to be 6.5×10^3 km and the magnetic field at this radius is 3.3×10^{-5} T, what must be the momentum of the proton? Ignore gravity in this problem.

4. In the Crab Nebula (the remnant of a supernova explosion), electrons of a momentum of up to about 10^{-16} kg·m/s orbit in a magnetic field of about 10^{-8} T. What is the orbital radius of such electrons? Note that it is necessary to use Eq. (30.6) for the calculation of the orbital radius.

5. At the Fermilab accelerator, protons of momentum 5.3×10^{-16} kg·m/s are held in a circular orbit of diameter 2.0 km by a vertical magnetic field. What is the strength of the magnetic field required for this?

6. What is the charge-to-mass ratio of a particle that orbits in a magnetic field of 0.200 T with an angular frequency of 3.51×10^{10} rad/s?

7. In a bubble chamber, a particle is observed to move in a circular orbit of radius 1.2 m. The magnetic field in the bubble chamber has a magnitude of 2.0 T. Assuming the charge of the particle is e, what is the magnitude of its momentum?

8. In a **mass spectrometer**, ions of a precisely selected speed are sent into a uniform magnetic field, where they move along a semicircle and then strike a photographic plate, making a mark (Fig. 30.26). A careful measurement of the distance of

the mark from the entrance slit gives us the diameter (and the radius) of the circular motion of the ion in the magnetic field, and from this radius the mass can be calculated by means of Eq. (30.5). Suppose that the magnetic field in the mass spectrometer is 0.050 T. Suppose that an ion of charge e and speed 4.0×10^4 m/s is found to strike the photographic plate at a distance of 0.332 m from the entrance slit. What is the mass of this ion?

9. Astrophysicists believe that the radio waves of 10^9 Hz reaching us from Jupiter are emitted by electrons of fairly low (nonrelativistic) energies orbiting in Jupiter's magnetic field. What must be the strength of this field if the cyclotron frequency is to be 1.0×10^9 Hz?

10. Figure 30.27 shows the tracks of an electron (charge $-e$) and an antielectron (charge $+e$) created in a bubble chamber. When the particles made these tracks, they were under the influence of a magnetic field of magnitude 1.0 T and of a direction perpendicular to and into the plane of the figure. What is the momentum of each particle? Assume that they are moving in the plane of the figure, and that this figure is $\frac{1}{10}$ natural size. Which is the track of the electron, and which is the track of the antielectron?

FIGURE 30.27 Tracks of an electron and an antielectron in a bubble chamber. The tracks spiral because the particles suffer a loss of energy as they pass through the liquid. For the purposes of Problem 10, concentrate on the initial portions of the tracks.

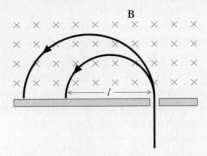

FIGURE 30.26 Paths of ions of different masses in a mass spectrometer. The crosses show the tails of the magnetic field vectors.

11. A beam containing the nuclei of two isotopes of carbon, ^{12}C (6 protons and 6 neutrons) and ^{14}C (6 protons and 8 neutrons), at the same momentum enters a magnetic field, which bends the trajectories of these two kinds of particles into circles of radii r_{12} and r_{14}, respectively. What is the ratio of the two radii?

†For help, see Online Concept Tutorial 33 at www.wwnorton.com/physics

12. In the Brookhaven AGS accelerator, protons are made to move around a circle of radius 128 m by the magnetic force exerted by a vertical magnetic field. The maximum field that the magnets of this accelerator can generate is 1.3 T.

 (a) Calculate the maximum permissible momentum of the protons.

 (b) Calculate the orbital frequency of such protons.

*13. You want to confine an electron of energy 3.0×10^4 eV by making it circle inside a solenoid of radius 10 cm under the influence of the force exerted by the magnetic field. The solenoid has 120 turns of wire per centimeter. What minimum current must you put through the wire if the electron is not to hit the wall of the solenoid?

*14. The Earth's magnetic field at the equator has a magnitude of 5.0×10^{-5} T; its direction is horizontal toward the north. Suppose that the Earth's atmospheric electric field is 100 V/m; its direction is vertically down. With what velocity (magnitude and direction) must we launch an electron if the electric force is to cancel the magnetic force?

*15. A high-speed electron is moving in a uniform magnetic field of 5.0×10^{-4} T. The electron has a velocity component of 6.0×10^6 m/s perpendicular to the magnetic field and a velocity component of 4.0×10^6 m/s parallel to the magnetic field.

 (a) Describe the motion of the electron qualitatively.

 (b) What are the cyclotron frequency and the corresponding period of the transverse motion?

 (c) How far does the electron advance parallel to the magnetic field in one such period?

*16. Suppose that a region with a uniform magnetic field **B** also has a uniform electric field **E** perpendicular to the magnetic field, an arrangement called **crossed fields**. Show that for a charged particle moving in such crossed fields in a direction perpendicular to both **E** and **B**, the electric force cancels the magnetic force, provided the particle has a speed

$$v = \frac{E}{B}$$

If the magnetic field is in the vertical upward direction and the electric field is in the northward direction, what must be the direction of motion of the charged particle to produce this cancellation? Crossed fields are used in **velocity selectors**, or velocity filters. In these devices, a beam of charged particles is aimed into a region of crossed fields, and this deflects particles out of the beam, except for those particles that have the critical velocity $v = E/B$.

*17. A velocity selector with crossed **E** and **B** fields (see Problem 16) is to be used to select alpha particles of energy 2.0×10^3 eV from a beam containing alpha particles of several energies. The electric field strength is 1.0×10^6 V/m. What must be the magnetic field strength? Alpha particles have charge $+2e$ and mass 4.0 u.

*18. In a mass spectrometer, a beam of ions is first made to pass through a velocity selector with crossed fields **E** and **B** (see

Problem 16). The selected ions are then made to enter a region of uniform magnetic field **B′**, where they move in arcs of circles (Fig. 30.28). The radii of these circles depend on the masses of the ions. Assume that an ion has a single charge e.

 (a) Show that in terms of E, B, $B′$, and the impact distance l marked in Fig. 30.28, the mass of the ion is

$$m = \frac{eBB'l}{2E}$$

 (b) Assume that in such a spectrometer, ions of the isotope ^{16}O impact at a distance of 29.20 cm and ions of a different isotope of oxygen impact at a distance of 32.86 cm. The mass of the ions of ^{16}O is 16.00 u. What is the mass of the other isotope?

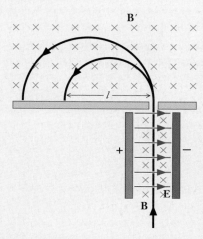

FIGURE 30.28 A mass spectrometer with a velocity selector.

*19. In a uniform magnetic field, an electron and a proton are orbiting in opposite directions along the same circle perpendicular to the magnetic field:

 (a) If the speed of the electron is 3.0×10^6 m/s, what must be the speed of the proton? Assume that the only force acting on the magnetic particles is the magnetic force.

 (b) If the particles are initially at the same point on the circle (moving in opposite directions), where on the circle will they meet again? Give the answer as an angle, in radians, measured along the circle.

30.2 Force on a Wire

20. A straight wire is placed in a uniform magnetic field; the wire makes an angle of 30° with the magnetic field. The wire carries a current of 6.0 A, and the magnetic field has a strength of 0.40 T. Calculate the force on a 10-cm segment of this wire. Show the direction of the force in a diagram.

21. The electric cable supplying an electric clothes dryer consists of two long straight wires separated by a distance of 1.2 cm. Opposite currents of 20 A flow on these wires. What is the magnetic force experienced by a 1.0-cm segment of wire due to the entire length of the other wire?

22. Two parallel cables of a high-voltage power line carry opposite currents of 1.8×10^3 A. The distance between the cables is 4.0 m. What is the magnetic force pushing on a 50-cm segment of one of these cables? Treat both cables as very long, straight wires.

*23. A closed loop of arbitrary (possibly three-dimensional) shape is placed near a very long, straight wire. Currents I and I', respectively, flow around the loop and along the straight wire. Prove that the force exerted by the straight wire on the loop must be perpendicular to the wire. (Hint: Use Newton's Third Law.)

*24. A straight segment of wire carries a current of 15 A from the point $\mathbf{r}_1 = 0.35\mathbf{i} + 0.50\mathbf{j}$ to the point $\mathbf{r}_2 = 0.50\mathbf{i} - 0.25\mathbf{j} + 0.40\mathbf{k}$, where distances are measured in meters. The wire segment is in a uniform magnetic field of 2.0 T parallel to the positive z axis. What is the magnetic force on the wire?

*25. A long wire carries current of 25 A parallel to the positive x axis, except for three of the four segments that follow the edges of a cube of side 0.25 m, as shown in Fig. 30.29. The wire is in a uniform magnetic field of 2.0 T directed parallel to the positive x axis. What is the net magnetic force on the wire?

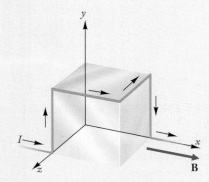

FIGURE 30.29 A long wire bent along four edges of a cube.

*26. Consider the long, straight wire and the U-shaped wire shown in Fig. 30.30. The wires lie in the same plane. The bottom of the U has a length l, and the sides of the U are very long. The wires are separated by a distance d, and they carry currents I and I', respectively. What is the force that the straight wire exerts on the U-shaped wire?

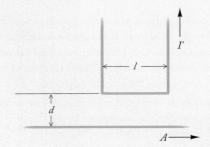

FIGURE 30.30 Straight wire and U-shaped wire.

*27. An electromagnetic launcher, or rail gun, consists of two parallel conducting rails across which is laid a conducting bar, which serves as a projectile. To launch this projectile, the rails are immersed in a magnetic field, and a current is sent through the rails and the bar (see Fig. 30.31). The magnetic force on the current in the bar then accelerates the bar (in actual rail guns of this kind, the magnetic field is itself produced by the current in the rails; but let us ignore this complication). Suppose that the magnetic field has a strength of 0.20 T and the bar has a length of 0.10 m and a mass of 0.20 kg. Ignore friction. What current must you send through the bar to give it an acceleration of 1.0×10^5 m/s²?

FIGURE 30.31 Electromagnetic launcher.

**28. A thin flexible wire carrying a current I hangs in a uniform magnetic field $\mathbf{B}$ (Fig. 30.32). A weight attached to one end of the wire provides a tension T. Within the magnetic field, the wire will adopt the shape of an arc of a circle.

(a) Show that the radius of this arc of a circle is $r = T/(BI)$.

(b) Show that if we remove the wire and launch a particle of charge $-q$ from the point P with a momentum $p = qT/I$ along the direction of the wire, it will move along the same arc of a circle. (This means that the wire can be used to simulate the orbit of the particle. Experimental physicists sometimes use such wires to check the orbits of particles through systems of magnets.)

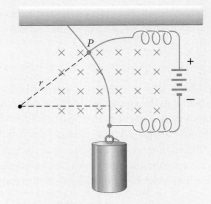

FIGURE 30.32 A wire with a weight hanging in a magnetic field.

30.3 Torque on a Loop

29. The coil in the mechanism of an ammeter is a rectangular loop, measuring 1.0 cm × 2.0 cm, with 120 turns of wire. The coil is immersed in a magnetic field of 0.010 T. What is the torque on this coil when its plane is parallel to the magnetic field and carries a current of 1.0×10^{-3} A?

30. A horizontal circular loop of wire of radius 20 cm carries a current of 25 A. At the location of the loop, the magnetic field of the Earth has a magnitude of 3.9×10^{-5} T and points down at an angle of 16° with the vertical. What is the magnitude of the torque that this magnetic field exerts on the loop?

31. The proton has a magnetic moment of 1.41×10^{-26} A·m².

    (a) If the proton is placed in a uniform magnetic field of 0.80 T, plot the torque as a function of the angle θ.

    (b) If the magnetic moment is initially oriented antiparallel to the field, how much energy will be released when the proton flips into the parallel orientation?

32. Consider the electric motor of Example 4. What is the average value of the torque over a complete rotation? If this motor rotates at the rate of 50 rev/s, what is the average horsepower that it delivers to its axle?

33. A circular coil consists of 60 turns of wire wound around the rim of a plastic disk of radius 2.0 cm. What is the magnetic moment of this coil if the current in the wire is 0.30 A?

*34. A pipe made of a superconducting material has a length of 0.30 m and a radius of 4.0 cm. A current of 4.0×10^3 A flows around the surface of the pipe; the current is uniformly distributed over the surface. What is the magnetic moment of this current distribution? (Hint: Treat the current distribution as a large number of rings stacked one on top of another.)

*35. An amount of charge Q is uniformly distributed over a disk of paper of radius R. The disk spins about its axis with angular velocity ω. Find the magnetic dipole moment of the disk. Sketch the lines of magnetic field and of electric field in the vicinity of the disk.

**36. A rectangular loop of wire of dimension 12 cm × 25 cm faces a long, straight wire. The two long sides of the loop are parallel to the wire, and the two short sides are perpendicular; the midpoint of the loop is 8.0 cm from the wire (Fig. 30.33). Currents of 95 A and 70 A flow in the straight wire and the loop, respectively.

    (a) What translational force does the straight wire exert on the loop?

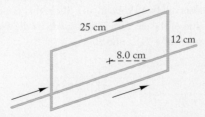

FIGURE 30.33 Long, straight wire and rectangular loop.

(b) What torque about an axis parallel to the straight wire and through the center of the loop does the straight wire exert on the loop?

**37. A compass needle is attached to an axle that permits it to turn freely in a horizontal plane so that only the horizontal component of the magnetic field affects its motion. The magnetic moment of the needle is 9.0×10^{-3} A·m², its moment of inertia is 2.0×10^{-8} kg·m², and the horizontal component of the Earth's magnetic field is 1.9×10^{-5} T.

    (a) What is the torque on the needle as a function of the angle θ between the needle and the north direction?

    (b) What is the frequency of small oscillations of the needle about the north direction? [Hint: Compare the equation of the rotational motion of the needle with Eq. (15.39) for the motion of a pendulum.]

30.4 Magnetism in Materials

38. Because of its relatively large susceptibility, the paramagnetic metal palladium is used as a calibration standard for magnetic measurements. At room temperature, palladium has a susceptibility of 7.9×10^{-4}. What is the relative permeability of palladium (its permeability in terms of μ_0)?

39. The **magnetization** M of a material is the magnetic field due to the material divided by μ_0; that is, $M = B_{\text{matter}}/\mu_0$. The magnetization also defines the magnetic moment per unit volume, or $M = $ [magnetic moment]/[volume]. If iron has a maximum internal field of $B_{\text{matter}} = 2.19$ T, calculate (a) the magnetization of iron, and (b) the magnetic moment per atom of iron (the density of iron atoms is 8.50×10^{28}/m³).

40. A certain long solenoid has 320 turns per meter and carries a current of 5.00 A. Determine the total field inside the solenoid when it is filled with (a) iron, which has a relative permeability of 200 under these conditions, and (b) permalloy, which has a relative permeability of 1500 under these conditions.

41. Some simple elemental superconductors (called type I) are perfect diamagnets, meaning that their internal magnetic field exactly cancels the applied field, or $B_{\text{matter}} = -B_{\text{external}}$. What are (a) the susceptibility and (b) the permeability of such superconductors?

*42. In so-called type I superconductors, surface currents fully shield the external magnetic field, ensuring zero net field in the interior of the superconductor; such perfect diamagnetism is known as the **Meissner effect**. A 10-cm-long superconducting cylindrical rod of indium metal has its axis parallel to an external field of 2.7×10^{-2} T; the material is a perfect diamagnet. What is the total current flowing around the rod? Such a current is confined to a thin surface layer with a thickness known as the **penetration depth**. For simplicity, assume that the surface current flowing around the rod is uniformly distributed over a penetration depth of 5.0×10^{-8} m. What is the current density in this layer?

30.5 The Hall Effect

43. A copper wire of diameter 3.0 mm is placed at right angles to a magnetic field of 2.0 T. The wire carries a current of 50 A. What is the Hall potential difference between opposite sides of the wire?

44. Repeat the preceding problem, with the assumption that the wire is placed at an angle of 60° with respect to the magnetic field.

45. A Hall sensor made from silicon is 0.30 mm thick and has a charge carrier density of $2.0 \times 10^{19}/m^3$. When it is placed in an unknown perpendicular magnetic field and carries a current of 5.0 mA along its length, a Hall voltage of 130 mV is measured across its width. What is the value of the magnetic field?

46. A 0.10-mm-thick strip of calcium metal is placed in a perpendicular magnetic field of 8.0 T. When a current of 100 mA is passed along its length, a Hall voltage of 5.4 μV is measured across its width. If the charge carriers in calcium are electrons, how many free electrons per calcium atom are there? (Calcium metal has 2.3×10^{28} atoms/m^3.)

47. A thin strip of rubidium metal, 65 μm thick and 1.0 mm wide, is placed in a field of 6.0 T. A current of 0.30 A is passed along its length. Rubidium has a density of 1.53 g/cm^3, a molar weight of 85.5 g/mole, and one conduction electron per atom.

   (a) Find the drift velocity of these conduction electrons.

   (b) What Hall voltage is measured across this strip?

48. A metal ribbon is 2.0 mm wide and 0.15 mm thick, and carries a current of 1.5 A along its length. In a perpendicular magnetic field of 2.0 T, a Hall voltage of 150 μV is measured across the width of the ribbon.

   (a) What is the number density of charge carriers in this metal?

   (b) What is the drift velocity of these electrons?

49. Related to the Hall voltage is a quantity known as the **Hall coefficient** R_H, which takes the value $R_H = -1/ne$ when the charge carriers are free electrons. What is the value of R_H for potassium? (Potassium has one conduction electron per atom, a density of 0.86 g/cm^3, and a molar weight of 39.1 g/mole.)

REVIEW PROBLEMS

50. An electromagnet produces a magnetic field of 0.60 T in the vertical downward direction. An electron of speed 6.0×10^6 m/s is moving through this magnetic field. What are the magnitude and the direction of the magnetic force on the electron if the electron is moving

   (a) Vertically downward?

   (b) Horizontally toward the east?

   (c) Horizontally toward the south?

   (d) Toward the south and downward at an angle of 30° from horizontal?

51. An electromagnet produces a magnetic field of 0.60 T in the vertical downward direction. A straight wire placed in this magnetic field carries a current of 3.0 A. Calculate the magnitude and direction of the magnetic force on a 0.10-m segment of this wire if the orientation is

   (a) Vertically downward

   (b) Horizontally toward the east

   (c) Horizontally toward the south

   (d) Toward the south and downward at an angle of 30° from horizontal

52. A piece of wire bent to form two perpendicular straight segments carries a current of 5.0 A, first along the x axis from $x = 0.50$ m to the origin, and then along the y axis from the origin to $y = -0.25$ m. The wire segment is in a uniform magnetic field of 1.5 T parallel to the positive z axis. What are the magnitude and direction of the magnetic force on the wire?

53. A proton of speed 4.0×10^5 m/s is seen to move in a circular orbit of radius 0.40 m in a magnetic field. What is the strength of the magnetic field that will give such circular motion? What is the frequency of the motion of the proton?

54. A circular loop of radius $r = 3.0$ cm has a total charge $Q = 5.0$ μC uniformly distributed around its circumference. The loop rotates about its axis with angular velocity $\omega = 6.3 \times 10^4$ rad/s. What is the magnetic moment of the loop?

*55. In 1897, **J. J. Thomson's** e/m **experiment** determined the charge-to-mass ratio of the electron. First, a beam of electrons of an unknown velocity v was deflected a distance y by being passed through a region of length L of uniform electric field E.

   (a) Show that this deflection distance is (see Section 23.4)

$$y = \frac{eEL^2}{2m_e v^2}$$

   (b) Then a magnetic field B was turned on perpendicular to the electric field and increased until no deflection was observed (see Problem 16). Show that the charge-to-mass ratio of the electron is then given by the experimental values of y, E, L, and B:

$$\frac{e}{m_e} = \frac{2yE}{B^2 L^2}$$

*56. A high-speed electron is moving in a uniform magnetic field of 5.0×10^{-4} T and a parallel uniform electric field of 2.0×10^3 V/m. The electron initially has a velocity component of 6.0×10^6 m/s perpendicular to the magnetic and electric fields and a velocity component of 4.0×10^6 m/s parallel to the magnetic and electric fields.

  (a) Describe the motion of the electron qualitatively.

  (b) What are the cyclotron frequency and the corresponding period of the transverse motion?

  (c) How far does the electron advance parallel to the magnetic and electric fields in the first such period? By how much does its parallel velocity increase in any such period?

*57. Two positively charged particles separated by a distance d, each with a charge q and a mass m, are initially moving with the same speed v in opposite directions perpendicular to the line joining them (see Fig. 30.34). A magnetic field applied perpendicularly to the plane of the page will bend the paths of the particles into circles. What strength of magnetic field is necessary to make them collide head-on midway between the two starting points? (Ignore the electric forces between the charges.)

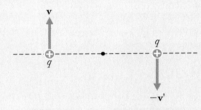

FIGURE 30.34 Two positively charge particles in a magnetic field.

*58. A rectangular loop of wire of dimensions 12 cm × 18 cm is near a long, straight wire. The rectangle and the straight wire lie in the same plane. One of the short sides of the rectangle is parallel to the straight wire and at a distance of 6.0 cm; the long sides are perpendicular to the straight wire (Fig. 30.35). A current of 40 A flows on the straight wire, and a current of 60 A flows around the loop. What are the magnitude and direction of the net magnetic force that the straight wire exerts on the loop?

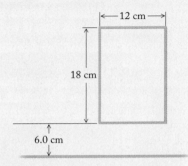

FIGURE 30.35 Long, straight wire and rectangular loop.

*59. A 25-turn circular coil has a mass of 0.050 kg. The coil is immersed in a uniform 0.20-T magnetic field, and is initially held so the plane of the coil is parallel to the magnetic field (see Fig. 30.36). The coil carries 5.0 A of current. When released, what will be the instantaneous angular acceleration of the coil about the horizontal axis?

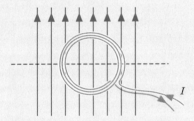

FIGURE 30.36 Circular coil in magnetic field.

Answers to Checkups

Checkup 30.1

1. The radius of circular motion for a charged particle in a magnetic field is given by Eq. (30.5): $r = mv/qB$. Since both electrons are in the same magnetic field, m, q, and B are the same, and the radius varies in proportion to the speed. Thus the second electron has twice the orbital radius of the first, or 30 cm.

2. Since the magnetic force is always perpendicular to the velocity $\mathbf{v} = d\mathbf{s}/dt$, it is always perpendicular to the displacement $d\mathbf{s}$. Thus the incremental amount of work, $dW = \mathbf{F} \cdot d\mathbf{s} = Fds \times \cos\theta$, is always zero, and so the magnetic field can do no work.

3. Since the cyclotron frequency is $f = qB/2\pi m$ [Eq. (30.7)], a smaller mass results in a larger cyclotron frequency for the electron.

4. Yes, any charged particle could be accelerated in a cyclotron. In a given magnetic field, a cyclotron used for electrons would operate at higher frequency (higher by the proton–electron mass ratio). Also, the electrons circulate in the opposite sense to protons (since the magnetic force is in the opposite direction for negative charges).

5. (E) Second, both same. As in Question 1, the larger speed corresponds to the larger radius, since $r = mv/qB$. Both have

the same orbital frequency, because the cyclotron frequency $f = qB/2\pi m$ does not depend on radius or speed.

Checkup 30.2

1. The force on them is repulsive. Recall from Eq. (29.5) that the magnetic force on a positive charge moving parallel to a current in a long wire is negative, or radially inward. For currents moving in antiparallel directions, the force is repulsive. This is also easily verified from the vector form $\mathbf{F} = I\mathbf{L} \times \mathbf{B}$.

2. If the wire is straight, the force will be zero when it is parallel to the magnetic field.

3. (B) To the left. From $\mathbf{F} = I\mathbf{L} \times \mathbf{B}$, we see from the right-hand rule that the cross product of an upward current with a field into the page in Fig. 30.12 gives a force that is to the left.

Checkup 30.3

1. For $\theta = 90°$, the direction of the cross product of $\boldsymbol{\mu}$ and $\mathbf{B}$ is out of the page; a torque out of the page implies that the loop will turn counterclockwise (right-hand rule). Similarly, for $\theta = -90°$, $\boldsymbol{\tau}$ is into the page; such a torque implies that the loop will turn clockwise.

2. From $\theta = 0°$, a small displacement results in a configuration with a torque that would return $\boldsymbol{\mu}$ toward $\theta = 0°$, and so $\theta = 0°$ is a stable equilibrium (apply the right-hand rule to Fig. 30.14). However, near $\theta = 180°$, a small displacement results in a torque that would also push $\boldsymbol{\mu}$ toward $\theta = 0°$; thus $\theta = 180°$ is a condition of unstable equilibrium.

3. Without a commutator, the loop would oscillate about its equilibrium orientation (where the coil is at right angles to the field, or the magnetic moment is parallel to the field); with friction, these oscillations would damp out and the magnetic moment would eventually remain parallel to the field.

4. The magnetic moment vector is always perpendicular to the plane of the loop, in the direction given by the right-hand rule: with the fingers wrapped around in the direction of the current, the thumb points upward in Fig. 30.13, and at an angle θ to the right of upward in Fig. 30.14.

5. From the plot in Fig. 30.17, or by setting $\sin\theta = 1$ in the result of part (a) of Example 4, we see that the maximum torque is 3.2 N·m. The torque has this magnitude when the angle $\theta = \pm 90°$, that is, when the plane of the coil is parallel to the magnetic field (magnetic moment perpendicular to the magnetic field). The torque is proportional to the magnetic moment, which is proportional to the number of turns of the coil, so tripling the number of turns to 120 would triple the torque to 9.6 N·m.

6. (D) $\mathbf{k}$. From the right-hand rule preceding Eq. (30.18), the magnetic moment is directed downward, in the $-\mathbf{j}$ direction. By the usual right-hand rule for cross products, the torque $\boldsymbol{\tau} = \boldsymbol{\mu} \times \mathbf{B}$ is then in the $+\mathbf{k}$ direction.

Checkup 30.4

1. Looking from above the ball, the charge rotates counterclockwise. Since this is negative charge, the current rotates in clockwise loops. By the right-hand rule, the field along the axis is thus pointing downward.

2. (B) Decreases. If the alignment of spins cannot increase, then B_{matter} remains constant. With increasing external field, the susceptibility $\chi = B_{\text{matter}}/B_{\text{external}}$ must then decrease, as must the permeability $\mu = \mu_0(1 + \chi)$.

Checkup 30.5

1. By Eq. (30.30), the Hall voltage varies inversely with thickness, so the second wire will have 10 times the Hall voltage of the first.

2. (C) Both $\mathbf{v}$ and $\mathbf{E}_\perp$. For positive charge carriers, the velocity is in the same direction as the current, and so is reversed from the case shown. The magnetic force on the moving positive charges is then in the same direction as previously (both q and $\mathbf{v}$ change sign); consequently, the buildup of charge on the sides of the conductor and the resulting transverse electric field are reversed.

Electromagnetic Induction

CONCEPTS IN CONTEXT

Concepts — in — Context

The coils of this large electric power generator, one of seventeen at Hoover Dam, surround an assembly of electromagnets. When the flow of water provides an external mechanical torque, the magnet assembly rotates, sweeping the magnetic field through the coils.

In this chapter we will see that such a relative motion between magnetic fields and coils converts mechanical work into electrical work by generating a voltage, or an electromotive force (emf), across the coils. We will consider such questions as:

? How does a simple generator produce an emf? (Section 31.2, page 1000)

? How does a rotating coil produce an alternating emf? (Example 4, page 1003)

? How can a rotating coil produce a steady, nonalternating emf? (Section 31.3 and Example 5, page 1009)

? How much magnetic energy is stored in the coils of a large generator such as shown in the photo? (Example 12, page 1015)

In this chapter we will discover that electric fields can be generated not only by charges, but also by changing magnetic fields or by changing magnetic flux. For instance, if we increase the currents in the windings of an electromagnet and thereby increase the strength of the magnetic field, this changing magnetic field generates an electric field, called an *induced electric field*. This kind of electric field exerts the usual electric forces on charges—in this regard the induced electric field does not differ from an ordinary electrostatic electric field.

However, the induced and the electrostatic electric fields differ in that the forces exerted by the latter are conservative, whereas the forces exerted by the former are nonconservative. We recall that, according to the requirement laid down in Chapter 8, what discriminates between conservative and nonconservative forces is the work done during a round trip. For conservative forces, such as the forces exerted by electrostatic electric fields, the work done during a round trip is zero; for nonconservative forces, such as the forces exerted by induced electric fields, the work done during a round trip is nonzero. This means that when a charge moves around a closed circuit, the induced electric fields deliver work to the charge, and they constitute a source of emf. Like the emf of a battery, this **induced emf** is capable of driving a current around the circuit. As we will see, a celebrated law formulated by Michael Faraday asserts that the magnitude of the induced emf is directly proportional to the rate of change of the magnetic flux intercepted by the circuit. Such induced emfs find practical application in electromagnetic generators, widely used for the generation of electric power.

We begin with a discussion of the induced emf produced by the motion of a conductor, such as a wire or a rod, in a constant magnetic field. As we will see, such a moving conductor sweeps across magnetic flux and thereby generates an induced emf, called a motional emf.

induced emf

31.1 MOTIONAL EMF

Suppose that we push a rod of metal with some velocity **v** through a uniform magnetic field, such as the magnetic field of a large electromagnet. The free electrons in the metal will then also have a velocity **v**, and they will experience a magnetic force. For motion perpendicular to the magnetic field, the magnitude of the magnetic force on an electron is evB [see Eq. (29.12)]. If the rod and the velocity **v** of the rod are perpendicular to each other and to the magnetic field, then the direction of this magnetic force on the free electrons is parallel to the rod, as indicated by the right-hand rule (see Fig. 31.1). The electrons will therefore flow along the rod, accumulating negative charge on the upper end and leaving positive charge on the lower end. The flow of charge will stop when the electric repulsion generated by the accumulated charges balances the magnetic force evB. However, if the ends of the rod are in sliding contact with a pair of long wires that provide a stationary return path, then the electrons will flow continuously around the circuit (see Fig. 31.2). Thus, the moving rod acts as a "pump of electricity," or a source of emf, which produces the same effect as a battery connected to the long wires. In Fig. 31.2, the upper end of the rod is the negative terminal of this source, and the lower end is the positive terminal.

The emf associated with the rod is defined as the work done by the driving force per unit positive charge that passes from the negative end of the rod to the positive. Since the magnitude of the force on an electron is evB, the magnitude of the force per unit

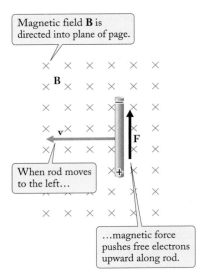

FIGURE 31.1 Conducting rod moving with velocity **v** through a uniform magnetic field. The magnetic field is directed into the plane of the page; the crosses indicate the tails of the magnetic field vectors. According to the right-hand rule, **v** × **B** is in the downward direction. The magnetic force −*e* **v** × **B** on a free electron within this rod is therefore in the upward direction.

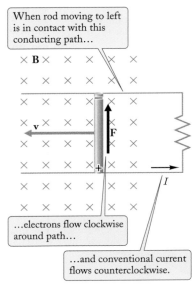

FIGURE 31.2 If the ends of the rod are in sliding contact with a pair of long wires, a current will flow around the circuit. This current can be detected by an ammeter.

charge is $evB/e = vB$. If the length of the rod is l, the work per unit charge equals the distance l times the force per unit charge, and therefore

$$\mathcal{E} = lvB \tag{31.1}$$

motional emf

This is called a **motional emf** because it is generated by the motion of the rod through the magnetic field.

EXAMPLE 1 Suppose you drop an aluminum rod of length 1.0 m out of a window at a place where the horizontal magnetic field of the Earth is 2.0×10^{-5} T. The rod is oriented horizontally, at right angles to the magnetic field (see Fig. 31.3). What is the induced emf between the ends of the rod when its instantaneous downward velocity reaches 12 m/s?

SOLUTION: The magnetic field of the Earth has both a horizontal and a vertical component. In our derivation of Eq. (31.1) for the motional emf, we assumed that the magnetic field **B** is perpendicular to the rod and to its velocity. For a vertically falling rod, as in Fig. 31.3, the horizontal component B_x of the Earth's magnetic field is perpendicular to the rod and to its velocity, and hence generates a motional emf. By contrast, the vertical component of the Earth's magnetic field exerts no force on the free electrons in the rod (remember that the magnetic force is zero when the velocity is parallel to the magnetic field), and therefore generates no motional emf.

With $B = B_x = 2.0 \times 10^{-5}$ T, (31.1) gives

$$\mathcal{E} = lvB_x = 1.0 \text{ m} \times 12 \text{ m/s} \times 2.0 \times 10^{-5} \text{ T} = 2.4 \times 10^{-4} \text{ V}$$

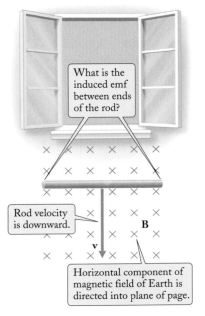

FIGURE 31.3 A rod has been dropped out of a window. The horizontal component of the magnetic field of the Earth is directed into the plane of the page.

electromagnetic flowmeter

A practical application of motional emfs is the **electromagnetic flowmeter** used to measure the speed of flow of a conducting liquid, such as detergent, liquid sodium, tomato pulp, beer, and so on. In this device, the liquid is made to flow in a nonconducting pipe placed at right angles to a magnetic field, and a motional emf of the magnitude given by Eq. (31.1) will then appear across the diameter of the column of liquid, because the diameter acts like a conducting rod. Electrodes inserted into the pipe on opposite sides pick up the emf, which can be measured on a sensitive voltmeter. The velocity of flow can then be calculated from the measured emf, the strength of the magnetic field, and the dimension of the pipe.

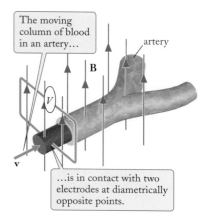

The moving column of blood in an artery…

…is in contact with two electrodes at diametrically opposite points.

FIGURE 31.4 Technique for measurement of the velocity of the flow of blood.

EXAMPLE 2   To measure the velocity of the flow of blood in the mesenteric artery in the abdomen of a dog, a researcher places the animal in a magnetic field of 3.0×10^{-2} T, inserts small electrodes through the wall of the artery from each side, and measures the emf with a voltmeter (see Fig. 31.4). The inner diameter of the artery is 0.30 cm and the measured emf is 1.8×10^{-6} V. What is the velocity of blood flow?

SOLUTION: According to Eq. (31.1), with $l = 0.30$ cm,

$$v = \frac{\mathcal{E}}{lB} = \frac{1.8 \times 10^{-6}\ \text{V}}{0.30 \times 10^{-2}\ \text{m} \times 3.0 \times 10^{-2}\ \text{T}} = 2.0 \times 10^{-2}\ \text{m/s}$$

It is instructive to reexamine the generation of the induced emf from the point of view of a reference frame in which the rod is at rest. In such a reference frame moving with the rod, the free charges have no velocity (except for perhaps a small drift velocity along the rod), and thus there is *no magnetic force along the rod*. Hence, the free charges must experience some other, nonmagnetic force along the rod that does work on them and supplies an emf. The question is then, what is this new kind of force in the moving reference frame that produces the same effect as the magnetic force in the stationary reference frame? Since the only kinds of force that act on electric charges are the magnetic force and electric force, the "new" kind of force must be electric. It must be due to a "new" kind of electric field, an electric field that exists in the moving reference frame, but not in the stationary reference frame (see Fig. 31.5).

The "new" electric field that exists in the reference frame of the rod moving through a magnetic field is called an **induced electric field**. We can determine the magnitude of the induced electric field from a consistency requirement: the "new" electric force $F' = qE'$ in the moving reference frame must coincide with the magnetic force $F = qvB$ in the stationary reference frame. This tells us that the electric field in the moving reference frame must have a magnitude

$$E' = vB \tag{31.2}$$

This electric field does work on the free charges and therefore represents a source of emf. The work done on a unit positive charge that passes through the rod is

$$\mathcal{E} = E'l \tag{31.3}$$

In view of Eq. (31.2), this value of the emf coincides with the value that we obtained in Eq. (31.1). Thus, the motional emf can be calculated either in a stationary reference frame or else in a moving reference frame; in the former case it arises from a magnetic field and in the latter from an electric field.

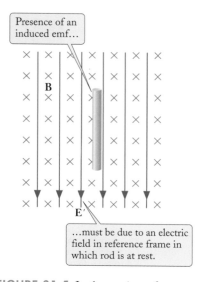

Presence of an induced emf…

B

E′

…must be due to an electric field in reference frame in which rod is at rest.

FIGURE 31.5 In the moving reference frame of the rod, there is a magnetic field **B** (blue crosses) and there is also an electric field **E′** (red lines).

Checkup 31.1

QUESTION 1: If the rod shown in Fig. 31.2 moves toward the right instead of the left, what is the direction of the current generated by the motional emf?

QUESTION 2: For the rod in Example 1, which end is the positive terminal for the motional emf? Suppose that instead of dropping the rod downward, we throw it upward. How does this change the emf?

QUESTION 3: When you run through the Earth's magnetic field, a motional emf will be generated from one side of your body to the other. In the Northern Hemisphere, the Earth's magnetic field is mainly downward (the horizontal component is small and can be ignored for the present purposes). If you run eastward, which side of your body will be the positive terminal for the induced emf? If you run westward? If you run northward?

QUESTION 4: What is the direction of the induced electric field in the reference frame of the rod shown in Fig. 31.3?

   (A) Up   (B) Into page   (C) Out of page   (D) To left   (E) To right

31.2 FARADAY'S LAW

Online
Concept
Tutorial

The quantity lvB appearing on the right side of Eq. (31.1) can be given an interesting interpretation in terms of the magnetic flux. As already mentioned in Section 29.2, the **magnetic flux** is defined in the same way as the electric flux. For any open or closed surface of area A, the *magnetic flux is defined as the integral of the normal component of the magnetic field over the area*:

$$\Phi_B = \int \mathbf{B} \cdot d\mathbf{A} = \int B_\perp \, dA \qquad (31.4)$$

magnetic flux

The magnetic flux is simply proportional to the number of magnetic field lines that are intercepted by the surface, lines crossing the surface in one direction being reckoned as positive, and lines in the opposite direction as negative. Note that for a flat surface in a constant magnetic field, the magnetic flux is simply

$$\Phi_B = \mathbf{B} \cdot \mathbf{A} = B_\perp A = BA \cos\theta \qquad (31.5)$$

where θ is the angle between the magnetic field and the perpendicular erected on the surface (see Fig. 31.6). The unit of magnetic flux is the product of the unit of magnetic field and the unit of area, that is, $\text{T} \cdot \text{m}^2$. In the SI system, this unit is called the **weber**:

$$1 \text{ weber} = 1 \text{ Wb} = 1 \text{ T} \cdot \text{m}^2$$

Now, consider a rod moving through a magnetic field at velocity **v**. In a time interval dt, the rod advances a distance $v\,dt$ and therefore sweeps through an area $l \times v\,dt$, perpendicular to the magnetic field (see Fig. 31.7). The product of this area and the magnetic field is the magnetic flux that the rod sweeps across; that is,

$$lv\,dt\,B = d\Phi_B$$

Thus,

$$lvB = \frac{d\Phi_B}{dt}$$

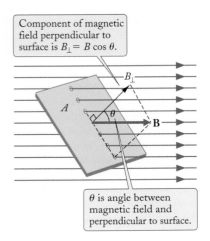

Component of magnetic field perpendicular to surface is $B_\perp = B \cos\theta$.

θ is angle between magnetic field and perpendicular to surface.

FIGURE 31.6 Flat rectangular surface in a uniform magnetic field.

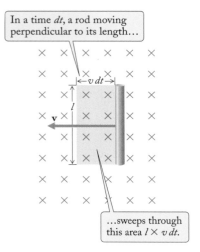

In a time dt, a rod moving perpendicular to its length…

$\leftarrow v\,dt \rightarrow$

l

$\mathbf{v}$

…sweeps through this area $l \times v\,dt$.

FIGURE 31.7 Rod moving through a magnetic field with speed v.

This tells us that the quantity lvB is the rate at which the rod sweeps across magnetic flux. We can then rewrite Eq. (31.1) as

$$\mathcal{E} = -\frac{d\Phi_B}{dt} \qquad (31.6)$$

The minus sign that we have inserted into Eq. (31.6) indicates how the polarity of the induced emf is related to the change of flux; we will discuss the determination of this polarity in the next section.

The advantage of this interpretation of the induced emf in terms of magnetic flux is that Eq. (31.6) is of more general validity than Eq. (31.1)—as we will see, Eq. (31.6) is a general law relating the induced emf to the rate at which the rod sweeps across flux, and it is valid regardless of how the sweeping of flux comes about. For instance, instead of moving the rod relative to the magnet that produces the magnetic field, we can hold the rod fixed and move the magnet. Flux will then sweep across the rod, and Eq. (31.6) suggests that an emf should be induced along the rod. Experiments do indeed confirm that in both cases the induced emf is exactly the same.

But there is another way in which magnetic flux will sweep across the rod: we can hold the rod fixed and increase or decrease the *strength* of the magnetic field. To understand why flux will be swept across the rod under these conditions, we must first take a look at what happens to the field lines of a current when the current changes. Figure 31.8a shows the field lines produced by a current on a long straight wire. If we increase the current, the magnetic field increases, that is, the number of field lines increases. Figure 31.8b shows the field lines of a stronger current. Where do the extra field lines come from? Obviously, the current has to make them. When the current increases, it makes new, small circles of field lines in its vicinity; meanwhile, the circles that already exist in Fig. 31.8a move outward, like ripples on a pond. Thus, the pattern shown in Fig. 31.8a gradually grows into the new pattern shown in Fig. 31.8b. Note that the pattern grows from the inside out.

Thus a change in the strength of the magnetic field involves moving field lines. If a *stationary* rod is located in the vicinity of the changing current, the moving field lines will sweep across the rod, that is, magnetic flux will sweep across the rod. Experiments show that such a sweeping of flux across a stationary rod induces an emf which is given by Eq. (31.6), the same formula as for the case of the sweeping of flux across a moving rod.

From the emf induced in a rod, we can infer the emf induced in a wire of arbitrary shape, since we can regard any such wire as consisting of short straight segments joined

MICHAEL FARADAY (1791–1867)

English physicist and chemist, Faraday's earliest research lay in chemistry, but he soon turned to research in electricity and magnetism, making contributions of the greatest significance. His discovery of electromagnetic induction was no accident, but arose from a systematic experimental investigation of whether magnetic fields can generate electric currents. Although Faraday was essentially an experimenter, with no formal training in mathematics, he made an important theoretical contribution by introducing the concept of field lines and by recognizing that electric and magnetic fields are physical entities.

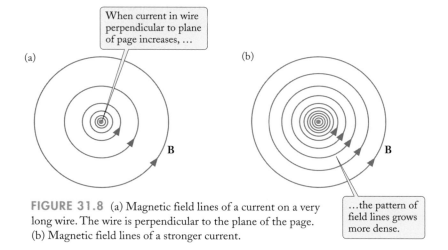

(a)

When current in wire perpendicular to plane of page increases, …

$\mathbf{B}$

(b)

…the pattern of field lines grows more dense.

$\mathbf{B}$

FIGURE 31.8 (a) Magnetic field lines of a current on a very long wire. The wire is perpendicular to the plane of the page. (b) Magnetic field lines of a stronger current.

one to another. The net emf induced between the two ends of the wire is then the sum of the emfs induced in all these short rods, and this net emf equals the net rate at which magnetic flux sweeps across the wire. More generally, we can make an assertion about the emf induced along an arbitrary moving or fixed mathematical path immersed in a constant or changing magnetic field. An emf will be induced between the ends of this path regardless of whether we place a rod or a wire along the path. Whenever a unit positive charge moves along this path, it will gain an amount $\mathcal{E}$ of energy from the induced electric field, regardless of whether the charge moves on a rod or through empty space. The rod or wire merely serves as a convenient conduit for the flow of charge. Of course, for the practical exploitation of the emf, we usually find it convenient to provide a rod, wire, or some other conductor along which the charge can flow, and we will then also have to provide a return path for the charge.

The general statement about the induced emf is known as **Faraday's Law of Induction**,

The induced emf along any moving or fixed path in a constant or changing magnetic field equals the rate at which magnetic flux sweeps across the path:

$$\mathcal{E} = -\frac{d\Phi_B}{dt} \tag{31.7}$$

For a *closed* path, or a closed circuit, Faraday's Law can be interpreted in terms of the flux intercepted by the surface within the path. For this purpose we imagine that the path is spanned (or closed off) by some smooth mathematical surface, such as shown in Fig. 31.9. Since the magnetic field lines are continuous (without sources or sinks), the exact shape of the surface is of no consequence; we can use any flat or curved surface that has the path as boundary—any such surface intercepts the same flux. If we move the path or if the magnetic field changes, flux will sweep across the path. But any flux that sweeps across the path represents a gain or a loss of flux intercepted by the surface, since any field line that moves across the path either moves into the surface or out of the surface. Therefore, the rate $d\Phi_B/dt$ at which flux sweeps across the closed path equals the rate of change of the flux intercepted by the surface, and we can interpret Eq. (31.7) as stating that the induced emf equals the rate of change of this intercepted flux. This means that for a closed path, Faraday's Law can be stated as follows:

The induced emf around a closed path in a magnetic field is equal to the rate of change of the magnetic flux intercepted by the surface within the path.

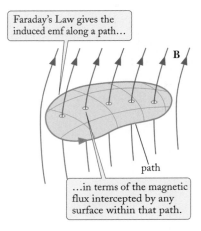

Faraday's Law gives the induced emf along a path…

B

path

…in terms of the magnetic flux intercepted by any surface within that path.

FIGURE 31.9 A closed path spanned by a surface.

Note that the magnetic flux intercepted by the surface within a closed path, say, a wire loop, can be made to change in a variety of ways (see Fig. 31.10). We can move the wire loop in or out of the magnetic field, or deform it so as to increase or decrease its area, or change its orientation, or change the strength of the magnetic field by changing the current in the magnet that produces the field—but in all these cases the emf induced around the loop is related to the change of flux by Eq. (31.7). For example, for a flat loop in a uniform field, Eqs. (31.7) and (31.5) give

$$\mathcal{E} = -\frac{d}{dt}(BA\cos\theta) \tag{31.8}$$

which makes it apparent that changing B, A, or θ will result in an induced emf. Examples of such induced emfs appear in the next section.

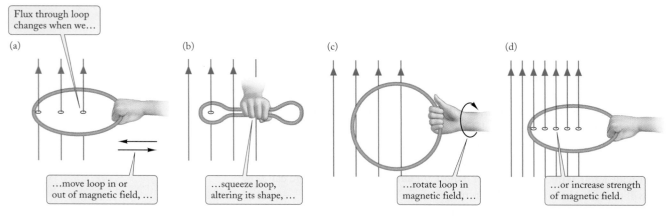

FIGURE 31.10 Several ways the flux through a loop can change.

Concepts
— *in* —
Context

The most important practical application of induction is the **electromagnetic generator**, widely used to generate the electric power that we need in our homes and factories. In its simplest form, the electromagnetic generator consists of a loop of wire that is made to rotate in the magnetic field of a magnet (see Fig. 31.11). When the loop is face on to the magnetic field, the flux intercepted by the surface within the loop is large; when the loop is edge on to the magnetic field, the flux is zero. Thus, the rotation of the loop about its axis leads to a rate of change of flux, which induces an emf and drives a current through the loop and through the external circuit connected to the ends of the loop.

In essence, the generator is an electric motor operating in "reverse." In a motor, we convert electric power into mechanical power—we send a current through a loop in a magnetic field, and we produce rotational motion. In a generator, we convert mechanical power into electric power—we rotate a loop in a magnetic field, and we produce an emf and a current in the loop and in the external circuit. Small or medium-size generators are driven by gasoline or diesel engines; large-size generators at electric power plants are driven by steam turbines or by hydraulic turbines.

There are many familiar, smaller-scale applications of Faraday's Law. For example, the **induction microphone**, or moving-coil microphone, consists of a flexible diaphragm with an attached small coil of wire, which is placed near a bar magnet (see Fig. 31.12). When a sound wave strikes the diaphragm, the pressure fluctuations move the diaphragm

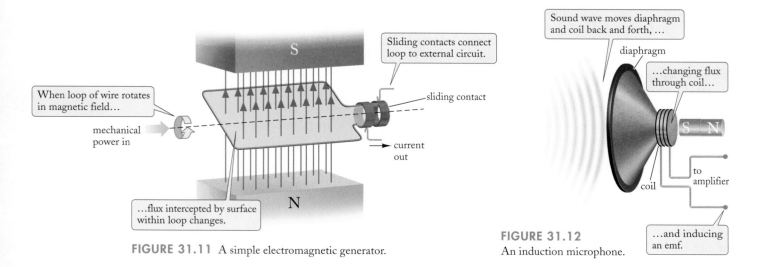

FIGURE 31.11 A simple electromagnetic generator.

FIGURE 31.12
An induction microphone.

and the coil back and forth, and this motion of the coil produces a changing flux and induces an emf. Thus, the microphone transforms pressure fluctuations into an electric signal, which can be amplified and sent to a loudspeaker. A loudspeaker is simply an induction microphone operating in reverse—a current fed into the coil experiences a force in the magnetic field, which pushes or pulls the coil of the bar magnet and the diaphragm back and forth, producing sound waves. Since the basic mechanism of the loudspeaker is the same as that of the induction microphone, it is actually possible to use a loudspeaker as a microphone—simply hook the loudspeaker into an amplifier and speak into it, and the loudspeaker will transform your sound signal into an electric signal and act as a (somewhat poor) microphone.

 ## Checkup 31.2

QUESTION 1: The magnetic flux through a closed surface is always zero. Why?

QUESTION 2: A closed path is immersed in a magnetic field. If we suddenly reverse the magnetic field (but leave its strength unchanged), will an emf be induced around the path?

QUESTION 3: You shoot a metallic arrow through a magnetic field. Will this generate an induced emf between the tip and the tail of the arrow?

QUESTION 4: Figure 31.13 shows several surfaces in a uniform magnetic field. Which has the largest magnetic flux?

(A) a   (B) b   (C) c

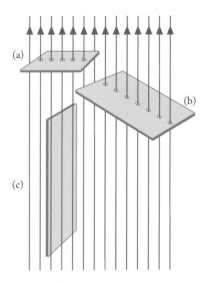

FIGURE 31.13 Several different surfaces in a uniform magnetic field.

31.3 SOME EXAMPLES; LENZ' LAW

Online *Concept* Tutorial

In this section we will look at some examples of the calculation of induced emfs. We begin by laying down a simple rule for finding the polarity of the induced emf, a rule known as **Lenz' Law:**

> *The induced current is always such as to oppose the change of flux (or the motion) that generated it.*

For instance, consider the rod moving in a magnetic field illustrated in Fig. 31.14. The rod slides on long parallel wires joined at the right, so as to form a closed circuit. The motion of the rod leads to an increasing magnetic flux due to the increasing area enclosed by the circuit. According to Lenz' Law, the extra flux contributed by the current induced in the rod and the wires must oppose this increase of flux; thus, the magnetic field of this induced current must be opposite to the original magnetic field—within the area enclosed by the circuit, the magnetic field of the current in the wire must be out of the plane of the page in Fig. 31.14, whereas the original magnetic field is into this plane. By the right-hand rule, the current must be counterclockwise around the circuit to produce such a magnetic field, in agreement with the direction indicated in Fig. 31.2.

Lenz' Law

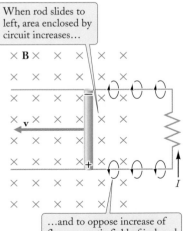

When rod slides to left, area enclosed by circuit increases…

…and to oppose increase of flux, magnetic field of induced current must be opposite to original magnetic field.

FIGURE 31.14 A rod moving in a uniform magnetic field is in sliding contact with parallel wires. The direction of the uniform magnetic field is perpendicularly into the plane of the page.

(Alternatively, we can deduce the direction of the current by examining the motion of the rod. The rod moves toward the left. According to Lenz' Law, the magnetic force on the current in the rod must oppose this motion, that is, the force must be toward the right. By the right-hand rule, if the magnetic force is to be toward the right, the current in the rod must be downward, which means the current must be counterclockwise around the circuit.)

It is important to understand that the induced current opposes the *change* in the flux, but does not necessarily oppose the flux itself. If the rod in Fig. 31.14 were moving toward the right instead of the left, then this motion would lead to a decreasing magnetic flux within the area enclosed by the circuit. The extra flux contributed by the induced current must oppose this decrease of flux; thus, the magnetic field of the induced current must be in the same direction as the original magnetic field, that is, into the plane of the page in Fig. 31.14. The current must then be clockwise around the circuit.

In practical applications, induced emfs are usually generated in coils consisting of many turns of wire, that is, many loops. In such a coil, each individual loop has a rate of change of flux and generates its own emf. Since the individual loops are connected in series, their emfs add together, and for a coil of N loops, the net emf is N times the emf of one loop:

Law of Induction for coil of N turns

$$\mathcal{E} = -N\frac{d\Phi_B}{dt} \qquad (31.9)$$

Alternatively, we can think of the flux through the N loops as crossing an area N times as large as a single loop. From this viewpoint, a single emf is generated, but the total flux is N times as large ($\Phi_{B,\text{total}} = N\Phi_B$), so we obtain the same result (31.9). It is sometimes helpful to be aware of these alternative points of view.

We now look at various examples of the calculation of induced emfs; we will examine changes in flux due to a changing magnetic field, a changing angle between the area and the magnetic field, and a changing area.

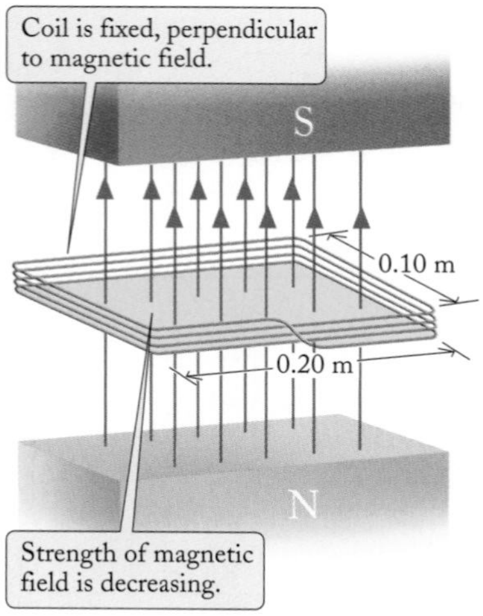

EXAMPLE 3   A rectangular coil of 150 loops forming a closed circuit measures 0.20 m × 0.10 m. The resistance of the coil is 5.0 Ω. The coil is placed between the poles of an electromagnet, face on to a magnetic field **B** (see Fig. 31.15). Suppose that when we switch the electromagnet off, the strength of the magnetic field decreases at the rate of 20 teslas per second. What is the induced emf in the coil? What is the direction of the induced current?

SOLUTION: The flux intercepted by each loop in the coil is $\Phi_B = BA$, where B is the magnetic field and A is the area of the loop. If only the magnetic field is changing, the rate of change of flux is

$$\frac{d\Phi_B}{dt} = \frac{dB}{dt}A$$

and Faraday's Law tells us that the total induced emf for the entire coil is

$$\mathcal{E} = -N\frac{d\Phi_B}{dt} = -N\frac{dB}{dt}A$$

$$= -150 \times (-20 \text{ T/s}) \times (0.20 \text{ m} \times 0.10 \text{ m}) = 60 \text{ V}$$

Coil is fixed, perpendicular to magnetic field.

S

0.10 m

0.20 m

N

Strength of magnetic field is decreasing.

FIGURE 31.15 A rectangular coil in a uniform, upward magnetic field.

According to Ohm's Law, the magnitude of the induced current is then

$$I = \frac{\mathcal{E}}{R} = \frac{60 \text{ V}}{5.0 \ \Omega} = 12 \text{ amperes}$$

To determine the direction of the induced current, we appeal to Lenz' Law. The magnetic field shown in Fig. 31.15 and its flux are decreasing. Hence, the current in the coil must provide a magnetic field that compensates (in part) for this decreasing magnetic flux. According to the right-hand rule, this requires a counterclockwise current around the loop (looking from above), so the magnetic field of this current is upward, in the direction of the original magnetic field.

COMMENTS: Note that for a current to flow, the coil needs to form a closed circuit. If there were a break in the coil, it would not form a closed circuit; an emf would still be generated, but no current could flow, just as in an ordinary open circuit. If instead of a coil, a solid conducting plate had been immersed in the changing field, the induced currents would flow in loops around the conductor. Such currents are called **eddy currents**, because they flow in loops. Eddy currents are sometimes desirable in a device; they remove motional energy and provide damping, for instance, in vibration-isolation devices. When damping is not desired and when eddy currents cause unwanted heating due to I^2R losses, they can be avoided by segmenting the conductor into separate pieces, so that no currents can flow from one piece to the next.

eddy current

EXAMPLE 4   An electromagnetic generator consist of a rectangular coil of N loops of wire that rotates about an axis perpendicular to a constant magnetic field **B**. Sliding contacts connect the coil to an external circuit (see Fig. 31.16). What emf does the coil deliver to the external circuit? The coil has an area A and rotates with an angular frequency ω.

Concepts
— in —
Context

SOLUTION: The magnetic flux varies as the coil rotates; it is maximum when the coil is oriented face on to the magnetic field, and mimimum when the coil is oriented parallel to the magnetic field. Instantaneously, the perpendicular to the coil makes an angle $\theta = \omega t$ with the magnetic field (see Fig. 31.16), and the component of the magnetic field perpendicular to the coil is $B_{\perp} = B \cos\theta = B \cos \omega t$. The instantaneous magnetic flux through the coil is then $\Phi_B = BA \cos \omega t$. To find

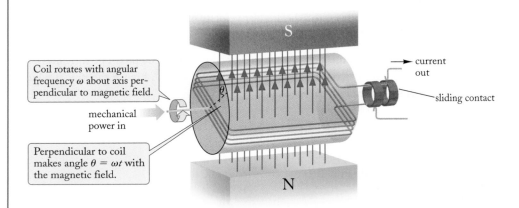

FIGURE 31.16 An electromagnetic generator.

the induced emf around each loop, we need to evaluate the rate of change of this flux, which means we need to evaluate the rate of change of $\cos \omega t$,

$$\frac{d}{dt} \cos \omega t = -\omega \sin \omega t$$

Thus, for the magnetic flux $\Phi_B = BA \cos \omega t$, the rate of change is

$$\frac{d\Phi_B}{dt} = -BA\omega \sin \omega t$$

According to Faraday's Law, the total induced emf around all N loops is

$$\mathcal{E} = -N\frac{d\Phi_B}{dt} = NBA\omega \sin \omega t \qquad (31.10)$$

At the instant shown in Fig. 31.16, the magnetic flux through the coil is increasing; by Lenz' Law, the induced current must therefore flow around the coil in a clockwise direction (viewed from above) so as to contribute an extra magnetic flux that opposes the increase of the original magnetic flux.

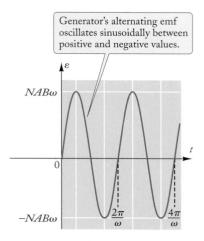

Generator's alternating emf oscillates sinusoidally between positive and negative values.

FIGURE 31.17 Emf of the generator; this is an alternating voltage (AC).

alternating emf (AC) and steady emf (DC)

Concepts — *in* — Context

The emf given by Eq. (31.10) is an **alternating emf**, or **AC voltage**, that oscillates sinusoidally between positive and negative values. A plot of this emf is shown in Fig. 31.17. Various methods can be used to produce a **steady emf**, or **DC voltage**, that does not alternate between positive and negative values but has a constant value. The most common method relies on a commutator or on an alternator. The **commutator** is simply a mechanical switch that reverses the connection between the generator and the external circuit every half cycle so that the emf delivered to the external circuit always has the same sign. The **alternator** is an electronic switch that uses diodes instead of mechanical devices to accomplish the same thing as a commutator. Another method relies on an entirely different kind of generator, called a **homopolar generator**. Instead of a rotating coil, this generator uses a rotating disk, which is equivalent to a rotating rod, as illustrated by the following example.

EXAMPLE 5 A straight rod of metal is rotating about one end on an axis parallel to a uniform magnetic field B (see Fig. 31.18a). The length of the rod is l, and the angular velocity of rotation is ω. What is the induced emf between the two ends of the rod?

SOLUTION: As the rod turns, it sweeps across magnetic flux. The rod takes a time $\Delta t = 2\pi/\omega$ to sweep out the circular area $\Delta A = \pi l^2$. Hence the rate at which it sweeps out area is

$$\frac{dA}{dt} = \frac{\Delta A}{\Delta t} = \frac{\pi l^2}{2\pi/\omega} = \frac{l^2 \omega}{2}$$

The rate at which it sweeps across magnetic flux is

$$\frac{d\Phi_B}{dt} = B\frac{dA}{dt} = \frac{Bl^2 \omega}{2}$$

According to Faraday's Law, the induced emf is then

$$\mathcal{E} = -\frac{Bl^2\omega}{2}$$

If the moving end of the rod is in sliding contact with a circular track connected to an external circuit (see Fig. 31.18b), the emf will drive a current around the circuit. By Lenz' Law, the induced current in the rod must be such that the magnetic force on the rod opposes the motion. Thus the magnetic force on the rod must oppose the counterclockwise motion of Fig. 31.18. By the right-hand rule for the magnetic force, this requires a current that flows from the moving end of the rod toward the center. In the external circuit, the current then flows from the center toward the moving end—the center acts as the positive terminal of the source of emf and the moving end as the negative terminal.

COMMENTS: For practical applications, homopolar generators are constructed with a rotating disk, rather than a rotating rod. This does not affect the emf of the generator, but it helps to reduce its internal resistance and mechanical stress. Homopolar generators are used in applications, such as electroplating, requiring a large current but only low voltage.

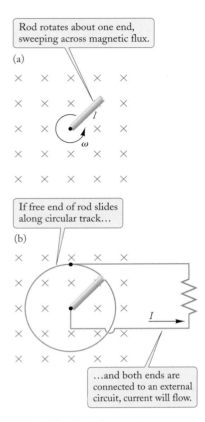

FIGURE 31.18 A homopolar generator. (a) Rod rotating in a uniform magnetic field. (b) Rotating rod connected to an external circuit.

For later use in Chapter 32, we need to reformulate Faraday's Law in terms of the induced electric field. As we saw in Section 31.1 in the example of the rod moving in a magnetic field, the induced emf for a moving straight path sweeping through a constant magnetic field is related to the induced electric field by

$$\mathcal{E} = E'l$$

More generally, the induced emf for an arbitrary moving or fixed path in a constant or changing magnetic field is related to the induced electric field (measured in the reference frame of the path) by

$$\mathcal{E} = \oint \mathbf{E} \cdot d\mathbf{s} = \oint E_{\parallel}\, ds \qquad (31.11)$$

where $E_{\parallel}$ is the component of the induced electric field tangent to the path and ds is a length element along the path. This permits us to express Faraday's Law as

$$\oint E_{\parallel}\, ds = -\frac{d\Phi_B}{dt} \qquad (31.12)$$

Faraday's Law in terms of induced electric field

This version of Faraday's Law can be used to calculate the induced electric field. Note that this form of Faraday's Law for the electric field due to a changing flux is mathematically similar to Ampère's Law, Eq. (29.18), for the magnetic field due to a current. In most of the problems of this chapter we will be concerned only with the induced emf, and the somewhat simpler version (31.7) of Faraday's Law will suffice. But sometimes we are interested in the induced electric field, as in the following example.

PROBLEM-SOLVING TECHNIQUES   FARADAY'S LAW; LENZ' LAW

• Before attempting to calculate an induced emf, ask yourself, where is there a changing magnetic flux? Since the flux in a uniform magnetic field can be written as $\Phi_B = BA \cos\theta$, the change of flux is readily obtained from a change in the magnetic field (Example 3), a change in orientation (Example 4), or a change in the position or shape of the path that encloses the flux (Example 5). Try to visualize the changes in the magnetic field in time or try to visualize the changes in the position, orientation, or shape of the path in time, as though you were watching a movie. Then evaluate the rate of change of the flux.

• When calculating the emf in a coil of N turns, remember that each loop in the coil generates its own emf, and the net emf of the coil is therefore proportional to N.

• The direction of the induced emf and the induced current can be determined by Lenz' Law, which can be applied either to the magnetic flux or to the motion of a conductor: the flux associated with the induced current opposes the original *change* of flux, or the magnetic force on the induced current opposes the original motion. Note that if the device that generates the emf is not actually connected to an external circuit (for instance, the rod in Fig. 31.18a), then it is helpful to imagine connecting it to an external circuit (as in Fig. 31.18b), so it becomes easier to imagine the direction of the induced current.

• The magnitude of the current that flows in an external circuit connected to the induced emf can be calculated in the usual way from the emf $\mathcal{E}$ and the resistance R, according to Ohm's Law, $I = \mathcal{E}/R$.

(a)

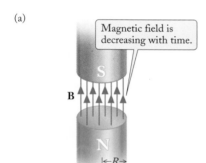

Magnetic field is decreasing with time.

B

(b)

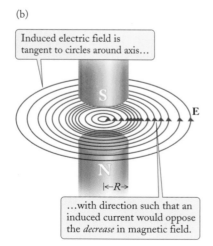

Induced electric field is tangent to circles around axis...

E

...with direction such that an induced current would oppose the *decrease* in magnetic field.

FIGURE 31.19 (a) Magnetic field **B** between the pole faces of an electromagnet. (b) Induced electric field.

EXAMPLE 6   The circular pole faces of the electromagnet shown in Fig. 31.19a have a radius $R = 0.10$ m. The magnetic field is suddenly shut off, decreasing (briefly) at a rate of 2.0×10^4 T/s. What is the induced electric field for $r \leq R$? For $r \geq R$? What is the numerical value of this field at $r = R$? Assume that there is only a magnetic field between the pole faces.

SOLUTION: By analogy with Ampère's Law (see Examples 4 and 5 of Chapter 29), the axial symmetry of the flux dictates that the direction of the induced electric field be tangent to circles around the axis (see Fig. 31.19b). Thus at any value of r, the left side of Faraday's Law in Eq. (31.12) reduces to

$$\oint E_{\parallel} \, ds = 2\pi r E_{\parallel}$$

For the right side of Faraday's Law, we must consider only the flux that crosses the area inside the path of radius r where we want to know the induced electric field. For $r \leq R$, this flux is

$$\Phi_B = BA = B\pi r^2$$

and its rate of change is

$$\frac{d\Phi_B}{dt} = \frac{dB}{dt} \pi r^2$$

Equating the two sides of Eq. (31.12) then gives us

$$2\pi r E_{\parallel} = -\frac{dB}{dt} \pi r^2$$

and we can solve for the electric field:

$$E_\parallel = -\frac{dB}{dt}\frac{r}{2} \quad \text{for} \quad r \leq R$$

Outside the pole faces, at any value of $r \geq R$ where we want to know the induced electric field, the magnetic flux is just the total flux between the pole faces of area $A = \pi R^2$, so

$$\frac{d\Phi_B}{dt} = \frac{dB}{dt}\pi R^2$$

This gives for Eq. (31.12)

$$2\pi r\, E_\parallel = -\frac{dB}{dt}\pi R^2$$

or

$$E_\parallel = -\frac{dB}{dt}\frac{R^2}{2r} \quad \text{for} \quad r \geq R$$

At $r = R$, either result gives

$$E_\parallel = -\frac{dB}{dt}\frac{R}{2}$$

$$= -(-2.0 \times 10^4 \text{ T/s}) \times \frac{0.10 \text{ m}}{2} = 1.0 \times 10^3 \text{ V/m}$$

COMMENT: Note that there is an induced electric field even in the region where there is no magnetic field! Such an induced electric field can be large enough to damage delicate electronic equipment near the magnet. In practice, it is advisable to increase and decrease magnetic fields slowly.

 Checkup 31.3

QUESTION 1: For each of the cases illustrated in Fig. 31.10, what is the direction of the induced current?

QUESTION 2: If the coil described in Example 3 has 300 loops instead of 150, how does this change the answers?

QUESTION 3: If the magnetic field in Example 3 is increasing rather than decreasing, how does this change the answer?

QUESTION 4: If the current in a long solenoid is increasing, what is the direction of the induced electric field inside and outside the solenoid?

QUESTION 5: A resistor is connected between two railroad tracks; a metal rod across the tracks is pushed toward the resistor (see Fig. 31.20). The Earth's magnetic field is directed into the page. What is the direction of the current, if any, in the resistor?

(A) To the right   (B) To the left   (C) No direction; current is zero

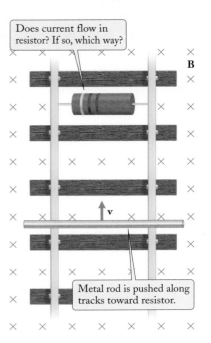

FIGURE 31.20 A metal rod and resistor in contact with railroad tracks.

PHYSICS IN PRACTICE   MAGNETIC LEVITATION

A popular science demonstration features the phenomenon of magnetic levitation, with a permanent magnet floating above a high-temperature superconductor in air (see Fig. 1). Of course, the like poles of two ordinary magnets repel, and so one will float above the other as shown in Fig. 2, but this configuration is not generally stable, and a guide is needed to hold the magnets in place. In the case of the superconductor, stability is readily achieved, even if the magnet is rotating; this behavior is exploited for the construction of high-performance superconducting bearings and gyroscopes.

The origin of superconducting magnetic levitation lies in Faraday's Law; we can understand this with the aid of Fig. 3. As a magnet is brought close to a superconductor, it attempts to increase the flux through the superconductor. But as discussed in Chapter 27, a superconductor has zero electrical resistance and can carry persistent currents. Thus any attempt to generate an emf in the material will result in an induced current sufficient to completely cancel the increase in flux. This simplest behavior, characteristic of type I superconductivity, reflects a property known as perfect diamagnetism: the material responds with an induced current and an internal field which exactly cancels the external field; in terms of the susceptibility discussed in Chapter 30, this means $\chi = -1$ (as mentioned in Section 30.4, ordinary materials can exhibit a much smaller amount of diamagnetism due to induced *atomic* currents). For the superconductor, levitation occurs because the atomic currents in the permanent magnet and the induced

FIGURE 1 A permanent magnet levitating above a high-temperature superconductor.

FIGURE 2 One permanent magnet "floats" above the other, due to their mutual repulsion.

31.4 INDUCTANCE

If a coil carrying a time-dependent current is near some other coil, then the changing magnetic field of the former can induce an emf in the latter. Thus, a time-dependent current in one coil can induce a current in another, nearby coil. For instance, consider the two coils in Fig. 31.21. The first of these coils is connected to some alternating source of emf, such as an AC generator, and it carries a time-dependent current. This coil therefore produces a time-dependent magnetic field $\mathbf{B}_1$. The changing magnetic flux Φ_{B1} through the second coil induces an emf in this coil. According to Faraday's Law, the emf in the second coil is

$$\mathcal{E}_2 = -\frac{d\Phi_{B1}}{dt} \tag{31.13}$$

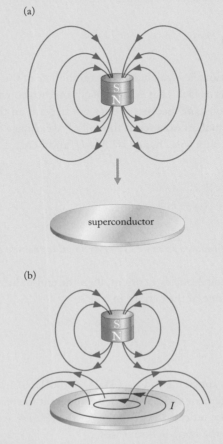

FIGURE 3 (a) A magnet is brought near a superconductor. (b) By Faraday's Law, currents are induced in the superconductor, which oppose the change in flux.

electric current in the superconductor are oppositely directed, and such currents repel one another.

Commercially useful superconductors do permit some flux to penetrate; these are called type II superconductors. Such penetrating flux can become stuck or "pinned" at defects in the superconductor, aiding in the stability of the supercurrents or the levitation. Even before high-temperature superconductivity, levitation phenomena using conventional type II materials had been exploited. Figure 4 shows a picture of a magnetically levitated train, which uses powerful superconducting magnets made of an alloy of niobium and titanium. Such trains can float at high speed, providing a comfortable and mechanically efficient means of transport. However, the expense of cooling the superconductor has so far been prohibitive, and such trains are not yet in common commercial use.

FIGURE 4 A superconductive magnetically levitated (maglev) train.

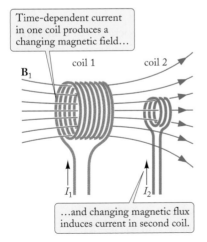

Time-dependent current in one coil produces a changing magnetic field...

coil 1   coil 2

$\mathbf{B}_1$

I_1 I_2

...and changing magnetic flux induces current in second coil.

FIGURE 31.21 Coil 1 creates a magnetic field $\mathbf{B}_1$. Some of the field lines pass through coil 2, and create a flux.

The flux Φ_{B1} is proportional to the strength of the magnetic field $\mathbf{B}_1$ in the second coil produced by the current I_1 in the first coil. This field strength is directly proportional to I_1 (an increase of the current by, say, a factor of 2 results in an increase of the magnetic field by the same factor; see Ampère's Law). Hence, the flux Φ_{B1} is also proportional to I_1. We can write the relationship between Φ_{B1} and I_1 as

mutual inductance M

$$\Phi_{B1} = MI_1 \tag{31.14}$$

where M is a constant of proportionality which depends on the size of the coils, their distance, and the number of turns in each; that is, M depends on the geometry of Fig. 31.21, but M is regarded as a "constant" because it does not depend on the current. The constant M is called the **mutual inductance** of the coils. If we change the current by some small amount dI_1, the flux will change by $M\,dI_1$, and hence Eq. (31.13) becomes

induced emf and inductance

$$\mathcal{E}_2 = -M\frac{dI_1}{dt} \tag{31.15}$$

This equation states that *the emf induced in coil 2 is proportional to the rate of change of the current in coil* 1.

The converse is also true: if coil 2 carries a current, then the emf induced in coil 1 is proportional to the rate of change of the current in coil 2:

$$\mathcal{E}_1 = -M\frac{dI_2}{dt} \tag{31.16}$$

The constants of proportionality appearing in Eqs. (31.15) and (31.16) are the same. Although we will accept this statement without proof, we note that the result is quite reasonable: the mutual inductance reflects the geometry of the *relative* arrangement of the coils, and that is of course the same in both cases.

The SI unit of inductance is called the **henry** (H):

henry (H)

$$1 \text{ henry} = 1 \text{ H} = 1 \text{ V·s/A} \tag{31.17}$$

Incidentally: The permeability constant μ_0 is commonly expressed in terms of this unit of inductance, so

$$\mu_0 = 4\pi \times 10^{-7} \text{ H/m} \tag{31.18}$$

transformer

Mutual inductance finds an important application in the operation of **transformers**, used to step up or step down the emf supplied by an AC generator. The transformer consists of two coils—the primary and the secondary—arranged close together. The emf supplied to the primary produces a current which induces an emf in the secondary. This induced emf will be larger or smaller than the original emf, depending on the number of turns in the coils (we will examine transformers in detail in Section 32.6).

EXAMPLE 7   To determine the mutual inductance experimentally, a physicist connects the first coil in Fig. 31.21 to an alternating source of emf and thereby produces a rate of change of the current of 40 A/s in this first coil. He finds that the induced emf measured across the second coil is -8.0×10^{-6} volt. What is the mutual inductance of the two coils?

SOLUTION: Solving Eq. (31.15) for the mutual inductance gives us

$$M = -\frac{\mathcal{E}_2}{dI_1/dt} = -\frac{(-8.0 \times 10^{-6} \text{ volt})}{40 \text{ A/s}} = 2.0 \times 10^{-7} \text{ H}$$

Note that for a given rate of change of current in the first coil, a larger induced voltage in the second coil implies a larger mutual inductance.

EXAMPLE 8   A long solenoid has n turns per unit length. A ring of wire of radius r is placed within the solenoid, perpendicular to the axis (see Fig. 31.22). What is the mutual inductance?

SOLUTION: If the current in the solenoid windings is I_1, the magnetic field is $B_1 = \mu_0 n I_1$ [see Eq. (29.21)], and the flux through the ring is

$$\Phi_{B1} = B_1 \times [\text{area}] = \mu_0 n I_1 \times \pi r^2$$

According to Eq. (31.14), the mutual inductance is

$$M = \frac{\Phi_{B1}}{I_1} = \mu_0 n \pi r^2 \qquad (31.19)$$

COMMENTS: Note again that the mutual inductance depends only on the geometry of the two elements, not on any particular values of B_1 or I_1. Note also that if the wire loop instead of the solenoid carried the current, the simple result (31.19) for the mutual inductance also gives us the ratio of the flux through the solenoid to the current I_2, even though that reversed arrangement would be much more difficult to calculate directly.

FIGURE 31.22 A ring of wire inside a solenoid.

A conductor by itself has a **self-inductance**. Consider a coil with a time-dependent current. The coil's own magnetic field (see Fig. 31.23) will then produce a time-dependent magnetic flux and, by Faraday's Law, an induced emf. This means that whenever the current is time-dependent, the coil will act back on the current and

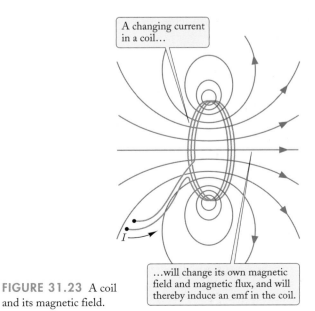

FIGURE 31.23 A coil and its magnetic field.

JOSEPH HENRY (1797–1878)
American experimental physicist. He made important improvements in electromagnets by winding coils of insulated wire around iron pole pieces, and invented an electromagnetic motor and a new, efficient telegraph. He discovered self-induction and investigated how currents in one circuit induce currents in another.

modify it (we will see how to calculate the resultant current in Section 32.3). For this reason the self-induced emf is called a **back emf**. From Lenz' Law, we immediately recognize that *the self-induced emf always acts in such a direction as to oppose the change in the current, that is, it attempts to maintain the current constant.*

The symbol L is used to distinguish the self-inductance from the mutual inductance M. In terms of the flux through the circuit, the definition of the self-inductance is of the same form as Eq. (31.14):

self-inductance L

$$\Phi_B = LI \tag{31.20}$$

and therefore the induced emf is

back emf and self-inductance

$$\mathcal{E} = -L \frac{dI}{dt} \tag{31.21}$$

EXAMPLE 9   A long solenoid has n turns per unit length and a radius R. What is its self-inductance per unit length?

SOLUTION: The magnetic field inside the solenoid is $B = \mu_0 nI$ [see Eq. (29.21)]. The number of loops in a length l is nl; each of these loops has a flux $\pi R^2 B$. Hence the flux through all the loops in a length l is

$$\Phi_B = \pi R^2 B \times [\text{number of turns}] = \pi R^2 B nl = \pi R^2 \mu_0 n^2 Il$$

and the self-inductance for the entire length l is

$$L = \Phi_B / I = \mu_0 n^2 \pi R^2 l \tag{31.22}$$

The self-inductance per unit length is therefore

self-inductance of solenoid

$$L/l = \mu_0 n^2 \pi R^2 \tag{31.23}$$

 ## Checkup 31.4

QUESTION 1: The two coils in Fig. 31.21 are face to face. How would the mutual inductance change if we were to move them farther apart? Bring them closer together? Turn one coil so it faces at a right angle to the other?

QUESTION 2: When a current in one coil changes at the rate of 10^3 A/s, an emf of 6 V is induced in another, nearby coil. If we next change the current in this other coil at the rate of 10^3 A/s, what will be the induced emf in the first coil?

QUESTION 3: Does the self-inductance of a coil depend on its size? Number of loops? Diameter of the wire?

QUESTION 4: Two solenoids have the same radius and length, but the second one has twice as many turns per unit length as the first one. What is the ratio of the self-inductance of the second to the self-inductance of the first?

(A) 1   (B) $\sqrt{2}$ (C) 2   (D) $2\sqrt{2}$ (E) 4

31.5 MAGNETIC ENERGY

Inductors store magnetic energy, just as capacitors store electric energy. When we connect an external source of emf to an inductor and start a current through the inductor, the back emf will oppose the increase of the current and the external emf must do work in order to overcome this opposition and establish the flow of current. This work is stored in the inductor, and it can be recovered by removing the external source of emf from the circuit and, at the same time, closing the resulting gap in the circuit (this can be done with a suitable switch). The current will then continue to flow for a while, at a gradually decreasing rate, because the inductor supplies a back emf which tends to maintain the current (opposes the decrease). Thus, the inductor delivers energy to the current while the current gradually decreases.

To calculate the amount of stored energy, we note that when the current i in the inductor increases at the rate di/dt, the back emf is

$$\mathcal{E} = -L\frac{di}{dt}$$

The inductor does work on the current at a rate given by the usual formula for the electric power [see Eq. (28.22)]:

$$P = \mathcal{E}i = -L\frac{di}{dt}i \tag{31.24}$$

Here the negative sign implies that the energy is delivered by the current to the inductor rather than vice versa. In a time dt, the energy U stored in the inductor therefore changes by an amount

$$dU = -P\,dt = -\mathcal{E}i\,dt = Li\,di \tag{31.25}$$

By integrating this from the initial value of the current ($i = 0$) to the final value ($i = I$), we find the final magnetic energy stored in the inductor:

$$U = \int dU = L\int_0^I i\,di = L\left(\frac{i^2}{2}\right)\Big|_0^I = L\left(\frac{I^2}{2} - 0\right)$$

that is,

$$U = \tfrac{1}{2}LI^2 \tag{31.26}$$

magnetic energy in inductor

Thus, *the total energy stored in the inductor is proportional to the square of the current.*

EXAMPLE 10 A solenoid has a radius of 2.0 cm; its winding has one turn of wire per millimeter. A current of 10 A flows through the winding. What is the amount of energy stored per unit length of the solenoid?

SOLUTION: According to Eq. (31.23), the inductance per unit length is

$$L/l = \mu_0 n^2 \pi R^2$$

and so from Eq. (31.26), the energy per unit length is

$$\frac{U}{l} = \frac{1}{2}\left(\frac{L}{l}\right)I^2 = \frac{1}{2}\mu_0 n^2 \pi R^2 I^2 \tag{31.27}$$

$$= \tfrac{1}{2} \times 1.26 \times 10^{-6} \text{ H/m} \times (1.0 \times 10^3/\text{m})^2 \times \pi \times (0.020 \text{ m})^2 \times (10 \text{ A})^2$$

$$= 7.9 \times 10^{-2} \text{ J/m}$$

The energy stored in a solenoid can be expressed in terms of the magnetic field. Consider a length l of the solenoid. From Eq. (31.27), the energy associated with this portion is

$$U = \tfrac{1}{2}\mu_0 n^2 \pi R^2 I^2 l \tag{31.28}$$

Since, for the solenoid, $B = \mu_0 nI$, we can write this as

$$U = \frac{1}{2\mu_0} B^2 \pi R^2 l \tag{31.29}$$

and since $\pi R^2 l$ is the volume filled with magnetic field, the energy is

$$U = \frac{1}{2\mu_0} B^2 \times [\text{volume}] \tag{31.30}$$

According to this equation, the quantity $(1/2\mu_0)B^2$ can be regarded as the magnetic energy per unit volume. Thus, the energy density u in the magnetic field is

energy density in magnetic field

$$[\text{energy density}] = u = \frac{1}{2\mu_0} B^2 \tag{31.31}$$

Although we have derived this equation only for the special case of a long solenoid, it turns out to be generally valid (in vacuum and nonmagnetic materials). The magnetic field, just as the electric field, stores energy. The magnetic energy density is proportional to B^2, just as the electric energy density is proportional to E^2 [compare Eq. (25.58)]. Accordingly, the magnetic energy is concentrated in the regions of space where the magnetic field is strong.

EXAMPLE 11 Near the surface, the Earth's magnetic field typically has a strength of 3.0×10^{-5} T and the Earth's atmospheric electric field typically has a strength of 100 V/m. What is the energy density in each field?

SOLUTION: The magnetic energy density is

$$\frac{1}{2\mu_0} B^2 = \frac{(3.0 \times 10^{-5}\text{T})^2}{2 \times 1.26 \times 10^{-6} \text{ H/m}} = 3.6 \times 10^{-4} \text{ J/m}^3$$

and the electric energy density is [see Eq. (25.58)]

$$\frac{\epsilon_0}{2} E^2 = \frac{8.85 \times 10^{-12} \text{ F/m}}{2} \times (100 \text{ V/m})^2 = 4.4 \times 10^{-8} \text{ J/m}^3$$

Thus, the magnetic energy density in our immediate environment is much larger than the electric energy density.

EXAMPLE 12 The coils of the giant generator shown in the chapter photo have a volume of approximately 45 m³. These stationary coils surround rotating electromagnets, which produce an average magnetic field within the coils of 0.35 T. What is the total magnetic energy stored in the coil volume?

SOLUTION: The total magnetic energy stored is the product of the magnetic energy density and the volume:

$$U = u \times [\text{volume}] = \frac{1}{2\mu_0}B^2 \times [\text{volume}]$$

$$= \frac{(0.35\text{ T})^2}{2 \times 1.26 \times 10^{-6}\text{ H/m}} \times 45\text{ m}^3 = 2.2 \times 10^6\text{ J}$$

 ## Checkup 31.5

QUESTION 1: If we compress a (long) solenoid, making its coils more tight, does the self-inductance change?

QUESTION 2: In the magnetic field of a long, straight wire, where is the magnetic energy density largest?

QUESTION 3: Two inductors, with $L = 10 \times 10^{-6}$ H and $L = 20 \times 10^{-6}$ H, respectively, carry identical electric currents. Which has the larger magnetic energy?

QUESTION 4: Consider the energy stored in the magnetic field of a solenoid. If you increase the current in the solenoid by a factor of 2, by what factor do you increase the magnetic energy?

(A) $\sqrt{2}$ (B) 2   (C) $2\sqrt{2}$ (D) 4

31.6 THE *RL* CIRCUIT

The *RL* circuit consists of a resistor and an inductor connected in series to a battery or some other source of emf. This circuit provides a good illustration of the effects of self-inductance on a current. Figure 31.24 is the schematic diagram for this circuit. In the diagram, the inductor is represented by a coiled line, reminiscent of a coil of wire. The inductance is L, the resistance is R, and the emf of the battery is $\mathcal{E}$. The inductor is supposed to be resistanceless; if the wire in the coils of the inductor has some resistance, then this resistance must be included in R.

We assume that the current in the circuit is initially zero, and that the battery is suddenly connected at the initial time $t = 0$. The current then starts to increase. But the self-inductance will generate an emf across the inductor, which, by Lenz' Law, opposes the increase of the current. Because of this self-induced emf, the current in the circuit cannot increase suddenly; it can increase only gradually. The self-induced emf continues to oppose the current and to restrain its growth until this current attains its final, steady value, $I = \mathcal{E}/R$. This steady, final value of the current is simply $\mathcal{E}/R$, as though the inductor were absent, because when the current becomes steady, the inductor ceases to contribute to the emf in the circuit, that is, it ceases to affect the circuit. Qualitatively, the gradual growth of current in an *RL* circuit is analogous to the increase of charge in an *RC* circuit. As we will see, the plot of the current vs. time

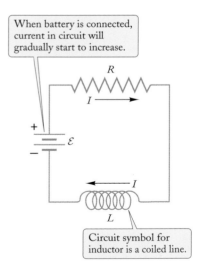

FIGURE 31.24 An *RL* circuit, consisting of a resistor and an inductor connected in series to a battery.

in the RL circuit is mathematically similar to the plot of the charge vs. time in the RC circuit.

For a mathematical treatment of the current in the circuit we turn to Kirchhoff's voltage rule: the sum of all the emfs and voltage drops around the circuit must be zero. The emf of the battery is $\mathcal{E}$. If the current at some instant is I, the voltage drop across the resistor is $-IR$ and the self-induced emf across the inductor is $-L\,(dI/dt)$. Hence

$$\mathcal{E} - IR - L\frac{dI}{dt} = 0 \tag{31.32}$$

At the initial time, $I = 0$ and Eq. (31.32) tells us that the initial potential drop occurs across the inductor, $\mathcal{E} = L(dI/dt)$. After a long time, the current is steady, so $dI/dt = 0$, and the entire potential drop occurs across the resistor, $\mathcal{E} = IR$. At intermediate times, both the resistor and the inductor contribute to the potential drop.

To calculate the current as a function of time, Eq. (31.32) may be solved in a manner identical to that used in Section 28.7 for the RC circuit. Following the technique used in Eqs. (28.28)–(28.33), such direct integration yields for the current

$$I = \frac{\mathcal{E}}{R}\left[1 - e^{-t/(L/R)}\right] \tag{31.33}$$

The quantity L/R that occurs in the exponential has the units of time, and is referred to as the **characteristic time** τ for the process of changing the current in the circuit:

characteristic time τ

$$\tau = \frac{L}{R} \tag{31.34}$$

Thus in terms of the characteristic time, the current is written

increase of current in RL circuit

$$I = \frac{\mathcal{E}}{R}\left(1 - e^{-t/\tau}\right) \tag{31.35}$$

Figure 31.25 is a plot of current vs. time according to this formula, showing the increase in the current from its initial value of zero to its final value of $\mathcal{E}/R$. At the characteristic time $t = \tau = L/R$, the current reaches

$$I = \frac{\mathcal{E}}{R}(1 - e^{-1}) = \frac{\mathcal{E}}{R}\left(1 - \frac{1}{2.718}\right) = \frac{\mathcal{E}}{R} \times 0.6321$$

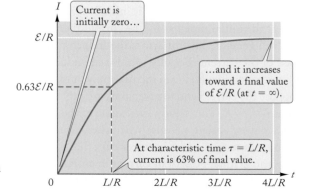

FIGURE 31.25 Growing current in the RL circuit as a function of time.

Thus, at the *characteristic time $\tau = L/R$, the current reaches approximately 63% of its final value.*

Although in the schematic diagram in Fig. 31.24 the inductance and the resistance are shown separated, this diagram can equally well represent a coil of resistive wire, such as the coil making up the windings of an electromagnet, which has both inductance and resistance. For purposes of calculation, the inductance and the resistance of such a coil may be regarded as placed in series, since each produces its own change of potential in the circuit. From our results for the *RL* circuit we therefore see that whenever an electromagnet is suddenly switched on, the current and the magnetic field in the magnet take a while to build up to steady values.

EXAMPLE 13   The windings of a large electromagnet have an inductance of 10.0 H and a resistance of 8.00 Ω. This electromagnet is connected to an external emf of 230 V. After the electromagnet is switched on, how long does the current take to build up to 63% of its final value? What is the final, steady value of the current that is attained after a fairly long time?

SOLUTION: The characteristic time for this electromagnet is

$$\tau = \frac{L}{R} = \frac{10.0\ \text{H}}{8.00\ \Omega} = 1.25\ \text{s}$$

This is the time required to reach 63% of the final value of the current.
    The final, steady value of the current is

$$I = \frac{\mathcal{E}}{R} = \frac{230\ \text{V}}{8.00\ \Omega} = 28.8\ \text{A}$$

This value of the current is attained after a long time, that is, a time long compared with 1.25 s.

Once the current has reached its final steady value $\mathcal{E}/R$, it will continue to flow without change as long as the battery continues to provide a steady emf $\mathcal{E}$. But if we remove the battery and, at the same time, close the resulting gap in the circuit (see Fig. 31.26), the inductor will try to maintain the current; however, the stored energy will gradually be dissipated by I^2R loss in the resistor. The result is a gradual decrease of the current from $\mathcal{E}/R$ to 0. The formula that describes the decrease of the current involves the same exponential function as in Eq. (31.33), with the same characteristic time $\tau = L/R$:

$$I = \frac{\mathcal{E}}{R} e^{-t/(L/R)} \qquad (31.36)$$

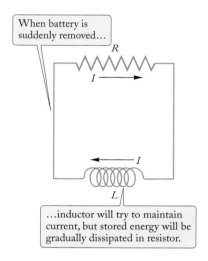

When battery is suddenly removed...

...inductor will try to maintain current, but stored energy will be gradually dissipated in resistor.

FIGURE 31.26 An *RL* circuit without a battery.

decrease of current in *RL* circuit

 Checkup 31.6

QUESTION 1: Suppose that we switch the electromagnet described in Example 13 off, by removing the emf and, and at the same time, closing the resulting gap with a wire of zero resistance. How long does it take for the current in the windings to fall to 37% of its initial value?

QUESTION 2: An *RL* circuit with a large inductance has a long relaxation time. Why does this make sense?

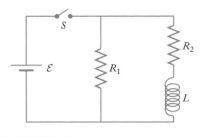

FIGURE 31.27 An *RL* circuit with two resistors.

QUESTION 3: An *RL* circuit with a low resistance has a long relaxation time. Why does this make sense?

QUESTION 4: According to Fig. 31.25, at what time does the current in the *RL* circuit reach one-half of the final value?

QUESTION 5: The switch *S* in Fig. 31.27 has been open for a long time. At $t = 0$ it is closed. What is the current through the inductor immediately after closing the switch?

(A) 0   (B) $\mathcal{E}/R_1$ (C) $\mathcal{E}/R_2$ (D) $\mathcal{E}/(R_1 + R_2)$ (E) $\mathcal{E}(R_1 + R_2)/R_1 R_2$

SUMMARY

PROBLEM-SOLVING TECHNIQUES Faraday's Law; Lenz' Law		**(page 1005)**
PHYSICS IN PRACTICE Magnetic Levitation		**(page 1008)**
MOTIONAL EMF IN ROD	$\mathcal{E} = lvB$	**(31.1)**
MAGNETIC FLUX	$\Phi_B = \int \mathbf{B} \cdot d\mathbf{A} = \int B_\perp \, dA$	**(31.4)**
For flat surface in uniform field	$\Phi_B = B_\perp A = BA \cos\theta$	**(31.5)**

FARADAY'S LAW OF INDUCTION

$$\mathcal{E} = -\frac{d\Phi_B}{dt}$$ **(31.7)**

For coil of N turns:

$$\mathcal{E} = -N\frac{d\Phi_B}{dt}$$ **(31.9)**

$d\Phi_B/dt$ can arise from change in magnetic field (Example 3), change in area (Example 5), or change in orientation of area (Example 4).

LENZ' LAW   Induced current opposes the *change* that produced it.

INDUCED ELECTRIC FIELD

$$\mathcal{E} = \oint \mathbf{E} \cdot d\mathbf{s} = \oint E_\parallel \, ds = -\frac{d\Phi_B}{dt}$$ **(31.12)**

For axial magnetic flux,

$$\oint E_\parallel \, ds = E_\parallel \times 2\pi r$$

SI UNIT OF INDUCTANCE $1 \text{ H} = 1 \text{ henry} = 1 \text{ volt-second/ampere}$ **(31.17)**

MUTUAL INDUCTANCE M	$\Phi_{B1} = MI_1$	**(31.14)**
	$\mathcal{E}_2 = -M\dfrac{dI_1}{dt}$	**(31.15)**
	$\mathcal{E}_1 = -M\dfrac{dI_2}{dt}$	**(31.16)**
SELF-INDUCTANCE L	$\Phi_B = LI$	**(31.20)**
	$\mathcal{E} = -L\dfrac{dI}{dt}$	**(31.21)**
SELF-INDUCTANCE OF SOLENOID	$L = \mu_0 n^2 \pi R^2 l$	**(31.23)**
MAGNETIC ENERGY IN INDUCTOR	$U = \tfrac{1}{2}LI^2$	**(31.26)**
ENERGY DENSITY IN MAGNETIC FIELD	$u = \dfrac{1}{2\mu_0}B^2$	**(31.31)**
CHARACTERISTIC TIME OF RL **CIRCUIT**	$\tau = L/R$	**(31.34)**
CURRENT IN RL **CIRCUIT**	$I = \dfrac{\mathcal{E}}{R}\left(1 - e^{-t/\tau}\right)$ (increasing current)	**(31.35)**
	$I = \dfrac{\mathcal{E}}{R}\,e^{-t/\tau}$ (decreasing current)	**(31.36)**

QUESTIONS FOR DISCUSSION

1. At the latitude of the United States, the magnetic field of the Earth has a downward component, larger than the northward component. Suppose that an airplane is flying due west in this magnetic field. Will there be an emf between its wingtips? Which wingtip will be positive? Will there be a flow of current?

2. What is the magnetic flux that the magnetic field of the Earth produces through the surface of the Earth?

3. Is Eq. (31.12) valid for an open path? Is Eq. (31.7) valid for a closed path?

4. A long straight wire carries a steady current. A square conducting loop is in the same plane as the wire. If we push the loop toward the wire, how is the direction of the current induced in the loop related to the direction of the current in the wire?

5. A **flip coil** serves to measure the strength of a magnetic field. It consists of a small coil of many turns connected to a sensitive ammeter. The coil is placed face on in the magnetic field and then suddenly flipped over. How does this indicate the presence of the magnetic field?

6. The **magneto** used in the ignition system of old automobile engines consists of a permanent magnet mounted on the flywheel of the engine. As the flywheel turns, the magnet passes by a stationary coil, which is connected to the spark plug. Explain how this device produces a spark.

7. A long straight wire carries a current that is increasing as a function of time. A rectangular loop is near the wire, in the same plane as the wire. How is the direction of the current induced in the loop related to the direction of the current in the wire?

8. Consider two adjacent rectangular circuits, in the same plane. If the current in one circuit is suddenly switched off, what is the direction of the current induced in the other circuit?

9. Figure 31.28 shows two coils of wire wound around a plastic cylinder. If the current in the left coil is made to increase, what is the direction of the current induced in the coil on the right?

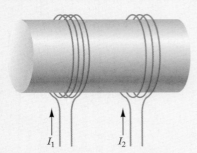

FIGURE 31.28 Two adjacent coils of wire.

10. A circular conducting ring is being pushed toward the north pole of a bar magnet. Describe the direction of the current induced in the ring.

11. Ganot's *Èléments de Physique*, a classic nineteenth-century textbook, states the following rules for the current induced in one loop face to face with another loop:

    I. The distance remaining the same, a continuous and constant current does not induce any current in an adjacent conductor.

    II. A current, at the moment of being closed, produces in an adjacent conductor an inverse current.

    III. A current, at the moment it ceases, produces a direct current.

    IV. A current which is removed, or whose strength diminishes, gives rise to a direct induced current.

    V. A current which is approached, or whose strength increases, gives rise to an inverse induced current.

    Explain these rules on the basis of Lenz' Law.

12. A sheet of aluminum is being pushed between the poles of a horseshoe magnet (Fig. 31.29). Describe the direction of flow

FIGURE 31.29
Sheet of aluminum and horseshoe magnet.

of the induced currents, or eddy currents, in the sheet. Explain why there is a strong friction force that opposes the motion of the sheet.

13. Some beam balances use a magnetic damping mechanism to stop excessive swinging of the beam. This mechanism consists of a small conducting plate attached to the beam and a small magnet mounted on a fixed support near this plate. How does this damp the motion of the beam? (Hint: See the preceding question.)

14. A bar magnet, oriented vertically, is dropped toward a flat horizontal copper plate. Explain why there will be a repulsive force between the bar magnet and the copper plate. Is this an elastic force, that is, will the bar magnet bounce if the magnet is very strong?

15. A conducting ring is falling toward a bar magnet (Fig. 31.30). Explain why there will be a repulsive force between the ring and the magnet. Explain why there will be no such force if the ring has a slot cut through it (see dashed lines in Fig. 31.30).

FIGURE 31.30 A conducting ring falling toward a bar magnet. The dashed lines indicate where a slot will be cut through the ring.

16. You have two coils, one of slightly smaller radius than the other. To achieve maximum mutual inductance, should you place these coils face to face or one inside the other?

17. Two circular coils are separated by some distance. Qualitatively, describe how the mutual inductance varies as a function of the orientation of the coils.

18. If a strong current flows in a circuit and you suddenly break the circuit (by opening a switch), a large spark is likely to jump across the switch. Explain.

19. What arguments can you give in favor of the view that the magnetic energy of an inductor is stored in the magnetic field, rather than in the current?

PROBLEMS

31.1 Motional Emf

1. An automobile travels at 88 km/h along a level road. The vertical downward component of the Earth's magnetic field is 5.8×10^{-5} T. What is the induced emf between the right and the left door handles separated by a distance of 1.8 m? Which side is positive and which is negative?

2. The DC-10 jet aircraft has a wingspan of 47 m. If such an aircraft is flying horizontally at 960 km/h at a place where the vertical component of the Earth's magnetic field is 6.0×10^{-5} T, what is the induced emf between its wingtips?

3. In order to detect the movement of water in the ocean, oceanographers sometimes rely on the motional emf generated by this movement of the water through the magnetic field of the Earth. Suppose that, at a place where the vertical magnetic field is 7.0×10^{-5} T, two electrodes are immersed in the water separated by a distance of 200 m measured perpendicularly to the movement of the water. If a sensitive voltmeter connected to the electrodes indicates a potential difference of 7.0×10^{-3} V, what is the speed of the water?

4. A runner runs northward at 9.0 m/s through the Earth's magnetic field, which, at her location, has a downward vertical component of 6.0×10^{-5} T. What is the motional emf between her right shoulder and her left shoulder separated by 50 cm?

5. The rate of flow of a conducting liquid can be measured with an electromagnetic flowmeter that detects the emf induced by the motion of the liquid in a magnetic field. Suppose that a plastic pipe of diameter 10 cm carries beer with a speed of 1.5 m/s. The pipe is in a transverse magnetic field of 1.5×10^{-2} T. What emf will be induced between the opposite sides of the column of liquid?

6. In an X-ray tube, an electron moves parallel to the horizontal component of the Earth's magnetic field with velocity 1.2×10^7 m/s. The vertical component of the Earth's magnetic field at the tube is 5.5×10^{-5} T. What is the induced electric field in the reference frame of the electron?

*7. Assume that the rod in Fig. 31.2 has a length of 0.86 m, the resistor has value 2.2 Ω, and a magnetic field of 8.0 T is directed into the page. The rod and rails have negligible resistance. The rod is pulled to the left at constant speed. What speed will produce a current of 1.5 A in the resistor? In what direction does the current flow? What pulling force must be applied to maintain a steady current?

*8. An automobile with a vertical radio antenna of length 75 cm travels at 80 km/h eastward. The Earth's magnetic field has a magnitude of 0.70×10^{-4} T and is directed 52° downward with respect to due north. What is the emf generated between the bottom and top of the antenna? Which end is at the higher potential?

†31.2 Faraday's Law
†31.3 Some Examples; Lenz' Law

9. A helicopter has blades of length 4.0 m rotating at 3.0 rev/s in a horizontal plane. If the vertical component of the Earth's magnetic field is 6.5×10^{-5} T, what is the induced emf between the tip of a blade and the hub?

10. The plane of a coil is initially at right angles to the direction of an applied magnetic field. By what angle must the coil be turned to *reverse* the direction of the flux and reduce the magnitude of the flux through the coil to 30% of its initial value?

11. In Idaho, the magnetic field of the Earth points downward at an angle of 69° below the horizontal. The strength of the magnetic field is 5.9×10^{-5} T. What is the magnetic flux through 1.0 m^2 of ground in Idaho?

12. An electromagnetic generator of the kind described in Example 4 has a coil of area 2.0×10^{-4} m^2 with 300 turns of wire. What is the amplitude of the alternating emf delivered by this coil when rotating at a rate of 2000 rev/min in a magnetic field of 0.020 T?

13. A homopolar generator consists of a metal disk rotating about a horizontal axis in a uniform horizontal magnetic field. The external circuit is connected to contact brushes touching the disk at the rim and at the axis. If the radius of the disk is 1.2 m and the strength of the magnetic field is 6.0×10^{-2} T, at what rate (rev/s) must you rotate the disk to obtain an emf of 6.0 V? (Hint: Any radius of the disk can be regarded as a conducting rod; see Example 5.)

14. Pulsars, or neutron stars, rotate at a fairly high speed, and they are surrounded by strong magnetic fields. The material in neutron stars is a good conductor, and hence a motional emf is induced between the center of the neutron star and the rim (this is similar to the emf induced in a rotating metallic rod; see Example 5). Suppose that a neutron star of radius 10 km rotates at the rate of 30 rev/s and that the magnetic field has a strength of 1.0×10^8 T. What is the emf induced between the center of the star and a point on its equator?

15. Large superconducting magnets are used in hospitals to produce pictures of the interior of the body by magnetic resonance imaging (MRI). For this purpose, the patient is shoved within the coils of the magnet, where the magnetic field is 1.5 T (see Fig. 31.31). Suppose that the patient is shoved into the magnetic field in a time of 10 s. Estimate the emf induced around the patient's trunk, 0.90 m in circumference. Should the patient be shoved into the magnetic field more slowly?

†For help, see Online Concept Tutorial 34 at www.wwnorton.com/physics

FIGURE 31.31 Large magnet used for MRI.

16. The bar magnet shown in Fig. 31.32 is thrust through the coil at a constant speed. Consider the following stages: (a) when the north pole approaches the coil, (b) when the magnet is centered in the coil, and (c) when the south pole leaves the right side of the coil. What is the direction of the current in the resistor at each of these stages?

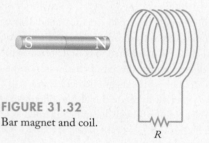

FIGURE 31.32
Bar magnet and coil.

17. A circular coil with 250 turns of 1.0-mm-diameter copper wire has an area of 0.35 m². A magnetic field directed at 30° with respect to the plane of the coil increases uniformly from zero to 5.5 T in 35 s. What is the emf induced across the coil during this time? If the ends of the coil are connected, what current flows? How much electrical energy is dissipated during this time?

*18. A circular loop of wire is placed in a magnetic field of 0.30 T while the free ends of the wire are attached to a 15-Ω resistor as shown in Fig. 31.33. When you squeeze the loop, the area of the loop is reduced at a constant rate from 200 to 100 cm² in 0.020 s. What are the magnitude and direction of the current in the resistor?

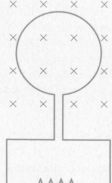

FIGURE 31.33 Circular loop in magnetic field.

*19. A compact disc (CD) is placed in a magnetic field of 1.5 T and rotates at 210 rev/min about an axis parallel to the field. What is the emf generated between a point on its outer track (radius 5.8 cm) and a point on its inner track (radius 2.3 cm)?

*20. A long rectangular conducting loop of width 25 cm is partially in a region of a horizontal magnetic field of 1.8 T perpendicular to the loop, as shown in Fig. 31.34. The mass of the loop is 12 g, and its resistance is 0.17 Ω. If the loop is released, what is its terminal velocity? Assume that the top of the loop stays in the magnetic field. (Hint: The terminal velocity occurs when the magnetic force on the induced current is equal in magnitude to the gravitational force.)

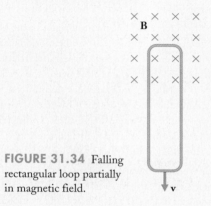

FIGURE 31.34 Falling rectangular loop partially in magnetic field.

*21. A rectangular loop of width $w = 0.30$ m and height $h = 0.50$ m is coplanar with and a distance $d = 0.10$ m from a long wire carrying a current of 2.5 A, as shown in Fig. 31.35. What is the flux through the loop? [Hint: Use Eq. (31.4) to sum the fluxes through vertical strips of area $dA = h\, dr$.]

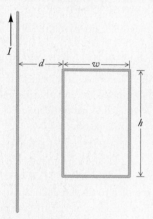

FIGURE 31.35 Long, straight wire and rectangular loop.

*22. A toroid has 150 turns on a rectangular cross section of height 4.0 cm and width 6.5 cm, as shown in Fig. 31.36. The inner radius of the toroid is $R = 7.0$ cm. The current in the wire of the toroid is $I = 2.0$ A. What is the flux through a circular path shown in the figure? [Hint: Use Eq. (31.4) to sum the fluxes through vertical strips of area $dA = h\, dr$.]

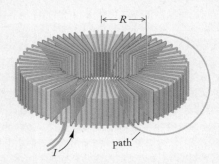

FIGURE 31.36 A toroid with rectangular cross section and a circular path.

*23. The current in a long solenoid of radius $R = 3.0$ cm is increasing uniformly at a rate of 1.5 A/s. The solenoid has 350 turns per meter. What is the induced electric field a distance r from the solenoid axis when (a) $r = 2.0$ cm and (b) $r = 4.0$ cm?

*24. The plane of a circular wire loop of radius 3.5 cm is initially perpendicular to a magnetic field of 8.2 T. The loop is then rotated 90°, so the magnetic field is parallel to the plane of the loop. How much charge flows past any point on the loop during this process? The resistance of the loop is 0.25 Ω.

*25. A metal disk with negligible resistance is 12 cm in diameter. The disk is rotated at 300 rev/s while in a 5.5-T magnetic field perpendicular to the disk. One end of a 33-Ω resistor is in contact with the center of the disk; the other end of the resistor is in contact with the edge of the disk. How much current flows in the resistor? What torque must be supplied to the disk to maintain this current?

*26. A very long train whose metal wheels are separated by a distance of 4 ft 9 in. is traveling at 80 mi/h on a level track. The vertical component of the magnetic field of the Earth is 6.2×10^{-5} T.

(a) What is the induced emf between the right wheels and the left wheels?

(b) The wheels are in contact with the rails, and the rails are connected by metal cross ties which close the circuit and permit a current to flow from one rail to the other. Since the number of cross ties is very large, their combined resistance is nearly zero; most of the resistance of the circuit is within the train. What is the current that flows in each axle of the train? The axles are cylindrical rods of iron of diameter 3 in. and length 4 ft 9 in.

(c) Calculate the power dissipated and the effective friction force on the train.

*27. A square loop of dimension 8.0 cm × 8.0 cm is made of copper wire of radius 1.0 mm. The loop is placed face on in a magnetic field which is increasing at the constant rate of 80 T/s. What induced current will flow around the loop? Draw a diagram showing the direction of the field and the induced current.

*28. A very long solenoid with 20 turns per centimeter of radius 5.0 cm is surrounded by a rectangular loop of copper wire. The rectangular loop measures 10 cm × 30 cm, and its wire has a radius of 0.050 cm. The resistivity of copper is 1.7×10^{-8} Ω·m. Find the induced current in the rectangular loop if the current in the solenoid is increasing at the rate of 5.0×10^4 A/s.

*29. A rectangular loop measuring 20 cm × 80 cm is made of heavy copper wire of radius 0.13 cm. Suppose you shove this loop, short side first, at a speed of 0.40 m/s into a magnetic field of 5.0×10^{-2} T. The rectangle is face on to the magnetic field, and the trailing short side remains outside the magnetic field (see Fig. 31.37). What induced current will flow around the loop?

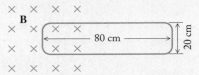

FIGURE 31.37 Rectangular loop partially in magnetic field.

*30. A washer (annulus) of aluminum is lying on top of a vertical solenoid (see Fig. 31.38). When the current in the solenoid is suddenly switched on, the washer flies upward. Carefully explain why the end of the solenoid exerts a repulsive force on the washer under these conditions. (Hint: Take into account that, at the end of the solenoid, the magnetic field lines spread out.)

FIGURE 31.38 Washer lying on top of a solenoid.

*31. The magnetic field of a betatron of radius 1.0 m has an amplitude of oscillation of 0.90 T and a frequency of 60 Hz. What is the amplitude of oscillation of the induced electric field at a radius of 0.80 m? At 1.5 m? What is the amplitude of oscillation of the induced emf around circular paths of each of these radii? (Consider the betatron to be a circular region of uniform magnetic field.)

**32. A square loop of dimension $l \times l$ is moving at a speed v toward a straight wire carrying a current I. The wire and the loop are in the same plane, and two of the sides of the loop are parallel to the wire. What is the induced emf of the loop as a function of the distance d between the wire and the nearest side of the loop?

**33. A circular coil of insulated wire has a radius of 9.0 cm and contains 60 turns of wire. The ends of the wire are connected in series with a 15-Ω resistor closing the circuit. The normal to the loop is initially parallel to a constant magnetic field of 5.0×10^{-2} T. If the loop is flipped over, so that the direction of the normal is reversed, a pulse of current will flow through the resistor. What amount of charge will flow through the resistor? Assume that the resistance of the wire is negligible compared with that of the resistor. (Hint: Suppose that the flipping takes a time Δt. What is the average rate of change of the magnetic flux? The average induced emf? The average current?)

31.4 Inductance

34. Two coils are arranged face to face, as in Fig. 31.21. Their mutual inductance is 2.0×10^{-2} H. The current in coil 1 oscillates sinusoidally with a frequency of 60 Hz and an amplitude of 12 A:

$$I_1 = 12 \sin(120\pi t)$$

where the current is measured in amperes and the time in seconds.

(a) What is the magnetic flux that this current generates in coil 2 at time $t = 0$?

(b) What is the induced emf that this current induces in coil 2 at time $t = 0$?

(c) What is the direction of the induced current in coil 2 at time $t = 0$, according to Lenz' Law? Assume that the positive direction for the current I_1 is as shown by the arrow in the figure.

35. A current of 15 A in a coil produces a magnetic flux of 0.10 T·m^2, or 0.10 Wb, through each of the turns of an adjacent coil of 60 turns. What is the mutual inductance?

36. A long solenoid has 400 turns per meter. A coil of radius 1.0 cm with 30 turns of insulated wire is placed inside the solenoid, its axis parallel to the axis of the solenoid. What is the mutual inductance? What emf will be induced around the coil if the current in the solenoid windings changes at the rate of 200 A/s?

37. A loop of wire carrying a current of 100 A generates a magnetic flux of 50 T·m^2, or 50 Wb, through the area bounded by the loop.

(a) What is the self-inductance of the loop?

(b) If the current is decreased at the rate of $dI/dt = 20$ A/s, what is the induced emf?

38. A long solenoid has 2000 turns per meter and a radius of 2.0 cm.

(a) What is the self-inductance for a 1.0-m segment of this solenoid?

(b) What back emf will this segment generate if the current in the solenoid is changing at the rate of 3.0×10^2 A/s?

39. A solenoid of self-inductance 2.2×10^{-3} H in which there is initially no current is suddenly connected in series with the

poles of a 24-V battery. What is the instantaneous initial rate of increase of the current in the solenoid?

40. A 7.5-mH inductor carries a time-dependent current given by $I = C_1 t - C_2 t^2$, where t is in seconds, $C_1 = 65$ A/s, and $C_2 = 25$ A/s^2. What is the induced emf across the inductor at $t = 1.0$ s? At $t = 2.0$ s?

41. A superconducting solenoid has an inductance of 25 H. If the current in the solenoid is to increase at a rate of 0.075 A/s, what emf must be supplied across the terminals of the solenoid?

42. In a fast digital circuit, the timing of signals is often limited by the inductance of circuit components. Suppose that a 5.0-V emf is suddenly applied to an effective inductance of 2.5 μH. How long does it take for the current in the inductor to reach 2.0 mA?

*43. A long solenoid of radius R has n turns per unit length. A circular coil of wire of radius R' with 200 turns surrounds the solenoid (Fig. 31.39). What is the mutual inductance? Does the shape of the coil of wire matter?

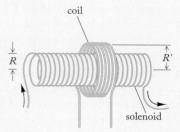

FIGURE 31.39 A long solenoid surrounded by a circular coil.

*44. Two long concentric solenoids of n_1 and n_2 turns per unit length have radii R_1 and R_2, respectively (Fig. 31.40). What is the mutual inductance per unit length of the solenoids? Assume $R_1 < R_2$.

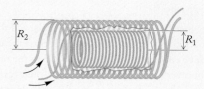

FIGURE 31.40 Two long concentric solenoids.

*45. Find the mutual inductance between the long wire and the rectangular loop described in Problem 21.

*46. An **induction furnace** exploits eddy currents to melt metals and is often used to produce alloys or to grow crystals of conducting materials. A particular induction furnace uses a 15-turn solenoid of length 25 cm and radius 3.0 cm. The current in the solenoid oscillates sinusoidally according to $I = I_{max} \sin \omega t$ where $I_{max} = 2.5$ A and $\omega = 1.2 \times 10^7$ rad/s. What is the maximum voltage induced across the solenoid? Neglect any resistance.

**47. Two inductors of self-inductance L_1 and L_2 are connected in parallel. The inductors are magnetically shielded from one another so that neither produces flux in the other. Show that the self-inductance of the combination is given by $1/L = 1/L_1 + 1/L_2$.

**48. Two inductors of self-inductance L_1 and L_2 are connected in series. The inductors are magnetically shielded from one another so that neither produces flux in the other. Show that the self-inductance of the combination is $L = L_1 + L_2$.

31.5 Magnetic Energy

49. A ring of thick wire has a self-inductance of 4.0×10^{-8} H. How much work must you do to establish a current of 25 A in this ring?

50. In a region of vacuum containing a magnetic field of 1.0 T and an electric field of 1.0 V/m, which field has the larger energy density?

51. What is the magnetic energy density at a point 3.0 mm from a long, thin wire carrying a current of 24 A?

52. The strongest magnetic field achieved in a laboratory is about 1.0×10^3 T. This field can be produced only for a short instant by compressing the magnetic field lines with an explosive device. What is the energy density in this field?

53. For each of the first six entries in Table 29.1, calculate the energy density in the magnetic field.

54. For a crude estimate of the energy in the Earth's magnetic field, pretend that this field has a strength of 5.0×10^{-5} T from the ground up to an altitude of 6.0×10^6 m above the ground. What is the total magnetic energy in this region?

55. A superconducting solenoid carries a current of 55 A, has an inductance of 35 H, and produces a magnetic field of 9.0 T. What energy is stored in the solenoid? What is the volume of the solenoid?

56. Suppose that the magnetic energy stored in an inductor is 2.0×10^{-3} J when the current in this inductor is 30 A. What will be the magnetic energy of the same inductor if the current is 60 A? 90 A?

57. The self-inductance of the aluminum cable of a high-voltage transmission line is 6.6×10^{-4} H per kilometer of length. If the cable carries a current of 800 A, what is the energy in the magnetic field of the cable, per kilometer?

*58. According to one proposal, the surplus energy from a power plant could be temporarily stored in the magnetic field within a very large toroid. If the strength of the magnetic field is 10 T, what volume of the magnetic field would we need to store 1.0×10^5 kW·h of energy? If the toroid has roughly the proportions of a doughnut, roughly what size would it have to be? (Hint: The volume of a toroid equals the cross-sectional area multiplied by 2π times the average radius.)

**59. Two long, straight, concentric tubes made of sheet metal carry equal currents in opposite directions. The inner tube has a radius of 1.5 cm and the outer tube a radius of 3.0 cm. The current on the surface of each is 120 A. What is the magnetic energy in a 1.0-m segment of these tubes?

31.6 The *RL* Circuit

60. An inductor with $L = 2.0$ H and a resistor with $R = 100\ \Omega$ are suddenly connected in series to a battery with $\mathcal{E} = 6.0$ V.

   (a) What is the current at $t = 0$? At $t = 0.010$ s? At $t = 0.020$ s?

   (b) What is the final, steady value of the current?

   (c) What is the rate of increase of the current at $t = 0$?

61. Consider the *RL* circuit shown in Fig. 31.26. At time $t = 0$, the circuit has an initial current $\mathcal{E}/R$. Apply Kirchhoff's rule to this circuit and find a differential equation for dI/dt. Show that Eq. (31.36) is a solution to this equation.

62. Design an *RL* circuit with an arrangement of switches so that the battery can be suddenly switched out of the circuit and the current in the inductor can be suddenly fed into the resistor. If the self-inductance of the inductor is 0.20 H, what resistance do you need to obtain a characteristic time of 10 s in your circuit?

63. A superconducting magnet has an inductance of 45 H and carries a current of 65 A. The terminals supplying current to the magnet are removed immediately after a resistor is connected across them. The current in the magnet drops to zero with a characteristic time of 12 s.

   (a) What is the value of the resistor?

   (b) What is the initial rate of power dissipation in the resistor?

   (c) What is the total energy dissipated in the resistor after a long time?

64. Initially, the switch shown in Fig. 31.41 is open and no currents flow in the circuit. At $t = 0$, the switch is closed.

   (a) What are the currents in the inductor, in resistor R_1, and in resistor R_2 immediately after the switch is closed?

   (b) What are those three currents after the switch has been closed for a long time?

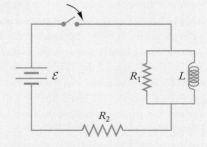

FIGURE 31.41 An *RL* circuit with two resistors.

65. *RL* circuits are sometimes used to generate high-voltage pulses. Consider the circuit shown in Fig. 31.42, where $\mathcal{E} = 12$ V, $R_1 = 1.5$ kΩ, and $R_2 = 6.0\ \Omega$. The switch has been in the upper position, connected to the battery for a long time.

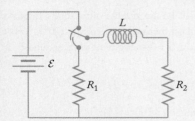

FIGURE 31.42 An RL circuit with two resistors.

(a) What is the current in the inductor?

(b) The switch is suddenly moved to the lower position. What is the voltage across the inductor immediately after the switch is moved to the lower position?

66. Suppose that the switch in Fig. 31.27 has been closed for a long time. At $t = 0$, it is opened. What is the current in the inductor for $t \geq 0$? What is the voltage across R_1 for $t > 0$? Across R_2?

67. Find the time constant of the circuit shown in Fig. 31.42 (a) when the switch is in the upper position and (b) when the switch is in the lower position.

68. An inductor is suddenly connected in series to a resistor with $R = 10\ \Omega$. The initial current in the inductor is 3.4 A. After a time of 6.0×10^{-2} s, the current has dropped to 1.7 A. What is the self-inductance?

*69. An RL circuit consists of two inductors of self-inductance $L_1 = 4.0$ H and $L_2 = 2.0$ H connected in parallel to each other, and connected in series to a resistor of 6.0 Ω and a battery of 3.0 V (Fig. 31.43). Assume that the inductors have no mutual inductance and no resistance.

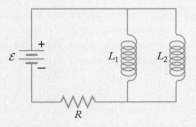

FIGURE 31.43 An RL circuit with two inductors.

(a) When the battery is suddenly connected, what is the initial rate of change of the current in each inductor?

(b) What is the final, steady current in the resistor? What are the final, steady currents in each inductor? (Hint: The currents in the inductors are inversely proportional to their inductances. Why?)

*70. What is the Joule heat dissipated by the current (31.36) in the resistor in the time interval $t = 0$ to $t = \infty$? Compare with the initial magnetic energy in the inductor.

*71. To measure the self-inductance and the internal resistance of an inductor, a physicist first connects the inductor across a 3.0-V battery. Under these conditions, the final, steady current in the inductor is 24 A. The physicist then suddenly short-circuits the inductor with a thick (resistanceless) wire placed across its terminals. The current then decreases from 24 A to 12 A in 0.22 s. What are the self-inductance and the internal resistance of the inductor?

REVIEW PROBLEMS

72. An experiment attempted with the Space Shuttle, but not completed because of mechanical difficulties, was designed to obtain electric power from the motional emf induced by the motion of the Space Shuttle through the Earth's magnetic field. While in orbit around the Earth at an altitude of 296 km at a speed of 7.7 km/s, a 20-km-long wire was to be stretched radially outward between the Space Shuttle and a small "tethered satellite." The magnetic field of the Earth at the altitude of the Space Shuttle is 2.7×10^{-5} T. Calculate the magnitude of the motional emf induced between the ends of such a 20-km wire in this magnetic field. Assume that the motion of the wire is at right angles to the magnetic field.

73. A girl uses a jump rope made of flexible wire. She holds the ends of the rope in her extended hands, so the rope has approximately the shape of a semicircle of radius 0.70 m, and

she whirls the rope at the rate of 1.0 revolution per second at a place where the magnetic field of the Earth is nearly vertical, of magnitude 5.0×10^{-5} T. The motion of the wire rope generates an alternating emf between the ends of the rope. What is the amplitude of oscillation of this alternating emf?

74. An electric generator consists of a rectangular coil of wire rotating about its longitudinal axis, which is perpendicular to a magnetic field of 2.00×10^{-2} T. The coil measures 10.0 cm × 20.0 cm and has 120 turns of wire. The ends of the wire are connected to an external circuit. At what speed (in rev/s) must you rotate this coil in order to induce an alternating emf of amplitude 12.0 V between the ends of the wire?

75. A circular current loop has a radius of 1.0 cm and carries a current of 3.0 A. What is the magnetic energy density at a point at the center of the loop?

76. A large toroid built in the early days of plasma research in the former Soviet Union had a major radius of 1.50 m and a minor radius of 0.40 m (Fig. 31.44). If the average magnetic field within such a toroid is 4.0 T, what is the magnetic energy? (Hint: The volume of a toroid equals the cross-sectional area multiplied by 2π times the average radius.)

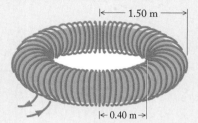

FIGURE 31.44 Giant toroid.

*77. Sharks have delicate sensors on their bodies that permit them to sense small differences in potential (Fig. 31.45). They can sense electrical disturbances created by other fish, and they can also sense the Earth's magnetic field and use this for navigation. Suppose that a shark is swimming horizontally at 25 km/h at a place where the magnetic field has a strength of 4.7×10^{-5} T and points down at an angle of 40° with the vertical. Treat the shark as a cylinder of diameter 30 cm. What is the largest induced emf between diametrically opposite points on the sides of the shark when heading north?

FIGURE 31.45 A shark.

*78. A metal rod of length l and mass m is free to slide, without friction, on two parallel metal tracks. The tracks are connected at one end so that they and the rod form a closed circuit (Fig. 31.46).

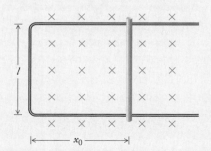

FIGURE 31.46 Rod sliding on parallel tracks. The crosses show the tails of the magnetic field vectors.

The rod has a resistance R, and the tracks have negligible resistance. A uniform magnetic field is perpendicular to the plane of this circuit. The magnetic field is increasing at a constant rate of dB/dt. Initially the magnetic field has strength B_0 and the rod is at rest at a distance x_0 from the connected end of the rails. Express the acceleration of the rod at this instant in terms of the given quantities.

*79. A boxcar of a train is 2.5 m wide, 9.5 m long, and 3.5 m high; it is made of sheet metal and is empty. The boxcar travels at 60 km/h on a level track at a place where the vertical component of the Earth's magnetic field is 6.2×10^{-5} T. Assume that the boxcar is not in electrical contact with the ground.

(a) What is the induced emf between the sides of the boxcar?

(b) Taking into account the electric field contributed by the charges that accumulate on the sides, what is the net electric field inside the boxcar (in the reference frame of the boxcar)?

(c) What is the surface charge density on each side? Treat the sides as two very large parallel plates.

*80. A 25-turn coil of wire has an area of 4.0×10^{-3} m^2 and is oriented perpendicularly to a magnetic field **B**. The coil has a resistance of 15 Ω. At what rate must the magnitude of **B** change for an induced current of 5.0 mA to appear in the coil?

*81. (a) A long solenoid has 300 turns of wire per meter and has a radius of 3.0 cm. If the current in the wire is increasing at the rate of 50 A/s, at what rate does the strength of the magnetic field in the solenoid increase?

(b) The solenoid is surrounded by a coil of wire with 120 turns (Fig. 31.47). The radius of this coil is 6.0 cm. What induced emf will be generated in this coil while the current in the solenoid is increasing?

(c) Suppose we replace the coil of radius 6.0 cm by a new coil of radius 8.0 cm. What will be the induced emf now?

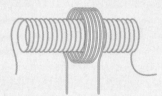

FIGURE 31.47 A long solenoid and a circular coil.

*82. The resistance of the coils of an electric motor is 2.0 Ω when the motor is not in operation (not rotating). When the motor is connected to 115 V and is rotating at full speed, it draws a current of 0.10 A. Deduce the back emf that the coils of the motor produce when rotating.

*83. The current in a 115-volt electric motor jumps from 4.0 A to 36 A when the motor coils are suddenly halted by a brake. What is the back emf when the motor is rotating at full speed?

*84. A current of 5.0 A flows through a cylindrical solenoid of 1500 turns. The solenoid is 40 cm long and has a diameter of 3.0 cm.

    (a) Find the magnetic field in the solenoid. Treat the solenoid as very long.

    (b) Find the energy density in the magnetic field, and find the magnetic energy stored in the space within the solenoid.

*85. An RL circuit with $L = 0.50$ H and $R = 0.025$ Ω is initially connected to a battery of 1.2 V. When the current reaches its maximum, steady value, the battery is suddenly switched out of the circuit and the current is switched into the resistor.

    (a) What is the maximum value of the current? At what time will the current in the inductor drop to 50% of its maximum value?

    (b) What is the maximum value of the energy stored in the inductor? What percentage of the energy remains when the current has dropped to 50% of its maximum value?

*86. A long straight wire carries a current that increases at a steady rate dI/dt.

    (a) What is the rate of increase of the magnetic field at a radial distance r?

    (b) The wire is in the same plane as the rectangular loop shown in Fig. 31.48. What is the induced emf around the loop? Assume that $a \ll r$, that is, the width of the loop is so small that the magnetic field and its rate of change are approximately the same at all points in the loop.

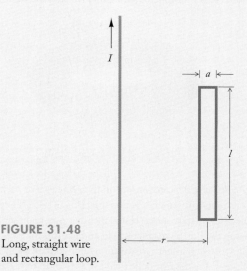

FIGURE 31.48
Long, straight wire and rectangular loop.

Answers to Checkups

Checkup 31.1

1. If the direction of the motion of the rod reverses, the force on the electrons reverses, so the current (the flow of positive charge) also reverses, and flows clockwise around the circuit.

2. Since we are looking in the direction of the horizontal magnetic field (see Fig. 31.3), $\mathbf{v} \times \mathbf{B}$ will point toward the right, so this is the end where positive charge will accumulate. Reversing the velocity (to upward) will reverse the force. making the left end positive.

3. No matter what horizontal direction your velocity is in, for a magnetic field pointing down, $\mathbf{v} \times \mathbf{B}$ will point toward your left side; thus, east, west, and north velocities yield a positive terminal on your north, south, and west side, respectively.

4. (E) To right. The induced electric field in the frame of reference moving with the rod points in the same direction as the magnetic force on a positive charge in the stationary reference frame; in Fig. 31.3, this is toward the right. For a conducting rod, this induced field causes charges to move until the sum of the induced field and the field due to the charges that accumulate at the ends of the rod is zero.

Checkup 31.2

1. Since magnetic field lines do not begin or end anywhere, but instead always form closed loops, any field line that enters a closed surface must also exit the surface, yielding zero net flux.

2. Yes. Reversing the magnetic flux requires it to change direction; an emf will be generated during the time of this sudden change. Since the emf is proportional to the rate of change, for a sudden change, a very large emf is generated during the change.

3. No. For a rod moving in a magnetic field, the induced emf is transverse to the motion, and so would only be generated between the sides of the arrow, not between the tip and the tail.

4. (B) b. The most field lines cross the area shown in Fig. 31.13b, and so this area intercepts the most magnetic flux, even though it is neither the most perpendicular to the magnetic field, nor the largest area.

Checkup 31.3

1. For case (a), if the loop is pulled to the right (out of the field region), the flux through the loop decreases; thus, to oppose

this change, the induced current will be counterclockwise, looking from above (by the right-hand rule). Conversely, if the loop is pushed into the field, the induced current will be clockwise. For case (b), squeezing the loop decreases its area and thus the flux through it, so the induced current will be counterclockwise, looking from above. For case (c), seen from what was the bottom as shown, the induced current will be clockwise to oppose the decrease in flux due to turning the loop. For the increasing field of case (d), the induced current will again be clockwise, seen from above, to oppose this change.

2. The induced emf is proportional to the number of loops, so it will also double to 120 V. However, the resistance will also double for twice as many turns, so the induced current will be the same, 12 A.

3. The induced emf and current will have the same magnitude, but to oppose the increase, they will have the opposite direction.

4. Like the case of the circular electromagnet (Example 6), the electric field will be circles around the axis. Note that for a circle outside the solenoid, there will be an induced electric field, even though there is no magnetic field, because the flux through part of the area bounded by the circle is changing.

5. (A) To the right. As the rod moves, the area of the loop formed by the rod, tracks, and resistor decreases. The induced current will oppose the *decrease* of flux and contribute flux into the page. Thus the current will flow clockwise around the loop. Note that the flux from the induced current opposes the *change* in flux, not the original flux.

Checkup 31.4

1. If we move the coils further apart, the magnetic field due to one will be smaller at the second. Thus the flux will be smaller, and so will the mutual inductance M. If we bring them closer together, the opposite is true: the flux will increase, and so will the mutual inductance. If one coil is at a right angle to the other, there will be no net flux crossing it, and so the mutual inductance will decrease to zero.

2. Since the mutual inductance is the same between a pair of coils regardless of which carries the current, the induced emf will be the same, 6 V.

3. According to Example 9, the self-inductance is inversely proportional to the length of a solenoid and directly proportional to the number of loops squared ($n^2l = N^2/l$). The diameter of the wire affects how closely a coil can be wound, so its self-

inductance can depend on the diameter of the wire; however, for the same number of turns per unit length, the wire diameter does not matter.

4. (E) 4. As in question 3 and Example 9, the self-inductance is proportional to the square of the number of turns, and so is 4 times as large for the second solenoid.

Checkup 31.5

1. From Eq. (31.22), the self-inductance, $L = \Phi_B/I = \mu_0 n^2 \pi R^2 l$, is proportional to $n^2l = N^2/l$ and so does change (increases) when the solenoid is compressed (when l decreases).

2. The magnetic energy density is largest where the magnetic field is the largest; this occurs at the surface of the wire (see Example 5 of Chapter 29).

3. Since the stored magnetic energy is proportional to the inductance, as in Eq. (31.26), the second inductor has the larger magnetic energy.

4. (D) 4. Since the stored energy is proportional to the square of the current [Eq. (31.26)], doubling the current increases the stored magnetic energy by a factor of 4.

Checkup 31.6

1. The decrease of the current in the electromagnet is governed by the exponential decay of Eq. (31.36); such a decay will reach 37% of its initial value in one characteristic time ($e^{-1} \approx 0.37$). Thus, from Example 13, the time will be $\tau = 1.25$ s.

2. A large inductance implies that the back emf, which opposes the change in current, is large. A large back emf tends to keep the current constant, implying a long relaxation time.

3. A low resistance implies a low "friction," so the current tends to keep going. For example, consider the decaying current of question 1, with some initial current I_0. The initial rate of energy dissipation will be $P = I_0^2 R$, so a lower resistance means that it will take longer for the stored energy in the inductor to dissipate in the resistor.

4. Estimating from the plot in Fig. 31.25, the current reaches one-half of its final value between $t \approx 0.6L/R$ and $0.7L/R$.

5. (A) 0. After the switch has been open for a long time, the current in the inductor must be zero, because there is no applied emf and any current in the right loop will have decayed to zero. Since the current in an inductor must change slowly, the current will still be zero immediately after closing the switch.

Alternating Current Circuits

CONCEPTS IN CONTEXT

Concepts
in
Context

Power companies prefer alternating currents to direct currents because of the ease with which alternating voltages can be stepped up or down by means of transformers, such as the ones shown here. Transformers make it possible to step up the output of a power plant to several hundred thousand volts, transmit the power along a high-voltage line that minimizes Joule heating losses, and finally step down the voltage to 230 volts AC or 115 volts AC just before consumer use in lighting and appliances.

As we learn about AC circuits, we will consider such questions as:

❓ For the "115-V AC" emf supplied at ordinary outlets in homes, what are the actual maximum and minimum emf? What is the relation between the resistance and the time-average power dissipated in a lightbulb or other resistive device connected to such an outlet? (Section 32.1, pages 1031 and 1033; Example 1, page 1034)

? How does a "dimmer" control the power delivered to a lightbulb? (Example 5, page 1049)

? How does a transformer step up or step down an alternating emf? An alternating current? (Section 32.6, page 1055; and Example 8, page 1056)

The current delivered by power companies to homes and factories is an oscillating, time-dependent current. It periodically flows forward and backward, cycling 60 times per second. This is called **alternating current**, or AC.

All the appliances connected to ordinary outlets in homes therefore involve circuits with oscillating currents. Furthermore, electronic devices—radios, TVs, cellular telephones, computers—involve a variety of circuits with oscillating currents of high frequency. Many of these circuits have a natural frequency of oscillation. Such circuits exhibit the phenomenon of resonance when the natural frequency matches the frequency of a signal applied to the circuit. For instance, in some radios, the tuning relies on an oscillating circuit whose frequency of oscillation is adjusted by means of a variable capacitor (attached to the tuning knob) so that it matches the frequency of the radio signal.

In this chapter, we will discuss how resistors, capacitors, and inductors behave in AC circuits, that is, circuits with oscillating currents. The calculations of currents in AC circuits are similar to the calculations in DC circuits. As we will see, the starting point for these calculations is again Kirchhoff's rule.

alternating current, or AC

32.1 RESISTOR CIRCUIT

In North America, the alternating emf supplied by the power companies to electric outlets in private homes is an alternating emf with an amplitude $\mathcal{E}_{max} = 163$ V and a frequency of 60 Hz. The mathematical formulas describing such an alternating emf are similar to those describing an oscillating particle (see Section 15.1). The angular frequency is $\omega = 2\pi f = 2\pi \times 60$ radians/s, and the time dependence of the emf is given by a cosine function

$$\mathcal{E} = \mathcal{E}_{max} \cos \omega t = 163 \text{ V} \times \cos(2\pi \times 60\ t) \tag{32.1}$$

Recall that the angular frequency $\omega = 2\pi \times 60$ s^{-1} ensures the correct periodicity of the cosine function: each time the time t advances $1/60$ second, the argument of the cosine increases by 2π, that is, one cycle. Figure 32.1 shows a plot of the emf as a

Concepts *in* Context

Online
Concept
Tutorial

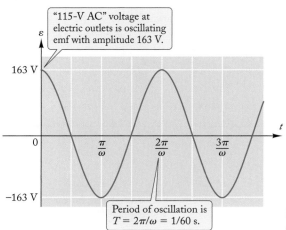

"115-V AC" voltage at electric outlets is oscillating emf with amplitude 163 V.

Period of oscillation is $T = 2\pi/\omega = 1/60$ s.

FIGURE 32.1 Emf supplied by an outlet as a function of time.

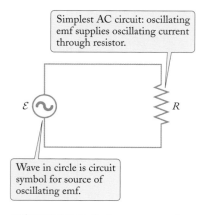

Simplest AC circuit: oscillating emf supplies oscillating current through resistor.

Wave in circle is circuit symbol for source of oscillating emf.

FIGURE 32.2 Resistor connected to a source of alternating emf.

current in resistor

FIGURE 32.3 The resistive wire in this toaster becomes orange-hot when connected to a wall outlet, a source of emf.

function of time. This kind of voltage is usually called "115 volts AC" for reasons that will become clear shortly. In many European countries, the amplitude of the AC voltage supplied to outlets in homes is twice as large; it is called "230 volts AC." An appliance designed for 115 V AC is likely to overheat and burn out if plugged into 230 V AC. To prevent this, the plugs for 115-V appliances and for 230-V are built in different shapes, so they cannot fit into the wrong kind of outlet.

The simplest possible AC circuit consists of a pure resistor connected to an oscillating source of emf. In the circuit diagram (see Fig. 32.2), the oscillating source of emf is symbolized by a wavy line enclosed in a circle. This circuit might represent an electric heater or an incandescent lamp plugged into an ordinary wall outlet (see Fig. 32.3).

According to Kirchhoff's rule, the sum of emfs and potential changes across resistors around any circuit must be zero. Although we first developed this law for DC circuits, it is equally valid for AC circuits. For the circuit shown in Fig. 32.2, this tells us that, at any instant of time,

$$\mathcal{E} - IR = 0 \tag{32.2}$$

from which we find

$$I = \frac{\mathcal{E}}{R} = \frac{\mathcal{E}_{max}\cos\omega t}{R} \tag{32.3}$$

Thus, *the current oscillates with the same frequency ω as the emf*, but whereas the amplitude of the emf is $\mathcal{E}_{max}$, the amplitude of the current is $\mathcal{E}_{max}/R$. Figure 32.4 compares plots of the emf and the current. The maxima and the minima of the current and the emf occur simultaneously, and the current is said to be **in phase** with the emf.

The instantaneous electric power dissipated in the resistor is the product of the instantaneous current and the emf, as in the case of a DC circuit [see Eq. (28.22)]:

$$P = \mathcal{E}I = \mathcal{E}_{max}\cos\omega t \times \frac{\mathcal{E}_{max}\cos\omega t}{R} = \frac{\mathcal{E}_{max}^2\cos^2\omega t}{R} \tag{32.4}$$

Although the emf and the current both are negative for one-half of each cycle, the power is always positive, because the emf and the current become negative simultaneously (the emf and the current are in phase), and therefore their product is always

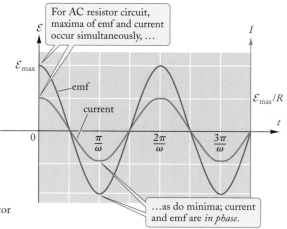

FIGURE 32.4 The emf (blue) and the current (red) in the resistor circuit as a function of time.

For AC resistor circuit, maxima of emf and current occur simultaneously, …

…as do minima; current and emf are *in phase*.

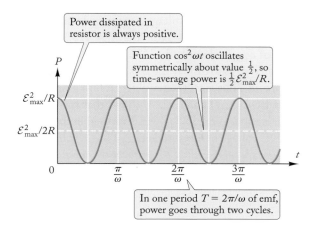

FIGURE 32.5 Instantaneous power dissipated in the resistor as a function of time. Note that the frequency of oscillation of the power is twice that of the emf.

positive. According to Eq. (32.4), the power oscillates between zero and a maximum value $\mathcal{E}^2_{max}/R$. Figure 32.5 is a plot of these oscillations.

For the operation of most appliances involving resistors, such as electric heaters or lightbulbs, we do not care about the oscillations of the power. We care only about the **time-average power** dissipated by the resistor, which tells us the average amount of heat or light produced by the heater or lightbulb. This time-average power can be obtained by averaging the quantity $\cos^2\omega t$ in Eq. (32.4) over one cycle of oscillation. To evaluate this average, note that the average of $\cos^2\omega t$ is the same as the average of $\sin^2\omega t$, because both of these quantities have the same number of ups and downs in one cycle. Using overbars to indicate averages, we therefore have

$$\overline{\cos^2\omega t} = \tfrac{1}{2}(\overline{\cos^2\omega t} + \overline{\cos^2\omega t}) = \tfrac{1}{2}(\overline{\cos^2\omega t} + \overline{\sin^2\omega t}) \qquad (32.5)$$

But $\cos^2\omega t + \sin^2\omega t = 1$, and hence Eq. (32.5) tells us that $\overline{\cos^2\omega t} = \tfrac{1}{2}$. This is also evident from Fig. 32.5, where we see that $\cos^2\omega t$ oscillates symmetrically about the value $\tfrac{1}{2}$. The time-average power is then

$$\overline{P} = \frac{\mathcal{E}^2_{max}}{2R} \qquad (32.6)$$

Concepts in Context

This is usually written in the form

$$\overline{P} = \frac{\mathcal{E}^2_{rms}}{R} \qquad (32.7)$$

power dissipated by resistor

where the quantity $\mathcal{E}_{rms}$, called the **root-mean-square voltage**, or rms voltage, is the maximum voltage divided by $\sqrt{2}$:

$$\mathcal{E}_{rms} = \frac{\mathcal{E}_{max}}{\sqrt{2}} \qquad (32.8)$$

root-mean-square voltage

Note that since $\mathcal{E}^2 = \mathcal{E}^2_{max}\cos^2\omega t$, the average value of $\mathcal{E}^2$ is $\tfrac{1}{2}\mathcal{E}^2_{max}$ and, according to Eq. (32.8), this equals $\mathcal{E}^2_{rms}$. Hence, the square of $\mathcal{E}_{rms}$ is the average, or the mean, of the square of $\mathcal{E}$.

In engineering practice, an AC voltage is usually described in terms of $\mathcal{E}_{rms}$. For example, if $\mathcal{E}_{max} = 163$ V, then $\mathcal{E}_{rms} = 163/\sqrt{2}$ V $= 115$ V. Thus an oscillating voltage with this value of $\mathcal{E}_{max}$ is described as "115 volts AC." Ordinary voltmeters and ammeters are calibrated so as to display the rms voltage and the rms current when connected to an AC circuit. Thus, a voltmeter plugged into a wall outlet will read 115 V.

Comparison of Eqs. (32.7) and (28.24) shows that the average AC power dissipated in the resistor is equal to the DC power dissipated in the same resistor when connected to a steady DC voltage of magnitude $\mathcal{E}_{rms}$. Thus, a voltage of 115 volts AC (*with* $\mathcal{E}_{max} = 163$ V) *delivers the same average power to the resistor as* 115 *volts DC.* This means that in any calculations involving the average electric power dissipated by resistors in such an AC circuit, we can pretend that we are dealing with a 115-volts DC circuit.

Concepts
— *in* —
Context

> **EXAMPLE 1**   A 115-V AC incandescent lightbulb is rated at 150 W. What is the resistance of this lightbulb (when at its operating temperature)?
>
> **SOLUTION:** We have $\mathcal{E}_{rms} = 115$ V and $\overline{P} = 150$ W. Hence Eq. (32.7) gives us
>
> $$R = \frac{\mathcal{E}_{rms}^2}{\overline{P}} = \frac{(115\text{ V})^2}{150\text{ W}} = 88\ \Omega \qquad (32.9)$$

When several appliances are plugged into the same outlet in a house, or into different outlets that are part of the same circuit in the house, all the appliances will be connected in parallel across the supplied emf of 115 V AC. In ordinary wall outlets for three-prong plugs, such as illustrated in Fig. 32.6, the small slot is at a potential of 115 V AC, the larger slot is at a potential of zero, and the round slot is also at a potential of zero. The two flat slots are connected to the two wires coming from the power station, and when an appliance is plugged into the outlet, the current flows into the appliance through one slot and out through the other. The round slot is connected to the ground, usually via a grounding plate or rod buried just outside the house. Under normal circumstances, this slot carries no current. It comes into play only when there is a failure in the electrical insulation of the appliance; then the ground slot permits the leaking currents to flow into the ground (instead of flowing through your body).

Note that the potentials indicated in Fig. 32.6 are open-circuit potentials, that is, they are the potentials before an appliance is plugged into the outlet. When an appliance drawing a heavy current—such as an electric heater or an air conditioner—is plugged into the outlet, the potential at the small flat slot may drop by several volts, and the potential at the large flat slot will rise by an equal amount. These changes in potential are due to the resistance of the wires connecting the outlets to the power station. The wires and the appliance form a series circuit, and the total potential drop of 115 V provided by the power station is distributed among these wires and the appliance in direct proportion to their resistances, as required by Ohm's Law.

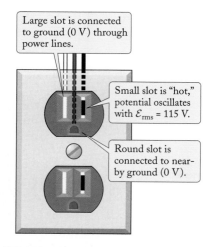

Large slot is connected to ground (0 V) through power lines.

Small slot is "hot," potential oscillates with $\mathcal{E}_{rms} = 115$ V.

Round slot is connected to nearby ground (0 V).

FIGURE 32.6 Standard power outlet for three-prong plug.

 Checkup 32.1

QUESTION 1: In Europe, outlets deliver electric power at 230 V AC. What is $\mathcal{E}_{max}$?

QUESTION 2: A lightbulb is connected to an AC outlet operating at 60 Hz. How many minima and how many maxima per second are there in the electric current? In the electric power?

QUESTION 3: Suppose you connect a lightbulb, by means of two wires, to the large slot and the round slot of the ordinary wall outlet illustrated in Fig. 32.6. Will the lightbulb shine?

QUESTION 4: During a "brownout," the oscillating emf supplied by power companies can drop significantly; suppose that the amplitude is $\mathcal{E}_{max} = 141$ V. What average power does this emf supply to a 100-Ω resistor?

(A) 100 W   (B) 141 W   (C) 200 W   (D) 282 W

32.2 CAPACITOR CIRCUIT

A capacitor connected to an alternating source of emf behaves very differently from a capacitor connected to a constant source of emf, such as a battery. When we connect the capacitor to the battery, there is an initial, brief surge of current that charges the capacitor. The current deposits positive charge on one plate of the capacitor, and negative charge on the other, and thereby increases the voltage across the capacitor. But once the voltage across the charged capacitor matches the emf, no further current flows, and the capacitor thereafter remains in equilibrium.

In contrast, when we connect a capacitor to an alternating source of emf, the current begins to deposit positive charges on one plate and negative charges on the other. But half a cycle later, the emf reverses, and so does the current. Thus, the current now removes the charges it deposited earlier, and begins to deposit reversed charges on the plates. This reversal of the current and of the charges repeats each half cycle. The current flows back and forth along the wires from the emf to the plates, and the signs of the charges on the plates alternate with the same frequency as the emf. Note that the capacitor permits this periodic back-and-forth flow of the current on the wires connected to its plates, even though no current can cross the gap between the plates.

Figure 32.7 shows the circuit diagram for a capacitor connected to our oscillating source of emf. When the instantaneous charge of the capacitor is Q, the instantaneous voltage across the capacitor is Q/C, and therefore Kirchhoff's rule gives

$$\mathcal{E} - \frac{Q}{C} = 0 \qquad (32.10)$$

With the expression (32.1) for our time-dependent $\mathcal{E}$, this yields

$$Q = C\mathcal{E} = C\mathcal{E}_{max}\cos\omega t \qquad (32.11)$$

Note that the charges on the capacitor plates oscillate between a positive value $C\mathcal{E}_{max}$ and a negative value $-C\mathcal{E}_{max}$; that is, the charges on the plates reverse every half cycle, in phase with the oscillating source of emf. The current in the circuit is the rate of change of the charge, $I = dQ/dt$. This means we must evaluate the rate of change of $\cos\omega t$:

$$\frac{d}{dt}\cos\omega t = -\omega\sin\omega t$$

Hence

$$I = \frac{dQ}{dt} = -\omega C\mathcal{E}_{max}\sin\omega t \qquad (32.12)$$

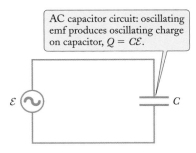

AC capacitor circuit: oscillating emf produces oscillating charge on capacitor, $Q = C\mathcal{E}$.

FIGURE 32.7 Capacitor connected to a source of alternating emf.

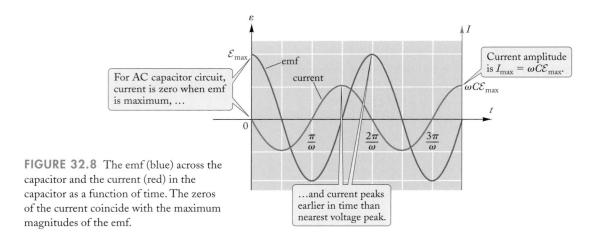

FIGURE 32.8 The emf (blue) across the capacitor and the current (red) in the capacitor as a function of time. The zeros of the current coincide with the maximum magnitudes of the emf.

According to this equation *the current oscillates sinusoidally with a frequency ω and with an amplitude $\omega C \mathcal{E}_{max}$*. Figure 32.8 shows plots of the emf and the current. Equation (32.12) tells us that the current has minimum magnitude (0) when the emf has maximum magnitude ($\mathcal{E}_{max}$) and that the current has a maximum magnitude ($\omega C \mathcal{E}_{max}$) when the emf has a minimum magnitude (0). Because of this, the current and the emf are said to be 90° **out of phase**. Because the current in a capacitive circuit peaks earlier than the time of the nearest voltage peak, we say that the *current leads the voltage* by 90°. The mnemonic *ICE* (ice) makes this easy to remember: in a capacitive circuit (C), the current I peaks before the voltage $\mathcal{E}$ (see Fig. 32.9). It is customary to write Eq. (32.12) as

current into capacitor

$$ I = -\frac{\mathcal{E}_{max}\sin\omega t}{X_C} \tag{32.13} $$

where

capacitive reactance

$$ X_C = \frac{1}{\omega C} \tag{32.14} $$

is called the **capacitive reactance**. The quantity X_C plays roughly the same role for a capacitor in an AC circuit as does the resistance R for a resistor [compare Eqs. (32.3) and (32.13)]. Note, however, that the reactance depends not only on the capacitance of the capacitor, but also on the frequency at which we are operating the circuit, whereas the resistance of a resistor does not depend on the frequency. The reactance is large if the frequency is low, and vice versa. The unit of reactance is the ohm, as it is for resistance.

From Eq. (32.12) or (32.13) we see that the current is small if the frequency ω is low. This was to be expected, since it means that the response of the capacitor to a low-frequency emf is nearly the same as its response to a constant emf (DC). As we saw in the discussion at the beginning of this section, for a constant emf, the capacitor blocks the flow of current.

The instantaneous power delivered to the capacitor is, again, the product of the instantaneous current, Eq. (32.12), and the instantaneous emf, Eq. (32.1):

$$ P = \mathcal{E}I = -\omega C \mathcal{E}_{max}^2 \cos\omega t \sin\omega t \tag{32.15} $$

FIGURE 32.9 The mnemonic "*ICE*" for an oscillating capacitor circuit.

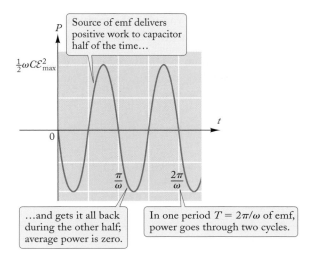

Source of emf delivers positive work to capacitor half of the time…

…and gets it all back during the other half; average power is zero.

In one period $T = 2\pi/\omega$ of emf, power goes through two cycles.

FIGURE 32.10 Instantaneous power delivered to the capacitor as a function of time.

The time dependence of this expression is contained in the factor $\cos \omega t \sin \omega t$. According to a standard trigonometric identity (see Appendix 3), this equals $\frac{1}{2} \sin 2\omega t$, which shows that the power oscillates at a frequency of 2ω. But the important point is that the average power delivered is zero—as Fig. 32.10 shows, within one cycle, there is as much positive power as negative power. The source of emf does work on the capacitor during parts of the cycle, but the capacitor does work on the source during other parts of the cycle, so **the average power is zero**. The ideal capacitor does not consume power, because it has no means of dissipating electric energy—it can only store charge and return it.

PHYSICS IN PRACTICE   FREQUENCY FILTER CIRCUITS

The selective blocking of low-frequency currents by a capacitor is exploited in stereo speaker systems. In order to ensure that sound waves diffract and spread to fill a room, stereo systems route high-frequency signals into small loudspeakers ("tweeters"); meanwhile, to attain adequate amplitudes, low-frequency signals are routed into large loudspeakers ("woofers"). To achieve this, a capacitor is connected in series with the tweeter (see Fig. 1). The capacitor has a large effective resistance (a large reactance) when the frequency is low. Thus, this capacitor acts as a filter that blocks low-frequency currents but permits the passage of high-frequency currents. As we will see in the next section, an inductor has the opposite effect of a capacitor. The inductor has a large effective resistance if the frequency is high. In the stereo speaker system, an inductor is connected in series with the woofer (see Fig. 2); this inductor acts as a filter that blocks high-frequency currents but permits the passage of low-frequency currents. Combinations of capacitors and inductors are used in many other kinds of electronic circuits that filter frequencies.

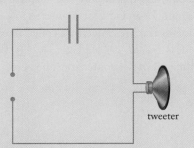

tweeter

FIGURE 1 Capacitor connected in series with tweeter.

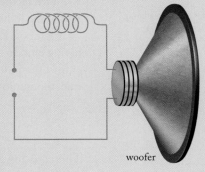

woofer

FIGURE 2 Inductor connected in series with woofer.

> | **EXAMPLE 2** | Suppose you plug the terminals of a 20-pF capacitor into a 115-V outlet. What is the maximum instantaneous current?

SOLUTION: The maximum instantaneous emf is $\mathcal{E}_{max} = 115\ \text{V} \times \sqrt{2} = 163\ \text{V}$. From Eq. (32.13) we see that the maximum instantaneous current occurs when $\sin \omega t = -1$, and it has the value

$$I_{max} = \frac{\mathcal{E}_{max}}{X_C} = \omega C \mathcal{E}_{max}$$

$$= (2\pi \times 60\ \text{Hz}) \times (20 \times 10^{-12}\ \text{F}) \times 163\ \text{V} = 1.2 \times 10^{-6}\ \text{A}$$

✔ Checkup 32.2

QUESTION 1: Does a capacitor connected to AC obey Ohm's Law?

QUESTION 2: For a capacitor with the emf and the current plotted in Fig. 32.8, make a plot of the electric energy in the capacitor as a function of time. At $t = 0$, is the energy maximum or minimum?

QUESTION 3: A capacitor has a capacitive reactance of $10^9\ \Omega$ at 60 Hz. What is the reactance at 600 Hz? 6000 Hz?

QUESTION 4: According to Eq. (32.12), the current in the capacitor circuit becomes very small if the frequency is low. Does this make sense?

QUESTION 5: Suppose that a capacitor is at first plugged into a 60-Hz, 115-V outlet. If you move the capacitor to a 60-Hz, 230-V outlet, by what factor is the new current amplitude related to the old one? By what factor is the new maximum power related to the old?

(A) 1, 4   (B) 1, 2   (C) 2, 4   (D) 2, 2   (E) 2, 1

Online Concept Tutorial 35

32.3 INDUCTOR CIRCUIT

An ideal inductor is a resistanceless coil of wire. We might expect that when such a coil is connected across a source of emf, a very large current will flow, since there is no resistance to oppose the current. However, for an alternating emf, the current in the inductor is limited by the self-inductance. As we know from Section 31.4, any increase of the current in the inductor generates a back emf, and this opposes the increase in the current.

Figure 32.11 shows the circuit diagram for an inductor connected to a source of alternating emf. In this diagram, the inductor is represented by a coiled line. The induced emf in the inductor, or the back emf, is $-L\ dI/dt$ [see Eq. (31.21)]. By Kirchhoff's rule, the sum of this back emf and the emf of the source must be zero:

$$\mathcal{E} - L\frac{dI}{dt} = 0 \tag{32.16}$$

This equation gives us

$$\frac{dI}{dt} = \frac{\mathcal{E}}{L} = \frac{\mathcal{E}_{max} \cos \omega t}{L} \tag{32.17}$$

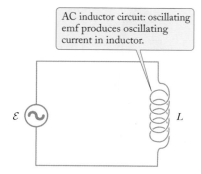

FIGURE 32.11 Inductor connected to a source of alternating emf.

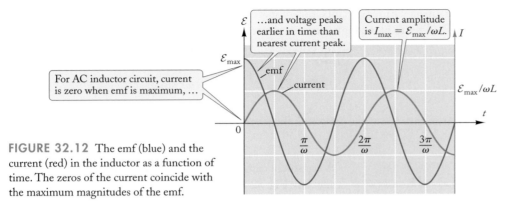

FIGURE 32.12 The emf (blue) and the current (red) in the inductor as a function of time. The zeros of the current coincide with the maximum magnitudes of the emf.

This tells us the rate of change of the current. To discover the current, we note that $\omega \cos \omega t$ is the rate of change of $\sin \omega t$:

$$\frac{d}{dt} \sin \omega t = \omega \cos \omega t$$

Thus, the quantity $\cos \omega t$ appearing in Eq. (32.17) is the rate of change of $(1/\omega) \sin \omega t$. But if the rates of change of two quantities are equal, then the two quantities can differ only by a constant. A constant current is consistent with Eq. (32.16), but it would not persist if there were some (small) resistance in the circuit. Ignoring any constant current, we find that the oscillating current must be

$$I = \frac{\mathcal{E}_{max} \sin \omega t}{\omega L} \tag{32.18}$$

This current oscillates sinusoidally with a frequency ω and an amplitude $\mathcal{E}_{max}/\omega L$. Figure 32.12 gives plots of the current and the emf. Again, comparison of these plots shows that the current is 90° out of phase with the emf—the current has minimum magnitude when the emf has maximum magnitude. Because the voltage in an inductive circuit peaks earlier than the time of the nearest current peak, we say that the *voltage leads the current* by 90°. The mnemonic $\mathcal{E}LI$ (Eli) makes this easy to remember: in an inductive circuit (L), the voltage $\mathcal{E}$ peaks before the current I (Fig. 32.13).

We can write Eq. (32.18) as

$$I = \frac{\mathcal{E}_{max} \sin \omega t}{X_L} \tag{32.19}$$

where

$$X_L = \omega L \tag{32.20}$$

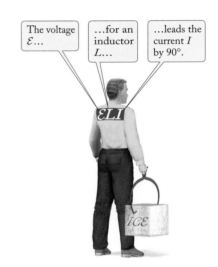

FIGURE 32.13 The mnemonic "$\mathcal{E}LI$ the $IC\mathcal{E}$ man" for oscillating inductor and capacitor circuits.

current in inductor

inductive reactance

is the **inductive reactance**. The inductive reactance depends on both the inductance and the frequency—the reactance is large if the frequency is high, and the reactance is small if the frequency is low. The unit of this reactance is, again, the ohm.

Note that according to Eq. (32.18) or (32.19), the amplitude of the current is small if the frequency is high, and the amplitude of the current is large if the frequency ω is low. This merely means that the response of the inductor to a low-frequency emf is close to its response to a steady emf—for a steady emf (DC), the inductor permits a very large current, since it has no resistance. As already mentioned in Physics in Practice:

Frequency Filter Circuits, this property of an inductor can be exploited in the design of a filter to block high-frequency currents.

The instantaneous power delivered to the inductor is

$$P = \mathcal{E}I = \frac{1}{\omega L}\mathcal{E}_{max}^2 \cos\omega t \, \sin\omega t \qquad (32.21)$$

As in the case of the capacitor, the average power is zero.

EXAMPLE 3   A resistor, a capacitor, and an inductor are connected in parallel (see Fig. 32.14) to a source of oscillating emf of frequency $\omega = 6.0 \times 10^3$ radians/s and amplitude 1.0×10^{-3} V. The resistance of the resistor is 200 Ω. (a) What is the maximum instantaneous current in the resistor? (b) If we want to make the maximum instantaneous currents in the capacitor and in the inductor equal to that in the resistor, what values of the capacitance and of the inductance must we select?

SOLUTION: (a) In a parallel connection, the source of emf supplies the same voltage across each of the circuit elements. Thus, the maximum instantaneous voltage across the resistor is 1.0×10^{-3} V, and the maximum instantaneous current in the resistor is, by Eq. (32.3),

$$I_{max} = \frac{\mathcal{E}_{max}}{R} = \frac{1.0 \times 10^{-3}\text{ V}}{200\ \Omega} = 5.0 \times 10^{-6}\text{ A}$$

(b) According to Eqs. (32.13) and (32.19), if the maximum instantaneous currents in the capacitor and the inductor are to match that in the resistor, their reactances must match the resistance:

$$X_C = X_L = 200\ \Omega$$

Thus,

$$\frac{1}{\omega C} = 200\ \Omega \qquad \text{and} \qquad \omega L = 200\ \Omega$$

which gives

$$C = \frac{1}{6.0 \times 10^3\text{ s}^{-1} \times 200\ \Omega} = 8.3 \times 10^{-7}\text{ F}$$

and

$$L = \frac{200\ \Omega}{6.0 \times 10^3\text{ s}^{-1}} = 3.3 \times 10^{-2}\text{ H}$$

COMMENTS: Note that with this choice of C and L, the maximum currents are equal at the frequency of 6.0×10^3 radians/s only. At any other frequency, this choice of C and L will not result in equal maximum currents. Also note that although the currents are equal in amplitude when $X_L = X_C$, they are 180° out of phase (the current in the capacitor leads the emf by 90° and the current in the inductor lags the emf by 90°), so the net instantaneous current in those two branches is zero and the net current is that in the resistor. Related behavior is discussed in more detail in Section 32.5.

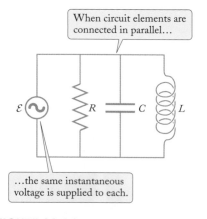

When circuit elements are connected in parallel...

...the same instantaneous voltage is supplied to each.

FIGURE 32.14 Resistor, capacitor, and inductor connected in parallel to a source of oscillating emf.

✔ Checkup 32.3

QUESTION 1: Does an inductor connected to AC obey Ohm's Law?

QUESTION 2: For an inductor with the emf and the current plotted in Fig. 32.12, make a plot of the magnetic energy in the inductor as a function of time. At $t = 0$ is the energy maximum or minimum?

QUESTION 3: An inductor has an inductive reactance of $10^3 \, \Omega$ at 60 Hz. What is the reactance at 600 Hz? 6000 Hz?

QUESTION 4: According to Eq. (32.18), the current in the inductor circuit becomes very large if the frequency is low. Does this make sense?

QUESTION 5: An inductor is connected to a 60-Hz, 115-V emf. Suppose that a piece of iron is inserted into the inductor so that the value of the inductance increases by a factor of 10. By what factor is the new current amplitude related to the old one? By what factor is the new inductive reactance related to the old?

(A) 10, 10   (B) 10, 100   (C) $\frac{1}{10}$, 1   (D) $\frac{1}{10}$, 10   (E) $\frac{1}{10}$, 100

32.4 FREELY OSCILLATING *LC* AND *RLC* CIRCUITS

In this section, we will examine the behavior of circuits when *no external source of alternating emf is present*. We have already examined such so-called free response or natural response for *RC* and *RL* circuits (Sections 28.7 and 31.6, respectively); we found that a smooth exponential function, with a characteristic time $\tau = RC$ or L/R, described the response for given initial conditions. In this section, we will consider *LC* and *RLC* circuits; we will find that such circuits have a natural tendency to oscillate.

An *LC* circuit consists of an ideal inductor and an ideal capacitor connected in series (see Fig. 32.15). The circuit has *no source of emf*; nevertheless, a current will flow in this circuit, provided that the capacitor is *initially charged*. The charge on one plate is then initially positive and that on the other plate negative. A current will begin to flow around the circuit from the positive plate to the negative. If the circuit had no inductance, the current would merely neutralize the charges on the plates, that is, the capacitor would discharge, and this would be the end of the current. But the inductor makes a difference: the inductor initially opposes the buildup of the current, but once the current has become established, the inductor will keep it going for some extra time. Hence *more* charge flows from one capacitor plate to the other than required for neutrality, and reversed charges accumulate on the capacitor plates. When the current finally stops, the capacitor will again be fully charged, with reversed charges. And then a reversed current will begin to flow, and so on. Thus, the charge sloshes back and forth around the circuit; because this sloshing is like water in a tank, an *LC* circuit is sometimes called a "tank circuit." If there is no resistance, these oscillations of the *LC* circuit continue forever.

The *LC* circuit is analogous to the mass–spring system of Section 15.2. *The inductor is analogous to the mass—it tends to keep the current constant and provides "inertia." The charged capacitor is analogous to the stretched spring—it tends to accelerate the charge and provides a "restoring force."*

The equation of motion of the *LC* circuit follows from Kirchhoff's rule: the sum of emfs and other voltage changes around the circuit must add to zero. Going around

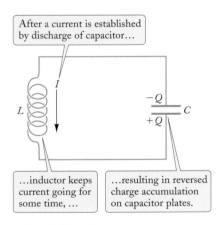

After a current is established by discharge of capacitor…

…inductor keeps current going for some time, …

…resulting in reversed charge accumulation on capacitor plates.

FIGURE 32.15 Inductor and capacitor connected in series.

the circuit in the direction of the arrow shown in Fig. 32.15, we find that the induced emf in the inductor (back emf) is

$$-L\frac{dI}{dt}$$

and the voltage across the capacitor is

$$-Q/C$$

Hence

$$-L\frac{dI}{dt} - \frac{Q}{C} = 0 \tag{32.22}$$

or, equivalently,

$$L\frac{dI}{dt} + \frac{Q}{C} = 0 \tag{32.23}$$

Note that here Q is reckoned as positive when the charge on the lower plate is positive, and I is reckoned as positive when the charge on the lower plate is increasing.

Equation (32.23) has exactly the same mathematical form as the equation of motion for the simple harmonic oscillator [see Eq. (15.18)],

$$m\frac{dv}{dt} + kx = 0$$

Comparing this with Eq. (32.23), we see that Q plays the role of x, whereas L replaces m and $1/C$ replaces k. The current ($I = dQ/dt$) plays the role of the velocity ($v = dx/dt$). Hence the solution of Eq. (32.23) can be found by recalling the solution for the simple harmonic oscillator [see Eq. (15.22)],

$$x = A\cos\left(\sqrt{\frac{k}{m}}\,t\right)$$

With the above replacements for x, m, and k, we immediately obtain

$$Q = Q_{max}\cos\left(\frac{1}{\sqrt{LC}}\,t\right) \tag{32.24}$$

where Q_{max} is the amount of charge on the positive plate at time $t = 0$. Likewise, from the equation for the velocity of the simple harmonic oscillator,

$$v = \frac{dx}{dt} = -\sqrt{\frac{k}{m}}\,A\sin\left(\sqrt{\frac{k}{m}}\,t\right)$$

we find that the current in the LC circuit must be

$$I = \frac{dQ}{dt} = -\frac{Q_{max}}{\sqrt{LC}}\sin\left(\frac{1}{\sqrt{LC}}\,t\right) \tag{32.25}$$

According to Eqs. (32.24) and (32.25), the charge and the current oscillate with a natural frequency, called the **resonant frequency** ω_0:

$$\omega_0 = \frac{1}{\sqrt{LC}} \tag{32.26}$$

Equation (32.25) also indicates that the current oscillates with amplitude

$$I_{max} = \frac{Q_{max}}{\sqrt{LC}} = \omega_0 Q_{max}$$

Figure 32.16 is a plot of the charge and the current in an *LC* circuit oscillating according to Eqs. (32.24) and (32.25).

The total stored enery in the *LC* circuit is the sum of the electric and magnetic energies [see Eqs. (26.27) and (31.26)]:

$$U = \frac{1}{2C}Q^2 + \frac{1}{2}LI^2 \tag{32.27}$$

Substituting Q from Eq. (32.24) and I from Eq. (32.25), we can see that the total energy remains constant, equal to the initial stored energy on the capacitor, although the total energy oscillates back and forth between electric and magnetic energy:

$$U = \frac{1}{2C} Q^2_{max} \cos^2 \omega_0 t + \frac{1}{2} L \frac{Q^2_{max}}{LC} \sin^2 \omega_0 t$$

$$= \frac{1}{2C} Q^2_{max} (\cos^2 \omega_0 t + \sin^2 \omega_0 t)$$

$$= \frac{1}{2C} Q^2_{max} \tag{32.28}$$

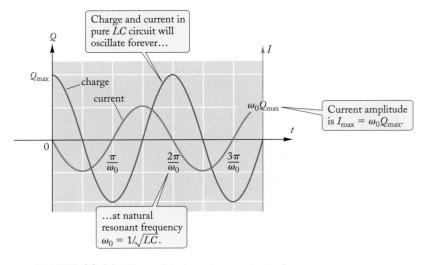

FIGURE 32.16 Charge (blue) on the capacitor and current (red) in the inductor as a function of time.

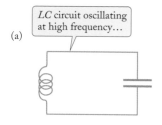

(a)

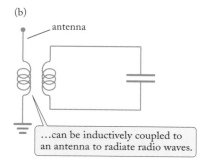

(b)

FIGURE 32.17 (a) An *LC* circuit in a radio. (b) The *LC* circuit is coupled to the antenna by the mutual inductance of the two inductors.

EXAMPLE 4   A primitive radio transmitter, like those used in the early days of "wireless telegraphy," consists of an *LC* circuit oscillating at high frequency (Fig. 32.17a). The circuit is inductively coupled to an antenna (Fig. 32.17b), so the oscillating current in the circuit induces an oscillating current on the antenna; the latter current then radiates radio waves (see Chapter 33). Suppose that the inductance in the circuit in Fig. 32.17a is 20 μH. What capacitance do we need if we want to produce oscillations of a frequency of 1.5×10^6 Hz?

SOLUTION: The angular frequency is $\omega_0 = 2\pi f = 2\pi \times 1.5 \times 10^6$ s^{-1}. Hence, from Eq. (32.26),

$$C = \frac{1}{\omega_0^2 L} = \frac{1}{(2\pi \times 1.5 \times 10^6 \text{ s}^{-1})^2 \times 20 \times 10^{-6} \text{ H}}$$

$$= 5.6 \times 10^{-10} \text{ F} = 560 \text{ pF}$$

A radio receiver employs much the same circuit as shown in Fig. 32.17b to pick up radio signals reaching the antenna. When a radio wave—consisting of oscillating electric and magnetic fields—reaches the antenna, it causes an oscillating current to flow along the antenna, which induces a current in the *LC* circuit. The current in the circuit will build up to a relatively large value if the frequency of the driving force supplied by the antenna matches the natural frequency of the circuit. To attain this resonance condition, the natural frequency of the circuit must be tuned to the frequency of the radio wave, which is done by adjusting the value of the capacitance. The capacitor in the radio circuit is a variable capacitor, whose capacitance can be controlled with a tuning knob (see Fig. 32.18). Turning this knob makes one of the sets of plates of the capacitor move parallel to the other, thereby changing the amount of overlap between the plates and the effective area of the capacitor.

If we add a resistor to an *LC* circuit (again without any external alternating emf), we obtain the *RLC* circuit shown in Fig. 32.19. We can readily analyze this circuit using Kirchhoff's voltage rule; we simply have to add the *IR* voltage drop across the resistor to our previous result (31.23):

$$L\frac{dI}{dt} + IR + \frac{Q}{C} = 0 \tag{32.29}$$

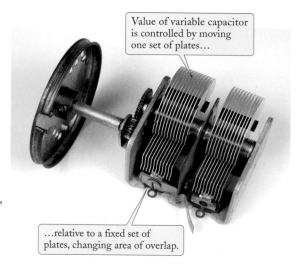

FIGURE 32.18 Adjustable capacitor for a radio. The capacitor consists of two sets of parallel, semicircular plates. The tuning knob controls the amount of overlap of the plates.

FIGURE 32.19 Resistor, inductor, and capacitor connected in series.

With $I = dQ/dt$, this equation describes a **damped harmonic oscillator**. In such a system, the charge and current tend to oscillate as they do in the freely oscillating *LC* circuit, but the amplitude of the oscillations decreases with time, in a manner equivalent to the mechanical oscillator with friction discussed in Section 15.5. For the *RLC* circuit, the decrease is due to electrical "friction": since the current flows through the resistor, the resulting Joule heating represents some energy loss, or damping, during each cycle of oscillation. If the resistance is small, these losses are small, and the oscillation amplitude decreases slowly, as shown in Fig. 32.20a; the circuit is said to be *underdamped*. If the resistance is too large, the current drops to zero before any oscillations can occur, as in Fig. 32.20b; the circuit is said to be *overdamped* (see Fig. 32.20b). Equation (32.29) can be solved by standard techniques for differential equations, but instead of attempting this here, we will only examine the behavior of the solution. The solution to Eq. (32.29) is an oscillating function multiplied by a decaying exponential function of time:

$$Q = Q_{max} e^{-Rt/2L} \cos \omega_d t \tag{32.30}$$

where the frequency of damped oscillations ω_d is somewhat smaller than the natural, undamped frequency ω_0, and is given by $\omega_d = \sqrt{\omega_0^2 - R^2/4L^2}$. For sufficiently small values of the resistance R (underdamped case), the oscillation frequency ω_d is almost equal to the natural frequency, $\omega_d \approx \omega_0$. This is the case plotted in Fig. 32.20a.

As mentioned in Section 15.5, the **quality factor**, or $\mathcal{Q}$, is a measure of how freely a system oscillates, and is defined by

$$\mathcal{Q} = -2\pi \frac{U}{\Delta U} \tag{32.31}$$

where U is the stored energy and ΔU is the energy change per cycle. Initially, the stored electrical energy is $Q_{max}^2/2C$. By Eq. (32.30), the electrical energy after a time t has amplitude

$$U = \frac{Q^2}{2C} = \frac{Q_{max}^2}{2C}(e^{-Rt/2L})^2 = \frac{Q_{max}^2}{2C}e^{-Rt/L} \tag{32.32}$$

For a small change in time Δt, the energy change is the product of the time derivative of U and Δt,

$$\Delta U = \frac{dU}{dt}\Delta t = \frac{Q_{max}^2}{2C}e^{-Rt/L} \times \left(-\frac{R}{L}\right) \times \Delta t = U \times \left(-\frac{R}{L}\right) \times \Delta t$$

Using this expression for ΔU and substituting $\Delta t = T \approx 2\pi/\omega_0$ for one cycle, we find that the quality factor (32.31) is

$$\mathcal{Q} = -2\pi \frac{U}{U \times (-R/L) \times (2\pi/\omega_0)}$$

or simply

$$\mathcal{Q} = \frac{\omega_0 L}{R} \tag{32.33}$$

Thus a small resistance results in a large $\mathcal{Q}$.

damped harmonic oscillator

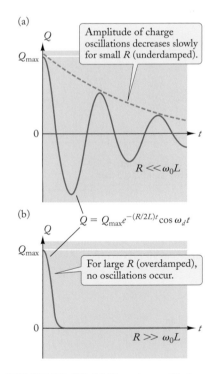

(a)

Q

Q_{max}

Amplitude of charge oscillations decreases slowly for small R (underdamped).

0

t

$R \ll \omega_0 L$

(b)

Q

$Q = Q_{max}e^{-(R/2L)t}\cos\omega_d t$

Q_{max}

For large R (overdamped), no oscillations occur.

0

t

$R \gg \omega_0 L$

FIGURE 32.20 (a) Decaying oscillation of the charge on the capacitor in an underdamped series *RLC* circuit. (b) Decay of charge on the capacitor in an overdamped *RLC* circuit.

quality factor $\mathcal{Q}$

From Eq. (32.30), the amplitude of the charge oscillations falls to $1/e$ of its initial value when $t = 2L/R$. If $T \approx 2\pi/\omega_0$ is the period of oscillation, the number of oscillations in the time t is

$$N = \frac{t}{T} = \frac{2L/R}{2\pi/\omega_0} = \frac{Q}{\pi} \tag{32.34}$$

Thus, as already mentioned in Section 15.5, the quality factor is a measure of how many oscillations occur before the amplitude of the system decreases appreciably. In simple electrical circuits, a Q of about 100 is often typical; using special techniques (for example, superconducting components), Q's on the order of 10^9 have been achieved.

✔ Checkup 32.4

QUESTION 1: Consider the LC circuit illustrated in Fig. 32.15. If you increase the separation between the plates, what happens to the natural frequency?

QUESTION 2: The circuit illustrated in Fig. 32.17a has a capacitor of 560 pF and a natural frequency of 1.5×10^6 Hz. If we replace this capacitor by two capacitors of 560 pF each connected in parallel, by what factor will the natural frequency change?

QUESTION 3: In an RLC circuit oscillating at resonance, the inductive reactance is 500 Ω and the resistance is 10 Ω. What is the Q of the circuit? After approximately how many oscillations will the amplitude of the charge oscillations fall to $1/e$ of its initial value?

QUESTION 4: In the circuit of Fig. 32.21, the switch has been in position 1 for a long time and the capacitor is initially uncharged. At $t = 0$, the switch is moved to position 2. What is Q_{max}, the maximum charge on the capacitor? What function describes the time dependence of the charge on the capacitor, Q/Q_{max}?

(A) $\dfrac{\mathcal{E}\sqrt{LC}}{R}$, $\cos\omega t$ (B) $\dfrac{\mathcal{E}\sqrt{LC}}{R}$, $\sin\omega t$

(C) $\dfrac{\mathcal{E}}{R\sqrt{LC}}$, $\cos\omega t$ (D) $\dfrac{\mathcal{E}}{R\sqrt{LC}}$, $\sin\omega t$

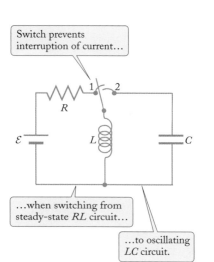

Switch prevents interruption of current…

…when switching from steady-state RL circuit…

…to oscillating LC circuit.

FIGURE 32.21 An RL circuit that can be switched to an LC circuit.

Online Concept Tutorial

forced oscillations

32.5 SERIES CIRCUITS WITH ALTERNATING EMF

We now consider oscillations in a circuit with more than one R, L, or C component connected in series to an external source of alternating emf, for instance, the RL circuit shown in Fig. 32.22. Such oscillations produced by an applied emf are called driven oscillations or **forced oscillations**. In contrast to the free oscillations, which proceed at the natural resonant frequency [see Eq. (32.26)], the forced oscillations proceed at the frequency of the applied emf. Recall that in a series circuit, the current is the same everywhere. The sum of the instantaneous voltages across the components, by Kirchhoff's voltage rule, must equal the applied emf $\mathcal{E}$:

$$\mathcal{E} = V_R + V_L \tag{32.35}$$

where V_R and V_L are the instantaneous voltages across the resistor and inductor, respectively. But as we saw above, the voltages across the resistor and inductor have different phases; the resistive voltage is in phase with the current, whereas the inductive voltage leads the current by 90°. Thus we need to know how to add oscillating voltages with different phases. For our circuit, the current may be assigned a reference phase of 0°:

$$I = I_{max} \cos \omega t \qquad (32.36)$$

so the voltages we need to add are

$$V_R = V_{max,R} \cos \omega t \qquad \text{and} \qquad V_L = V_{max,L} \cos(\omega t + 90°) \qquad (32.37)$$

These voltages can be summed using trigonometric identities, but that becomes cumbersome for more complicated circuits. A simpler way to sum sinusoidal functions relies on a geometrical construction known as a **phasor**, which is a vector rotating in two dimensions (the two dimensions are fictitious; they have nothing to do with the x and y dimensions of real space). To understand this method, first note from Fig. 32.23 than *an oscillating function, such as $V_R = V_{max,R} \cos \omega t$, can be represented as the x component of a rotating vector.* The rotating vector shown has length equal to the amplitude $V_{max,R}$ of the oscillating voltage across the resistor, and at any instant makes an angle ωt with respect to the x axis, where ω is the frequency of the applied emf. As is evident in Fig. 32.23, the x component is then indeed the instantaneous value V_R, and will vary from $+V_{max,R}$ to $-V_{max,R}$ as the vector rotates (as the oscillations occur).

Similarly, the oscillating voltage V_L can be represented by the x component of the phasor shown in Fig. 32.24; the only difference is that this vector makes an extra angle of 90° with respect to the x axis. Since V_R and V_L oscillate at the same frequency, they will always have the same relative angle of 90° between them, as was already evident from the functional form of Eq. (32.37).

In order to sum the two voltages, we need to sum the x components of the two voltage phasors shown in Fig. 32.24. This is easy to do if we first perform the vector sum of the two phasors, and *then* take the x component of the sum. Figure 32.25a shows both phasors; for simplicity, the phasors are shown at time $t = 0$. Figure 32.25b shows the vector sum, performed by putting the tail of one phasor on the tip of the

phasor

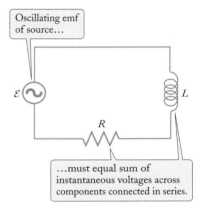

FIGURE 32.22 A series RL circuit with oscillating emf.

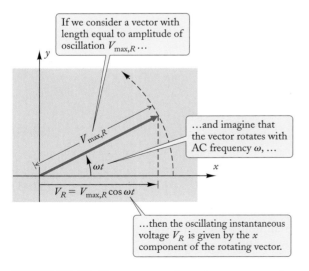

FIGURE 32.23 Phasor representing voltage across a resistor.

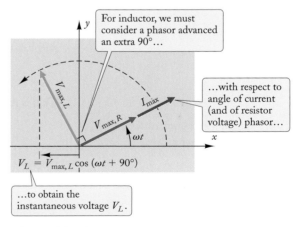

FIGURE 32.24 Three phasors representing series current, voltage across the inductor, and voltage across the resistor. Compared with the current and the resistor voltage, the phase of the inductor voltage leads by 90°.

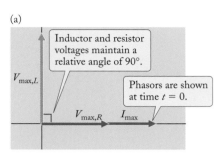

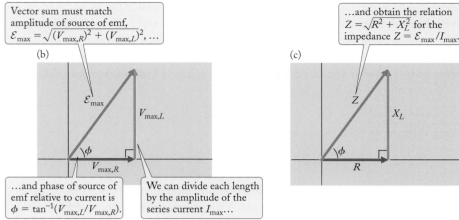

FIGURE 32.25 Addition of oscillating voltages using phasors. (a) Phasors for resistor (blue) and inductor (green) voltages. (b) The total voltage $\mathcal{E}$ can be represented by a phasor (red) that is the vector sum of the resistor and inductor phasors. (c) Impedance Z related to resistance R and inductive reactance X_L.

other. From the right triangle, the sum then has amplitude $\sqrt{(V_{max,R})^2 + (V_{max,L})^2}$, and this must match the amplitude of the applied emf:

$$\mathcal{E}_{max} = \sqrt{(V_{max,R})^2 + (V_{max,L})^2} \tag{32.38}$$

We also see that the applied emf has a phase (with respect to the current I and V_R) of

$$\phi = \tan^{-1}\left(\frac{V_{max,L}}{V_{max,R}}\right) \tag{32.39}$$

The instantaneous source voltage is the x component of the vector sum:

$$\mathcal{E} = \mathcal{E}_{max}\cos(\omega t + \phi) \tag{32.40}$$

In an RL circuit, the source voltage still leads the current, but by an angle less than 90°.

Recall from Sections 32.1–32.3 that the amplitudes of the current and voltage across each individual circuit component are related by

$$V_{max,R} = I_{max}R \quad \text{and} \quad V_{max,L} = I_{max}X_L \tag{32.41}$$

Thus Eq. (32.39) may be rewritten as

$$\phi = \tan^{-1}\left(\frac{X_L}{R}\right) \tag{32.42}$$

and Eq. (32.38) as

$$\mathcal{E}_{max} = \sqrt{(I_{max}R)^2 + (I_{max}X_L)^2} = I_{max}\sqrt{R^2 + X_L^2} \tag{32.43}$$

This is usually expressed as

amplitudes of emf and current related to impedance Z

$$\mathcal{E}_{max} = I_{max}Z \tag{32.44}$$

where

$$Z = \sqrt{R^2 + X_L^2} \tag{32.45}$$

is called the **impedance** of the RL circuit. The impedance measures the total opposition to the flow of alternating current from both the resistance R and the reactance X_L = ωL. The relation (32.45) can be easily reconstructed from Fig. 32.25b by dividing the magnitude of each phasor by the common current amplitude I_{max}. This gives the geometry of Fig. 32.25c.

EXAMPLE 5   A light dimmer is an application of a series *RL* circuit (see Fig. 32.26). Assume for simplicity that the resistance of the light-bulb is constant with, say, $R = 100 \; \Omega$. For a dimmer, the bulb is connected in series with a variable inductance L. The circuit is connected to an ordinary emf of 115 V AC oscillating at 60 Hz. What range of L is needed if it is desired to vary the average power from 30 W to 100 W?

SOLUTION: The current in the circuit is given by Eq. (32.44),

$$I_{max} = \frac{\mathcal{E}_{max}}{Z} = \frac{\mathcal{E}_{max}}{\sqrt{R^2 + (\omega L)^2}}$$

Thus the average power dissipated in the resistor is

$$\overline{P} = \frac{1}{2} I_{max}^2 R = \frac{1}{2} \frac{\mathcal{E}_{max}^2 R}{R^2 + (\omega L)^2}$$

Rearranging to solve for the inductance L, we find

$$L = \frac{1}{\omega} \sqrt{\frac{\mathcal{E}_{max}^2 R}{2\overline{P}} - R^2}$$

Inserting the values for $\omega = 2\pi f = 377 \text{ s}^{-1}, \mathcal{E}_{max} = \sqrt{2}\mathcal{E}_{rms} = 163$ V, $R = 100 \; \Omega$, and $\overline{P} = 100$ W, we obtain

$$L = \frac{1}{377 \text{ s}^{-1}} \sqrt{\frac{(163 \text{ V})^2 \times 100 \; \Omega}{2 \times 100 \text{ W}} - (100 \; \Omega)^2} = 0.15 \text{ H}$$

With $\overline{P} = 30$ W, the same calculation yields $L = 0.49$ H.

Thus a variable inductor with a range $L = 0.15 - 0.49$ H would limit the power delivered to the lightbulb to the desired range.

COMMENTS: A lightbulb could instead be dimmed with a variable resistor, but this would add unnecessary Joule heating. The advantage of using a reactive circuit component to reduce the current is that it does not waste energy; no power is dissipated in an ideal inductor or capacitor. Many common household dimmers work on a much different principle: a semiconductor device is used to turn off the applied emf for a variable fraction of each 60-Hz cycle; such devices typically emit a hefty amount of electromagnetic noise, due to the large induced emf's associated with the sudden changes in current.

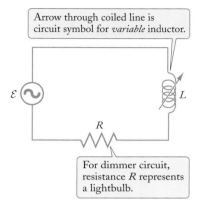

Arrow through coiled line is circuit symbol for *variable* inductor.

For dimmer circuit, resistance R represents a lightbulb.

FIGURE 32.26 An *RL* light dimmer circuit with a variable inductance.

An *RC* circuit can be analyzed in the same way as an *RL* circuit; the only difference is that the phase angle is then negative, since the capacitive voltage lags the current. The resulting relations are identical to those found above if we replace X_L with $-X_C$.

More interesting is the behavior of a series *RLC* circuit, as illustrated in Fig. 32.27. The appropriate phasors are shown in Fig. 32.28a; the vector addition is readily performed with the aid of Fig. 32.28b. Since the inductive and capacitive voltages differ in phase by 180°, the corresponding phasors directly oppose one another. The right triangle in Fig. 32.28b immediately provides the relations

$$\mathcal{E}_{max} = \sqrt{(V_{max,R})^2 + (V_{max,L} - V_{max,C})^2} \tag{32.46}$$

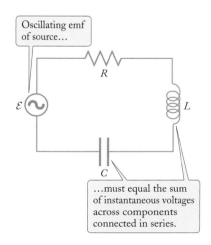

Oscillating emf of source...

...must equal the sum of instantaneous voltages across components connected in series.

FIGURE 32.27 A series *RLC* circuit with oscillating emf.

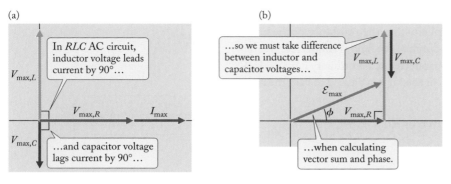
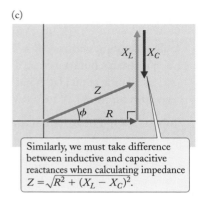

FIGURE 32.28 Addition of oscillating voltages using phasors. (a) Phasors representing resistive and reactive voltages. (b) Source voltage $\mathcal{E}$ is represented by a phasor that is the vector sum of the resistive and reactive voltages. (c) By dividing each voltage amplitude in (b) by the current amplitude I_{max}, the impedance Z is related to the resistance R and the reactances X_L and X_C.

and

$$\phi = \tan^{-1}\left(\frac{V_{max,L} - V_{max,C}}{V_{max,R}}\right) \tag{32.47}$$

As above, if we divide each voltage in Eqs. (32.46) and (32.47) by the current amplitude, we obtain the impedance $Z = \mathcal{E}_{max}/I_{max}$ of the circuit,

impedance Z of RLC circuit

$$Z = \sqrt{R^2 + (X_L - X_C)^2} \tag{32.48}$$

and the phase of the source of emf,

phase ϕ of RLC circuit

$$\phi = \tan^{-1}\left(\frac{X_L - X_C}{R}\right) \tag{32.49}$$

Notice that for the series RLC circuit, the phase will be positive if the inductive reactance dominates (at high frequency, since $X_L = \omega L$), and the phase will be negative if the capacitive reactance dominates [at low frequency, since $X_C = 1/(\omega C)$].

If we substitute $X_L = \omega L$ and $X_C = 1/(\omega C)$ into Eq. (32.48), we can examine the behavior of the impedance Z for different frequencies of the applied emf:

$$Z = \sqrt{R^2 + \left(\omega L - \frac{1}{\omega C}\right)^2} \tag{32.50}$$

For given values of R, L, and C, the impedance will thus vary with the frequency, and will have its minimum value when the term in parentheses is zero, that is, when the two reactances cancel. Thus the minimum value

$$Z_0 = R \tag{32.51}$$

occurs at one particular frequency ω_0, when

$$\omega_0 L - \frac{1}{\omega_0 C} = 0$$

or

$$\omega_0 = \frac{1}{\sqrt{LC}} \tag{32.52}$$

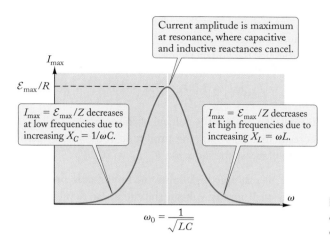

FIGURE 32.29 Current in a series RLC circuit as a function of the frequency of the oscillating emf.

This is called the *resonant frequency* and is the same as the natural resonant frequency of free oscillations of an LC circuit discussed in the previous section. Since the impedance is a minimum at resonance, the current amplitude

$$I_{max} = \frac{\mathcal{E}_{max}}{Z} = \frac{\mathcal{E}_{max}}{\sqrt{R^2 + [\omega L - (1/\omega C)]^2}} \tag{32.53}$$

current amplitude in *RLC* circuit

and the average power $\bar{P} = \frac{1}{2}I_{max}^2 R$ will both be maximum at resonance. Figure 32.29 is a plot of the current amplitude (32.53) as a function of frequency. Notice the characteristic peak of this resonance curve. The current is large at ω_0 but approaches zero for small and large frequencies. Such behavior is what permits the selection of a particular frequency by the tuning circuit of a radio, as discussed in the previous section.

EXAMPLE 6   A series RLC circuit has an applied AC voltage of amplitude $\mathcal{E}_{max} = 1.0$ V, which oscillates at a frequency equal to the resonant frequency of the circuit. If $R = 0.50\ \Omega$, $L = 20$ mH, and $C = 2.0\ \mu$F, what is the resonant frequency (in hertz)? What is the amplitude of the current? What is the amplitude of the voltage across the resistor? Across the inductor? Across the capacitor?

SOLUTION: The resonant angular frequency is given by Eq. (32.52):

$$\omega_0 = \frac{1}{\sqrt{LC}} = \frac{1}{\sqrt{20 \times 10^{-3}\ \text{H} \times 2.0 \times 10^{-6}\ \text{F}}} = 5.0 \times 10^3\ \text{radians/s}$$

The resonant frequency in hertz is

$$f_0 = \frac{\omega_0}{2\pi} = \frac{5.0 \times 10^3\ \text{radians/s}}{2\pi\ \text{radians/cycle}} = 800\ \text{cycles/s} = 800\ \text{Hz}$$

At resonance, $X_L = X_C$ and $Z = \sqrt{R^2 + (X_L - X_C)^2} = R$, so the amplitude of the current is

$$I_{max} = \frac{\mathcal{E}_{max}}{Z} = \frac{\mathcal{E}_{max}}{R} = \frac{1.0\ \text{V}}{0.50\ \Omega} = 2.0\ \text{A}$$

The amplitude of the voltage across the resistor is

$$V_{max,R} = I_{max}R = 2.0\ \text{A} \times 0.50\ \Omega = 1.0\ \text{V}$$

At resonance, this is the same as the voltage of the source of emf, which it must be, since $Z = R$ at resonance.

The amplitude of the voltage across the inductor is

$$V_{max,L} = I_{max}X_L = I_{max}\omega L = 2.0 \text{ A} \times 5.0 \times 10^3 \text{ s}^{-1} \times 20 \times 10^{-3} \text{ H} = 200 \text{ V}$$

This is 200 times larger than the voltage of the source!

Since at resonance $X_L = X_C$, the amplitude of the voltage across the capacitor is the same as that across the inductor, $V_{max,C} = 200$ V.

COMMENTS: This example highlights the remarkable properties of a resonant circuit: even for a small applied voltage, the circuit can develop large voltages across the reactive components, the inductor and capacitor. In reality, this does not happen instantly; after connecting the external emf, it takes a number of cycles for the amplitudes to build up to the values calculated above (the inverse process to the decaying oscillations of the previous section). Thereafter, the amplitudes maintain these steady values.

In Section 32.4, we discussed the quality factor, or Q, of a resonant circuit in terms of freely decaying oscillations. For the forced oscillations examined here, we can employ the same definition of the Q. The energy stored in the circuit is the magnetic energy when the current is a maximum, $\frac{1}{2}LI_{max}^2$. The energy dissipated per cycle is the average power times the time for one cycle, $\overline{P} \times T = (I_{max}^2 R/2) \times (2\pi/\omega_0)$. Thus the Q is

$$Q = -2\pi \frac{U}{\Delta U} = -2\pi \frac{\frac{1}{2}LI_{max}^2}{(-I_{max}^2 R/2) \times (2\pi/\omega_0)} = \frac{\omega_0 L}{R}$$

as obtained previously in Eq. (32.33). For forced oscillations, we can see that the Q also measures by what factor the reactive voltage amplitude at resonance is increased compared with the applied voltage:

$$Q = \frac{\omega_0 L}{R} = \frac{X_L}{R} = \frac{I_{max}X_L}{I_{max}R} = \frac{V_{max,L}}{V_{max,R}} \tag{32.54}$$

or, since $V_{max,R} = \mathcal{E}_{max}$ at resonance,

$$V_{max,L} = Q\mathcal{E}_{max}$$

Thus the resonant circuit multiplies the same voltage by a factor of Q.

The Q is also a measure of the sharpness of the resonance peak; it is common to plot the power dissipated as a function of frequency as in Fig. 32.30. If $\Delta\omega$ is the full width of the power peak at half the maximum value, it can be shown (Problem 54) that the above value of the Q is equivalent to

$$Q = \frac{\omega_0}{\Delta\omega} \tag{32.55}$$

Thus a circuit with a large quality factor has a narrow resonance curve (small $\Delta\omega$, see Fig. 32.30) and a circuit with a low quality factor has a broad resonance curve (see also Fig. 15.22).

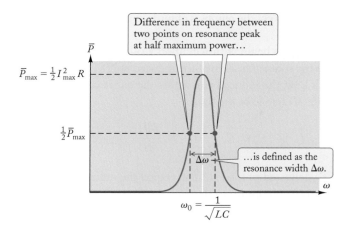

FIGURE 32.30 Average power in a series RLC circuit as a function of the frequency of the oscillating emf.

Difference in frequency between two points on resonance peak at half maximum power…

…is defined as the resonance width $\Delta\omega$.

Checkup 32.5

QUESTION 1: An AC circuit consists of a fixed-amplitude emf, a resistor, and one other component. When the frequency increases, the current increases. What is the other component?

QUESTION 2: An AC emf with amplitude 1.0 V is applied to a series RLC circuit. The frequency is varied and the current is observed to have a maximum value of 0.20 A when the inductive reactance is $X_L = 50\ \Omega$. What is the value of R? What is the quality factor of the circuit?

QUESTION 3: What is the Q of the RLC circuit of Example 6?

QUESTION 4: A series RLC circuit is connected to an emf of amplitude 2.0 V. At the resonant frequency, the amplitude of the voltage across the inductor is 50 V. What is the amplitude of the voltage across the resistor?

   (A) 102 V   (B) 98 V   (C) 52 V   (D) 48 V   (E) 2.0 V

32.6 THE TRANSFORMER

A transformer consists of two coils arranged in such a way that (almost) all the magnetic field lines generated by one of them pass through the other. This can be achieved by winding both coils on a common iron core (see Fig. 32.31). As we saw in Section 30.4, the iron increases the strength of the magnetic field in its interior by a large factor. Since the field is much stronger in the iron than outside, most of the field lines have to stay inside the iron; this means that *the iron tends to keep the field lines together and acts as a conduit for the field lines from one coil to the other*.

Each coil is part of a separate electric circuit (see Fig. 32.32). The **primary** circuit has a source of alternating emf, and the **secondary** circuit has a lightbulb, or an electric heater, or some other "load" that consumes electric power. The alternating current in the primary circuit induces an alternating emf in the secondary circuit. We will show that the emf $\mathcal{E}_2$ produced in the secondary circuit is related as follows to the emf $\mathcal{E}_1$ in the primary circuit:

$$\mathcal{E}_2 = \mathcal{E}_1 \frac{N_2}{N_1} \qquad (32.56)$$

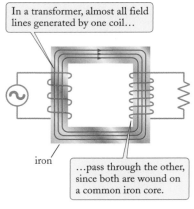

In a transformer, almost all field lines generated by one coil…

…pass through the other, since both are wound on a common iron core.

iron

FIGURE 32.31 A transformer.

primary circuit and secondary circuit

emfs in transformer

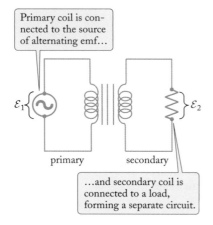

Primary coil is connected to the source of alternating emf…

primary   secondary

…and secondary coil is connected to a load, forming a separate circuit.

FIGURE 32.32 Circuit diagram for the transformer. The parallel lines represent a solid iron core.

step-up transformer and step-down transformer

where N_1 and N_2 are, respectively, the numbers of turns in the primary and the secondary coils.

To prove Eq. (32.56), we begin with Kirchhoff's rule as it applies to the primary circuit: the emf $\mathcal{E}_1$ of the source must equal the induced emf $\mathcal{E}_{1,\text{ind}}$ across the primary coil. But by Faraday's Law, the induced emf across the entire coil equals the number of loops times the rate of change of the flux:

$$\mathcal{E}_1 = \mathcal{E}_{1,\text{ind}} = -N_1 \frac{d\Phi_1}{dt} \qquad (32.57)$$

Likewise, the emf $\mathcal{E}_2$ delivered to the load must equal the induced emf $\mathcal{E}_{2,\text{ind}}$, which, again, equals the number of loops times the rate of change of the flux in that coil:

$$\mathcal{E}_2 = \mathcal{E}_{2,\text{ind}} = -N_2 \frac{d\Phi_2}{dt} \qquad (32.58)$$

Since the same numbers of magnetic field lines pass through both coils, we know that

$$\frac{d\Phi_1}{dt} = \frac{d\Phi_2}{dt} \qquad (32.59)$$

Thus, the ratio of Eqs. (32.58) and (32.57) is

$$\frac{\mathcal{E}_2}{\mathcal{E}_1} = \frac{N_2}{N_1} \qquad (32.60)$$

which is equivalent to Eq. (32.56).

If $N_2 > N_1$, we have a **step-up transformer**, and if $N_2 < N_1$, we have a **step-down transformer**.

EXAMPLE 7   Doorbells and buzzers usually are designed for 12 volts AC and they are powered by small transformers that step down 115 volts AC to 12 volts AC. Suppose that such a transformer has a primary winding with 1500 turns. How many turns are there in the secondary winding?

SOLUTION: Equation (32.56) applies to the instantaneous voltages. It is therefore also valid for the rms voltages. With the appropriate numerical values, Eq. (32.56) gives

$$N_2 = N_1 \frac{\mathcal{E}_2}{\mathcal{E}_1} = 1500 \text{ turns} \times \frac{12 \text{ V}}{115 \text{ V}} = 157 \text{ turns}$$

As long as the secondary circuit is open and carries no current ($I_2 = 0$), *an ideal transformer does not consume electric power*. Under these conditions, the primary circuit consists of nothing but the source of emf and an inductor—it is a pure L circuit. In such a circuit, the power delivered by the source of emf to the inductor averages to zero (see Section 32.3).

If the secondary circuit is closed—that is, if it is connected to some external load— a current will flow ($I_2 \neq 0$). This current contributes to the magnetic flux in the transformer and induces a current in the primary circuit. The current in the latter is then different from that in a pure L circuit, and the power will *not* average to zero over a cycle.

In an ideal transformer, the electric power that the primary circuit takes from the source of emf exactly matches the power that the secondary circuit delivers to the external load. Since the power is the product of the current and the emf, we can express the equality of these instantaneous powers as

$$I_1 \mathcal{E}_1 = I_2 \mathcal{E}_2 \tag{32.61}$$

where the currents and emfs are also taken to be instantaneous values. Comparison with Eq. (32.60) makes it evident that the currents are in inverse proportion to the corresponding ratio of turns of the coils:

$$\frac{I_1}{I_2} = \frac{N_2}{N_1} \tag{32.62}$$

Good transformers approach the ideal condition (32.61) of conservation of electric energy fairly closely: about 99% of the power supplied to the input terminals emerges at the output terminals; the difference is essentially all lost as heat in the iron core and the windings. Such high efficiency is attained by avoiding eddy-current losses (see Section 31.3), by means of an iron core constructed from thin layers separated by insulation. For the immense transformers at power plants (chapter photo), even the loss of a tiny fraction of the power transmitted requires the use of special cooling to carry away the waste heat. Power-plant transformers are typically immersed in oil, and circulation of this liquid carries the heat to large-area metal surfaces; these radiator tubes are air-cooled with arrays of fans (see Fig. 32.33).

Transformers play a large role in our electric technology. As we saw in Example 8 of Chapter 28, transmission lines for electric power operate much more efficiently at high voltage, since this reduces the Joule heating losses caused by the resistance of the wires. To take advantage of this high efficiency, power lines are made to operate

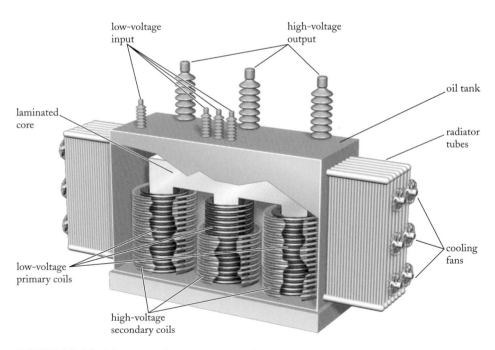

FIGURE 32.33 A large transformer at a power plant.

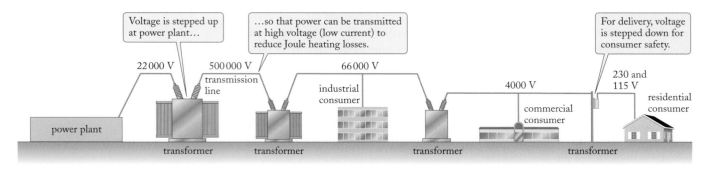

FIGURE 32.34 Typical voltage transformations during the transmission of electric power. The small transformers used to step the power down to 115 V for residential use are often attached to telephone poles.

at several hundred kilovolts. The voltage must be stepped up to this value at the power plant, and for safety's sake, it must be stepped down just before it reaches the consumer (see Fig. 32.34). For these operations, large banks of transformers are needed at both ends. Transformers are also used in many electric and electronic devices, such as TV tubes and X-ray machines, which require high voltages to accelerate beams of electrons.

Concepts
— in —
Context

EXAMPLE 8 A single generator of a large power plant delivers an electric power of 130 MW at 22 kilovolts AC. For transmission, this voltage is stepped up to 500 kV by a transformer. What is the rms current delivered by the generator? What is the rms current in the transmission line? By what factor is the energy loss (Joule heating) of the transmission line reduced by using the transformer? Assume that the transformer does not waste any power, and that the loads are purely resistive.

SOLUTION: The power delivered and the AC voltage refer to the average and rms values, respectively. For a resistive load, the average power is the product of the rms current and voltage, so the rms current delivered by the generator is

$$I_1 = \frac{\overline{P}}{\mathcal{E}_1} = \frac{130 \times 10^6 \text{ W}}{22 \times 10^3 \text{ V}} = 5.9 \times 10^3 \text{ A}$$

The same power is delivered to the transmission line, so the current there is reduced to

$$I_2 = \frac{\overline{P}}{\mathcal{E}_2} = \frac{130 \times 10^6 \text{ W}}{500 \times 10^3 \text{ V}} = 260 \text{ A}$$

Alternatively, we can note that the transformer currents are in inverse ratio to the voltages [Eq. (32.61)]:

$$I_2 = \frac{\mathcal{E}_1}{\mathcal{E}_2}I_1 = \frac{22 \times 10^3 \text{ V}}{500 \times 10^3 \text{ V}} \times 5.9 \times 10^3 \text{ A} = 260 \text{ A}$$

The energy loss by Joule heating in the transmission line is equal to I^2R; for a given transmission line, the resistance R is fixed, so the energy loss is reduced by the square of the current ratio,

$$\frac{I_1^2}{I_2^2} = \frac{(5.9 \times 10^3 \text{ A})^2}{(260 \text{ A})^2} = 5.2 \times 10^2$$

Energy losses are reduced by more than a factor of 500! For a 2.0-Ω transmission line, this is the difference between the loss of more than half of the power ($I_1^2 R = (5.9 \times 10^3 \text{ A})^2 \times 2.0 \ \Omega = 70 \text{ MW}$, or 54%) and an almost negligible loss (0.1%). If the power were transmitted at an even lower voltage, as in Example 8 of Chapter 28, the energy loss would become much more extreme.

 ## Checkup 32.6

QUESTION 1: If we want to use a transformer to step up 115 V AC to 230 V AC, by what factor must the number of coils in the secondary differ from that in the primary?

QUESTION 2: Suppose that the transformer described in Example 7 delivers some power to doorbell. Is the current larger in the primary or in the secondary?

QUESTION 3: Suppose that you have two transformers, one with a turns ratio of 2 and one with a turns ratio of 5. Each may be used to step up or to step down a voltage. The two are to be connected one after the other. What are possible values for the ratio of input to output voltages for the pair?

    (A) 10, 0.1   (B) 10, 5, 2.5   (C) 25, 4   (D) 10, 5, 2, 0.1   (E) 10, 2.5, 0.4, 0.1

SUMMARY

PHYSICS IN PRACTICE   Frequency Filter Circuits   **(page 1037)**

CURRENT IN RESISTOR CIRCUIT	$I = \dfrac{\mathcal{E}_{\max} \cos \omega t}{R}$	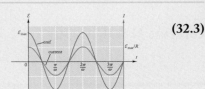 (32.3)
RMS (ROOT-MEAN-SQUARE) VOLTAGE	$\mathcal{E}_{\text{rms}} = \dfrac{\mathcal{E}_{\max}}{\sqrt{2}}$	(32.8)
AVERAGE AC POWER DISSIPATED BY RESISTOR	$\overline{P} = \dfrac{\mathcal{E}_{\max}^2}{2R} = \dfrac{\mathcal{E}_{\text{rms}}^2}{R}$	(32.6, 32.7)
REACTANCE OF CAPACITOR CIRCUIT	$X_C = \dfrac{1}{\omega C}$	(32.14)
CURRENT IN CAPACITOR CIRCUIT	$I = -\dfrac{\mathcal{E}_{\max} \sin \omega t}{X_C}$	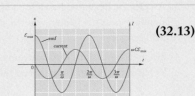 (32.13)

| REACTANCE OF INDUCTOR CIRCUIT | $X_L = \omega L$ | (32.20) |

| CURRENT IN INDUCTOR CIRCUIT | $I = \dfrac{\mathcal{E}_{max} \sin \omega t}{X_L}$ | (32.19) |

| NATURAL FREQUENCY OF *LC* CIRCUIT; RESONANT FREQUENCY | $\omega_0 = \dfrac{1}{\sqrt{LC}}$ | (32.26) |

| SLOWLY DECAYING OSCILLATIONS IN *RLC* CIRCUIT   (damped harmonic oscillator) | $Q = Q_{max} e^{-Rt/2L} \cos \omega_d t$ | (32.30) |

| *Q* OF *RLC* CIRCUIT | $Q = \dfrac{\omega_0 L}{R}$ | |

| IMPEDANCE IN SERIES *RLC* CIRCUIT | $Z = \sqrt{R^2 + (X_L - X_C)^2}$ | (32.48) |

| IMPEDANCE RELATED TO AMPLITUDES OF EMF AND CURRENT | $Z = \dfrac{\mathcal{E}_{max}}{I_{max}}$ | (32.44) |

| PHASE ANGLE OF EMF WITH RESPECT TO CURRENT IN A SERIES *RLC* CIRCUIT | $\phi = \tan^{-1}\left(\dfrac{X_L - X_C}{R}\right)$ | (32.49) |

| VOLTAGE AMPLITUDES IN A SERIES *RLC* CIRCUIT | $V_{max,R} = I_{max} R \quad V_{max,L} = I_{max} X_L \quad V_{max,C} = I_{max} X_C$ | (32.41) |

| EMFs OF TRANSFORMER | $\mathcal{E}_2 = \mathcal{E}_1 \dfrac{N_2}{N_1}$ | (32.56) |

| CURRENTS IN TRANSFORMER | $I_2 = I_1 \dfrac{N_1}{N_2}$ | (32.62) |

QUESTIONS FOR DISCUSSION

1. In most European countries, the voltage available at outlets in homes is 230 V AC. What is the actual amplitude of oscillation of this voltage?

2. You can perceive the 120-Hz flicker (two peaks of intensity per AC cycle) in a fluorescent light tube (by sweeping your eye quickly across the tube), but you cannot perceive any such flicker in an incandescent lightbulb. Explain.

3. An electric heater operates from 60-Hz, 115-V AC. When standing near the heater, why do we not feel 120 heat pulses per second?

4. Some electric motors operate only on DC, others only on AC. What is the difference between these motors?

5. If you connect a capacitor across a 115-V outlet, does any current flow through the connecting wires? Through the space between the capacitor plates? Does the outlet deliver instaneous electric power? Average electric power?

6. It is sometimes said that a capacitor becomes a short circuit at high frequencies, and that an inductor becomes an open circuit at high frequencies. Explain.

7. Can you blow a fuse by connecting a very large capacitor across an ordinary 115-V outlet?

8. How could you use an *LC* circuit to measure the capacitance of a capacitor?

9. Consider a series RLC circuit. Can the voltage across the capacitor ever be larger than $\mathcal{E}_{max}$? Across the inductor? Across the resistor? (Hint: Inspect the phasor diagram.)

10. Roughly plot the phase angle ϕ given by Eq. (32.49) as a function of ω. What is the phase angle at resonance?

11. If you use an AC voltmeter to measure the driving emf and the voltages across the inductor, the capacitor, and the resistor in a series RLC circuit, you will find that the emf is greater than or equal to the voltage across the resistor, but smaller than the sum of the voltages across the inductor, capacitor, and resistor. Explain.

12. What is the impedance of a series RLC circuit at resonance?

13. If we substitute $R = 0$ in Eq. (32.53), we obtain the equation for the current in a driven LC circuit. What must we substitute to obtain the equation for a driven RC circuit? A driven RL circuit?

14. Show that the equation for the current in a series RLC circuit [Eq. (32.53)] includes Eq. (32.3), (32.13), and (32.19) as special cases. What are the individual impedances of a resistor, a capacitor, and an inductor?

15. Why can we not use a transformer to step up the voltage supplied by a battery?

16. Does an electric motor absorb more electric power when working against a mechanical load than when running freely?

PROBLEMS

†32.1 Resistor Circuit

1. An electric heater (Fig. 32.35) plugged into a 115-V AC outlet uses an average electric power of 1200 W.

    (a) What are the rms current and the maximum instantaneous current through the heater?

    (b) What are the maximum instantaneous power and the minimum instantaneous power?

FIGURE 32.35 An electric heater.

2. An immersible heating element used to boil water consumes an (average) electric power of 400 W when connected to a source of 115 volts AC. Suppose that you connect this heating element to a source of 115 volts DC. What power will it consume?

3. A high-voltage power line operates on an rms voltage of 230 000 volts AC and delivers an rms current of 740 A to a resistive load.

    (a) What are the maximum instantaneous voltage and current?

    (b) What are the maximum instantaneous power and the average power delivered?

4. The GG-1 electric locomotive develops 4600 hp; it runs on an rms AC voltage of 1100 V.

    (a) What rms current does this locomotive draw?

    (b) Why is it advantageous to supply the electric power for locomotives at high voltage (and fairly low current)?

5. An electric heater operating with a 115-V AC power supply delivers 1200 W of heat.

    (a) What is the rms current through this heater?

    (b) What is the maximum instantaneous current?

    (c) What is the resistance of this heater?

6. A frequency generator delivers an average power of 2.0 W into a load of 50 Ω. What is the rms voltage? The rms current?

7. In a test of an AM radio station transmitter, an AC voltage amplitude of 2.2 kV is measured across an oil-cooled resistive load of 50 Ω. What is the rms current? What is the average power delivered?

*8. A square loop of nichrome wire with resistance 11 Ω and area 0.15 m² is rotated in a magnetic field of 2.0 T. For what frequency of rotation will the average power dissipated in the loop be 1.0 W? 100 W?

†For help, see Online Concept Tutorial 35 at www.wwnorton.com/physics

†32.2 Capacitor Circuit

9. A capacitor of 2.2 μF is connected to a generator supplying 12 V AC. What is the rms current if the angular frequency of the generator is 1.2×10^3 radians/s? What is the rms current if the angular frequency of the generator is 2.4×10^3 radians/s?

10. A capacitor has a capacitance of 0.15 μF. What is the reactance when it is connected to a circuit operating at a $\omega = 6.0 \times 10^3$ radians/s? At $\omega = 1.8 \times 10^4$ radians/s?

11. A capacitor has a reactance of 3.0×10^6 Ω at a frequency of 2.0 kHz. What is its reactance at 3.0 kHz? At 4.0 kHz?

12. A capacitor of 600 pF is connected to an audio generator of adjustable frequency. At what frequency is the reactance of this capacitor 2.0×10^5 Ω?

13. A capacitor with $C = 8.0 \times 10^{-7}$ F is connected to an oscillating source of emf. This source provides an emf $\mathcal{E} = \mathcal{E}_{max} \cos \omega t$, with $\mathcal{E}_{max} = 0.20$ V and $\omega = 6.0 \times 10^3$ radians/s.

    (a) What is the reactance of the capacitor?

    (b) What is the maximum current in the circuit?

    (c) What is the current at time $t = 0$? At time $t = \pi/4\omega$?

14. A capacitor of 0.40 μF is connected to an AC source of amplitude 12 V and angular frequency 3.0×10^3 radians/s. What is the instantaneous current in the circuit when the instantaneous voltage is 12 V? What is the instantaneous current one-quarter cycle later? One-half cycle later? Three-quarters cycle later?

15. To measure the capacitance of a capacitor, a physicist connects it to an oscillating source of emf of angular frequency 2.0×10^3 radians/s and of amplitude 4.0×10^{-3} V. She finds that the maximum instantaneous current in the capacitor is 8.0×10^{-9} A. What is the capacitance?

16. We want to generate a maximum current of 5.0×10^{-5} A in a circuit consisting of a capacitor of 2.0 pF connected to a source of oscillating emf of angular frequency 4.0×10^4 radians/s. What amplitude of oscillation of the emf do we need?

17. An emf $\mathcal{E} = \mathcal{E}_{max} \cos \omega t$ with $\omega = 6.0 \times 10^2$ radians/s is applied to a 2.0-μF capacitor. Calculate the instantaneous power delivered to the capacitor at $t = 1.0$ ms, 2.0 ms, 3.0 ms, and 4.0 ms.

18. A capacitor, used as a sensor for measuring the level of liquid nitrogen in a container, has a capacitance of 0.30 nF when the container is empty and 0.43 nF when it is full and the capacitor is immersed in liquid nitrogen. If an emf with an amplitude of 15 V and angular frequency of 6.0×10^4 radians/s is applied to the sensor, what is the amplitude of the current that will flow when the container is empty? When full?

*19. A 47-Ω resistor and a capacitor are to be connected in parallel to a source of emf oscillating with amplitude $\mathcal{E}_{max} = 6.2$ V at a frequency of 60 Hz. For what value of the capacitance C will the amplitude of the current into the capacitor equal the ampli-

tude of the current through the resistor? (Hint: For circuit elements connected in parallel, the instantaneous voltage across each is the same.)

*20. A circuit consists of three capacitors of 0.80 μF each, connected in parallel to the terminals of a generator that delivers 24 V AC at an angular frequency of 1.8×10^3 radians/s (see Fig. 32.36). What is the net rms current in this circuit?

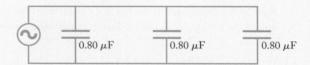

FIGURE 32.36 Three capacitors in parallel connected to a source of emf.

†32.3 Inductor Circuit

21. What is the rate of change of the current in a coil with an inductance of 250 μH that has an instantaneous voltage of 25 V across it?

22. An inductor has an inductance of 0.25 H. What is the reactance when it is connected to a circuit operating at $\omega = 6.0 \times 10^3$ radians/s? At $\omega = 1.8 \times 10^4$ radians/s?

23. An inductor of 0.20 H is connected to an audio generator of adjustable frequency. At what frequency is the reactance of this inductor 2.0×10^3 Ω?

24. A 1.0-km segment of the aluminum cable of a high-voltage transmission line has a resistance of 7.3×10^{-2} Ω, a capacitance of 4.7×10^{-8} F, and a self-inductance of 6.6×10^{-4} H. At the standard frequency of 60 Hz, what are the capacitive and the inductive reactances of this segment? Compare the reactances X_C and X_L and the resistance R. Which of these quantities is largest? Smallest?

25. An inductor of 0.30 H is connected to a generator supplying 12 V AC. What is the rms current if the angular frequency of the generator is 1.2×10^3 radians/s? What is the rms current if the angular frequency of the generator is 2.4×10^3 radians/s?

26. We want to generate a maximum current of 5.0×10^{-5} A in an inductor of 1.0×10^{-2} H connected to a source of oscillating emf of angular frequency 4.0×10^4 radians/s. What amplitude of oscillation of the emf do we need?

27. The primary winding of a transformer has an inductance of 6.2 H. What rms current will flow in this winding if it is connected to an outlet supplying 115 V AC?

28. To measure the inductance of an inductor, an experimenter connects it to an oscillating source of emf of angular frequency 3.0×10^3 radians/s and of amplitude 8.0×10^{-3} V. He finds that the maximum instantaneous current in the inductor is 2.0×10^{-7} A. What is the inductance?

†For help, see Online Concept Tutorial 35 at www.wwnorton.com/physics

†For help, see Online Concept Tutorial 35 at www.wwnorton.com/physics

29. The voltage across a superconducting solenoid of inductance 35 H may not exceed 12 V without malfunction. What is the largest permitted rate of change of current through the solenoid? If an alternating current with amplitude 25 A is to used, what is the maximum frequency at which this current may oscillate?

30. An AM radio antenna has the shape of a coil with inductance 0.90 mH. A radio wave with frequency 1.2 MHz induces a current of amplitude 5.2 nA in the coil. What is the amplitude of the induced voltage across the coil?

31. To prevent radio frequency interference in electronic equipment, inductors are often placed wherever the transmission of only low-frequency signals is desired. If such an **rf choke** has an inductance of 47 μH, what reactance does it present to a 60-Hz power supply current? To a 100-MHz radio signal?

*32. An inductor with $L = 4.0 \times 10^{-2}$ H is connected to a source of alternating emf. This source provides an emf $\mathcal{E} = \mathcal{E}_{max} \times \cos \omega t$, with $\mathcal{E}_{max} = 0.20$ V and $\omega = 6.0 \times 10^3$ radians/s.

(a) What is the reactance of the inductor?

(b) What is the maximum current in the circuit?

(c) What is the current at time $t = 0$? At time $t = \pi/4\omega$?

*33. A 2.0-μF capacitor and a 3.0-mH inductor are connected in parallel with an AC power supply. At what frequency will the inductive reactance be 3 times the capacitive reactance?

*34. Consider the circuit shown in Fig. 32.37. The emf of the source is of the form $\mathcal{E}_{max} \cos \omega t$. In terms of this emf and the capacitance C and the inductance L, find the instantaneous currents through the capacitor and the inductor. Find the instantaneous current and the instantaneous power delivered by the source of emf.

FIGURE 32.37 Inductor and capacitor connected to a source of emf.

*35. In Example 3 we examined the current in a circuit containing a resistor, a capacitor, and an inductor connected in parallel to a source of oscillating emf of an angular frequency 6.0×10^3 radians/s and amplitude 1.0×10^{-3} V. How would the answers to this example change if we reduced the angular frequency to 3.0×10^3 radians/s?

*36. Suppose that in Example 3 the emf is of the form $\mathcal{E} = \mathcal{E}_{max} \times \cos \omega t$. What is the net instantaneous current in the circuit at $t = 0$? At $t = \pi/2\omega$? At $t = \pi/\omega$?

*37. An inductor of 4.5×10^{-2} H and a capacitor of 0.25 μF are connected in parallel to a source of alternating emf of frequency ω. For what value of ω will the rms currents in the inductor and the capacitor be of equal magnitudes?

*38. An inductor is built in the shape of a solenoid of radius 0.20 cm, length 4.0 cm, with 1000 turns. This inductor is connected to a generator supplying an rms voltage of 1.2×10^{-4} V AC at an angular frequency of 9.0×10^3 radians/s.

(a) What is the inductance? Assume that the solenoid can be treated as very long.

(b) What is the rms current flowing through the inductor?

(c) What is the rms energy stored in the inductor?

32.4 Freely Oscillating *LC* and *RLC* Circuits

39. What is the natural frequency for an *LC* circuit consisting of a 2.2×10^{-6} F capacitor and an 8.0×10^{-2} H inductor?

40. You want to construct an *LC* circuit of natural frequency 8.0×10^3 Hz. You have available a capacitor of 0.20 μF. What inductor do you need?

41. What is the natural frequency of oscillation of the circuit shown in Fig. 32.38? The capacitances are 2.4×10^{-5} F each and the inductance is 1.2×10^{-3} H.

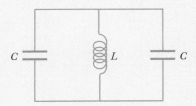

FIGURE 32.38 Two equal capacitors connected to an inductor.

42. A radio receiver contains an *LC* circuit whose natural frequency of oscillation can be adjusted, or tuned, to match the frequency of incoming radio waves. The adjustment is made by means of a variable capacitor. Suppose that the inductance of the circuit is 15 μH. Over what range of capacitances must the capacitor be adjustable if the frequencies of oscillation of the circuit are to span the range from 530 kHz to 1600 kHz?

43. In Fig. 32.39, the emf has value $\mathcal{E} = 12$ V and the capacitance is $C = 22$ μF. After the switch S in the circuit has been in position 1 for a long time, it is moved to position 2. The current in the right loop then oscillates with amplitude 3.5 mA. What is the frequency of oscillation? What is the value of the inductance L?

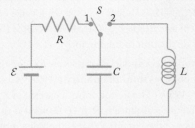

FIGURE 32.39 Switch-selected *RC* and *LC* circuits.

44. Consider an underdamped series RLC circuit (Fig. 32.19) with $R = 0.75\ \Omega$, $L = 25\ \mu H$, and $C = 100\ nF$. What is the value of the quality factor Q for this circuit? If the capacitor has an initial charge Q_{max}, how many oscillations will occur before the charge amplitude is reduced to $(1/e) \times Q_{max} \approx 0.37Q_{max}$?

45. An underdamped series RLC circuit with $R = 2.2 \times 10^{-3}\ \Omega$ is to have an oscillation frequency of 3.5×10^6 Hz and a quality factor Q of 750. What must be the values of the capacitance C and the inductance L?

*46. The circuit of Fig. 32.40a is oscillating with the switch S closed. The graph of current vs. time is shown in Fig. 32.40b.

    (a) At time t_1 the switch S is suddenly opened. Is the frequency of oscillation increased, decreased, or unchanged? In the space on the right in Fig. 32.40b sketch a rough, qualitative graph of current for times after t_1.

    (b) At time t_2 the switch S is closed. Sketch the graph of current after this time.

(a)

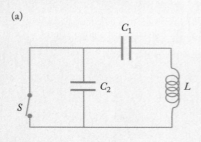

(b)

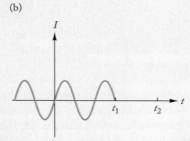

FIGURE 32.40 (a) Inductor and two capacitors in a circuit. The switch S is initially closed. (b) Current in the circuit as a function of time. At $t = t_1$, the switch S is opened.

†32.5 Series Circuits with Alternating Emf

47. A series RC circuit is connected to an alternating 60-Hz emf with $\mathcal{E}_{max} = 163$ V. If $R = 50\ \Omega$, what value of C will result in a current amplitude $I_{max} = 0.50$ A? What is the average rate of power dissipation in this circuit?

48. A motor may be modeled as a series RL circuit. A particular motor operates from an ordinary 60-Hz wall outlet with $\mathcal{E}_{rms} = 115$ V. The motor draws a current of $I_{rms} = 2.5$ A. If the effective resistance is $R = 16\ \Omega$, what is the inductive reactance of the circuit? What is the value of the inductance?

†For help, see Online Concept Tutorial 35 at www.wwnorton.com/physics

49. In a series AC circuit, average power is dissipated only in the resistor. That power is $\bar{P} = \frac{1}{2}I_{max}^2 R$.

    (a) Use Eq. (32.44) to verify that the average power can be written

$$\bar{P} = \tfrac{1}{2}\mathcal{E}_{max}I_{max}\frac{R}{Z}$$

    (b) Use the phasor diagram of Fig. 32.28c to rewrite this as

$$\bar{P} = \tfrac{1}{2}\mathcal{E}_{max}I_{max}\cos\phi$$

    (c) The quantity $\cos\phi$ is called the **power factor**; it is the factor by which the average power is reduced from that of a purely resistive circuit. A certain series circuit has $\mathcal{E}_{max} = 167$ V, $I_{max} = 14$ A, and dissipates an average power of 750 W. What is the power factor? What is the phase angle? Can we tell whether the circuit contains a capacitor, an inductor, or both?

50. A series RLC circuit is connected to a source of alternating emf as shown in Fig. 32.41. The components have values $R = 5.2\ \Omega$, $L = 36$ mH, and $C = 0.41\ \mu F$. The amplitude of the emf of the source is 100 V. Initially, the frequency of the emf is tuned to resonance.

    (a) What is the resonant frequency?

    (b) At resonance, what is the amplitude of the voltage across the inductor?

    (c) At resonance, what is the amplitude of the voltage between points 1 and 2 of Fig. 32.41?

    (d) The frequency is changed to 4.7×10^5 radians per second. Now what is the amplitude of the voltage between points 1 and 2 of Fig. 32.41?

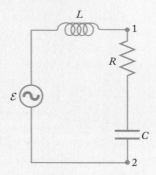

FIGURE 32.41 A series RLC circuit.

51. A series RLC circuit with $R = 3.0\ \Omega$, $C = 20$ pF, and a resonant frequency of 17 MHz is connected to an alternating emf with $\mathcal{E}_{max} = 12$ V.

    (a) What is the amplitude of the voltage across the inductor at resonance?

    (b) What is the amplitude of the voltage across the resistor at resonance?

    (c) What is the average power dissipated in this circuit?

    (d) What is the quality factor Q of this circuit?

52. A series RC circuit is often used as a **high-pass filter**. Consider a resistor R and capacitor C connected in series to an oscillating source of emf with amplitude $\mathcal{E}_{max}$. In terms of $\mathcal{E}_{max}$, what is the amplitude of the voltage across the resistor for a low frequency, when $\omega RC = 1/10$? For a high frequency, when $\omega RC = 10$? At what frequency is the amplitude of the voltage across the resistor equal to $\mathcal{E}_{max}/\sqrt{2}$?

53. A series RL circuit is often used as a **low-pass filter**. Consider a resistor R and inductor L connected in series to an oscillating source of emf with amplitude $\mathcal{E}_{max}$. In terms of $\mathcal{E}_{max}$, what is the amplitude of the voltage across the resistor for a low frequency, when $\omega L/R = 1/10$? For a high frequency, when $\omega L/R = 10$? At what frequency is the amplitude of the voltage across the resistor equal to $\mathcal{E}_{max}/\sqrt{2}$?

*54. The average power dissipated in a series RLC circuit is $\overline{P} = \frac{1}{2}I_{max}^2 R$, where I_{max} is given by Eq. (32.53). We can define the **width of resonance** in terms of the half-power points $\overline{P}_{1/2}$ of the resonance curve; these have half the peak value, or $\overline{P}_{1/2} = \frac{1}{2}(\frac{1}{2}\mathcal{E}_{max}^2/R)$. By setting $\overline{P} = \overline{P}_{1/2}$, obtain an equation for the values of ω at the half-power points. Solve the resulting quadratic equation and show that the two roots are separated by

$$\Delta\omega = \frac{R}{L}$$

Thus we can verify the equivalence of Eqs. (32.33) and (32.55), $Q = \omega_0 L/R = \omega_0/\Delta\omega$.

*55. Consider that an emf is connected to a **parallel RLC circuit**. In this case, the voltage V_{max} is the same across the source and each element, but the net current is the sum of the individual currents, so a phasor diagram for the *currents* can be drawn instead. Draw this diagram. By dividing the magnitude of each current phasor by the common value of V_{max}, obtain a phasor diagram that relates impedance, reactances, and resistance. Use the geometry of the diagram to show that

$$Z = \left[\frac{1}{R^2} + \left(\frac{1}{X_L} - \frac{1}{X_C}\right)^2\right]^{-1/2}$$

56. Consider a *parallel RLC* circuit (see Problem 55) with $R = 5.0\ \Omega$, $L = 30\ \mu H$, and $C = 25$ nF. An alternating emf with $\mathcal{E}_{max} = 15$ V is applied. Determine the total current in the circuit when

   (a) It operates at the resonant frequency.

   (b) It operates at half the resonant frequency.

32.6 The Transformer

57. A transformer used to step up 115 V to 5000 V has a primary coil of 100 turns. What must be the number of turns in the secondary coil?

58. The primary winding of a transformer has 1200 turns and the secondary winding has 80 turns. If the emf supplied to the primary is 115 V AC, what is the emf delivered by the secondary?

59. Consider each of the four transformers illustrated in Fig. 32.34; assume that the last of these delivers 115 V AC. What is the ratio N_1/N_2 of the numbers of turns in the primary and secondary windings of each of these transformers?

60. How much current is flowing in the 1200-V secondary of a transformer if the primary has a voltage of 12 V and a current of 3.0 A?

61. The generators of a large power plant deliver an electric power of 2000 MW at 22 kilovolts AC. For transmission, this rms voltage is stepped up to 400 kV by a transformer. What is the rms current delivered by the generators? What is the rms current in the transmission line? Assume that the transformer does not waste any power.

62. A transformer operating on a primary voltage of 115 volts AC delivers a secondary voltage of 6.0 volts AC to a small electric buzzer (see Fig. 32.42). If the current in the secondary circuit is 3.0 A, what is the rms current in the primary circuit? Assume that no electric power is lost in the transformer.

FIGURE 32.42 Small transformers used for buzzers and bells.

63. In a transformer, the ratio of turns of primary and secondary is 10:1. If the primary voltage is 120 V and the secondary current is 3.0 A, how much power is being absorbed in the primary circuit of the transformer?

64. The largest transformer ever built handles a power of 1.50×10^9 W. This transformer is used to step down 765 kVAC to 345 kVAC. What is the rms current in the primary? What is the current in the secondary? Assume that no electric power is lost by the transformer.

65. The secondary coils of many transformers have *taps*; these are extra connections to an intermediate part of the secondary coil, so that, in addition to the full secondary voltage, a fixed fraction of the secondary voltage is also available. Consider a transformer with a 115-V AC primary and a 12-V AC secondary that has a **center tap**. The 12-V AC secondary voltage is applied to a load that dissipates an average power of 4.0 W, and the center tap (6.0 V AC) to a load of average power 12 W. What is the current in the primary?

66. A **variac** is a variable transformer, usually with a sliding contact that can be moved along the coil to select some integer number of the total turns in the secondary coil. If the 115-V AC primary of a variac has 200 turns and the secondary has 250 turns, what is the largest rms voltage available at the secondary? What is the smallest increment by which the secondary rms voltage may be changed?

REVIEW PROBLEMS

67. A resistor of 100 Ω is connected to an AC source with an amplitude of 163 V. What is the instantaneous current in the resistor when the instantaneous voltage is 163 V? What is the instantaneous current one-quarter cycle later? One-half cycle later? Three-quarters cycle later?

68. A parallel-plate capacitor has plates of area $0.30 \ m^2$ separated by an air gap of 0.0020 cm. This capacitor is connected to a generator supplying an rms voltage of 12 V AC at an angular frequency of 9.0×10^3 radians/s.

   (a) What is the capacitance?

   (b) What is the rms current flowing into the plates?

   (c) What is the rms charge on the capacitor?

   (d) What is the rms energy stored in the capacitor?

*69. A circuit consists of a resistor connected in series to a battery; the resistance is 5.0 Ω and the emf of the battery is 12 V. The wires (of negligible resistance) connecting these circuit elements are laid out along a square of 20 cm × 20 cm (Fig. 32.43). The entire circuit is placed face on in an oscillating magnetic field. The instantaneous value of the magnetic field is

$$B = B_0 \sin \omega t$$

with $B_0 = 0.15$ T and $\omega = 360$ radians/s.

   (a) Find the instantaneous current in the resistor.

   (b) Find the average power dissipated in the resistor.

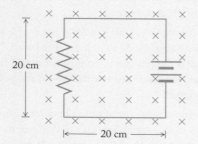

FIGURE 32.43 Circuit placed in magnetic field.

*70. A capacitor with $C = 4.0 \times 10^{-8}$ F is connected to an oscillating source of emf. The source provides an emf $\mathcal{E} = \mathcal{E}_{max} \times \sin \omega t$, with $\mathcal{E}_{max} = 0.80$ V and $\omega = 6.0 \times 10^3$ radians/s.

   (a) What is the reactance of the capacitor?

   (b) What is the maximum current in the circuit?

   (c) What is the current at time $t = 0$? At time $t = \pi/4\omega$?

*71. A circuit consists of two capacitors of 6.0×10^{-8} F and 9.0×10^{-8} F connected in series to an oscillating source of emf (Fig. 32.44). This source delivers a cosinusoidal emf $\mathcal{E} = 1.8 \times \cos (120 \ \pi t)$, where $\mathcal{E}$ is in volts and t in seconds.

   (a) Find the charge on each capacitor as a function of time.

   (b) At what time is the charge on the capacitors maximum? At what time minimum?

   (c) What is the maximum energy in the capacitors? What is the time-average energy?

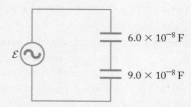

FIGURE 32.44 Two capacitors in series connected to a source of emf.

72. Three capacitors, with $C_1 = 5.0 \times 10^{-8}$ F, $C_2 = 2.0 \times 10^{-8}$ F, and $C_3 = 1.0 \times 10^{-8}$ F, are connected to an oscillating source of emf as shown in Fig. 32.45. The source supplies an emf $\mathcal{E} = 2.0 \cos(120\pi t)$, where $\mathcal{E}$ is in volts and t in seconds.

   (a) Find the instantaneous current supplied by the source of emf. What is the maximum value of this current?

   (b) Find the instantaneous current that flows through each capacitor, and find the instantaneous potential difference across each capacitor.

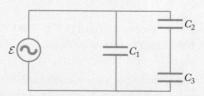

FIGURE 32.45 Three capacitors connected to a source of emf.

73. An inductor of 3.0 mH is connected to an AC source of amplitude 8.0 V and angular frequency 3.0×10^3 radians/s. What is the instantaneous current in the inductor when the instantaneous voltage is 8.0 V? What is the instantaneous current one-quarter cycle later? One-half cycle later? Three-quarters cycle later?

*74. An inductor of 1.6×10^{-3} H is connected to a source of alternating emf. The current in the inductor is $I = I_0 \sin \omega t$, with $I_0 = 180$ A and $\omega = 120\pi$ radians/s.

(a) What is the potential difference across the inductor at time $t = 0$? At time $t = 1/240$ s?

(b) What is the energy in the inductor at time $t = 0$? At time $t = 1/240$ s?

(c) What is the instantaneous power delivered by the source of emf to the inductor at time $t = 0$? At time $t = 1/240$ s?

75. An inductor and a resistor are connected in parallel to a battery and to a source of oscillating emf as shown in Fig. 32.46. The inductance is $L = 5.0 \times 10^{-2}$ H, the resistance is $R = 2000 \ \Omega$, the emf of the battery is 3.0 V, and the emf of the oscillating source is $\mathcal{E} = 1.5 \cos(6000\pi t)$, where $\mathcal{E}$ is in volts and t in seconds. At the initial time $t = 0$, the current in the inductor is zero.

(a) From this initial condition, calculate the instantaneous current in the inductor and in the resistor for times after $t = 0$.

(b) Calculate the instantaneous power delivered to the inductor and to the resistor.

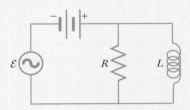

FIGURE 32.46 Inductor and resistor connected to a battery and a source of oscillating emf.

*76. What is the resonant frequency of the circuit shown in Fig. 32.47? The inductance is $L = 1.5$ H, and the capacitances are $C_1 = 20 \ \mu\text{F}$ and $C_2 = 10 \ \mu\text{F}$.

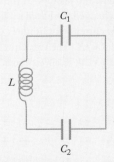

FIGURE 32.47 Two capacitors connected to an inductor.

*77. An LC circuit has an inductance of 5.0×10^{-2} H and a capacitance of 5.0×10^{-6} F. At $t = 0$, the capacitor is fully charged so $Q_{max} = 1.2 \times 10^{-4}$ C. What is the energy in this circuit? At what time, after $t = 0$, will the energy be purely magnetic? At what *later* time will it be purely electric?

78. In a series RL circuit (see Fig. 32.22), a 1.2-kHz oscillating emf has amplitude $\mathcal{E}_{max} = 24$ V and the current amplitude is $I_{max} = 0.45$ A. The ratio L/R is 5.0×10^{-3} s. What is the value of the resistor?

79. A series RLC circuit (see Fig. 32.41) has $\mathcal{E}_{max} = 15$ V, $R = 6.0 \ \Omega$, $L = 55$ mH, and $C = 0.25 \ \mu\text{F}$.

(a) What is the resonant frequency (in Hz)?

(b) Calculate the amplitude of the voltage across the capacitor at resonance.

(c) What is the quality factor Q for this circuit?

80. The primary winding of a transformer has 140 turns and the secondary has 2200 turns. This transformer is connected to a second, identical transformer, so the output of the first transformer serves as input for the second transformer. If the primary of the first transformer is connected to a source of 115 V AC, what is the voltage delivered by the secondary of the second transformer?

81. As illustrated in Fig. 32.34, transformers are used by electric companies to step up or down the voltage from the power plant to your home. If you use a current of 5.0 A in your vacuum cleaner at 115 V AC, how much of a current does this require at the power plant (at 22 000 V)? In the transmission line (at 500 000 V)?

82. A power station feeds 1.0×10^8 W of electric power at 760 kV into a transmission line. Suppose that 10% of this power is lost in Joule heat in the transmission line. What percentage of the power would be lost if the power station were to feed 340 kV into the transmission line instead of 760 kV, other things being equal?

83. A transformer consists of two concentric coils of thin insulated wire of low resistance. One coil has 800 turns and the other has 200 turns. The second coil is wound tightly around the first, so the two have nearly the same radius. If the first coil is connected to a source of oscillating emf supplying a voltage of amplitude 60 V and frequency 1500 Hz, what will be the voltage across the terminals of the second coil?

Answers to Checkups

Checkup 32.1

1. Since this is twice the rms voltage of 115 V AC, the maximum emf will also be doubled to 2×163 V = 326 V. Alternatively, one can use Eq. (32.8) to obtain $\mathcal{E}_{max} = \sqrt{2}\mathcal{E}_{rms} = \sqrt{2} \times 230$ V = 325 V.

2. By the usual definitions, the current is maximum at its most positive value (I_{max}) and is minimum at its most negative value ($-I_{max}$). Thus, for a frequency of 60 Hz, there are 60 minima and 60 maxima per second. The power is always positive for a resistor circuit ($P = I^2R$), and so varies between zero and P_{max} twice per current cycle (once for positive current, once for negative current); thus the power has 120 minima and 120 maxima per second.

3. Both the large slot and the round slot are nearly at a potential of zero volts, so no appreciable current will flow in the lightbulb and it will not shine.

4. (A) 100 W. An emf with a 141-V amplitude has a root-mean-square voltage $\mathcal{E}_{rms} = \mathcal{E}_{max}/\sqrt{2} = 141$ V$/\sqrt{2} = 100$ V, and thus an average power $\overline{P} = \mathcal{E}_{rms}^2/R = (100$ V$)^2/(100 \ \Omega) = 100$ W.

Checkup 32.2

1. No—instantaneously, the current is zero when the voltage is maximum, clearly different from Ohm's Law. However, the *amplitudes* of the current and voltage obey a relationship similar to Ohm's Law: $I_{max} = \mathcal{E}_{max}/X_C$, where the capacitive reactance $X_C = 1/(\omega C)$ provides a frequency-dependent opposition to AC.

2. The stored energy varies with the square of the instantaneous emf: $U = \frac{1}{2}C\mathcal{E}^2$ [see Eq. (26.27)]. Since $\mathcal{E} = \mathcal{E}_{max} \cos\omega t$, a plot similar to Fig. 32.5 is obtained, except that the maximum value, occurring at $t = 0, \pi/\omega, 2\pi/\omega$, etc., is an energy with value $U_{max} = \frac{1}{2}C\mathcal{E}_{max}^2$.

3. The capacitive reactance is inversely proportional to frequency, $X_C = 1/(\omega C)$, and so is 10 times less, or $10^8 \ \Omega$, at 600 Hz. Similarly, it is $10^7 \ \Omega$ at 6000 Hz.

4. Yes—low frequency means almost static conditions (DC corresponds to zero frequency), where the charge on the capacitor plates does not change at any appreciable rate, so there is essentially no current.

5. (C) 2, 4. For a given capacitor at fixed frequency, the current is proportional to the voltage amplitude, $I_{max} = \mathcal{E}_{max}/X_C$, and so increases a factor of 2. For a capacitor, the maximum power is proportional to the square of the voltage amplitude, and so the new maximum power is related to the old by a factor of 4.

Checkup 32.3

1. No—instantaneously, the current is zero when the voltage is maximum, clearly different from Ohm's Law. However, the *amplitudes* of the current and voltage obey a relationship similar to Ohm's Law: $I_{max} = \mathcal{E}_{max}/X_L$, where the inductive reactance $X_L = \omega L$ provides a frequency-dependent opposition to AC.

2. The stored energy varies with the square of the instantaneous current: $U = LI^2/2$ [see Eq. (31.26)]. Since, from Eq. (32.18), $I = [\mathcal{E}_{max}/(\omega L)]\sin\omega t$, a $\sin^2\omega t$ plot is obtained for the energy (similar to the $\cos^2\omega t$ plot of Fig. 32.5), with a minimum value of zero at $t = 0$ (and at $t = \pi/\omega, 2\pi/\omega$, etc.) and maximum values of $U_{max} = \mathcal{E}_{max}^2/(2\omega^2L)$.

3. The inductive reactance is proportional to the frequency: $X_L = \omega L = 2\pi fL$. Thus at 10 times the frequency (600 Hz), the inductive reactance will be 10 times as large, or $10^4 \ \Omega$. Similarly, it is $10^5 \ \Omega$ at 6000 Hz.

4. Yes—low frequency means almost static conditions, so there is no back emf to oppose the driving voltage; since there is no resistance, the current will be large.

5. (D) $\frac{1}{10}$; 10. Increasing the inductance by a factor of 10 increases the inductive reactance by the same factor ($X_L = \omega L$), and so *decreases* the current by a factor of 10 ($I_{max} = \mathcal{E}_{max}/X_L$).

Checkup 32.4

1. Increasing the plate separation decreases the capacitance and so, by Eq. (32.26), increases the resonant frequency.

2. Connecting two identical capacitors in parallel increases the capacitance by a factor of 2; this decreases $\omega_0 = 1/\sqrt{LC}$ by a factor of $\sqrt{2}$ [Eq. (32.26)].

3. From Eq. (32.33), $Q = \omega_0 L/R = X_L/R = 500 \ \Omega/10 \ \Omega = 50$. From Eq. (32.34), one characteristic time for the decaying oscillation amplitude will include $N = Q/\pi = 50/\pi \approx 16$ cycles.

4. (B) $\mathcal{E}\sqrt{LC}/R$; $\sin\omega t$. With the switch in position 1 for a long time, the current in the left loop becomes steady at $I = \mathcal{E}/R$. When switched to position 2, the current in the inductor is initially unchanged, and then oscillates with time. By Eq. (32.25), this current amplitude is $Q_{max}/\sqrt{LC}$ which implies $Q_{max} = \mathcal{E}\sqrt{LC}/R$. The capacitor is initially uncharged, that is, $Q = 0$ at $t = 0$, so $\sin\omega t$ provides the time dependence of the charge.

Checkup 32.5

1. If the current increases, the impedance must have decreased. Only a capacitive reactance decreases with frequency $[X_C = 1/(\omega C)]$, so if there is only one other component, it must be a capacitor.

2. The current is maximum at resonance, when $I_{max} = \mathcal{E}_{max}/R$, so $R = \mathcal{E}_{max}/I_{max} = 1.0 \text{ V}/0.20 \text{ A} = 5.0 \text{ } \Omega$. By Eq. (32.54), the quality factor is $Q = X_L/R = 50 \text{ } \Omega/5.0 \text{ } \Omega = 10$.

3. There are several ways to calculate the Q; for example, Eq. (32.54) may be used: $Q = V_{max,L}/V_{max,R} = 200 \text{ V}/1.0 \text{ V} = 200$.

4. (E) 2.0 V. At resonance, the voltages across the inductor and capacitor have equal amplitudes, but are 180° out of phase and sum to zero. Thus, as discussed in Example 6, the amplitude of the voltage across the resistor is the same as the emf; that is, at ω_0, $V_{max,R} = \mathcal{E}_{max} = 2.0 \text{ V}$.

Checkup 32.6

1. The ratio of emfs is the same as the ratio of turns in the coils, so the number of turns in the secondary needs to be a factor of 2 greater than the number in the primary.

2. For a step-down transformer, the smaller voltage requires a larger current in the secondary, since the power delivered to the load equals the power drawn from the source [Eq. (32.61)].

3. (E) 10, 2.5, 0.4, 0.1. The transformer with turns ratio 2 will provide a voltage ratio of 2 or $\frac{1}{2}$ depending on whether it is used as a step-up or step-down transformer. Similarly, the other transformer can provide a voltage ratio of 5 or $\frac{1}{5}$. Connected one after the other, we have four possible combinations: $2 \times 5 = 10$, $\frac{1}{2} \times 5 = 2.5$, $\frac{1}{5} \times 2 = 0.4$, and $\frac{1}{5} \times \frac{1}{2} = 0.1$.

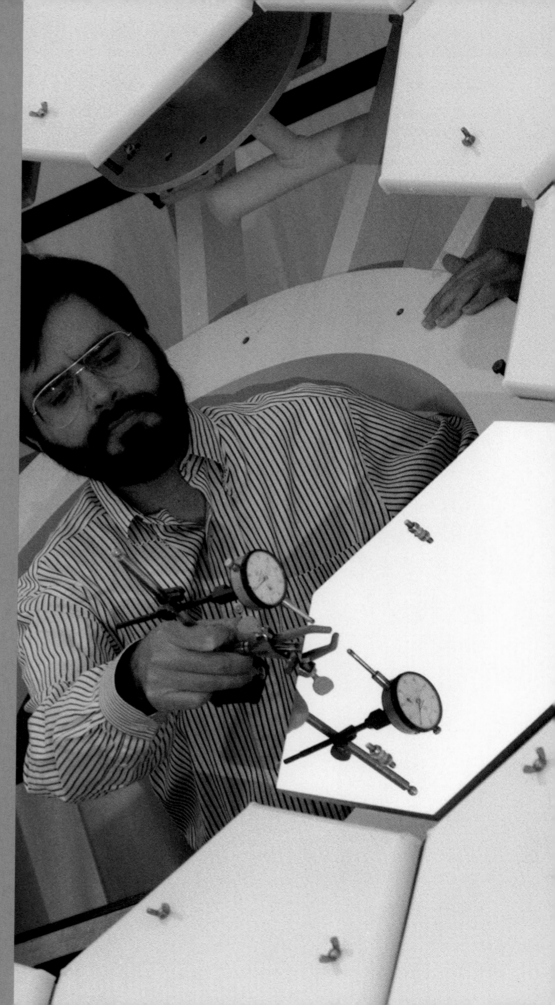

PART

5

Waves and Optics

CONTENTS

Each segment in this 5.5-meter diameter array of hexagonal mirrors has a spherical surface, so the combination of all 144 segments forms an overall approximation of a parabolic surface. A refined optics manufacturing process permits these aluminum segments to attain the smoothness and reflectivity of glass mirrors.

CHAPTER

33

Electromagnetic Waves

CONCEPTS IN CONTEXT

Concepts
— *in* —
Context

Electromagnetic waves are found throughout our environment. The antennas on the tower in the photo emit and receive electromagnetic waves; even the light coming from the photo to your eyes consists of electromagnetic waves.

As we learn about electromagnetic waves, we will consider such questions as:

❓ How are the electric and magnetic fields of an electromagnetic wave related? (Section 33.3 and Example 3, page 1081; and Example 8, page 1098)

❓ How does an antenna generate electromagnetic waves? What determines the size of an antenna? (Section 33.4 and Example 5, page 1088)

? In what way is a periodic electromagnetic wave altered for AM and FM radio communications? (Physics in Practice: AM and FM Radio, page 1089)

? How does the strength of an electromagnetic wave decrease with distance from the source? (Example 6 and Section 33.5, page 1094)

We already know that a changing magnetic field induces an electric field, described by Faraday's Law of induction. In this chapter, we will discover that the converse is also true: a changing electric field induces a magnetic field. The law describing this induction effect of electric fields was formulated by James Clerk Maxwell, who thereby achieved a wide-ranging unification of all the laws of electricity and magnetism. These laws became known as Maxwell's equations. The next three chapters of this book are, in essence, nothing but applications of Maxwell's equations.

The mutual induction of electric and magnetic fields gives rise to the phenomenon of self-supporting electromagnetic oscillations in empty space. If, initially, there exists an oscillating electric field, it will induce a magnetic field, and this will induce a new electric field, and so on. Thus, these fields can perpetuate each other. An oscillating charge or current is needed to get the fields started, but after this initiation the fields continue on their own. These self-supporting oscillations are **electromagnetic waves**, either traveling waves or standing waves. Electromagnetic waves are similar to mechanical waves, such as waves on a string or sound waves in air, in that the disturbance at one point causes a disturbance at a neighboring point, and this causes a disturbance at the next neighboring point, etc. But electromagnetic waves differ from mechanical waves in that they propagate through empty space—they propagate without any material medium. Among such electromagnetic waves are radio waves, microwaves, light, and X rays. All of these kinds of waves are qualitatively the same; they consist of oscillating electric and magnetic fields. The only difference between a radio wave and light is in their wavelength and frequency—the radio wave has a much longer wavelength and smaller frequency than light. In this chapter we will examine the properties of electromagnetic waves, and see how they are generated by accelerated electric charges.

33.1 INDUCTION OF MAGNETIC FIELDS; MAXWELL'S EQUATIONS

As we saw in Chapter 31 [Eq. (31.12)], Faraday's Law for the electric field induced by a changing magnetic field can be expressed in the form

$$\oint E_{\parallel} \, ds = -\frac{d\Phi_B}{dt} \tag{33.1}$$

where $E_{\parallel}$ is the component of the induced electric field parallel to a closed path (see Fig. 33.1), ds is the magnitude of a displacement along the path, and Φ_B is the magnetic flux intercepted by the surface within the path.

The magnetic field induced by a changing electric field is described by an equation analogous to Faraday's Law. On the left side of the equation that describes this new kind of induction there appears the induced magnetic field (see Fig. 33.2), and on the right side there appears the rate of change of the electric flux intercepted by the surface within the path:

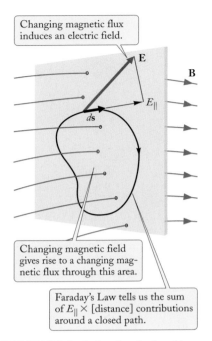

Changing magnetic flux induces an electric field.

Changing magnetic field gives rise to a changing magnetic flux through this area.

Faraday's Law tells us the sum of $E_{\parallel} \times$ [distance] contributions around a closed path.

FIGURE 33.1 A closed path placed in a changing magnetic field. $E_{\parallel}$ is the component of the induced electric field tangent to the path.

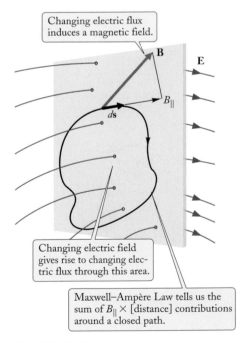

Changing electric flux induces a magnetic field.

Changing electric field gives rise to changing electric flux through this area.

Maxwell–Ampère Law tells us the sum of $B_{\parallel} \times$ [distance] contributions around a closed path.

FIGURE 33.2 A closed path placed in a changing electric field. $B_{\parallel}$ is the component of the induced magnetic field tangent to the path.

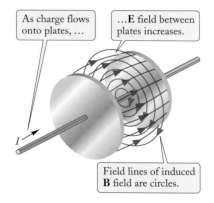

FIGURE 33.3 A capacitor is being charged by wires connected to its plates. In the space between the plates there is an increasing electric field (red) and an increasing electric flux. The lines of the induced magnetic field (blue) are circles.

(a)

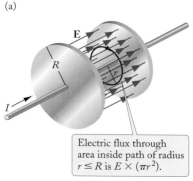

(b)

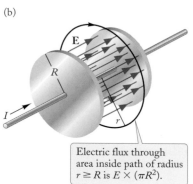

FIGURE 33.4 (a) A path of radius $r \leq R$ for the determination of the induced magnetic field between two capacitor plates. (b) A path of radius $r \geq R$ for the determination of the induced magnetic field outside the capacitor plates.

$$\oint B_{\parallel} \, ds = \mu_0 \epsilon_0 \frac{d\Phi_E}{dt} \qquad (33.2)$$

Apart from the minus sign in Eq. (33.1) and the constant $\mu_0\epsilon_0$ in Eq. (33.2), these two equations are "mirror images" of each other. Equation (33.1) states that a changing magnetic flux induces an electric field; Eq. (33.2) states that *a changing electric flux induces a magnetic field*. Thus, these equations can be obtained from each other by exchanging electric and magnetic fields. This indicates a certain symmetry in the effects of electric and magnetic fields on one another. The induction of magnetic fields by a changing electric field was first proposed by Maxwell on purely theoretical considerations; the experimental demonstration of this kind of induction came only later.

For a simple example of a system with a rate of change of electric flux and an induced magnetic field, consider a capacitor that is being charged by currents flowing into its plates (see Fig. 33.3). In the space between the plates there is then an increasing electric field, and there is increasing electric flux. If the plates of the capacitor are circular, as shown in Fig. 33.3, the induced magnetic field lines will have the shape of concentric circles, just like the field lines surrounding a long wire carrying a current. The strength of the magnetic field can be calculated in the manner familiar from the calculations with Ampère's Law in Chapter 29: take a path that follows one of the circular field lines and evaluate each side of Eq. (33.2) for this choice of path. According to Eq. (33.2), we then find that the strength of the induced magnetic field is directly proportional to the rate of change of the electric flux, or to the rate of change of the electric field between the plates.

EXAMPLE 1   If the capacitor plates in Fig. 33.3 have radius R and the electric field is increasing at a rate dE/dt, find an expression for the induced magnetic field for $r \leq R$. Do the same for $r \geq R$. What is the current in the wire?

SOLUTION: By symmetry, we can see that the magnetic field lines will be tangent to circles around the capacitor axis, as shown in Fig. 33.3. The magnetic field must also have a constant magnitude along a given circular path of some radius r. Thus, similar to what was done with Ampère's Law in Section 29.3, the left side of Eq. (33.2) for any such circular path can be rewritten

$$\oint B_{\parallel} \, ds = B \times 2\pi r$$

For $r \leq R$ the electric field is constant and perpendicular to the surface of area $A = \pi r^2$ (see Fig. 33.4a). Thus we can write the rate of change of the electric flux as

$$\frac{d\Phi_E}{dt} = A \frac{d}{dt} E = \pi r^2 \frac{dE}{dt} \qquad (r \leq R)$$

Inserting these relations in Eq. (33.2) gives

$$B \times 2\pi r = \mu_0 \epsilon_0 \pi r^2 \frac{dE}{dt}$$

or, solving for the magnetic field,

$$B = \frac{\mu_0 \epsilon_0}{2} r \frac{dE}{dt} \qquad (r \leq R)$$

If $r \geq R$, then the electric field intercepts only part of the surface within a path of radius r. As shown in Fig. 33.4b, this is the part with radius R. Hence

$$\frac{d\Phi_E}{dt} = A\frac{d}{dt}E = \pi R^2 \frac{dE}{dt} \quad (r \geq R)$$

By equating the two sides of Eq. (33.2) we have

$$B \times 2\pi r = \mu_0 \epsilon_0 \pi R^2 \frac{dE}{dt}$$

and we obtain for the magnetic field

$$B = \frac{\mu_0 \epsilon_0}{2}\frac{R^2}{r}\frac{dE}{dt} \quad (r \geq R)$$

The current can be determined from the known behavior of a capacitor. The electric field between the plates is $E = \sigma/\epsilon_0$, where $\sigma = Q/(\pi R^2)$ is the charge density on a plate [Eq. (23.13)]. The charge on the plates is $Q = \pi R^2 \sigma = \epsilon_0 \pi R^2 E$, and the current I is

$$I = \frac{dQ}{dt} = \epsilon_0 \pi R^2 \frac{dE}{dt}$$

Notice that if we compare the last two equations, we can conclude

$$B = \frac{\mu_0 I}{2\pi r}$$

which is a result familiar from Ampère's Law [see Eq. (29.9)]. Thus, outside the capacitor, the magnetic field induced by the total changing electric flux is the same as that produced by the current causing that changing flux (but the magnetic field is weaker than this at points inside the capacitor).

Although the magnetic fields of currents (described by Ampère's Law) and the magnetic fields induced by electric fields [described by Eq. (33.2)] can have different sources, the equations describing these fields both involve the tangential component of the field. Hence Ampère's Law and Eq. (33.2) can be combined into one single equation, the **Maxwell–Ampère Law**:

$$\oint \mathbf{B} \cdot d\mathbf{s} = \mu_0 I + \mu_0 \epsilon_0 \frac{d\Phi_E}{dt} \quad \text{or} \quad \oint B_\| \, ds = \mu_0 I + \mu_0 \epsilon_0 \frac{d\Phi_E}{dt} \quad (33.3)$$

Maxwell-Ampère Law

This equation covers all conceivable cases of production of magnetic fields, by currents, by changing electric fields, or by any combination of both. For example, if the space between the capacitor plates in Fig. 33.3 is filled with a slightly conducting material (a leaky capacitor), then there will be both an electric current and an electric flux in this space, and both will contribute to the magnetic field.

The right side of Eq. (33.3) can be written as $\mu_0(I + \epsilon_0 \, d\Phi_E/dt)$, which shows that the quantity $\epsilon_0 \, d\Phi_E/dt$ has the same effect for the production of magnetic fields as the ordinary electric current I. Accordingly, the quantity $\epsilon_0 \, d\Phi_E/dt$ is called the **displacement current** (although there is a good reason for the word *current*, there is no good reason for the word *displacement*).

displacement current

Note that in the case of the capacitor plates illustrated in Fig. 33.3, the displacement current between the plates is large whenever the ordinary current flowing on the external wires connected to the plates is large. As we saw in Example 1, the ordinary current deposits charge on the plates of the capacitor, and whenever the ordinary current is large, the rate of change of the charge is large, and so is the rate of change of the electric field between the plates, and so is the rate of change of the electric flux. Since they both produce the same magnetic field, it is evident from Example 1 that the magnitude of the total displacement current between the plates is equal to the magnitude of the ordinary current flowing on the external wires. *Thus, the displacement current between the capacitor plates can be viewed as a "continuation" of the ordinary current by a corresponding rate of change of electric flux.*

Equation (33.3) is Maxwell's modification of Ampère's Law. The great importance of this equation lies in its general validity—Maxwell boldly proposed that this equation is valid not only for the electric and magnetic fields associated with capacitors, wires, and other such devices, but also for the fields associated with electromagnetic waves.

Equation (33.3) is the last of the fundamental laws that we need for a complete mathematical description of the behavior of electric and magnetic fields. There are four fundamental laws, as follows:

Gauss' Law for electricity [Eq. (24.10)],

$$\oint \mathbf{E} \cdot d\mathbf{A} = \frac{Q_{\text{inside}}}{\epsilon_0} \quad \text{or} \quad \oint E_\perp \, dA = \frac{Q_{\text{inside}}}{\epsilon_0} \tag{33.4}$$

Gauss' Law for magnetism [Eq. (29.15)],

$$\oint \mathbf{B} \cdot d\mathbf{A} = 0 \quad \text{or} \quad \oint B_\perp \, dA = 0 \tag{33.5}$$

Maxwell's equations

Faraday's Law [Eqs. (31.11) and (31.12)],

$$\oint \mathbf{E} \cdot d\mathbf{s} = -\frac{d\Phi_B}{dt} \quad \text{or} \quad \oint E_\parallel \, ds = -\frac{d\Phi_B}{dt} \tag{33.6}$$

and the Maxwell–Ampère Law [Eq. (33.3)],

$$\oint \mathbf{B} \cdot d\mathbf{s} = \mu_0 I + \mu_0 \epsilon_0 \frac{d\Phi_E}{dt} \quad \text{or} \quad \oint B_\parallel \, ds = \mu_0 I + \mu_0 \epsilon_0 \frac{d\Phi_E}{dt} \tag{33.7}$$

The physical basis for each of these four laws may be briefly summarized as follows:

- Gauss' Law is based on Coulomb's Law describing the forces of attraction and repulsion between stationary charges.
- Gauss' Law for magnetism asserts that there are no sources or sinks of magnetic field lines.
- Faraday's Law describes the induction of an electric field by a changing magnetic flux.
- The Maxwell–Ampère Law is based on the law of magnetic force between moving charges and it also contains the induction of a magnetic field by a changing electric flux.

Taken as a whole, the four laws (33.4)–(33.7) are known as **Maxwell's equations**, because Maxwell supplied the missing link between the magnetic and the electric fields [Eq. (33.3)] and thereby placed the capstone on electromagnetic theory. Maxwell recognized that these equations imply a dynamic interplay between electric and magnetic fields, an interplay that couples and unifies electric and magnetic phenomena.

Maxwell's equations provide a complete description of the interactions among charges, currents, electric fields, and magnetic fields. All the properties of the fields can be deduced by mathematical manipulation of these equations. If the distribution of charges and currents is given, then these equations uniquely determine the corresponding fields. Even more important, Maxwell's equations uniquely determine the time evolution of the fields, starting from a given initial condition for these fields—if we know the fields at an initial time, we can calculate them at any later time. Thus, these equations accomplish for the dynamics of electromagnetic fields what Newton's equations of motion accomplish for the dynamics of particles.

Although the experimental foundation on which we based the development of Maxwell's equations was restricted to charges at rest or charges in uniform motion, these equations also govern the fields of accelerated charges and the fields of light and radio waves. In the remaining sections of this chapter we will calculate the fields of electromagnetic waves from our equations and we will see that the results are in agreement with the observed properties of light and radio waves.

 ## Checkup 33.1

QUESTION 1: If a current of 2 A is charging a capacitor of 400 pF, what is the total displacement current between the capacitor plates? What if the capacitance is 800 pF?

QUESTION 2: In some region of space, where there is no electric charge and no current, there are electric and magnetic field lines that form closed loops. Can such electric and magnetic fields be static?

QUESTION 3: Do the field lines of an induced electric field start and stop on electric charges?

QUESTION 4: Figure 33.3 shows a current charging a capacitor. Suppose that at some time the current stops and the capacitor remains with some constant charge. After this time, is there an electric field between the plates? A magnetic field?

QUESTION 5: Consider a constant current I flowing into a capacitor, as in Fig. 33.4. At what radial distance from the center of the capacitor is the magnetic field the largest?

(A) $r = 0$ (B) $r = R/2$ (C) $r = R$ (D) $r = 2R$ (E) $r = \infty$

33.2 THE ELECTROMAGNETIC WAVE PULSE

We started our study of electricity and magnetism with the electric field of a point charge at rest—this is the Coulomb field given by Eq. (23.2). Later, we dealt with the fields of a current, that is, charges in motion with uniform velocity—in addition to the Coulomb field, such moving charges have a magnetic field. Now we will investigate the fields of a charge with *accelerated motion*. We will find that in this case there are extra electric and magnetic fields that spread outward from the position of the charge, like ripples on a pond in which a stone has been dropped, and carry away energy and momentum. These spreading fields are called **radiation fields**, or **wave fields**.

radiation fields

We can gain some insight into the fields spreading out from an accelerated charge by the following argument. Consider a charge that is initially at rest, then is quickly accelerated for some short time interval, and then continues to move with a constant final velocity. The initial electric field lines originate at the initial position of the charge. But the field lines at some later time, after the acceleration ceases, must originate on the new position of the charge. The field lines cannot change from their initial configuration to their final configuration instantaneously; rather a disturbance must travel outward from the position of the charge and gradually change the field lines. The disturbance takes the form of a kink connecting the old and the new field lines.

The disturbance travels at some speed c (we will establish below that the speed of the disturbance is actually the speed of light, and our notation anticipates this result). Figure 33.5 shows the situation at some time after the acceleration has ended. Suppose that the acceleration lasts from time $t = 0$ to time $t = \tau$. Then the disturbance of the electric field lines begins at time $t = 0$ and ends at time $t = \tau$. The leading edge of the disturbance (outer edge of kink) travels outward from the initial position of the charge and in a time t it reaches out to a distances ct. Beyond the sphere of radius ct, the electric field is still the old field with field lines centered on the initial position of the charge.

The disturbance ceases as soon as the acceleration ceases. The field in the vicinity of the uniformly moving charge then settles into the new radial configuration centered on the new position of the charge. The trailing edge of the disturbance (inner edge of kink) marking the cessation of the acceleration travels outward from the position that the charge has at the time $t = \tau$, when the acceleration ceases; and by some later time t it reaches out to a distance $c(t - \tau)$. Within the sphere of radius $c(t - \tau)$, the electric field is the new radial field centered on the new position of the uniformly moving charge.

The disturbance produced by the accelerated charge is confined to the space between the larger and the smaller spheres in Fig. 33.5. The field lines in this zone must connect the lines of the new field of the uniformly moving charge with the lines of the old field of the stationary charge. As shown in Fig. 33.5, each connecting segment of field line makes an angle with the radial line to the position of the charge. Thus, *the electric field in the zone of the kink has both a radial component and a tangential, or trans-*

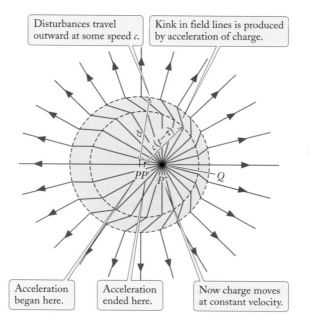

Disturbances travel outward at some speed c.

Kink in field lines is produced by acceleration of charge.

Acceleration began here.

Acceleration ended here.

Now charge moves at constant velocity.

FIGURE 33.5 Electric field lines of a charge that has suffered an acceleration. Here P'' is the present position of the charge, P is the initial position of the charge, and P' is an intermediate position. Between P and P' the charge suffered a constant acceleration. Between P' and P'' the charge moved at constant velocity. The outer dashed sphere (outer edge of the kink) has radius ct and is centered on P; the inner dashed sphere (inner edge of kink) has radius $c(t - \tau)$ and is centered on P'.

(a)

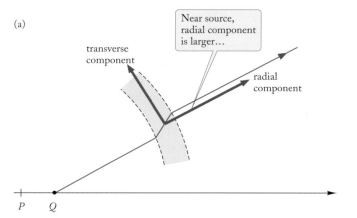

(b)

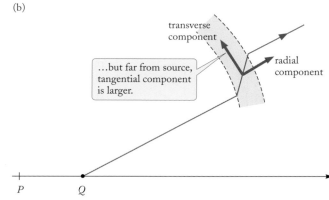

FIGURE 33.6 (a) One of the field lines of Fig. 33.5. In the zone of the kink the electric field has both a radial and a transverse component. (b) The same field line at a later time, when the particle has traveled further to the right and the kink has traveled farther outward. The kink is more pronounced because the radial component of the electric field has decreased more than the transverse component (the radial component decreases in proportion to $1/r^2$; the transverse component decreases in proportion to $1/r$).

verse, component. This transverse component is the radiation field, or the wave field, of the accelerated charge. From Fig. 33.5, we see that this transverse component is maximum for directions perpendicular to the original acceleration, and it is zero along the direction of the original acceleration.

The traveling kink is a single wave pulse, analogous to the single wave pulse we might produce on a stretched string by flicking its end just once. If we want to produce a periodic electromagnetic wave, we have to accelerate the charge back and forth periodically, and thereby generate a succession of kinks in the field lines. This picture of an electromagnetic wave as a succession of kinks in the field lines explains how such waves are able to travel in a vacuum, a puzzle that greatly worried the physicists of the nineteenth century. The electromagnetic wave does not really travel in a vacuum—it is a disturbance in the electric field of the charge, that is, it travels in the electric field of the charge. The space surrounding the charge is not really a vacuum; the space is filled with the electric field of the charge, and this electric field is the medium in which the electromagnetic wave propagates.

As the kink travels outward, the angle between the radial direction and the direction of the electric field in the zone of the kink increases; thus, the kink becomes more pronounced and the electric field becomes more transverse; Fig. 33.6 illustrates this for one field line. Thus, when the kink reaches a large distance from the charge, the electric field in the zone of the kink will be almost entirely transverse. Such an electric field at right angles to the direction of propagation is a characteristic feature of electromagnetic waves.

The transverse kink propagates in the outward direction as a spherical wave pulse, but the strength of this pulse depends on direction. Often we will be interested in the behavior of the wave over only a relatively small interval of distances, an interval small compared with the radius of the spherical wave (see Fig. 33.7). For example, we might be interested in the wave that the moving charges on the antenna of a distant radio transmitter produce in the room in which we are sitting. If the length of the room is, say, 5 m and the distance to the transmitter is 20 km, then the portion of the spherical wave pulse that passes through the room will look flat and uniform, and we can approximate the spherical wave pulse by a plane wave pulse, with plane wave fronts and with straight electric field lines at right angles to the direction of propagation.

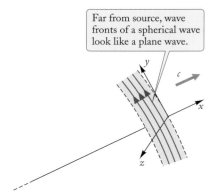

FIGURE 33.7 A small portion of the spherical wave pulse can be regarded as a plane wave pulse. The electric field in this wave pulse is almost perpendicular to the direction of propagation.

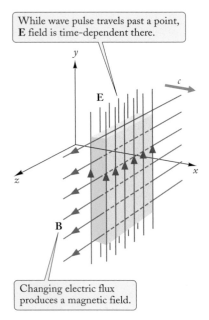

While wave pulse travels past a point, **E** field is time-dependent there.

Changing electric flux produces a magnetic field.

FIGURE 33.8 Electric field lines of a plane wave pulse consisting of a region of uniform electric field **E**. The wave front, or front surface of the pulse, is a flat plane. The magnetic field lines are at right angles to the electric field lines. The entire pattern of field lines travels in the x direction with the wave speed c.

right-hand rule for electromagnetic waves

Figure 33.8 shows such a flat wave pulse propagating in the x direction; the transverse electric field, at right angles to the direction of propagation, is represented by the straight field lines parallel to the y direction. For the sake of simplicity, we will assume for now that the electric field in the wave pulse is of constant magnitude (the magnitude of the electric field in the wave pulse depends on the acceleration of the charge that generated the pulse—constant magnitude of the electric field requires constant magnitude of the acceleration).

Since the wave pulse sweeps through the room at the speed c, which, as we will see, is the speed of light, it lasts only some short time at any one point of the room. This means the electric field of the wave pulse is time-dependent: initially it is zero, then it suddenly increases to some constant value E when the leading edge of the pulse arrives, then it remains constant for a while, and finally it drops to zero when the rear edge of the pulse passes.

Besides the transverse electric field, *the wave pulse is also endowed with a transverse magnetic field, at right angles to the electric field and to the direction of propagation* (see Fig. 33.8). This magnetic field arises by induction, from the time-dependent electric field and the time-dependent electric flux. The induced magnetic field can be calculated from the time-dependent electric field by means of the Maxwell–Ampère Law. In the next section, we will relate the magnitude of the induced magnetic field to that of the electric field; here, let us try to understand why its direction is perpendicular to the electric field. For this purpose, consider a small stationary rectangle perpendicular to the electric field. In Fig. 33.9, this rectangle is in the x–z plane. This rectangle registers a change of electric flux when an edge of the wave pulse sweeps across it; thus there is a rate of change of flux, or a displacement current, associated with each edge of the wave pulse. In Fig. 33.9, the displacement current is parallel to the changing vertical electric field lines. Since the effect of such a displacement current is the same as that of a real current, the magnetic field produced by the changing electric flux is the same as that of a vertical plane of real current at the pulse edge. According to the usual right-hand rule for current and magnetic field, such a current produces a magnetic field perpendicular to the current; thus, the magnetic fields contributed by the two edges are in the z direction, and so is the net magnetic field shown in Fig. 33.9. Note that the correct directions of the magnetic field and the electric field are related by a simple **right-hand rule for electromagnetic waves**: if the fingers of the

FIGURE 33.9 A small stationary rectangle (green) perpendicular to the electric field lines registers a rate of change of electric flux at the instant the front edge of the wave pulse sweeps across the rectangle, and again at the instant the rear edge of the wave pulse sweeps across the rectangle. Hence there is a sheet of displacement current at the front edge of the wave pulse, and another sheet of displacement current at the rear edge. The black lines indicate the flow of the displacement currents. The blue lines indicate the magnetic field produced by these currents.

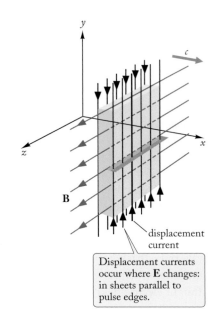

displacement current

Displacement currents occur where **E** changes: in sheets parallel to pulse edges.

right hand are placed along the direction of the electric field and curl toward the magnetic field, the thumb will point along the direction of propagation of the wave (see Fig. 33.10). In vector notation, the propagation is in the direction of the vector cross product, **E × B**.

Like the electric field of the wave pulse, the magnetic field is time-dependent: initially it is zero, then it suddenly increases to some constant value B when the leading edge of the pulse arrives, then it remains constant for a while, and then it drops to zero when the rear edge of the pulse passes. According to Faraday's Law, this time-dependent magnetic field and its time-dependent magnetic flux give rise to an induced electric field. By means of Faraday's Law, it can be verified that this induced electric field coincides with the original electric field E of the wave pulse. Thus, the electric and magnetic fields of the wave pulse mutually induce each other—the electric field induces the magnetic field, and the magnetic field induces the original electric field. These fields thereby become self-supporting, and they continue to exist and to propagate independently of what happens to the charge that initially generated them.

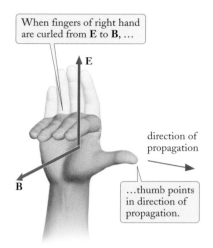

FIGURE 33.10 Right-hand rule for electromagnetic waves. The fingers of the right hand are placed along the direction of the electric field and curl toward the magnetic field; the thumb then points along the direction of propagation of the wave.

 Checkup 33.2

QUESTION 1: For each of the following wave pulses, what is the direction of the magnetic field?

   (A) Wave pulse is traveling eastward and electric field is southward.

   (B) Wave pulse is traveling westward and electric field is southward.

   (C) Wave pulse is traveling upward and electric field is southward.

QUESTION 2: For each of the following wave pulses, what is the direction of the electric field?

   (A) Wave pulse is traveling eastward and magnetic field is southward.

   (B) Wave pulse is traveling westward and magnetic field is southward.

   (C) Wave pulse is traveling upward and magnetic field is southward.

QUESTION 3: Which of Maxwell's equations determine how the electric and magnetic fields of the wave pulse induce each other?

QUESTION 4: A wave pulse is traveling southward and the electric field is downward. What is the direction of the magnetic field?

   (A) Upward   (B) Northward   (C) Eastward   (D) Westward

33.3 PLANE WAVES; POLARIZATION

Online *Concept* Tutorial

In this section, we will concentrate on periodic electromagnetic waves, produced by accelerating a charge back and forth repeatedly. *A periodic electromagnetic wave consists of alternating zones of positive and negative electric field and magnetic field.* The positive and negative electric fields define the wave crests and wave troughs of the wave. The wave is characterized by a frequency and a wavelength.

As described in the preceding section, the electric and magnetic fields of the electromagnetic wave mutually induce each other, and the wave thereby becomes self-supporting. The wave is initially generated by some accelerated charges; but once it has been generated, it propagates on its own, independent of the accelerated charges that generated it.

By a detailed theoretical analysis (Section 33.6), based on this mutual induction of the electric and magnetic fields according to the Maxwell–Ampère Law and Faraday's Law, we can establish that the speed c of the electromagnetic wave is given by the expression

speed of electromagnetic wave

$$c = \frac{1}{\sqrt{\mu_0 \epsilon_0}} \tag{33.8}$$

This result was first obtained by Maxwell. Numerically, this theoretical expression for the speed of electromagnetic waves yields

$$c = \frac{1}{\sqrt{\mu_0 \epsilon_0}} = \frac{1}{\sqrt{1.26 \times 10^{-6} \text{ H/m} \times 8.55 \times 10^{-12} \text{ F/m}}}$$

$$= 3.00 \times 10^8 \text{ m/s} \tag{33.9}$$

This value coincides with the measured value of the speed of light in vacuum, a coincidence that led Maxwell to propose that light waves are electromagnetic waves, consisting of self-supporting, mutually induced electric and magnetic fields.

Maxwell's theory of the propagation of electromagnetic waves and of their generation by accelerated charges received direct experimental confirmation at the hands of Heinrich Hertz, who generated the first artificial radio waves by means of sparks triggered in a gap in a high-frequency LC circuit.

The speed c of the electromagnetic wave equals the speed of light. We recall from Chapter 16 that the **wavelength** λ of the wave is the distance between one wave crest and the next, and the **frequency** f is the number of wave crests arriving at some fixed point per second. As for any wave, the product of the wavelength and the frequency of an electromagnetic wave equals the speed of the wave, that is, the speed of light:

wavelength λ and frequency f

wavelength, frequency, and speed of electromagnetic wave

$$\lambda f = c \tag{33.10}$$

EXAMPLE 2 The wavelength of green light is 5.50×10^{-7} m. What is the frequency of this kind of light?

SOLUTION: From Eq. (33.10),

$$f = \frac{c}{\lambda} = \frac{3.00 \times 10^8 \text{ m/s}}{5.50 \times 10^{-7} \text{ m}} = 5.45 \times 10^{14} \text{ Hz}$$

The most precise modern method for the determination of the speed of light relies on separate measurements of the wavelength and the frequency of the light emitted by a stabilized laser. The speed can then be evaluated as the product of these measured quantities, as in Eq. (33.10). The value of the speed obtained by these means is

$$c = 299\ 792\ 458 \text{ m/s} \tag{33.11}$$

As stated in Chapter 1, this value of the speed of light was adopted as a standard of speed in 1983, and it is now used as the basis for the definition of the meter.[1]

HEINRICH RUDOLF HERTZ (1857–1894) *German physicist, and professor at Bonn. He supplied the first experimental evidence for the electromagnetic waves predicted by Maxwell's theory. Hertz generated these waves by means of an electric spark, measured their speed and wavelength, and established their similarity to light waves in the phenomena of reflection, refraction, and polarization.*

---

[1]According to modern practice, Eq. (33.9) is then used to calculate the value of ϵ_0 from the value of the speed of light and the value of μ_0.

A periodic wave may be regarded as a succession of positive and negative wave pulses. Therefore the arrangement of the electric and magnetic fields in the periodic electromagnetic wave is the same as in the electromagnetic wave pulse, discussed in Section 33.2. *The directions of the electric and magnetic fields are perpendicular to the direction of propagation and they are perpendicular to each other.* The direction of the magnetic field is related to the direction of the electric field by the right-hand rule stated in the preceding section: if you place the fingers of the right hand along the direction of **E** and curl them toward the direction of **B**, the thumb will point in the direction of propagation (see Fig. 33.10).

The theoretical analysis of the mutual induction of the electric and magnetic fields in the wave (Section 33.6) shows that the magnitude of the magnetic field differs from that of the electric field by a factor of c:

$$B = \frac{1}{c}E \qquad (33.12)$$

 magnetic field of wave

Thus, the magnetic field of the wave is large wherever the electric field is large; that is, the wave crests (and wave troughs) of the magnetic field coincide with the wave crests (and wave troughs) of the electric field.

Figure 33.11a shows the electric and magnetic field lines for such a periodic wave, with the direction of propagation parallel to the x axis, the direction of the electric field parallel to the y axis, and the direction of the magnetic field parallel to the z axis. Figure 33.11b gives plots of the strengths of the electric and magnetic field in a harmonic wave, that is, a wave with a sinusoidal dependence on position. Figure 33.11c displays the wave fronts, or the locations of the wave crests. The maximum strength of the electric field of such a wave, or the strength of the electric field at the wave crests, is called the **amplitude** of the wave, usually designated by E_0. Note that the electric and magnetic fields of the wave are always in phase—the wave crests of the magnetic field are always at the same position as the wave crests of the electric field (but the direction of the magnetic field is perpendicular to the electric field).

amplitude

EXAMPLE 3   The wave reaching a point at some distance from a radio transmitter has an electric field with an amplitude of 2.0×10^{-3} V/m. What is the amplitude of the magnetic field?

SOLUTION: By Eq. (33.12), the amplitude of the magnetic field is

$$B_0 = \frac{E_0}{c} = \frac{2.0 \times 10^{-3} \text{ V/m}}{3.0 \times 10^8 \text{ m/s}} = 6.7 \times 10^{-12} \frac{\text{V·s}}{\text{m}^2}$$

The units V·s/m² on the right side do not look like the unit of the magnetic field, but we can recognize that the units are correct if we perform some substitutions and cancellations:

$$1 \frac{\text{V·s}}{\text{m}^2} = 1 \frac{(\text{J/C})·\text{s}}{\text{m}^2} = 1 \frac{\text{N·m·s}}{\text{C·m}^2} = 1 \frac{\text{N·s}}{\text{C·m}} = 1 \text{ T}$$

Here the last step used the definition of the tesla in Section 29.2, $1 \text{ T} = 1 \text{ N·s/(C·m)}$.

The wave shown in Fig. 33.11 is a **plane wave**, with uniform electric and magnetic fields over any plane perpendicular to the direction of propagation, such as the planes shown in Fig. 33.11c. Such perpendicular planes, drawn at the location of the wave crests, where the electric field is maximum, are the wave fronts of the wave.

plane wave

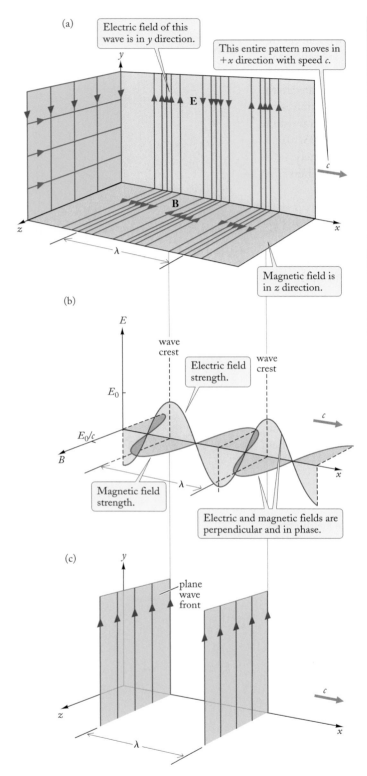

FIGURE 33.11 (a) Electric (red) and magnetic (blue) field lines for a plane wave traveling toward the right, shown at one instant of time. The electric field is vertical, and the magnetic field is horizontal. Only the electric field lines in the x–y plane and the magnetic field lines in the x–z plane are shown. There are many more electric and magnetic field lines parallel to those shown. These lines fill slabs perpendicular to the x axis. (b) Plot of the strengths of the electric and magnetic fields as a function of x. (c) Two of the wave fronts of the plane wave.

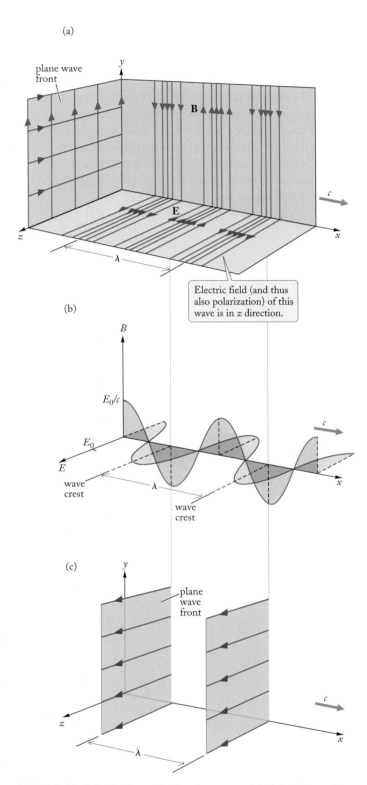

FIGURE 33.12 (a) Electric (red) and magnetic (blue) field lines of another plane wave traveling toward the right, shown at one instant of time. The electric field is horizontal, and the magnetic field is vertical. (b) Plot of the strengths of the electric and magnetic fields as a function of x. (c) Two of the wave fronts of the plane wave.

The direction of the electric field is called the direction of **polarization** of the wave. Thus, the wave shown in Fig. 33.11 is polarized in the y direction (the electric field is in the $\pm y$ directions). By contrast, the wave shown in Fig. 33.12 is polarized in the z direction. To generate a wave polarized in the y direction, we need to accelerate an electric charge back and forth along the y direction; to generate a wave polarized in the z direction, we need to accelerate the charge back and forth along the z direction. It is of course also possible to generate waves polarized in some intermediate direction, say, at $45°$ to the y and z axes; but such waves are a superposition of those shown in Figs. 33.11 and 33.12, and they therefore are nothing essentially new. Hence, electromagnetic waves (of a given direction of propagation) have only **two independent directions of polarization**. In this regard, electromagnetic waves are analogous to transverse waves on a string; if the string is stretched horizontally, then there are only two independent transverse waves, since we can either shake the string up and down (vertical polarization) or right and left (horizontal polarization). Schematically, the direction of polarization of a light wave is often indicated by a double-headed arrow, which indicates the direction of the positive and negative electric fields of the wave (see Fig. 33.13).

Although an individual light wave, like any other kind of electromagnetic wave, is always polarized in some direction or another, the light beams produced by ordinary light sources—the Sun, a lightbulb, a candle—do not exhibit any noticeable polarization. Such "unpolarized" light consists of a superposition of a very large number of plane waves with random directions of polarization (see Fig. 33.14a). Hence, on the average, there is no polarization in the beam. Note that, in Fig. 33.14a, any of the plane waves of an intermediate direction of polarization can be regarded as a superposition of waves of horizontal and of vertical polarizations; hence we can equally well represent an unpolarized light beam as a random mixture of horizontally and vertically polarized waves (see Fig. 33.14b).

polarization

two independent directions of polarization

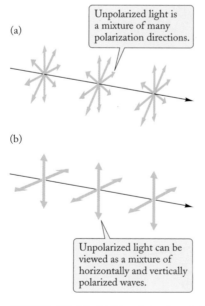

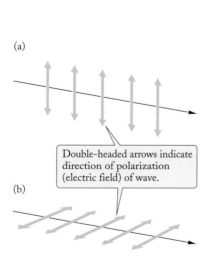

FIGURE 33.13 Schematic representation of the direction of polarization of a light wave. (a) Vertical polarization. (b) Horizontal polarization.

FIGURE 33.14 (a) Unpolarized sunlight consists of a superposition of many plane waves with random directions of polarization. (b) Such light can be represented as a random mixture of horizontally and vertically polarized waves.

polarizer

Unpolarized light can be given a polarization by passing it through a **polarizer** that permits the passage of only the electric field component parallel to the preferential direction in the sheet and absorbs the electric field component perpendicular to the preferential direction. For a wave on a string, the analog of a polarizer is a slot in a plate (see Fig. 33.15). The slot readily permits the passage of a wave with polarization parallel to this slot, but blocks a wave with polarization perpendicular to it. For a light wave, the most commonly employed polarizer is a sheet of Polaroid, which contains long chains of organic molecules arranged parallel to each other. The preferential direction that permits passage of the electric field of a wave is *perpendicular* to the direction of alignment of these molecules. An analogous polarizer for microwaves, or radio waves of short wavelength, can be constructed out of a number of thin conducting rods or wires arranged parallel to each other (see Fig. 33.16). The preferential direction of polarization that permits the passage of the electric field of a wave is then *perpendicular* to the direction of the wires, because the wires have very little effect on a perpendicular electric field. In contrast, an electric field *parallel* to the wires causes strong currents to flow along the wires, which both reflect the wave and dissipate its energy. The preferential

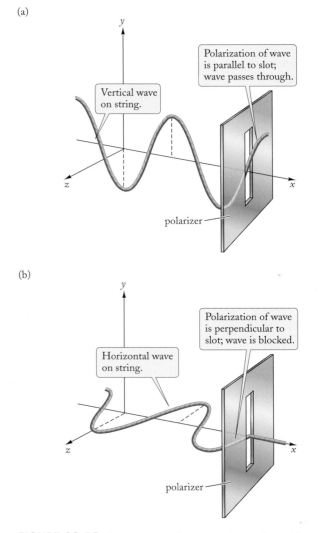

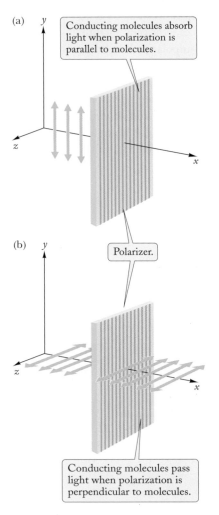

FIGURE 33.15 A wave on a string approaches a plate with a vertical slot. (a) If the direction of polarization of the wave on the string is vertical, the slot permits passage of the wave. (b) If the direction of polarization is horizontal, the slot blocks the wave.

FIGURE 33.16 An array of vertical wires or conducting molecules (a) blocks the passage of a microwave of vertical polarization but (b) permits the passage of a microwave of horizontal polarization.

direction for the passage of waves in Fig. 33.16 is contrary to our intuition—on the basis of the analogy with a wave on a string, we would expect that the slots between the wires in Fig. 33.16a permit the passage of a vertically polarized wave. But in this regard the analogy with the wave on a string misleads us. The wave on a string and the electromagnetic wave respond to the array of wires in a different way.

Polaroid is widely used in sunglasses. On the average, unpolarized sunlight consists of an equal mixture of both directions of polarization, parallel to the preferential direction of the Polaroid and perpendicular. Hence the Polaroid will absorb half the sunlight (actually somewhat more, since the Polaroid in sunglasses is slightly tinted). But the important advantage of Polaroid sunglasses over ordinary sunglasses is that they strongly attenuate the glare of reflected sunlight, such as the sunlight reflected by water or by a road. In distinction to ordinary sunlight, the reflected sunlight is somewhat polarized in the horizontal direction (see Section 34.3); since Polaroid sunglasses have their preferential direction in the vertical plane (unless you tilt your head!), they block the horizontally polarized light.

A variety of interesting experiments can be performed with two or more polarizers arranged in tandem. For instance, Fig. 33.17 shows a simple arrangement of two polarizers. Unpolarized light is incident on the first polarizer, which selects waves of vertical polarization and allows them to pass; the light emerging from this first polarizer is therefore vertically polarized. When this light is incident on the second polarizer, or **analyzer**, its electric field vector makes an angle ϕ with the preferential direction. We can regard a light wave with such an electric field as a superposition of two light waves, whose electric fields are, respectively, parallel and perpendicular to the preferential direction of the analyzer. The analyzer permits the passage of the former wave but blocks the latter. If the amplitude of the wave incident on the analyzer is E_0, then the amplitude of the parallel wave is

$$E_0' = E_0 \cos\phi \qquad (33.13)$$

analyzer

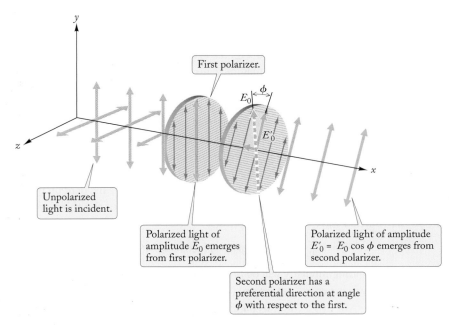

First polarizer.

E_0

ϕ

E_0'

Unpolarized light is incident.

Polarized light of amplitude E_0 emerges from first polarizer.

Second polarizer has a preferential direction at angle ϕ with respect to the first.

Polarized light of amplitude $E_0' = E_0 \cos\phi$ emerges from second polarizer.

FIGURE 33.17 Two polarizers arranged in tandem. The first polarizer has its preferential direction oriented vertically as indicated by the green arrows; the second (the analyzer) has its preferential direction inclined at an angle ϕ.

Since the intensity of a light wave, like that of any other kind of wave, is proportional to the square of the amplitude,[2] *the intensity I of the transmitted wave emerging from the analyzer is smaller than the intensity I_0 of the incident polarized wave by a factor* $\cos^2 \phi$:

$$[\text{transmitted intensity}] = [\text{incident intensity}] \times \cos^2 \phi \qquad (33.14)$$

or

$$I = I_0 \cos^2 \phi \qquad (33.15)$$

This relation between the intensity incident on the analyzer and the intensity transmitted by the analyzer is called the **Law of Malus**. Note that if $\phi = 90°$, the transmitted intensity is zero; that is, such "crossed" polarizers block the light completely.

This blocking of light by crossed polarizers can be readily demonstrated by means of two Polaroid sunglasses (see Fig. 33.18). However, such sunglasses cannot be used for a quantitative test of Eq. (33.14), because the reduction of intensity of light by the sunglasses is caused not entirely by polarization but also by the tint of the glass.

EXAMPLE 4   Suppose that the preferential direction of the second polarizer makes an angle of 30° with the preferential direction of the first. If unpolarized light is incident on the first polarizer from the left, what fraction of this incident light will pass through both polarizers and emerge on the right?

SOLUTION: In unpolarized light, on the average, one-half of the light waves are polarized in the vertical direction and one-half in the horizontal direction. Hence, one-half of the light will be able to pass through the first polarizer. According to Eq. (33.14), a fraction $\cos^2 \phi = \cos^2 30° = 0.75$ of this light will then pass through the second polarizer. Hence the fraction of the light that passes through both polarizers is $0.5 \times 0.75 = 0.375$.

A remarkable demonstration of the behavior of polarizers is illustrated in Fig. 33.19. In Fig. 33.19a, two polarizers are crossed, so that none of the light incident on the first is transmitted through the second. However, if another polarizer is inserted *between* the crossed polarizers, as in Fig. 33.19b, some intensity is then transmitted through all three! Such polarizer "magic" is no more than a consequence of Malus' Law. To understand this, let us assume that the preferential direction of the first polarizer is vertical. After the first polarizer, a light wave of amplitude E_0 is polarized vertically; if the preferential direction of the inserted polarizer makes an angle ϕ with the vertical, then the emerging wave has amplitude $E_0' = E_0 \cos \phi$ and direction of polarization at the angle ϕ, and when this wave is incident on the third polarizer, some of it will be able to pass through. Since the first and third polarizers are crossed, the preferential direction of the third polarizer is horizontal. Thus the wave incident on the third polarizer makes an angle $90° - \phi$ with preferential direction of the third polarizer, and the transmitted wave will have amplitude

$$E_0'' = E_0' \cos(90° - \phi) = E_0 \cos \phi \cos(90° - \phi)$$

and the corresponding intensity will be proportional to $\cos^2 \phi \cos^2(90° - \phi)$. The transmitted intensity will in general not be zero (unless the preferential direction of the inserted polarizer is at $\phi = 0°$ or $90°$, that is, parallel to that of one of the crossed polarizers).

Law of Malus

FIGURE 33.18 The light that passes through the upper sunglasses becomes polarized in the vertical direction, because this is the preferential direction of the Polaroid in the sunglasses. If the lower sunglasses were oriented parallel to the upper sunglasses ($\phi = 0$), the polarized light could pass and reach us. But if the upper sunglasses are oriented perpendicular to the lower sunglasses ($\phi = 90°$), as shown, the polarized light is blocked.

ÉTIENNE MALUS (MALÜS) (1775–1812) *French army engineer and physicist, noted for his mathematical and experimental investigations in optics and in double refraction. He discovered the polarization of light by reflection while looking though a calcite crystal at sunlight reflected by the windows of the Luxembourg Palace in Paris.*

---

[2]For a precise definition of the intensity of a light wave, we just examine the energy flux in the wave (that is, the energy carried by the wave per square meter per second); we will deal with this precise definition of the intensity in Section 33.5.

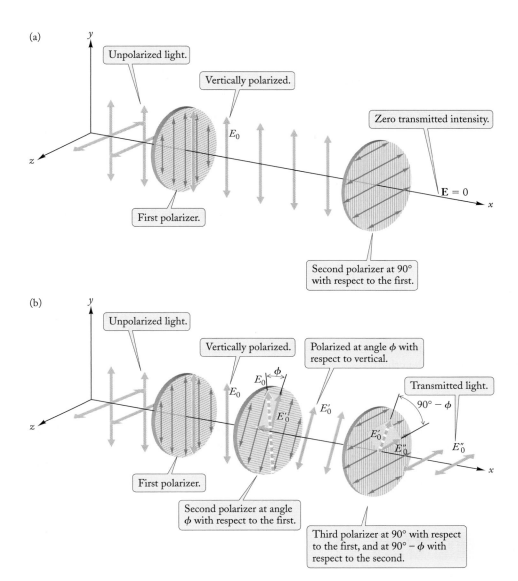

(a)

FIGURE 33.19 (a) Two crossed polarizers (perpendicular preferential directions, as indicated by the green arrows); zero intensity is transmitted through such a system. (b) Two crossed polarizers with another polarizer in between. As long as the pass direction of the second polarizer is not parallel to one of the crossed polarizers, some intensity is transmitted through the system.

✔ Checkup 33.3

QUESTION 1: For each of the wave pulses described in Checkup questions 1 and 2 of Section 33.2, what is the direction of polarization?

QUESTION 2: Suppose the y and z axes are vertical and horizontal, respectively, as in Figs. 33.8 and 33.9. An electromagnetic wave traveling in the $+x$ direction is polarized at 45° from the vertical. This wave can be regarded as a superposition of the two waves illustrated in Figs. 33.11 and 33.12. If the amplitude of the wave is E_0, what are the amplitudes of the two waves in the superposition?

QUESTION 3: If the preferential direction of a polarizer is inclined at an angle of 45° to the direction of polarization of an incident wave, by what factor is the transmitted intensity reduced? What if the angle is 60°?

QUESTION 4: Unpolarized light of intensity I_0 is incident upon the first of two polarizers. The second polarizer has its preferential direction at 60° with respect to the first one. What is the final transmitted intensity?

    (A) $I_0/8$ (B) $I_0/4$ (C) $3I_0/8$ (D) $\sqrt{3}I_0/4$.   (E) $3I_0/4$

33.4 THE GENERATION OF ELECTROMAGNETIC WAVES

Concepts
— in —
Context

FIGURE 33.20 Antennas of radio stations atop tall masts.

As discussed in some detail in Section 33.2, an accelerated point charge creates an electromagnetic wave pulse that spreads outward from the charge. In essence, this wave pulse is a disturbance of the familiar electric and magnetic fields with which we began our study of electricity and magnetism. As long as the charge moves with uniform velocity, these fields accompany the charge—they move as though they were rigidly attached to the charge. But if the charge is forced to accelerate, then parts of the fields break away—they become independent of the charge and they travel outward as an electromagnetic wave pulse. If the charge moves back and forth with periodic motion and periodic acceleration, then it will radiate a periodic wave, with a frequency equal to that of the motion of the charge. The wavelength of the wave is related to its frequency by Eq. (33.10),

$$\lambda = \frac{c}{f}$$

For instance, the electric charges on the antennas of FM radio stations (Fig. 33.20) typically oscillate back and forth with a frequency of 1.0×10^8 Hz (or 100 MHz); correspondingly, the wavelength of the radiation emitted by these accelerated charges has a wavelength of

$$\lambda = \frac{c}{f} = \frac{3.0 \times 10^8 \text{ m/s}}{1.0 \times 10^8 \text{ s}^{-1}} = 3.0 \text{ m} \tag{33.16}$$

long waves, medium waves, and short waves

TV waves

The oscillations of the charges on the antenna are produced by means of a resonating LC circuit coupled to the antenna by a mutual inductance (see Fig. 32.17). In essence, this is the method used to generate **long waves, medium waves** (including AM), and **short waves** (including FM), as well as **TV waves**. Such radio waves span a wavelength range from 10^5 m to a few centimeters.

Concepts
— in —
Context

EXAMPLE 5   The acceleration of charges up and down an efficient antenna is provided by a resonant standing wave, with the travel time up and down the antenna matched to one period of the oscillation. However, the speed v of the wave in the conductor is somewhat slower than the speed of light; for a typical antenna, $v = 0.68c$. How long should such an antenna be for an AM radio station broadcasting at a frequency of 1.0 MHz? For a cellular telephone operating around 1.0 GHz?

SOLUTION: The travel time up and down the antenna is one oscillation period, or T. For an antenna of length l, the distance up and down is $2l$, and this distance must equal the velocity of the wave in the antenna times the travel time:

$$2l = v \times T$$

The period is the inverse of the frequency f, so the length of the antenna is

$$l = \frac{v}{2f}$$

We are told $v = 0.68c$ for a typical antenna. For an antenna transmitting AM radio waves at 1.0 MHz, the length is

$$l = \frac{v}{2f} = \frac{0.68 \times 3.00 \times 10^8 \text{ m/s}}{2 \times 1.0 \times 10^6 \text{ Hz}} = 102 \text{ m}$$

PHYSICS IN PRACTICE   AM AND FM RADIO

Concepts
—*in*—
Context

The frequencies of AM or FM radio waves indicated on the dials of radio receivers refer to the frequency of the carrier wave, which is the steady, periodic wave emitted by the radio station during moments of silence, when there is no audio signal (no voice signal and no musical signal). AM radio waves span a frequency range from 550 kHz to 1.7 MHz, and FM radio waves a range from 88 MHz to 108 MHz. AM and FM stations use different methods for imprinting the audio signal on the carrier wave. In the AM (*A*mplitude *M*odulation) method, the *amplitude* of the carrier wave is altered in accordance with the amplitude of the audio signal to be transmitted; thus, the carrier amplitude is increased or decreased when the amplitude of the audio signal increases or decreases (see Fig. 1). In the FM (*F*requency *M*odulation) method, the *frequency* of the carrier wave is altered in accordance with the amplitude of the audio signal; thus, the carrier frequency is increased or decreased when the amplitude of the audio signal increases or decreases (see Fig. 2), but the amplitude of the carrier wave is kept constant. FM attains a higher fidelity than AM because it is quite insensitive to disturbances in the strength of the carrier wave. Any such disturbances in the strength of the FM carrier wave have no effect on its frequency, and hence do not alter the audio signal imprinted on the wave. By contrast, any disturbance in the strength of an AM carrier wave leads to a distortion of the audio signal. FM broadcasts also have higher fidelity because FM stations are permitted to broadcast a wider range of frequencies: In order to avoid overcrowding of the broadcast bands, AM stations are limited by law to imprinting audio signals of no more than 10 kHz on the carrier, which is smaller than the frequency range of human hearing. The modulation may reach up to 20 kHz for FM stations, closer to the full range of human hearing.

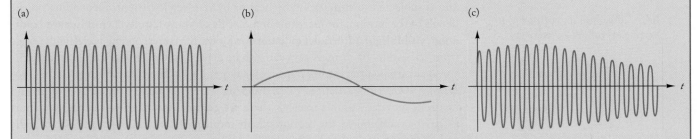

FIGURE 1 (a) Carrier radio wave. (b) Audio, or sound, signal. (c) AM radio wave. The amplitude of the wave is modulated according to the audio signal.

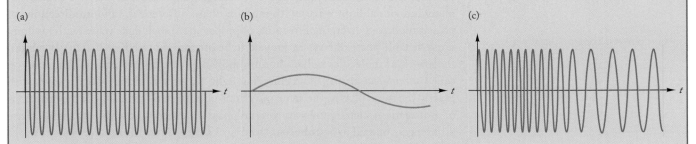

FIGURE 2 (a) Carrier radio wave. (b) Audio signal. (c) FM radio wave. The frequency of the wave is modulated according to the audio signal.

Similarly, for a cellular telephone operating at 1.0 GHz, the length is

$$l = \frac{0.68 \times 3.00 \times 10^8 \text{ m/s}}{2 \times 1.0 \times 10^9 \text{ Hz}} = 0.102 \text{ m} = 10 \text{ cm}$$

This is why an AM radio transmitter antenna is very long, whereas a cellular telephone antenna is quite short. Incidentally: Instead of using a long wire to intercept the electric field, portable AM radio receivers use a small coil of wire to intercept the changing magnetic flux due to the electromagnetic wave.

microwaves

Waves of a wavelength shorter than that of radio waves, called **microwaves**, are best generated by a resonating electromagnetic cavity, consisting of an empty metallic can in which a standing electromagnetic wave is set up by an electron beam passing through, much as a standing sound wave is set up in an organ pipe by a stream of air passing over the blowhole. The antenna that radiates the microwaves is simply a horn attached to the electromagnetic cavity by a metallic pipe, or waveguide, which permits the waves to spill out into space (see Fig. 33.21). This method can be used to generate waves as short as about a millimeter. Shorter wavelengths cannot be generated with currents oscillating in macroscopic laboratory equipment; however, short wavelengths can be easily generated by electrons oscillating within molecules and atoms subjected to stimulation by heat or by an electric current. Depending on the details of the motion, the electrons in molecules and in atoms will emit **infrared radiation**, **visible light**, **ultraviolet radiation**, or **X rays**; the corresponding wavelengths range from 10^{-3} m to 10^{-11} m. X rays can also be generated by the acceleration that high-speed electrons suffer during impact on a target; this is **Bremsstrahlung** (German for *braking radiation*). Radiations of even shorter wavelengths are emitted by protons and neutrons moving within a nucleus; these are **gamma rays**, with wavelengths as short as 10^{-13} m. Of course, the motion of subatomic particles and their emission of radiation cannot be calculated by classical mechanics or classical electricity and magnetism; such calculations require quantum mechanics.

infrared radiation, visible light, ultraviolet radiation, and X rays

Bremsstrahlung

gamma rays

In an ordinary light source, such as an incandescent lightbulb, the individual atoms or molecules radiate independently. The emerging light consists of a superposition of many individual light waves with random phase differences, random directions of polarization, and diverging directions of propagation; such light waves with random, unpredictable phase differences are said to be **incoherent** (see Fig. 33.22a). In a **laser**, the atoms or molecules radiate in unison, by a quantum-mechanical phenomenon called *stimulated emission*. The emerging light is a superposition of light waves with exactly the same phases, the same directions of polarization, and the same directions of propagation; such light waves with no phase differences, or with predictable phase differences, are said to be **coherent** (see Fig. 33.22b). Since the individual light waves in this kind of light combine constructively, the light emerging from the laser is very intense, and it also forms a very narrow, sharp beam.

incoherent and coherent light waves

Another important mechanism for the generation of electromagnetic waves is **cyclotron emission**. This involves high-speed electrons undergoing centripetal acceleration while spiraling in a magnetic field (see Section 30.1). Depending on the speed of the electron and the strength of the magnetic field, the radiation may consist of radio waves, X rays, or anything in between (including visible light). Most of the radio waves reaching us from stars, pulsars, and radio galaxies are generated by this process.

cyclotron emission

Figure 33.23 displays the **electromagnetic spectrum**, the wavelength and the frequency bands of electromagnetic radiation. The bands overlap to some extent, because the names assigned to the different ranges of wavelengths depend not only on the

electromagnetic spectrum

FIGURE 33.21 (a) Horns of microwave antennas. (b) Microwave oven.

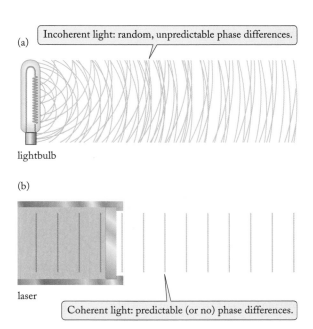

FIGURE 33.22 (a) Wave fronts of incoherent light waves emitted by a lightbulb. (b) Wave fronts of coherent light waves emitted by a laser.

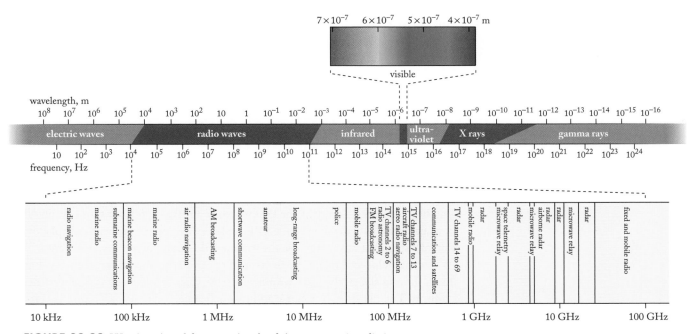

FIGURE 33.23 Wavelength and frequency bands of electromagnetic radiation.

value of the wavelength, but also on the method used to generate or detect the radiation. For example, radiation of a wavelength of a tenth of a millimeter will be called a radio wave (microwave) if detected by a radio receiver, but it will be called infrared radiation if detected by a heat sensor.

The wavelengths of visible light range from about 7×10^{-7} m to 4×10^{-7} m. The wavelength of light is usually expressed in nanometers (1 nm = 10^{-9} m), in terms

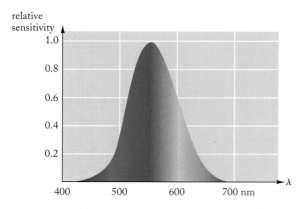

FIGURE 33.24 Sensitivity of the human eye at different wavelengths and colors in bright light. The sensitivity is maximum at about 550 nm (yellow green light) and decreases to about 1% of this maximum at 690 nm (deep red) and at 430 nm (violet). In dim light, the sensitivity changes; it becomes maximum at about 500 nm (blue green).

of which *the range of wavelengths of visible light extends from* 400 nm *to* 700 nm. This is the range of wavelengths over which our eyes are sensitive to light. We perceive different wavelengths within the visible region as having different colors. Figure 33.24 is a plot of the sensitivity of the human eye and shows how colors are correlated with wavelength.

Incidentally: Our eyes are almost completely insensitive to the polarization of light waves. We can detect the polarization only with special equipment, such as Polaroid sunglasses. Radio and TV antennas are, of course, very sensitive to the direction of polarization of radio waves, and they must have the proper orientation to pick up a signal. For instance, a simple antenna in the form of a straight wire or rod is sensitive to a radio wave polarized in the same direction as the antenna, but insensitive to a radio wave polarized at right angles to the antenna. Some TV antennas are equipped with an electric motor so that they can be easily rotated to the optimal orientation for each TV station you want to receive.

 Checkup 33.4

QUESTION 1: Why do FM radio waves give higher fidelity than AM?

QUESTION 2: What is the color of light of wavelength 650 nm? 550 nm?

QUESTION 3: What is the wavelength of yellow light?

QUESTION 4: Arrange in order of increasing wavelength: AM, FM, radar, TV.

QUESTION 5: Which of the following objects is closest in size to the wavelength of visible light?

    (A) An atom ($\sim 10^{-10}$ m)   (B) A virus ($\sim 10^{-6}$ m)   (C) A dime (~ 1 cm)
    (D) A person (~ 2 m)   (E) A football field (~ 100 m)

33.5 ENERGY OF A WAVE

The electric and magnetic fields of an electromagnetic wave contain energy. As the wave moves along, so does this energy—the wave transports energy. For instance, the light waves of sunlight transport energy from the surface of the Sun to the Earth, and we can readily perceive this energy by the heat we feel when we expose our skin to sunlight.

Let us calculate the flow of energy in a plane wave, or an approximately plane portion of a spherical wave. Suppose the wave propagates in the positive x direction, with its electric fields and magnetic fields in the y and z directions, respectively. Figure 33.25 shows a slab of the wave within which the electric and magnetic field are nearly uniform. The densities of electric and magnetic energy are $(\epsilon_0/2)E^2$ and $(1/2\mu_0)B^2$ [see Eqs. (25.58) and (31.31)], so the energy density u in the wave is

energy density in electromagnetic wave

$$u = \frac{\epsilon_0}{2}E^2 + \frac{1}{2\mu_0}B^2 \qquad (33.17)$$

If the thickness of the wave slab is dx and the frontal area is A (see Fig. 33.25), then the volume of the slab is $A\,dx$ and the total amount of energy in the slab is the energy density times the volume:

$$dU = \left(\frac{\epsilon_0}{2}E^2 + \frac{1}{2\mu_0}B^2\right) \times A\,dx \qquad (33.18)$$

Since $B = E/c$ and $\epsilon_0 = 1/(\mu_0 c^2)$, we can write this energy as

$$dU = \left[\frac{1}{2\mu_0 c^2} E^2 + \frac{1}{2\mu_0} \left(\frac{E}{c} \right)^2 \right] \times A \, dx \qquad (33.19)$$

$$= \frac{1}{\mu_0 c^2} E^2 A \, dx \qquad (33.20)$$

Note that the two terms on the right side of Eq. (33.19) are equal, that is, *the electric and magnetic energy densities in an electromagnetic wave are equal.*

The forward speed of the wave slab is c, and hence the amount of energy dU moves out of the (stationary) volume $A \, dx$ in a time $dt = dx/c$. The rate of flow of energy, or the power, is therefore

$$\frac{dU}{dt} = \frac{dU}{dx/c} = \frac{1}{\mu_0 c} A E^2 \qquad (33.21)$$

In Fig. 33.25, the energy flows toward the right, along the direction of propagation of the wave. The **energy flux in the wave** is defined as the rate of energy flow per unit frontal area, that is, $(1/A)dU/dt$. The rate of energy flow is the power; hence the energy flux is the power carried per unit area of wave front. The units for this energy flux are watts per square meter (W/m^2). According to Eq. (33.21), the energy flux for our plane wave is

$$\frac{1}{A} \frac{dU}{dt} = \frac{1}{\mu_0 c} E^2$$

This energy flux is called the **Poynting flux**, usually designated by S:

$$S = \frac{1}{\mu_0 c} E^2 \qquad (33.22)$$

energy flux in wave

This energy flux provides us with a precise measure of the instantaneous energy flux of the wave. A wave of high intensity—such as the wave produced by a powerful laser—is a wave of large energy flux.

Since an electromagnetic wave propagates in the $\mathbf{E} \times \mathbf{B}$ direction and $\mathbf{E}$ and $\mathbf{B}$ are perpendicular, the Poynting flux is sometimes expressed as a vector of the form (again using $B = E/c$)

$$\mathbf{S} = \frac{1}{\mu_0} \mathbf{E} \times \mathbf{B} \qquad (33.23)$$

Poynting vector S

Thus the **Poynting vector S** has a magnitude equal to the instantaneous energy flux of the wave and points in the direction of propagation.

Since the electric field oscillates in time, so does the energy flux. For a harmonic wave, such as shown in Fig. 33.11b, with a sinusoidal dependence on time, the average of the square of the electric field is one-half of the square of the amplitude E_0 of the electric field, that is, $\overline{E^2} = E_0^2/2$ (compare the calculation of the average of the square of the emf in Section 32.1). Hence the **time-average energy flux** in the wave is

$$\overline{S} = \frac{1}{2\mu_0 c} E_0^2 \qquad (33.24)$$

time average energy flux in wave

The time-average energy flux is the **intensity** of the wave.

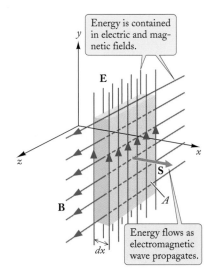

FIGURE 33.25 A slab of electric (red) and magnetic (blue) fields in a plane wave propagating toward the right. The slab of thickness dx and frontal area A moves with the wave.

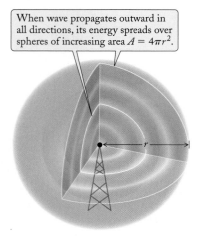

When wave propagates outward in all directions, its energy spreads over spheres of increasing area $A = 4\pi r^2$.

FIGURE 33.26 The energy of the radio wave is spread out over a sphere of radius $r = 8.5$ km.

EXAMPLE 6   At a distance of 8.5 km from a radio transmitter, the amplitude of the oscillating electric field in the radio wave is $E_0 = 0.13$ V/m. What is the time-average energy flux there? What is the total power radiated by the radio transmitter? Assume that this transmitter radiates uniformly in all directions.

SOLUTION: From Eq. (33.24), the time-average energy flux is

$$\overline{S} = \frac{1}{2\mu_0 c} E_0^2 = \frac{(0.13 \text{ V/m})^2}{2 \times 1.26 \times 10^{-6} \text{ H/m} \times 3.00 \times 10^8 \text{ m/s}}$$

$$= 2.2 \times 10^{-5} \text{ W/m}^2$$

To obtain the total power, we must multiply the calculated power per unit area (the energy flux, or intensity) by the area over which the radio wave has spread. This is the area of a sphere of radius r, namely, an area $A = 4\pi r^2$, with $r = 8.5$ km (see Fig. 33.26). Therefore the total power is

$$\overline{P} = [\text{area}] \times [\text{energy flux}] = 4\pi r^2 \overline{S}$$

$$= 4\pi \times (8.5 \times 10^3 \text{ m})^2 \times 2.2 \times 10^{-5} \text{ W/m}^2 \qquad (33.25)$$

$$= 2.0 \times 10^4 \text{ W} = 20 \text{ kW}$$

Note that, according to Eq. (33.25), the energy flux $\overline{S}$ for a spherical wave spreading out from a source in all directions is inversely proportional to the square of the distance:

$$\overline{S} = \frac{\overline{P}}{4\pi r^2} \qquad (33.26)$$

We must take this dependence of the flux on distance into account whenever we want to investigate the spreading of the wave over a large range of distances (but we can ignore this dependence over a small range of distances, where the wave can be approximated as a plane wave of constant energy flux).

A wave carries energy, so we might expect that it also carries momentum. Maxwell showed this to be the case. He showed that when a body absorbs an electromagnetic wave with total energy U, the momentum p transferred to the body is given by

$$p = \frac{U}{c} \qquad (33.26)$$

From this expression for the momentum, we can deduce that the wave exerts a pressure on the absorbing body. Since the pressure is the force per unit area, or the change in momentum per unit time per unit area, the pressure exerted by the wave on the absorbing body is

$$[\text{pressure}] = \frac{F}{A} = \frac{1}{A}\frac{dp}{dt} = \frac{1}{A} \times \frac{1}{c}\frac{dU}{dt}$$

or, in terms of the energy flux $S = (1/A)\, dU/dt$,

$$[\text{pressure}] = \frac{S}{c} \qquad (\text{absorption}) \qquad (33.27)$$

If a body reflects rather than absorbs, then the wave is sent back from whence it came; it thus leaves with a momentum opposite to its original value. Thus, for a reflecting body, the momentum transferred and pressure exerted are twice as large as for an absorbing body:

$$[\text{pressure}] = \frac{2S}{c} \quad \text{(reflection)} \qquad (33.28)$$

EXAMPLE 7   Near the Earth's orbit, the energy flux from the Sun is approximately 1400 W/m² at normal incidence. One proposed form of space propulsion is the "solar sail," driven by sunlight. What is the pressure due to sunlight at normal incidence on a perfectly reflecting sail? What is the force due to this pressure on a sail of area 25 m²? If such a sail is quite thin, so it has a mass of only 10 grams, what is the Sun's gravitational force on the sail? If released from rest, what will be the speed of the sail after one day? Assume that the sail is released far from the Earth and that the force is essentially constant over one day.

SOLUTION: As mentioned just above, for reflection, the pressure is given by Eq. (33.28):

$$[\text{pressure}] = \frac{2S}{c} = \frac{2 \times 1400 \text{ W/m}^2}{3.00 \times 10^8 \text{ m/s}}$$

$$= 9.3 \times 10^{-6} \text{ N/m}^2$$

The force due to this pressure on an area of 25 m² has magnitude

$$F_{\text{pressure}} = [\text{pressure}] \times [\text{area}] = 9.3 \times 10^{-6} \text{ N/m}^2 \times 25 \text{ m}^2 = 2.3 \times 10^{-4} \text{ N}$$

and is directed away from the Sun.

For a mass of 0.010 kg, and using data listed inside the book cover, the gravitational force due to the Sun has magnitude

$$F_{\text{gravitational}} = \frac{GM_S m}{r^2}$$

$$= \frac{6.67 \times 10^{-11} \text{ N}\cdot\text{m}^2/\text{kg}^2 \times 1.99 \times 10^{30} \text{ kg} \times 0.010 \text{ kg}}{(1.50 \times 10^{11} \text{ m})^2}$$

$$= 5.9 \times 10^{-5} \text{ N}$$

and is directed toward the Sun. This is less than the pressure force, so the net force is directed away from the Sun.

The speed of the sail is the acceleration multiplied by the time, $v = at$. The acceleration is

$$a = \frac{F_{\text{net}}}{m} = \frac{F_{\text{pressure}} - F_{\text{gravitational}}}{m}$$

$$= \frac{2.3 \times 10^{-4} \text{ N} - 5.9 \times 10^{-5} \text{ N}}{0.010 \text{ kg}} = 1.7 \times 10^{-2} \text{ m/s}^2$$

After one day, the velocity will be

$$v = at = 1.7 \times 10^{-2} \text{ m/s}^2 \times 24 \times 60 \times 60 \text{ s}$$

$$= 1.5 \times 10^3 \text{ m/s} = 1.5 \text{ km/s}$$

Although large speeds could thus be reached for a sail by itself, the speed will be much lower, or the gravitational force will dominate, if a spacecraft of appreciable mass is attached to the sail (unless we manage to "miniaturize" the spacecraft, so its mass is not much more than that of the sail).

✔ Checkup 33.5

QUESTION 1: Given that a light wave and an ultraviolet wave have the same amplitude, which has the higher electric and magnetic energy density?

QUESTION 2: What is the distance between adjacent maxima in the energy density of an electromagnetic wave of wavelength λ?

QUESTION 3: A wave has an amplitude $E_0 = 1.0 \times 10^{-3}$ V/m; another wave has an amplitude $E_0 = 2.0 \times 10^{-3}$ V/m. What is the ratio of the energy densities in these two waves?

QUESTION 4: If the amplitude of the electric field in a radio wave is $E_0 = 0.13$ V/m at a distance of 8.5 km (as in Example 6), what will be the amplitude when this wave spreads to a distance of 17 km? What will be the time-average energy flux?

QUESTION 5: Suppose that a plane wave strikes a polarizer whose preferential direction is at 45° to the direction of polarization of the wave. What fraction of the energy of the wave is transmitted? What happens to the energy that is not transmitted?

QUESTION 6: An electromagnetic wave is absorbed by a body. Another wave with twice the amplitude of the first is reflected by a different body. By what factor do the pressures exerted on the bodies differ?

QUESTION 7: A lightbulb emits 125 W of average power uniformly in all directions. A book is placed 1.0 m away from the bulb. The book has an area of 0.10 m² perpendicular to the light and is perfectly absorbing. What is the average energy flux (intensity) reaching the book?

(A) 1.0 W/m²   (B) 10 W/m²   (C) 12.5 W/m²
(D) 125 W/m²   (E) 1250 W/m²

33.6 THE WAVE EQUATION

We saw in Section 33.2 how the changing electric field in the wave front of a pulse induces a magnetic field (and vice versa); after that, we examined the behavior of sinusoidal electromagnetic waves. Here, we will see that such sinusoidal waves are solutions of a *wave equation* that we can obtain directly from Faraday's Law and the Maxwell–Ampère Law. We will also see from our analysis that the amplitudes of the fields are indeed related by $B_0 = E_0/c$, as stated previously in Eq. (33.12).

Consider a region of space in which there are perpendicular electric and magnetic fields, but no electric charges or currents. As shown in Fig. 33.27, the electric field **E** is in the y direction and the magnetic field **B** is in the z direction. Suppose these fields vary with position in the x direction. We first evaluate Faraday's Law

$$\oint E_{\parallel} \, ds = -\frac{d\Phi_B}{dt} \tag{33.29}$$

for the small vertical rectangular path shown in Fig. 33.27, which has a width dx and a height dy. In the integration of $E_{\parallel}$ around this path, only the vertical segments contribute, since $\mathbf{E}$ is perpendicular to the horizontal segments. The electric field is E along the left segment and $E + dE$ along the right segment, so the path integral is $-E\,dy + (E + dE)\,dy = dE\,dy$. The magnetic flux through the rectangle is $\Phi_B = B\,dx\,dy$, and therefore Eq. (33.29) becomes

$$dE \, dy = -\frac{dB}{dt} \, dx \, dy$$

or

$$\frac{dE}{dx} = -\frac{dB}{dt} \tag{33.30}$$

Since E and B are functions of x and t, the derivatives are actually partial derivatives (see Appendix 4.2), and the proper form of Eq. (33.30) is

$$\frac{\partial E}{\partial x} = -\frac{\partial B}{\partial t} \tag{33.31}$$

If we perform the same type of analysis by applying the Maxwell–Ampère Law (with $I = 0$) to the horizontal rectangle in Fig. 33.27, we obtain a similar equation:

$$-\frac{\partial B}{\partial x} = \mu_0 \epsilon_0 \frac{\partial E}{\partial t} \tag{33.32}$$

Equations (33.31) and (33.32) can be combined by taking the partial derivative of (33.31) with respect to x. For the left side, we have

$$\frac{\partial}{\partial x} \frac{\partial E}{\partial x} = \frac{\partial^2 E}{\partial x^2}$$

We do the same for the right side of Eq. (33.31), then change the order of derivatives and substitute from Eq. (33.32) as follows:

$$-\frac{\partial}{\partial x} \frac{\partial B}{\partial t} = -\frac{\partial}{\partial t} \frac{\partial B}{\partial x} = +\frac{\partial}{\partial t}\left(\mu_0 \epsilon_0 \frac{\partial E}{\partial t} \right) = \mu_0 \epsilon_0 \frac{\partial^2 E}{\partial t^2}$$

Equating these two sides, we obtain

$$\frac{\partial^2 E}{\partial x^2} = \mu_0 \epsilon_0 \frac{\partial^2 E}{\partial t^2} \tag{33.33}$$

wave equation

Equation (33.33) is an instance of an important differential equation known as the **wave equation**. Similar wave equations arise in other areas of physics. The study of the solutions of such wave equations plays a large role in microwave and radio communications, radar, acoustics, sonar, and oceanography, as well as in quantum mechanics and high-energy physics.

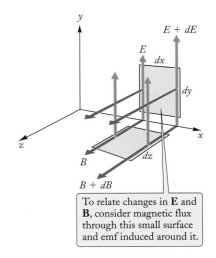

FIGURE 33.27 Electric and magnetic fields in a plane wave; the magnitudes of the field vary as a function of x. Evaluation of Faraday's Law around the vertical tan rectangle (and the Maxwell–Ampère Law around the horizontal one) provide relations between changes in E and B.

To relate changes in **E** and **B**, consider magnetic flux through this small surface and emf induced around it.

Among the simplest solutions of Eq. (33.33) are harmonic traveling waves. We can write these in a form familiar from Ch. 16 [see Eq. (16.8)]:

electric field of harmonic traveling wave

$$E = E_0 \cos[k(x - vt)] \qquad (33.34)$$

where v is the speed of the wave and k is the wave number, which is related to the wavelength by $k = 2\pi/\lambda$ [see Eq. (16.7)]. If we substitute this tentative solution into Eq. (33.33) and take the space and time derivatives, we obtain

$$-k^2 \cos[k(x - vt)] = -\mu_0 \epsilon_0 k^2 v^2 \cos[k(x - vt)]$$

From this we see that the wave equation is satisfied provided that $\mu_0 \epsilon_0 v^2 = 1$, or

$$v = \frac{1}{\sqrt{\mu_0 \epsilon_0}} \qquad (33.35)$$

This is Maxwell's result for the speed c of electromagnetic waves [see Eq. (33.8)].

Finally, we can derive the relation between E and B. Similar to the analysis above, Eqs. (33.31) and (33.32) can be combined to show that the magnetic field B also satisfies the wave equation. These equations also require that the traveling wave solution for the magnetic field be of the same form (33.34) as for the electric field,

magnetic field of harmonic traveling wave

$$B = B_0 \cos[k(x - vt)] \qquad (33.36)$$

Substituting the solutions (33.34) and (33.36) into Eq. (33.31), we have

$$\frac{\partial}{\partial x} E_0 \cos[k(x - vt)] = -\frac{\partial}{\partial t} B_0 \cos[k(x - vt)]$$

or

$$-kE_0 \sin[k(x - vt)] = -kvB_0 \sin[k(x - vt)] \qquad (33.37)$$

Solving for B_0, we obtain

$$B_0 = \frac{1}{v} E_0 \qquad (33.38)$$

Since the speed of electromagnetic waves is $v = c$, we see that Eq. (33.38) gives the same relation between E and B presented earlier [see Eq. (33.12)].

Concepts in Context

EXAMPLE 8   An AM radio signal consists of a plane wave traveling in the $+x$ direction with frequency $f = 1750$ kHz; its oscillating magnetic field is in the $\pm z$ direction and has amplitude 5.0×10^{-10} T. Write down the traveling wave forms (33.34) and (33.36) for the electric and magnetic fields for this wave.

SOLUTION: To express the traveling wave in the forms (33.34) and (33.36), we need the amplitude and wave number. Since B_0 and f are given, we can calculate both the electric field amplitude

$$E_0 = cB_0 = 3.00 \times 10^8 \text{ m/s} \times 5.0 \times 10^{-10} \text{ T} = 0.15 \text{ V/m}$$

and the wave number

$$k = \frac{2\pi}{\lambda} = \frac{2\pi f}{c} = \frac{2\pi \times 1750 \times 10^3 \text{ s}^{-1}}{3.00 \times 10^8 \text{ m/s}} = 3.67 \times 10^{-2} \text{ m}^{-1}$$

Since the $+x$ direction of propagation is in the $\mathbf{E} \times \mathbf{B}$ direction, for $\mathbf{B}$ in the $+z$ direction, $\mathbf{E}$ must be in the $+y$ direction. Thus the traveling waves may be written in the forms (33.34) and (33.36):

$$E = E_y = (0.15 \text{ V/m}) \times \cos\{(3.67 \times 10^{-2} \text{ m}^{-1})[x - (3.00 \times 10^8 \text{ m/s})t]\}$$

$$B = B_z = (5.0 \times 10^{-10} \text{ T}) \times \cos\{(3.67 \times 10^{-2} \text{ m}^{-1})[x - (3.00 \times 10^8 \text{ m/s})t]\}$$

 ## Checkup 33.6

QUESTION 1: One electromagnetic wave has a wave number k that is twice as large as that of a second electromagnetic wave. How do their wavelengths compare? Their speeds? Their frequencies?

QUESTION 2: The electric field of a traveling electromagnetic wave varies periodically with a wavelength of 300 m. What is the frequency of the corresponding periodic magnetic field?

QUESTION 3: Is the function $E = E_0 x^2/L^2$ a solution of the wave equation? (E_0 and L are constants.)

QUESTION 4: Which of the following functions is *not* a solution of the wave equation? (The quantities k and A and the velocity v are constants.)

(A) $\sin[k(x - vt)]$ (B) $\cos^2[k(x - vt)]$ (C) $A(x + vt)$
(D) $A(x - vt)^2$ (E) $A(x^2 - v^2t^2)$

SUMMARY

PHYSICS IN PRACTICE   AM and FM Radio   **(page 1089)**

MAXWELL'S EQUATIONS

Gauss' Law

$$\oint \mathbf{E} \cdot d\mathbf{A} = \oint E_\perp \, dA = \frac{Q_{\text{inside}}}{\epsilon_0} \tag{33.4}$$

Gauss' Law for magnetism

$$\oint \mathbf{B} \cdot d\mathbf{A} = \oint B_\perp \, dA = 0 \tag{33.5}$$

Faraday's Law

$$\oint \mathbf{E} \cdot d\mathbf{s} = \oint E_\parallel \, ds = -\frac{d\Phi_B}{dt} \tag{33.6}$$

Maxwell–Ampère Law

$$\oint \mathbf{B} \cdot d\mathbf{s} = \oint B_\parallel \, ds = \mu_0 I + \mu_0 \epsilon_0 \frac{d\Phi_E}{dt} \tag{33.7}$$

DISPLACEMENT CURRENT

$$I_{\text{displacement}} = \epsilon_0 \frac{d\Phi_E}{dt}$$

INDUCED MAGNETIC FIELD   For axial, cylindrically symmetric electric flux,

$$B \times 2\pi r = \mu_0 \epsilon_0 \frac{d}{dt} E \times [\text{area}]$$

(Example 1)

SPEED OF ELECTROMAGNETIC WAVE

$$c = \frac{1}{\sqrt{\mu_0 \epsilon_0}}$$

(33.8)

WAVELENGTH λ Distance for one wave oscillation.

FREQUENCY f Number of wave oscillations per second.

WAVELENGTH/FREQUENCY SPEED RELATION

$$\lambda f = c$$

(33.10)

MAGNETIC FIELD OF WAVE

$$B = \frac{1}{c} E$$

(33.12)

TRANSMITTED INTENSITY THROUGH A POLARIZER
For an unpolarized incident intensity I_0:

$$I = \tfrac{1}{2} I_0$$

(Example 4)

For a polarized incident intensity I_0 (Malus' Law):

$$I = I_0 \cos^2 \phi$$

(33.15)

ENERGY DENSITY IN ELECTROMAGNETIC WAVE

$$u = \frac{\epsilon_0}{2} E^2 + \frac{1}{2\mu_0} B^2 = \epsilon_0 E^2$$

(33.17)

TIME-AVERAGE ENERGY FLUX (INTENSITY) IN PLANE WAVE

$$\overline{S} = \frac{1}{2\mu_0 c} E_0^2$$

(33.24)

POYNTING VECTOR S (propagation direction)

$$\mathbf{S} = \frac{1}{\mu_0} \mathbf{E} \times \mathbf{B}$$

(33.23)

ENERGY FLUX S IN TERMS OF POWER P FOR SPHERICAL WAVE	$\bar{S} = \dfrac{\bar{P}}{4\pi r^2}$ 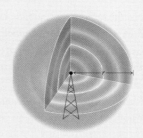	**(33.26)**

PRESSURE OF ELECTROMAGNETIC WAVE

$$[\text{pressure}] = \begin{cases} \dfrac{S}{c} & (\text{absorption}) \\[2ex] \dfrac{2S}{c} & (\text{reflection}) \end{cases} \qquad \textbf{(33.27, 33.28)}$$

WAVE EQUATION

$$\frac{\partial^2 E}{\partial x^2} = \mu_0 \epsilon_0 \frac{\partial^2 E}{\partial t^2} \qquad \textbf{(33.33)}$$

HARMONIC TRAVELING WAVE SOLUTIONS OF THE WAVE EQUATION

$$E = E_0 \cos[k(x - vt)] \qquad \textbf{(33.34)}$$
$$B = B_0 \cos[k(x - vt)] \qquad \textbf{(33.36)}$$

WAVE NUMBER

$$k = \frac{2\pi}{\lambda}$$

QUESTIONS FOR DISCUSSION

1. The displacement current between the plates of a capacitor has the same magnitude as the conduction current in the wires connected to the capacitor and yet the magnetic field produced by the former current near and within the capacitor is much weaker than that produced by the latter current near and within the wire. Explain.

2. Consider the electric field of a single positive electric charge moving at constant velocity. What is the direction of the displacement current intercepted by a circular area perpendicular to the velocity in front of the charge? Behind the charge?

3. Which of Maxwell's equations permits us to deduce the electric Coulomb field of a static charge? Which of Maxwell's equations permits us to deduce the magnetic field of a charge moving with uniform velocity?

4. Suppose that there exist magnetic monopoles, that is, positive and negative magnetic charges that act as sources and sinks of magnetic field lines, analogous to positive and negative electric charges. Which of Maxwell's equations would have to be modified to take into account such monopoles? Qualitatively, what are the required modifications?

5. In practice, the electromagnetic oscillations in a cavity are damped by "frictional" losses. How does this "friction" arise?

6. Efficient waveguides are manufactured out of a very good conductor, such as copper, sometimes with a silver lining. Why is high conductivity essential for high efficiency?

7. Is the transverse electric radiation field E_0 a conservative field?

8. Since the speed of light has now been adopted as the standard of speed and has been assigned the value $2.997\ 924\ 58 \times 10^8$ m/s *by definition*, why is it still meaningful to compare this number with Maxwell's theoretical prediction for the speed of light?

9. Describe the direction of polarization of the radiation field of an accelerated charge at a few typical points in the space surrounding the charge. Verify that the direction of polarization is never perpendicular to the direction of acceleration.

10. Figure 33.24 is a plot of the sensitivity of the human eye as a function of the wavelength of light. The sensitivity is maximum at about 550 nm and drops to about 1% at 690 nm and at 430 nm. Suppose that the sensitivity of your eye were constant over the entire interval of wavelengths shown in Fig. 33.24. How would this alter your visual perception of some of the things you see in your everyday life?

11. An atom radiates visible light of a wavelength several thousand times longer than the size of the atom. How is this possible?

12. Why does the radio reception fade in the receiver of your automobile when you enter a tunnel?

13. Shortwave radio waves are reflected by the ionosphere of the Earth; this makes them very useful for long-range communications. Explain.

14. In the seventeenth century, the Danish astronomer Ole Roemer noticed that the orbital periods of the moons of Jupiter, as observed from the Earth, exhibit some systematic irregularities; the periods are slightly longer when the Earth is moving away from Jupiter, and slightly shorter when the Earth is moving toward Jupiter. Roemer attributed this apparent irregularity to the finite speed of propagation of light, and exploited it to make the first determination of the speed of light. Explain how one can use the lengthening or the shortening of the observed period to deduce the speed of light. (This shift of period may be regarded as the earliest discovery of a Doppler shift.)

15. It has been proposed that we could eliminate the glare of the headlights of approaching automobiles by covering the windshields and the headlights with sheets of Polaroid. What orientation should we pick for the sheets of Polaroid installed on windshields and on headlights so that the light of every approaching automobile is blocked out, but our own light is not?

16. Suppose you are given a sheet of Polaroid that has no marking identifying its preferential direction. You have available a beam of unpolarized light. How can you determine the preferential direction of the sheet? (Hint: Cut the sheet in two and place one sheet behind the other rotated by 90° around the axis of the beam, so the light is completely blocked. What will happen if you now rotated one sheet about a *transverse axis*, that is, an axis perpendicular to the beam?)

17. The scattered light reaching you from the blue sky is (partially) polarized: if you look straight up, the direction of polarization is perpendicular to the direction of the Sun. How does this polarization arise? [Hint: Consider a beam of (unpolarized) sunlight passing overhead. The electric field of the light waves in this beam accelerates electrons in the molecules of air, and the radiation emitted by these electrons constitutes the scattered light of the sky. Since the direction of acceleration is perpendicular to the incident beam, what can you say about the polarization of the radiation emitted downward, toward you?]

18. A charged particle moving around a circular orbit at uniform speed has a centripetal acceleration and therefore produces a radiation field. However, a uniform current flowing around the circular loop does *not* produce a radiation field. Is this a contradiction? Explain.

PROBLEMS

33.1 Induction of Magnetic Fields; Maxwell's Equations

1. A parallel-plate capacitor of plate area A is being charged by a current I flowing into its plates via external wires. At one instant, the charge on the capacitor plates is Q.

    (a) Assume that the electric field between the plates is uniform. Show that the electric field between the plates at this instant is $E = Q/\epsilon_0 A$.

    (b) Show that the electric flux crossing the mathematical midplane between the plate surfaces is Q/ϵ_0.

    (c) What is the displacement current of this capacitor?

    (d) Show that this displacement current matches the ordinary current I flowing into the plates.

2. Consider the baglike surface shown in Fig. 33.28. If the radius of the mouth of this surface is one-half the radius of the capacitor plates, what is the electric current intercepted by this surface? What is the displacement current intercepted by this surface? What is the sum of the intercepted electric current and displacement current?

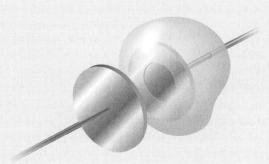

FIGURE 33.28 A parallel-plate capacitor with a baglike surface.

3. A parallel-plate capacitor is being charged by a current of 4.0 A.

    (a) What is the displacement current between its plates?

    (b) What is the rate of change of the electric flux intercepted by each plate?

4. A parallel-plate capacitor consists of circular plates of radius 0.30 m separated by a distance of 0.20 cm. The voltage applied to the capacitor is made to increase at a steady rate of 2.0×10^3 V/s. Assume that the electric charge distributes itself uniformly over the plates, and ignore the fringing effects.

   (a) What is the rate of increase of the electric field between the plates?

   (b) What is the displacement current between the plates within a radius of 0.15 m? Within a radius of 0.30 m?

   (c) What is the magnetic field between the plates at a radius of 0.15 m? At 0.30 m?

5. A parallel-plate capacitor has circular plates of radius 20 cm and a uniform electric field between the plates. The capacitor is being charged at a rate of 0.10 A.

   (a) What is the net displacement current between the plates?

   (b) What is the displacement current between the plates within the radial interval of $0 \le r \le 5.0$ cm?

6. The electric field in a parallel-plate capacitor with circular plates of radius $R = 5.0$ cm is changing with time t according to $E = Ct^2$, where $C = 5.0 \times 10^4$ V/m·s². What is the induced magnetic field at $t = 0.50$ s and $r = 1.0$ cm from the axis? What is the induced magnetic field at $t = 2.0$ s and $r = 6.0$ cm from the axis?

*7. Suppose that the parallel-plate capacitor discussed in Section 33.1 (see Fig. 33.3) is filled with a slab of dielectric with a dielectric constant κ. Assume that, as in the case of the empty capacitor, the displacement current can be viewed as a continuation of the ordinary current. Show that this implies that the displacement current in this filled capacitor must be

$$\kappa\epsilon_0 \frac{d\Phi_E}{dt}$$

and Maxwell's modification of Ampère's Law must be

$$\oint B_\parallel \, ds = \mu_0 I + \kappa\mu_0\epsilon_0 \frac{d\Phi_E}{dt}$$

*8. An emf $\mathcal{E}_{max} \sin\omega t$, with $\mathcal{E}_{max} = 0.50$ V and $\omega = 4.0 \times 10^3$ radians/s, is applied to the terminals of a capacitor with $C = 2.0$ pF. What is the displacement current between the capacitor plates?

*9. The space between the plates of a leaky capacitor is filled with a material of resistance 5.0×10^5 Ω. The capacitor has a capacitance of 2.0×10^{-6} F; its plates are circular, with a radius of 30 cm; and its electric field is uniform. At time $t = 0$, the initial voltage across the capacitor is zero.

   (a) What is the displacement current if we increase the voltage at the steady rate of 1.0×10^3 V/s?

   (b) At what time will the real current leaking through the capacitor equal the displacement current?

   (c) What is the displacement current between the plates within a radius of 20 cm? What is the magnitude of the

magnetic field between the plates at radius $r = 20$ cm at $t = 0$? At $t = 1.0$ s? At $t = 2.0$ s?

*10. A parallel-plate capacitor with circular plates of radius 25 cm separated by a distance of 0.15 cm is connected to a source of alternating emf. The voltage across the plates oscillates with an amplitude of 5000 V and a frequency of 60 Hz. What is the amplitude of the magnetic field between the plates at a distance of 20 cm from the axis of the capacitor?

*11. A long, straight wire of radius R and resistivity ρ carries a current which changes direction at $t = 0$ and can be described by the function $I = Ct$, where the rate C is a constant.

   (a) What is the (time-dependent) electric field inside the wire?

   (b) At $t = 0$, what is the induced magnetic field at a distance r from the axis for $r \le R$?

   (c) What is it for $r \ge R$?

*12. A long solenoid of length l and N turns is wound with wire of cross-sectional area A and resistivity ρ. The wire carries a time-dependent current which changes direction at $t = 0$ and can be described by the function $I = Ct$, where the rate C is a constant. At $t = 0$, what is the induced magnetic field inside the solenoid?

*13. Write Maxwell's equations for a medium of dielectric constant κ.

**14. Suppose that there exist magnetic monopoles, that is, positive and negative magnetic charges that act as sources and sinks of magnetic field lines analogous to positive and negative electric charges. The magnetic field generated by a magnetic charge q_m is an inverse-square field, $B = q_m/4\pi r^2$. Write a new set of Maxwell's equations that take into account the magnetic charge. Be careful with the constants ϵ_0 and μ_0 and with the signs.

33.2 The Electromagnetic Wave Pulse

15. In a collision with an atom, an electron suddenly stops. Describe the directions of the electric and magnetic radiation fields at some distance from the electron at right angles to the acceleration.

16. Draw a diagram analogous to Fig. 33.5, showing the electric field lines of a charge that has suffered an acceleration toward the left (that is, an acceleration opposite to that involved in Fig. 33.5).

17. Suppose that a charge is initially at rest, then is accelerated for a short time interval from $t = 0$ to $t = \tau$, then is decelerated from $t = \tau$ to $t = 2\tau$, and then remains at rest after $t = 2\tau$. Carefully draw a diagram analogous to Fig. 33.5, showing the electric field lines of such a charge, at some time after the deceleration has ended.

18. Figure 33.8 shows the electric and magnetic field lines for a flat wave pulse propagating in the positive x direction. Draw an analogous diagram for a flat wave pulse propagating in the negative x direction.

19. An electromagnetic pulse travels upward from the surface of the Earth. The electric field throughout the pulse points northward. Determine the direction of the magnetic field of the pulse.

20. A sheet of charge in the y–z plane at $x = 0$ is briefly accelerated in the $+y$ direction. As a result, a planar electromagnetic pulse travels outward in the $+x$ and $-x$ directions. Consider a point on the negative x axis. What is the direction of the electric field as the pulse passes that point? The direction of the magnetic field?

*21. Consider the plane wave pulse shown in Figs. 33.8 and 33.9. If the magnitude of the electric field in this pulse is 4.0×10^{-3} V/m, what is the magnitude of the displacement current flowing along the front surface of the pulse per meter of length measured perpendicularly to the current?

†33.3 Plane Waves; Polarization

22. A light-year is defined as the distance that light travels in one year. Calculate how much this is in meters.

23. Laser range finders used by surveyors (see Fig. 33.29) determine the distance traveled to a reflecting target by means of a pulse of laser light, which travels from the range finder to the target and back. The distance is automatically calculated from the travel time of this pulse. If such a range finder is to determine a distance of 100 m to within 1 cm, what is the maximum permitted error in the measurement of the travel time?

FIGURE 33.29 A laser range finder.

24. When the American astronauts on the Moon were in conversation with Mission Control on the Earth, there was a noticeable delay between questions and answers. What is the round-trip travel time for a radio signal from Earth to Moon and back? The distance to the Moon is 3.8×10^8 m.

25. The mean distance from the Earth to the Sun is 1.50×10^{11} m. How long does it take for light to travel from the Sun to the Earth?

†For help, see Online Concept Tutorial 36 at www.wwnorton.com/physics

26. A transmitter in Los Angeles sends a live broadcast to a listener in Texas; the electromagnetic wave travels 2000 km. How long does it take to travel?

27. A severe limitation on the speed of computation of large electronic computers is imposed by the speed of light, because the electric signals on the connecting wires within the computer are electromagnetic waves ("guided waves"), which travel at a speed roughly equal to the speed of light. If the computer measures about 1.0 m across, what is the minimum travel time required for a typical signal sent from one end of the computer to another? What is the maximum number of signals that can be sent back and forth (sequentially) per second? Is there any way to avoid the limitations imposed by the travel time of signals?

28. (a) One type of antenna for a radio receiver consists of a short piece of straight wire; when the electric field of a radio wave strikes this wire it makes currents flow along it, which are detected and amplified by the receiver. Suppose that the electric field of a radio wave is vertical. What must be the orientation of the wire for maximum sensitivity?

    (b) Another type of antenna consists of a circular loop; when the magnetic field of a radio wave strikes this loop it induces a current around it. Suppose that the magnetic field of a radio wave is horizontal. What must be the orientation of the loop for maximum sensitivity?

29. A plane electromagnetic wave travels in the eastward direction. At one instant the electric field at a given point has a magnitude of 0.60 V/m and points down. What are the magnitude and direction of the magnetic field at this instant? Draw a diagram showing the electric field, the magnetic field, and the direction of propagation.

30. Suppose that an unpolarized light beam is incident from the left on the arrangement of two polarizers illustrated in Fig. 33.17. If the intensity of the light emerging on the right is 30% of the incident intensity, what must be the angle between the preferential directions of the polarizers?

31. The preferential directions of two adjacent sheets of Polaroid make an angle of 45°. A beam of polarized light, whose direction of polarization coincides with the preferential direction of the *second* sheet, is incident on the *first* sheet. By what factor is the intensity of the transmitted beam emerging from the second sheet reduced compared with the intensity of the incident beam? Assume that the sheets act as ideal polarizing filters.

32. Two sheets of Polaroid are placed on top of one another. Unpolarized light is perpendicularly incident on the sheets. By what factor is the intensity of the emerging light reduced (relative to the incident light) if the preferential directions of the sheets differ by an angle of 30°? 45°? 60°?

33. Unpolarized light is incident upon the first of two polarizers. The second polarizer has its preferential direction at 80° with respect to the first one. What percentage of the original incident intensity is transmitted?

34. Unpolarized light of intensity I_0 is incident upon the first of three polarizers. The first polarizer has its preferential direc-

tion vertical, the second polarizer has its preferential direction at 60° with respect to vertical, and the third has its preferential direction at 20° with respect to vertical. What is the final transmitted intensity?   0

35. Polarized light is incident upon the first of two polarizers with its plane of polarization making an angle of 45° with respect to the pass direction of this polarizer. The intensity transmitted through the second polarizer is 30% of the original incident intensity. By what angle do the preferential directions of the two polarizers differ?

*36. (a) Consider two polarizers. The second has its preferential direction at 45° with respect to the first. Polarized light strikes the first polarizer at 45° with respect to its preferential direction. Show that the transmitted intensity is 25% of the original wave. (b) Next, consider three polarizers, each oriented at 90°/3 = 30° with respect to the previous one. Polarized light strikes the first polarizer at 30° with respect to its preferential direction. Calculate the percentage of the incident intensity that is transmitted. (c) Finally, consider a large number N of polarizers, each oriented at 90°/N with respect to the former. What is the percentage of the incident intensity that is transmitted through all N polarizers for $N = 90$? For $N \rightarrow \infty$?

*37. An electromagnetic wave traveling along the x axis consists of the following superposition of two waves polarized along the y and z directions, respectively:

$$\mathbf{E} = \mathbf{j}E_0 \sin\left(\omega t - \frac{\omega x}{c}\right) + \mathbf{k}E_0 \cos\left(\omega t - \frac{\omega x}{c}\right)$$

where $\omega = kc$. This electromagnetic wave is said to be **circularly polarized**.

(a) Show that the magnitude of the electric field is E_0 at all points of space at all times.

(b) Consider the point $x = y = z = 0$. What is the angle between $\mathbf{E}$ and the z axis at time $t = 0$? $t = \pi/2\omega$? $t = \pi/\omega$? $t = 3\pi/2\omega$? Draw a diagram showing the y and z axes and the direction of $\mathbf{E}$ at these times. In a few words, describe the behavior of $\mathbf{E}$ as a function of time.

*38. If the preferential directions of two adjacent sheets of Polaroid are at right angles, they will completely block a light beam. However, if you insert a third sheet of Polaroid between the other two, then some light will pass through (see Fig. 33.19). A formula for the dependence of the intensity of the transmitted light as a function of the angle that the preferential direction of the inserted sheet makes with that of the first sheet can be obtained from the discussion at the end of Section 33.3. For what orientation of the inserted sheet is the transmitted intensity maximum?

*39. Two sheets of Polaroid are arranged as polarizer and analyzer. Suppose that the preferential direction of the second sheet is rotated by an angle ϕ about the direction of incidence and then rotated by an angle α about the vertical direction (see Fig. 33.30). If unpolarized light of intensity I_0 is incident from the left, what is the intensity of the light emerging on the right?

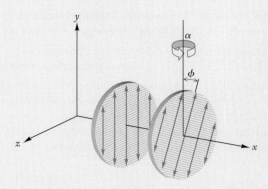

FIGURE 33.30 Two sheets of Polaroid.

33.4 The Generation of Electromagnetic Waves

40. At many coastal locations, radio stations of the National Weather Service transmit continuous weather reports at a frequency of 162.5 MHz. What is the wavelength of these transmissions?

41. The shortest microwaves have a wavelength of about 1.0 mm. What is the frequency of such a wave?

42. Radio station WWV of the National Institute of Standards and Technology, Fort Collins, Colorado, transmits precise time signals at radio frequencies of 2.5, 5.0, 10, 15, and 20 MHz. What are the wavelengths of these transmissions?

43. An ordinary radio receiver, such as found in homes across the country, has an AM dial and an FM dial (Fig. 33.31). The AM dial covers a range from 530 to 1700 kHz and the FM dial a range from 88 to 108 MHz. What is the range of wavelengths for AM? For FM?

FIGURE 33.31 An AM–FM radio.

44. A proton in cyclotron motion moves in a circle in a magnetic field of 8.0 T. Find the wavelength of electromagnetic waves produced at the cyclotron frequency.

45. Find the wavelengths of waves of the following frequencies: (a) 2.0×10^{18} Hz. (b) 3.0×10^{10} Hz. (c) 60 Hz. What part of the electromagnetic spectrum corresponds to each of these frequencies?

46. Hydrogen atoms in interstellar clouds of gas emit radio waves of wavelength 21 cm. What is the frequency of these waves?

47. A radar antenna emits radio waves of a frequency of 1.1×10^{10} Hz. What is the wavelength of these waves?

48. Figure 33.24 gives the sensitivity of the human eye as a function of the wavelength of light. For what wavelength is the sensitivity maximum? For what wavelengths is the sensitivity one-half the maximum? One-quarter of the maximum? What color corresponds to each of these wavelengths?

33.5 Energy of a Wave

49. At one point in an electromagnetic wave, the instantaneous electric field has a magnitude of 80 V/m. What is the energy density?

50. At a distance of several kilometers from a radio transmitter, the electric field of the emitted radio wave has a magnitude of 0.12 V/m at one instant of time. What is the energy density in this electric field? What is the energy density in the magnetic field of the radio wave?

51. A radio wave has an instantaneous magnetic field of 2.0×10^{-10} T. What is the magnitude of the instantaneous Poynting flux?

52. The average energy flux of sunlight incident on the top of the Earth's atmosphere is 1.4×10^3 W/m². What are the corresponding amplitudes of oscillation of the electric and magnetic fields?

53. A plane electromagnetic wave travels in the northward direction. At one instant, the electric field at a given point has a magnitude of 0.50 V/m and is in the eastward direction. What are the magnitude and direction of the magnetic field at the given point? What are the magnitude and direction of the Poynting vector?

54. A laser used as a torch, to cut plates of metal (Fig. 33.32), produces a light beam with an average energy flux of 1.0×10^9 W/m². What are the magnitudes of the rms electric and magnetic fields in such a light beam?

FIGURE 33.32 Laser used as torch.

55. Starlight arriving at the Earth from the star Capella has an rms energy flux of 1.2×10^{-8} W/m². The distance of this star is 4.3×10^{17} m. Calculate the power radiated by this star.

56. The beam of light produced by a small laser is cylindrical, of diameter 2.5 mm. The rms power that the laser feeds into this beam is 1.2 W. Calculate the rms values of the electric and magnetic fields.

57. The Sun emits radiation uniformly in all directions. At the Earth, at a distance of 1.5×10^{11} m, the energy flux of sunlight is 1.4×10^3 W/m². Calculate the power radiated by the Sun.

58. The beam of a powerful laser has a diameter of 0.20 cm and carries a power of 6.0 kW. What is the time-average Poynting flux in this beam? What are the rms values of the electric and magnetic fields?

59. A TV transmitter emits a spherical wave, that is, a wave spreading out uniformly in all directions. At a distance of 5.0 km from the transmitter, the amplitude of the wave is 0.22 V/m. What is the magnitude of the time-average Poynting flux at this distance? What is the time-average power emitted by the transmitter?

60. A silicon solar cell (Fig. 33.33) of frontal area 13 cm² delivers 0.20 A at 0.45 V when exposed to full sunlight of energy flux 1.0×10^3 W/m². What is the efficiency for conversion of light energy into electric energy?

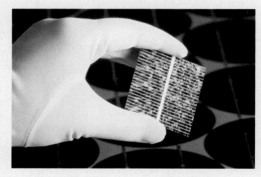

FIGURE 33.33 Solar cells.

61. At a distance of 6.0 km from a radio transmitter, the amplitude of the electric radiation field of the emitted radio wave is $E_0 = 0.13$ V/m. What will be the amplitude of the radio wave when it reaches a distance of 12.0 km? A distance of 18.0 km?

62. A point source, with a power output of 100 W, emits electromagnetic waves uniformly in all directions. At what distance from the source is the amplitude of the electric field 2.0 V/m?

63. A small diode laser pointer emits 5.0 mW of average power in a beam that has a diameter of 1.0 mm. What is the intensity of the beam? What is the amplitude of the electric field in the beam? How much energy is contained in a beam of length 10 m at a given instant?

64. A pulsed laser used for evaporation ("ablation") in materials science has a total energy of 0.50 J in a pulse of ultraviolet light that is 2.0×10^{-8} s long. The beam is 2.0 mm in diameter. What is the power delivered during the pulse? What is the intensity in the pulsed beam? What is the electric field amplitude during the pulse?

65. A point source of light delivers 75 W of average power uniformly in all directions. A circular mirror with a radius of 5.0 cm is placed 2.0 m from the source. The light strikes the mirror (essentially) perpendicularly; the mirror is perfectly reflecting. What is the intensity of light at the mirror? What is the force on the mirror?

66. A radio transmitter broadcasts uniformly in all directions with a power output of 50 kW. What is the time-average Poynting flux 50 km from the transmitter? What is the amplitude of the electric field there?

67. A solar energy "farm" produces 1.0 MW of power for a small town. The solar cells used convert incoming solar energy to electricity with 25% efficiency. Assume that the average solar flux is 500 W/m^2. What total area of solar cells is required?

*68. A magnifying glass of diameter 10 cm focuses sunlight into a spot of diameter 0.50 cm. The energy flux in the sunlight incident on the lens is 0.10 W/cm^2.

(a) What is the energy flux in the focal spot? Assume that all points in the spot receive the same flux.

(b) Will newspaper ignite when placed at the focal spot? Assume that the flux required for ignition is 2.0 W/cm^2.

*69. Binoculars are usually marked with their magnification and lens size (Fig. 33.34). For instance, 7 × 50 binoculars magnify angles by a factor of 7 and their collecting lenses have an aperture of diameter 50 mm. Your pupil, when dark-adapted, has an aperture of diameter 7.0 mm. When observing a distant pointlike light source at night, by what factor do these binoculars increase the energy flux penetrating your eye? Neglect reflection of light by the lenses.

FIGURE 33.34 Binoculars.

*70. A radio receiver has a sensitivity of 2.0×10^{-4} V/m. At what maximum distance from a radio transmitter emitting a time-average power of 10 kW will this radio receiver still be able to detect a signal? Assume that the transmitter radiates uniformly in all directions.

*71. At normal incidence, the intensity of sunlight at the Earth is 1400 W/m^2. Assume for simplicity that all of the incident energy that strikes the Earth is absorbed. What is the force on the Earth due to radiation pressure from the Sun? Compare this with the gravitational force on the Earth due to the Sun.

*72. A Poynting flux can also be defined for static fields. Consider a long, straight wire of radius r, length l, and resistance R. The wire carries a constant current I uniformly distributed over its cross section. In terms of the given quantities, determine the electric field throughout the wire, the magnetic field at the surface of the wire, and the Poynting flux at a point on the surface of the wire. In what direction is the Poynting vector?

**73. Two plane wave pulses of the kind described in Section 33.3 are traveling in opposite directions. Their polarizations are parallel and the magnitudes of their electric fields are 2.0×10^{-3} V/m.

(a) What are the electric energy density and the magnetic energy density in each pulse?

(b) Suppose that at one instant the two pulses overlap. What are the magnitudes of the electric field and the magnetic field in this superposition?

(c) What are the electric energy density and the magnetic energy density in this superposition?

33.6 The Wave Equation

74. Show that the electric field $E_y = E_0 \sin(kx - \omega t)$ is a solution to the wave equation, Eq. (33.33). In what direction does this wave travel? Write an expression for the magnetic field of this wave.

75. The magnetic field of an electromagnetic wave is described by $B_x = B_0 \cos(kz + \omega t)$. If the frequency of this wave is 3.75 MHz, what is the wavelength? What are the values of k and ω? In what direction does this wave travel? Write an expression for the electric field of this wave.

*76. Show in general that any function f which is a function of the variable $(x - ct)$ is a solution to the wave equation, Eq. (33.33). Then verify this explicitly by checking that the function $E = E_0 \exp[-k^2(x - ct)^2]$ satisfies the wave equation.

*77. Proceeding as in Eqs. (33.29)–(33.31), apply the Maxwell–Ampère law to the horizontal rectangle in Fig. 33.27 to obtain Eq. (33.32).

REVIEW PROBLEMS

78. A capacitor has circular plates, such as illustrated in Fig. 33.3, of radius 0.15 m. Between these plates there is a uniform electric field. Suppose that this electric field is increasing at the rate of 3.8×10^{13} V/m per second. What is the displacement current between the plates? What is the magnitude of the magnetic field at the edge of the capacitor, halfway between the plates?

*79. A parallel-plate capacitor has circular plates of area A separated by a distance d. A thin straight wire of length d lies along the axis of the capacitor and connects the two plates (Fig. 33.35); this wire has a resistance R. The exterior terminals of the plates are connected to a source of alternating emf with a voltage $V = V_0 \sin \omega t$.

(a) What is the current in the thin wire?

(b) What is the displacement current through the capacitor?

(c) What is the current arriving at the outside terminals of the capacitor?

(d) What is the magnetic field between the capacitor plates at a distance r from the axis? Assume that r is less than the radius of the plates.

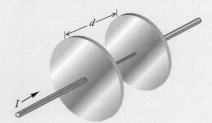

FIGURE 33.35 Parallel-plate capacitor with a thin wire connecting the inside faces of the plates.

80. Sketch the electric field lines for a positive charge that is initially moving at uniform velocity, suddenly stops, and remains stopped.

81. Radio station WWV at Fort Collins, Colorado, continually transmits time signals at several shortwave frequencies. These short waves are reflected by the ionosphere and also by the ground. The waves bounce back and forth between the ionosphere and the ground, and they can therefore roughly follow the curvature of the Earth's surface and travel all the way around the Earth. Suppose that a navigator on a ship in the south of the Indian Ocean, halfway around the Earth, listens to the WWV signals. If he sets his chronometer according to what he hears, roughly how late will he be?

82. Linearly polarized light is incident on a polarizer whose preferential direction is inclined at an angle of 20° relative to the plane of polarization of the incident light. What fraction of the incident light is transmitted? Repeat for an angle of 40°, and repeat for an angle of 60°.

83. A light wave of intensity 0.50 W/m^2 is polarized in the vertical plane. This wave is incident on a polarizer whose preferential direction is inclined at 60° to the vertical, and then on a second polarizer whose preferential direction is vertical. What is the intensity of light that emerges from the second polarizer?

*84. An electromagnetic wave has the form

$$\mathbf{E} = \mathbf{i}E_0 \sin\left(\omega t + \frac{\omega z}{c}\right) + \mathbf{j}2E_0 \sin\left(\omega t + \frac{\omega z}{c}\right)$$

(a) What is the direction of propagation of this wave?

(b) What is the direction of polarization, that is, what angle does the direction of polarization make with the x, y, and z axes?

(c) Write down the formula for the magnetic field of this wave as a function of space and time.

85. An electromagnetic wave is traveling vertically upward. The instantaneous electric field at some point in this wave is eastward, of magnitude 150 V/m. What are the magnitude and the direction of the instantaneous magnetic field?

86. For radio communication with submerged submarines, the U.S. Navy uses ELF (extremely low frequency) radio waves of wavelength 4000 km; such waves can penetrate for some distance below water. What is the frequency of such waves?

87. The beam of light produced by a small laser is cylindrical, of diameter 2.5 mm. The rms power that the laser feeds into this beam is 1.2 W. Calculate the rms values of the electric and magnetic fields.

*88. According to a proposed scheme, solar energy is to be collected by a large power station on a satellite orbiting the Earth. The energy is them to be transmitted down to the surface of the Earth as a beam of microwaves. At the surface of the Earth, the beam is to have a width of about 10 km × 10 km and is to carry a power of 5.0×10^9 W.

(a) What would be the time-average Poynting flux in this beam?

(b) What would be the amplitude of the electric and magnetic fields in the beam?

89. A radio transmitter emits a time-average power of 5.0 kW in the form of a radio wave with uniform intensity in all directions. What are the amplitudes of the electric and magnetic fields of this radio wave at a distance of 10 km from the transmitter?

*90. A steady current of 12 A flows in a copper wire of radius 0.13 cm.

(a) What is the longitudinal electric field in the wire?

(b) What is the magnetic field at the surface of the wire?

(c) What is the magnitude of the radial Poynting flux at the surface of the wire?

(d) Consider a 1.0-m segment of this wire. According to the Poynting flux, what amount of power flows into this piece of wire from the surrounding space?

(e) Show that the power calculated in part (d) coincides with the power of the Joule heat developed in the 1.0-m segment of wire.

91. The quasar 3C 273 is at a distance of 2.8×10^9 light-years from the Earth. The flux of radio waves reaching the Earth from this quasar is 4.1×10^{-25} W/m^2 in a frequency interval of 1.0 Hz around 1410 Hz (comparable amounts of flux are found at other radio frequencies). What is the power in this 1.0-Hz-wide frequency interval captured by the Arecibo radio telescope, a dish of diameter 300 m? What is the power emitted by the quasar? Assume the quasar radiates uniformly in all directions.

92. In the United States, the accepted standard[3] for the safe maximum level of continuous whole-body exposure to microwave radiation is 10 milliwatts/cm^2.

[3] It is of interest that in Russia, where many experiments on the effects of microwaves on the human body have been performed, the accepted standard is much lower, 10 microwatts/cm^2.

(a) For this energy flux, what are the corresponding amplitudes of oscillation of the electric and magnetic fields?

(b) Suppose that a man of frontal area 1.0 m^2 completely absorbs microwaves with an intensity of 10 milliwatts/cm^2 incident on this area and that the microwave energy is converted to heat within his body. What is the rate (in calories per second) at which his body develops heat?

*93. At night, the naked, dark-adapted eye can see a star provided the energy reaching the eye is 8.8×10^{-11} W/m^2.

(a) Under these conditions, how many watts of power enter the eye? The diameter of the dark-adapted pupil is 7.0 mm.

(b) Assume that in our neighborhood there are, on the average, 3.5×10^{-3} stars per cubic light-year and that each of these emits the same amount of light as the Sun (3.9×10^{26} W). If so, how far would the faintest visible star be? How many stars could we see in the sky with the naked eye?

Answers to Checkups

Checkup 33.1

1. As discussed, the total displacement current $\epsilon_0 \, d\Phi_E/dt$ is equivalent to the ordinary current charging the plates, and so is 2 A in either case.

2. No; if the electric field line forms a closed loop, $\oint E_\parallel \, ds \neq 0$ along this line, and by Eq. (33.6), the magnetic flux and the magnetic field must be time-dependent. Similarly, if there are no ordinary currents within the closed loop formed by the magnetic field, then Eq. (33.7) implies that such a magnetic field was caused by a time-dependent electric field.

3. No—such field lines form continuous closed loops in space, and thus do not originate on charges.

4. There is an electric field due to the static charges residing on the plates. However, if the amount of charge (and thus the electric field and electric flux) is not changing, then there is no magnetic field between the plates.

5. (C) $r = R$. According to the results of Example 1, for $r \leq R$, the induced magnetic field increases with r; for $r \geq R$, the induced magnetic field decreases with r. The maximum occurs at $r = R$.

Checkup 33.2

1. To determine the direction of the magnetic field, apply the right-hand rule: the direction of $\mathbf{E} \times \mathbf{B}$ must give the direction of propagation of the wave. Thus for case (A), if the electric field is southward, the magnetic field must be downward

for $\mathbf{E} \times \mathbf{B}$ to be eastward. Similarly, for case (B), for a southward $\mathbf{E}$ crossed with $\mathbf{B}$ to give a westward propagation, $\mathbf{B}$ must be upward. For case (C), $\mathbf{B}$ must be eastward for $\mathbf{E} \times \mathbf{B}$ to be upward.

2. Similar to question 1, we need the direction of the E vector which, when crossed with $\mathbf{B}$, gives a vector with the given propagation direction. For case (A), $\mathbf{E}$ must be upward (curling the fingers from upward to southward, the thumb gives the required eastward direction of propagation). Similarly, for case (B), the electric field must be downward. For case (C), a westward $\mathbf{E}$ crossed with a southward $\mathbf{B}$ gives an upward direction of propagation.

3. The induction of the electric field by the changing magnetic field is governed by Faraday's Law, $\oint E_\parallel \, ds = -d\Phi_B/dt$. The induced magnetic field is determined by the Maxwell–Ampère Law; for the wave pulse, there is zero real current, so only the displacement-current term contributes, $\oint B_\parallel \, ds = \mu_0 \epsilon_0 \, d\Phi_E/dt$.

4. (C) Eastward. To determine the direction of the magnetic field, apply the right-hand rule: the direction of $\mathbf{E} \times \mathbf{B}$ must give the direction of propagation of the wave. Thus if the electric field is downward, the magnetic field must be eastward for $\mathbf{E} \times \mathbf{B}$ to be southward.

Checkup 33.3

1. The direction of polarization is parallel to the direction of the electric field of the wave or wave pulse; thus, for question 1 of the preceding section, all the cases have southward polariza-

tions. For question 2, the electric fields (and thus the directions of polarization) were in the (A) upward, (B) downward, and (C) westward directions.

2. If the wave of amplitude E_0 is polarized at 45° with respect to the y axis, then the components are equal: $E_y = E_0 \cos 45° = E_0/\sqrt{2}$ and $E_x = E_0 \sin 45° = E_0/\sqrt{2}$.

3. From Malus' Law, the transmitted intensity will be the incident value multiplied by a factor of $\cos^2 \phi$. In the first case, this is $\cos^2 45° = \frac{1}{2}$ (intensity reduced by a factor of 2); in the second case we have $\cos^2 60° = \frac{1}{4}$ (intensity reduced by a factor of 4).

4. (A) $I_0/8$. For unpolarized light, a single polarizer reduces the intensity to one-half of the incident value (see Example 4). The second polarizer reduces the polarized light intensity by a factor of $\cos^2 \phi = \cos^2(60°) = (1/2)^2 = 1/4$. So the final intensity is $(1/2)I_0 \times (1/4) = I_0/8$.

Checkup 33.4

1. For FM radio waves, the signal is imprinted as variations in frequency, and so is not affected by changes in wave amplitude, whereas AM signals fluctuate with noise and any other amplitude variations (for example, while driving under an overpass); also, regulations permit FM broadcast signals to have a frequency range (20 kHz) twice the AM bandwidth.

2. From Fig. 33.23 or 33.24, we see that light of wavelength 650 nm is red, while light of wavelength 550 nm is green or yellow-green.

3. From Figs. 33.23 and 33.24, we perceive the color yellow over a range of wavelengths near 575–600 nm.

4. By familiarizing ourselves with Fig. 33.23, we see that radar has the shortest wavelength (around 1 cm); most TV stations come next (around 1 m); FM is next (around 3 m); while AM has much longer wavelengths (hundreds of meters).

5. (B) a virus ($\sim 10^{-6}$ m). Visible light has wavelengths in the 400–700 nm range, or about 0.4–0.7 $\times$ 10^{-6} m.

Checkup 33.5

1. The energy densities depend only on the amplitude of the wave, as in Eq. (33.17), so both are equal.

2. The energy density is a maximum whenever the magnitude of the field is a maximum. This happens twice each cycle, when $E = \pm E_0$. Thus the distance between energy density maxima is $\lambda/2$.

3. The energy density is proportional to the square of the amplitude; since the amplitudes have a ratio of 1 to 2, the energy densities have a ratio of 1 to 4.

4. Assuming, as in Example 6, that the wave spreads uniformly in all directions, then for twice the distance the power is spread over an area 4 times as large. Thus the intensity (the time-average energy flux) is one-fourth as large, or $(2.2 \times 10^{-5} \text{ W/m}^2)/4 = 5.5 \times 10^{-6} \text{ W/m}^2$. The intensity is proportional to the square of the amplitude, and one-fourth the intensity implies one-half the amplitude, or $(1/2) \times 0.13 \text{ V/m} = 0.065 \text{ V/m}$.

5. By Malus' Law, Eq. 33.15, the fraction of the intensity transmitted is $\cos^2 45° = \frac{1}{2}$. The energy that is not transmitted is absorbed by the polarizer, usually by inducing currents which dissipate the energy as heat.

6. The pressure is proportional to the energy flux, which is proportional to the amplitude squared, so a wave with twice the amplitude exerts 4 times the pressure. However, reflection also produces twice as much pressure as absorption, so overall, the second wave exerts 8 times the pressure of the first.

7. (B) 10 W/m^2. The intensity is the power per unit area at the distance of the book; the 125 W of power is spread over an area $A = 4\pi r^2$ at $r = 1.0$ m, and so the intensity is $I = \bar{P}/A = (125 \text{ W})/[4\pi(1.0 \text{ m})^2] = 10 \text{ W/m}^2$.

Checkup 33.6

1. Since the wave number is proportional to the reciprocal of the wavelength, the first wave will have a wavelength half as large as the second wave. Their speeds will be identical, both equal to c, the speed of light. For equal speeds, frequency is proportional to wave number ($\omega = ck$, or $f = c/\lambda$), so the first wave will have twice the frequency of the second.

2. Both the electric and magnetic fields vary the same way, and both have frequencies given by $f = c/\lambda = (3 \times 10^8 \text{ m/s})/300 \text{ m} = 1 \times 10^6 \text{ s}^{-1} = 1$ MHz.

3. No. This function has no time dependence, and so has no traveling wave character. Mathematically, the second derivative with respect to x is a constant, $2E_0/L^2$, whereas the second derivative with respect to time is zero. Thus this function cannot satisfy Eq. (33.33).

4. (E) $A(x^2 - v^2t^2)$. Each of the other expressions is a function of $(x - vt)$ or $(x + vt)$ which automatically provides a traveling solution; for each function f except (E), direct substitution verifies that $\partial^2 f/\partial x^2 = (1/v^2) \, \partial^2 f/\partial t^2$ is satisfied.

Reflection, Refraction, and Optics

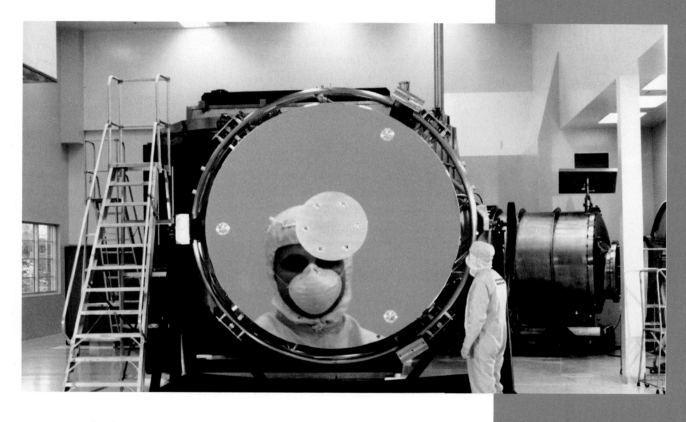

CONCEPTS IN CONTEXT

Concepts *in* Context

The Hubble Space Telescope uses this large concave mirror, with a radius of curvature of 11 m, to image distant galaxies.

With the laws of geometric optics presented in this chapter, we can answer a variety of questions about images formed by the mirror:

❓ How far from the mirror does the image of a distant galaxy form? (Example 7, page 1129)

❓ The technician is 4.0 m from the mirror. Where is his image? (Example 10, page 1134)

❓ The image of the technician is large and upright. How do we calculate such image properties? (Example 13, page 1140)

❓ How do we measure the magnification of a telescope? (Example 16, page 1151)

reflection

refraction

So far we have examined the propagation of light waves and other electromagnetic waves only in a vacuum. There, a plane wave will simply propagate in a fixed direction at the constant speed c. But if the wave encounters the surface of a region filled with matter—a sheet of metal, a pane of glass, or a layer of water—then the wave will interact with the matter and can suffer a change in speed, direction, intensity, and polarization. One part of the wave will be **reflected** by the surface; that is, it will bounce off. The other part will be **refracted**; that is, it will penetrate into the matter-filled region and continue to propagate, with some change of speed and of direction. You can notice this partial reflection of light by a water surface when you look at the light from streetlamps or other bright sources reflected by the surface of a calm dark pond; the reflected light is not as strong as the direct light. You can notice the refraction of light by looking at your fingers through a drinking glass filled with water; your fingers appear distorted, because the rays of light traveling from your fingers to your eye suffer a change of direction when they enter the drinking glass, and they suffer another change of direction when they leave the drinking glass.

Since light is an electromagnetic wave, the changes in speed, direction, intensity, and polarization of the light wave can be calculated from Maxwell's equations, taking into account the motion of the electric charges and the flow of current caused by the action of the wave on the matter. But the complete calculation of all the changes in the wave is rather complicated, and furthermore, Maxwell's equations often tell us more than we want to know. For instance, if a light wave encounters a water surface, we may wish to compute the angle at which it penetrates the water, but we do not always need to know the changes in intensity or polarization.

In this chapter we will see that much can be learned by considering merely one aspect of Maxwell's equations, namely, their implications for the speed of light in matter-filled regions. We will see that in such regions the speed of light is reduced, and that this change of the speed of light leads to the change of direction of propagation during refraction, when the wave penetrates from vacuum into a matter-filled region or when the wave penetrates from one matter-filled region into another.

geometric optics

The laws of reflection and refraction are the basis for **geometric optics**, which we will study in this chapter. Geometric optics relies on the assumption that *light propagates in a fixed direction, along a straight line (rectilinearly), while in a uniform medium; and suffers changes of direction only when it encounters the surface separating two different media.* You can observe this straight-line propagation in the beam of a strong searchlight aimed at the dark sky, or in beams of sunlight piercing through holes in a cloud (Fig. 34.1). But the most dramatic demonstration of straight-line propagation is provided by laser beams, which look like fine straight lines in space (Fig. 34.2).

FIGURE 34.1 Rays of sunlight.

FIGURE 34.2 Beams of laser light.

34.1 HUYGENS' CONSTRUCTION

Online
Concept
Tutorial

The propagation of a light wave or any other electromagnetic wave can be conveniently described by means of the **wave fronts**, or wave crests—that is, the points at which the electric field of the wave has maximum strength at some instant of time. For example, Fig. 34.3 shows the instantaneous wave fronts of the radio wave emitted by a radio station. The interval between one wave front and the next is one wavelength. With the passing of time, each of these wave fronts spreads in the outward direction.

The rule governing the propagation of wave fronts is **Huygens' Construction**:

> *To find the change of position of a wave front in a small time interval Δt, draw many small spheres of radius [wave speed] $\times \Delta t$ with centers on the old wave fronts. The new wave front is the surface that touches the leading edges of these small spheres.*

The small spheres employed in this construction are called **wavelets.** Figure 34.4 shows how Huygens' Construction applies to the propagation of the spherical wave fronts in Fig. 34.3. The wave speed in this example is simply c, and hence the radius of the wavelets is $c\,\Delta t$. Erecting wavelets of this radius on the old wave front, we find the new wave front that touches the outer edges of these wavelets; since, in this example, all the wavelets have the same radius, the new wave front is concentric with the old wave front.

Figure 34.5 shows a similar construction for the case of the propagation of a plane radio wave or a plane light wave. When we erect the wavelets on the plane wave front, the result is another plane wave front, parallel to the first. This means that the plane continues to propagate in the same direction, without deviating to one side or another. A light beam, such as the light beam from a laser, consists of plane wave fronts, and Huygens' Construction therefore accounts for the straight-line propagation of such a light beam.

Huygens' Construction applies not only to the propagation of light waves in a vacuum, but also to their propagation in any transparent material, such as air, glass, or water. As we will see in the following sections, this construction permits us to derive

wave front

Huygens' Construction

wavelet

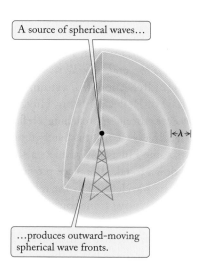

FIGURE 34.3 Spherical wave fronts at one instant of time. At a later time, each of these wave fronts will have moved outward by some distance.

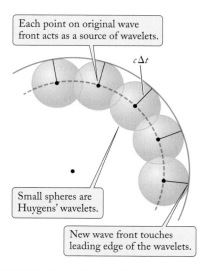

FIGURE 34.4 Huygens' Construction for the propagation of a spherical wave front. The inner (blue dashed) arc shows the wave front at time t; the outer (blue solid) arc shows the propagated wave front at time $t + \Delta t$.

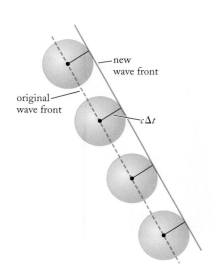

FIGURE 34.5 Huygens' Construction for the propagation of a plane wave front.

the laws of reflection and refraction. Although our emphasis will be on the propagation of light, Huygens' Construction is a general feature of wave propagation—it applies just as well to sound waves and water waves. Apart from differences arising from the different speeds of these waves, the laws of reflection and refraction for all such waves are essentially the same, and many of our results for light waves can be readily generalized to such other waves.

✔ Checkup 34.1

QUESTION 1: Suppose that in Fig. 34.4 the light is propagating *inward*, that is, toward the center of the circle. If the position of the wave front at time t is along the blue dashed arc, what is the propagated wave front at time $t + \Delta t$?

QUESTION 2: Consider the wave front of a water wave on the surface of a pond generated by a brick dropped into the pond with its flat side down. The wave front at the initial instant is then rectangular. According to Huygens' Construction, what is the shape of the wave front a short time Δt later?

QUESTION 3: When a thin stick of length L is dropped onto water (with its length parallel to the water surface), the impact produces a water wave that spreads outward. The wave front soon after the initial instant is long and narrow, like the stick. What is the length-to-width ratio of the wave front when it has traveled a distance L from the stick? When it has traveled a distance much greater than L?

(A) $2, \frac{3}{2}$ (B) $1, 1$ (C) $3, 2$ (D) $\frac{3}{2}, 1$ (E) $3, 1$

34.2 REFLECTION

Online
Concept
Tutorial

When a light wave encounters the surface of a transparent material—such as the surface of a pane of glass, or the surface of a pond—part of the wave penetrates the surface and part is reflected. When a light wave encounters the surface of a very smoothly polished metal—such as the silvered surface of a mirror—almost all of the wave is reflected. In this section we will deal with the reflected part of the wave; in the next section we will deal with the part of the wave that penetrates from one transparent material into the other.

The **Law of Reflection** for a wave incident on a flat surface at an angle has been known since ancient times: *the angle of incidence equals the angle of reflection*. Figure 34.6 gives an experimental demonstration of this Law of Reflection with a strong light beam shining down on a mirror. To derive this Law of Reflection from Huygens' Construction, we begin with Fig. 34.7a, which shows wave fronts approaching a reflecting surface; at the instant shown, one edge of the leading wave front is barely touching the surface at the point P. Figure 34.7b shows some Huygens' wavelets a short time later, when the second wave front has moved down to take the place previously occupied by the leading wave front. The portions of the wavelets below the reflecting surface have been omitted as irrelevant. The new leading wave front constructed on these wavelets touches the surface at the point P'. Obviously, to the right of the point P', the new wave front is simply parallel to the old wave front, that is, this part of the wave has not yet been reflected. To find the new wave front to the left of the point P', we draw a straight line that starts at P' and is tangent to the wavelet centered on P. This straight line represents the part of the wave front that has already been reflected. To see that the incident wave front and the reflected wave front make the same angle

FIGURE 34.6 Reflection of a beam of light by a mirror. The angle of incidence equals the angle of reflection.

(a)

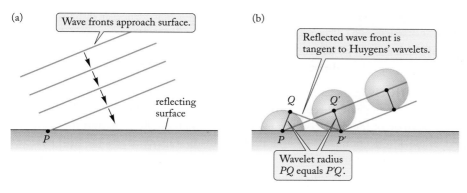

(b)

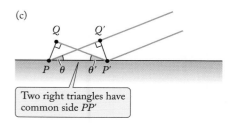

(c)

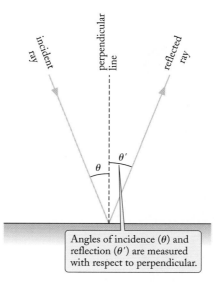

FIGURE 34.7 (a) Wave front approaching a reflecting surface. The leading wave front barely touches the reflecting surface. (b) Huygens' wavelets erected on the leading wave front of part (a). (c) The incident wave front PQ' makes an angle θ with the reflecting surface; the reflected wave front QP' makes an angle θ' with this surface.

with the reflecting surface, we appeal to Fig. 34.7c. The right triangles $PQ'P'$ and $P'QP$ are identical, because they have a common long side (PP') and their short sides (PQ and $P'Q'$) are equal. Hence the angles θ and θ' are equal.

The direction of propagation of a wave is commonly described by the **rays** of the wave. These *rays are lines perpendicular to the wave fronts.* For example, Fig. 34.8 shows the rays associated with the incident and the reflected wave fronts.

The angle θ (or θ') between the wave front and the reflecting surface is equal to the angle between the ray and the perpendicular to the surface. The angles θ and θ' are called the **angles of incidence and of reflection** (see Fig. 34.9). Thus, from the Huygens' Construction, we have deduced that the angle of incidence equals the angle of reflection. This is the Law of Reflection,

$$\theta_{\text{incident}} = \theta_{\text{reflected}} \tag{34.1}$$

Note that the incident ray, the reflected ray, and the perpendicular to the surface all lie in the same plane.

ray

angles of incidence and of reflection

Law of Reflection

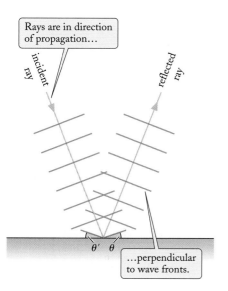

FIGURE 34.8 The rays of the wave are perpendicular to the wave fronts.

FIGURE 34.9 The angle of incidence θ and the angle of reflection θ'. These angles are the same as in Fig. 34.8.

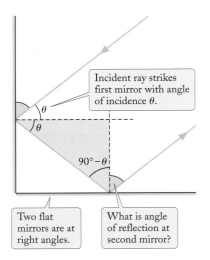

FIGURE 34.10 Two flat mirrors arranged in a corner. The blue angles indicate the directions of the first incident ray and the final reflected ray relative to the surface of the first mirror.

EXAMPLE 1 Two flat mirrors are arranged in a corner, at right angles (see Fig. 34.10). A ray of light strikes the first mirror with an angle of incidence θ. The reflected ray then strikes the second mirror. At what angle does the second reflected ray emerge?

SOLUTION: In Fig. 34.10, the angle of incidence of the ray on the first mirror is θ, and the angle of reflection is the same. Inspecting the colored triangle, we see that the angle of incidence on the second mirror is then $90° - \theta$, and therefore the angle of reflection is also $90° - \theta$. Hence the first incident and the final reflected rays both make an angle of $90° - \theta$ with respect to the surface of the first mirror. These rays are therefore parallel.

COMMENTS: The corner mirrors return the ray in the direction from which it came. This property of reflection by a corner is also valid for a three-dimensional corner, consisting of three mirrors arranged at right angles, called a corner reflector. When a ray is incident on such a corner reflector at some angle, it is returned in the direction from which it came after reflection on the three mirrors. Reflectors on automobiles and bicycles make use of an array of many small corner reflectors (see Fig. 34.11).

When light from some source strikes a flat mirror, the reflection of the light leads to formation of an image of the source. Figure 34.12 shows a point source of light and the rays emerging from it; the figure also shows the reflected rays. If we extrapolate the reflected rays to the far side of the mirror, we find that they *all appear to come from a point source of light placed beyond the mirror*. This apparent point source is the **image**. To an eye looking into the mirror, the image looks like the original source—the eye perceives the mirror image as existing in the space beyond the mirror. But this mirror image is an illusion; the light does not come from beyond the mirror. This kind of illusory image that gives the impression that light rays emerge from where they do not is called a **virtual image**.

image and virtual image

FIGURE 34.11 A reflector on the rear of a bicycle. Each of the small elements is a corner reflector.

Many small corner mirrors return light to direction from which it came.

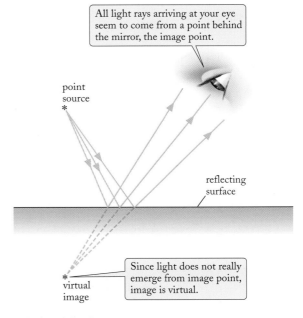

FIGURE 34.12 Rays emerging from a point source are reflected by a mirror. The extrapolated rays (dashed) appear to come from a point source beyond the mirror.

If, instead of a single luminous point, our light source consists of an extended object—such as a book, a hand, or a face—then the mirror image will also be an extended object. If the object is illuminated by sunlight or by a lamp, each point on the surface of the object scatters light and acts as a luminous point. Each such point forms an image in the mirror, and the net result is a mirror image of the entire object. Note that the mirror image of an object is a mirror-reversed object. For instance, Fig. 34.13 shows some written letters and their mirror images. This reversal is commonly referred to as a reversal of left to right. However, it is more accurately described as a reversal of front to back—mirror writing is ordinary writing seen from behind. And the mirror image of, say, a hand facing north is a hand facing south (the reversal is not an ordinary "about face," but involves passing the front of the hand through its back, thereby converting a right hand into a left hand, and vice versa; see Fig. 34.14).

 ## Checkup 34.2

QUESTION 1: If a ray makes an angle of 20° with the surface of a mirror, what is the angle of incidence?

QUESTION 2: If the angle of incidence on a flat mirror is 0°, what is the angle of reflection? What happens to the ray?

QUESTION 3: Consider Fig. 34.6. The image of the flashlight is not visible in the photograph. Where is the image?

QUESTION 4: You stand 2 m from a door mirror. What is the distance from you to your image?

QUESTION 5: You run toward a door mirror at 8 m/s. What is the speed of your image relative to you?

QUESTION 6: What is the length of a mirror needed for a person of height h to see his or her reflection from head to toe?

(A) $2h$ (B) $\frac{3}{2}h$ (C) h (D) $\frac{2}{3}h$ (E) $\frac{1}{2}h$

 Online Concept Tutorial 37

34.3 REFRACTION

The speed of light in a transparent material—such as air, water, or glass—differs from the speed of light in vacuum. We can recognize this immediately by recalling the theoretical formula for the speed of light derived from Maxwell's equations, Eq. (33.8):

$$c = \frac{1}{\sqrt{\epsilon_0 \mu_0}} \tag{34.2}$$

We know from Chapter 26 that in a material with given dielectric characteristics, the quantity ϵ_0 in Maxwell's equations must be replaced by $\kappa\epsilon_0$, where κ is the dielectric constant of the material [see Eq. (26.24)].[1] Hence, the formula (34.2) for the speed of light likewise must be replaced by a new formula for the speed,

$$v = \frac{1}{\sqrt{\kappa\epsilon_0 \mu_0}} \tag{34.3}$$

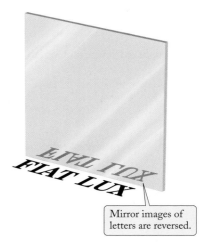

Mirror images of letters are reversed.

FIGURE 34.13 Some letters and their images in a mirror.

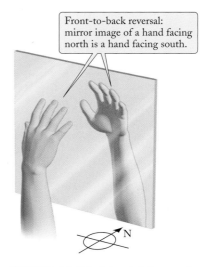

Front-to-back reversal: mirror image of a hand facing north is a hand facing south.

FIGURE 34.14 A hand facing north and its image in a mirror. The north side of the hand (the palm) becomes the south side in the mirror.

---

[1] In a material with magnetic properties, μ_0 must also be multiplied by a factor of the relative magnetic permeability (see Section 30.4); but this factor is very near 1 except in ferromagnetic materials. We will therefore ignore any correction factor of μ_0.

This is usually written as

$$v = \frac{c}{n} \qquad (34.4)$$

where $c = 1/\sqrt{\epsilon_0 \mu_0}$ is the standard speed of light in vacuum and

$$n = \sqrt{\kappa} \qquad (34.5)$$

The quantity n is called the **index of refraction** of the material. The index of refraction of any material is larger than 1; according to Eq. (34.4), *the speed of light in the material is less than the speed of light in vacuum.* In connection with Eq. (34.5), it is important to keep in mind that the value of the dielectric constant depends on the frequency of the electric field. Hence, the values of the dielectric constants from Table 26.1 cannot be inserted into Eq. (34.5), because the former values apply only to static fields, whereas we are now concerned with the high-frequency fields of a light wave.

Table 34.1 gives the values of the indices of refraction of a few materials. For instance, water has $n = 1.33$, and the speed of light in water is

$$v = \frac{c}{n} = \frac{c}{1.33} = \frac{3.00 \times 10^8 \text{ m/s}}{1.33} = 2.26 \times 10^8 \text{ m/s}$$

The values in the table apply to light waves of medium frequency (yellow-green light). The index of refraction is slightly larger for blue light and slightly smaller for red light; we will deal with this complication later in this section. Note that the index of refraction of air is very close to 1; consequently, in most calculations we can ignore the distinction between air and vacuum.

With the wave speed $v = c/n$, the relation between frequency and wavelength becomes [see Eq. (33.10)]

$$\lambda f = v = \frac{c}{n}$$

or

$$\lambda = \frac{1}{n}\frac{c}{f} \qquad (34.6)$$

Since c/f is the wavelength that the wave would have in vacuum [see Eq. (33.10)], we can also write Eq. (34.6) as

$$\lambda = \frac{\lambda_{\text{vac}}}{n} \qquad (34.7)$$

Thus the wavelength of an electromagnetic wave is shorter in a material than in empty space. For example, if a wave penetrates from vacuum or from air into water, where $n = 1.33$, its speed is reduced by a factor of 1.33, from 3.00×10^8 m/s to 2.26×10^8 m/s, but its frequency remains constant. Consequently, Eq. (34.6) shows that its wavelength will be reduced by a factor of 1.33 (see Fig. 34.15). The fact that the frequency of the wave remains constant can be understood in terms of the atomic mechanism underlying the interaction of the wave with the material. When the wave strikes the water surface, it shakes the electrons of the water molecules; this acceleration of electric charges produces extra waves, which combine with the original wave and result in a sloweddown wave. The frequency of the combined, slowed wave is the same as that of the orig-

TABLE 34.1	INDICES OF REFRACTION OF SOME MATERIALS[a]
MATERIAL	*n*
Air, 1 atm, 0°C	1.000 29
1 atm, 15°C	1.000 28
1 atm, 30°C	1.000 26
Water	1.33
Ethyl alcohol	1.36
Castor oil	1.48
Quartz, fused	1.46
Glass, crown	1.52
light flint	1.58
heavy flint	1.66

[a]For light of wavelength $\approx$ 550 nm.

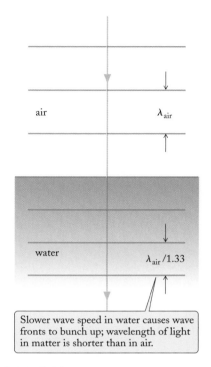

Slower wave speed in water causes wave fronts to bunch up; wavelength of light in matter is shorter than in air.

FIGURE 34.15 Change of wavelength of a light wave as it penetrates from air into water. The wavelength in water is shorter by a factor of $n = 1.33$.

inal wave, because the shaking of the electrons proceeds at the original frequency, and the extra waves produced by the electrons are therefore also of the same frequency as the original wave. But the slower speed results in a decreased wavelength, like the decreased spacing of cars when encountering a slowdown in traffic on a highway.

EXAMPLE 2   A light wave of wavelength 550 nm in vacuum enters a plate of glass of index of refraction $n = 1.52$. What is the speed of the light in the glass? What is the wavelength of the light in the glass? What is the frequency of the wave in the glass?

SOLUTION: The speed of light in the glass is

$$v = \frac{c}{n} = \frac{3.00 \times 10^8 \text{ m/s}}{1.52} = 1.97 \times 10^8 \text{ m/s}$$

The wavelength in the glass is

$$\lambda = \frac{\lambda_{\text{vac}}}{n} = \frac{550 \text{ nm}}{1.52} = 362 \text{ nm}$$

The frequency in the glass is the same as the frequency in vacuum,

$$f = \frac{c}{\lambda_{\text{vac}}} = \frac{3.00 \times 10^8 \text{ m/s}}{550 \times 10^{-9} \text{ m}} = 5.45 \times 10^{14} \text{ Hz}$$

Alternatively, we can calculate this frequency from the speed and the wavelength in the glass, with the same result:

$$f = \frac{v}{\lambda} = \frac{1.97 \times 10^8 \text{ m/s}}{362 \times 10^{-9} \text{ m}} = 5.45 \times 10^{14} \text{ Hz}$$

Incidentally: Equation (34.6) does not imply that a light source changes color when immersed in water. The *color* we perceive depends on the *frequency* of the light reaching our eyes, and this frequency is independent of whether the light source, our eyes, or both are immersed in air or water.

When a wave strikes the surface of a transparent material, part of it is reflected and part of it penetrates into the material. In the preceding section we investigated the direction of propagation of the reflected wave; now let us investigate the penetrating wave. Again, we will use Huygens' Construction to find out what the wave does when its strikes the surface of the material. In vacuum the speed of light is c; in the material it is c/n. Figure 34.16a shows the approaching wave fronts at one instant of time. The left edge of one wave front barely touches the surface. Figure 34.16a also shows the Huygens' wavelets that determine the position of this wave front at a later time; only the forward portions of these wavelets are relevant. Above the surface, the wavelet has a radius $c\,\Delta t$; below the surface, in the material, the wavelet has a smaller radius $(c/n)\Delta t$. The reduction of the speed of propagation of one side of the wave front causes the wave front to swing around, changing its direction of advance. This is analogous to the change of the direction of advance of a row of marching soldiers when the soldiers on one edge slow down by taking short steps while the soldiers on the other edge take longer steps.

refraction

This change of direction is called **refraction**. To determine the angle of refraction when light passes from one material into another, we need to examine the triangles in Fig. 34.16b. The right triangles $PP'Q$ and $PP'Q'$ have the long side PP' in common. In terms of the length PP' of this common side, the sines of the angles between the wave fronts and the surface are

$$\sin\theta = \frac{c\,\Delta t}{PP'} \qquad (34.8)$$

and

$$\sin\theta' = \frac{(c/n)\Delta t}{PP'} \qquad (34.9)$$

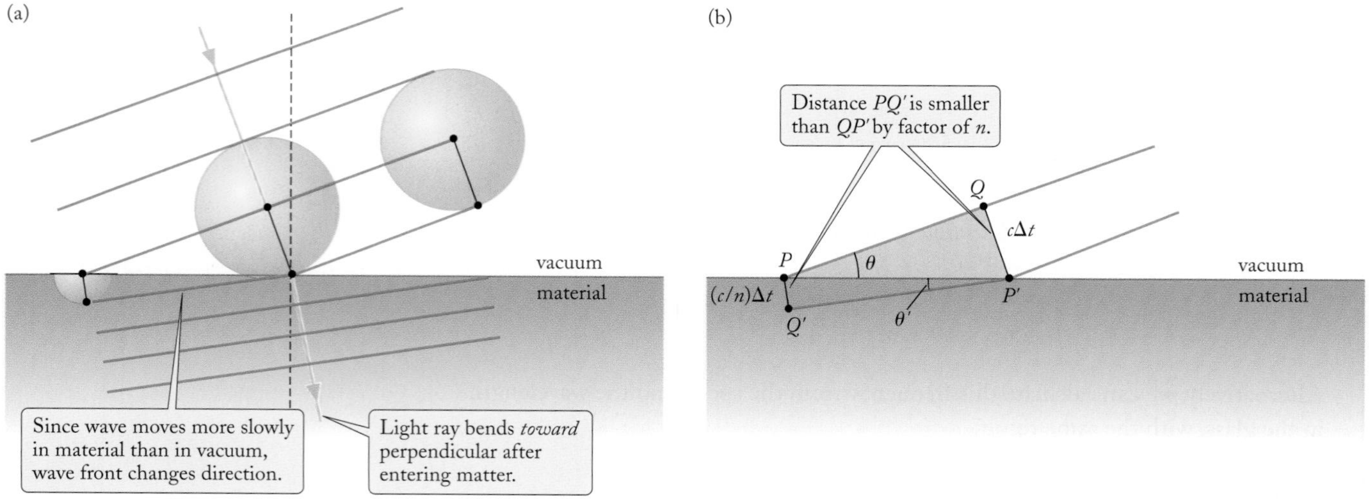

(a)

Since wave moves more slowly in material than in vacuum, wave front changes direction.

Light ray bends *toward* perpendicular after entering matter.

(b)

Distance PQ' is smaller than QP' by factor of n.

FIGURE 34.16 (a) Huygens' wavelets erected on a wave from whose edge barely touches the dielectric surface. (b) The incident wave front makes an angle θ with the dielectric surface; the refracted wave front makes an angle θ'.

The ratio of the sines is therefore

$$\frac{\sin\theta}{\sin\theta'} = \frac{c}{c/n}$$

from which

$$\sin\theta = n\sin\theta' \qquad (34.10)$$

This equation describes the change of direction of the wave upon penetration into a material. Equation (34.10) is called the **Law of Refraction**, or **Snell's Law**. The angle θ is the angle of incidence, and θ' is the angle of refraction. It is usually more convenient to measure these angles between the two rays and the perpendicular to the surface; all three lie in the same plane. Figure 34.17 shows the incident and refracted rays, and the angles of incidence and refraction. As can be seen in this figure, the ray in the material is bent toward the perpendicular (θ' is smaller than θ).

Note that we can also use the formula (34.10) for a ray of light that emerges from the material into vacuum. The formula is valid for the ray shown in Fig. 34.17 and also for the ray propagating in the reverse direction, provided we always assume that θ is the angle in vacuum and θ' is the angle in the material, regardless of the direction of propagation.

Our formula (34.10) describing refraction at the interface between a vacuum and a material is a special case of a general formula describing refraction at the interface of two different material media. If the indices of refraction are n_1 and n_2 and the angles between the rays and the perpendicular are θ_1 and θ_2, respectively, then

$$n_1\sin\theta_1 = n_2\sin\theta_2 \qquad (34.11)$$

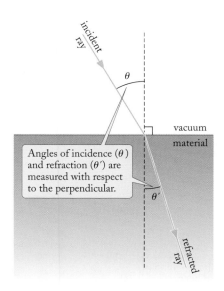

FIGURE 34.17 The angle of incidence θ and the angle of refraction θ'.

Law of Refraction (Snell's Law)

EXAMPLE 3   A ray of light enters a thick plate of glass of index of refraction $n = 1.52$ at an angle of incidence of 45° (see Fig. 34.18). (a) What is the angle of refraction of the ray at the upper surface of the glass? (b) When the ray reaches the lower surface of the glass, it is refracted again and it emerges into air. What is the angle at which it emerges?

SOLUTION: (a) With $n = 1.52$ and $\theta = 45°$, the Law of Refraction gives us

$$\sin 45° = 1.52\sin\theta' \qquad (34.12)$$

from which

$$\sin\theta' = \frac{\sin 45°}{1.52} = \frac{0.707}{1.52} = 0.465 \qquad (34.13)$$

With our calculator we then find that the angle of refraction is

$$\theta' = 28°$$

(b) Within the glass, the ray travels at 28° with respect to the perpendicular, and since the two surfaces of a plate of glass are parallel, this will be the angle of incidence at the lower surface. Hence refraction at the lower surface (where the ray proceeds from glass into air) is simply the reverse of the refraction at the upper surface (where the ray had proceeded from air into glass). The angle at which the ray emerges is therefore 45°. Note that the final ray, although shifted, is parallel to the initial ray.

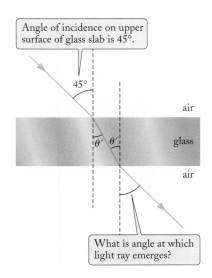

FIGURE 34.18 A ray of light passes through a thick plate of glass.

COMMENTS: This result can also be obtained from the Law of Refraction. We can apply Eq. (34.10) to a ray emerging from glass, provided we assume that θ' in this formula is the angle in the glass and θ is the angle in air; thus

$$\sin\theta = 1.52\,\sin 28° \qquad (34.14)$$

This again leads to the result $\theta = 45°$.

| **EXAMPLE 4** | A small, shiny fish is in the water 1.0 m below the surface. Where does an angler looking downward into the water see |

the fish; that is, where is the image of the fish? Assume that the angler's eye is almost directly above the fish.

SOLUTION: Figure 34.19 shows two light rays from the fish to the eye of the angler. The first light ray is perpendicular to the surface and is not bent. The second light ray is bent away from the perpendicular when it emerges from the water. With the assumption that θ' is the angle in water and θ the angle in air, Eq. (34.10) can be applied to this ray emerging from the water. Since the angles are small when the fish is viewed from almost directly above, we can use the familiar approximation that the sine of the angle is approximately equal to the angle expressed in radians that is, $\sin\theta \approx \theta$ and $\sin\theta' \approx \theta'$. Hence Eq. (34.10) becomes approximately

$$\theta \approx n\theta' = 1.33\,\theta'$$

where we have used the value $n = 1.33$ for water from Table 34.1. When the eye extrapolates the refracted ray back into the water, it seems to intersect the vertical ray at the point P', above the point P. Hence the image is above the fish. The image distance OP' and the fish distance OP are related as follows (see Fig. 34.19):

$$OP\tan\theta' = OP'\tan\theta$$

For small angles, we can make the approximations $\tan\theta \approx \theta$ and $\tan\theta' \approx \theta'$. Hence

$$\frac{OP'}{OP} \approx \frac{\theta'}{\theta} \approx \frac{1}{1.33}$$

This shows that the image depth is smaller than the fish depth by a factor of 1.33, the index of refraction. For instance, if $OP = 1.0$ m, then $OP' = 0.75$ m. The fish seems to be nearer to the surface than it is (if the angler wants to touch the fish, she must immerse her hand down deeper than where the fish *seems* to be). This smaller **apparent depth** occurs for any flat interface:

$$[\text{apparent depth}] = \frac{[\text{actual depth}]}{n} \qquad (34.15)$$

COMMENTS: The apparent bending that you perceive when you look at a straight rod partially immersed in water results from this apparent shrinking of the vertical distance. All the immersed portions of the rod seem to be nearer the water surface than they are, and the rod seems to have a kink at the place where it enters the water (see Fig. 34.20).

When the eye looking into the water is not nearly above the rod, that is, when the angles θ and θ' are not small, there is additional distortion and an exaggeration of the apparent bending.

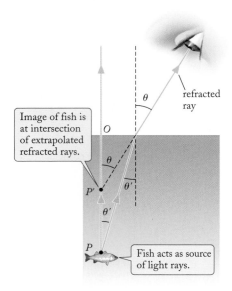

Image of fish is at intersection of extrapolated refracted rays.

Fish acts as source of light rays.

FIGURE 34.19 A small, shiny fish as a source of light. Note that the direction of propagation of the light is opposite to that shown in Fig. 34.17. This does not affect the validity of Eq. (34.10). The extrapolated ray (dashed) appears to come from the point P'. The angles shown are exaggerated for clarity; the location of the image of the fish is independent of angle only for small angles; that is, only when viewed from almost directly above.

FIGURE 34.20 A drinking straw partially immersed in water appears bent.

For a ray attempting to leave water, there is a critical angle beyond which refraction is impossible. As the light ray emerges into air, it is bent *away* from the vertical; in the extreme case, it is bent so much that it lies almost along the water surface (see Fig. 34.21); this extreme case corresponds to $\theta = 90°$ in Eq. (34.10):

$$n \sin\theta' = \sin 90° = 1$$

The critical angle for this extreme form of refraction is therefore given by

$$\sin\theta_{\text{crit}} = \frac{1}{n} \qquad (34.16)$$

For water, with $n = 1.33$, we then find

$$\theta_{\text{crit}} = 49°$$

If a light ray strikes a water surface from below with an angle of incidence larger than this, refraction is impossible. The only alternative is reflection—the water surface behaves as a perfect mirror. This phenomenon is called **total internal reflection**. It can occur whenever the index of refraction of the medium containing the light ray is larger than the index of refraction of the adjacent medium.

Total internal reflection has many important practical applications in optics. For instance, in periscopes the light is reflected down the tube by internal reflection in a prism (see Fig. 34.22); because the reflection is total, without any loss of light, this gives a much better image than reflection in a mirror. Such reflecting prisms are also used in binocular telescopes to reflect the path of the rays of light back and forth, and increase the effective length of the telescope.

critical angle for total internal reflection

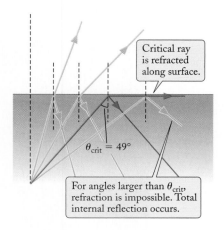

Critical ray is refracted along surface.

$\theta_{\text{crit}} = 49°$

For angles larger than θ_{crit}, refraction is impossible. Total internal reflection occurs.

FIGURE 34.21 A ray approaching a water surface from below with an angle of incidence $\theta = 49°$ is refracted along the water surface.

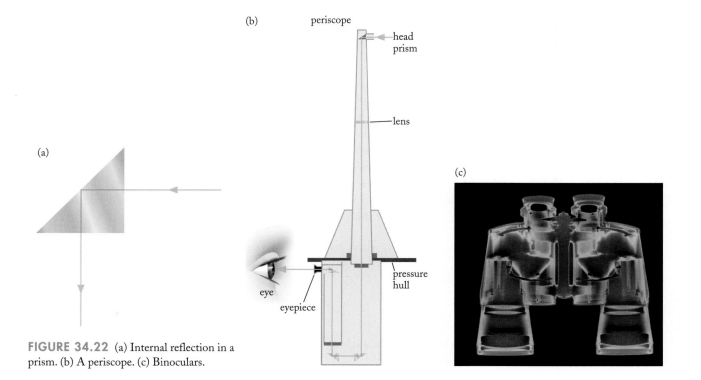

(b) periscope

head prism

lens

eye

eyepiece

pressure hull

(a)

(c)

FIGURE 34.22 (a) Internal reflection in a prism. (b) A periscope. (c) Binoculars.

PHYSICS IN PRACTICE   OPTICAL FIBERS

In an optical fiber, light moves along a thin rod made of transparent material; the light zigzags back and forth between the walls of the rod, undergoing a sequence of total internal reflections—the fiber acts as a pipe for light (see Fig. 1). Such optical fibers are being used to replace telephone and other communication cables. The electrical pulses normally carried on a wire cable are converted into pulses of infrared laser light, which can be transmitted in an optical fiber. The efficiency of optical telephone lines is very high because an optical fiber can carry many telephone conversations simultaneously; in modern telephone systems, single optical fibers are being used to carry several hundred telephone conversations simultaneously.

Optical fibers also find application in flexible endoscopes, used by physicians to examine the interior of the intestine and the stomach. This device is a bundle of optical fibers, sometimes over a meter in length, one end of which is inserted into the patients stomach or intestine while the physician looks into the other end. A small lightbulb at the leading end provides illumination (see Fig. 2).

FIGURE 1 Internal reflection in an optical fiber. Light enters the optical fiber at the bottom, travels around several loops of the fiber, and emerges at the top.

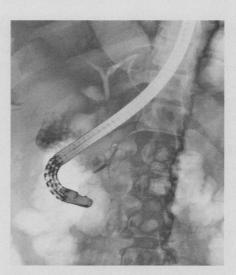

FIGURE 2 Endoscope inside patient.

As we saw in Chapter 33, ordinary light is unpolarized. An experimental observation, which can be derived by detailed analysis of Maxwell's equations, is that *reflection causes partial to complete polarization of light*. In the special case where the reflected and refracted rays are perpendicular to each other, the reflected ray is completely polarized, with the plane of polarization parallel to the reflecting surface, as shown in Fig. 34.23. In this case, the angle between the surface and the refracted ray is the same as the angles of incidence and reflection, since both are complements of the angle labeled ϕ. For incidence at this polarizing angle θ_p, Snell's Law implies

$$\sin\theta_p = n \sin\theta' = n \sin(90° - \theta_p) = n \cos\theta_p$$

which immediately gives

Brewster's Law

$$\tan\theta_p = n \tag{34.17}$$

This relation for the angle of incidence for complete polarizaton by reflection is known as **Brewster's Law**.

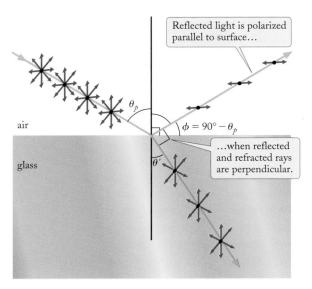

FIGURE 34.23 Polarization by reflection. The incident light can have any polarization, but when incidence is at Brewster's angle, the reflected light is polarized in a plane parallel to the surface.

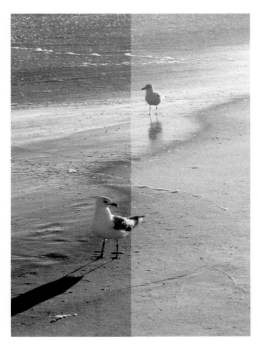

FIGURE 34.24 A view through polarizing sunglasses (left) reduces reflected glare compared with unpolarized sunglasses (right).

Recall from Section 33.3 that the glare of reflected light, with its partial or complete horizontal polarization, can be attenuated by using polarizing sunglasses, which only pass light polarized in the vertical plane (see Fig. 34.24).

In most materials, the index of refraction depends somewhat on the wavelength of the light. Usually, the index of refraction increases as the wavelength decreases. For instance, Fig. 34.25 displays a plot of the index of refraction of light in water over a range of wavelengths (the wavelengths plotted along the horizontal axis of Fig. 34.25 are measured in air, before the light penetrates the water). When a ray of light containing several wavelengths, or colors, is refracted by a material with an index of refraction that depends on wavelength, the refracted rays of different colors will emerge at somewhat different angles. The separation of a ray by refraction into distinct rays of different colors is called **dispersion**.

dispersion

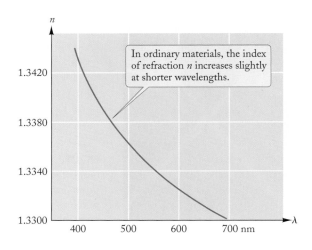

FIGURE 34.25 Index of refraction of light in water as a function of wavelength. The index of refraction varies by about 1% over the range of visible wavelengths.

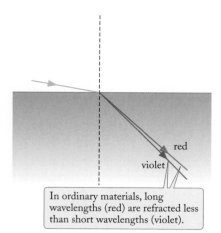

In ordinary materials, long wavelengths (red) are refracted less than short wavelengths (violet).

FIGURE 34.26 Refraction of red and of violet light in water. The ray of violet light is refracted more than the ray of red light. The difference between the angles of the refracted rays has been exaggerated for the sake of clarity.

EXAMPLE 5 The index of refraction for red light in water is 1.330, and for violet light it is 1.342. Suppose that a ray of light approaches a water surface with an angle of incidence of 80.0°. What are the angles of refraction for red light and for violet light?

SOLUTION: For $n = 1.330$, Eq. (34.10) yields

$$\sin\theta' = \frac{\sin\theta}{n} = \frac{\sin 80.0°}{1.330} = 0.740$$

for which our calculator gives us $\theta' = 47.7°$.

For $n = 1.342$, Eq. (34.10) yields

$$\sin\theta' = \frac{\sin 80°}{1.342} = 0.734$$

and $\theta' = 47.2°$. The violet light is bent more toward the vertical than the red light (see Fig. 34.26). For $\theta = 80°$, the difference in the angles of refraction 0.5°. Thus, refraction in water separates light rays according to colors. A beautiful demonstration of this effect is found in rainbows, which are produced by the refraction of sunlight in water droplets (sese Problem 47).

prism

A **prism** is the traditional device employed for the separation of light rays into their constituent colors. The basic mechanism is the same as discussed in Example 5: the glass in the prism has slightly different indices of refraction for light of different wavelengths, and hence it bends rays of different colors by different amounts (see Fig. 34.27). In passing through a prism, light is refracted twice, first when it enters the glass, and then when it leaves the glass. Under normal operating conditions, a good prism will introduce a difference of several degrees between the angular directions of the emerging red and and violet rays.

spectrum

The pattern of colors produced by the analysis of light by means of a prism is called the **spectrum** of the light. The white light emitted by the Sun has a continuous spectrum consisting of a blend of all the colors (see Fig. 34.28). The colored light emitted by the atoms of a chemical element in an electric discharge tube, such as a neon tube, has a discrete spectrum consisting of just a few sharp colors. Each of these discrete

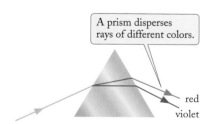

A prism disperses rays of different colors.

red
violet

FIGURE 34.27 Refraction of red and of violet light by a prism.

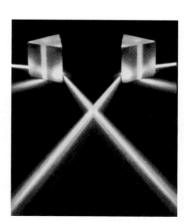

FIGURE 34.28 Analysis of a beam of white light by means of a prism reveals a continuous spectrum of colors, from red to orange, yellow, green, blue, and violet.

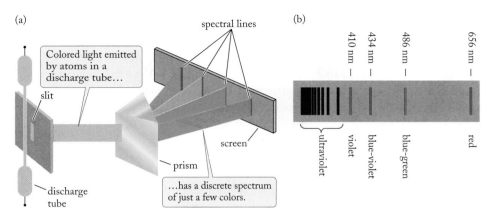

FIGURE 34.29 (a) Arrangement for the analysis of light emitted by the atoms of a chemical element. Each discrete color forms a spectral line on the screen or photographic plate. (b) The spectral lines of hydrogen. (A color print of these and other spectral lines is shown on page 1289.)

colors is essentially pure—it is light of a single wavelength. For example, hydrogen atoms emit the following discrete colors: red, blue green, blue violet, and violet. Such discrete colors are called **spectral lines**; in Chapter 38 we will see how such spectral lines arise from the quantum mechanics of the atom. Figure 34.29 shows the spectral lines of hydrogen as displayed by a prism illuminated with light from a fine slit. Each of the lines in the spectrum is a separate image of the slit made by a separate color after refraction by the prism.

spectral lines

✔️ Checkup 34.3

QUESTION 1: A ray of light enters a slab of glass with an angle of incidence of 0°. What is the angle of refraction? Does the answer depend on the index of refraction?

QUESTION 2: Qualitatively, why is the index of refraction of cold air larger than that of hot air?

QUESTION 3: If the wavelength of light in vacuum is 600 nm, what is the wavelength when this light enters water?

QUESTION 4: In each case, state whether a light ray passing from the first medium to the second will be bent closer to the perpendicular or farther away: (a) air to water; (b) water to glass; and (c) quartz to ethyl alcohol.

QUESTION 5: The critical angle for a light ray in water is 49°. What is the final direction of a light ray incident on a water surface from below with an angle barely more than 49°? Barely less than 49°?

QUESTION 6: A ray of light penetrates from a medium of index of refraction 1.3 into a medium with index of refraction 1.6. Can this ray suffer total internal reflection?

QUESTION 7: Light within a material with index of refraction $n = 2.0$ is incident upon an interface with air; the incident ray makes an angle θ with the perpendicular. Which of the following is the full range of angles of incidence for which total internal reflection occurs?

    (A) $0 < \theta < 30°$ (B) $0 < \theta < 60°$ (C) $30° < \theta < 60°$

    (D) $30° < \theta < 90°$ (E) $60° < \theta < 90°$

34.4 SPHERICAL MIRRORS

concave mirror

The Law of Reflection is also valid for a wave incident on a small portion of a curved reflecting surface, since such a portion can be approximated by a flat, tangent surface. Figure 34.30 shows a **concave** mirror with a surface curved like the inner surface of a sphere. When a plane wave is incident on this mirror, each portion of the wave is reflected according to the Law of Reflection, with equal angles of incidence and reflection. However, the angles of incidence for different portions of the wave are different; for instance, the angle of incidence is 0° for the portion of the wave that strikes the exact center of the mirror (this portion of the wave is reflected back on itself), but the angle of incidence is more than 0° for the portions of the wave that strike above or below the center. Since the angles of reflection for different portions of the wave are different, the plane wave incident on the spherical mirror becomes a wave with curved wave fronts.

focal point

As can be seen in Fig. 34.30, the reflected wave fronts converge toward a point, the **focal point** of the mirror. We can describe the direction of propagation of the wave fronts by rays; Fig. 34.31 shows the incident and reflected rays. For incident rays parallel to the axis, the reflected rays intersect at the focal point.

The focal point of the spherical mirror is halfway between the mirror and the center of curvature of the spherical surface. For a proof, we use Fig. 34.32, which shows the path of a single ray of light. The focal point F is the intersection of the ray with the axial line CA. To find the distance FA from the focal point to the mirror, called the **focal length**, we begin with the observation that in the isosceles triangle CFQ the length CF equals FQ. Under the assumption that angle θ is small (equivalently, that the incident ray is near the axial line), the length FQ is approximately equal to FA. Hence

$$CF = FA \qquad (34.18)$$

and thus the point F is halfway between the mirror (A) and the center of curvature (C). This means the focal length is one-half of the radius of curvature of the spherical surface. Designating the focal length by f and the radius of curvature by R, we can write

focal length

$$f = \tfrac{1}{2}R \qquad (34.19)$$

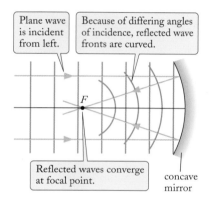

FIGURE 34.30 A concave spherical mirror focuses an incident plane wave to a point.

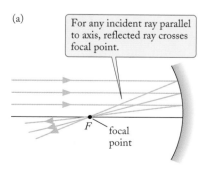

(a)

FIGURE 34.31 (a) Reflection of parallel rays by a concave spherical mirror. (b) Rays and concave mirror.

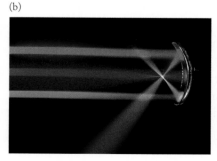

(b)

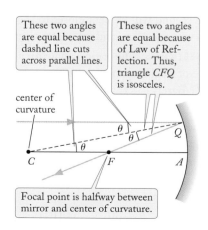

FIGURE 34.32 Reflection of a single ray. The geometry shown implies $f = R/2$.

EXAMPLE 6   A dentist uses a concave mirror to throw light into the patient's mouth (see Fig. 34.33). If a parallel beam of light supplied by a lamp is to be concentrated at a distance of 25 cm from the mirror, what should be the radius of curvature of the mirror?

SOLUTION: For strong concentration of the light, we want the focal point of the mirror to be at the distance of 25 cm from the mirror. According to Eq. (34.19), the radius of curvature of the mirror must be

$$R = 2f = 2 \times 25 \text{ cm} = 50 \text{ cm}$$

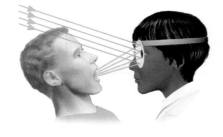

FIGURE 34.33 Mirror used to throw light into the patient's mouth.

EXAMPLE 7   Consider the Hubble Space Telescope mirror described at the beginning of the chapter, with a radius of curvature of 11 m. How far from the mirror does the image of a distant galaxy form?

SOLUTION: We know that parallel rays, such as those from a distant source, converge at the focal point. Thus from Eq. (34.19), the image will form at a distance from the mirror equal to half the radius of curvature,

$$f = \frac{R}{2} = \frac{11 \text{ m}}{2} = 5.5 \text{ m}$$

Concepts
— in —
Context

Figure 34.34 shows a **convex** spherical mirror. This kind of mirror is also curved like the surface of a sphere; but, in contrast to the concave mirror, the reflecting surface is the outer surface of the sphere. Parallel rays incident on this mirror diverge upon reflection. If we extrapolate the divergent rays to the far side of the mirror, they all seem to come from a single point, the focal point of the convex mirror. An argument similar to that given above demonstrates that the focal length is again one-half of the radius of the spherical surface:

convex mirror

$$f = -\tfrac{1}{2} R \tag{34.20}$$

A negative sign has been inserted in Eq. (34.20) to indicate that the focal point is on the far side of the spherical surface for a convex mirror. This negative sign will prove useful in algebraic calculations with mirrors (see below).

Both concave and convex mirrors will form images of light sources placed in front of them. Figure 34.35 shows a point source of light *P* in front of a concave mirror. To find the position of the image, we must trace some rays of light and find where they intersect. The rays that are easiest to trace are the **principal rays** shown in Fig. 34.35:

1. The first of these rays (*PC* in Fig. 34.35a) passes through the center of curvature of the mirror. It then strikes the mirror perpendicularly and is therefore reflected back on itself.
2. The second ray (*PQ* in Fig. 34.35b) begins parallel to the axial line and thus passes through the focal point after being reflected.
3. The third ray (*PF* in Fig. 34.35c) passes through the focal point. Upon reflection, this ray emerges parallel to the axial line (this behavior of the third ray becomes obvious if we recognize that the third ray is merely the reverse of a ray that arrives at the mirror from a direction parallel to the axial line).
4. The fourth ray (*PA* in Fig. 34.35d) is reflected symmetrically with respect to the axial line, so the angles θ and θ' are equal.

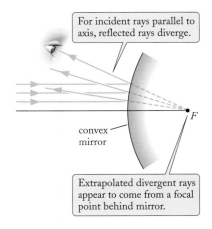

For incident rays parallel to axis, reflected rays diverge.

convex mirror

F

Extrapolated divergent rays appear to come from a focal point behind mirror.

FIGURE 34.34 Reflection of parallel rays by a convex mirror.

(a)

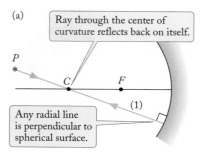

Ray through the center of curvature reflects back on itself.

Any radial line is perpendicular to spherical surface.

(b)

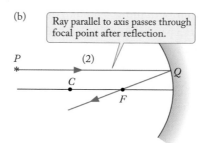

Ray parallel to axis passes through focal point after reflection.

(c)

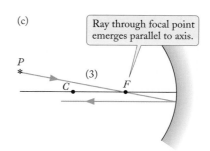

Ray through focal point emerges parallel to axis.

(d)

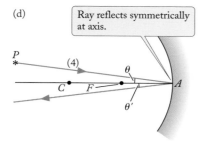

Ray reflects symmetrically at axis.

(e)

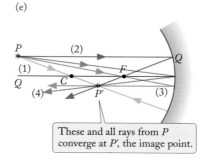

These and all rays from P converge at P', the image point.

FIGURE 34.35 Concave mirror. (a) Ray from P through center of curvature. (b) Ray from P parallel to axial line passes through focus. (c) Ray from P through focus emerges parallel to axial line. (d) Ray from P striking mirror at the axis has angles of incidence and reflection symmetric about the axis. (e) The rays (1), (2), (3), and (4) intersect at the image P'.

object

All these rays, and any other rays originating from the point source P that also strike the mirror, come together at P' (see Fig. 34.35e). This point P' is the image of the point source P. Note that to locate the image, two out of the four rays mentioned above are already sufficient—the other two are redundant but serve as useful checks.

The point source, or any source of light in a mirror or lens system, is known as the **object**. If the source of light is an extended object, then we must find the image of each of its points. For instance, a luminous object in the shape of an arrow has an image as shown in Fig. 34.36. We can easily verify this by drawing the rays that emerge from, say, the tip of the arrow, the midpoint of the arrow, the tail of the arrow, and so on. All the rays emerging from the tip of the arrow intersect at the tip of the image; all the rays emerging from the midpoint of the arrow intersect at the midpoint of the image; and so on. However, for finding the position of the image, it often suffices to draw the rays emerging from one or two points of the object. For instance, in Fig. 34.36, we need only the rays emerging from the tip, since we know that the tail of the arrow and the image of the tail both sit on the axial line.

The ray-tracing technique summarized in Fig. 34.36 is a graphical method for finding the image of an object. This method also applies to convex mirrors; an example of this is shown in Fig. 34.37. Notice from Fig. 34.37 that for a convex mirror, a

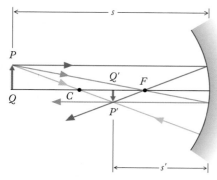

FIGURE 34.36 An object PQ in the shape of an arrow and its image $P'Q'$ formed by a concave mirror.

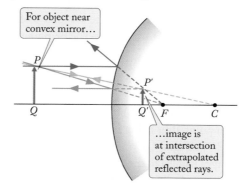

For object near convex mirror…

…image is at intersection of extrapolated reflected rays.

FIGURE 34.37 An object PQ and its image $P'Q'$ formed by a convex mirror.

ray parallel to the axis is reflected as if it came from the focal point, a ray heading toward the center of curvature is reflected back on itself; and a ray heading toward the focal point is reflected parallel to the axial line.

Although ray tracing is useful to locate the image roughly and to gain a qualitative understanding of how the positions of object, mirror, and image are related, its precision is limited. For higher precision, we must calculate the position of the image algebraically by means of the **mirror equation**,

$$\frac{1}{s} + \frac{1}{s'} = \frac{1}{f} \qquad (34.21)$$

mirror equation

where s is the distance from the object to the mirror and s' is the distance from the image to the mirror, measured along the axial line, or optic axis (see Fig. 34.36). The sign conventions for these distances are as summarized in the box Problem-Solving Techniques: Sign Conventions for Mirrors (see also Fig. 34.38).

For a proof of Eq. (34.21), we make use of Fig. 34.39, which shows an object PQ, its image $P'Q'$, and two rays. The ray PCP' passes through the center of curvature of the spherical surface and is reflected on itself; the ray PAP' strikes the center of the mirror and is reflected symmetrically with respect to the axial line so the angles θ and θ' are equal. The triangles PQA and $P'Q'A$ are similar; that is, they have the same angles. Hence their corresponding sides must be in proportion:

$$\frac{PQ}{P'Q'} = \frac{s}{s'} \qquad (34.22)$$

The triangles PQC and $P'Q'C$ are also similar; hence

$$\frac{PQ}{P'Q'} = \frac{QC}{Q'C} \qquad (34.23)$$

or, since $QC = s - R$ and $Q'C = R - s'$,

$$\frac{PQ}{P'Q'} = \frac{s - R}{R - s'} \qquad (34.24)$$

Combining Eqs. (34.22) and (34.24), we find

$$\frac{s}{s'} = \frac{s - R}{R - s'} \qquad (34.25)$$

We can rearrange this equation as

$$\frac{R - s'}{s'} = \frac{s - R}{s}$$

or

$$\frac{R}{s'} - 1 = 1 - \frac{R}{s}$$

If we divide both sides of this by R and shift terms from one side of the equation to the other, we obtain

$$\frac{1}{s} + \frac{1}{s'} = \frac{2}{R} \qquad (34.26)$$

Since $f = R/2$ [see Eq. (34.19)], this is the same as Eq. (34.21).

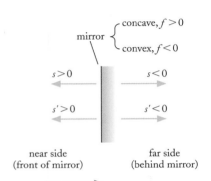

FIGURE 34.38 Summary of sign conventions for spherical mirrors.

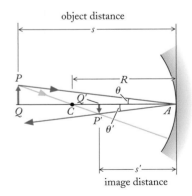

FIGURE 34.39 The geometry used to derive the mirror equation. The angles θ and θ' are equal; hence the right triangles PQA and $P'Q'A$ are similar. The right triangles PQC and $P'Q'C$ are also similar, since they have identical angles at C.

PROBLEM-SOLVING TECHNIQUES | SIGN CONVENTIONS FOR MIRRORS

The object distance s or image distance s' is positive if the object or image is in front of the mirror; the distance s or s' is negative if the object or image is behind the mirror (the object can be behind the mirror if what serves as "object" for the mirror is actually an image produced by another mirror or a lens). As mentioned in the text, f is positive for a concave mirror, negative for a convex mirror. These sign conventions for spherical mirrors are summarized in Fig. 34.38.

EXAMPLE 8 A candle is placed 41.0 cm in front of a convex spherical mirror of radius of curvature 60.0 cm (see Fig. 34.40). Where is the image?

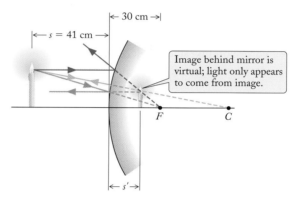

Image behind mirror is virtual; light only appears to come from image.

FIGURE 34.40 A candle and its image in a convex mirror. This is a *virtual* image; it lies behind the mirror, and it is upright.

SOLUTION: The focal length is half the radius of curvature, and is negative for a convex mirror [see Eq. (34.20)]. With $s = 41.0$ cm and $f = -R/2 = -30.0$ cm, Eq. (34.21) gives

$$\frac{1}{41.0 \text{ cm}} + \frac{1}{s'} = -\frac{1}{30.0 \text{ cm}}$$

or

$$\frac{1}{s'} = -\frac{1}{30.0 \text{ cm}} - \frac{1}{41.0 \text{ cm}} = \frac{-41.0 - 30.0}{41.0 \times 30.0} \frac{1}{\text{cm}}$$

from which

$$s' = -17.3 \text{ cm}$$

The negative value of s' indicates that the image is behind the mirror.

EXAMPLE 9 Suppose that the same candle is placed 41.0 cm in front of a concave spherical mirror of radius of curvature 60.0 cm (see Fig. 34.41). Where is the image?

(a)

|←— s = 41 cm —→|

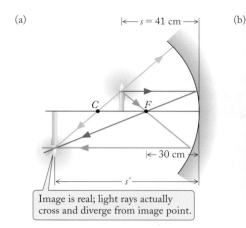

|← 30 cm →|

s'

Image is real; light rays actually
cross and diverge from image point.

(b)

FIGURE 34.41 (a) A candle and its image in a concave mirror. (b) This is a *real* image; it lies in front of the mirror, and it is inverted.

SOLUTION: With $s = 41.0$ cm and now $f = +30.0$ cm, Eq. (34.21) gives

$$\frac{1}{41.0 \text{ cm}} + \frac{1}{s'} = \frac{1}{30.0 \text{ cm}}$$

or

$$\frac{1}{s'} = \frac{1}{30.0 \text{ cm}} - \frac{1}{41.0 \text{ cm}} = \frac{41.0 - 30.0}{41.0 \times 30.0} \frac{1}{\text{cm}}$$

from which

$$s' = 112 \text{ cm}$$

The positive value of s' indicates that the image is on the same side of the mirror as the object.

The image in Example 9 is a **real image**. The real image is in front of the mirror; if we insert a sheet of paper at its location, the image will be projected on the sheet of paper, and we can see it sharply on this paper. Even without a sheet of paper, we can see the image if we look toward the mirror from the far left of Fig. 34.41; the image then gives the visual impression of a ghostly replica of the object floating in midair. *The light rays that we see when we look toward the mirror not only seem to come from the real image, but they actually do.* As Fig. 34.41 shows, the light rays pass through the real image and diverge from it, just as they diverge from the object. In contrast, the image in Example 8 is a virtual image, similar to the virtual images that we examined in our discussion of flat mirrors. The virtual image is behind the mirror. Light rays do not actually pass through a virtual image; they merely *seem* to come from such an image (see Fig. 34.40).

In both the case of a real image and the case of a virtual image, the geometry of the emerging light rays is as though the image were an object that acts as the source of the rays. This property of images allows us to analyze optical systems consisting of more than one element, for instance, two mirrors, or one mirror and one lens, or two lenses, etc. To find the final image produced by such a compound system, find the image produced by the first element, then treat this image as a new object for the second element, and find the new image that the second element produces of this new object.

real image

PROBLEM-SOLVING TECHNIQUES   IMAGES OF SPHERICAL MIRRORS

1 Begin the solution of a problem involving images in spherical mirrors by drawing a clear, careful diagram showing the mirror, the object, and relevant distances. Draw all the relevant distances to scale, or at least draw all the distances in roughly correct proportions. Represent the object by a heavy arrow, as in several of the diagrams of this section.

2 Trace two or three of the four principal rays that start from the tip of the heavy arrow, and find where they intersect, forming an image. Any two of the four principal rays are sufficient to determine the intersection; but it is sound practice to draw a third ray to serve as a check. Figures 34.35, 34.40, and 34.41 can be used as ray-tracing guides.

3 If the rays do not intersect in front of the mirror after reflection, extrapolate them with dotted lines, and find where the extrapolated rays intersect behind the mirror. If the rays intersect in front of the mirror, the image is real (and inverted); if the rays intersect behind the mirror, the image is virtual (and upright). An object placed in front of a convex mirror always has a virtual image; an object placed in front of a concave mirror can have either a virtual or a real image, depending on whether the object is within or beyond the focal length, respectively.

4 If you have drawn all distances to scale, the ray diagram provides you with a graphical solution of the image problem. But even if you have drawn the distances only in rough proportion, the ray diagram will tell you on which side of the mirror the image is located, and how its distance compares with other relevant distances.

5 Use the mirror equation (34.21) only after you know what to expect on the basis of the ray diagram. Remember to be careful about the sign conventions for s, s', and f; these sign conventions are summarized in Fig. 34.38 and the box on sign conventions on page 1132.

6 To decide whether the image is real or virtual, inspect the ray diagram. Ask yourself, is the image an actual point of intersection of rays (real), or is it only a point of intersection of rays extrapolated backward (virtual)? The sign of the image distance s' can be used to verify your conclusion.

Concepts in Context

EXAMPLE 10 Consider the Hubble Space Telescope mirror described at the beginning of the chapter. The technician is 4.0 m from the mirror, which has a radius of curvature of 11 m. Where is the image of the technician?

SOLUTION: The focal length is $f = R/2 = 5.5$ m, as we found in Example 7. Thus the mirror equation tells us

$$\frac{1}{s'} = \frac{1}{f} - \frac{1}{s} = \frac{1}{5.5\ \text{m}} - \frac{1}{4.0\ \text{m}} = -0.068\frac{1}{\text{m}}$$

from which we obtain

$$s' = -15\ \text{m}$$

Thus the image is virtual and 15 m behind the mirror. The appropriate ray diagram is shown in Fig. 34.42.

COMMENT: This arrangement, in comparison with Figs. 34.39 and 34.41, highlights a general property of concave mirrors: when the object is beyond the focal length (as in Figs. 34.39 and 34.41), the image is real and inverted, but when the object is closer to the mirror than the focal point (as in Fig. 34.42 and the chapter photo), the image is virtual and upright.

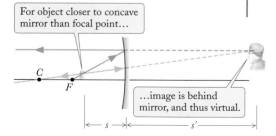

FIGURE 34.42 Ray diagram corresponding to the chapter photo.

Checkup 34.4

QUESTION 1: A fly is at a distance of 60 cm from a concave spherical mirror of radius 30 cm. Is the image behind the mirror or in front? Is it real or virtual? Is it upright or inverted?

QUESTION 2: A fly is at a distance of 10 cm from a convex spherical mirror of radius 30 cm. Is the image behind the mirror or in front? Is it real or virtual? Is it upright or inverted?

QUESTION 3: Can a convex mirror produce a real image?

QUESTION 4: Could the dentist in Example 6 use a convex mirror to concentrate light in the patient's mouth?

QUESTION 5: Consider a concave mirror, with an object closer to the mirror than the focal point. Which of the following is true? The image is:

(A) Virtual and upright, with $s' < 0$ (B) Virtual and inverted, with $s' < 0$

(C) Real and inverted, with $s' > 0$ (D) Real and upright, with $s' < 0$

(E) Virtual and upright, with $s' > 0$

34.5 THIN LENSES

Online
Concept
Tutorial

A lens made of a refracting material with two convex spherical surfaces will focus the parallel rays in a beam of light to a point (see Fig. 34.43). For a thin lens in vacuum or in air, the focal length f is given by the **lens maker's formula**:

$$\frac{1}{f} = (n - 1)\left(\frac{1}{R_1} + \frac{1}{R_2}\right) \tag{34.27}$$

lens maker's formula

where n is the index of refraction of the material of the lens and where R_1 and R_2 are the radii of curvature of the two spherical surfaces making up the lens. This equation is based on the assumption that the lens is thin (its thickness is small compared with R_1 and R_2) and that the incident rays are near the axial line.

Equation (34.27) can be verified by tracing rays through the lens, taking into account their refraction at the two curved surfaces. We will not perform this tedious calculation here (see Problems 81–82). We only note that the focusing depends on the

(a)

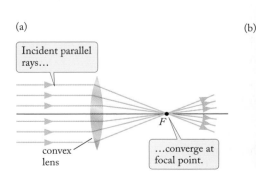

(b)

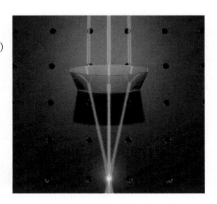

FIGURE 34.43 (a) Refraction of parallel rays by a convex lens. The focal length is positive. (b) Rays passing through a convex lens.

(a)

(b)

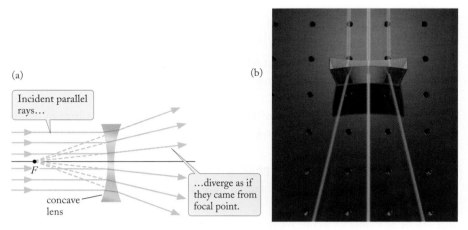

Incident parallel rays...

concave lens

...diverge as if they came from focal point.

F

FIGURE 34.44 (a) Refraction of parallel rays by a concave lens. The focal length is negative. (b) Rays passing through a concave lens.

curvature of the lens surface, because of which parallel rays at some distance from the axial line strike the lens surface with larger angles of incidence than rays near the axial line, and consequently they are bent more sharply toward the axial line. This is of course exactly what is required to make these distant rays cross the axial line at the same point (the focus) as the near rays.

Equation (34.27) may also be applied to a lens with two concave surfaces (see Fig. 34.44). In this case, the radii R_1 and R_2 must be reckoned as *negative*, and the focal distance f is then also negative. The meaning of a negative focal length is the same as in the case of mirrors: parallel rays incident on the lens diverge when they emerge from the other side of the lens, and the focal point is where the extrapolated rays appear to intersect (see Fig. 34.44). Furthermore, Eq. (34.27) can be applied to a concave–convex lens, such as shown in Fig. 34.45. Whether such a lens produces net convergence or divergence depends on whether the positive radius (convex) or the negative radius (concave) is smaller; for instance, the lens of Fig. 34.45a produces convergence, because the convex radius of curvature is smaller; and the lens of Fig. 34.45b produces divergence, because the concave radius of curvature is smaller.

Note that *a lens has two focal points at equal distances to the right and the left of the lens*. The point on the right of a converging lens is the focus of a parallel beam coming from the left and conversely, the point on the left is the focus of a parallel beam coming

(a)

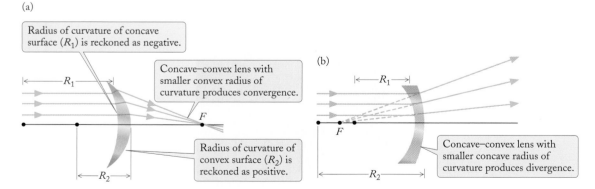

Radius of curvature of concave surface (R_1) is reckoned as negative.

R_1

Concave–convex lens with smaller convex radius of curvature produces convergence.

F

Radius of curvature of convex surface (R_2) is reckoned as positive.

R_2

(b)

R_1

F

Concave–convex lens with smaller concave radius of curvature produces divergence.

R_2

FIGURE 34.45 Refraction of rays by concave–convex lenses. (a) For this lens, the concave surface (left surface) has a larger radius of curvature than the convex surface (right surface). For this lens, the sum $1/R_1 + 1/R_2$ in Eq. (34.27) is positive. (b) For this lens, the concave surface (left surface) has a smaller radius of curvature than the convex surface (right surface). For this lens, the sum $1/R_1 + 1/R_2$ in Eq. 34.27 is negative.

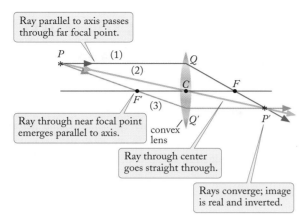

FIGURE 34.46 A point source of light *P* in front of a convex lens. The rays (1), (2), and (3) intersect at the image *P′*.

from the right. Moreover, for any thin lens, concave, convex, or concave–convex, the left and right focal distances are equal.

To find the image of an object placed near the lens, we can use a ray-tracing technique similar to that used for mirrors. Figure 34.46 shows the three *principal rays* that are easy to trace:

1. The first ray (*PQP′* in Fig. 34.46) starts parallel to the axial line. This ray is deflected by the lens so it passes through the far focal point *F*.
2. The second ray (*PCP′* in Fig. 34.46) passes through the center of the lens. This ray continues straight through the lens, without deflection (actually, the ray suffers a deflection when it enters the lens and another deflection when it leaves the lens; but for a thin lens, these deflections are opposite and their effects cancel).
3. The third ray (*PQ′P′* in Fig. 34.46) passes through the near focal point *F′*. That ray is deflected by the lens so it emerges parallel to the axial line (this becomes obvious if we recognize that the third ray is merely the reverse of a ray that arrives at the lens from the right, parallel to the axial line).

All these rays intersect at the point *P′*, the image point. As in the case of mirrors, two of the above three rays are already sufficient to locate the image. And, of course, the same ray-tracing technique can be applied to find the virtual image formed by a diverging lens. Figure 34.47 gives an example of such ray tracing for a concave lens. Here, the ray parallel to the axis diverges as if it came from the near focal point, and a ray heading for the far focal point emerges parallel to the axis. Note that in Figs. 34.46 and 34.47 (and other figures hereafter) the deflections of rays by a lens are indicated

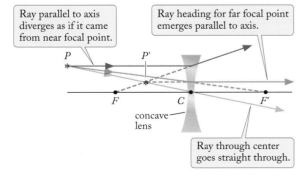

FIGURE 34.47 A point source of light *P* in front of a concave lens, and the image *P′*.

PROBLEM-SOLVING TECHNIQUES   SIGN CONVENTIONS FOR LENSES

The object distance s is positive if the object is on the near side of the lens and negative if it is on the far side; the image distance s' is positive if the image is on the far side of the lens and negative if it is on the near side. In this context, the "near" side is the side from which the light rays are incident on the lens (left side in the preceding figures), and the "far" side is the other side (right side in the preceding figures). The object distance can be negative if what serves as "object" for the lens is actually an image on the far side produced by a previous lens, sometimes known as a "virtual object." The sign of f is positive for a convex (converging) lens, and negative for a concave (diverging) lens. These sign conventions for lenses are summarized in Fig. 34.48.

schematically as occurring in the mid plane of the lens, and not at the lens surfaces. This is a convenient simplification for thin lenses.

The equation to be used for the algebraic calculation of the image distances is the same as Eq. (34.21),

thin-lens equation

$$\frac{1}{s} + \frac{1}{s'} = \frac{1}{f} \tag{34.28}$$

also known as the **thin-lens equation**, but the sign conventions are slightly different, as summarized in the box Problem-Solving Techniques: Sign Conventions for Lenses (see also Fig. 34.48).

Although the derivation of the lens equation (34.28) can be based on a geometric argument similar to that used for the mirror formula (34.21), we can bypass this labor by a trick. We begin by noting that a concave mirror is equivalent to one-half of a convex lens placed directly in front of a flat mirror. If the concave mirror and the (entire) convex lens have the same focal length, then the two arrangements shown in Figs. 34.49a and b also have this same focal length. Each ray in Fig. 34.49b has to pass through the half lens twice, once before reflection by the flat mirror and once after; the deflection angle imposed on the ray by the half lens is then the same as the deflection angle imposed by the entire lens, and it is therefore also the same as the deflection angle imposed by the concave mirror (see, for instance, the deflection angles marked in Figs. 34.49a, b, c for the blue ray that passes through the focal point). Thus, the arrangements shown in Figs. 34.49a and b have exactly the same optical properties—in both cases the image distances are the same. If we now remove the flat mirror in Fig. 34.49a and replace the half lens by the entire lens, the image distances will remain the same, but the image will form on the opposite side of the lens (see Fig. 34.49c), that is, the

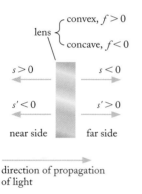

FIGURE 34.48 Summary of sign conventions for lenses.

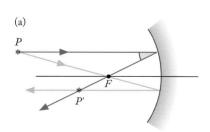   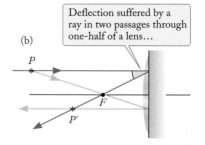   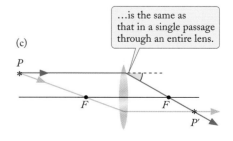

FIGURE 34.49 (a) A point source of light and its image formed by a concave mirror. (b) A similar image is formed by one-half of a convex lens placed in front of a flat mirror. (c) A similar image is also formed by the entire lens, but the image is now on the other side of the lens.

sign of the image distance will be reversed. Consequently, the same equation (34.21) must apply to the concave mirror and to the convex lens; the only difference is that the sign of the image distance is reversed. This reversal of sign has already been taken into account in our description of the sign conventions associated with Eqs. (34.21) and (34.28)—for the mirror, s' is taken as positive if the image is on the *near* side of the mirror, whereas for the lens, s' is taken as positive if it is on the *far* side of the lens. There is, of course, a similar correspondence between a convex mirror and a concave lens.

EXAMPLE 11   A convex lens of focal length 25 cm is placed at a distance of 10 cm from a printed page. What is the image distance? How much larger is the image of the page than the page?

SOLUTION: With $s = 10$ cm and $f = 25$ cm, Eq. (34.28) gives

$$\frac{1}{10 \text{ cm}} + \frac{1}{s'} = \frac{1}{25 \text{ cm}}$$

from which

$$\frac{1}{s'} = \frac{1}{25 \text{ cm}} - \frac{1}{10 \text{ cm}} = \frac{10 - 25}{25 \times 10} \frac{1}{\text{cm}}$$

and

$$s' = -17 \text{ cm}$$

The negative sign indicates that the image is on the near side of the lens (see Fig. 34.50). The image is virtual (for an observer looking at the lens from the right, light rays *appear* to diverge from the image, but they actually do not).

Since the triangles $P'Q'C$ and PQC are similar, the sizes of image and object are in the ratio

$$\frac{P'Q'}{PQ} = \frac{Q'C}{QC} = -\frac{s'}{s}$$

$$= \frac{17 \text{ cm}}{10 \text{ cm}} = 1.7$$

(34.29)

that is, the image is larger than the object by a factor of 1.7. This is the principle involved in the magnifying glass.

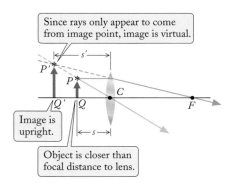

FIGURE 34.50 An object and its image for a convex lens.

EXAMPLE 12   A concave lens of focal length 25 cm is placed at a distance of 10 cm from the same printed page. What is the image distance? How much larger is the image of the page than the page?

SOLUTION: With $s = 10$ cm and, for a concave (diverging) lens, $f = -25$ cm, Eq. (34.28) gives

$$\frac{1}{10 \text{ cm}} + \frac{1}{s'} = -\frac{1}{25 \text{ cm}}$$

from which

$$\frac{1}{s'} = -\frac{1}{25 \text{ cm}} - \frac{1}{10 \text{ cm}} = \frac{-10 - 25}{25 \times 10} \frac{1}{\text{cm}}$$

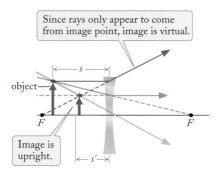

Since rays only appear to come from image point, image is virtual.

object

F   F

Image is upright.

$\leftarrow s' \rightarrow$

FIGURE 34.51 An object and its image for a concave lens.

and

$$s' = -7.1 \text{ cm}$$

The negative sign again indicates that the image is on the near side of the lens (compared with the object; see Fig. 34.51). The image is virtual.

As in the preceding example, the sizes of image and object are in the ratio

$$-\frac{s'}{s} = \frac{7.1 \text{ cm}}{10 \text{ cm}} = 0.71$$

that is, the image is smaller than the object by a factor of 0.71.

Equation (34.29) is a general result for the magnification produced by a single lens or by a mirror. The size of the image always equals the size of the object multiplied by a factor $-s'/s$. This factor is called the **linear magnification** M:

linear magnification M

$$M = -\frac{s'}{s} \tag{34.30}$$

Thus a magnification $M = 1.7$, as in Example 11, means that the image is 1.7 times the size of the object (and 1.7 times as far from the lens).

The sign of M tells us something about the character of the image. For a single lens, if M is positive, then, as in Examples 11 and 12, the image is upright and virtual (s' is negative). If M is negative, then the image is inverted and real (s' is positive). The sign of M tells us the same information for a mirror.

Concepts
— in —
Context

EXAMPLE 13   The image of the technician in the chapter photo is quite large. From the data of Example 10, find the magnification of this image.

SOLUTION: We know the object distance $s = 4.0$ m and the image distance $s' = -15$ m. Thus the magnification is

$$M = -\frac{s'}{s} = -\frac{-15 \text{ m}}{4.0 \text{ m}} = +3.8$$

Thus the image is 3.8 times as large as the object. Also, the positive sign of the magnification for a single mirror tells us that the image is upright and virtual (s' is negative).

For a multiple-lens system, we use the image of the first lens as the object for the second lens, and so on. In this case, the overall magnification is simply the product of the individual lens magnifications; for example, the magnification of a two-lens system is

$$M = M_1 \times M_2 = \left(-\frac{s'_1}{s_1}\right) \times \left(-\frac{s'_2}{s_2}\right) \tag{34.31}$$

For multiple lenses, the overall sign of the magnification still tells us whether the final image is upright with respect to the original object (M positive), or inverted (M negative). However, whether the final image is virtual or real cannot be determined from the sign of the magnification alone.

EXAMPLE 14   A bird at a (large) distance of 100 m (1.00×10^4 cm) is viewed through a two-lens system; the light first travels through a convex lens with $f_1 = 100$ cm, and then through a concave lens, placed 90.0 cm behind the first lens, as shown in Fig. 34.52a. The concave lens has a focal length $f_2 = -8.0$ cm. Locate the final image and determine the overall linear magnification of this system.

SOLUTION: We begin by finding the image produced by the first lens; we will then use this as the object for the second lens. For the first lens, with $f_1 = 100$ cm, the object distance is $s_1 = 100$ m $= 1.00 \times 10^4$ cm. We can solve the thin-lens equation (34.28) for the image distance s_1':

$$\frac{1}{s_1'} = \frac{1}{f_1} - \frac{1}{s_1} = \frac{1}{100 \text{ cm}} - \frac{1}{1.00 \times 10^4 \text{ cm}} = 9.90 \times 10^{-3} \frac{1}{\text{cm}}$$

which gives

$$s_1' = 101 \text{ cm}$$

Thus, by itself, the first lens would form a real, inverted image, as depicted in the ray diagram of Fig. 34.52b. Now we use this image as the object for the second lens; since the lenses are separated by 90.0 cm, our image (101 cm from the first lens) is 11 cm beyond the second lens. Since it is *beyond* the second lens, the object distance is negative, $s_2 = -11$ cm (see Fig. 34.48 and the box Problem-Solving Techniques: Sign Conventions for Lenses). With $f_2 = -8.0$ cm, we can again apply the thin-lens equation to obtain the image distance from the second lens:

$$\frac{1}{s_2'} = \frac{1}{f_2} - \frac{1}{s_2} = -\frac{1}{8.0 \text{ cm}} - \frac{1}{-11 \text{ cm}} = -0.034 \frac{1}{\text{cm}}$$

from which we obtain

$$s_2' = -29 \text{ cm}$$

The final image distance is negative, so it is 29 cm before (to the left of) the second lens, and the final image is virtual, as shown in Fig. 34.52c.

As shown in Fig. 34.52c, the final image is upright compared with the original object (inverted with respect to the first image). We can verify the orientation by examining the magnification. The magnification of the first lens is

$$M_1 = -\frac{s_1'}{s_1} = -\frac{101 \text{ cm}}{1.00 \times 10^4 \text{ cm}} = -0.0101$$

Thus the first image is inverted and about $1/100$ the size of the object. The magnification of the second lens is

$$M_2 = -\frac{s_2'}{s_2} = -\frac{-29 \text{ cm}}{-11 \text{ cm}} = -2.6$$

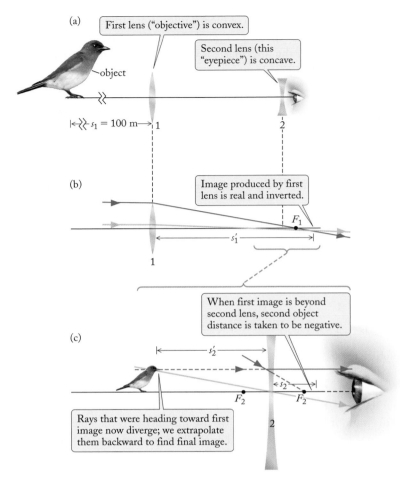

(a)

First lens ("objective") is convex.

Second lens (this "eyepiece") is concave.

object

$\mid\!\!\longleftarrow\!\!\Large\gg\normalsize s_1 = 100\ \text{m}\longrightarrow\mid$ 1

2

(b)

Image produced by first lens is real and inverted.

F_1

s_1'

1

When first image is beyond second lens, second object distance is taken to be negative.

(c)

s_2'

$\mid\!\!\ll s_2\!\!\longrightarrow\!\mid$

F_2 F_2

Rays that were heading toward first image now diverge; we extrapolate them backward to find final image.

2

FIGURE 34.52 (a) A two-lens system. (b) The image produced by the first lens is real and inverted. (c) Expanded view with the second lens in place; the final image is virtual and upright.

Thus the second lens inverts the first image and enlarges it 2.6 times. The overall magnification is the product of the two:

$$M = M_1 M_2 = (-0.0101) \times (-2.6) = +0.026$$

The overall magnification is positive, so the final image is indeed upright.

COMMENTS: Although the final image is much smaller than the object $(1/0.026 \approx 40$ times smaller), it is much, much closer to the eye than the object $[(100\ \text{m})/(0.29\ \text{m}) \approx 300$ times closer], and to the eye the image appears very large, so this lens system acts as a telescope. Thus, for distant objects, the ordinary magnification is not always the best way to characterize a lens system (for example, for a telescope to be useful, an image of the Moon need only appear larger than the Moon appears, but need not be larger than the Moon itself!). In the next section, we will introduce a different measure of magnification, the angular magnification. The lens arrangement of this example is similar to a **Galilean telescope**, often preferred for convenient use because the image remains upright.

Galilean telescope

PROBLEM-SOLVING TECHNIQUES   IMAGES OF LENSES

The procedure for finding the images produced by lenses is similar to the procedure we learned for mirrors:

1   Begin the solution of the problem by drawing a clear, careful diagram showing the lens, the object, and relevant distances. Draw all the relevant distances to scale, or at least draw all the distances in roughly correct proportions. Represent the object by a heavy arrow.

2   Trace the two or three principal rays that start from the tip of the heavy arrow, and find where they intersect, forming an image. Two of the three principal rays are sufficient to determine the intersection; but it is sound practice to draw all three, so the third ray can serve as a check (see Example 12).

3   If the rays do not intersect after passing through the lens, extrapolate them backward with dotted lines, and find where the extrapolated rays intersect on the near side of the lens. If the rays do intersect on the far side of the lens, the image is real (and inverted); if extrapolated rays intersect on the near side of the lens, the image is virtual (and upright). An object placed in front of a concave lens always has a virtual image; an object placed in front of a convex lens can have either a virtual or a real image, depending on whether the object is within or beyond the focal length, respectively.

4   If you have drawn all distances to scale, the ray diagram provides you with a graphical solution of the image

problem. But even if you have drawn the distances only in rough proportion, the ray diagram will tell you on which side of the lens the image is located, and how its distance compares with other relevant distances.

5   Use the lens equation (34.28) only after you know what to expect on the basis of the ray diagram. Remember to be careful about the sign conventions for s, s', and f; these sign conventions are summarized in Fig. 34.48.

6   To decide whether the image is real or virtual, inspect the ray diagram. Ask yourself, Is the image an actual point of intersection of rays, or is it only a point of intersection of rays extrapolated backward?

7   For a multiple-lens system, proceed in sequence, beginning from the source of light, using the image distance from the first lens and the separation between lenses to determine the object distance for the second lens. If the image from the first lens is on the far side of the second lens, then the object distance for the second lens is negative (as in Example 14 and in the sign conventions of Fig. 34.48).

8   The relative size of the image and object can be calculated from the magnification of Eq. (34.30), or, for a multiple-lens system, from the product of such magnifications, as in Eq. (34.31).

✔ Checkup 34.5

QUESTION 1: A thin plano–convex lens (one flat side, one convex side) has a focal length of 15 cm when the flat face is aimed toward the light source. If we turn this lens around, so its rounded face is toward the light source, what will be its focal length?

QUESTION 2: A fly is at a distance of 20 cm in front of a convex lens of focal length 30 cm. Is the image on the near side or the far side? Is it real or virtual? Is it upright or inverted?

QUESTION 3. A fly is at a distance of 10 cm in front of a concave lens of focal length 30 cm. Is the image on the near side or on the far side? Is it real or virtual? Is it upright or inverted?

QUESTION 4: Can a concave lens form a real image of a real object?

QUESTION 5: A thin lens, like the one in Fig. 34.45, has one convex surface with radius of curvature 2.0 m and one concave surface with radius of curvature 1.0 m. If the lens is made of glass of index of refraction 1.5, what is its focal length?

(A) 4.0 m   (B) −2.0 m   (C) 1.3 m   (D) 2.0 m   (E) −4.0 m

QUESTION 6: Two converging lenses have the same focal length f. An object is a distance $s_1 = 2f$ from the first lens, and the lenses are separated by a distance $5f$. With respect to the upright object, the final image of the second lens is

(A) Upright and real   (B) Upright and virtual
(C) Inverted and real   (D) Inverted and virtual

34.6 OPTICAL INSTRUMENTS

Optical instruments—cameras, magnifiers, microscopes, and telescopes—employ lenses and mirrors to form images in different ways. Most of the lenses in high-quality optical systems are compound lenses, that is, combinations of several simple lenses chosen in such a way that the optical deficiencies of one lens are canceled by those of another. Such deficiencies include **chromatic aberration**, where rays of different colors are refracted differently by the glass of the lens, and are therefore focused differently. Blurring of images is also caused by **spherical aberration**, where parallel rays that are far from the axis of a lens or mirror miss the focal point. Only parabolic surfaces truly focus parallel rays; spherical surfaces are merely conveniently fabricated approximations of parabolic surfaces. However, in the following discussion of optical instruments, we will ignore these complications, and we will schematically represent the compound lenses in the instrument by single thin lenses of suitable focal lengths.

chromatic aberration

spherical aberration

The Photographic Camera

The lens of the camera focuses a real image of the object on the photographic film or digital sensor and thereby imprints the image on the film or sensor (see Fig. 34.53). The distance between the lens and the film is adjusted in focusing, so a properly focused image always falls on the film, regardless of the object distance. A shutter controls the exposure time during which light is admitted to the camera. For a given exposure time, the amount of light entering the camera is proportional to the area of the lens. The size of the camera lens is commonly labeled by the **f number**, which is defined as the ratio of the focal length of the lens to its diameter. For instance, a lens of focal length 55 mm and a diameter of 32 mm has an f number of 55/32, or 1.7. Reducing the diameter of the lens by one-half would double the f number to 3.4. It would also reduce the area of the lens by a factor of 4, and for a given exposure time, this would reduce the amount of light entering the camera by a factor of 4. If we want to compensate for the reduced lens diameter and keep constant the amount of light reaching the film, we would have to increase the exposure time by a factor of 4. A lens of small f number is said to be "fast," because it collects sufficient light for a photograph in a short exposure time.

Good cameras have an adjustable iris diaphragm that can be used to block part of the area of the lens and thereby alter its effective f number. If the diaphragm is closed down so only a small central portion of the lens remains unblocked, the camera will require a long exposure time, but it will have a large **depth of field**—it simultaneously forms sharp images for objects spanning a large range of object distances. This is so because the rays emitted by any object then enter the camera at small angles, and they therefore intersect the image at small angles; thus the rays will stay close together on the photographic film, even if the position of the image does not fall exactly on the

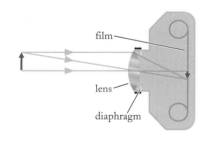

film

lens

diaphragm

FIGURE 34.53 A photographic camera.

depth of field

film (see Fig. 34.54). Some camera lens systems have a "zoom" lens, permitting the focal length to be increased (other cameras permit substitution of a lens with longer focal length); this feature provides a larger image for a given object distance.

The Eye

In principle, the eye is similar to a camera. The lens of the eye forms a real image on the retina, a delicate membrane packed with light-receptor cells, which send nerve impulses to the brain. Figure 34.55 shows a horizontal section through the human eye. The diameter of the eyeball is typically 2.3 cm.

The cornea and the aqueous humor act as a lens; they provide most of the refraction for rays entering the eye. The crystalline lens merely provides the fine adjustments of focal length required to make the image of an object fall on the retina, regardless of the object distance. The crystalline lens is flexible and its focal length is adjusted by the ciliary muscles; this adjustment is called accommodation. If the eye is viewing a distant object, the muscles are relaxed and the lens is fairly flat, with a long focal length. If the eye is viewing a nearby object, the muscles are contracted and the lens is more rounded, with a short focal length.

The shortest attainable focal length determines the shortest distance at which an object can be placed from the eye and still be seen sharply. This shortest distance is called the **near point** of the eye. For a normal young adult, the near point is typically 25 cm (although nearsighted persons can view much closer objects). With advancing age the lens loses its flexibility and the near point recedes; for instance, at an age of 60 years, the near point is typically around 200 cm.

The iris serves the same function as the diaphragm of the camera—it controls the size of the pupil and the amount of light admitted to the eye. The pupil contracts to a diameter of 2 mm in bright light, and expands to 7 mm in faint light, a change of area by a factor of about 12.

The receptor cells lining the retina are of two kinds: about 120 million rods and 6 million cones. The rods are more sensitive to light than the cones, but they do not discriminate among colors. The cones are blind at low light levels, but they give us color perception. See page xlvii of the Prelude for images of the retina and of red cells.

The cones also give us high-acuity vision, because some 2000 of them are densely packed into a small spot, the fovea, located near the center of the retina. However, the density of rod cells is maximum at retinal locations about 20° away from the fovea. For this reason, very faint light is best perceived peripherally, by looking to one side of the light source, instead of staring directly at it.

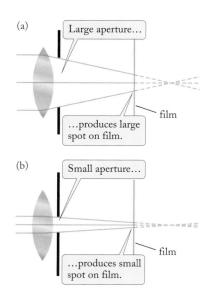

FIGURE 34.54 (a) When the aperture is large, the rays are spread over a large angle. If the focal point is beyond the film, the rays make a large spot on the film. (b) When the aperture is small, the rays are confined to a narrow range of angles. Even if the focal point is beyond the film, the rays make only a small spot on the film.

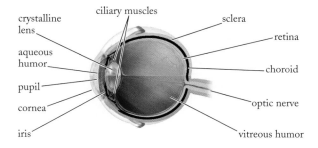

FIGURE 34.55 The human eye (natural size). The space between the cornea and the crystalline lens is filled with a transparent jelly, the aqueous humor. The main body of the eye is also filled with a transparent jelly, the vitreous humor. The index of refraction of the humors is 1.34, nearly the same as for water. The index of refraction of the crystalline lens is 1.44. The sclera is the thick white outer casing of the eye. The choroid is a pigmented black membrane, like the black paint inside cameras.

FIGURE 34.56 Hold the book at a distance of half a meter. Close your left eye and steadily stare at the cross. If you gradually bring the book closer to your eye, the dot will disappear when its image falls on your blind spot; the dot will reappear when the image moves beyond your blind spot.

The bundle of nerve fibers connecting the receptor cells to the brain leaves the eye at a place somewhat to one side of the fovea. This is called the blind spot, because it lacks receptor cells. You are normally not aware of this blind spot, because the brain fills in the visual picture from information acquired during eye movements; but you can demonstrate the existence of the blind spot by means of Fig. 34.56.

The three most common optical defects of the eye are nearsightedness, farsightedness, and astigmatism. *In a nearsighted (myopic) eye, the focal length is excessively short,* even when the ciliary muscles are completely relaxed. Thus, parallel rays from a distant object come to a focus in front of the retina (see Fig. 34.57a) and fail to form a sharp image on the retina—vision of distant objects is blurred. This condition can be corrected by eyeglasses with divergent lenses (see Fig. 34.57b).

In a farsighted (hyperopic) eye, the focal length is excessively long, even when the ciliary muscles are fully contracted (in other words, the near point of the eye is farther away than normal). Hence rays from nearby objects converge toward an image beyond the retina (see Fig. 34.58a) and fail to form a sharp image on the retina—vision of nearby objects is blurred. The condition can be corrected by eyeglasses with convergent lenses (see Fig. 34.58b). In old age, both nearsightedness and farsightedness often occur simultaneously, through loss of flexibility of the crystalline lens and the weakening of the ciliary muscles, with a consequent failure of the eye to accommodate over the full range of distances (presbyopia). The correction then requires bifocal lenses, with a lower convergent portion for near vision and an upper divergent portion for far vision.

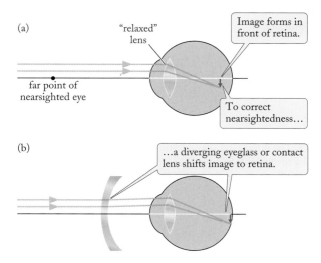

FIGURE 34.57 (a) Nearsighted eye forms an image in front of the retina. (b) Nearsighted eye with corrective lens.

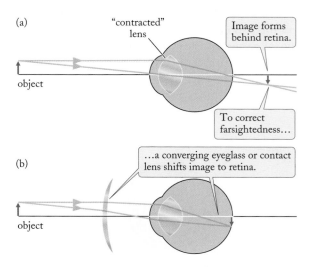

FIGURE 34.58 (a) Farsighted eye forms an image beyond the retina. (b) Farsighted eye with corrective lens.

| **EXAMPLE 15** | A farsighted person has a near point of 200 cm. What kind of glasses does this person need to be able to read a newspaper held at 25 cm from the eyes? |

SOLUTION: The object distance is 25 cm (we ignore the small difference between distances measured from the eye and from the glasses). Since the nearest point at which the person can see distinctly is 200 cm away, the lens must form an image of the newspaper at this distance—that is, $s' = -200$ cm. The lens equation then tells us

$$\frac{1}{f} = \frac{1}{s} + \frac{1}{s'} = \frac{1}{25 \text{ cm}} + \frac{1}{-200 \text{ cm}} = \frac{8-1}{200} \frac{1}{\text{cm}}$$

and

$$f = 29 \text{ cm}$$

Incidentally: Eyeglass prescriptions are specified in **diopters**, which is the inverse of the focal length in meters; thus for the farsighted person of this example, reading glasses of strength $1/0.29 = +3.4$ diopters would permit reading at 25 cm. Inexpensive reading glasses of various strengths are available at drug or convenience stores.

Astigmatism is an inability to focus simultaneously light rays arriving in different planes, for instance, light rays arriving in the vertical plane and light rays arriving in the horizontal plane. This is caused by a slight horizontal or vertical flattening of the cornea—instead of being curved spherically, the cornea is slightly out of round, with more curvature in one direction than in the other. Although astigmatism is quite common, it is often so slight as to be unnoticeable. Figure 34.59 provides a simple test for astigmatism. Correction of this condition requires a lens with a cylindrical surface, which focuses rays of light in the, say, vertical plane but does not deflect rays of light in the horizontal plane.

The Magnifier

In order to see fine detail with the naked eye, we must bring the object very close to the eye, so the angular size of the object is large and, correspondingly, the image on the retina is large (see Fig. 34.60). This means we want to bring the object to the near point, at a typical distance of 25 cm for the eye of a young adult. To see finer detail, we need a magnifier. This consists of a strong convergent lens placed adjacent to the eye, as shown in Fig. 34.61.[2] Such a lens permits us to bring the object closer to the eye than 25 cm, and thereby increase the size of the image on the retina.

The angular magnification of the magnifier is defined as the ratio of the angular size of the image produced by the magnifier at the largest possible distance (as in Fig. 34.62) to the angular size of the object seen by the naked eye at the standard distance of 25 cm (as in Fig. 34.60). If the magnifier is to form an image at a large distance, we must make the extrapolated rays from the image nearly parallel, which requires that we place the object very near the focal distance of the magnifier. The extrapolated rays (shown dashed in Fig. 34.62) are then nearly parallel, and they intersect far

diopters

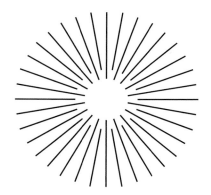

FIGURE 34.59 Close one eye and focus the other on the central ends of the radial lines. If some of the lines appear less sharp or less black than others, your eye is astigmatic. (To detect mild astigmatism, you may find it helpful to perform this test on each eye and notice the difference, if any.)

---

[2] Note that such a magnifier is not the same thing as a magnifying glass. In common use, the magnifying glass is placed at an appreciable distance from the eye, near the object to be magnified, because this maximizes the magnification. A magnifying glass can be regarded as a magnifier (in the technical sense of the word) only if it is placed near the eye.

from the lens. In the extreme case, we place the object at the focal distance f; the extrapolated rays are then parallel. The intersection of these rays is then at infinite distance, and the image is said to be at infinity.

To evaluate the magnification, we need to compare the angles θ and θ' in Figs. 34.60 and 34.62. Since the angle θ in Fig. 34.60 is small, we can use the familiar formula that this angle (in radians) approximately equals the tangent of the angle, or the size h divided by the distance of 25 cm:

$$\theta = \frac{h}{25 \text{ cm}}$$

Likewise, the angle θ' in Fig. 34.62 is small and approximately equals the size h divided by the focal distance f:

$$\theta' = \frac{h}{f}$$

Taking the ratio of these two angles, we find the angular magnification of the magnifier,

angular magnification of magnifier

$$[\text{angular magnification}] = \frac{\theta'}{\theta} = \frac{25 \text{ cm}}{f} \tag{34.32}$$

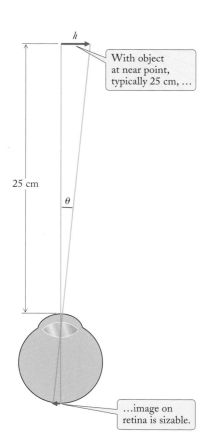

FIGURE 34.60 The angular size of the object determines the size of the image on the retina. Here the object has been placed at a distance of 25 cm from the eye.

With object at near point, typically 25 cm, …

…image on retina is sizable.

FIGURE 34.61 A magnifier used by a jeweler.

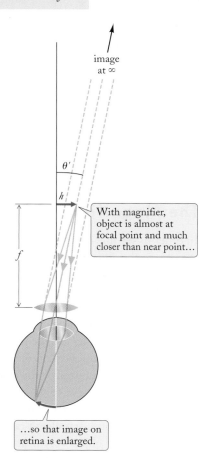

image at ∞

With magnifier, object is almost at focal point and much closer than near point…

…so that image on retina is enlarged.

FIGURE 34.62 The magnifier is adjacent to the eye. The object has been placed at a distance slightly shorter than the focal distance, so that the eye sees the image nearly at infinity.

This tells us the magnification relative to that of the typical naked eye; that is, it tells us how much better the magnifier is than the naked eye. For example, a magnifier with $f = 5$ cm has an angular magnification of $(25 \text{ cm})/(5 \text{ cm}) = 5$. Note that this result is valid only under the assumption that the magnifier is placed adjacent to the eye, and that the object is placed near the focus of the magnifier, so the image is at infinity. The second of these assumptions is not crucial—if the object is placed somewhat closer than the focus, the magnification will be changed only slightly. But the first assumption is crucial—if the magnifier is placed at some appreciable distance from the eye, then the magnification will be quite different.

The Microscope

The microscope consists of two lenses: the objective and the ocular, or eyepiece. Both of these lenses have very short focal lengths. The objective is placed near the object, and it forms a real, magnified image of the object. This image serves as object for the ocular, which acts as a magnifier and forms a virtual image at infinity (see Fig. 34.63). Thus both the objective and the ocular contribute to the magnification of the microscope. The net angular magnification of the microscope is the angular magnification of the ocular multiplied by the magnification of the objective. The angular magnification of the ocular is given by Eq. (34.32) with $f = f_{\text{ocular}}$; and the magnification of the objective is given by Eq. (34.30), where s and s' are, respectively, the object and image distances for the objective (these distances are marked in Fig. 34.63). If we combine these two magnifications, and if we ignore the sign in Eq. (34.30), we find that the net angular magnification of the microscope is

$$[\text{angular magnification}] = \frac{25 \text{ cm}}{f_{\text{ocular}}} \times \frac{s'}{s} \qquad (34.33)$$

angular magnification of microscope

Note that, as in the case of the magnifier, this tells us the magnification relative to the (typical) naked eye. Good microscopes operate at magnifications of up to 1400. Although higher magnifications can be achieved, this serves little purpose because diffraction of the light (see the next chapter) at the objective limits the details that can be resolved. To overcome this limitation, we need to use waves of shorter wavelength than light waves, such as the electron waves used in electron microscopes, which will be discussed in Chapter 38.

The Telescope

A simple astronomical telescope consists of an objective of very long focal length and an ocular of short focal length. These two lenses are separated by a distance (nearly) equal to the sum of their individual focal lengths, so their focal points coincide (see Fig. 34.64). The objective forms a real image FP' of a distant object. This image serves as object for the ocular, which forms a magnified virtual image at infinity (Fig. 34.64).

To find the angular magnification produced by this telescope, we begin by noting that the lens equation with object distance $s = \infty$ applied to the objective gives

$$\frac{1}{\infty} + \frac{1}{s'} = \frac{1}{f_{\text{objective}}}$$

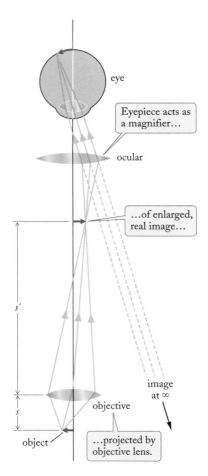

eye

Eyepiece acts as a magnifier...

ocular

...of enlarged, real image...

s'

image at ∞

s

objective

object

...projected by objective lens.

FIGURE 34.63 Arrangement of lenses in a microscope. The object to be magnified is below the objective, and the eye is just above the ocular. The ocular forms a virtual image at infinity, and the lens of the eye focuses on the retina the parallel rays emerging from the ocular.

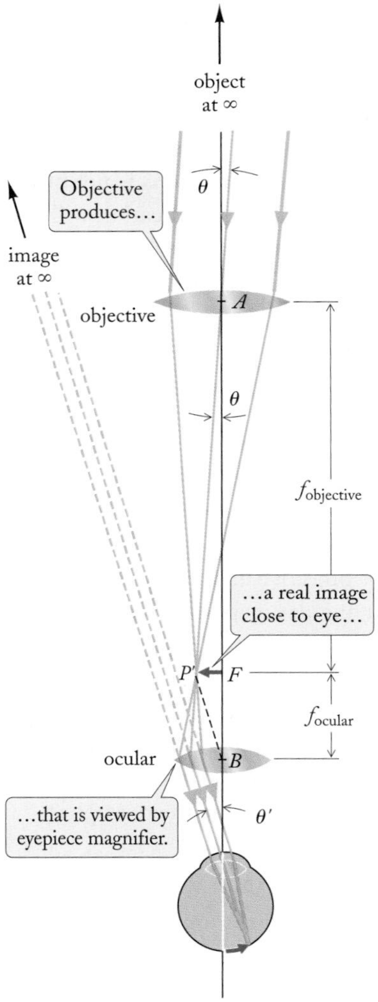

FIGURE 34.64 An astronomical telescope. The object (a star) is at a large distance above the telescope. The observer's eye is below the ocular. The ocular forms a virtual image at infinity, and the lens of the eye focuses on the retina the parallel rays emerging from the ocular.

But $1/\infty = 0$, and hence

$$s' = f_{\text{objective}} \tag{34.34}$$

This means that the first image is at the focal distance of the objective. Accordingly, in Fig. 34.64, the image FP' is shown located at the focal point F. We can now use the geometric relationships contained in this figure. The angular magnification is the ratio of the angles θ and θ' that represent, respectively, the angular sizes of the object and the final image viewed by the eye. Since both these angles are small, their values (in radians) are approximately the transverse distance FP' divided by the distance to the appropriate lens:

$$\frac{\theta'}{\theta} = \frac{FP'/FB}{FP'/FA} = \frac{FA}{FB} = \frac{f_{\text{objective}}}{f_{\text{ocular}}}$$

and so the angular magnification is

angular magnification of telescope

$$[\text{angular magnification}] = \frac{\theta'}{\theta} = \frac{f_{\text{objective}}}{f_{\text{ocular}}} \tag{34.35}$$

For example, an astronomical telescope with $f_{\text{objective}} = 120$ cm and $f_{\text{ocular}} = 2.5$ cm has an angular magnification of $(120 \text{ cm})/(2.5 \text{ cm}) = 48$.

Many astronomical telescopes are reflecting telescopes in which a concave mirror plays the role of the objective. The mirror forms a real image that serves as object for the ocular. Of course, the ocular (or a small mirror diverting light toward the ocular) must be placed in front of the large concave mirror, blocking out some of the light. Figure 34.65 shows an arrangement of concave mirror and ocular. The formula for the angular magnification of such a mirror telescope is, again, Eq. (34.35).

Concepts
— *in* —
Context

| **EXAMPLE 16** | Suppose that an ocular with focal length $f = 1.5$ cm is used with the Hubble telescope mirror of the chapter photo, discussed in Example 7. What is the system's angular magnification? |

SOLUTION: The angular magnification is given by Eq. (34.35); thus, using $f_{objective} = 5.5$ m from Example 7 and $f_{ocular} = 1.5$ cm,

$$[\text{angular magnification}] = \frac{f_{objective}}{f_{ocular}} = \frac{5.5 \text{ m}}{1.5 \times 10^{-2} \text{ m}} = 370$$

COMMENT: In addition to a large focal length and a large magnification, a good astronomical telescope must have other attributes. In order to view faint stars, a telescope must collect as much light as possible, and thus must have a large diameter. A large diameter is also needed to better resolve adjacent distant objects (to be discussed in Section 35.6, Diffraction by a Circular Aperture).

Because large mirrors of good quality are easier to manufacture than large lenses of good quality, the largest astronomical telescopes all use mirrors. For instance, the telescope on Mount Palomar (see Fig. 34.66a) uses a mirror of diameter 5.1 m and a focal length of 16.8 m. The Hobby–Eberly telescope in Texas and the Keck telescope on the summit of Mauna Kea in Hawaii are the largest reflectors; the former uses an interlocking array of 91 precisely aligned hexagonal mirrors (see Fig. 34.66b), acting as a single mirror of diameter 10.5 m and focal length 13.1 m.

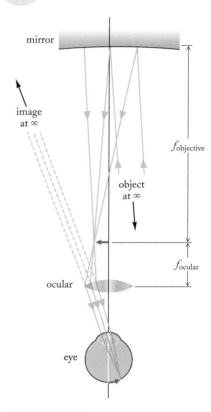

FIGURE 34.65 A reflecting telescope. In this diagram, the object has been placed below, to facilitate comparison with Fig. 34.64.

(a)

(b)

FIGURE 34.66 (a) The 200-in. telescope on Mount Palomar in California. The mirror can be seen under the observer cage. (b) The Hobby–Eberly telescope. The large reflector is made up of many hexagonal mirrors.

Checkup 34.6

QUESTION 1: You want to use a camera to make a photograph in dim light. Do you need a camera of small f number or large f number?

QUESTION 2: A farsighted person has difficulty reading in normal light, but can still read well in very bright light. Explain.

QUESTION 3: Consider the microscope illustrated in Fig. 34.63.

(A) Does the objective produce a linear magnification of the object? An angular magnification?

(B) Does the ocular produce a linear magnification? An angular magnification?

(C) Is the image produced by the objective real or virtual? Upright or inverted?

(D) Is the image produced by the ocular real or virtual? Upright or inverted?

QUESTION 4: Consider the telescope illustrated in Fig. 34.64.

(A) Does the objective produce a linear magnification of the object? An angular magnification?

(B) Does the ocular produce a linear magnification? An angular magnification?

(C) Is the image produced by the objective real or virtual? Upright or inverted?

(D) Is the image produced by the ocular real or virtual? Upright or inverted?

QUESTION 5: The arrangement of the lenses in the microscope and the telescope is quite similar. What is the main difference?

QUESTION 6: For which instrument(s) does the angular magnification depend on the (somewhat arbitrary) typical distance of the near point of vision, 25 cm?

(A) Microscope only   (B) Magnifier only   (C) Telescope only

(D) Microscope and magnifier only   (E) Microscope, magnifier, and telescope

SUMMARY

WAVELET   Sphere with radius [wave speed] $\times \Delta t$ centered on old wave front.

HUYGENS' CONSTRUCTION   The new wave front is the surface that touches the wavelets erected on the old wave front.

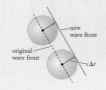

LAW OF REFLECTION

$$\theta_{\text{incident}} = \theta_{\text{reflected}}$$

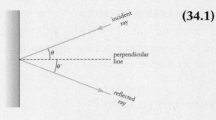

(34.1)

INDEX OF REFRACTION
(v is the speed of light in the material.)

$$n = \frac{c}{v}$$

(34.4)

WAVELENGTH IN MATERIAL

$$\lambda = \frac{\lambda_{\text{vac}}}{n}$$

(34.7)

INDEX OF REFRACTION
(in terms of dielectric constant)

$$n = \sqrt{\kappa}$$

(34.5)

LAW OF REFRACTION (SNELL'S LAW)
where angles are measured with respect to a
perpendicular to the surface.

$$n_1 \sin\theta_1 = n_2 \sin\theta_2$$

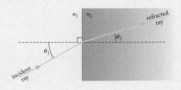

(34.11)

APPARENT DEPTH (for a flat surface, when
viewing from almost directly above)

$$[\text{apparent depth}] = \frac{[\text{actual depth}]}{n}$$

(34.15)

**CRITICAL ANGLE FOR TOTAL INTERNAL
REFLECTION**

$$\sin\theta_{\text{crit}} = \frac{1}{n}$$

(34.16)

**COMPLETE POLARIZATION BY REFLECTION
(BREWSTER'S LAW)**

$$\tan\theta_p = n$$

(34.17)

DISPERSION Separation of colors of refracted
light due to a wavelength-dependent index of
refraction.

FOCAL LENGTH OF SPHERICAL MIRROR (f is
positive for concave mirror, negative for convex;
R is radius of curvature of mirror surface.)

$$f = \pm\tfrac{1}{2}R$$

(34.19, 34.20)

MIRROR EQUATION (s is the object distance; s'
is the image distance. s or s' is positive if object or
image is in front of mirror, negative if behind.)

$$\frac{1}{s} + \frac{1}{s'} = \frac{1}{f}$$

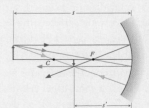

(34.21)

LENS MAKER'S FORMULA   (f is positive for a converging lens, negative for a diverging lens; R_1 or R_2 is positive for a convex surface, negative for a concave surface.)

$$\frac{1}{f} = (n - 1)\left(\frac{1}{R_1} + \frac{1}{R_2}\right)$$ (34.27)

THIN-LENS EQUATION   [s is positive if object is on near side of lens, negative if on far side (relative to the source of light); s' is positive if image is on far side of lens, negative if on near side.]

$$\frac{1}{s} + \frac{1}{s'} = \frac{1}{f}$$ (34.28)

LINEAR MAGNIFICATION
(for a lens or a mirror)

$$M = -\frac{s'}{s}$$ (34.30)

LINEAR MAGNIFICATION OF A SYSTEM OF LENSES OR MIRRORS   (M_1 is the magnification of the first lens or mirror, etc.)

$$M = M_1 \times M_2 \times \cdots$$ (34.37)

---

f NUMBER   Ratio of focal length of lens to diameter of aperture.

DEPTH OF FIELD   Range of object distances over which a sharp image forms.

NEAR POINT   The shortest distance for sharp vision, typically 25 cm.

ANGULAR MAGNIFICATION OF MAGNIFIER

$$\frac{25 \text{ cm}}{f}$$ (34.32)

ANGULAR MAGNIFICATION OF MICROSCOPE

$$\frac{25 \text{ cm}}{f_{\text{ocular}}} \times \frac{s'}{s}$$ (34.33)

ANGULAR MAGNIFICATION OF TELESCOPE

$$\frac{f_{\text{objective}}}{f_{\text{ocular}}}$$ (34.35)

QUESTIONS FOR DISCUSSION

1. When light is incident on a smooth surface—a glass surface, a painted surface, a water surface—the reflection is strongest if the angle of incidence is near 90° (grazing incidence). Can Huygens' Construction explain this?

2. In celestial navigation, the navigator measures the angle between the Sun, or some other celestial body, and the horizon with a sextant. If the navigator is on dry land, where the horizon is not visible, he can measure instead the angle between the Sun and its reflection in a pan full of water, and divide this angle by 2. Explain.

3. Suppose we release a short flash of light in the space between two parallel mirrors placed face to face. Why does this light flash not travel back and forth between the two mirrors forever?

4. To measure the index of refraction of some small transparent chips of plastic of irregular shape, a physicist places the chips in a glass beaker full of water. She then gradually adds sucrose to the water until the chips suddenly become nearly invisible. Explain why a measurement of the index of refraction of the sucrose solution then yields the index of refraction of the chips of plastic.

5. At sunset, the image of the Sun remains visible for some time after the actual position of the Sun has sunk below the horizon. Explain.

6. After a navigator measures the angle between the Sun and the horizon with a sextant, he must make a correction for the refraction of sunlight by the atmosphere of the Earth. Does this refraction increase or decrease the apparent angle between the Sun and the horizon?

7. Artists are notorious for making mistakes when drawing or painting mirror images. What is wrong with the position and orientation of the image shown in the mirror in Fig. 34.67?

FIGURE 34.67 A detail from a medieval tapestry at the Musée de l'Hôtel de Cluny, Paris. The head of the unicorn is reflected in the mirror. What is wrong with this image?

8. Two parallel mirrors are face to face. Describe what you see if you stand between these mirrors.

9. Figure 34.68 shows spots of sunlight on a wall in the shade of a tree. The spots were made by sunlight that has passed through very small gaps between the leaves of the tree. Explain why all the spots are round and of nearly the same size, even though the gaps are of irregular shape and size. (Hint: The Sun is round.)

FIGURE 34.68 Spots of sunlight on a wall.

10. At amusement parks you find mirrors that make you look very short and fat or very tall and thin. What kinds of mirrors achieve these effects?

11. Two mirrors are arranged at right angles to one another. If an object is near the mirrors, what is the largest number of images of that object that could be visible?

12. Store owners often install convex mirrors at strategic locations in their stores to supervise the customers (Fig. 34.69). What is the advantage of a convex mirror over a flat mirror?

FIGURE 34.69 Convex mirror used to supervise customers in a store.

13. You look toward a lens or a mirror, and you see the image of an object. How can you tell whether this image is real or virtual?

14. Hand mirrors are sometimes concave, but never convex. Why?

15. If you place a book in front of a concave mirror, will it show you mirror writing? Does the answer depend on the distance of the book from the mirror?

16. How could you make a lens that focuses sound waves?

17. If you place a small lightbulb at the focus of a convex lens and look toward the lens from the other side, what will you see?

18. Consider Fig. 34.46. How do we know that the ray $PQ'P'$ emerges parallel to the axis of the lens?

19. Suppose you place a magnifying glass against a flat mirror and look into the glass. What do you see if your face is very near the glass? If it is not very near?

20. Lenses, like mirrors, suffer from aberration. Consider rays incident on a convex lens, parallel to the axis. If the point of incidence is far from the axis, would you expect the ray to pass in front of the focus or behind?

21. Figure 34.70a shows a large **Fresnel lens** used in the lantern of a lighthouse. The lens consists of annular segments, each with a curved surface similar to the curved surface of an ordinary lens (see Fig. 34.70b). Why is this arrangement better than a single curved surface?

22. A convex lens is made of plastic of index of refraction 1.2. If you immerse this lens in water, will it produce convergence or divergence of incident parallel rays?

23. Binoculars use prisms to reflect the light back and forth (Fig. 34.71). What is the purpose of these prisms, and what is their advantage over mirrors?

(a)

(b)

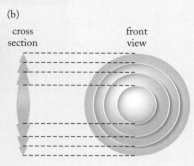

cross section   front view

FIGURE 34.70 (a) Fresnel lens of the lighthouse at Point Reyes, California. (b) Two views of a Fresnel lens.

FIGURE 34.71 Arrangement of prisms in a binocular.

PROBLEMS

†34.2 Reflection

1. According to a (questionable) story, Archimedes set fire to the Roman ships besieging Syracuse by focusing the light of the Sun on them with mirrors. Suppose that Archimedes used flat mirrors. How many flat mirrors must simultaneously reflect sunlight at a piece of canvas if it is to catch fire? The energy flux of the sunlight at the surface of the Earth is 0.10 W/cm^2, and the energy flux required for ignition of canvas is 4.0 W/cm^2. Assume that the mirrors reflect the sunlight without loss.

2. Two flat mirrors are joined along their edges so they make an angle of 90° with each other (see Fig. 34.72). A cat sits in the angle formed by the mirrors. Draw a diagram showing the location and orientation of all the images of the cat in this system of mirrors.

FIGURE 34.72 Cat between two mirrors at an angle of 90°.

3. A ray of light strikes a mirror at an angle of θ with respect to the perpendicular. At what angle with respect to the mirror surface does it reflect from the mirror?

4. If you wish to take a picture of yourself in a plane mirror, for what distance should the camera focus be set if you and the camera are 1.2 m from the mirror?

5. A ray of light is incident on a flat mirror at an angle θ. If we then rotate the mirror by 10° (see Fig. 34.73) while keeping the incident ray fixed, by what angle will we rotate the reflected ray?

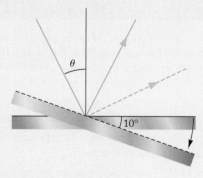

FIGURE 34.73 A flat mirror, before and after rotation.

†For help, see Online Concept Tutorial 37 at www.wwnorton.com/physics

*6. Two mirrors are arranged so they make an angle of 60° with each other. If a hummingbird hovers near the vertex where the mirrors meet, what is the number of images of itself that the hummingbird can see?

*7. Two mirrors are arranged so they make an angle of 45° with each other. If you stand near the vertex where the two mirrors meet, what is the number of images of yourself that you can see?

*8. A vertical mirror, oriented toward the Sun, throws a rectangular patch of sunlight on the floor in front of the mirror. The size of the mirror is 0.50 m × 0.50 m, and its bottom rests on the floor. If the Sun is 50° above the horizon, what is the size of the patch of sunlight on the floor?

*9. What are the minimum length and minimum width of a mirror hanging on a wall such that you can see your entire body when standing in front of it? Assume that your height is 1.80 m, your width is 0.50 m, and your eyes are separated by 0.06 m.

†34.3 Refraction

10. What is the speed of light in crown glass? Light flint glass? Heavy flint glass?

11. A helium–neon laser, producing the red light commonly used in demonstrations, has a wavelength of 633 nm in air. What is the frequency of the light in air? What is the wavelength of such light in crown glass? What is the frequency in the glass? What is the speed of the light in the glass?

12. The speed of light in diamond is 1.24×10^8 m/s. Calculate the index of refraction of diamond.

13. By what factor does the speed of light change when it goes from castor oil to ethyl alcohol?

14. A ray of sunlight approaches the surface of a smooth pond at an angle of incidence of 40°. What is the angle of refraction?

15. The speed of sound in air is 340 m/s, and in water it is 1500 m/s. If a sound wave in air approaches a water surface with an angle of incidence of 10°, what will be the angle of refraction?

16. Make a plot of the angle of incidence vs. the angle of refraction for light rays incident on a water surface. What is the maximum angle of refraction?

17. A ship's navigator observes the position of the Sun with his sextant and measures that the Sun is 39.000° away from the vertical. Taking into account the refraction of the Sun's light by air, what is the true angular position of the Sun? For the purpose of this problem, assume that the Earth is flat and that the atmosphere can be regarded as a flat, transparent plate of uniform density and index of refraction of 1.0003.

18. A pebble is at the bottom of a 2.0-m-deep swimming pool. What is the apparent depth of the pebble when viewed from outside the water, directly above the pebble?

³ In all problems, assume the index of refraction of air is exactly 1, unless otherwise stated.

† For help, see Online Concept Tutorial 37 at www.wwnorton.com/physics

19. White light is incident on one of the materials in Table 34.1 at angle of 30.0°. If the blue wavelengths are refracted at an angle of 19.9° and the red wavelengths at an angle of 20.2°, what is the material?

20. A coin is embedded in a plastic holder with index of refraction 1.58. If the coin is at the center of a 10-cm-thick block of plastic, what is its apparent depth when viewed straight on?

21. In a glass ($n = 1.62$) fish tank filled with water, a ray of light strikes the inner surface of the glass tank at an angle of incidence of 30.0°. It then travels through the glass and out into the air. With what angle with respect to the perpendicular does it exit the glass?

22. Light of wavelength 550 nm in air strikes a surface of a clear material. The light is incident at 45° with respect to the normal; the ray refracts in the material at 28° with respect to the normal. What is the wavelength of light in the material?

23. A scuba diver is underwater; from there, the Sun appears to be at 30° from the vertical. At what actual angle is the Sun located (with respect to directly overhead)?

24. Laser scientists can now produce pulses of light in air almost as short as 1×10^{-15} s = 1 femtosecond. Such pulses are made up of different colors of light. Suppose such a pulse is normally incident on a slab of crown glass that is 1.00 mm thick. Dispersion in the glass can cause appreciable spreading of a short pulse. The index of refraction of blue light is 1.530, and the index of refraction of red light is 1.510. How much sooner does the red part of the pulse get to the other side of the glass compared with the blue part of the pulse?

25. If diamond has an index of refraction of 2.42, what is the critical angle for total internal reflection? (This relatively small angle facilitates total internal reflection, making the diamond more briliant.)

26. One of the angles of a prism is 90°. The prism is made of glass of index of refraction 1.58. If a ray of light enters this prism at an angle of incidence of 70°, as shown in Fig. 34.74, what is the angle at which this ray emerges?

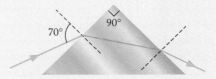

FIGURE 34.74 A prism with a 90° angle.

27. A prism of flint glass has an acute angle of 30.0°. A ray of white light is perpendicularly incident on one face of this prism (see Fig. 34.75). What are the angles at which red, green, and violet light emerge? The indices of refraction for red (660 nm), green (550 nm), and violet (400 nm) light are 1.650, 1.660, and 1.690, respectively.

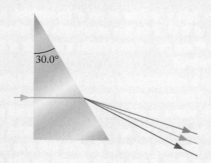

FIGURE 34.75 A 30° prism used to analyze a beam of white light.

28. An optical fiber is made of a thin strand of glass of index of refraction 1.50. If a light ray is to remain trapped within this fiber, what is the largest angle it can make with the surface of the fiber?

29. (a) A transparent medium of index of refraction n_1 adjoins a transparent medium of index of refraction n_2. Assuming $n_1 > n_2$, show that the critical angle for total internal reflection of a ray attempting to leave the first medium is

$$\sin\theta_{\text{crit}} = \frac{n_2}{n_1}$$

   (b) A layer of kerosene ($n_2 = 1.20$) floats on a water surface. In this case, what is the critical angle for total internal reflection for a light ray attempting to leave the water?

30. To determine the index of refraction of a liquid, an experimenter sends a ray of light through the liquid from below and measures the critical angle for total internal reflection at the upper surface. She finds that this angle is 56°. What is the index of refraction?

31. What is the critical angle for a ray of light to be internally reflected inside a plate of crown glass? What if this plate is immersed in water? (See Problem 29.)

32. Use the formula derived from Snell's Law in Problem 29 to consider total internal reflection at an interface with a material other than air. If a light pipe is made by coating a cylinder of crown glass with light flint glass, what is the critical angle for total internal reflection completely inside the crown glass? (Such coatings avoid problems with scratches on the outer surface.) Are there angles for which light in the crown glass will refract out into the flint glass but will still be totally internally reflected in the flint glass?

33. A *mirage* is caused by a thin layer of hot air just above the ground, and its appearance is due to total internal reflection. When looking from far away, an observer sees a reflecting surface on the ground (which has the appearance of a water surface), coming from rays with an angle of reflection of 89.4° or more. If the cooler air has an index of refraction $n = 1.000\,26$,

what is the index of refraction of the thin layer of hot air? (Hint: Use the result of Problem 29.)

34. When unpolarized light is incident on sapphire at an angle of 60.5°, the reflected light is completely polarized. What is the index of refraction of sapphire?

35. At what angle above the horizon is the Sun when its light, reflected from a calm swimming pool, is linearly polarized?

36. The critical angle for total internal reflection in cubic zirconia (in air) is 27.0°. What is the polarizing angle for reflection from cubic zirconia?

*37. A ray of light strikes a plate of window glass of index of refraction 1.5 and thickness 2.0 mm with an angle of incidence of 50°. Find the transverse displacement between the transmitted ray and the extrapolation of the incident ray.

*38. A point source of light is placed above a thick plate of glass of index of refraction n (Fig. 34.76). The distance from the source to the upper surface of the plate is l, and the thickness of the plate is d. A ray of light from the source may suffer either a single reflection at the upper surface, or a single reflection at the lower surface, or multiple alternating reflections at the lower and upper surfaces. Thus, each ray splits into several rays, giving rise to multiple images. In terms of l and d, find the distance of the first, second, and third images below the upper surface of the plate. Assume that the angle of incidence of the ray is small.

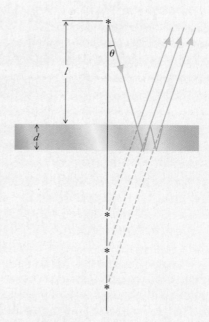

FIGURE 34.76 Ray of light reflected and refracted by a thick plate of glass.

*39. Light polarized in the plane of incidence (see Fig. 34.77) is incident upon water from air. There is no reflected ray. What is the angle of incidence?

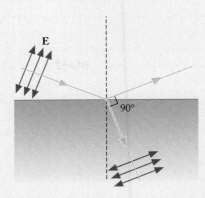

FIGURE 34.77 The light wave is polarized in the plane of the page. The refracted ray is perpendicular to the direction expected for a reflected ray.

*40. Because of refraction in the plate of glass, an object viewed through an ordinary window will seem somewhat nearer than its actual distance.

(a) Consider the rays that strike the glass at a small angle of incidence (nearly normal). Show that if an object is at a distance *l* from a window, the image is at a distance that is shorter by an amount $\Delta l = (1 - 1/n)d$, where *n* is the index of refraction and *d* is the thickness of the glass.

(b) What change in distance does this formula give if the windowpane is ordinary glass with $d = 2.0$ mm and $n = 1.5$?

(c) What change in distance does this formula give if the windowpane is heavy plate glass with $d = 8.0$ mm and $n = 1.5$?

*41. The highly reflective paint used on highway signs contains small glass beads, which reverse the direction of propagation of a light ray, throwing it back toward the source of light (see Fig. 34.78a). The reversal of direction will occur only for those

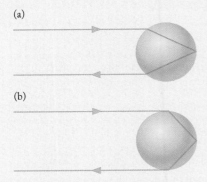

(a)

(b)

FIGURE 34.78 (a) Path of a light ray that reverses direction in a transparent sphere. (b) Path of a light ray that penetrates at maximum distance from the axis of a sphere.

light rays incident on the sphere of glass at a selected distance from the axis, a distance that depends on the index of refraction of the glass. Show that if the reversal of direction of an off-axis ray is to occur at all, the index of refraction of the glass must be in the range $1 < n \leq \sqrt{2}$. (Hint: Consider the light ray shown in Fig. 34.78b; what index of refraction does the reversal of this light ray require?)

*42. A signal rocket explodes at a height of 200 m above a ship on the surface of a smooth lake. The explosion sends out sound waves in all directions. Since the speed of sound in water (1500 m/s) is larger than the speed of sound in air (340 m/s), a sound wave can suffer total reflection at a water surface if it strikes at a sufficiently large angle of incidence. At what minimum distance from the ship will a sound wave from the explosion suffer total reflection?

*43. A layer of oil, of index of refraction n', floats on a surface of water. A ray of light coming from below attempts to pass from the water to the oil and from there to the air above. What is the maximum angle of incidence of the ray on the water–oil surface that will permit the ultimate escape of the ray into the air? Does your answer depend on n'? (Hint: Use the formula derived in Problem 29.)

*44. A seagull sits on the (smooth) surface of the sea. A shark swims toward the seagull at a constant depth of 5.0 m. How close (measured horizontally) can the shark approach before the seagull can see it?

*45. To discover the percentage of sucrose (cane sugar) in an aqueous solution, a chemist determines the index of refraction of the solution very precisely and then finds the percentage in a table giving the dependence of index of refraction on sucrose concentration. The chemist determines the index of refraction by immersing a glass prism in the sucrose and measuring the critical angle for total internal reflection of a light ray inside the glass prism.

(a) Suppose that with a prism of index of refraction 1.6640 the critical angle is 57.295°. Use the result of Problem 29 to find the index of refraction of the sucrose solution.

(b) Use the following table, interpolating if necessary, to find the concentration of sucrose to four significant figures:

CONCENTRATION	n
40.00%	1.3997
40.10	1.3999
40.20	1.4001
40.30	1.4003

**46. A prism of glass of index of refraction 1.50 has angles of 45.0°, 45.0°, and 90.0°. A ray of light is incident on one of the short faces at an angle θ (see Fig. 34.79). What is the maximum value of θ for which the ray of light will suffer total internal reflection at the long face?

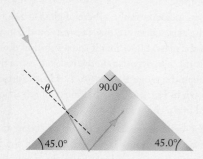

FIGURE 34.79 Total internal reflection in a prism.

**47. Rainbows are produced by the refraction of sunlight by drops of water. Figure 34.80 shows a ray of light entering a spherical drop of water. The ray is refracted at A, reflected at B, and refracted at C. The angle of incidence at A (between the ray and the normal to the surface) is θ, and the angle of refraction is θ'.

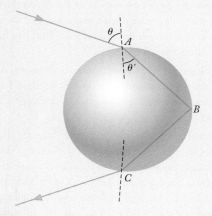

FIGURE 34.80 Path of a light ray in a drop of water.

(a) By geometry, show that the angles of incidence and reflection at B coincide with θ', and that the angles of incidence and refraction at C coincide with θ' and θ, respectively.

(b) Show that the angular deflection of the ray from its path is $\theta - \theta'$ at A, $\pi - 2\theta'$ at B, and $\theta - \theta'$ at C. These angles are all in radians, and they are measured clockwise from the incident path at each point.

(c) The total angular deflection of the ray by the raindrop is $\Delta = 2(\theta - \theta') + \pi - 2\theta'$. A rainbow will form when all the rays within an infinitesimal range $d\theta$ of angles of incidence suffer the same angular deflection, that is, when the derivative $d\Delta/d\theta = 2 - 4d\theta'/d\theta$ is zero. If this condition is satisfied, the rays sent back by the raindrops are concentrated, producing a bright zone in the sky. Show that the critical angle θ_c, at which $d\Delta/d\theta = 0$, is given by

$$\cos^2\theta_c = \tfrac{1}{3}(n^2 - 1)$$

where n is the index of refraction of water.

(d) The index of refraction for red light in water is 1.330. Find θ_c, and find Δ in degrees and minutes of arc. Draw a diagram showing a red ray coming from the Sun, hitting the drop, and reaching the eye of a rainbow watcher.

(e) The index of refraction for violet light in water is 1.342. Find θ_c and find Δ. On top of the preceding diagram, draw a violet ray coming from the Sun, hitting the drop (at a different point), and reaching the eye of the rainbow watcher. Will the watcher see the red color above or below the violet?

†34.4 Spherical Mirrors

48. In the headlamp of an automobile, a lightbulb placed in front of a concave spherical mirror generates a parallel beam of light. If the distance between the lightbulb and the mirror is 4.0 cm, what must the radius of curvature of the mirror be?

49. A lightbulb is placed at a distance of 15.0 cm in front of a convex mirror of radius of curvature 40.0 cm. Where is the image of the lightbulb? Draw a diagram showing the light-bulb, the mirror, and the image. Show three of the principal rays.

50. At what distance from a concave mirror of radius R must you place an object if the image is to be at the same position as the object?

51. A concave mirror has a radius of curvature R. If you want to form a real image, within what range of distances from the mirror must you place the object? If you want to form a virtual image, within what range of distances must you place the object?

52. A candle is placed at a distance of 15.0 cm in front of a concave mirror of radius 40.0 cm. Where is the image of the candle? Draw a diagram showing the candle, the mirror, and the image. Show three of the principal rays.

53. A candy cane is 15.0 cm from a reflecting sphere (a holiday ornament) that is 6.0 cm in diameter. What is the focal length? Where is the image of the candy cane?

54. A concave mirror reflects an image of the Moon 6.0 cm from the mirror. If a coin is placed 8.0 cm from the same mirror, where will its image be?

55. A convex mirror has a focal length of -25.0 cm. What is the object distance if the image is 10.0 cm behind the mirror?

56. A soupspoon is a spherical mirror on either side. When looking at one side of the spoon, a woman sees an image of her eye 7.0 cm from the spoon, on the near side. Without changing positions, she flips the spoon over. Now the image of her eye is 2.0 cm on the far side of the spoon. What is the radius of curvature of the spoon? How far is the woman's eye from the spoon?

†For help, see Online Concept Tutorial 38 at www.wwnorton.com/physics

57. A hand mirror is to show a (virtual) image of your face magnified 1.5 times when held at a distance of 20 cm from your face. What must be the radius of curvature of a spherical mirror that will serve the purpose? Must it be concave or convex? (Hint: Note from the similar triangles in Fig. 34.42 that this magnification is equal to the magnitude of the ratio of the image to object distances.)

58. The surface of a highly polished doorknob of brass has a radius of curvature of 4.5 cm. If you hold this doorknob 15 cm away from your face, where is the image you see? By what factor does the size of the image differ from the size of your face? (Hunt: Note from the similar triangles in Fig. 34.37 that the magnification is equal to the magnitude of the ratio of the image to object distances.)

*59. A concave mirror of radius of curvature 30.0 cm faces a second concave mirror, of radius of curvature 24.0 cm. The distance between the mirrors is 80.0 cm, and their axes coincide. A lightbulb is suspended between the mirrors, at a distance of 20.0 cm from the first mirror.

    (a) Where does the first mirror form an image of the lightbulb?

    (b) Where does the second mirror form an image of the lightbulb?

*60. A concave mirror of radius of curvature 60.0 cm faces a convex mirror of the same radius of curvature. The distance between the mirrors is 50.0 cm, and their axes coincide. A candle is held between the mirrors, at a distance of 10.0 cm from the convex mirror. Consider rays of light that reflect first off the concave mirror and then off the convex mirror. Where do these rays form an image?

†34.5 Thin Lenses

61. The crystalline lens of a human eye has two convex surfaces with radii of curvature of 10 mm and 6.0 mm. The index of refraction of its material is 1.45. Treating it as a thin lens, calculate its focal length when removed from the eye and placed in air.

62. A thin lens of flint glass with $n = 1.58$ has one concave surface of radius of curvature 15 cm and one flat surface

    (a) What is the focal length of this lens?

    (b) If you place this lens at a distance of 40 cm from a candle, where will you find the image of the candle?

63. A thin, symmetric convex lens of crown glass with index of refraction $n = 1.52$ is to have a focal length of 20 cm. What are the correct radii of curvature of the spherical surfaces of the lens?

64. A double-convex lens has a focal length of 16 cm, and one of its surfaces has a radius of curvature of 27 cm. The lens glass has an index of refraction of 1.66. What is the radius of curvature of the other surface?

65. A double-concave lens has radii of curvature of 16 cm and 12 cm; the lens glass has an index of refraction of 1.58. What is the focal length of the lens?

66. A contact lens has one convex surface of radius of curvature 2.5 cm and one concave surface of radius of curvature 3.0 cm. It is made from a plastic of index of refraction 1.43. What is the focal length of the contact lens?

67. An object is located 15 cm from a concave lens of focal length −20 cm. What is the image distance? Is the image real or virtual? Upright or inverted? Larger or smaller than the object? What is the magnification?

68. The Moon subtends an angle of 0.52° when viewed from Earth. If a lens with a focal length of 150 mm is used to image the Moon, what is the height of the Moon's image?

69. A slide projector has a lens of focal length 13.0 cm. The lens is at a distance of 2.0 m from the screen. What must be the distance from the slide to the lens if a sharp image of the slide is to be seen on the screen? What is the magnification of the slide projector?

70. The convex lens of a magnifying glass has a focal length of 20 cm. At what distance from a postage stamp must you hold this lens to produce a (virtual) stamp twice as large as the stamp?

71. If you place a convex lens of focal length 18 cm at a distance of 30 cm from a small lightbulb, where will you find the image of the lightbulb? Is this a real or a virtual image? Is it upright or inverted?

72. The lens maker's formula, Eq. (34.27), follows from the Law of Refraction at the air–lens surfaces. Argue that if a thin lens of index of refraction n is immersed in a fluid (for example, water) of index of refraction n_0, then the lens maker's formula must be modified as follows:

$$\frac{1}{f} = \left(\frac{n}{n_0} - 1 \right) \left(\frac{1}{R_1} + \frac{1}{R_2} \right)$$

(Hint: In the Law of Refraction, only the ratio of the indices of refraction is relevant.)

*73. Show that if two thin lenses of focal lengths f_1 and f_2 are placed next to one another (in contact; see Fig. 34.81), the net focal length f is given by

$$\frac{1}{f} = \frac{1}{f_1} + \frac{1}{f_2}$$

FIGURE 34.81 Two thin lenses in contact.

*74. A convex lens of focal length 25 cm is at a distance of 60 cm from a concave mirror of focal length 20 cm. A lightbulb is 80 cm from the lens (Fig. 34.82).

(a) Where does the lens form an image of the lightbulb?

(b) Where does the mirror form an image of the first image?

(c) Where does the lens form the (final) image of the second image?

FIGURE 34.82 Lightbulb, convex lens, and concave mirror.

*75. Two lenses, one concave and one convex, have focal lengths of magnitude 30 cm. The lenses are separated by a distance of 10 cm. A candle is 20 cm from the convex lens (Fig. 34.83).

(a) Where does the convex lens form an image?

(a) Where does the concave lens form an image of this image?

FIGURE 34.83 Candle, convex lens, and concave lens.

*76. An object is a distance $\frac{3}{2} f$ from a convex lens of focal length f. A second, identical lens is a distance $d = f$ beyond the first. Where is the final image? What is the overall magnification of the two-lens system?

*77. Prove that the lens equation can be put in the form $xx' = f^2$, where x and x' are the object and image distances measured from the focal points (see Fig. 34.84). This is called the **Newtonian form** of the lens equation.

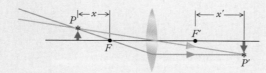

FIGURE 34.84 Convex lens and distances x and x'.

*78. To measure the focal length of a thin concave lens, a physicist places this lens adjacent to a thin convex lens of focal length 12.0 cm. She finds that when the joined lenses face the Sun, the rays of sunlight are focused on a spot 22.0 cm beyond the lenses. What focal length for the concave lens can she deduce from this?

*79. A lightbulb is 15.0 cm in front of a convex mirror of radius of curvature 10.0 cm. A convex lens of focal length 25.0 cm is 5.0 cm beyond the lightbulb (Fig. 34.85). Where do you see the image of the lightbulb if you look through the convex lens at the mirror?

FIGURE 34.85 Lightbulb, convex mirror, and convex lens.

*80. Two convex lenses of focal lengths 40 cm and 60 cm are separated by a distance of 10 cm. Where is the focal point of this system, measured from the center of the lens system, for a parallel beam of light incident from the left? For a beam incident from the right?

*81. Consider **refraction by a spherical surface** between two transparent materials. For instance, Fig. 34.86 shows a sphere of glass of index of refraction n_2 immersed in a medium of index of refraction n_1. For a convex glass surface and $n_2 > n_1$, a real image will form inside the glass if a source is placed sufficiently far outside the glass. For rays making small angles with the axis, show that the equation relating the object distance s and image distance s' shown in Fig. 34.86 is

$$\frac{n_1}{s} + \frac{n_2}{s'} = \frac{n_2 - n_1}{R}$$

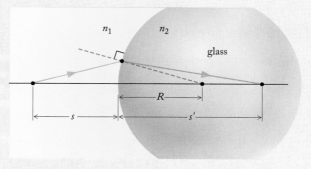

FIGURE 34.86 Spherical refracting surface.

*82. Use the result of Problem 81 to derive the **lens maker's formula** and the **thin-lens equation**, Eqs. (34.27) and (34.28), by considering refraction at two successive convex spherical glass surfaces in air. Treat the image of the first surface as the object for the second surface. For a thin lens, neglect the distance between the surfaces. Recall that the focal length is the image distance for an object at infinity.

**83. The "crystal" ball of a fortune-teller has a diameter of 15.0 cm. It is made of glass of index of refraction 1.60. At what distance from the center will this sphere bring an incident beam of sunlight to a focus? (Hint: Use the result of Problem 81 twice.)

**84. A Fresnel lens consists of a large number of segments each of which is an annular portion of an ordinary lens. Fresnel lenses of diameters more than 1 m are commonly used in the lamps of lighthouses, where ordinary lenses without segments would be much too thick and too heavy (see Fig. 34.70). Consider one of these annular segments in a plano-convex Fresnel lens (Fig. 34.87); for the sake of simplicity, assume that the width of the segment is infinitesimal. Derive a formula for the angle of inclination θ of the surface of this segment in terms of the index of refraction n, the distance r of the segment from the axis of the lens, and the focal length of the lens.

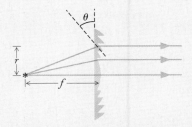

FIGURE 34.87 A ray passing through one of the annular segments of a Fresnel lens.

34.6 Optical Instruments

85. The lens of a camera (Fig. 34.88) has a focal length of 50 mm. The distance of the lens from the film is adjustable over a range from 50 mm to 62 mm. Over what range of object distances (measured from the lens) is this camera capable of producing sharp pictures?

FIGURE 34.88 Lens of a camera.

86. A miniature Minox camera has a lens of focal length 15 mm. This camera can be focused on an object as close as 20 cm, or as far away as infinity. What must be the distance from the lens to the film if the camera is set for 20 cm? What if the camera is set for infinity?

87. The light meter of a 35-mm camera with a lens of f number 1.7 indicates that the correct exposure time for a photograph is $\frac{1}{250}$ s. If the iris diaphragm is closed down so the f number becomes 8, what will be the correct exposure time?

88. Pretend that the cornea, the crystalline lens, and the fluid of the human eye act together as a single thin lens placed at a distance of 2.2 cm from the retina (Fig. 34.89). The lens is deformable; it can change its focal length by changing its shape.

   (a) What must be the focal length if the eye is viewing an object at a very large distance?

   (b) What must be the focal length if the eye is viewing an object at a distance of 25 cm?

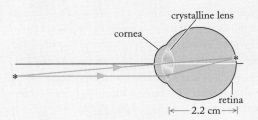

FIGURE 34.89 Lens of human eye forms image on retina.

89. In a nearsighted eye the (relaxed) lens has an abnormally short focal length, and consequently the eye fails to form an image of a distant object on the retina. This defect can be corrected with a contact lens. Pretend that both the lens of the eye and the contact lens are thin lenses so that the formula given in Problem 73 applies.

   (a) Suppose that the focal length of the eye is 2.00 cm. What must be the focal length of the contact lens if it is to increase the net focal length to 2.20 cm? Should it be convergent or divergent?

   (b) The radius of curvature of the side of the contact lens next to the eye should be -0.80 cm so as to fit tightly on the cornea. What must be the radius of curvature of the other side? The contact lens is made of plastic with index of refraction 1.33.

90. Would the contact lens of Problem 66 be prescribed for a nearsighted person or for a farsighted person? Express the strength of this lens in diopters.

91. Equation (34.32) gives the angular magnification for a magnifier if the object is so placed that the image is at infinity. Show that if the object is so placed that the image is at a distance of 25 cm, then the angular magnification is $1 + (25 \text{ cm})/f$.

92. A microscope has an objective of focal length 4.0 mm. This lens forms an image at a distance of 224 mm from the lens. If we want to attain a net angular magnification of 550, what choice must we make for the angular magnification of the ocular?

93. A telescope has an objective of focal length 160 cm and an ocular of focal length 2.5 cm. If you look into the *objective* (that is, into the wrong end) of this telescope, you will see distant objects *reduced* in size. By what factor will the angular size of objects be reduced?

94. An amateur astronomer uses a telescope with an objective of focal length 90.0 cm and an ocular of focal length 1.25 cm. What is the angular magnification of this telescope?

95. The large reflecting telescope on Palomar Mountain has a minor of focal length 1680 cm (see Fig. 34.66a). If this telescope is operated with an ocular focal length of 1.25 cm what is the angular magnification?

96. It has been proposed that the six 1.8-m diameter mirrors in the Multiple-Mirror Telescope (Fig. 34.90) be replaced by a single 6.5-m diameter mirror, which would fit into the existing mount. By what factor would this replacement enhance the amount of light collected by the telescope?

97. The telescopes built by Galileo Galilei consisted of a convex objective lens and a concave ocular lens. The lenses are arranged so the focal point of the objective coincides with the focal point of the ocular (similar to Fig. 34.64). Explain how this telescope produces an angular magnification. Show that

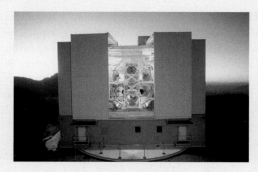

FIGURE 34.90 The Multiple-Mirror Telescope on Mt. Hopkins, Arizona.

the eye sees an upright image. (Hint: The concave lens placed *before* the point F in Fig. 34.64 has the same effect as the convex lens placed *beyond* the point F in Fig. 34.64—the lens produces an image at infinity. Is this image upright or inverted?)

REVIEW PROBLEMS

98. Two flat mirrors are arranged at an angle of 145° with respect to each other (see Fig. 34.91). A ray of light strikes one of the mirrors at an angle of incidence of 60°, it is reflected, and it then strikes the other mirror. What is the angle of reflection at the second mirror?

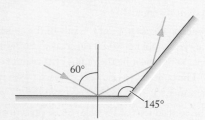

FIGURE 34.91 Two adjacent flat mirrors an angle.

99. A ray of light strikes a plate of window glass of index of refraction 1.5 with an angle of incidence of 50°.

   (a) What is the angle of refraction?

   (b) At what angle does the ray of light emerge on the other side of the glass?

100. The bottom corner of a swimming pool is 2.0 m below the surface of the water. If a ray of light travels upward from the corner and reaches the water surface at an angle of 40° with the vertical, what will be the angle the ray makes with the vertical when it emerges from the water? If your eye intercepts this ray, how deep will the corner of the swimming pool seem to you?

101. A prism has three equal angles of 60°. The prism is made of flint glass of index of refraction 1.66. At what angle should a ray of light enter this prism so it is refracted symmetrically (see Fig. 34.92)?

FIGURE 34.92 A symmetric prism.

102. A prism of glass has two angles of 45.0° and one of 90.0°. To determine the index of refraction of the glass in this prism, an experimenter aims a beam of light perpendicularly into a face of this prism and measures the angle at which the ray emerges on the far side (see Fig. 34.93). If this angle is 72.0°, what is the index of refraction?

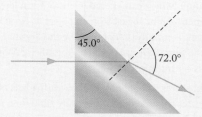

FIGURE 34.93 A 45° prism.

*103. The bottom half of a beaker of depth 20.0 cm is filled with water ($n = 1.33$), and the top half is filled with oil ($n = 1.48$). If you look into this beaker from above, how far below the upper surface of the oil does the bottom of the beaker seem to be?

104. The walls of a (filled) aquarium are made of glass with an index of refraction of 1.5. If a ray strikes the glass from the inside at an angle of incidence of 45°, what is the angle at which it emerges into the air?

105. A prism is to be used for total internal reflection of a ray of light that is perpendicularly incident on one of its faces (see Fig. 34.94). The index of refraction of the glass is 1.60. For what range of vertex angles α of this prism will this total internal reflection occur?

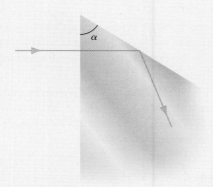

FIGURE 34.94 Total internal reflection in a prism.

106. A swimming pool has a depth of 2.0 m. A waterproof electric light is installed in the bottom of the pool which illuminates the pool at night. When looking down on the pool, you see that this light produces an illuminated circle on the surface, but it leaves the rest of the surface dark. What is the radius of this illuminated circle?

107. You place an object in front of a concave mirror, at a distance smaller than the focal length. Is the image real or virtual? Upright or inverted? Magnified or reduced? What if the distance is greater than the focal length? What if the mirror is convex?

108. The outside rearview mirror on an automobile is convex (Fig. 34.95). When you look into this mirror, you notice that the image of a distant truck or some other distant object is about 1.0 m beyond the mirror. What is the radius of curvature of the mirror?

FIGURE 34.95 Rearview mirror of an automobile.

109. In a grocery store, a convex mirror with $R = 4.0$ m is used for surveillance. Suppose a person of height 1.7 m is 12 m from such a mirror. Where is the image? What is the size of the image?

*110. A concave mirror of radius of curvature 60 cm faces a convex mirror of the same radius of curvature. The distance between the mirrors is 50 cm, and their axes coincide. A candle is held between the mirrors, at a distance of 10 cm from the convex mirror. Consider rays of light that reflect first off the concave mirror and then off the convex mirror. Where do these rays form an image?

111. You place an object in front of a convex lens at a distance smaller than the focal length. Is the image real or virtual? Upright or inverted? Magnified or reduced? What if the distance is greater than the focal length? What if the lens is concave?

112. A candle is placed at a distance of 50 cm in front of a concave lens of focal length 30 cm. Where is the image of the candle? Draw a diagram showing the candle, the lens, and the image. Show three principal rays.

113. A lightbulb is placed at a distance of 15 cm in front of a convex lens of focal length 30 cm. Where is the image of the lightbulb? Draw a diagram showing the lightbulb, the lens, and the image. Show the three principal rays.

*114. Three identical convex lenses of focal length 60 cm each are separated by distances of 12 cm from one to the next. If a small light bulb is placed 20 cm in front of the left lens, where will the final image be? What is the magnification of the final image?

115. A microscope has an objective of focal length 1.9 mm and an ocular of focal length 25 mm. The distance between these lenses is 180 mm.

(a) At what distance from the objective must the object be placed so that the ocular forms an image at infinity, as shown in Fig. 34.63?

(b) What is the net angular magnification of this microscope?

Answers to Checkups

Checkup 34.1

1. If we draw a set of identical spherical wavelets centered on different points on the blue dashed arc (Huygens' Construction), their leading edges (when traveling inward) form a sphere of smaller radius, precisely the reverse of the outward wave.

2. If we draw a set of small, circular wavelets centered on different points around the rectangular edge of the brick, we find a nearly rectangular shape, but with rounded corners, for the propagated wave front.

3. (D) $\frac{3}{2}$; 1. When the wave front moves a distance L out from each tip, it will have a total length $3L$ (including the length L of the stick); out from the sides, its width will be only $2L$ (the stick is thin). Thus, after traveling a distance L, the length-to-width ratio will be 3/2. Much farther out, the wave front will be nearly circular (for example, after traveling $50L$, it will be $101L$ long and $100L$ wide): the length-to-width ratio approaches 1.

Checkup 34.2

1. Since the angle of incidence is defined as the angle that the ray makes with the *perpendicular to the surface*, a ray making an angle of 20° with the surface has an angle of incidence of 70°.

2. The law of reflection tells us that the angle of reflection is also 0°. Thus both the incident and reflected rays are perpendicular to the surface, and the ray returns back on itself.

3. Similar to the image illustrated schematically in Fig. 34.12, the image of the flashlight in Fig. 34.6 is located below the mirror, at a distance equal to the distance the flashlight is above the mirror.

4. As illustrated in Fig. 34.12, a (virtual) image is the same distance behind a mirror as you are in front: thus, if you are 2 m from the mirror, your image is 4 m away from you.

5. Your image approaches the back of the mirror at the same rate that you approach the front (8 m/s); thus, relative to you, the image travels at 16 m/s.

6. (E) $\frac{1}{2} h$. Since the incident and reflected rays from the foot to the eye make the same angle, and similarly for the rays from the top of the head to the eye, you can see from head to toe in a mirror (appropriately positioned) of only half your height, $\frac{1}{2} h$.

Checkup 34.3

1. The refracted ray is also along the perpendicular, with an angle of refraction of 0° (from Snell's Law). For this one angle, the answer is independent of the index of refraction.

2. Cold air is denser than warm air, and the denser air has a larger dielectric constant, and hence a larger index of refraction [see Eq. (34.5)].

3. The wavelength will be $(600 \text{ nm})/1.33 \approx 450$ nm.

4. The ray will be bent closer to the perpendicular when it enters a medium of greater index of refraction. Thus, from the data of Table 34.1, we can determine that the ray will be bent (a) closer to the perpendicular ($n = 1$ to $n = 1.33$); (b) closer to the perpendicular ($n = 1.33$ to $n = 1.52$–1.66); (c) farther away from the perpendicular ($n = 1.46$ to $n = 1.36$).

5. If the angle is slightly more than the critical angle of 49°, there will be no refracted ray, and only the ray reflected back downward at 49° will occur. If the angle is slightly less than the critical angle of 49°, then the refracted ray will be just above the surface (nearly parallel to it); of course, part of the ray will still be reflected downward with an angle of reflection equal to the angle of incidence.

6. No; for this ray the angle of refraction is always less than the angle of incidence, and hence the angle of refraction can never attain 90°. For this example, the angle of refraction will always satisfy $\sin\theta < 1.3/1.6$, or $\theta < 55°$.

7. (D) $30° < \theta < 90°$. The critical angle is given by $\sin\theta_{\text{crit}} = 1/n$. So here, $\sin\theta_{\text{crit}} = 1/(2.0)$, or $\theta_{\text{crit}} = 30°$. Total internal reflection will occur for all angles greater than this, up to the maximum angle of incidence, 90°.

Checkup 34.4

1. As in Example 9 and Fig. 34.41, an object further away from a concave mirror than the focal point produces a real, inverted image in front of the mirror.

2. For convex mirrors (like flat mirrors), the image is always behind the mirror, and is thus virtual. As in the ray diagram of Fig. 34.40, such images are upright. You should verify that the closer object distance here does not change the nature of the ray diagram.

3. As in Example 8 and Fig. 34.40, and question 2 above, a convex mirror always produces a virtual image. Thus the answer is no.

4. No. The convex mirror does not bring the rays together, but instead always causes them to diverge.

5. (A) Virtual and upright, with $s' < 0$. This situation was addressed in Example 9 and Fig. 34.42.

Checkup 34.5

1. The focal length of any thin lens does not depend on which side the object is on, so its focal length is still 15 cm.

2. This is the "magnifier" situation of Fig. 34.50, where the object is closer to the convex lens than the focal length of the lens. The image is on the near side (the same side as the object); the image is upright and virtual.

3. For a single concave (diverging) lens, the image is always on the same side as the object; it is always upright and virtual (as in Fig. 34.47).

4. No. For a single concave (diverging) lens (a negative f), the thin lens equation implies that the image distance s' will always be negative for a positive object distance s, implying a virtual, upright image.

5. (E) -4.0 m. For the lens maker's equation (34.27), we use $+2.0$ m for the convex radius and -1.0 m for the concave radius, giving $1/f = (n-1)(1/R_1 + 1/R_2) = (1.5 - 1)[1/(2.0 \text{ m}) + 1/(-1.0 \text{ m})] = -0.25/\text{m}$, so $f = -4.0$ m.

6. (A) Upright and real. The image produced by the first lens is real and inverted; the thin lens equation puts it at a distance $2f$ beyond the first lens. This is the object for the second lens; since the lenses are separated by $5f$, the second object distance is $3f$. Since this is outside the focal length f, a real image forms which is reinverted, or upright with respect to the original object.

Checkup 34.6

1. To get more light into the camera, you need a larger-diameter aperture. Since the f number is inversely proportional to the aperture diameter, you need a small f number.

2. In bright light, the pupil gets smaller. As in a camera, this reduces the amount of light entering the eye, but increases the depth of field: because of the smaller aperture, light rays enter the eye at small angles, and they stay close together on the retina and form a fairly small spot, even when the position of the image does not fall exactly on the retina.

3. (A) The arrangement is such that the object distance is smaller than the image distance, so the objective does produce a linear magnification. Both object and image subtend the same angle, and the object may be at or beyond the typical near point of 25 cm, so no ordinary angular magnification occurs.

(B) The ocular produces a large linear magnification (in principle, infinite, since the image is usually at infinity). The ocular produces an angular magnification, the same as a magnifier; see Eq. (34.32).

(C) From Fig. 34.63, the first image, formed by the objective, is real and inverted.

(D) Again from Fig. 34.63, the final image produced by the ocular is virtual and inverted (with respect to the original object).

4. (A) The objective produces no magnification (or, more accurately, it produces a magnification of zero, since the image distance is finite, and the object is essentially at infinity). As depicted in Fig. 34.64, the object and image subtend the same angle, so there is no angular magnification [see answer 3(A), above].

(B) As in the case of the microscope of question 2, the ocular does produce a large (in principle, infinite) linear magnification, and does produce an angular magnification, given by Eq. (34.32).

(C) From Fig. 34.64, the first image, formed by the objective, is real and inverted.

(D) Again from Fig. 34.64, the final image produced by the ocular is virtual and inverted (with respect to the original object).

5. The microscope uses an objective lens of small focal length to form an enlarged image of a nearby object; the telescope uses an objective lens of large focal length to form an image of a distant object at its focal point.

6. (D) Microscope and magnifier only. As discussed above, it is only for enlarging near objects that the near point of vision plays a role [compare Eqs. (34.32), (34.33), and (34.35)].

Interference and Diffraction

CONCEPTS IN CONTEXT

Concepts
— *in* —
Context

Everyday objects, such as these compact discs (CDs), can display spectacular rainbows of color. Each "groove" or track on the CD spreads, or diffracts, a reflected light wave, and waves from different tracks combine to produce the observed patterns.

With the concepts developed in this chapter, we can consider several questions about such interference patterns:

? What do the colors tell us about the source of light? (Section 35.4, page 1185)

? What does the pattern tell us about the spacing between tracks? (Example 5, page 1185)

? How does the pattern depend on the size of a track? (Example 8, page 1195)

Geometric optics relies on the assumption that the sizes of the mirrors or lenses and the separations between them are much larger than the wavelength of light. Under these conditions, we can adequately describe the propagation of light by rays that are rectilinear except when reflected or when refracted by the interface between different media. Thus, in geometric optics the wave properties of light do not show up explicitly (although these wave properties enter into the derivations of the laws of reflection and refraction).

The absence of any explicit wave effects during the propagation of light through a system of mirrors and lenses is related to the small wavelength of light. In comparison with the sizes of typical mirrors and lenses, the wavelength of light is extremely small (about 5×10^{-7} m). We cannot detect this small wavelength and the small-scale wave behavior of light by means of ordinary mirrors and lenses for the same reason that we cannot detect ripples on the surface of the ocean by means of a super-tanker—such instruments are too crude. If we want to detect ripples on the surface of the ocean, we need to use small pieces of flotsam, preferably of a size comparable with the wavelength of the ripples. Likewise, to detect the wave behavior of light, we need to let the light interact with a body or an obstacle of a size comparable with a wavelength. In that case, light will display its wave properties by the phenomena of **interference** and **diffraction**. As mentioned in Chapters 16 and 17, *interference is the constructive or destructive superposition of two or more waves meeting at one place; diffraction is the bending and spreading of waves around obstacles.* **Wave optics**, or physical optics, deals with interference and diffraction of light, and other phenomena that directly involve the wave properties of light.

wave optics

In this chapter we will begin by examining the interference of light waves and other electromagnetic waves. Like all electric and magnetic fields, the fields of electromagnetic waves obey the Superposition Principle: if two waves meet at some point, the resultant electric or magnetic field is simply the vector sum of the individual fields. If two waves of equal amplitude meet crest to crest, they combine and produce a wave of double amplitude; if they meet crest to trough, they cancel and give a wave of zero amplitude. The former case is called *constructive interference* and the latter *destructive interference.* We will encounter both cases of interference in the following sections.

35.1 THIN FILMS

The interference between light waves is difficult to observe. For instance, the simplest case of interference is that between two waves of the same frequency traveling in opposite directions, giving a standing wave, with nodes and antinodes. Such a standing wave can be set up by shining light perpendicularly at a mirror, so the incident wave interferes with the reflected wave. However, since the wavelength of light is so small, our eyes cannot perceive the individual nodes and antinodes—we only see the average intensity, without noticeable interference effects.

Spectacular interference effects may become visible when light waves with a range of frequencies and wavelengths are reflected by a thin film, such as a thin film of oil floating on water (see Fig. 35.1) or a thin film of soapy water on the surface of a soap bubble. When the incident waves enter the oil film (see Fig. 35.2), they will set up a multitude of reflected waves due to reflection at the upper surface, reflection at the lower surface, and multiple zigzags between the surfaces. The reflected waves that re-emerge all travel in the same direction, and they can interfere destructively or constructively in the region of space above the film. Note that we are now interested only

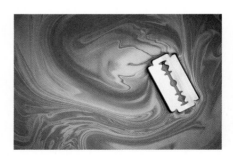

FIGURE 35.1 Colored fringes in the light reflected by a thin film of oil floating on water.

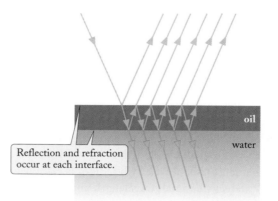

FIGURE 35.2 Incident ray and multiple reflected and refracted rays produced by a thin film.

Reflection and refraction occur at each interface.

THOMAS YOUNG (1773–1829)

English physicist, physician, and Egyptologist. Young worked on a wide variety of scientific problems, ranging from the structure of the eye and the mechanism of vision to the decipherment of the Rosetta stone. He revived the wave theory of light and recognised that interference phenomena provide proof of the wave properties of light.

in the interference between the several reflected waves—the incident wave plays no direct role in this. The beautiful shimmering color displays seen in oil slicks and on soap bubbles arise from such interference effects. Different portions of an oil or soap film usually have different thicknesses, and they therefore give constructive interference for different wavelengths. The result is a pattern of bright colored bands, or colored **fringes**.

To find the conditions for constructive and destructive interference between the waves reflected by a thin film, let us make the simplifying assumption that the direction of propagation of the incident wave is nearly perpendicular to the surface of the film and that this wave has a given frequency (and wavelength). The two most intense waves are those that experience only one reflection: the wave that reflects only at the upper surface and the wave that reflects only at the lower surface. Figure 35.3 shows the rays corresponding to these two waves. Let us *first assume that the reflection affects the phases of both waves in the same way*, which is true for the air–oil and oil–water interfaces here. Under what conditions will these waves interfere constructively in the space above the film? Obviously, the wave that is reflected from the lower surface has to travel an extra distance to emerge from the film. If the thickness of the film is *d*, and if the direction of propagation is nearly perpendicular to the film, then the extra distance the wave has to travel is 2*d*, that is, down through the film and then back up.

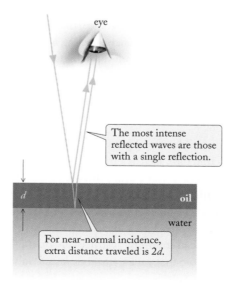

eye

The most intense reflected waves are those with a single reflection.

For near-normal incidence, extra distance traveled is 2*d*.

FIGURE 35.3 Incident ray and reflected rays for nearly perpendicular incidence and reflection. The wave reflected at the lower surface has to travel an extra distance of approximately *d* while propagating down and *d* while propagating up, that is, an extra distance 2*d*.

Provided that this extra distance is equal to one, two, three, etc., wavelengths, the wave reflected from the lower surface will meet crest to crest with the wave reflected from the upper surface (see Fig. 35.4). The condition for constructive interference is therefore

$$2d = \lambda, \ 2\lambda, \ 3\lambda, \ldots \tag{35.1}$$

constructive interference for thin film

Likewise, if the extra distance is equal to one half wavelength, three half wavelengths, five half wavelengths, etc., the wave reflected from the lower surface will meet crest to trough with the wave reflected from the upper surface (see Fig. 35.5). The condition for destructive interference is therefore

$$2d = \tfrac{1}{2}\lambda, \ \tfrac{3}{2}\lambda, \ \tfrac{5}{2}\lambda, \ldots \tag{35.2}$$

destructive interference for thin film

Thus, depending on the thickness of the film and on the wavelength of the light, the reflected wave can be either very strong or very weak. Note that the wavelength λ in Eqs. (35.1) and (35.2) is the wavelength of the light within the film; the wavelength outside the film will differ from this by a factor equal to the film's index of refraction n. If $\lambda_{air} \approx \lambda_{vacuum}$ is the wavelength in the air outside the film, recall from Eq. (34.7) that the wavelength λ inside the material of the film is smaller:

$$\lambda = \frac{\lambda_{air}}{n} \tag{35.3}$$

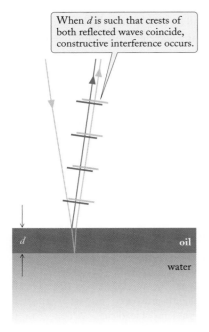

When d is such that crests of both reflected waves coincide, constructive interference occurs.

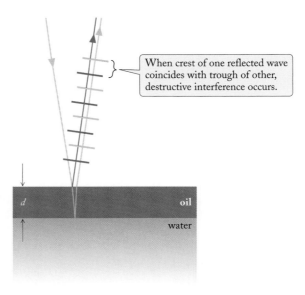

When crest of one reflected wave coincides with trough of other, destructive interference occurs.

FIGURE 35.4 The wave (orange lines) reflected from the lower surface meets crest to crest with the wave (blue lines) reflected from the upper surface.

FIGURE 35.5 The wave (orange lines) reflected from the lower surface meets crest to trough with the wave (blue lines) reflected from the upper surface.

EXAMPLE 1　A film of kerosene 450 nm thick floats on water. White light, a mixture of all visible colors, is vertically incident on this film. Which of the wavelengths contained in the white light will give maximum intensity upon reflection? Which will give minimum intensity? The index of refraction of kerosene is 1.20.

SOLUTION: According to Eq. (35.1), for maximum intensity (constructive interference), we need

$$\lambda = 2d, \frac{2d}{2}, \frac{2d}{3}, \ldots$$
$$= 900 \text{ nm}, 450 \text{ nm}, 300 \text{ nm}, \ldots$$

These are the wavelengths in kerosene; to obtain the wavelengths in air, we must multiply by the index of refraction of kerosene, $n = 1.20$. We are interested only in wavelengths in the visible range, approximately 400–700 nm. Of the resulting wavelengths, the only one in the visible region is $1.20 \times 450 \text{ nm} = 540 \text{ nm}$.

For minimum intensity (destructive interference) we need, according to Eq. (35.2),

$$\lambda = 4d, \frac{4}{3}d, \frac{4}{5}d, \ldots$$
$$= 1800 \text{ nm}, 600 \text{ nm}, 360 \text{ nm}, \ldots$$

Upon multiplication by 1.20, we find that the only wavelength in air in the visible region is $1.20 \times 360 \text{ nm} = 432 \text{ nm}$.

COMMENTS: A wavelength of 540 nm corresponds to a yellow green color. The film will therefore be seen to have this color in reflected light.

If the oil film is thick, no interference fringes will be visible. In a thick film, the distance from the top of the film to the bottom is many wavelengths. Thus, even a very small change in the angle at which the ray of light enters the film results in a change of travel distance by many wavelengths. The interference maxima and minima for rays of different angles therefore overlap, and they tend to average out. The rays emitted by an ordinary source, such as the Sun or a lamp, and the rays entering your eye always include a range of different angles; thus, you see only an average intensity when observing such rays reflected by a thick film, and you do not see interference fringes (Fig. 35.6).

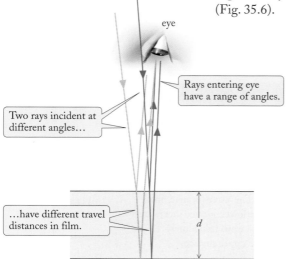

eye

Rays entering eye have a range of angles.

Two rays incident at different angles...

...have different travel distances in film.

d

FIGURE 35.6 Two rays arriving at slightly different angles at a thick film. For the ray with the smaller angle of incidence (blue), the travel distance through the film and back is, say, 5000 wavelengths; for the ray with the larger angle of incidence (orange), the travel distance is, say, 5000.5 wavelengths. The former ray gives constructive interference; the latter gives destructive interference. The interference effects tend to average out.

We can also use Eqs. (35.1) and (35.2) for a soap film suspended in air, instead of an oil film floating on water; but we must take into account a minor complication that can be derived from analysis of reflection of electromagnetic waves according to Maxwell's equations. *Whenever an electromagnetic wave propagating in one medium is reflected at the surface of another medium of larger index of refraction, the wave suffers a reversal of its electric field, that is, a change of phase by* 180° (this change of phase upon reflection is similar to that discussed in Section 16.4 for a wave on a string; see Fig. 16.14). In the case of the oil film floating on water, the wave reflected at the upper (or outer) surface of the film suffers such a change of phase, and so does the wave reflected at the lower (or inner) surface; the net result is that the relative phase between the two waves is not altered, and the extra phase changes introduced by the process of reflection can be ignored, as in Example 1. However, in the case of a soap film, or some other film, suspended in air, only the wave reflected at the outer surface suffers the extra change of phase. Thus, besides the phase difference introduced by the thickness of the film, there is an extra phase difference of 180° between the two waves. This means that the waves reflected from the outer and the inner surfaces of the soap film surfaces will be out of phase if the distance $2d$ is equal to one, two, three, etc., wavelengths—and that Eq. (35.1) is now the condition for destructive interference. Conversely, Eq. (35.2) is now the condition for constructive interference.

Similar interference effects can also arise in a narrow gap between two adjacent glass surfaces. Figure 35.7 shows a photograph of the interference fringes produced by the thin film of air between a flat glass plate and a lens of large radius of curvature. The convex surface of the lens is in contact with the plate at the center, but leaves a gradually widening gap for increasing distances from the center (Fig. 35.8). At the bright fringes, the width of the gap is such as to give constructive interference of the reflected light. At the dark fringes, the width is such as to give destructive interference. The rings in Fig. 35.7 are called **Newton's rings**.

Thin films are of great importance in the manufacture of optical instruments. The lenses of high-quality instruments are often coated with a thin film of dielectric material, so that undesirable reflections of light are eliminated and more light can enter the instrument. With one thin film, we can achieve destructive interference at only one wavelength; but with several layers of thin films, we can achieve destructive interference at several wavelengths.

Newton's rings

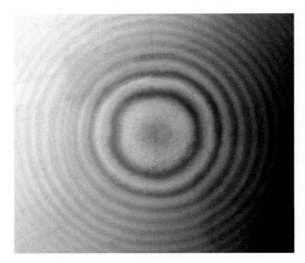

FIGURE 35.7 Fringes of constructive and destructive interference seen in the light reflected in the gap between a flat glass plate and a spherical lens in contact.

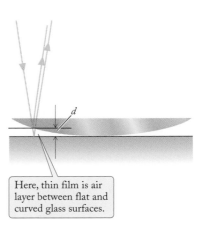

Here, thin film is air layer between flat and curved glass surfaces.

FIGURE 35.8 Rays reflected at the top and the bottom of the air gap between the flat glass plate and the lens.

PROBLEM-SOLVING TECHNIQUES   THIN-FILM INTERFERENCE

When examining the interference of light rays reflected in or at thin films, keep in mind that:

- The condition for destructive or constructive interference involves the wave length λ within the film, which is smaller than the wavelength in air by a factor equal to the index of refraction n of the film: $\lambda = \lambda_{air}/n$.

- Whenever the wave reflects on the surface of a medium of higher index of refraction, it suffers an extra phase change of 180°. If neither or both reflections occur at interfaces with a medium of higher index of refraction, there is no net extra effect. However, if only one of the reflections occurs at an interface with a medium of higher index of refraction, then the extra phase change of 180° reverses the conditions for constructive and destructive interference.

 Checkup 35.1

QUESTION 1: Why is the central spot in Newton's rings (see Fig. 35.7) dark?

QUESTION 2: If the light in Example 1 is incident on the kerosene film from below (from within the water), are Eqs. (35.1) and (35.2) still valid for constructive and destructive interference? What is different?

QUESTION 3: A "nonreflective" coating is sometimes put on glass lenses to increase the light entering an optical system. For a wavelength of 500 nm in air, what is the smallest coating thickness that could be used to minimize reflection? The index of refraction of the coating is 1.25, and that of the glass is 1.50.

(A) 100 nm   (B) 125 nm   (C) 156 nm   (D) 200 nm   (E) 250 nm

35.2 THE MICHELSON INTERFEROMETER

The Michelson interferometer takes advantage of the interference between two light waves to accomplish an extremely precise comparison between two lengths. Figure 35.9 is a schematic diagram of the essential parts of such an interferometer. The apparatus consists of two arms at the ends of which are mounted mirrors M_1 and M_2. Light waves of a given frequency (and wavelength) from a monochromatic source S fall on a semitransparent mirror M (a half-silvered mirror). This mirror splits the light wave into two parts: one part continues straight ahead and reaches mirror M_1, the other part is reflected and reaches mirror M_2. These mirrors reflect the waves back toward the central mirror M, and upon reflection or transmission by this mirror, the waves emerge from the interferometer. When they emerge, they interfere constructively or destructively. Suppose that the lengths MM_1 and MM_2 differ by d. Then one of the waves must travel an extra distance $2d$, and the condition for the constructive interference is

$$2d = 0, \ \lambda, \ 2\lambda, \ldots \tag{35.4}$$

constructive and destructive interference in interferometer

and for destructive interference it is

$$2d = \tfrac{1}{2}\lambda, \ \tfrac{3}{2}\lambda, \ \tfrac{5}{2}\lambda, \ldots \tag{35.5}$$

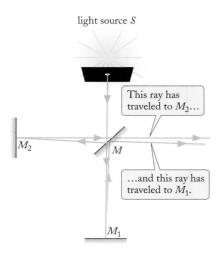

light source S

This ray has traveled to M_2...

...and this ray has traveled to M_1.

M_2

M

M_1

FIGURE 35.9 Paths of rays in a Michelson interferometer. For the sake of clarity, rays are shown reaching the mirrors M_1 and M_2 with a small angle between them; they are actually parallel, and they overlap.

ALBERT ABRAHAM MICHELSON (1852–1931) *American experimental physicist, professor at the Case School of Applied Science and at the University of Chicago. He made precise measurements of the speed of light and used interferometric methods to determine the wavelength of spectral lines in terms of the standard meter. He first performed the "Michelson–Morley" experiment on his own in 1881 and then repeated it several times in collaboration with E. W. Morley, with increasing accuracy. Michelson received the Nobel Prize in 1907.*

To achieve this interference, the mirrors M_1 and M_2 must be very precisely aligned, so that the image of M_1 seen in M is exactly parallel to M_2. The alignment can be achieved by means of adjusting screws on the backs of the mirrors.

The mirror M_1 is usually mounted on a carriage that can be driven along a track by means of an accurately machined screw. If mirror M_1 is slowly moved inward or outward, the interference of the emerging waves will change back and forth between constructive and destructive whenever the mirror is displaced by $\frac{1}{4}$ wavelength—and the intensity of the emerging light will change back and forth between maxima and minima, which are seen as bright and dark fringes. Thus, the displacement of the mirror can be measured very precisely by counting fringes and fractions of a fringe. This measurement expresses the displacement in terms of the wavelength of the light. Modern interferometers used for length measurements, such as that illustrated in Fig. 35.10, are designed to count fringes automatically with a photoelectric sensor. Laboratory interferometers can determine displacements as small as one-millionth of a wavelength. A large-scale instrument, the Laser Interferometer Gravitational Observatory (LIGO), is designed to measure displacements smaller than one-billionth of a wavelength!

Interferometers have played an important role in establishing that the speed of light is independent of the motion of the Earth. This was demonstrated by the famous **Michelson–Morley experiment**, first performed in 1881. The idea behind this experiment is as follows. If light were to propagate in an interplanetary medium in a manner analogous to the propagation of sound in air, then we would expect that the motion of the Earth toward or away from a light wave propagating through the "stationary" interplanetary medium would affect the speed of the light relative to the Earth, just

FIGURE 35.10 Special interferometer used at the International Bureau of Weights and Measures for the comparison of lengths. The interferometer automatically counts the wavelengths when the carriage moves through a given length.

(a)

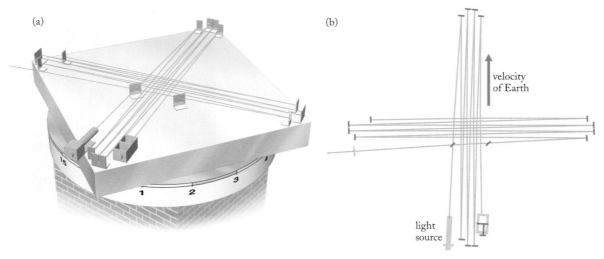

(b)

velocity
of Earth

light
source

FIGURE 35.11 (a) The interferometer used in the experiment of Michelson and Morley. (b) The many mirrors reflect the light beams back and forth several times, increasing the path length of the light. The vertical arm is parallel to the direction of motion of the Earth and the horizontal arm is perpendicular.

as the motion of a train toward or away from a sound wave propagating through the stationary air affects the speed of the sound relative to the train. Such an alteration of the speed of light could be detected with an interferometer by orienting one of the arms parallel to the direction of motion of the Earth and the other arm perpendicular. A difference in the speeds of light c along the arms would entail a difference in the corresponding wavelengths ($\lambda = c/f$) and alter the conditions (35.4) and (35.5) for bright and dark fringes. The easiest way to detect the speed difference is by rotating the interferometer, so the arm that has been parallel to the motion becomes perpendicular and vice versa. If the speeds were different in the two directions, this rotation could shift the fringes from bright to dark or vice versa.

Michelson and Morley found no observable fringe shift in their experiment. However, a small fringe shift could have been masked by experimental errors. Taking into account possible experimental errors, Michelson and Morley cautiously concluded that the effect of the motion of the Earth on the speed of light was no larger than 5 km/s. This value is substantially less than the speed of the Earth around the Sun ($\approx$30 km/s) and proved beyond all reasonable doubt that the propagation of light through space is *not* analogous to the propagation of sound through air. The demonstration that light did not propagate through some stationary interplanetary medium was fundamental to the development of the theory of Special Relativity, discussed in Chapter 36. Figure 35.11 shows Michelson and Morley's interferometer.

✔ Checkup 35.2

QUESTION 1: Since the light rays in Fig. 35.9 reflect on a medium of larger index of refraction, they suffer phase changes of 180°. Does this alter the conditions (35.4) and (35.5)?

QUESTION 2: Since the light rays in Fig. 35.9 pass through the glass of mirror M, they suffer an extra phase change because the index of refraction of the glass differs from that of air. Does this alter the conditions (35.4) and (35.5)?

QUESTION 3: Using a screw mechanism, one mirror of a Michelson interferometer is moved. If the wavelength of the light used is 500 nm, and if one full turn of the screw results in the observation of 400 bright fringes, through what distance was the mirror moved?

(A) 0.05 mm   (B) 0.10 mm   (C) 0.16 mm   (D) 0.20 mm   (E) 0.40 mm

35.3 INTERFERENCE FROM TWO SLITS

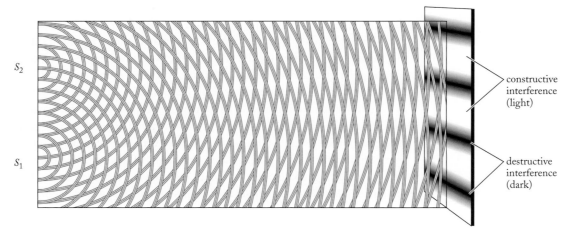

Online
Concept
Tutorial
39

A very clear experimental demonstration of interference effects in light can be performed with two small light sources placed near each other. The light waves spread out from the sources, run into one another, and interfere constructively or destructively, giving rise to a pattern of bright and dark zones (see Fig. 35.12). These interference effects were first discovered by Thomas Young around 1800, and, in conjunction with diffraction effects (see Section 35.5), they settled a long-standing controversy regarding the nature of light. Before Young performed his experiments, some physicists—among them Newton—had argued in favor of a particle nature of light; they had conjectured that light is a stream of particles. But Young's demonstration of interference gave unmistakable evidence in favor of waves. However, the question of what these waves were made of was not settled until much later, when Maxwell formulated his equations and established that light is a wave consisting of electric and magnetic fields.

If the interference pattern produced by two light sources is to remain stationary in space, the two sources must be **coherent**, that is, *they must emit waves of the same frequency* and the same phase (a *constant* phase difference is also acceptable; what is crucial is that the phase difference must not fluctuate in time). Such coherent sources are easily manufactured by aiming a monochromatic wave with plane or spherical wave fronts at an opaque plate with two narrow slits or holes; the waves diverging from the

coherent light

FIGURE 35.12 This diagram by Thomas Young illustrates how the interference of light waves emerging from two small light sources S_1 and S_2 produces a pattern of light and dark zones. Along the directions where the wave crests of the two waves coincide, constructive interference occurs. Along intermediate directions, the wave crests of one wave fall on the wave troughs of the other, giving destructive interference. If viewed from along the edge of the figure, the pattern is more obvious. If the light is intercepted by a screen, these directions of constructive and destructive interference give rise to bright and dark bands, or interference fringes.

two slits are then coherent because they arise from a single original wave (see Fig. 35.13). An extended object, such as an ordinary lightbulb, can be used to illuminate the slits provided it is placed at a *very large* distance. The lightbulb will then effectively act as a point source, illuminating the slits with light waves that consist of a succession of nearly plane wave fronts.[1]

Note that if we use two *separate* lightbulbs as sources, then we will not see any interference pattern. The light from two separate lightbulbs is **incoherent**; it consists of a mixture of a large number of light waves with variable phase differences. We can understand the distinction between coherent and incoherent light sources by means of an analogy with water waves. If we simultaneously drop two large, identical stones into a pond, they act as coherent sources of water waves. The waves from these two sources will interfere constructively at some places, and destructively at others (we can observe the interference effects of the waves by means of bits of cork scattered on the water; on the midline between the sources, the waves interfere constructively, and the bits of cork bob up and down with twice the amplitude they would have had if only one stone had been dropped; at some other places, where the waves interfere destructively, the bits of cork remain at rest). But if, instead of dropping two large stones, we dribble two handfuls of small pebbles into the pond, then the two sources are incoherent. There will then be no distinguishable interference pattern on the surface of the pond—all the bits of cork will bob up and down, more or less at random, although on the average the energy of their bobbing motion will be twice as large as if we had dropped a single handful of pebbles.

In modern practice, a laser is often used to illuminate the two slits, because it provides a very intense plane wave, which makes even faint interference and diffraction effects visible. If the slits in the plate are very narrow, less than one wavelength in width, then the portions of the wave passing through them will diverge from the slits, spreading out in all directions, as illustrated in Fig. 35.13 (the spreading of the wave is a diffraction effect; we will study this in some detail in Section 35.5). This means that the two slits act as two pointlike, coherent sources.

To find the zones of constructive and destructive interference, or the zones of maximum and minimum light intensity, in the space beyond the slits, we need to examine the path difference between the rays from each of the slits. Figure 35.14 shows the light wave incident on the plate from the left and the rays QP and $Q'P$ leading from the slits to the point P on the right. The waves emerging from the slits and reaching P interfere constructively if the difference between the lengths QP and $Q'P$ is zero, or one wavelength, or two wavelengths, etc.; and they interfere destructively if this difference is one half wavelength, or three half wavelengths, five half wavelengths, etc.

We can obtain a simple formula for the angular positions of the maxima and minima if we make the additional assumption that the point P is at a very large distance from the plate (to be precise, QP is assumed to be very large compared with the distance between the slits, $d = QQ'$). If so, then the rays QP and $Q'P$ are nearly parallel, as illustrated in Fig. 35.15. The difference between the lengths QP and $Q'P$ is then the distance $Q'R$; this is the path difference between the two waves arriving at P. This path

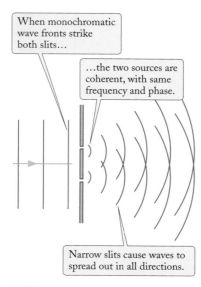

incoherent light

When monochromatic wave fronts strike both slits...

...the two sources are coherent, with same frequency and phase.

Narrow slits cause waves to spread out in all directions.

FIGURE 35.13 A plane light wave strikes a plate with two very narrow slits. The slits act as two coherent light sources. Light waves diverge from the two slits and interfere.

---

[1] If the lightbulb is placed too close to the slits, then light waves arrive at the slits from several directions at once, and this tends to wash out the interference pattern because waves of different directions of incidence will produce overlapping zones of brightness and darkness. For convenience, the lightbulb is sometimes placed fairly near the slits; but it must then be covered with a shield perforated by a single pinhole, so the emerging light comes from just one portion on the surface of the lightbulb; this again makes the light source into a point source.

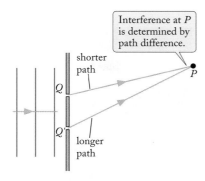

FIGURE 35.14 The waves reaching the point P have different path lengths QP and $Q'P$.

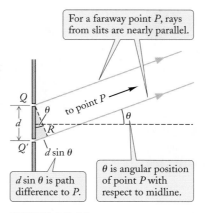

FIGURE 35.15 If P is far away, QP and $Q'P$ are nearly parallel. The lengths QP and $Q'P$ then differ by $d \sin\theta$.

difference $Q'R$ is the short side of the right triangle $QQ'R$. Since the distance d between the slits is the hypotenuse of this triangle, we can express $Q'R$ in terms of d and the angle θ:

$$[\text{path difference}] = Q'R = d \sin\theta \qquad (35.6)$$

The angle θ in the triangle equals the angle between the line toward P and the perpendicular midline of the plate (see Fig. 35.15); thus, θ represents the angular position of the point P.

Our condition for constructive interference is then that the path difference $Q'R$ equals zero, one, two, etc., wavelengths:

$$d \sin\theta = 0, \ \lambda, \ 2\lambda, \ \ldots \qquad (35.7)$$

maxima of interference pattern

or

$$d \sin\theta = m\lambda \quad \text{where } m = 0, \ 1, \ 2, \ \ldots \qquad (35.8)$$

Likewise, the condition for destructive interference is

$$d \sin\theta = \tfrac{1}{2}\lambda, \ \tfrac{3}{2}\lambda, \ \tfrac{5}{2}\lambda, \ \ldots \qquad (35.9)$$

minima of interference pattern

For a distant point, these equations give the angular positions of the maxima and the minima of the interference pattern.

The zones of high intensity have the shape of beams fanning out in the space beyond the slits. The central beam (corresponding to $\sin\theta = 0$, or $\theta = 0$) is called the *central maximum*, or the *zero-order maximum*. The beams on each side of this (corresponding to $d \sin\theta = \lambda$, where θ is measured either up or down) are called the *first-order maxima*, etc. Thus, the integer m in Eq. (35.8) is called the *order number*. The beams of high intensity are separated by lines of zero intensity; these are nodal lines, analogous to the nodal points produced by destructive interference in the standing wave on a string (see Section 16.4). The lines of zero intensity between the zero-order maximum and the first-order maxima (corresponding to $d \sin\theta = \tfrac{1}{2}\lambda$) are called the *first-order minima*, etc.

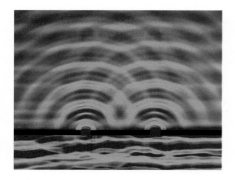

FIGURE 35.16 Interference between water waves spreading out from two coherent pointlike sources in a ripple tank.

(a)

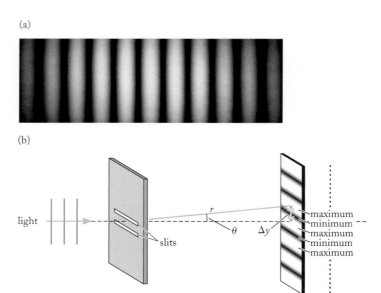

(b)

FIGURE 35.17 (a) A photographic film placed beyond two illuminated narrow slits records a regular pattern of bright and dark fringes. (b) Placement of the photographic film beyond the slits. For a small angle θ, the vertical distance Δy is approximately $\Delta y = r\theta$.

Figure 35.16 is a photograph of such a pattern of fanlike beams produced by the interference of water waves spreading out from two pointlike sources [the formulas (35.7) and (35.9) apply to water waves and to any other kinds of waves]. With light waves, it is difficult to photograph the entire pattern of beams at once; instead we must be content with the photograph in Fig. 35.17, which shows the pattern of bright and dark fringes recorded on a photographic film that intercepts the beam at some fixed distance beyond the slits.

EXAMPLE 2   Two narrow slits separated by a distance of 0.12 mm are illuminated with light of wavelength 589 nm from a sodium lamp. What is the angular position of the first-order maximum in the interference pattern? If the light is intercepted by a photographic film placed 2.0 m beyond the slits, what is the distance on the film between this maximum and the central maximum?

SOLUTION: With $d = 1.2 \times 10^{-4}$ m and $\lambda = 5.89 \times 10^{-7}$ m, Eq. (35.7) gives

$$\sin\theta = \frac{\lambda}{d} = \frac{5.89 \times 10^{-7}\,\text{m}}{1.2 \times 10^{-4}\,\text{m}} = 4.9 \times 10^{-3}$$

We recall that, for small angles, the sine (or the tangent) of the angle is approximately equal to the angle in radians. Thus,

$$\theta \approx 4.9 \times 10^{-3}\ \text{radian}$$

The distance between the points with $\theta = 0$ and $\theta = 4.9 \times 10^{-3}$ radian on the photographic film is approximately equal to the radial distance r from the slit to the film (see Fig. 35.17b) multiplied by the angle of 4.9×10^{-3} radian:

$$\Delta y \approx r\theta$$

$$= 2.0\ \text{m} \times 4.9 \times 10^{-3} = 9.8 \times 10^{-3}\ \text{m} = 0.98\ \text{mm}$$

Thus, the maxima are separated by nearly 1 cm.

PROBLEM-SOLVING TECHNIQUES	TWO-SLIT INTERFERENCE

- In general, the path difference between the two waves arriving at the observation point P is the difference in the length of the two long sides of the triangle in Fig. 35.14. This path difference can always be calculated by trigonometry. But if the point P is at a very large distance compared with the separation d of the sources, then the path difference takes the simple form $d \sin \theta$.

- Whenever the angular position θ of the observation point P is small, then we can use the approximation $\sin \theta \approx \theta$. Furthermore, the transverse displacement of the observation point from the midline is then approximately $\Delta y \approx r\theta$.

We can calculate the intensity distribution as a function of angle by examining the superposition of the waves contributed by the two slits. The wave propagating outward from each of the slits is of the form [compare Eq. (33.34)]

$$E \propto \cos\left(\omega t - \frac{2\pi r}{\lambda}\right)$$

where ω is the frequency of the wave and r is the distance from the slit to the point P. For the wave emerging from the upper slit, the distance r is

$$r_1 = r_0 - \frac{d}{2}\sin\theta$$

where r_0 is the distance measured from the midpoint between the two slits (see Fig. 35.18). And for the wave emerging from the lower slit, the distance r is

$$r_2 = r_0 + \frac{d}{2}\sin\theta$$

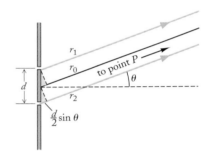

FIGURE 35.18 Rays from the slits to the point P.

At our faraway point P, each wave has essentially the same amplitude E_0. Hence the electric fields contributed by the upper and the lower slits are, respectively,

$$E_1 = E_0 \cos\left(\omega t - \frac{2\pi r_0}{\lambda} + \frac{\pi d}{\lambda}\sin\theta\right)$$

and

$$E_2 = E_0 \cos\left(\omega t - \frac{2\pi r_0}{\lambda} - \frac{\pi d}{\lambda}\sin\theta\right)$$

The superposition of these two electric fields then gives

$$E = E_1 + E_2$$
$$= E_0\left[\cos\left(\omega t - \frac{2\pi r_0}{\lambda} + \frac{\pi d}{\lambda}\sin\theta\right) + \cos\left(\omega t - \frac{2\pi r_0}{\lambda} - \frac{\pi d}{\lambda}\sin\theta\right)\right]$$

With the trigonometric identity $\cos(\alpha + \beta) + \cos(\alpha - \beta) = 2\cos\alpha\cos\beta$, (see Appendix 14) this becomes

$$E = 2E_0 \cos\left(\omega t - \frac{2\pi r_0}{\lambda}\right)\cos\left(\frac{\pi d}{\lambda}\sin\theta\right) \qquad (35.10)$$

The factor $\cos(\omega t - 2\pi r_0/\lambda)$ is the usual oscillating function of space and time, characteristic of a wave propagating outward. The factor $2E_0 \cos[(\pi d/\lambda) \sin \theta]$ is the amplitude of the wave and depends on the position angle θ.

To determine the interference pattern, we are interested in the intensity as a function of position. The intensity of a wave is proportional to E^2. Thus the maximum intensity of the summed wave (35.10) is 4 times the intensity of the wave arriving from either slit. Since we are interested only in the time-average intensity, we can replace the factor $\cos^2(\omega t - 2\pi r_0/\lambda)$ from Eq. (35.10) by its time-average value $\frac{1}{2}$. The dependence of the intensity I on the position angle is then given by the remaining factor:

intensity for two-slit interference pattern

$$I = I_{max} \cos^2\left(\frac{\pi d}{\lambda} \sin\theta\right) \tag{35.11}$$

where I_{max} is the intensity at the center of the interference pattern, at $\theta = 0$. This formula describes the intensity distribution produced by the two slits. Figure 35.19 is a plot of the intensity as a function of the angle θ. The plot represents the intensity of light found at different angles at some constant distance r_0 from the midpoint of the slits. As expected, the intensity is maximum when $(d/\lambda) \sin \theta = 0, 1, 2, 3$, etc., and it is minimum (zero) when $(d/\lambda) \sin \theta = \frac{1}{2}, \frac{3}{2}, \frac{5}{2}$, etc.

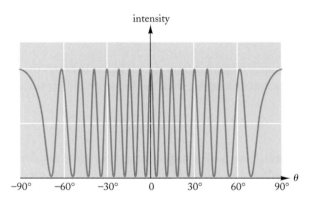

FIGURE 35.19 Intensity as a function of the angle θ for light arriving at a screen placed beyond two slits. The distance between the slits is $d = 8\lambda$.

EXAMPLE 3   Consider the intensity in the two-slit interference pattern on a distant screen at an angle midway between a maximum and a minimum. How does this intensity compare with the intensity at the maximum?

SOLUTION: For the usual experimental arrangement, we can assume, for a distant screen, that the angle θ is small, so the sine of the angle is approximately equal to the angle. The angle midway between the central maximum and the first-order minimum then has $d \times \theta = \frac{1}{4}\lambda$, which is midway between the values $d \times \theta = 0$ (central maximum) and $d \times \theta = \frac{1}{2}\lambda$ (first minimum). According to Eq. (35.11), the intensity at this position is

$$[\text{intensity}] \propto \cos^2\left(\frac{\pi}{\lambda} d \sin\theta\right) \approx \cos^2\left(\frac{\pi}{\lambda} d \times \theta\right) = \cos^2\left(\frac{\pi}{\lambda} \frac{\lambda}{4}\right) = \cos^2\frac{\pi}{4} = \frac{1}{2}$$

Thus, the intensity at this point is one-half the intensity at the central maximum.

✔ Checkup 35.3

QUESTION 1: If, in Example 2, we *decrease* the distance between the slits, does the distance between the first maximum and the central maximum decrease or increase?

QUESTION 2: Suppose that the distance d between the slits equals λ. Describe the intensity distribution for this case.

QUESTION 3: If the first-order maximum is to be at $\theta = 45°$, what wavelength is required for a given slit separation d?

QUESTION 4: Consider a double-slit experiment with $d = 5\lambda$. At what angular position do the third-order minima occur?

    (A) 0°   (B) 30°   (C) 45°   (D) 60°   (E) 90°

QUESTION 5: In a double-slit experiment, let I_{max} be the intensity at the peak of the central maximum. If the path difference to a point on the screen is $\lambda/6$, what is the intensity at that point?

    (A) 0   (B) $\frac{1}{4}I_{max}$ (C) $\frac{1}{2}I_{max}$ (D) $\frac{3}{4}I_{max}$ (E) I_{max}

35.4 INTERFERENCE FROM MULTIPLE SLITS

Online
Concept
Tutorial

It is easy to see that Eq. (35.8) also applies to the case of a plane wave incident on multiple slits. For instance, Fig. 35.20 shows an opaque plate with three evenly spaced slits. If waves emerging from *adjacent* slits interfere constructively, then the waves emerging from *all three* slits interfere constructively. Hence the condition for maximum intensity is the same as Eq. (35.8),

$$d\sin\theta = m\lambda \qquad \text{where } m = 0,\ 1,\ 2,\dots \qquad (35.12)$$

and this same equation also applies to more than three slits. However, the condition for minimum intensity is *not* Eq. (35.9). For instance, in the case of three slits, destructive interference of waves from adjacent slits will lead to cancellation of the wave originating from one pair of slits, but the wave from the remaining slit will not experience cancellation. Thus, the condition for minima is somewhat more complicated; a technique to determine the detailed behavior between the maxima given by Eq. (35.12) is discussed later in this section. It turns out that between the maxima given by Eq. (35.12), called **principal maxima**, there can be several minima and several **secondary maxima**. Figure 35.21 is a photograph illustrating the strong principal maxima and the weaker secondary maxima of the diffraction pattern of three slits. We will consider secondary maxima in more detail at the end of this section. For now, we will concentrate our attention on the principal maxima, because the secondary maxima are quite weak, especially if the number of slits is large.

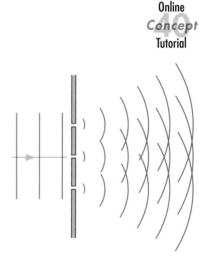

FIGURE 35.20 A plane wave strikes a plate with three very narrow slits.

principal maxima

secondary maxima

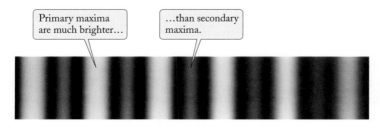

Primary maxima are much brighter…

…than secondary maxima.

FIGURE 35.21 A photographic film placed beyond the slits shows the strong principal maxima and the weaker secondary maxima between them.

diffraction gratings

Arrangements of large numbers of slits are called **diffraction gratings** or, simply, **gratings**. When the number of slits is large, the principal maxima become very sharply peaked and well separated. Gratings are commonly used in spectroscopy laboratories to analyze light into its colors. If a light beam containing several wavelengths passes though such a grating, the maxima for these different wavelengths will form beams at different angles—according to Eq. (35.12), the angle θ increases with wavelength, and hence the beams of long-wavelength light will be found at larger angles than the beams of short-wavelength light. Thus, the grating separates the light according to color and produces a spectrum in much the same way that a prism does. There is, however, one important difference between the spectra formed by a prism and by a grating: in the prism the long-wavelength light (red) experiences the least deflection; in the grating the long-wavelength light experiences the most deflection.

The grating will produce one complete spectrum for each of the possible alternatives listed for the right side in Eq. (35.12)—that is, each principal maximum, except the central maximum, gets spread out by color. These spectra are called, in turn, the first-order spectrum [$m = 1$, $d \sin \theta = \lambda$ in Eq. (35.12)], the second-order spectrum ($m = 2$, $d \sin \theta = 2\lambda$), and so on. Sometimes these spectra overlap; for example, the red end of the second-order spectrum may show up at the same angle as the blue end of the third-order spectrum (see Fig. 35.22).

FIGURE 35.22 First-, second-, and third-order spectra ($m = 1$, $m = 2$, and $m = 3$) of hydrogen light produced by a system of N slits. The lines correspond to the principal maxima; the secondary maxima are weak and can be ignored. Each spectrum consists of violet, blue violet, blue, and red spectral lines (compare Fig. 34.29b). The pattern of spectral lines for negative values of θ (negative m) is similar.

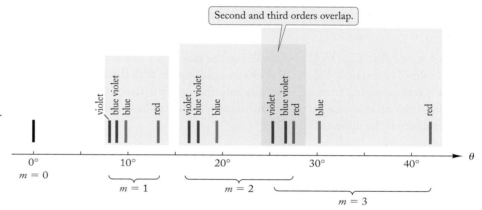

When atoms of iron are heated, they emit numerous spectral lines, among which are spectral lines of wavelengths $\lambda = 500.57\,\text{nm}$ and $\lambda = 500.61\,\text{nm}$. Suppose we analyze the spectrum of iron with a grating with $d = 4.0 \times 10^{-6}$ m. What is the angular separation between the maxima for these wavelengths in the first-order spectrum produced by this grating?

EXAMPLE 4

SOLUTION: The first-order maxima occur near

$$\sin \theta = \frac{\lambda}{d} = \frac{500.6 \times 10^{-9}\,\text{m}}{4.0 \times 10^{-6}\,\text{m}} \approx 0.13$$

This is a sufficiently small value that the sine of the angle is approximately equal to the angle in radians. Hence the equation $d \sin \theta = \lambda$ for the angular position of the maxima in the first-order spectrum becomes

$$d \times \theta \approx \lambda$$

Thus, the angle is proportional to the wavelength, and the difference in angle is therefore proportional to the difference in wavelength:

$$d\,\Delta\theta = \Delta\lambda$$

With $\Delta\lambda = 500.61$ nm $- 500.57$ nm $= 0.04$ nm, we find

$$\Delta\theta = \frac{\Delta\lambda}{d} = \frac{0.04 \times 10^{-9} \text{ m}}{4.0 \times 10^{-6} \text{ m}} = 1 \times 10^{-5} \text{ radian}$$

The best gratings have a very large number of slits, because this produces sharp, narrow beams and enhances the ability of the grating to discriminate between spectral lines of nearly identical wavelengths. Since it is difficult to cut a large number of slits in an opaque plate, gratings are usually manufactured by cutting fine parallel grooves in a glass or metal surface with a diamond stylus guided by a special ruling machine. When illuminated by a plane light wave, the edges of the grooves act as coherent light sources, in much the same way as slits. High-quality gratings used in spectroscopy have 100 000 or more grooves with distances of about 10^{-6} m between them.

You can verify the effects of a grating on light by a simple experiment with a CD or an LP record; the former is illustrated in the chapter photo. The tracks or grooves on these disks make them behave like crude **reflection gratings**. You can see spectral colors with these crude gratings if you let them reflect the light from some narrow light source, such as sunlight coming through a crack between windows or light from the edge of a bright lamp. Compact discs have tighter grooves than LP records, and they give a more pronounced spread of the spectral colors. When you look across the surface of such a disk, you see light from a range of angles, and thus light of many colors, at the same time. Also, the grooves are curved, and so you see several patterns in different directions. The left CD in the chapter photo shows a continuous spectrum of colors, such as in sunlight or from the hot filament of a lightbulb, whereas the right CD predominantly shows a discrete spectrum, characteristic of the individual atoms of a particular element, in this case hydrogen in a gas discharge tube.

reflection gratings

EXAMPLE 5   The patterns on the CDs in the photo at the beginning of the chapter result from interference. For the CD on the right side, the light is normally incident on the CD and the third-order blue interference maximum ($\lambda = 486$ nm) is seen when viewing at $60°$ from the normal. What is the effective slit separation d of the compact-disc grating?

SOLUTION: Since the light is normally incident, it behaves like a wave normally incident on slits. The third-order interference maximum is then given by Eq. (35.12) with $m = 3$,

$$d \sin\theta = 3\lambda$$

so the slit spacing is

$$d = \frac{3\lambda}{\sin\theta} = \frac{3 \times 486 \text{ nm}}{\sin 60°} = 1.7 \times 10^{-6} \text{ m} = 1.7 \ \mu\text{m}$$

COMMENT: In science and engineering practice, interference phenomena are often used to determine spacings, as in this example. Thus, an X-ray diffractometer or an electron diffraction device each obtain an interference pattern by reflecting a wave from a "grating" of atoms (in a crystal or on a surface, respectively); the measured angles of the interference maxima then determine atomic spacings that are too small to see.

Radio telescopes, consisting of a regularly spaced array of antennas, act as "gratings" for radio waves. The same formula (35.12) for the angular directions of the maxima applies to these as to ordinary gratings used with light. Of course, there is a difference between the operation of a radio telescope and that of an ordinary grating: in the former the radio waves from a distant source *enter* the antennas and interfere constructively or destructively within the radio receiver, whereas in the latter the light waves *emerge* from the slits and travel to a distant point where they interfere. Nevertheless, the condition for, say, maximum intensity can still be expressed in terms of the direction of the entering or emerging wave by Eq. (35.12), because this condition hinges only on the phase relationships among the waves, and these relationships are independent of the direction of propagation (for instance, in Fig. 35.15, if the waves traveling away from the slits in some direction reach a distant point in phase, then waves traveling in the reverse direction toward the slits from this point will arrive at the slits in phase).

EXAMPLE 6   One branch of the VLA (Very Large Array) radio telescope at Socorro, New Mexico, has nine antennas arranged on a straight line with distances of 2.0 km between one antenna and the next (see Fig. 35.23). These antennas are all connected to a single radio receiver by waveguides of equal length; the receiver then registers a maximum intensity if the radio waves incident on all antennas are in phase. Suppose that radio waves of wavelength 21 cm from a pointlike quasar in the sky strike this radio telescope. When the quasar is at the zenith, the intensity in the radio receiver is maximum. What angular displacement of the quasar from the zenith will also result in maximum intensity?

SOLUTION: Let us assume (and later verify) that the desired angular displacement may be small, so that we can use the approximation that for small angles, the sine of the angle equals the angle in radians. Thus, Eq. (35.12) gives, for the maximum nearest the central maximum,

$$d \times \theta = \lambda \tag{35.13}$$

Hence

$$\theta = \frac{\lambda}{d} = \frac{0.21 \text{ m}}{2.0 \times 10^3 \text{ m}} = 1.1 \times 10^{-4} \text{ radian}$$

The angular displacement at the first maximum is indeed small, so our assumption was justified.

FIGURE 35.23 The VLA radio telescope. The distance between the antennas can be adjusted by rolling them along tracks. Here the antennas are shown close together.

For a quantitative analysis of the interference due to multiple slits, we can use the technique of **phasors**, previously introduced in Chapter 32. A phasor is a rotating vector in a fictitious two-dimensional space, useful because any oscillating function can be represented as the x component of a rotating vector. The rotating vector shown in Fig. 35.24 has length equal to the amplitude E_0 of the wave disturbance at some point, and at any instant makes an angle ωt with respect to the x axis. As is evident in Fig. 35.24, the x component then indeed equals the instantaneous value of $E = E_0 \cos \omega t$. The utility of the technique can be seen in Fig. 35.25, where two waves of amplitude E_0 with a phase difference δ between them are summed vectorially to determine the amplitude of the resultant wave. To do this, we merely plot the corresponding phasors tail to tip, with a relative angle equal to the phase difference. From the geometry of Fig. 35.25, we see that the resultant wave has amplitude

$$E = 2E_0 \cos\left(\frac{\delta}{2}\right) \tag{35.14}$$

For $\delta = 0$, the resultant is $E = 2E_0$, which corresponds to constructive interference. As the phase difference δ increases, the resultant goes through a series of minima ($\delta = \pi, 3\pi, 5\pi, \ldots$) and maxima ($\delta = 2\pi, 4\pi, 6\pi, \ldots$) as the second phasor alternately cancels or adds to the first phasor. We can apply this to the double-slit experiment. The phase difference between the two waves expressed as a fraction of 2π is the same as the path difference expressed as a fraction of λ:

$$\frac{\delta}{2\pi} = \frac{d \sin\theta}{\lambda} \tag{35.15}$$

It is easy to verify that with this phase difference δ, Eq. (35.14) gives the same intensity as a function of the position angle θ as obtained previously in Eq. (35.11). Thus either the algebraic result (35.11) or the geometric result (35.14) gives the two-slit intensity pattern with comparable ease.

For multiple, nonidentical, or other complex slit arrangements, the phasor technique can readily provide insight into the locations of the minima and the primary and secondary maxima, as well as the detailed intensity pattern. For example, consider interference from three identical slits, discussed above (see Figs. 35.20 and 35.21). In this case we sum three phasors, as shown in Figs. 35.26a–f. For equally spaced slits, the first and second phasors have the same phase difference δ between them as the second

phasor

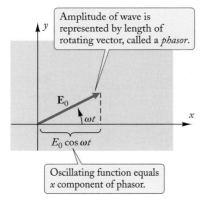

FIGURE 35.24 A rotating vector, or phasor, has magnitude equal to the amplitude of the wave it represents.

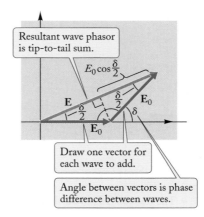

FIGURE 35.25 To add two waves of amplitude E_0, the corresponding phasors are drawn tip to tail and the angle between them is set equal to the phase difference of the waves.

and third phasors [each difference given by Eq. (35.15)]. As we increase our angular position θ, the phase difference δ increases, as usual. As shown in Fig. 35.26, the net amplitude E begins at $3E_0$ (Fig. 35.26a) and decreases smoothly with increasing phase difference until $E = 0$ at $\delta = 2\pi/3$ (Fig. 35.26d), where the phasors form an equilateral triangle. Beyond $\delta = 2\pi/3$, the intensity of the three-slit pattern then increases smoothly until $E = E_0$ at $\delta = \pi$ (Fig. 35.26f), a secondary maximum. Then the sequence of resultant amplitudes continues in reverse, through another minimum at $\delta = 4\pi/3$, until the first-order principal maximum ($E = 3E_0$) is reached at $\delta = 2\pi$; thereafter, the entire sequence repeats for every 2π increase of phase difference between adjacent slits. The resulting intensity pattern (proportional to the square of the amplitude) is shown in Fig. 35.27a (a photograph of this pattern is shown in Fig. 35.21). For the three-slit pattern, there is one secondary maximum between the primary maxima; in general, there will be one more secondary maximum for every additional slit as in Fig. 35.27b.

The same technique of examining the resultant amplitude as the phase difference between adjacent slits increases can be applied to any number of slits, even to unequal slits, which produce waves of different amplitudes. Similarly, for slits that are not evenly spaced, the technique can be applied by maintaining the phase difference between different adjacent pairs of phasors in direct proportion to the spacing between the corresponding pairs of adjacent slits.

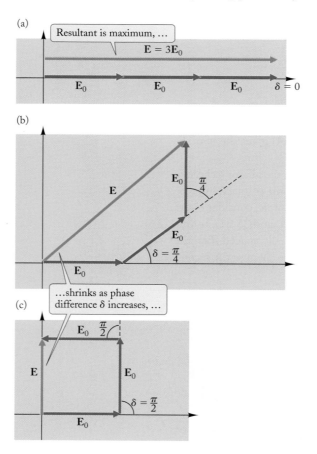

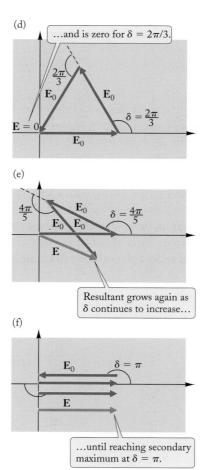

FIGURE 35.26 Phasor diagrams for three identical evenly spaced slits. The phase difference $\delta = (2\pi/\lambda)\, d \sin\theta$ increases as we consider the resultant for increasing angular position θ. The resultant amplitude decreases from a primary maximum to zero (a–d) and then increases from zero to a secondary maximum (d–f).

(a)

Between principal maxima, for three slits...

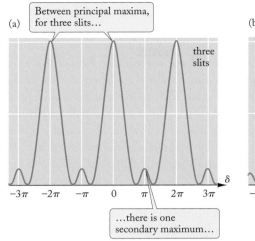

three slits

-3π -2π $-\pi$ 0 π 2π 3π

...there is one secondary maximum...

(b)

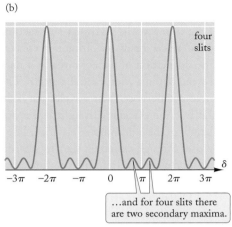

four slits

-3π -2π $-\pi$ 0 π 2π 3π

...and for four slits there are two secondary maxima.

FIGURE 35.27 Intensity as a function of the phase difference δ between adjacent slits for light arriving at a screen placed beyond (a) three slits, and (b) four slits. For three slits, there is one secondary maximum between principal maxima; for four slits, there are two. (Note that these plots can be converted into plots of intensity as a function of the angle θ by using $\delta = (2\pi/\lambda)\sin\theta$.)

For three slits, we saw that the first interference minimum is at $\delta = 2\pi/3$, where the phasors form an equilateral triangle (see Fig. 35.26d). This corresponds to a path difference between adjacent slits of $d\sin\theta = \lambda/3$. Similarly, for any number N of identical, evenly spaced slits, the first minimum occurs when the phasors form an N-sided regular polygon. For N slits, the angle θ of the first minimum is then given by

$$d\sin\theta = \frac{\lambda}{N} \qquad (35.16)$$

first minimum for N slits

Thus for large N, such as with a grating, the principal maxima are extremely narrow.

 ## Checkup 35.4

QUESTION 1: In Fig. 35.22, a red spectral line appears between a blue and a blue violet line. How can this be?

QUESTION 2: In Example 4 we calculated the angular separation between two spectral lines in the first-order spectrum produced by a grating. Would the separation between these spectral lines be larger or smaller in the second-order spectrum?

QUESTION 3: Consider the three-slit phasor diagrams of Fig. 35.26. What is the amplitude of the resultant when the phase difference between adjacent slits is $\pi/2$ (90°)?

QUESTION 4: Consider two slits, but now the first is narrow and produces a wave of amplitude E_0 and the second is wider and produces a wave of amplitude $2E_0$. Let I_0 be the intensity due to the first slit alone. Using phasors, answer the following. What is the intensity of each maximum? What is the intensity at each minimum (it is not zero)? What is the intensity half way between a maximum and a minimum?

QUESTION 5: For five identical, evenly spaced slits, the first interference minimum occurs when the phase difference δ between adjacent slits is equal to:

(A) $\pi/10$ (B) $\pi/5$ (C) $2\pi/5$ (D) $\pi/2$ (E) 2π

QUESTION 6: For six identical, evenly spaced slits, how many secondary interference maxima occur between each adjacent pair of primary maxima?

(A) 1   (B) 2   (C) 3   (D) 4   (E) 5

35.5 DIFFRACTION BY A SINGLE SLIT

Like any wave, light displays diffraction effects—it bends and spreads around obstacles. We can easily see diffraction effects in water waves. For instance, Fig. 35.28 shows the diffraction of water waves by a narrow aperture; behind the aperture, the waves spread out in a wide, fanlike pattern. The diffraction of these water waves is quite pronounced because their wavelength is fairly large; it is about as large as the aperture. Diffraction effects with light waves are much harder to observe, because the wavelength of the light waves is so small. The diffraction becomes observable only when light passes through extremely narrow slits, or when it strikes extremely small opaque obstacles, or when we examine the fine detail at the boundaries of shadows produced by sharp edges (Fig. 35.29).

Diffraction effects with light were first noticed in the seventeenth century. Nevertheless, Newton and some of his contemporaries held to the belief that light is of a corpuscular nature, and consists of a stream of particles. The importance of diffraction effects in establishing the wave nature of light was not fully appreciated until 1818, when Augustin Fresnel mathematically formulated the theory of diffraction and demonstrated that diffraction is a general and distinctive characteristic of waves. The brilliant success of this theory finally convinced physicists that light is a wave. In this section we will discuss the simple case of diffraction by a single narrow slit.

Figure 35.30 shows a plane light wave approaching a slit in an opaque plate. To find the light distribution in the space beyond the slit, we use the following prescription, called the **Huygens–Fresnel Principle**:

Huygens–Fresnel Principle

Pretend that each point of the wave front reaching the slit can be regarded as a point source of light emitting a spherical wave; the net wave in the region beyond the slit is simply the superposition of all these waves.

This prescription has some obvious similarities to Huygens' Construction (Section 34.1). However, the latter is merely a geometric construction for finding the successive positions of the wave fronts and does not yield any information about the distribution of intensity in different directions, whereas the aim of our new prescription is precisely the calculation of this distribution of intensity. Although the new prescription has a strong intuitive appeal, it it is not all that easy to justify rigorously. Nevertheless, it gives the right answer, or essentially the right answer; this can be

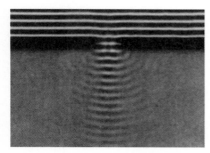

FIGURE 35.28 Diffraction of water waves in a ripple tank. Beyond the narrow aperture, the waves spread out in a fanlike pattern.

FIGURE 35.29 Diffraction of light around a razor blade. The diffracted light generates a complex pattern of fine fringes at the edges of the shadow. This photograph was prepared by illuminating the razor blade with a distant light source and throwing its shadow on a distant screen.

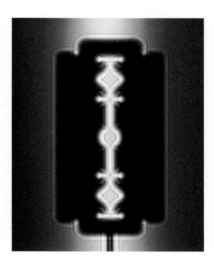

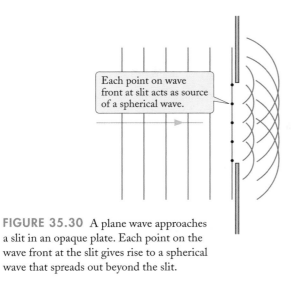

FIGURE 35.30 A plane wave approaches a slit in an opaque plate. Each point on the wave front at the slit gives rise to a spherical wave that spreads out beyond the slit.

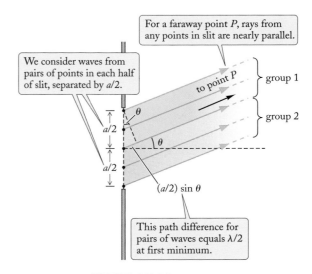

FIGURE 35.31 Rays from points in the slit to the point *P*. The path difference between the uppermost ray and the middle ray is $(a/2)\sin\theta$.

shown by a somewhat more sophisticated mathematical argument based on Maxwell's equations.

We can use the Huygens–Fresnel Principle to find the positions of the minima in the diffraction pattern of the single slit. Figure 35.31 shows some rays associated with waves that spread out from the points of a wave front at the slit. All the rays lead to a point *P* in the space beyond the slit. As in our calculation for the case of two slits, we will assume that *P* is very far away, so all the rays are nearly parallel. We can divide these rays into two equal groups: those that come from the upper part of the slit, and those that come from the lower. Rays from the first group have a shorter distance to travel to the point *P* than rays from the second group. Consider the uppermost ray from the first group and the uppermost ray from the second group. The path difference between these is $(a/2)\sin\theta$, where *a* is the width of the slit (see Fig. 35.31). If this path difference equals $\frac{1}{2}\lambda$, the two rays will interfere destructively. Furthermore, pairs of rays that originate at an equal distance below each of the two uppermost rays in the two groups will also interfere destructively. This establishes that all the waves cancel in pairs when

$$\frac{a}{2}\sin\theta = \frac{1}{2}\lambda$$

or

$$a\sin\theta = \lambda \tag{35.17}$$

This is the condition for the first single-slit diffraction minimum.

To find the next minimum, we divide the rays into four equal groups and consider the uppermost rays from the first and second groups (see Fig. 35.32). These rays and other pairs of rays will interfere destructively if their path difference equals $\frac{1}{2}\lambda$:

$$\frac{a}{4}\sin\theta = \frac{1}{2}\lambda \tag{35.18}$$

or

$$a\sin\theta = 2\lambda \tag{35.19}$$

AUGUSTIN FRESNEL (frenel) (1788–1827) *French physicist and engineer. His brilliant experimental and mathematical investigations firmly established the wave theory of light. Fresnel was commissioner of lighthouses, for which he designed large segmented lenses (Fresnel lenses) as a replacement for systems of mirrors.*

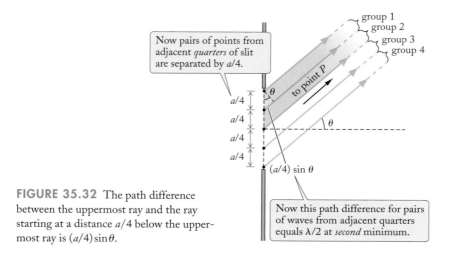

Now pairs of points from adjacent *quarters* of slit are separated by $a/4$.

$a/4$

$a/4$

$a/4$

$a/4$

$(a/4)\sin\theta$

to point P

group 1
group 2
group 3
group 4

Now this path difference for pairs of waves from adjacent quarters equals $\lambda/2$ at *second* minimum.

FIGURE 35.32 The path difference between the uppermost ray and the ray starting at a distance $a/4$ below the uppermost ray is $(a/4)\sin\theta$.

By continuing this argument, we find a general condition for single-slit diffraction minima:

minima for diffraction pattern

$$a\sin\theta = \lambda,\ 2\lambda,\ 3\lambda,\ \ldots \qquad (35.20)$$

As regards the maxima, there is of course a strong central maximum, at $\theta = 0$, where the path lengths from all of the points in the slit to a distant point on the axis are the same. The secondary maxima cannot be found by any simple argument; we can calculate their locations and heights from the intensity relation derived later in this section. Roughly, their position is halfway between the successive minima. On a photographic film placed at some distance, the maxima and minima show up as a pattern of bright and dark fringes (see Fig. 35.33). The central maximum is much brighter than the first secondary maxima, which are much brighter than the next secondary maxima, and so on.

FIGURE 35.33 A photographic film placed beyond an illuminated single slit records a strong central maximum and successively weaker secondary maxima.

EXAMPLE 7   Equation (35.20) applies not only to light waves, but also to radio waves and other waves. Suppose that radio waves from a TV transmitter with a wavelength of 0.80 m strike the wall of a large building. In this wall there is a very wide window with a height of 1.4 m (a horizontal slit; see Fig. 35.34). The wall is opaque to radio waves, and the window is transparent. What is the angular width of the central maximum of the diffraction pattern formed by the waves inside the building?

SOLUTION: According to Eq. (35.20), the first minimum is at an angle such that

$$a\sin\theta = \lambda$$

which yields

$$\sin\theta = \frac{\lambda}{a} = \frac{0.80\text{ m}}{1.4\text{ m}} = 0.57$$

Our calculator tells us that the angle θ is then 35°. The central maximum extends from $-35°$ to $+35°$; that is, the width is 70°.

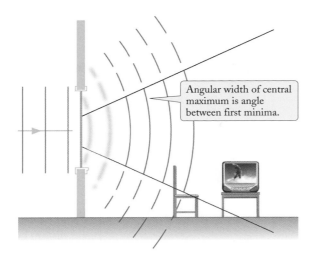

Angular width of central maximum is angle between first minima.

FIGURE 35.34 Diffraction of TV waves at a window.

For a given wavelength, the relation $a \sin\theta = \lambda$ implies that the angle at which the first minimum occurs *increases* when the slit width *a decreases*. Thus a wide slit produces a narrow intensity peak on a distant screen, but for a narrow slit, the pattern spreads out over the screen. This is the fundamental diffraction phenomenon: as the size of the opening or obstruction becomes small compared with a wavelength, the bending and spreading of the waves increases. The angular half-width of the central peak is given by the formula for the first minimum, as in Example 7:

$$\sin\theta = \frac{\lambda}{a} \tag{35.21}$$

The quantity $\sin\theta$ cannot be greater than 1; if $a < \lambda$, then there are no minima, that is, the central peak is so wide that it covers all angles.

To calculate the complete intensity distribution as a function of angle, we must evaluate the superposition of all the waves propagating outward from all the points within the slit (see Fig. 35.35). We divide the slit into a large number k of point sources and sum the electric fields the waves contribute at a distant point P, as previously. Each wave from each point source can be represented by a phasor of length E_0. Figure 35.36a shows the phasor diagram for the central maximum, where all the phasors are in phase and the resultant field is $E_{max} = kE_0$. Figure 35.36b shows the situation away from the central maximum, where the total phase difference between the first and last phasors is ϕ. Because successive points in the slit contribute waves of the same amplitude and equal phase increments, their phasors form a polygonal arc, which we can approximate as an arc of a circle of radius R. The length of the arc is the same total length, E_{max}, as in Fig. 35.36a; the length of the resultant is E. Because the radial lines labeled R are perpendicular to the circular arc, the geometry of Fig. 35.36b shows that the arc subtends the phase angle ϕ. Since angle is defined as arc length over radius, we have

$$\phi = \frac{E_{max}}{R} \tag{35.22}$$

From either of the right triangles formed by the bisector of the angle ϕ we have

$$\sin\left(\frac{\phi}{2}\right) = \frac{E/2}{R} \tag{35.23}$$

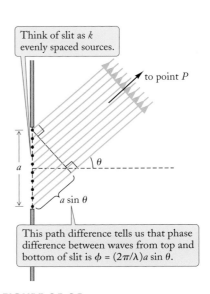

Think of slit as k evenly spaced sources.

to point P

This path difference tells us that phase difference between waves from top and bottom of slit is $\phi = (2\pi/\lambda)a \sin\theta$.

FIGURE 35.35 A single slit considered to be a large number k of point sources.

(a)

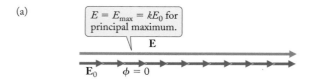

FIGURE 35.36 Phasor diagram for a single slit. (a) At the central maximum, all k phasors are in phase, and the net amplitude is $E_{max} = kE_0$. (b) At other positions, the phasors differ by equal, small phase increments. The k phasors form a polygonal arc of k sides, which we can approximate as an arc of a circle. The central angle of this arc equals the phase difference between the first and last phasors.

(b)

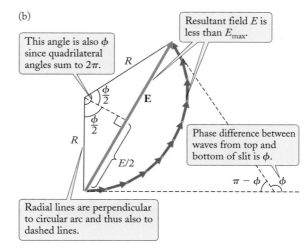

Substituting $R = E_{max}/\phi$ from Eq. (35.22) and solving for E gives us

$$E = E_{max} \frac{\sin(\phi/2)}{\phi/2} \tag{35.24}$$

Since intensity is proportional to the square of the amplitude, this yields an intensity

intensity for single-slit diffraction pattern

$$I = I_{max}\left[\frac{\sin(\phi/2)}{\phi/2}\right]^2 \tag{35.25}$$

where I_{max} is the intensity of the central maximum and, as previously [compare Eq. (35.15)], the phase difference ϕ between waves from the top and bottom of the slit is $2\pi/\lambda$ times the path difference $a\sin\theta$ of Fig. (35.35):

$$\phi = \frac{2\pi}{\lambda} a\sin\theta \tag{35.26}$$

Fraunhofer diffraction pattern

Equation (35.25) is the formula for the intensity pattern for a single slit, known as the **Fraunhofer diffraction pattern**. The intensity is plotted as a function of the angle θ toward a distant screen in Fig. 35.37, for the special case of $a = 3\lambda$. As discussed above, the width of the pattern will spread for a narrower slit, and it will shrink for a wider slit.

We previously showed that the intensity goes to zero at the *minima*, where $a\sin\theta = m\lambda$, or where $\phi = 2\pi, 4\pi, 6\pi, \ldots$. The exact location and intensity of the secondary *maxima* may be obtained using differential calculus, by setting the derivative of Eq. (35.25) equal to zero. However, the maxima are fairly close to halfway between the minima. Thus the maxima essentially occur where $\phi = 3\pi, 5\pi, \ldots$. For example, the intensity of the first secondary maximum is approximately

$$I = I_{max}\left[\frac{\sin(3\pi/2)}{3\pi/2}\right]^2 = \frac{4}{9\pi^2}I_{max} \approx 0.045 I_{max}$$

or less than 5% of the intensity of the central maximum.

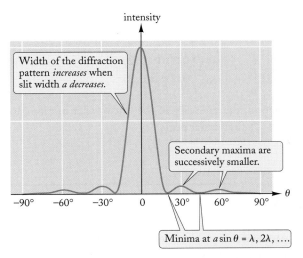

FIGURE 35.37 Intensity as a function of the angle for light arriving at a screen placed behind a narrow slit of width $a = 3\lambda$.

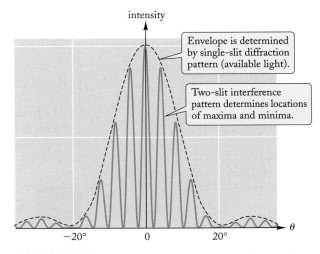

FIGURE 35.38 Intensity as a function of angle for two slits with slit width $a = 3\lambda$ and slit separation $d = 5a$, showing both single-slit diffraction and two-slit interference effects.

Finally, we point out that the single-slit diffraction pattern determines where on a screen light is available to interfere when more than one slit is present. To obtain the full intensity as function of angle for, say, a two-slit interference pattern, we need only to replace the constant I_{max} in the two-slit interference pattern of Eq. (35.11) by the position-dependent single-slit diffraction intensity I of Eq. (35.25). For example, Fig. 35.38 plots the intensity of such a two-slit interference pattern, where the slit spacing d is 5 times the individual slit width a and $a = 3\lambda$. As is evident, there are many two-slit interference maxima within each single-slit diffraction maximum. Also, the fifth-order two-slit interference maxima are missing, since they coincide with the first minima of the single-slit diffraction pattern; there is no *available light* at these points to interfere.

EXAMPLE 8   The compact discs in the chapter photo exhibit interference maxima from several orders. Suppose that the effective slit width for a CD track is $a = 800$ nm. What is the ratio of intensities of the third- and first-order interference maxima for the blue light ($\lambda = 486$ nm)? (See also Example 5.)

Concepts
— in —
Context

SOLUTION: In Example 5, we saw that the third-order interference maximum for blue light was at $\theta = 60°$. The phase angle ϕ for the single-slit pattern at the third-order interference maximum is, from Eq. (35.26),

$$\phi_3 = 2\pi\,\frac{a}{\lambda}\,\sin\theta = 2\pi \times \frac{800\ \text{nm}}{486\ \text{nm}} \times \sin 60° = 9.0 \text{ radians}$$

The first-order interference maximum occurs at a value of $\sin\theta$ that is one-third as large, so the single-slit diffraction phase angle at the first-order interference maximum is

$$\phi_1 = \tfrac{1}{3}\,\phi_3 = \tfrac{1}{3} \times 9.0 \text{ radians} = 3.0 \text{ radians}$$

We then can obtain the ratio of the intensities of these two peaks by evaluating Eq. (35.25) for ϕ_1 and for ϕ_3 and taking the ratio:

$$\frac{I_3}{I_1} = \left[\frac{\sin(\phi_3/2)/(\phi_3/2)}{\sin(\phi_1/2)/(\phi_1/2)} \right]^2 = \left[\frac{\sin(9.0/2)/(9.0/2)}{\sin(3.0/2)/(3.0/2)} \right]^2 = 0.11$$

Thus, due to single-slit diffraction, the intensity of the third-order interference peak is about one-tenth that of the first-order peak.

 Checkup 35.5

QUESTION 1: You aim a light beam at a narrow slit, and you aim a particle beam at a similar narrow slit. How do the behaviors of these beams differ if you gradually decrease the width of the slit?

QUESTION 2: Suppose that the width a of a single slit equals $\sqrt{2}\lambda$. Describe the intensity distribution for this case.

QUESTION 3: If $a \gg \lambda$, then the angle of the first minimum [Eq. (35.21)] is very small. Does this mean that light passing through a large aperture will be focused to a narrow spot?

QUESTION 4: Estimate the angle θ (in radians) of the first minimum in a single-slit diffraction experiment. The slit width is $a = 0.1$ mm; the wavelength is $\lambda = 500$ nm.

(A) 5×10^{-3} (B) 1×10^{-3} (C) 5×10^{-4}
(D) 1×10^{-4} (E) 5×10^{-5}

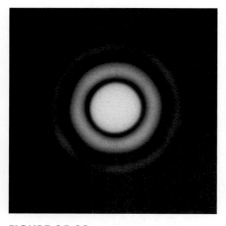

FIGURE 35.39 A photographic film placed beyond a circular aperture records a strong central maximum and annular secondary maxima.

35.6 DIFFRACTION BY A CIRCULAR APERTURE; RAYLEIGH'S CRITERION

The diffraction of light by a circular aperture is in principle no different from the diffraction by a slit (a very long rectangular aperture). To find the maxima and minima in the diffraction pattern of the light emerging from such a circular aperture, we must sum the waves originating from all points of a wave front at the circular aperture. Because we must sum waves from points throughout a circle and not just across a slit, this calculation is fairly complicated, and we will not attempt it. Figure 35.39 is a photograph that displays the central maximum and the bright and dark fringes generated by diffraction at a circular aperture. The angular position of the first minimum is given by the simple formula

first minimum for circular aperture

$$\sin\theta = 1.22 \frac{\lambda}{a} \tag{35.27}$$

where a is the diameter of the circular aperture. Note the similarity with Eq. (35.17).

Many optical instruments—telescopes, microscopes, cameras, etc.—have circular apertures, and these will diffract light. For instance, the objective lens of an astronomical telescope will act like a circular opening in a plate; the parallel wave fronts arriving from some distant star will experience diffraction effects at this aperture and produce an intensity distribution such as that shown in Fig. 35.39. The image of the star as seen through this telescope will then not be a bright point, but a disk surrounded by concentric rings, as in Fig. 35.39. For example, as seen through a telescope with an objective lens 6 cm in diameter, stars look like small disks, about 2×10^{-5} radian (or 4 seconds of arc) across.

This spreading out of the image puts a limit on the detail that can be perceived through the telescope. If two stars are very close together, their images tend to merge, and it may be impossible to tell them apart. Figures 35.40a–d show the images produced by a pair of pointlike light sources upon diffraction by a circular aperture. In the first of these figures, the angular separation of the sources is so small that the two images look like one image. In the second figure, the angular separation is large enough to give a clear indication of the existence of two separate images.

The angular separation of the sources in Fig. 35.40b is such that the central maximum of the diffraction pattern of one source coincides with the minimum of the diffraction pattern of the other source. Since we are now dealing with small angles, we can approximate sin θ by θ, and Eq. (35.26) then tells us that the angular position of the first minimum, or the angular separation between the sources in Fig. 35.40b, is

$$\theta = 1.22 \frac{\lambda}{a} \qquad (35.28)$$

We will regard this as the critical angle that decides whether the two sources are clearly distinguishable: the telescope (or other optical instrument) can resolve the sources if their angular separation is larger than that in Eq. (35.28), and it cannot resolve them if the separation is smaller. This is **Rayleigh's criterion**.

EXAMPLE 9

The star ζ (zeta) Orionis is a binary star; that is, it consists of two stars very close together. The angular separation of the stars is 2.8 seconds of arc. Can the stars be resolved with a telescope having an objective lens 6.0 cm in diameter? Assume that the wavelength of the starlight is 550 nm.

SOLUTION: According to Rayleigh's criterion, a telescope of this aperture can resolve stars as close as

$$\theta = 1.22 \frac{\lambda}{a} = 1.22 \times \frac{550 \times 10^{-9} \text{ m}}{0.060 \text{ m}}$$

$$= 1.1 \times 10^{-5} \text{ radian} \times \frac{360 \text{ degrees}}{2\pi \text{ radians}} \times \frac{3600 \text{ seconds of arc}}{1 \text{ degree}}$$

$$= 2.3 \text{ seconds of arc}$$

Since the angular separation of the stars is 2.8 seconds of arc, the telescope can resolve the double stars.

(a)

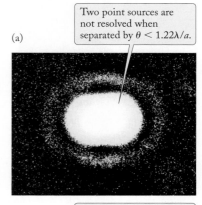

Two point sources are not resolved when separated by $\theta < 1.22\lambda/a$.

(b)

Rayleigh's criterion: two point sources are barely resolved when separated by $\theta = 1.22\lambda/a$.

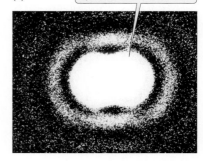

(c)

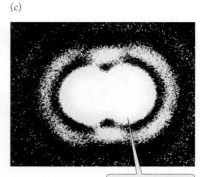

Two point sources are well resolved when $\theta > 1.22\lambda/a$.

(d)

FIGURE 35.40 When the light waves from two pointlike sources arrive at a circular aperture simultaneously, each set of light waves will produce its own diffraction pattern. If the angular separation between the two sources is small, the diffraction patterns overlap. (a) Here the angular separation is small. (b–d) The angular separation is progressively larger.

JOHN WILLIAM STRUTT, 3RD BARON RAYLEIGH (1842–1919)
English physicist, professor at Cambridge and at the Royal Institution. He is best known for his extensive mathematical investigations of sound and of light. He also investigated the behavior of gases at high densities, and he discovered argon; for this he was awarded the Nobel Prize in 1904.

The ability of a telescope to resolve stars or other objects of small angular separation improves with the size of the telescope—a telescope of 30 cm diameter can resolve angular separations as small as 0.5 second of arc. However, beyond 30 cm, the resolution of an Earth-bound telescope does not improve further with size, because fluctuations in the density of the atmosphere introduce irregularities in the path of the light coming down from the sky; this smears out the light and prevents the telescope from reaching its full potential.

The Hubble Space Telescope, launched in 1990, is in orbit above the atmosphere of the Earth, and it is not affected by atmospheric fluctuations. This telescope (see Fig. 35.41) has an aperture of of 2.4 m, and it achieves an angular resolution of about 0.1 second of arc, close to the limit set by Rayleigh's criterion. Thus, the space telescope achieves higher angular resolution than any Earth-bound optical telescope. Furthermore, it is able to observe at ultraviolet and infrared wavelengths, which are blocked by the atmosphere.

For Earth-bound radio telescopes, the atmosphere poses no problem—radio waves do not suffer from the effects of atmospheric density fluctuations. Hence, with increasing size, the resolution of a radio telescope improves indefinitely. Figure 35.42 shows the large radio telescope at Arecibo, Puerto Rico. The concave "mirror" of this telescope has an aperture of 300 m and a radius of curvature which is also 300 m. The shortest wavelength at which this telescope has been operated is 4.0 cm. For this wavelength, Rayleigh's criterion gives a limiting angular resolution

$$\theta = 1.22 \frac{\lambda}{a} = 1.22 \times \frac{0.040 \text{ m}}{300 \text{ m}} = 1.6 \times 10^{-4} \text{ radian}$$

which is about 30 seconds of arc.

FIGURE 35.41 The Hubble Space Telescope.

FIGURE 35.42 The Arecibo radio telescope.

Checkup 35.6

QUESTION 1: If we apply a criterion similar to Rayleigh's to a narrow slit, instead of a circular aperture, what critical angle for the resolution of two images do we obtain?

QUESTION 2: The Arecibo radio telescope has an aperture 125 times as large as the Hubble Space Telescope. However, its resolution is less than that of Hubble. Why?

QUESTION 3: Two dots of red ink on a page are separated by 0.2 mm. Suppose your eye has an aperture of 0.5 cm. Of the following, what is the largest distance at which it is possible to resolve the dots, assuming otherwise perfect eyesight?

   (A) 3 cm   (B) 10 cm   (C) 30 cm   (D) 100 cm   (E) 300 cm

SUMMARY

PROBLEM-SOLVING TECHNIQUES Thin-Film Interference	**(page 1174)**

PROBLEM-SOLVING TECHNIQUES Two-Slit Interference	**(page 1181)**

CONSTRUCTIVE INTERFERENCE Waves meet crest to crest, making a total wave of larger amplitude.

DESTRUCTIVE INTERFERENCE Waves meet crest to trough, making a total wave of smaller or zero amplitude.

THIN-FILM INTERFERENCE where $\lambda = \dfrac{\lambda_{air}}{n}$, with film thickness d, for both waves reflecting at an interface with a material of higher index of refraction:
(When only one of the waves reflects at an interface with a material of higher index of refraction, the conditions are reversed.)

Maxima: $2d = \lambda,\ 2\lambda,\ 3\lambda, \ldots$

Minima: $2d = \frac{1}{2}\lambda,\ \frac{3}{2}\lambda,\ \frac{5}{2}\lambda, \ldots$

(35.1, 35.2, 35.3)

TWO-SLIT INTERFERENCE PATTERN with slit separation d, where θ is the angle with respect to the midline. For small angles, the transverse displacement from the midline at distance r is $\Delta y \approx r\theta$.

Maxima: $d \sin\theta = 0,\ \lambda,\ 2\lambda, \ldots$

Minima: $d \sin\theta = \frac{1}{2}\lambda,\ \frac{3}{2}\lambda,\ \frac{5}{2}\lambda, \ldots$

(35.7, 35.9)

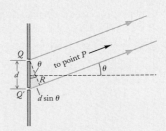

INTENSITY FOR TWO-SLIT INTERFERENCE PATTERN

$$I = I_{max} \cos^2\left(\frac{\pi d}{\lambda} \sin\theta\right)$$

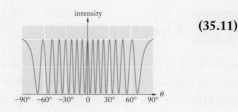

(35.11)

MULTIPLE-SLIT INTERFERENCE PATTERN (N slits)

Principal maxima: $d \sin\theta = m\lambda$ where $m = 0, 1, 2, \ldots$

First minima: $d \sin\theta = \dfrac{\lambda}{N}$ (35.12, 35.16)

PHASOR DIAGRAMS   To obtain the resultant wave, add vectors representing the amplitudes and phases of the interfering waves tip to tail.

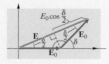

MINIMA FOR SINGLE SLIT DIFFRACTION PATTERN with slit width a

$$a \sin\theta = \lambda, \ 2\lambda, \ 3\lambda, \ldots$$

(35.20)

INTENSITY FOR SINGLE-SLIT DIFFRACTION PATTERN

where $\phi = \dfrac{2\pi}{\lambda} a \sin\theta$

$$I = I_{max}\left[\frac{\sin(\phi/2)}{\phi/2}\right]^2$$

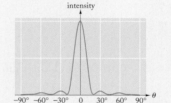

(35.25)

(35.26)

FIRST MINIMUM FOR CIRCULAR APERTURE with diameter a

$$\sin\theta = 1.22\frac{\lambda}{a}$$

(35.27)

RAYLEIGH'S CRITERION

$$\theta = 1.22\frac{\lambda}{a}$$

(35.28)

QUESTIONS FOR DISCUSSION

1. The light wave inside the tube of a laser is a standing wave, reflected at both ends of the tube. If the distance between the ends is L, what condition must the wavelength of the standing wave satisfy?

2. Would you expect sound waves reflected by a wall to produce a standing wave? How could you test this experimentally?

3. The radar wave emitted by a stationary police radar unit is reflected by an approaching automobile. Does this set up a standing wave?

4. When two waves interfere destructively at one place, what happens to their energy?

5. When two coherent waves of equal intensity interfere constructively at one place, the energy density at that place becomes 4 times as large as the energy density of each individual wave. Is this a violation of the Law of Conservation of Energy?

6. Suppose that N waves of equal intensity meet at one place. If the waves are coherent, the net intensity is N^2 times the intensity of each individual wave. If the waves are incoherent, the

(average) intensity is N times the intensity of each individual wave. Explain, and give an example of each case.

7. When light strikes a windowpane, some rays will be reflected back and forth between the two glass surfaces. Why do these reflected rays not produce visible colored interference fringes?

8. Suppose that a lens is covered with an antireflective coating that eliminates the reflection of perpendicularly incident light of some given color. Will this coating also eliminate the reflection of this light incident at an angle?

9. Is it possible to cover the surface of an aircraft with an antireflective coating so that it does not reflect radar waves of wavelength, say, 5 cm?

10. Explain how Newton's rings may be used as a sensitive test of the rotation symmetry of a lens.

11. Two flat plates of glass are in contact at one edge and separated by a thin spacer at the other edge (Fig. 35.43a). Explain why we see parallel interference fringes in the reflected light if we illuminate these plates from above (Fig. 35.43b).

(a)

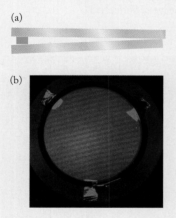

(b)

FIGURE 35.43 (a) Two flat plates of glass, separated by a thin wedge of air. (b) Interference fringes.

12. In the experiment that sought to test the dependence of the speed of light on the motion of the Earth, Michelson and Morley used an interferometer with very long arms (about 11 m, obtained by multiple reflections back and forth between sets of mirrors). Why does this make the instrument more sensitive?

13. If you stand next to your radio receiver, your body will sometimes affect the reception. Why?

14. Consider (1) sunlight, (2) sunlight passed through a monochromatic filter selecting one wavelength, (3) light from a neon tube, (4) light from a laser, (5) starlight passed through a monochromatic filter, and (6) radio waves emitted by a radio station. Which of these kinds of light or electromagnetic radiation are sufficiently coherent to give rise to an interference pattern when used to illuminate two slits such as shown in Fig. 35.13?

15. When a pair of stereo loudspeakers are installed, the terminals of the loudspeakers should be connected to the amplifier in the same way, so that the loudspeakers are in phase. What would happen to sound waves of long wavelength if the loudspeakers were out of phase?

16. What happens to the interference pattern plotted in Fig. 35.19 if $d < \lambda$?

17. Suppose we cover one of the slits in Fig. 35.13 with a glass plate. Since the wavelength in the glass is shorter than in air, the path in the glass includes more wavelengths than the path in air, and the phases of the two waves emerging from the slits will be different. If the phase difference between these waves is exactly 180°, how does this change the location of maxima and minima?

18. Consider the interference pattern shown in Fig. 35.19. How would this pattern change if the entire interference apparatus were immersed in water?

19. Suppose that a plane wave is incident at an angle on a plate with two narrow slits (Fig. 35.44). In what direction will we then find the central maximum?

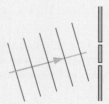

FIGURE 35.44 Plane wave obliquely incident on a plate with two slits.

20. Several radio antennas are arranged at regular intervals along a straight line; the antennas are connected to the same radio transmitter, so they radiate coherently. How can this array be used to concentrate the radio emission in a selected direction?

21. When a crystal is illuminated with X rays, each atom acts as a pointlike source of scattered X rays. The typical separation between adjacent atoms in a crystal is 0.1 nm. Roughly what must be the wavelength of the X rays if they are to exibit distinct interference effects?

22. Gratings used for analyzing light are often called diffraction gratings, but it would be more accurate to call them interference gratings. Why?

23. Consider the diffraction pattern shown in Fig. 35.37. How would this pattern change if the entire apparatus were immersed in water?

24. Figure 17.21 shows the diffraction of water waves at the entrance of a harbor. Could this diffraction be eliminated by making the entrance smaller?

25. In the center of the shadow of a disk or sphere there is a small bright spot, called the **Poisson spot** (Fig. 35.45). This spot is very faint near the disk but becomes more noticeable at large distances. Qualitatively, explain why the diffraction of the light waves around the edges gives rise to this spot.

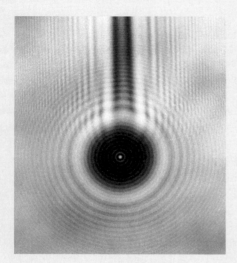

FIGURE 35.45 The Poisson spot.

26. Besides good angular resolution, what other advantage does a telescope of large aperture have over a telescope of small aperture?

27. Spy satellites use cameras with lenses of very large diameter, 30 cm or more. Why are such large diameters necessary?

28. In order to beam a sound wave sharply in one direction with a megaphone, should the horn of the megaphone have a small aperture or a large aperture?

29. The manufacturers of the Questar telescope claim that this telescope can distinguish two stars even if their separation is somewhat smaller than that specified by Rayleigh's criterion. Why is this possible?

30. The maximum useful magnification of an optical microscope is determined by diffraction effects at the objective lens. Explain.

31. Other things being equal, how much resolution can you gain by operating an optical microscope with blue light instead of red light? Why can you not operate an ordinary microscope with ultraviolet light?

32. The picture of atoms in Fig. 35.46 was taken with an electron microscope using electron waves of extremely short wavelength. Roughly, how short must the wavelength be to make individual atoms visible?

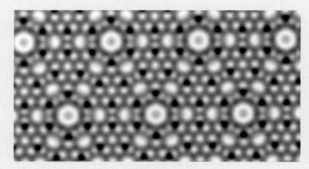

FIGURE 35.46 Atoms in a crystal.

33. The compound eye of insects consists of a large number of small eyes, or ommatidia (Fig. 35.47). Each ommatidium is typically 0.03 mm across; it does not form an image, but merely acts as a sensor of the intensity of light arriving from a narrow cone of directions. What are some of the advantages and disadvantages of such a compound eye as compared with the camera eye of vertebrates?

FIGURE 35.47 Compound eyes of a fly.

34. The corona formed by diffraction of sunlight by small droplets of water in a cloud consists of a bright ring surrounding the Sun. What is the color of the outer edge of the corona? The inner edge?

35. Antennas used for the transmission of microwaves in communication links consist of metal dishes. What factors determine the size of these dishes?

36. Very-long-baseline interferometry (VLBI) with two or more radio telescopes placed on different parts of the Earth has been used for precise determination of the rate of continental drift and also the rate of rotation of the Earth (sidereal day). Explain how data collected by radio telescopes monitoring a fixed radio source in the sky can be used for such purposes.

PROBLEMS

35.1 Thin Films

1. In 1887, Hertz gave the first direct experimental demonstration of the existence of the electromagnetic waves that had been predicted by Maxwell's theory. For this experiment, Hertz placed a high-frequency spark generator at some distance in front of a large vertical zinc plate. The wave emitted by the spark reflected off the zinc plate, setting up a standing wave. Hertz found that when the generator was operating at a frequency of about 4.0×10^7 cycles per second, the distance between a node and an antinode in the standing-wave pattern was about 2.0 m. What value of the speed of propagation of the electromagnetic disturbances could he deduce from this?

2. The wall of a soap bubble floating in air has a thickness of 400 nm. If sunlight strikes the wall perpendicularly, what colors seen in the reflected light will be strongly enhanced? The index of refraction of the soap film is 1.35.

3. A thin film with index of refraction $n_{film} = 1.38$ is used to coat a lens, which has $n_{lens} = 1.50$. What is the thinnest film that will highly reflect near-ultraviolet light of wavelength 350 nm? What is the thinnest film that will minimize reflection of 700-nm light?

4. Similar to the arrangement in Fig. 35.43, two flat plates of glass are in contact at one edge and a human hair is used as a spacer at the other edge. Green light of wavelength 546 nm is normally incident from above, and 76 dark fringes appear across the plate. What is the thickness of the hair?

5. A Newton's ring arrangement (see Figs. 35.7 and 35.8) exhibits its 25th dark ring out at the edge of the lens when light of wavelength 590 nm is used; that ring has radius 6.0 cm. What is the radius of curvature of the convex surface of the plano–convex lens? (Hint: You may use the formula of Problem 35.13.)

6. In a Newton's ring experiment (see Figs. 35.7 and 35.8), light of wavelength 546 nm is used. When the air space between the plano–convex lens and the flat glass plate is filled with liquid, the radius of the fifth dark ring decreases from 2.0 cm to 1.5 cm. What is the index of refraction of the liquid? (Hint: Use the formula of Problem 35.13.)

7. A coating is used to maximize the light entering a silicon solar cell. If the indices of refraction of silicon and the coating are 3.5 and 1.6 respectively, what minimum coating thickness can be used for 500-nm light?

*8. The tube of a laser has a mirror at each end (Fig. 35.48). A standing electromagnetic wave fills the space between these mirrors. This wave must satisfy the boundary condition that its electric field is zero at all times at the position of the mirrors.

   (a) Show that the distance d between the mirrors and the wavelength λ of the wave must be related by

$$d = \left(\frac{m+1}{2}\right)\lambda \quad m = 0, 1, 2, \dots$$

   (b) A He–Ne laser with $d = 30.00$ cm operates at $\lambda = 633$ nm. What is the value of m? How many minima and how many maxima does the standing wave have between the two mirrors?

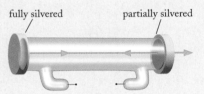

FIGURE 35.48 Tube of a laser with mirrors at its ends.

*9. Consider the light *transmitted* by a thin film. For a light wave with a direction of propagation perpendicular to the surface of the film, what is the condition for constructive interference between the direct wave and the wave that is reflected once by the lower surface and once by the upper surface? Would you expect that this condition for constructive interference in transmission coincides with the condition of destructive interference in reflection?

*10. A thin oil slick, of index of refraction 1.25, floats on water. When a beam of white light strikes this film vertically, the only colors seen enhanced in the reflected beam are red (at a wavelength of 675 nm in air) and violet (at 450 nm). From this information, deduce the thickness of the oil slick.

*11. Two flat parallel plates of glass are separated by thin spacers so as to leave a gap of width d (Fig. 35.49). If light of wavelength λ is normally incident on these plates, what is the condition for constructive interference between the rays reflected by the lower surface of the top plate and the upper surface of the bottom plate?

FIGURE 35.49 Two parallel flat plates of glass.

*12. A layer of oil of thickness 200 nm floats on top of a layer of water of thickness 400 nm resting on a flat metallic mirror. The index of refraction of the oil is 1.24, and that of the water is 1.33. A beam of light is normally incident on these layers. What must be the wavelength of the beam if the light reflected by the top surface of the oil is to interfere destructively with the light reflected by the mirror?

**13. A lens with one flat surface and one convex surface rests on a flat plate of glass (Fig. 35.50). A light ray normally incident on the lens will be partially reflected by the curved surface of the lens and partially reflected by the flat plate of glass. The interference between these reflected rays will be constructive or destructive, depending on the height of the air gap between the lens and the plate. The interference gives rise to the pattern of Newton's rings, shown in Fig. 35.7.

(a) Why is the center of the pattern dark?

(b) Show that the radius of the mth dark ring is

$$r = \sqrt{m\lambda R - m^2\lambda^2/4}$$

where λ is the wavelength of the light and R is the radius of curvature of the convex surface of the lens.

(c) What is the radius of the first dark ring if $\lambda = 500$ nm and $R = 3.0$ m?

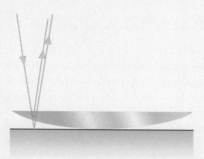

FIGURE 35.50 Lens resting on a flat glass plate.

35.2 The Michelson Interferometer

14. In a Michelson interferometer such as shown in Fig. 35.9, one mirror is moved 0.450 mm, and 1422 interference fringes are counted. What wavelength of light was used?

15. In a Michelson interferometer such as shown in Fig. 35.9, light of wavelength 546 nm is used. When one mirror is moved, 930 interference fringes are counted. Through what distance was the mirror moved?

*16. The **Fabry–Perot interferometer** consists of two parallel half-silvered mirrors. A ray of light entering the space between the mirrors may pass straight through, or be reflected once or several times by each mirror (Fig. 35.51). Show that the condition for constructive interference of the emerging light (at large distance from the mirrors) is

$$2d\cos\theta = 0,\ \lambda,\ 2\lambda,\ldots$$

where d is the distance between the mirrors, λ the wavelength of light, and θ the angle of incidence of the light.

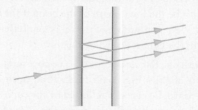

FIGURE 35.51 Fabry–Perot interferometer.

*17. The interferometer of the International Bureau of Weights and Measures can count 19 000 bright fringes (maxima) per second. To achieve this count rate, what must be the speed of motion of the moving mirror (mirror M_1 in Fig. 35.9)? Assume that the interferometer operates with krypton light of wavelength $\lambda = 605.8$ nm.

*18. An **etalon** consists of two mirrors held a fixed distance apart by means of a rigid support (Fig. 35.52a). The distance between the two mirrors can serve as a standard of length. To measure the distance in terms of wavelengths of light, the etalon is installed in a Michelson interferometer, replacing the fixed mirror M_2 (Fig. 35.52b). For a start, the distances MM_1 and MM_3 are made exactly equal; then the mirror M_1 is slowly moved outward, producing a sequence of interference maxima and minima until the distances MM_1 and MM_4 are exactly equal.[2] Suppose that you operate a Michelson interferometer with krypton light of wavelength 605.7802 nm and that you count 36 484.8 interference maxima while moving the mirror M_1 from its initial position to its final position. To within six significant figures, what is the length of the etalon?

(a)

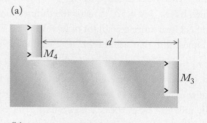

(b)

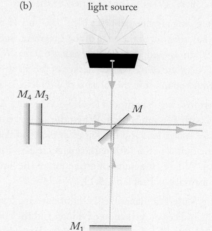

FIGURE 35.52 (a) An etalon. (b) The etalon (M_3, M_4) installed in the Michelson interferometer.

[2] Exact equality of the lengths of the arms can be verified by checking for constructive interference with *white* light. For white light, containing many wavelengths, constructive interference is possible only if the lengths are equal, that is, $d = 0$ in Eq. (35.4).

*19. The index of refraction of air for monochromatic light of some given wavelength can be determined very precisely with a Michelson interferometer one of whose arms (MM_1 in Fig. 35.9) is in air while the other (MM_2) is in a vacuum tank. With the vacuum tank filled with air, the arms are first adjusted so they are exactly equal. Then the vacuum tank is slowly evacuated, while at the same time the length of the arm is gradually increased so as to maintain constructive interference (this increase of length maintains a fixed number of wavelengths in the arm). Show that the index of refraction of air can be expressed in terms of the fractional change of length:

$$n - 1 = \frac{\Delta L}{L}$$

Suppose that for light of wavelength 580 nm, an interferometer with arms of length 40.00 cm requires a readjustment of 0.011 06 cm to maintain constructive interference while the vacuum tank is being evacuated. What is the index of refraction of air at this wavelength?

†35.3 Interference from Two Slits

20. Microwaves of wavelength 2.0 cm from a radio transmitter are aimed at two narrow parallel slits in an aluminum plate. The slits are separated by a distance of 5.0 cm. At what angles at some distance beyond the plate will we find maxima in the interference pattern?

21. Suppose that the two slits, the film, and the light source described in Example 2 are immersed in water. What will now be the distance between the central maximum and the first lateral maximum?

22. Laser light of wavelength 633 nm is incident upon two narrow slits separated by 0.15 mm. On a screen 4.00 m beyond the slits, how far apart are the interference fringes?

23. In a double-slit experiment, green light of wavelength 546 nm is used. On a screen 1.50 m away, what is the distance between the central maximum and the first minimum if the slits are separated by 0.22 mm?

24. Two radio antennas transmit electromagnetic waves of the same frequency and phase. If their separation is $\lambda/2$ what is the angular position of the first interference minimum?

25. Two narrow slits, separated by 0.10 mm, are illuminated with light of wavelength 633 nm. On a screen 1.20 m away, what is the intensity, relative to the maximum, at a point 1.5 cm from the central maximum?

26. A piece of aluminum foil with two narrow slits is being illuminated with red light of wavelength 694.3 nm from a laser. This yields a row of evenly spaced bright bands on a screen placed

3.00 m beyond the slits. The interval between the bright bands is 1.40 cm. What is the distance between the two slits?

27. Consider the water waves shown in Fig. 35.16. With a ruler, measure the wavelength of the waves and the distance between the sources; with a protractor, measure the angular positions of the nodal lines (minima). Do these measured quantities satisfy Eq. (35.9)?

28. For the most detailed mapping of the sky, radio astronomers use radio telescopes placed on different continents, with (straight-line) separation of several thousand kilometers. Suppose that two such radio telescopes are separated by an east–west distance of 5000 km, and that they are connected to a single radio receiver tuned to a wavelength of 21 cm. If a source is symmetrically located above these radio telescopes, the radio waves reaching the receiver will be in phase, and the receiver will register maximum intensity. What westward angular displacement of the source from this location will, again, result in maximum intensity?

29. An interferometric radio telescope consists of two antennas separated by a distance of 1.00 km. The two antennas feed their signals into a common receiver tuned to a frequency of 2300 MHz. The receiver will detect a maximum (constructive interference) if the wave sent out by a radio source in the sky arrives at the two antennas with the same phase. What possible angular positions of the radio source will result in such a maximum? Reckon the angular position from the vertical line erected at the midpoint of the antennas. Treat the antennas as pointlike.

30. Light of wavelength 694.3 nm from a ruby laser is incident on two narrow parallel slits cut in a thin sheet of metal. The slits are separated by a distance of 0.11 mm. A screen is placed 1.5 m beyond the slits. Find the intensity, relative to the central maximum, at a point on the screen 1.2 cm to one side of the central maximum.

31. In a double-slit experiment, the intensity at the peak of the central maximum is I_{max}. If the slit spacing is 12 times the wavelength of light used, what is the intensity at an angular position 1.0° away from the peak of the central maximum? At 2.0°? At 3.0°?

32. In a double-slit experiment, the intensity at the peak of the central maximum is I_{max}. What is the intensity when the path difference to a point on the screen is $\lambda/3$?

*33. In a double-slit experiment such as that in Fig. 35.17, the slit spacing is 0.20 mm, the distance to the screen is 3.00 m, and light of wavelength 633 nm is used. If a thin sheet of polystyrene ($n = 1.49$) is placed over the bottom slit only, the entire interference pattern moves downward on the screen by 4.0 mm. What is the thickness of the polystyrene sheet?

34. A 1.0-kHz sound wave is normally incident on two narrow windows, separated by 2.0 m. If a listener is standing 5.0 m beyond the windows on the midline, how far must she move laterally to be standing at the first interference minimum? Take the speed of sound to be 340 m/s.

†For help, see Online Concept Tutorial 39 at www.wwnorton.com/physics

*35. Light of wavelength λ is obliquely incident on a pair of narrow slits separated by a distance d. The angle of incidence of the light on the slits is ϕ (Fig. 35.53).

(a) Show that the diffracted light emerging at an angle θ interferes constructively if

$$|d \sin \theta - d \sin \phi| = 0, \ \lambda, \ 2\lambda, \dots$$

and destructively if

$$|d \sin \theta - d \sin \phi| = \tfrac{1}{2}\lambda, \ \tfrac{3}{2}\lambda, \ \tfrac{5}{2}\lambda, \dots$$

(b) Show that if θ is small, the angular separation between the interference maxima and minima is independent of the angle ϕ.

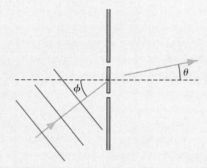

FIGURE 35.53 Plane wave obliquely incident on a pair of slits in a plate.

*36. A device called **Lloyd's mirror** produces interference between a ray reaching a vertical screen directly and a ray reaching the screen after reflection by a horizontal mirror. Show that in terms of the distances z, z_0, and l defined in Fig. 35.54, the condition for constructive interference is

$$\sqrt{l^2 + (z + z_0)^2} - \sqrt{l^2 + (z - z_0)^2} = \tfrac{1}{2}\lambda, \ \tfrac{3}{2}\lambda, \ \tfrac{5}{2}\lambda, \dots$$

and that for destructive interference is

$$\sqrt{l^2 + (z + z_0)^2} - \sqrt{l^2 + (z - z_0)^2} = 0, \ \lambda, \ 2\lambda, \dots$$

Note that in this calculation you must take into account the reversal, or change in phase, of the wave during reflection.

FIGURE 35.54 Lloyd's mirror.

**37. Two radio beacons transmit waves of the same phases and frequencies. The transmitters are on the x axis, at $\pm x_0$. Show that the interference is constructive at those points of the x–y plane satisfying the condition

$$\sqrt{(x + x_0)^2 + y^2} - \sqrt{(x - x_0)^2 + y^2} = m\lambda$$

where $m = 0, 1, 2, \dots$. Show that for a given nonzero value of m, this is the equation of a hyperbola. (To show this, use either graphic methods to plot the curve or else your knowledge of analytic geometry.)

**38. In the first application of interferometric methods in radio astronomy, Australian astronomers observed the interference between a radio wave arriving at their antenna on a direct path from the Sun and on a path involving the reflection on the surface of the sea (Fig. 35.55). Assuming that the radio waves have a frequency of 6.0×10^7 Hz and that the radio receiver is at a height of 25 m above the level of the sea, what is the least angle of the source above the horizon that will give destructive interference of the waves at the receiver?

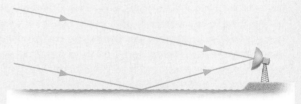

FIGURE 35.55 Direct and reflected radio waves.

†35.4 Interference from Multiple Slits

39. A grating has 5000 lines per centimeter. What are the angular positions of the principal maxima produced by this grating when illuminated with light of wavelength 650 nm?

40. Sodium light with wavelengths 588.99 nm and 589.59 nm is incident on a grating with 5500 lines per centimeter. A screen is placed 3.0 m beyond the grating. What is the distance between the two spectral lines in the first-order spectrum on the screen? In the second-order spectrum?

41. A good grating cut in speculum metal has 5900 lines per centimeter. If this grating is illuminated with white light ranging over wavelengths from 400 nm to 700 nm, it will produce a spectrum ranging over some interval of angles. From what angle to what angle does the first-order spectrum extend? The second-order spectrum? The third-order spectrum? Do these angular intervals overlap? Is the third-order spectrum complete?

42. Six identical, evenly spaced slits are illuminated by light of wavelength 633 nm. What is the angular width of the central maximum if the slit spacing is $d = 0.25$ mm?

---

†For help, see Online Concept Tutorial 40 at www.wwnorton.com/physics

43. A grating has 40 000 evenly spaced lines over a width of 8.0 cm, illuminated by a beam of monochromatic light. What is the angular width of the central maximum when light of wavelength 550 nm is used?

44. Use the phasor diagrams of Fig. 35.26 to calculate the intensity, relative to a principal maximum, of the three-slit pattern for each of the values of the phase difference δ shown in the figure.

*45. A thin curtain of fine batiste consists of vertical and horizontal threads of cotton forming a net that has a regular array of square holes. While looking through this curtain at the red (670 nm) taillight of an automobile, a physicist notices that the taillight appears as a multiple vertical array of images (an array of principal maxima). The angular separation between adjacent images is 2.0×10^{-3} radian. From this information deduce the vertical spacing between the threads in the batiste curtain.

*46. Consider four evenly spaced identical slits. Use phasor diagrams to calculate the intensity, relative to a principal maximum, when the phase difference between adjacent slits is 0, $\pi/6$, $\pi/4$, $\pi/3$, $\pi/2$, $2\pi/3$, $3\pi/4$, $5\pi/6$, and π. Plot the resulting intensity pattern and compare your plot with Fig. 35.27b.

*47. Three evenly spaced slits are illuminated by monochromatic light. The middle slit is wider than the other two, so the intensity due to the middle slit alone is 4 times the intensity due to either of the top or bottom slits alone. Use phasor diagrams to find the intensity when the phase difference between adjacent slits is 0, $\pi/4$, $\pi/2$, $3\pi/4$, and π. Plot these values of intensity as a function of the phase angle.

*48. Three identical narrow slits are illuminated by monochromatic light. The top and center slits are separated by a distance d; the center and bottom slits are separated by a distance $2d$. Use phasor diagrams to find the intensity on a distant screen when the phase difference between the top and center slits is 0, $\pi/4$, $\pi/3$, $\pi/2$, $2\pi/3$, $3\pi/4$, and π. Make a plot of intensity vs. phase angle.

*49. A **reflection grating** consists of a metal plate in which a large number of closely spaced parallel grooves have been cut.

(a) Show that if light of wavelength λ is obliquely incident on this grating at an angle ϕ (see Fig. 35.56), the light returned by the grating at an angle θ interferes constructively if

$$|d \sin\theta - d \sin\phi| = 0, \ \lambda, \ 2\lambda, \ldots$$

(b) For what angle of incidence ϕ will the direction of the emerging light overlap with the direction of the incident light?

(c) For what angle θ does the reflection grating give constructive interference with white light, consisting of a mixture of many wavelengths?

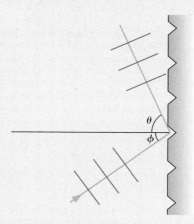

FIGURE 35.56 A reflection grating.

†35.5 Diffraction by a Single Slit

50. Consider the diffraction pattern shown in Fig. 35.33. How would this pattern change if the slit, the light source, and the film were immersed in water?

51. Light of wavelength 632.8 nm from a He–Ne laser illuminates a single slit of width 0.10 mm. What is the width of the central maximum formed on a screen placed 2.0 m beyond the slit?

52. Consider the water waves shown in Fig. 35.57. With a ruler, measure the wavelength of the waves and the length of the gap; with a protractor, measure the angular position of the two nodal lines (minima). Check whether the quantities satisfy Eq. (35.20).

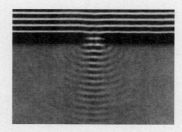

FIGURE 35.57 Diffraction of water waves by an aperture.

53. Light from a 633-nm laser illuminates a single slit. If the full width of the central diffraction maximum on a screen 1.00 m away is 3.0 cm, what is the slit width?

54. A sound wave of frequency 820 Hz passes through a doorway of width 1.0 m. What are the angular directions of the minima of the diffraction pattern? Assume the velocity of sound is 331 m/s.

†For help, see Online Concept Tutorial 39 at www.wwnorton.com/physics

55. Water waves of wavelength 20 m approach a harbor entrance 50 m across at right angles to their path. What is the angular width of the central beam of diffracted waves beyond the entrance?

56. You want to prepare a photograph, such as Fig. 35.33, showing diffraction by a single slit. Suppose you use light of wavelength 577 nm from a mercury lamp and you place your photographic film 2.0 m behind the slit. If the width of the central maximum is to be at least 0.50 cm on your film, what width of the slit do you require?

57. Suppose you want to detect diffraction fringes in the intensity pattern produced by *sunlight* passing through a fine slit cut into the blinds covering a window. You have available a filter that blocks all wavelengths except 550 nm, and you place this filter over the slit.

    (a) If the width of the slit is 0.10 mm, will you be able to detect diffraction fringes? Take into account that the Sun's angular diameter is about $\frac{1}{2}°$.

    (b) If the width of the slit is 0.010 mm?

    (c) If you remove the filter?

58. Relative to the intensity of the central maximum, what is the intensity of the single-slit diffraction pattern at an angular position halfway between the peak of the central maximum and the first minimum? (You may assume small angles.)

59. Relative to the intensity of the central maximum, what is the intensity of the single-slit diffraction pattern at an angular position halfway between the first and the second minimum (that is, near the first secondary maximum)? (You may assume small angles.)

60. Figure 35.58 shows a combined diffraction–interference pattern. How many slits were illuminated? The distance between adjacent slits is 0.12 mm. Deduce the width of each slit. Deduce the wavelength of the light.

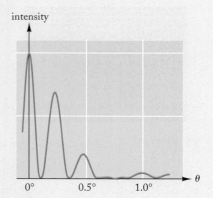

FIGURE 35.58 Combined interference and diffraction pattern.

*61. Two narrow slits of width a are separated by a center-to-center distance d. Suppose that the ratio of d to a is an integer, $d/a = m$.

    (a) Show that in the diffraction pattern produced by this arrangement of slits, the mth interference maximum (cor-

responding to $d\sin\theta = m\lambda$) is suppressed because of coincidence with a diffraction minimum. Show that this is also true for the $2m$th, $3m$th, etc., interference maxima.

    (b) How many interference maxima are there between one diffraction minimum on one side and the next on the same side?

*62. Figure 35.59 shows the intensity curve for the interference pattern produced by three very narrow slits separated by distances of 12λ. Suppose the three very thin slits are replaced by three new slits of width 6λ. This will change the intensity curve.

    (a) On top of Fig. 35.59 (or on a copy of Fig. 35.59), plot the intensity curve of the diffraction pattern produced by *one* of these new slits.

    (b) By suitably combining these two curves, obtain the complete intensity curve for the system of the three new slits.

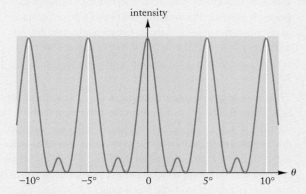

FIGURE 35.59 Interference pattern due to three narrow slits.

*63. Find the actual location of the first secondary maximum in the single-slit diffraction pattern by differentiating Eq. (35.25) and solving for the phase angle to at least three significant figures iteratively (by trial and error). Express your result as a multiple of 3π, the phase angle midway between the first two minima.

*64. Light of wavelength λ is obliquely incident on a slit of width a. The angle of incidence of the light on the slit is ϕ (Fig. 35.60).

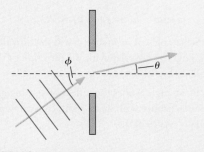

FIGURE 35.60 Plane wave obliquely incident on a single slit.

(a) Show that the diffracted light emerging at an angle θ interferes destructively if

$$a \sin\theta - a \sin\phi = \lambda, \ 2\lambda, \ 3\lambda, \ldots$$

(b) Show that, for small values of θ, the angular separation between the directions of destructive interference is independent of ϕ.

35.6 Diffraction by a Circular Aperture; Rayleigh's Criterion

65. The impressionist painter Georges Seurat used small dots of paint, which form continuous images when viewed from a distance. How small must two adjacent green dots ($\lambda = 550$ nm) be so that a human eye, with a 6.0-mm aperture, can barely resolve them when viewed from 2.0 m (and thus not resolve them from larger distances)?

66. When the eye looks at a star (a point of light), diffraction at the pupil spreads the image of the star on the retina into a small disk.

 (a) When opened to maximum size, the pupil of a human eye has a diameter of 7.0 mm. Assuming the starlight has a wavelength of 550 nm, what is the angular size of the image on the retina?

 (b) The distance from the pupil to the retina is 23 mm. What is the linear size of the image of the star?

 (c) At the midpoint on the retina (fovea), there are 150 000 light-sensitive cells (rods) per mm^2. How many of these cells are illuminated when the eye looks at a star?

67. A sailor uses a speaking trumpet to concentrate his voice into a beam. The opening at the front end of the speaking trumpet has a diameter of 25 cm. If the sailor emits a sound of wavelength 15 cm (this is roughly the wavelength a man emits when yelling "eeeee . . ."), what is the angular width of the central maximum of the beam of sound?

68. The antenna of a small radar transmitter operating at 1.5×10^{10} Hz consists of a circular dish of diameter 1.0 m. What is the angular width of the central maximum of the radar beam? What is the linear width at a distance of 5.0 km from the transmitter?

69. The Hubble Space Telescope placed into orbit above the atomsphere of the Earth has an aperture of 2.4 m. According to Rayleigh's criterion, what angular resolution can this telescope achieve with visible light of wavelength 550 nm? With ultraviolet light of wavelength 120 nm? How much better is this than the angular resolution of 0.5 seconds of arc achieved by telescopes on the surface of the Earth?

70. The radio telescope at Jodrell Bank (England) is a dish with a circular aperture of diameter 76 m. What angular resolution can this radio telescope achieve when operating at a wavelength of 21 cm?

*71. According to newspaper reports, a photographic camera on a Blackbird reconnaissance jet flying at an altitude of 27 km can distinguish detail on the ground as small as the size of a person.

 (a) Roughly, what angular resolution does this require?

 (b) According to Rayleigh's criterion, what minimum diameter must the lens of the camera have?

*72. For an optically perfect lens, the size of the focal spot is limited only by diffraction effects. Suppose that a lens of diameter 10 cm and focal length 18 cm is illuminated with parallel light of wavelength 550 nm. What is the angular width of the central maximum in the diffraction pattern? What is the corresponding linear width at the focal distance?

*73. Kalahari bushmen are said to be able to see the four brightest moons of Jupiter with the naked eye. According to Rayleigh's criterion, what must be the minimum separation between two small light sources placed at the distance of Jupiter if they are to be resolved by the human eye? Compare this with the separations between these four moons of Jupiter. Does the limit on the resolving power of the human eye prevent *you* from seeing the moons? The (average) distance to Jupiter is 6.3×10^8 km; the separations between the moons are typically 4.0×10^5 km. Assume that the diameter of the pupil of the eye is 7.0 mm and that the wavelength of the light is 550 nm.

*74. Mars has a radius of 3400 km; when Mars is at its closest to the Earth, its distance is 7.8×10^7 km. Calculate the angular size of Mars as seen from Earth. Estimate the diameter of the objective lens of the telescope of smallest size that will permit you to tell that Mars has a disk, that is, that the image of Mars is wider than the image of a star.

*75. (a) According to Rayleigh's criterion, what is the angular resolution that the human eye can achieve for light of wavelength 550 nm? The fully distended pupil of the human eye has a diameter of 7.0 mm.

 (b) Even during steady fixation, the eye has a spontaneous tremor that swings it through angles of 20 or 30 seconds of arc. Compare this angular tremor with the angular resolution that you found in part (a). Would the elimination of the tremor greatly improve the acuity of the eye?

*76. At night, on a long stretch of straight road in Nevada, a truck driver sees the distant headlights of another truck. How close must he be to the other truck in order for his eyes to resolve two headlights? Assume that the pupils of the truck driver have a diameter of 5.0 mm, that the headlights are separated by 1.8 m, and that the light has a wavelength of 550 nm.

*77. Some spy satellites carry cameras with lenses 30 cm in diameter and with a focal length of 2.4 m.

 (a) What is the angular resolution of the camera according to Rayleigh's criterion? Assume that the wavelength of light is 550 nm.

(b) If such a satellite looks down on the Earth from a height of 150 km, what is the distance between two points on the ground that the camera can barely resolve?

(c) The lens projects images of the two points on a film at the focal plane of the lens. What is the distance between the two images projected on the film?

*78. "7 × 50" binoculars magnify angles by a factor of 7.0, and their objective lenses have an aperture of 50 mm diameter.

(a) According to Rayleigh's criterion, what is the intrinsic angular resolution of these binoculars? Assume that the light has a wavelength of 550 nm.

(b) At best, the pupil of your eye has an aperture of 7.0 mm diameter. Compare the angular resolution of your eye divided by a factor of 7.0 with the intrinsic angular resolution of the binoculars. Which of the two numbers determines the actual angular resolution you can achieve while looking through the binoculars?

*79. When exposed to strong light, the pupil of the eye of a cat narrows to a fine slit, about 0.30 mm across. Suppose that the cat is looking at two white mice 20 m away and separated by a distance of 5.0 cm. Can the cat distinguish one mouse from the other? Assume that the wavelength of light is 550 nm.

REVIEW PROBLEMS

*80. A lens made of flint glass with an index of refraction of 1.61 is to be coated with a thin layer of magnesium fluoride with an index of refraction of 1.38 in order to reduce reflection.

(a) How thick should the layer be so as to give destructive interference for the perpendicular reflection of light of wavelength 550 nm seen in air?

(b) Does your choice of thickness permit *constructive* interference for the reflection of light of some other wavelength in the visible spectrum? (If it does, you ought to make a better choice.)

*81. When you look at the point on a spherical soap bubble nearest your eye (where the rays of light reaching your eye are perpendicular to the bubble's surface), you see strongly reflected red light, of a wavelength of 650 nm. What can you conclude about the thickness of the wall of the soap bubble? The index of refraction of the soapy water is 1.35.

*82. When you look at a spherical soap bubble, under the condition described in the preceding problem, you see strongly reflected light of a wavelength of 650 nm. Suppose you now inflate the bubble (without adding liquid) until the strongly reflected light becomes blue violet, of a wavelength of 430 nm. By what factor must you inflate the diameter of the bubble to accomplish this? The index of refraction of the soapy water is 1.35.

83. When a beam of monochromatic light is incident on two narrow slits separated by a distance 0.15 mm, the angle between the central beam and the third lateral maximum in the interference pattern is 0.52°. What is the wavelength of the light?

84. The red line in the spectrum of hydrogen has a wavelength of 656.3 nm; the blue violet line in this spectrum has a wavelength of 434.2 nm. If hydrogen light falls on a grating of 6000 slits per centimeter, what will be the angular separation (in degrees) of these two spectral lines as seen in the first-order spectrum?

*85. Two radio beacons are located on an east–west line and separated by a distance of 6.0 km. The radio beacons emit synchronous (in-phase) sinusoidal waves of a frequency of 1.0×10^5 Hz. The navigator of a ship wants to determine his position relative to the radio beacons. His radio receiver indicates zero signal strength at the position of the ship. What are the possible angular bearings of the ship relative to the radio beacons? Assume that the distance between the ship and the radio beacons is much larger than 6 km.

86. Crystal structures are most often determined by the diffraction of X rays from atoms. Consider atomic planes in a crystal as shown in Fig. 35.61. The incident and reflected X rays make an angle θ with the atomic planes. Show that the reflected rays will interfere constructively provided that

$$2d \sin\theta = m\lambda \qquad m = 1,\ 2,\ 3,\ \ldots$$

where d is the distance between planes. This result was first derived by W. L. Bragg and is known as the **Bragg law**; the interference peaks are known as *Bragg reflections*. Note that this relation has solutions only for $\lambda \le 2d$; thus, for atomic spacings, X rays must be used instead of visible light.

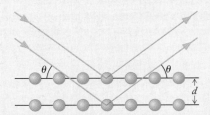

FIGURE 35.61 Reflection of X rays from two planes of atoms in a crystal.

*87. The radio waves from a transmitter to a receiver may follow either a direct path or else an indirect path involving a reflection on the ground (Fig. 35.62). This can lead to destructive interference of the two waves and a consequent fading of the radio signal at certain locations. Suppose that a transmitter and a receiver operating at a wavelength λ are at a height h on tall buildings with bare ground between. The distance from the emitter to the receiver is d.

(a) Show that the condition for destructive interference is

$$\sqrt{d^2 + 4h^2} - d = \lambda, \ 2\lambda, \ 3\lambda, \ldots$$

Note that in this calculation you must take into account the reversal, or change of phase, of the wave during reflection on the ground.

(b) If $h = 60$ m and $d = 2300$ m, what wavelengths will lead to destructive interference? What is the lowest-frequency radio wave for which destructive interference will occur?

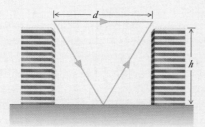

FIGURE 35.62 Antennas on two buildings.

88. Interference effects can be detected with sound waves passing through a picket fence. Suppose that the fence consists of vertical boards or rods separated by a distance of 20 cm. Suppose you stand on one side of the fence and a friend stands on the other side of the fence, at some reasonably large distance at an angle of 30° away from the direction perpendicular to the fence. If you clap your hands, producing white noise, what wavelengths and frequencies will your friend hear strongly?

89. A vehicle approaches at night from a large distance. Only resolving a single source, at first you think it is a motorcycle. But as it gets closer, you can distinguish two headlights. If the headlights are separated by 1.3 m and the pupil of your eye has a diameter of 7.0 mm, how far away was the vehicle when the headlights became distinguishable? Assume $\lambda = 580$ nm and otherwise perfect conditions.

*90. Two radio beacons emit waves of frequency 2.0×10^5 Hz. The beacons are on a north–south line, separated by a distance of 3.0 km. The southern beacon emits waves $\frac{1}{4}$ cycle later than the northern beacon. Find the angular directions for constructive interference. Reckon the angles relative to the east–west line, and assume that the distance between the beacons and the point of observation is much larger than 3 km.

91. The beam of a ruby laser, with light of wavelength 693 nm, is aimed at a slit of width 0.050 mm cut in an aluminum sheet. The diffracted light is intercepted by a screen placed 2.5 m beyond the slit. How far from the centerline in this screen will the first-order minimum be found? The second-order minimum?

*92. A slit of width 0.11 mm cut in a sheet of metal is illuminated with light of wavelength 577 nm from a mercury lamp. A screen is placed 4.0 m beyond the slit.

(a) Find the width of the central maximum in the diffraction pattern on the screen—that is, find the distance between the first minimum on the left and on the right.

(b) Find the width of the secondary maximum—that is, find the distance between the first minimum and the second minimum on the same side.

93. According to a recent proposal, solar energy is to be collected by a large power station on an artificial satellite orbiting the Earth at an altitude of 35 000 km. The power is to be beamed down to the surface of the Earth in the form of microwaves. If the microwaves have a wavelength of 10 cm and if the antenna emitting the microwaves is 1.5 km in diameter, what is the angular width of the central beam emerging from this antenna? What will be the transverse dimension of the beam when it reaches the surface of the Earth?

94. Rumor has it that a photographic camera on a spy satellite can read the license plate of an automobile on the ground.

(a) If the altitude of the satellite is 160 km, roughly what angular resolution does the camera need to read a license plate? Assume that the reading requires a linear resolution of about 5.0 cm.

(b) To attain the angular resolution, what must be the diameter of the aperture of the camera? Assume that the wavelength of light is 550 nm.

*95. A microwave antenna used to relay communication signals has the shape of a circular dish of diameter 1.5 m. The antenna emits waves with $\lambda = 4.0$ cm.

(a) What is the width of the central maximum of the beam of this antenna at a distance of 30 km?

(b) The power emitted by the antenna is 1.5×10^3 W. What is the energy flux directly in front of the antenna? What is the energy flux at a distance of 30 km? Assume that the power is evenly distributed over the width of the central beam.

Answers to Checkups

Checkup 35.1

1. At the center, there is no effective path difference (the flat and curved glass surfaces are essentially touching). The light reflected at the lower (flat) glass surface (a medium of higher index of refraction than air) suffers a 180° phase change. The light reflected from the upper (curved) glass surface (above the air gap) suffers no change in phase; thus these two waves are 180° out of phase and destructively interfere, providing a dark central spot.

2. Equations (35.1) and (35.2) are still valid. When the light was incident from above, the waves reflected from the two interfaces ($n_{air} < n_{oil}$ and $n_{oil} < n_{water}$) each suffer a 180° phase change; when the light is incident from below, neither one does. Thus, in either case, there is no relative phase change due to reflection, and the interference is determined solely by the path difference, as in Eqs. (35.1) and (35.2).

3. (A) 100 nm. For minimal reflection and the smallest thickness, we use the first destructive interference condition of Eq. (35.2) [and the relation $\lambda = \lambda_{air}/n$ of Eq. (35.3)], $d = \lambda/4 = \lambda_{air}/(4n) = (500 \text{ nm})/(4 \times 1.25) = 100$ nm.

Checkup 35.2

1. No; each ray is reflected twice (at an air-to-silver or glass-to-silver interface), and so each suffers a total phase change of 360°. Thus neither has any overall phase change, and there is no relative phase change.

2. Yes—the ray from mirror M_1 passes through the glass of mirror M three times, and the ray from mirror M_2 passes through it once, so there is a relative phase change due to the extra optical path difference. A correction for this extra path difference is provided by a *compensatory*, a slab of glass visible in Fig. 35.11. In practice, the path lengths in the Michelson interferometer need not be known exactly, since it is only the shift in fringes upon changing the interferometer orientation that needs to be determined.

3. (B) 0.10 mm. Because the light travels to and from the mirror, the path difference is twice the mirror motion, so for 400 fringes Eq. (35.4) gives $2d = 400\lambda$ or $d = 200\lambda = 200 \times 500$ nm = 0.10 mm.

Checkup 35.3

1. Since the product $d\sin\theta = \lambda$ is constant for the first maximum, decreasing the distance d between the slits results in an increase in the angle θ and thus an increase in the distance between the central and the first maximum.

2. For $d = \lambda$, Eq. (35.11) implies intensity maxima when $\sin\theta = 0$ or 1, so there is only a central maximum (and a maximum at $\theta = 90°$, which is at the border of the field of view). There will be one minimum in between for $\sin\theta = \frac{1}{2}$, that is, for $\theta = 30°$.

3. To have the first maximum at 45°, Eq. (35.8) or (35.11) requires that $\lambda = d\sin 45° = d/\sqrt{2}$.

4. (B) 30°. As in Eq. (35.9), the minima are at $d\sin\theta = \frac{1}{2}\lambda, \frac{3}{2}\lambda, \frac{5}{2}\lambda, \dots$. Thus the third-order minima satisfy $d\sin\theta = \frac{5}{2}\lambda$. With the given $d = 5\lambda$, this mean $5\lambda\sin\theta = \frac{5}{2}\lambda$, or $\sin\theta = \frac{1}{2}$. The angle whose sine is $\frac{1}{2}$ is 30°; the third-order minima occur at 30° on either side of the central maximum.

5. (D) $\frac{3}{4}I_{max}$. The intensity is given by Eq. (35.11); for a path difference as small as $\lambda/6$, the point in question is on the central maximum, slightly away from the peak. Since the path difference equals $d\sin\theta$, the intensity is $I = I_{max}\cos^2(\pi d\sin\theta/\lambda) = I_{max}\cos^2(\pi/6) = I_{max}(\sqrt{3}/2)^2 = \frac{3}{4}I_{max}$.

Checkup 35.4

1. The red line belongs to the second-order spectrum ($m = 2$ in $m\lambda = d\sin\theta$); the blue and blue violet lines belong to the third-order spectrum ($m = 3$). Since the wavelength λ is shorter for blue light, the value of 3λ for blue wavelengths occurs at the same θ as the value of 2λ for the longer red wavelengths; the orders overlap.

2. Since the mth-order spectrum is at $d\sin\theta = m\lambda$, each line of the second-order spectrum will be at approximately twice the angle of the first (approximately, since $\sin\theta \approx \theta$ becomes less accurate in the second order); thus, the angular separation will be twice as large in the second-order spectrum as in the first-order spectrum.

3. When the phase difference is $\pi/2$, the three phasors of amplitude E_0 form three sides of a square as in Fig. 35.26c; the resultant is the fourth side, which also has ampliude E_0.

4. At a maximum, the two phasors point the same way, so the amplitudes add to give a total amplitude of $3E_0$; thus, since intensity is proportional to the square of the amplitude, the intensity at the maxima is $9I_0$. At a minimum, the two phasors point in opposite directions, and so sum to give a total amplitude of E_0, corresponding to a minimum intensity of I_0. Halfway between a maximum and a minimum, the phasors point at right angles, so the net amplitude is $\sqrt{2^2 + 1^2}E_0 = \sqrt{5}E_0$, and so the intensity there is $5I_0$.

5. (C) $2\pi/5$. For identical, evenly spaced slits, the first minimum occurs when the five phasors close to form a pentagon; in this case, each is oriented at a phase angle $2\pi/5$ with respect to the previous one. Alternatively, one can see this result by noting that a full 2π of phase difference must be achieved by the five slits.

6. (D) 4. For two slits there are no secondary maxima, for three slits there is one, and so on, so that for six slits there are four secondary maxima. In general, for N slits, there are $N - 2$ secondary maxima.

Checkup 35.5

1. If you decrease the size of the narrow slit (comparable to the wavelength of light), the width of the illuminated zone beyond the slit grows wider for the light beam (a diffraction pattern), but narrower for the particle beam (an ordinary shadow). (We will see in Chapter 37 that particles also have wave properties; however, the corresponding wavelengths are usually much smaller than the wavelength of light and play no role unless the slit is *extremely* narrow.)

2. There will, of course, be a central maximum; the value $a = \sqrt{2}\lambda$ places the first minimum ($a \sin\theta = \lambda$) at $\sin\theta = 1/\sqrt{2}$, or $\theta = 45°$. Beyond that the intensity will increase and be close to the first secondary maximum at $\theta = 90°$, however, there will be no other maxima or minima.

3. No, light passing through a large aperture will form an ordinary shadow; the angle $\theta = \lambda/a$ still gives the direction of the first diffraction minimum at a sufficiently *large* distance from the slit.

4. (A) 5×10^{-3}. The first minimum occurs at $\sin\theta = \lambda/a = (500 \times 10^{-9}\text{ m})/(1 \times 10^{-4}\text{ m}) = 5 \times 10^{-3}$. Since this is small, we may use $\sin\theta \approx \theta$ to obtain $\theta \approx 5 \times 10^{-3}$.

Checkup 35.6

1. For a single slit, the first diffraction minimum of one image will coincide with the central diffraction peak of the other when $\theta = \lambda/a$, as in Eq. (35.21), where we assume small angles (as is always the case when resolving distant objects).

2. Although a is larger, the wavelength of radio waves is much larger than that of light ($4\text{ cm}/500\text{ nm} = 8 \times 10^4$). Thus, even though the aperture is 125 times larger, the resolution is hundreds of times poorer.

3. (D) 100 cm. The dots subtend a small angle $\theta = y/r$, where $y = 0.2$ mm is the dot separation and r is the distance. Thus, using Rayleigh's criterion, $y/r = \theta = 1.22\lambda/a$, with $\lambda \approx 600$ nm, we can solve for the largest resolving distance $r = ya/(1.22\lambda) = 140$ cm. Thus the dots will be resolved at 100 cm, but not at 300 cm.

Relativity, Quanta, and Particles

CONTENTS

The solar cells in the 73-meter long photovoltaic arrays on the International Space Station convert solar energy into electrical power. The arrays are fitted with gimbals that angle the arrays toward the Sun at all times so as to maximize the power supplied to the Space Station.

The Theory of Special Relativity

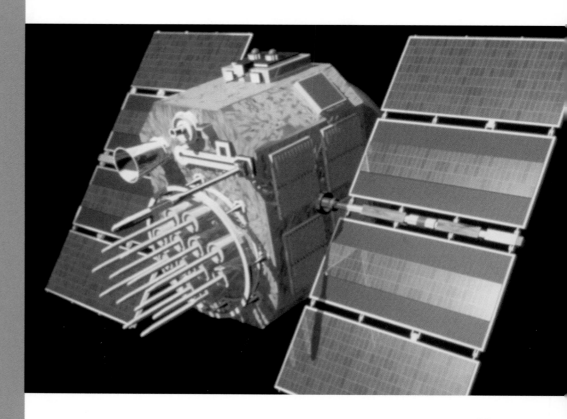

CONCEPTS IN CONTEXT

Concepts
— in —
Context

Determination of latitude and longitude by means of radio signals from Global Positioning System (GPS) satellites, such as the one shown here, requires measurement of the travel time from several satellites to the relevant point on the ground. GPS satellites are in relative motion with respect to the Earth's surface, but the speed of their radio signals is not affected by this motion.

In our study of the theory of Special Relativity we will study the propagation of light in different reference frames in relative motion, and we can then ask:

? How are distances calculated from GPS signals, and how do we know that the speed of light is unaffected by the motion of the satellite or by the translational motion of the Earth? (Section 36.1, page 1219)

? Relative to clocks on the surface of the Earth, clocks on GPS satellites are in motion at a somewhat high speed. How does this affect the rate of the clocks? (Example 2, page 1227)

? How does the motion of a GPS satellite relative to the Earth affect the frequency of the radio signal? (Example 3, page 1228)

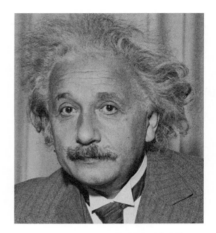

ALBERT EINSTEIN (1879–1955)
German (and Swiss, and American) theoretical physicist, professor at Zurich, at Berlin, and at the Institute for Advanced Study at Princeton. Einstein was the most celebrated physicist of the twentieth century. He formulated the theory of Special Relativity in 1905 and the theory of General Relativity in 1916. Einstein also made incisive contributions to modern quantum theory, for which he received the Nobel Prize in 1921. Einstein spent the last years of his life in an unsuccessful quest for a unified theory of forces that was supposed to incorporate gravity and electricity in a single set of equations.

As we saw in Chapter 5, Newton's laws of motion are equally valid in every inertial reference frame. All inertial reference frames are unaccelerated, but they can differ in their uniform translational motion. Since Newton's laws make no distinction between different inertial reference frames, no mechanical experiment can detect a uniform translational motion of one inertial reference frame by itself. Such motion can be detected only as a relative motion of one reference frame with respect to another reference frame. This is the Newtonian principle of relativity. For a concrete illustration of this principle, consider the reference frame of a cruise ship moving steadily away from the shore, without acceleration, and consider the reference frame of the shore. Both of these reference frames are inertial, and the behavior of a ball used in a game of tennis on the ship is not different from the behavior of a similar ball on shore—balls on the ship and on the shore accelerate in the same way when subjected to forces, and they obey the same laws of conservation of energy, conservation of momentum, etc.[1] Thus, experiments with such balls aboard the ship will not reveal the uniform motion of the ship relative to the shore. To detect this motion, the crew of the ship must take sightings of points on the shore or use some other navigational technique that fixes the position and velocity of the ship relative to the shore. Hence, in regard to mechanical experiments, uniform translational motion of our inertial reference frame is always *relative motion—it can be detected only as motion of our reference frame with respect to another reference frame.* There is no such thing as absolute motion.

The question naturally arises whether the relativity of motion indicated by mechanical experiments also applies to electric, magnetic, optical, and other experiments. Do any of these experiments permit us to detect an absolute motion or our reference frame? In 1905, Albert Einstein proposed that no experiment of any kind should ever permit us to detect such motion. He postulated a principle of relativity for *all* the laws of physics. This postulate serves as the foundation for Einstein's theory of Special Relativity, widely regarded as one of the greatest achievements of twentieth-century physics. The theory of relativity requires a drastic revision of our concepts of space and time, and it also requires a drastic revision of Newton's laws. *At high speeds—near the speed of light— particles obey new, relativistic laws which are quite different from Newton's laws.* However, at low speeds—small compared with the speed of light—the differences between Einstein's and Newton's theories are usually undetectable. Hence Newton's laws are adequate for describing the behavior of particles and of other bodies at the relatively low speeds we encounter in our everyday experience.

Before we deal with the details of Einstein's theory of relativity, we will briefly describe why nonmechanical experiments—and, especially, experiments with light— might be expected to detect absolute motion, which mechanical experiments cannot detect.

---

[1] This assumes the ship moves steadily. If the ship lurches forward, or pitches, or rolls, it ceases to be an inertial reference frame, and ball will behave in a manner "inconsistent" with Newton's laws.

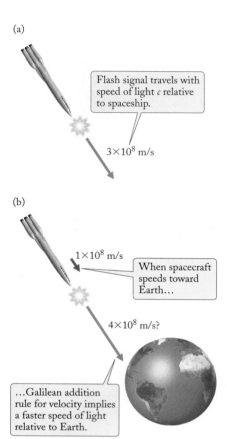

FIGURE 36.1 Velocities according to the Galilean addition rule in (a) reference frame of a spaceship and (b) reference frame of the Earth. A speed of light different from 3×10^8 m/s is at odds with Maxwell's equations.

ether

36.1 THE SPEED OF LIGHT; THE ETHER

Since the laws of mechanics are the same in all inertial reference frames, it might seem quite natural to assume that the laws of electricity, magnetism, and optics are also the same in all inertial reference frames. But this assumption immediately leads to a paradox concerning the speed of light. As we know from Chapter 33, light is an oscillating electric and magnetic disturbance propagating through space, and Maxwell's equations permit us to deduce that the speed of propagation of this disturbance must always be[2] 3.00×10^8 m/s. The trouble with this deduction is that, according to the Galilean addition rule for velocity [Eq. (4.53)], the speed of light ought *not* to be the same in all reference frames. For instance, imagine that an alien spaceship approaching the Earth with a speed of, say, 1.00×10^8 m/s flashes a light signal toward the Earth. If this signal has a speed of 3.00×10^8 m/s in the reference frame of the spaceship, then the Galilean addition rule tells us that it ought to have a speed of 4.00×10^8 m/s in the reference frame of the Earth (see Fig. 36.1).

To resolve this paradox, either we must give up the notion that the laws of electricity and magnetism (and the values of the speed of light) are the same in all inertial reference frames, or else we must give up the Galilean addition rule for velocities. Both alternatives are unpleasant: the former means that we must abandon all hope for a principle of relativity embracing electricity and magnetism, and the latter means that together with the Galilean addition rule we must abandon the transformation rule for position vectors measured in different reference frames [Eq. (4.50)] as well as the intuitively "obvious" notions of absolute time and length from which these rules are derived.

Since the failure of a relativity principle embracing electricity and magnetism might seem to be the lesser of two evils, let us first explore this alternative. Let us assume that there exists a preferred inertial reference frame in which the laws of electricity and magnetism take their simplest form, that is, the form expressed in Maxwell's equations, Eqs. (33.4)–(33.7). In this reference frame, the speed of light has its standard value of $c = 3.00 \times 10^8$ m/s, whereas in any other reference frame it is larger or smaller according to the Galilean addition rule. The propagation of light is then analogous to the propagation of sound. There exists a preferred reference frame in which the equations for the propagation of sound waves in, say, air take their simplest form: the reference frame in which the air is at rest. In this reference frame, sound has its standard speed of 331 m/s. In any other reference frame, the equations for the propagation of sound waves are more complicated, but the speed of propagation can always be obtained directly from the Galilean addition rule. For instance, if a wind of 40 m/s (a hurricane) is blowing over the surface of the Earth, then sound waves have a speed of 331 m/s relative to the air, but their speed relative to the ground depends on direction—downwind the speed is 371 m/s, whereas upwind it is 291 m/s.

This analogy between the propagation of light and of sound suggests that there might exist some pervasive medium whose oscillations bring about the propagation of light, just as the oscillations of air bring about the propagation of sound. Presumably this ghostly medium fills all of space, even the interplanetary and interstellar space, which is normally regarded as a vacuum. The physicists of the nineteenth century called this hypothetical medium the **ether**, and they attempted to describe light waves as oscillations of the ether, analogous to sound waves as oscillations of the air. The preferred reference frame in which light has its standard speed is then the reference frame in which the ether is at rest. The existence of such a preferred reference frame would

[2]The exact value of the speed of light is 2.997 924 58 $\times 10^8$ m/s, but throughout this chapter we will round this off to 3.00×10^8 m/s.

imply that velocity is absolute—the ether frame would set an absolute standard of rest, and the velocity of any body could always be referred to this frame. For instance, instead of describing the velocity of the Earth relative to some other material body, such as the Sun, we could always describe its velocity relative to the ether.

Presumably the Earth has some nonzero velocity relative to the ether. Even if the Earth were at rest in the ether at one instant, this condition could not last, since the Earth continually changes its motion as it orbits around the Sun. The motion of the ether past the Earth was called the *ether wind* by the nineteenth-century physicists (Fig. 36.2). If the Sun is at rest in the ether, then the ether wind would have a velocity opposite to the velocity of the Earth around the Sun—about 30 km/s; if the Sun is in (steady) motion, then the ether wind would vary with the seasons—it would reach a maximum when the orbital motion of the Earth is parallel to the motion of the Sun, and a minimum when antiparallel.

Experimenters attempted to detect this ether wind by its effects on the propagation of light. A light wave in a laboratory on the Earth would have a greater speed when moving downwind and a smaller speed when moving upwind or across the wind. If the speed of the ether wind "blowing" through the laboratory is V, then the speed of light in this laboratory would be $c + V$ for a light signal with downwind motion, $c - V$ for upwind motion, and $\sqrt{c^2 - V^2}$ for motion perpendicular to the wind (see Fig. 36.3). With a value of 300 000 km/s for c and a value of approximately 30 km/s for V, the increase or decrease of the speed of light amounts to only about 1 part in 10 000, and a very sensitive apparatus is required for the detection of this small change.

In a famous experiment first performed in 1881 and often repeated thereafter, A. A. Michelson and E. W. Morley attempted to detect small changes in the speed of light by means of an interferometer. The results of their experiment were negative. As discussed in Section 35.2, the sensitivity of the original experiment of Michelson and Morley was such that an ether wind of 5 km/s would have produced a detectable effect. Since the expected wind is about 30 km/s, the experimental result contradicts the ether theory of the propagation of light. Later, more refined versions of the experiment established that if there were an ether wind, its speed would certainly have to be less than 3 m/s. The experimental evidence therefore establishes conclusively that the motion of the Earth has no effect on the propagation of light. As the Earth moves around the Sun, its velocity is first in one direction, then in another, and another; and the Earth is first in one inertial reference frame, then in another, and another. But all these inertial reference frames appear to be completely equivalent in regard to the propagation of light. There is no preferred reference frame. There is no ether.

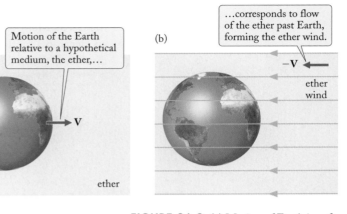

FIGURE 36.2 (a) Motion of Earth in reference frame of ether. (b) Motion of ether in reference frame of Earth.

Concepts — *in* — **Context**

Michelson-Morley experiment

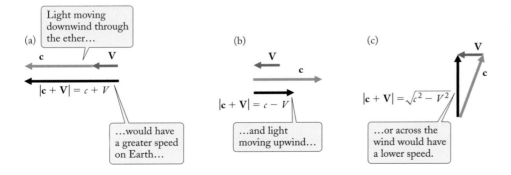

FIGURE 36.3 The velocity of light is **c** relative to the ether, and the velocity of the ether is **V** relative to the laboratory. The velocity of light relative to the laboratory is then the vector sum **c** + **V**. (a) If **c** and **V** are parallel, the magnitude of the vector sum is $c + V$. (b) If **c** and **V** are antiparallel, the magnitude of the vector sum is $c - V$. (c) If **c** and **V** form this right triangle, the magnitude of the vector sum is $\sqrt{c^2 - V^2}$.

Today, we take it for granted that the speed of light is unaffected by the motion of the Earth. Many experiments and instruments rely on the fact that the travel time for an electromagnetic signal between an emitter and a receiver depends only on the distance between the two; the speed of light, and thus the travel time, does not depend on the motion of the emitter or of the receiver. For example, accurate determination of a position on the ground using the Global Positioning System (GPS) is achieved by precisely measuring the travel time of signals from several satellites. When a GPS receiver calculates the distances to several satellites, it assumes that the speed of light is unaffected by the motion of the satellites and by the translational motion of the Earth.

✔ Checkup 36.1

QUESTION 1: You are in an open automobile (a convertible) traveling at 80 km/h. Does the speed of sound waves (relative to you) depend on direction?

QUESTION 2: At the equator of the Earth, the rotational speed is 0.46 km/s. Was the original Michelson–Morley experiment capable of detecting the corresponding ether wind?

QUESTION 3: If Michelson and Morley had detected an ether wind of 30 km/s opposite to the motion of the Earth around the Sun, what could they have concluded about the absolute motion of the Earth? The Sun?

QUESTION 4: In 1887, after observations lasting a few days, a null result was obtained in the Michelson–Morley experiment with 5 km/s sensitivity. To be sure that there was no ether, Michelson and Morley had to repeat the experiment

    (A) With a different color of light   (B) At 1 km/s resolution
    (C) At a different time of year

Online
Concept
Tutorial

36.2 EINSTEIN'S PRINCIPLE OF RELATIVITY

Neither the laws of mechanics nor the laws for the propagation of light reveal any intrinsic distinction between different inertial reference frames. This motivated Einstein to take a bold step and to propose a general hypothesis concerning *all* the laws of physics. This hypothesis is the **Principle of Relativity**:

Principle of Relativity

All the laws of physics are the same in all inertial reference frames.

In addition, Einstein proposed the **Principle of the Universality of the Speed of Light**:

Principle of Universality of Speed of Light

The speed of light (in vacuum) is the same in all inertial reference frames; it always has the value $c = 3.00 \times 10^8$ m/s.

These deceptively simple principles form the foundation of the theory of Special Relativity. As we pointed out in the preceding section, the universality of the speed of light conflicts with the Galilean addition rule for velocity. We will therefore have to discard this rule, and we will also have to discard the transformation rule for position vectors or coordinates on which it is based (see Section 4.6).

The universality of the speed of light also requires that we give up some of our intuitive, everyday notions of space and time. Obviously, the fact that a light signal always has a speed of 3.00×10^8 m/s, regardless of how hard we try to move toward it or away from it in a fast aircraft or spaceship, does violence to our intuition. This strange behavior of light is only possible because of a strange behavior of length and time in relativistic physics. As we will see later in this chapter, neither length nor time is absolute—they both depend on the reference frame in which they are measured, and they suffer contraction or dilation when the reference frame changes.

Before we can inquire into the consequences of Einstein's two principles, we must carefully describe the construction of reference frames and the synchronization of their clocks. A reference frame is a coordinate grid erected around some given origin and a set of clocks (see Fig. 36.4), which can be used to determine the space and time coordinates of any **event**. In relativity, as in everyday life, an event is an occurrence that happens at one point of space at one point of time (for example, a climber stepping onto the summit of Mt. McKinley). An event is therefore represented by one point of space and time. The space coordinates of the event are directly given by the coordinate grid intersections at the event. The time coordinate of the event is the time registered by the clock at the event. Different choices of reference frame result in different space coordinates and different time coordinates for the event, and in Section 36.5 we will see how these different coordinates in different reference frames are related.

event

The clocks of any chosen reference frame must be synchronized with each other and with the master clock sitting at its origin of coordinates. Einstein proposed that this synchronization can be accomplished by sending out a flash of light from a point exactly midway between the clock at the origin and the other clock (see Fig. 36.5). The two clocks are synchronized if both show exactly the same time when the light from the midpoint reaches them. Note that this synchronization procedure hinges on the universality of the speed of light. If the speed of light were not a universal constant, but were dependent on the reference frame and on the direction of propagation (say, faster toward the right in Fig. 36.5 and slower toward the left), then we could not achieve synchronization by the simple procedure with a flash of light from the midpoint.

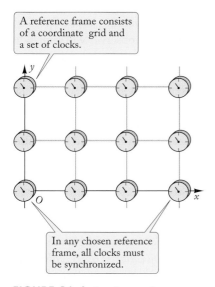

A reference frame consists of a coordinate grid and a set of clocks.

In any chosen reference frame, all clocks must be synchronized.

FIGURE 36.4 A reference frame.

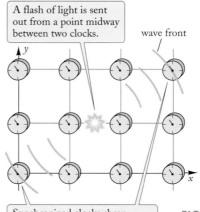

A flash of light is sent out from a point midway between two clocks.

wave front

Synchronized clocks show exactly the same time when light from midpoint reaches them.

FIGURE 36.5 Synchronization procedure for a pair of clocks. The leading wave front reaches the clocks at the lower left corner and the upper right corner simultaneously.

One immediate consequence of our synchronization procedure is that *simultaneity is relative, that is, the simultaneity of two events depends on the reference frame.* Einstein illustrated this with the following concrete example. Suppose that a train is traveling at high speed along a straight track and two bolts of lightning strike the front end and the rear end of the train, leaving scorch marks on the train and on the ground (the existence of these marks helps us to remember exactly where the lightning struck in each reference frame). Suppose that these two strokes of lightning are exactly simultaneous in the reference frame of the Earth. Then in the reference frame of the train, the two strokes of lightning will not be simultaneous—as judged by the clocks on the train, the stroke at the front end of the train occurs slightly earlier than the stroke at the rear end.

To see how this difference between the two reference frames comes about, let us apply our procedure for testing simultaneity. Suppose that in the reference frame of the Earth, an observer stands near the track at the midpoint between the two scorch marks the lightning made on the ground (see Fig. 36.6); she will then receive flashes of light from the lightning at her left and her right at the same instant. Thus, this observer will confirm that in the reference frame of the Earth, the lightning was simultaneous.

In the reference frame of the train, an observer can likewise test for simultaneity by placing himself exactly at the midpoint between the scorch marks the lightning made at the front and rear ends of the train (see Fig. 36.7) and waiting for the arrival of the flashes of light. Will he receive the flashes of light from the front and the rear of the train at the same instant? We can answer this by examining the motion of this observer and the propagation of the flashes of light in the reference frame of the ground (see Fig. 36.8). This observer is traveling toward the flash of light approaching him from the front end of the train, and he is traveling away from the flash of light trying to catch up with him from the back end of the train. Thus, this observer will encounter the flash of light from the front end before the flash of light from the rear end can catch up. In the reference frame of the ground, this delay between the flashes of light seen by the observer on the train is attributed to his motion toward one flash and away from the other. But in the reference frame of the train, the observer cannot attribute the delay to such a difference in motion—the light flashes from the front and the rear of the train originated at exactly equal distances from him, and, according to the Principle of Universality of the Speed of Light, they traveled at equal speeds. Hence this observer will conclude that the stroke of lightning at the front end of the train occurred earlier and the stroke of lightning at the rear end occurred later.

Although this qualitative argument shows that simultaneity depends on the reference frame, it does not tell us by how much. A quantitative calculation shows that for a train traveling at ordinary speed the delay is insignificant, 10^{-13} s or less. But the delay increases with the speed and it also increases with the distance between the flashes of lightning. For instance, consider a fast spaceship traveling by the Earth at 90% of the speed of light, and consider two flashes of lightning at two points on the Earth separated by a fairly large distance, say, Boston and New York, separated by a distance of 300 km. If these flashes are simultaneous in the reference frame of the Earth, they will differ by 0.001 s in the reference frame of the spaceship.

If simultaneity is relative, then *the synchronization of clocks is also relative.* In the reference frame of the Earth, all the clocks of this reference frame are synchronized, that is, the hands of these clocks reach the, say, noon position simultaneously. But when observed from the reference frame of

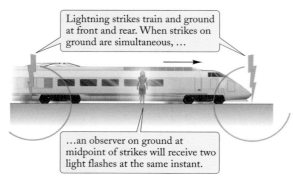

FIGURE 36.6 Observer on the ground at the midpoint between the scorch marks on the track watches for the arrival of flashes of light.

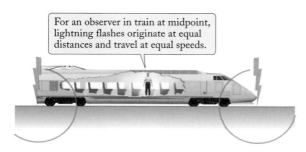

FIGURE 36.7 Observer in the train at the midpoint between the front and rear ends, where lightning has made scorch marks.

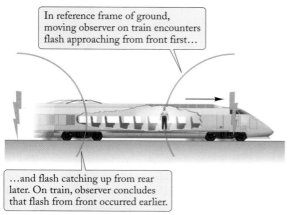

FIGURE 36.8 Observer on a train moving toward the right.

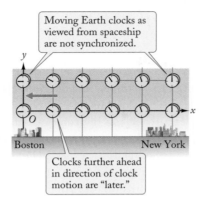

FIGURE 36.9 Clocks of the reference frame of the Earth as observed at one instant of spaceship time.

Moving Earth clocks as viewed from spaceship are not synchronized.

Boston   New York

Clocks further ahead in direction of clock motion are "later."

Clocks on moving spaceship as viewed from Earth are not synchronized.

Clocks further ahead in direction of clock motion are "later."

FIGURE 36.10 Clocks of the reference frame of the spaceship as observed at one instant of Earth time.

the train or the spaceship, the clocks in the Earth's reference frame that are further ahead in the direction of clock motion show earlier times. Consequently, they reach the noon position later than those on the right side—they are "late" in the same way as the lightning of the left side is late. This can be seen in Fig. 36.9, which shows the clocks belonging to the reference frame of the Earth as observed at one instant of time from a fast spaceship traveling in the direction from Boston to New York.

The effect is symmetric. In the reference frame of the spaceship, all the clocks onboard are synchronized. But, as observed from the reference frame of the Earth, the clocks on the front part of the spaceship are late. Figure 36.10 shows the clocks belonging to the reference frame of the spaceship as observed at one instant from the Earth. Note that in Fig. 36.9 we are viewing the reference frame of the Earth moving past the spaceship, and in Fig. 36.10 we are viewing the reference frame of the spaceship moving past the Earth. In either case, *the clocks on the leading edge of the reference frame are late.*

The relativity of synchronization is a direct consequence of the universality of the speed of light, since our procedure for testing simultaneity depends crucially on the speed of light. The failure of an absolute synchronization valid for all reference frames implies that there exists no absolute time. Each reference frame has its own way of reckoning time—*time is relative.* Instead of the single absolute time coordinate t we used in Newtonian physics, we now have to use a separate time coordinate for each individual reference frame.

 Checkup 36.2

QUESTION 1: A spaceship approaches the Earth at 1.00×10^8 m/s and sends a light signal toward the Earth, as in Fig. 36.1. According to Einstein, what is the speed of this light signal relative to the Earth?

QUESTION 2: Figure 36.10 shows the clocks of the reference frame of a spaceship in motion relative to the Earth. The clocks at the front of the spaceship are late. Does this mean that the crew of the spaceship find that their clocks are out of synchronization?

QUESTION 3: An earthquake occurs in San Francisco, and simultaneously (in Earth time) another earthquake occurs in New York. These earthquakes are not simultaneous as seen in the reference frame of a fast spaceship traveling westward in the direction from New York to San Francisco. Which is late?

QUESTION 4: A satellite, moving away from you at speed v, emits a pulse of radio waves in your direction. The speed of the waves relative to you is

(A) $c + v$ (B) $c - v$ (C) $\sqrt{c^2 - v^2}$ (D) c (E) v

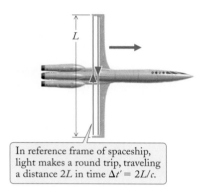

In reference frame of spaceship, light makes a round trip, traveling a distance 2L in time $\Delta t' = 2L/c$.

FIGURE 36.11 Spaceship with a "racetrack" for a light pulse.

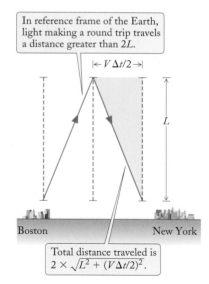

In reference frame of the Earth, light making a round trip travels a distance greater than 2L.

$|\leftarrow V\Delta t/2 \rightarrow|$

L

Boston   New York

Total distance traveled is $2 \times \sqrt{L^2 + (V\Delta t/2)^2}$.

FIGURE 36.12 The trajectory of the light pulse as observed from the Earth. The light pulse has both a vertical motion (up or down) and also a horizontal motion (toward the right).

36.3 TIME DILATION

The relativity of time shows up not only in the synchronization of clocks, but also in the rate of clocks. When experimenters on Earth observe a clock onboard the moving spaceship, they find that it suffers **time dilation** relative to the clocks on Earth: the rate of the moving clock is slow compared with the rate of identically manufactured clocks at rest on Earth. To see how this comes about, imagine that experimenters in the spaceship set up a racetrack of length L perpendicular to the direction of motion of the spaceship (see Fig. 36.11). If the experimenters in the spaceship use one of their clocks to measure the time of flight $\Delta t'$ of a light signal that goes from one end of the track to the other and returns to its starting point, they will then find that the light signal takes a time of $\Delta t' = 2L/c$ to complete the round trip.

But experimenters on the Earth see that the light signal has concurrent vertical and horizontal motions (see Fig. 36.12). For the experimenters on Earth, the light signal travels a total distance *larger* than 2L. Since the speed must still be the standard speed of light c, they will find that according to their clocks the light signal now takes a time Δt *longer* than $2L/c$ to complete the trip. Thus, a given time interval $\Delta t'$ registered by a clock on the spaceship is registered as a longer time interval Δt by the clocks on the Earth. This means that the clock on the spaceship runs slow when judged by the clocks on the Earth. Note that the experiment involves *one* clock on the spaceship (the clock at the starting point of the track), but several clocks on the Earth, because the light signal does not return to the point at which it started on Earth, and the observers on Earth will have to use one (stationary) clock at the starting point and another (stationary) clock at the end point to measure the time of flight.

For a quantitative evaluation of the time dilation, we note that in Fig. 36.12 the upward portion of the path of the light signal is the hypotenuse of a right triangle of sides L and $V \Delta t/2$, where V is the speed of the spaceship relative to the Earth. The total length of the path that the light signal has to cover in the reference frame of the Earth is therefore

$$2 \times \sqrt{L^2 + (V\Delta t/2)^2} \tag{36.1}$$

The time taken to cover this distance is

$$\Delta t = \frac{2 \times \sqrt{L^2 + (V\Delta t/2)^2}}{c} \tag{36.2}$$

If we square both sides of this equation, we obtain

$$(\Delta t)^2 = \frac{4L^2 + V^2(\Delta t)^2}{c^2} \tag{36.3}$$

which we can solve for $(\Delta t)^2$ and then for Δt:

$$(\Delta t)^2 = \frac{4L^2/c^2}{1 - V^2/c^2}$$

and

$$\Delta t = \frac{2L/c}{\sqrt{1 - V^2/c^2}} \tag{36.4}$$

Since, in the reference frame of the spaceship, $2L/c = \Delta t'$, this gives us

$$\Delta t = \frac{\Delta t'}{\sqrt{1 - V^2/c^2}} \qquad \begin{cases} \text{clock at rest in spaceship registers } \Delta t' \\ \text{clocks in Earth reference frame measure } \Delta t \end{cases} \qquad (36.5)$$

time dilation

This is the time-dilation formula. It shows that the time registered by the clocks on the Earth is longer than the time registered by the clock on the spaceship by a factor of $1/\sqrt{1 - V^2/c^2}$, that is, the clock on the spaceship runs slow when measured with the clocks on the Earth. Figure 36.13 is a plot of the time-dilation factor $1/\sqrt{1 - V^2/c^2}$ for speeds in the range from 0 to c. At low speeds the time-dilation effect is insignificant, but at speeds near c, it becomes quite large.

The slowing down of the rate of lapse of time applies to all physical processes—atomic, nuclear, biological, etc. Thus, the astronauts on the spaceship will perform all their tasks in "slow motion," and they will age slower than normal when measured with the clocks on the Earth. But they themselves will be unaware of this. From their point of view, they are in an inertial reference frame in which the usual laws of physics are valid, and the physical processes in their reference frame proceed at the normal rate, without any indication of anything unusual.

The time-dilation effect is symmetric: as measured by the clocks on the spaceship, a clock on the Earth runs slow by the same factor:

$$\Delta t' = \frac{\Delta t}{\sqrt{1 - V^2/c^2}} \qquad \begin{cases} \text{clock at rest on Earth registers } \Delta t \\ \text{clocks in spaceship reference frame measure } \Delta t' \end{cases} \qquad (36.6)$$

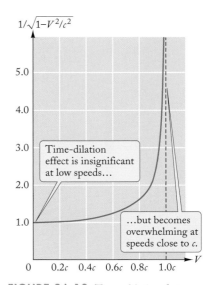

FIGURE 36.13 Time-dilation factor as a function of speed V.

The derivation of Eq. (36.6) can be based on an argument similar to that given above, with a racetrack for light at rest on the Earth. Incidentally: in these arguments we have implicitly assumed that the length of the racetrack is unaffected by the motion of the spaceship or the Earth, that is, we have assumed that the length is absolute. As we will see in the next section, this is true for lengths perpendicular to the direction of motion, although it is not true for lengths along the direction of motion.

EXAMPLE 1   Very drastic time-dilation effects have been observed in the decay of short-lived elementary particles. For instance, a muon particle (see Chapter 41) usually decays in about 2.2×10^{-6} s; but if it is moving at high speed through the laboratory, then the internal processes that produce the decay will slow down as judged by the clocks in our laboratory, and the muon lives a longer time. In accurate experiments performed at the European Organization for Nuclear Research (CERN) accelerator near Geneva, muons with a speed of 99.94% of the speed of light were found to have a lifetime 29 times as long as the lifetime of muons at rest. Is this dilation of the lifetime in agreement with Eq. (36.5)?

SOLUTION: We can regard the muon as a clock at rest in the reference frame of an (imaginary) spaceship with a speed of $V = 0.9994c$. Over the lifetime of the muon, this clock registers a time interval of $\Delta t' = 2.2 \times 10^{-6}$ s. However, the experimenters in the laboratory see the spaceship moving at $V = 0.9994c$, and over the lifetime of the muon they see their own clocks register a larger time Δt than the time $\Delta t'$ registered by the moving clock. According to Eq. (36.5), the times Δt and $\Delta t'$ differ by the time-dilation factor $1/\sqrt{1 - V^2/c^2}$, that is,

$$\frac{1}{\sqrt{1 - V^2/c^2}} = \frac{1}{\sqrt{1 - (0.9994c)^2/c^2}} = \frac{1}{\sqrt{1 - (0.9994)^2}} = \frac{1}{0.035} = 29$$

This time-dilation factor is in agreement with the experimental result.

At everyday speeds, the time-dilation effect is extremely small. For example, consider a clock aboard an airplane traveling at 300 m/s over the ground. The time-dilation factor is then

$$\frac{1}{\sqrt{1 - V^2/c^2}} = \frac{1}{\sqrt{1 - (300)^2/(3.00 \times 10^8)^2}} = \frac{1}{\sqrt{1 - 10^{-12}}} \qquad (36.7)$$

Evaluation of this gives 1.000 000 000 000 5, which means that a clock in the airplane will slow down by only 5 parts in 10^{13}! However, detection of such a small change is not beyond the reach of modern atomic clocks. In a notable experiment, scientists from the National Institute of Standards and Technology placed portable atomic clocks aboard a commercial airliner and kept them flying for several days, making a complete round trip around the world. Before and after the trip, the clocks were compared with an identical clock that was kept on the ground. The flying clocks were found to have lost time—in one instance, the total time lost because of the motion of the clock was about 10^{-7} s.

twin paradox

The time-dilation effect leads to the famous **twin paradox**, which we can state as follows: a pair of identical twins, Terra and Stella, celebrate their, say, twentieth birthday on Earth. Then Stella boards a spaceship that carries her at a speed of $V = 0.99c$ to Proxima Centauri, 4 light-years away; the spaceship immediately turns around and brings Stella back to Earth. According to the clocks on Earth, this trip takes about 4 years each way, so Terra's age will be 28 years when the twins meet again. But Stella has benefited from time dilation—relative to the reference frame of the Earth, the spaceship clocks run slow by a factor

$$\frac{1}{\sqrt{1 - V^2/c^2}} = \frac{1}{\sqrt{1 - (0.99)^2}} = \frac{1}{0.14} \qquad (36.8)$$

Hence 8 years of travel registered by the Earth clocks amount to only $8 \times 0.14 = 1.1$ years according to the spaceship clocks, and Stella's biological age on return will be only 21 years. Stella will be younger than Terra.

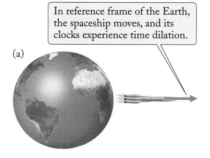

(a)

In reference frame of the Earth, the spaceship moves, and its clocks experience time dilation.

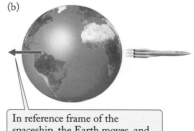

(b)

In reference frame of the spaceship, the Earth moves, and its clocks experience time dilation.

FIGURE 36.14 Time-dilation effect symmetry.

The paradox arises when we examine the elapsed times from the point of view of the reference frame of the spaceship. In this reference frame, the Earth is moving away from the spacecraft (see Fig. 36.14). Hence in this reference frame, the Earth clocks run slow—and Terra should be younger than Stella.

The resolution of this paradox hinges on the fact that our time-dilation formula is valid only if the time of a moving clock is measured from the point of view of an *inertial* reference frame. The reference frame of the Earth is (approximately) inertial, and therefore our calculation of the time dilation of the spaceship clocks is valid. But the reference frame of the spaceship is not inertial—the spaceship must decelerate when it reaches Proxima Centauri, stop, and then accelerate toward the Earth. If the reference frame is not inertial, the Principle of Relativity does not apply. Therefore, we cannot use the simple time-dilation formula to find the time dilation of the Earth clocks from the point of view of the spaceship reference frame. The "paradox" results from the misuse of this formula.

A detailed analysis of the behavior of the Earth clocks from the point of view of the spaceship reference frame establishes that the Earth clocks indeed do also run slow as long as the spaceship is moving with uniform velocity, but that the Earth clocks run *fast* when the spaceship is undergoing its acceleration to turn around at Proxima Centauri. The time that the Earth clocks gain during the accelerated portions of the trip more than compensates for the time they lose during the other portions of the trip. This confirms that Stella will be younger than Terra, even from the point of view of the spaceship reference frame.

EXAMPLE 2   The speed of a GPS satellite relative to a point on the Earth's surface is typically $V = 3.9 \times 10^3$ m/s. Assume that the clock on a GPS satellite is synchronized with the clocks of the Earth's reference frame at one instant. By how much do they differ 1.0 hour later? What is the corresponding distance error for a radio signal? In this calculation, ignore the rotation of the Earth and treat the satellite motion the same as motion with uniform velocity.

Concepts — in — Context

SOLUTION: A time interval Δt measured on Earth will differ from the interval $\Delta t'$ measured on the GPS satellite by the time-dilation factor:

$$\Delta t = \frac{\Delta t'}{\sqrt{1 - V^2/c^2}}$$

We wish to find the difference $\Delta t - \Delta t'$ when $\Delta t' = 1.0$ hour. Since the speed of the satellite is much less than the speed of light, we can simplify the time-dilation factor using the expansion $(1 + x)^n \approx 1 + nx$ for small x. Here, with $n = -\frac{1}{2}$ and $x = -V^2/c^2$, we have

$$\frac{1}{\sqrt{1 - V^2/c^2}} = \left(1 - \frac{V^2}{c^2}\right)^{-1/2} \approx 1 + \frac{1}{2}\frac{V^2}{c^2} \tag{36.9}$$

Inserting this into the time-dilation expression, we have

$$\Delta t = \Delta t' \times \left(1 + \frac{1}{2}\frac{V^2}{c^2}\right)$$

or

$$\Delta t - \Delta t' = \Delta t' \times \frac{1}{2}\frac{V^2}{c^2} \tag{36.10}$$

With $\Delta t' = 1.0$ h $= 3600$ s and $V = 3.9 \times 10^3$ m/s, we find that the clocks will differ by

$$\Delta t - \Delta t' = 3600 \text{ s} \times \frac{1}{2} \times \left(\frac{3.9 \times 10^3 \text{ m/s}}{3.00 \times 10^8 \text{ m/s}} \right)^2 = 3.0 \times 10^{-7} \text{ s}$$

If such a timing error were uncorrected and the satellite clock were used to determine the distance of a point on the Earth by light travel time, then the error in this distance would be

$$\Delta x = c \times (\Delta t - \Delta t') = 3.00 \times 10^8 \text{ m/s} \times 3.0 \times 10^{-7} \text{ s} = 90 \text{ m} \quad (36.11)$$

Since the accuracy of GPS positioning is required to be much better than that, corrections for the time dilation must be applied, in addition to many more mundane corrections, for example, those due to refraction and the decrease in the speed of radio waves in the atmosphere.

Finally, we point out that the time dilation of relativistic physics affects the **Doppler shift** of the frequency of light waves whenever the emitter is in motion relative to the receiver. Since we want to find the frequency detected by the *receiver*, we consider the reference frame of the receiver, and we pretend that this reference frame serves as the "medium" in which the light propagates. In Newtonian physics the Doppler shift for an emitter moving at speed V through a medium in which the emitted wave has a speed c is simply $f' = f/(1 \pm V/c)$, where f is the frequency radiated by the emitter, f' is the frequency detected by the receiver, and the positive sign applies if the emitter is receding, the negative sign if approaching [see the derivation of Eq. (17.17) for the case of sound waves]. In relativistic physics, the Doppler shift must also include the time dilation of the emitter. The time dilation of the period of the waves corresponds to a *reduction* in the detected frequency by the additional factor $\sqrt{1 - V^2/c^2}$. Including this factor, we have

$$f' = \frac{\sqrt{1 - V^2/c^2}}{1 \pm V/c} f \quad (36.12)$$

where now V is to be interpreted as the speed of the emitter relative to the receiver. Since $1 - V^2/c^2 = (1 - V/c)(1 + V/c)$, Eq. (36.12) can be simplified by appropriate cancellations:

relativistic Doppler shift

$$f' = \sqrt{\frac{1 - V/c}{1 + V/c}} f \qquad \text{for receding emitter}$$

$$\qquad (36.13)$$

$$f' = \sqrt{\frac{1 + V/c}{1 - V/c}} f \qquad \text{for approaching emitter}$$

Concepts — *in* — Context

EXAMPLE 3   A GPS satellite transmits radio (microwave) signals at a frequency of 1.575 GHz. Assume for simplicity that a GPS satellite is directly approaching your location on the Earth's surface with a relative speed of 3.9×10^3 m/s. By what factor is the frequency that you detect on Earth increased due to the ordinary Doppler shift of Newtonian physics? Due to the time dilation of relativistic physics?

SOLUTION: For an approaching emitter, the Newtonian result for the upward-shifted frequency that you receive is given solely by the denominator in Eq. (36.12) [see also Eq. (17.17)]:

$$f' = \frac{1}{1 - V/c} f$$

Taking the ratio of received to emitted frequencies and inserting the values, we obtain

$$\frac{f'}{f} = \frac{1}{1 - (3.9 \times 10^3 \text{ m/s})/(3.00 \times 10^8 \text{ m/s})} = 1.000\,013$$

This is an upward shift of 13 parts per million, or of about $(1.3 \times 10^{-5}) \times (1.575$ GHz$) = 20$ kHz.

The time dilation factor is

$$\sqrt{1 - \frac{V^2}{c^2}} \approx 1 - \frac{1}{2}\frac{V^2}{c^2}$$

which shifts the received frequency *downward* by a factor that differs from unity by

$$\frac{1}{2}\frac{V^2}{c^2} = \frac{1}{2} \times \frac{(3.9 \times 10^3 \text{ m/s})^2}{(3.00 \times 10^8 \text{ m/s})^2} = 8.4 \times 10^{-11}$$

or less than a tenth of a part per *billion*. Thus for such a "low" speed, the time dilation contribution to the frequency shift is completely negligible compared with the ordinary Doppler effect. We already used this fact when we examined police radar guns in Example 17.7.

 Checkup 36.3

QUESTION 1: Distinguish between the relativity of the synchronization of clocks and the relativity of the rates of clocks.

QUESTION 2: Consider the plot of the time-dilation factor given in Fig. 36.13. If we increase the speed of a clock by a factor of 2, do we increase the time-dilation factor by a factor of 2, or by less, or by more?

QUESTION 3: An astronaut aims a beam of light from a green laser pointer at you. If you approach the astronaut at $V = 0.10c$, do you detect blue light or yellow light?

QUESTION 4: A spaceship moves away from Earth at high speed. How do experimenters on the Earth measure a clock in the spaceship to be running? How do those in the spaceship measure a clock on the Earth to be running?

    (A) Slow; fast   (B) Fast; slow   (C) Slow; slow   (D) Fast; fast

36.4 LENGTH CONTRACTION

In the preceding sections we have seen that time is relative—both the synchronization of clocks and the rate of clocks depend on the reference frame. Now we will see that length is also relative. A measuring rod, or any other body, onboard the spaceship suffers **length contraction** along the direction of motion. The length of the moving measuring rod will be short when compared with the length of an identically manufactured measuring rod at rest on the Earth. The reason for this is that the length measurement of a moving body depends on simultaneity, and since simultaneity is relative, so is length.

Suppose that the spaceship, traveling from Boston toward New York, carries a measuring rod that has a length of, say, 300 km in the reference frame of the spaceship (see Fig. 36.15). To measure the length of this rod in the reference frame of the Earth, we station observers in the vicinity of New York and Boston with instructions to ascertain the positions of the front and the rear ends of the measuring rod at one instant of time, say, at noon.

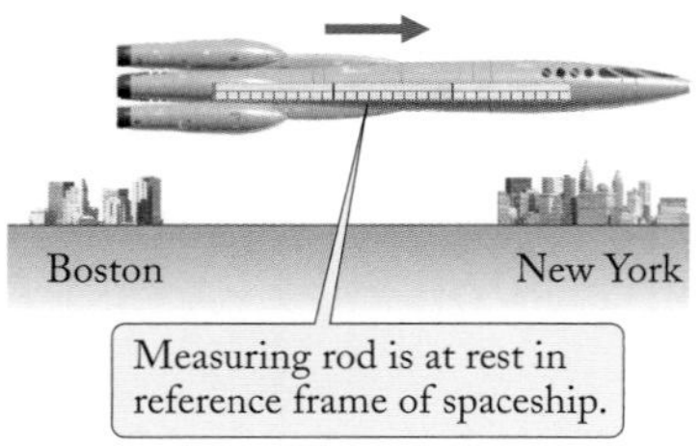

But when the observers on the Earth do this, the observers on the spaceship will claim that the position measurements were not done simultaneously, and that the observers in the vicinity of Boston measured the position of the rear end at a later time. In the extra time, the rear end moves an extra distance to the right, and hence the distance between the positions measured for the rear and the front ends will be reduced. From the point of view of the observers on the spaceship, it is therefore immediately obvious that the length measured by the observers on the Earth will be *short*. Figures 36.16 and 36.17 show the reference frame of the spaceship moving past the Earth and the reference frame of the Earth moving past the spaceship, respectively. In these figures the length contraction has been included (it was left out in Figs. 36.9 and 36.10). As illustrated in these figures, the length-contraction effect is symmetric: a body at rest in the spaceship will suffer contraction when measured in the reference frame of the Earth, and a body at rest on the Earth will suffer contraction when measured in the reference frame of the spaceship.

We can obtain a formula for the length contraction by exploiting the formula for the time dilation. Consider a rod of length L' at rest in the spaceship. According to the spaceship clocks, an observer on the Earth takes a time $\Delta t' = L'/V$ to travel from one end of the rod to the other. Taking time dilation into account, we then conclude that the observer on the Earth will judge that this takes a shorter time of only

$$\Delta t = \sqrt{1 - \frac{V^2}{c^2}}\, \Delta t' = \sqrt{1 - \frac{V^2}{c^2}}\, \frac{L'}{V} \tag{36.14}$$

However, the observer on the Earth cannot attribute this reduction of the travel time to a slowing of his clock—in his own reference frame his clock runs at a normal rate. Instead, he must attribute the reduction of travel time to a contraction of length of the rod. Measured in the reference frame of the Earth, the rod in the spaceship has some contracted length L, and it moves at speed V. Hence, the rod passes by a point on the Earth in a time L/V. This time must agree with the time Δt calculated in Eq. (36.14), so

$$\frac{L}{V} = \sqrt{1 - \frac{V^2}{c^2}}\, \frac{L'}{V} \tag{36.15}$$

or

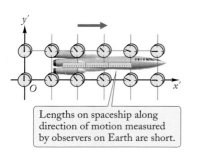

FIGURE 36.15 Spaceship with a measuring rod oriented along the direction of its motion.

Measuring rod is at rest in reference frame of spaceship.

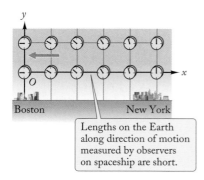

Lengths on spaceship along direction of motion measured by observers on Earth are short.

FIGURE 36.16 Reference frame of the spaceship as observed at one instant of Earth time (including length contraction; this length contraction was not included in Fig. 36.10).

Lengths on the Earth along direction of motion measured by observers on spaceship are short.

FIGURE 36.17 Reference frame of the Earth as observed at one instant of spaceship time (including length contraction; this length contraction was not included in Fig. 36.9).

$$L = \sqrt{1 - \frac{V^2}{c^2}} L' \quad \text{for a rod } L' \text{ at rest in spaceship} \qquad (36.16)$$

This is the formula for length contraction. According to this formula, the length of a rod or any body in motion relative to a reference frame is shortened by a factor of $\sqrt{1 - V^2/c^2}$. Figure 36.18 is a plot of the length-contraction factor $\sqrt{1 - V^2/c^2}$ as a function of speed.

As already mentioned, this contraction effect is symetric: if the rod is at rest on the Earth and is measured in the reference frame of the spaceship, the formula for the length contraction is

$$L' = \sqrt{1 - \frac{V^2}{c^2}} L \quad \text{for a rod } L \text{ at rest on Earth} \qquad (36.17)$$

The length contraction has not been tested directly by experiment. There is no practical method for a high-precision measurement of the length of a fast-moving body. Our best bet might be high-speed photography, but this is nowhere near accurate enough, since the contraction is extremely small even at the highest speeds that we can impart to a macroscopic body. Note, however, that the experimental evidence for time dilation can be regarded as indirect evidence for length contraction, since, as we saw above, the former implies the latter.

The contraction effect applies only to lengths along the direction of motion of the body. Lengths perpendicular to the direction of motion are unaffected. The proof of this is by contradiction: imagine that we have two identically manufactured pieces of pipe, one at rest on the Earth, one at rest on the spaceship (see Fig. 36.19). If the motion of the spaceship relative to Earth were to bring about a transverse contraction of the spaceship pipe, then, by symmetry, the motion of the Earth relative to the spaceship would bring about a contraction of the Earth pipe. These contraction effects are contradictory, since in one case the spaceship pipe would fit inside the Earth pipe, and in the other case it would fit outside.

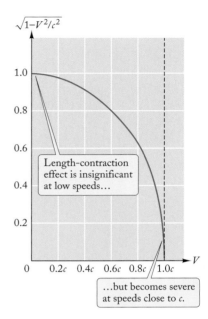

FIGURE 36.18 Length-contraction factor as a function of speed V.

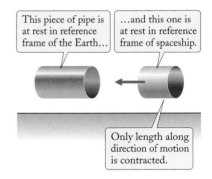

FIGURE 36.19 Two identical pieces of pipe.

EXAMPLE 4

A proton is passing by the Earth at a speed of 0.50c. In the reference frame of the proton, what is the length of the diameter of the Earth in a direction parallel to that of the motion of the proton? By how much is this shorter than the diameter in the reference frame of the Earth?

SOLUTION: We can regard the reference frame of the proton as the reference frame of a spaceship, and we can regard the diameter of the Earth as a rod at rest in the reference frame of the Earth. Relative to the proton, or the spaceship, the Earth has a speed $V = 0.50c$. In the reference frame of the Earth, the diameter has the familiar value $L = 1.3 \times 10^7$ m. But in the reference frame of the proton, or the spaceship, the Earth has a speed of $V = 0.50c$, and the rod is observed to have a shorter length L'. According to Eq. (36.16), the lengths L and L' differ by the length contraction factor $\sqrt{1 - V^2/c^2}$, that is,

$$\sqrt{1 - V^2/c^2} = \sqrt{1 - (0.50c)^2/c^2} = \sqrt{1 - 0.25} = 0.87$$

Since the diameter of the Earth in its own reference frame is $L = 1.3 \times 10^7$ m, the length L' in the reference frame of the proton is

$$L' = 0.87 \times L = 0.87 \times 1.3 \times 10^7 \text{ m} = 1.1 \times 10^7 \text{ m}$$

This is shorter than 1.3×10^7 m by 2×10^6 m, or 2000 km!

COMMENTS: The dimensions of the Earth perpendicular to the direction of motion do not contract. This implies that, in the reference frame of the proton, the Earth is not a sphere, but an ellipsoid.

From the length contraction of a three-dimensional body we can deduce the volume contraction. The volume of the Earth, which is calculated by taking a product of the dimension parallel to the motion and the two dimensions perpendicular to the motion, will be contracted by just one factor of $\sqrt{1 - V^2/c^2}$, that is, a factor of 0.87 in the case of Example 4.

 Checkup 36.4

QUESTION 1: A cannonball is perfectly round in its own reference frame. Describe the shape of the cannonball in a reference frame relative to which it is moving at a high speed, say, 0.5c.

QUESTION 2: Could we use the argument based on the two identical pieces of pipe (see Fig. 36.19) to prove that lengths *along* the direction of motion are not affected? Why, or Why not?

QUESTION 3: In view of the length contraction, does the density of a material depend on its speed?

QUESTION 4: A track is 100 m long. A particle moves parallel to the track with speed V such that $1 - V^2/c^2 = 0.25$. In the reference frame of the particle, the length of the track is

(A) 25 m   (B) 50 m   (C) 100 m   (D) 200 m   (E) 400 m

36.5 THE LORENTZ TRANSFORMATIONS AND THE COMBINATION OF VELOCITIES

Suppose that we measure the space and time coordinates of an event—such as the impact of lightning on a point on the ground—in the reference frame of the Earth and also in the reference frame of a moving spaceship. We will then obtain different values of these coordinates in the Earth and in the spaceship reference frames, but these different values of the coordinates are related by transformation formulas. In Einstein's physics, the transformation formulas for the coordinates are fairly complicated, because they are designed so as to keep the speed of light the same in all reference frames, and they incorporate the length contraction and the time dilation. Before dealing with these complicated formulas, let us examine the much simpler transformation formulas for coordinates in Newton's physics, where there is no length contraction and no time dilation.

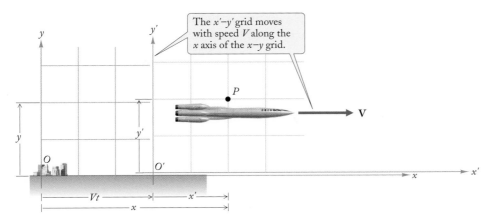

FIGURE 36.20 The coordinate grid x–y belonging to the reference frame of the Earth and the coordinate grid x′–y′ belonging to the reference frame of the spaceship. The coordinates of the point P are x, y in the first grid and x′, y′ in the second grid.

Figure 36.20 shows the coordinate grids x–y and x′–y′ of the first (Earth) and the second (spaceship) reference frames. The second reference frame is moving with velocity V along the x axis of the first reference frame. We assume that at time $t = 0$, the origins of the two coordinate grids coincide; at time t, the distance between the origins is then Vt. The coordinates of the point P are x, y in the first reference frame and x′, y′ in the second reference frame. By inspection of Fig. 36.20, we see that the distance x equals the sum of the distances x′ and Vt:

$$x = x' + Vt \tag{36.18}$$

Hence

$$x' = x - Vt \tag{36.19}$$

Furthermore, the distance y equals the distance y′

$$y' = y \tag{36.20}$$

Equations (36.19) and (36.20) are the transformation equations for the x and y coordinates in Newton's physics. These two equations are merely the x and y components of the general vector equation $\mathbf{r}' = \mathbf{r} - \mathbf{V}t$ for the transformation of the position vector we found in Chapter 4 [see Eq. (4.50)]. We could have obtained our equations for the transformation of the x and y coordinates from the general vector equation; but it is just as easy to re-derive these results by inspection of Fig. 36.20. Note that although Fig. 36.20 makes the equations for the transformation of the x and y coordinates seem self-evident, these equations hinge on the absolute character of length in Newton's physics. Absolute length means that the observers in the two reference frames agree on the measurement of any length or any distance between two points. If the observers disagreed on the values of the distances x or x′—for example, if one observer claimed that the distance x was 3.0 m and the other observer claimed that this distance x was contracted to 2.5 m—then Eq. (36.18) would not be valid. The left side of Eq. (36.18) is a distance defined at one instant in the first reference frame, whereas the right side is a sum of a distance x′ defined at one instant in the second reference frame and a distance (Vt) defined at one instant in the first reference frame, and such a sum makes no sense unless the observers agree on the values of these distances.

Furthermore, in Newton's physics time is absolute. This means that the times registered by the clocks in the two reference frames are always equal,

$$t' = t \tag{36.21}$$

Taken together, Eqs. (36.19)–(36.21) are called the **Galilean transformation equations**; they relate the space and time coordinates in one reference frame to those in the other.

From these equations we can deduce the Galilean addition rule for the components of the velocity. For instance, if the x coordinate changes by dx in a time dt, then Eqs. (36.19) and (36.21) give us

$$dx' = dx - V\,dt \tag{36.22}$$

$$dt' = dt \tag{36.23}$$

Dividing these two equations side by side, we obtain

$$\frac{dx'}{dt'} = \frac{dx}{dt} - V \tag{36.24}$$

Here, dx'/dt' is the x velocity of the particle, light signal, or whatever, measured in the second reference frame; and dx/dt is the x velocity measured in the first reference frame. Hence Eq. (36.24) says

$$v'_x = v_x - V \tag{36.25}$$

This, of course, is simply the Galilean addition rule for the x component of the velocity [see Eq. (4.53)].

In Einstein's physics, the Galilean formulas for the transformation of the coordinates and for the addition of velocities must be replaced by more complicated formulas, designed in such a way as to keep the speed of light the same in all reference frames. The transformation equations that accomplish this trick are called the **Lorentz transformations**. If the new reference frame moves, again, with velocity V along the x axis of the first reference frame, and if the origins coincide at time $t = 0$, the Lorentz transformations take the form

$$x' = \frac{x - Vt}{\sqrt{1 - V^2/c^2}} \tag{36.26}$$

$$y' = y \tag{36.27}$$

$$t' = \frac{t - Vx/c^2}{\sqrt{1 - V^2/c^2}} \tag{36.28}$$

These equations cannot be obtained by simple inspection of Fig. 36.20, because the distances displayed in this figure are not absolute in Einstein's physics, and they cannot simply be added as in Newton's physics.

We will not bother with a formal derivation of the Lorentz transformation equations, because we can achieve a clear understanding of the various terms and factors in these equations by comparing them with the Galilean transformation equations. Equation (36.26) differs from the Galilean equation (36.19) only by the factor $1/\sqrt{1 - V^2/c^2}$; this factor represents the length contraction. Equation (36.27) is identical to the Galilean equation, because lengths perpendicular to the direction of motion remain unchanged. And Eq. (36.28) differs from the Galilean equation in two ways: it contains an extra factor $1/\sqrt{1 - V^2/c^2}$ representing the time dilation, and it contains an extra term $-(Vx/c^2)/\sqrt{1 - V^2/c^2}$ representing the relativity of synchronization discussed in Section 36.2. We already gave quantitative treatments of the time dilation and the length contraction, and we therefore do not need to reexamine these here. But we did not yet give a quantitative treatment of the relativity of synchronization, and we will now examine this, to justify the presence of the extra term in Eq. (36.28).

Suppose that observers in the Earth frame send a light signal in the positive x direction from the origin O (with coordinate $x = 0$) to some point P (with coordinate $x > 0$). The signal leaves the origin O at time $t = 0$ and arrives at the point P at time t. According to the observers in the Earth frame, the arrival time is $t = x/c$, since the light signal travels at speed c. We want to know the arrival time t', as seen by the observers in the spaceship frame. For these observers, the light signal moves in the positive x' direction at speed c, while simultaneously the Earth frame and the point P move in the negative x' direction at speed V. Hence the "closing speed" of the light signal and the point P is $c + V$ (this closing speed is larger than c, but that is not objectionable; it merely reflects the fact that if the target and light signal are both moving toward each other, they will meet sooner than if the target is at rest). To calculate the time of arrival of the light signal at the point P, the spaceship observers have to divide the length OP by the closing speed $c + V$. But since the length OP is a moving length, these observers must take the length contraction into account: the length between O and P is not x, but is $x\sqrt{1 - V^2/c^2}$. Accordingly, the observers in the spaceship frame find that the arrival time of the light signal is

$$t' = \frac{x\sqrt{1 - V^2/c^2}}{c + V}$$

HENDRIK ANTOON LORENTZ
(1853–1928) *Dutch theoretical physicist, professor at Leiden. He investigated the relationship between electricity, magnetism, and mechanics. In order to explain the observed effect of magnetic fields on emitters of light (Zeeman effect), he postulated the existence of electric charges in the atom, for which he was awarded the Nobel Prize in 1902. He derived the Lorentz transformation equations by examining Maxwell's equations, but he was not aware that this leads to a new concept of space and time.*

Before we compare this with the value of t' given by the Lorentz transformation (36.28), let us multiply the numerator and denominator by $\sqrt{1 - V^2/c^2}$ and rearrange the result:

$$t' = \frac{x\sqrt{1 - V^2/c^2}}{c + V} \times \frac{\sqrt{1 - V^2/c^2}}{\sqrt{1 - V^2/c^2}} = \frac{x(1 - V^2/c^2)}{c(1 + V/c)\sqrt{1 - V^2/c^2}}$$

$$= \frac{x(1 + V/c)(1 - V/c)}{c(1 + V/c)\sqrt{1 - V^2/c^2}}$$

$$= \frac{x/c - Vx/c^2}{\sqrt{1 - V^2/c^2}} \qquad (36.29)$$

Here, the term x/c in the numerator is simply the time t that the light signal takes to arrive according to observers in the Earth frame. The term $-Vx/c^2$ in the numerator represents the relativity of synchronization. Comparing Eqs. (36.28) and (36.29), (with $x/c = t$), we see they agree exactly—and this agreement provides the justification of the extra term in Eq. (36.28).

EXAMPLE 5   A measuring rod of length $\Delta x'$ is at rest along the x' axis of the spaceship frame, which is moving in the positive x' direction with speed V relative to the Earth frame. What is the length of the measuring rod in the Earth frame according to the Lorentz transformation equations?

SOLUTION: We begin with Eq. (36.26) written as an equation for differences:

$$\Delta x' = \frac{\Delta x - V \Delta t}{\sqrt{1 - V^2/c^2}} \tag{36.30}$$

In the Earth frame, the length of the rod is measured at one instant t of time, so $\Delta t = 0$. Hence Eq. (36.30) reduces to

$$\Delta x' = \frac{\Delta x}{\sqrt{1 - V^2/c^2}}$$

which gives

$$\Delta x = \Delta x' \times \sqrt{1 - \frac{V^2}{c^2}}$$

This is the expected length-contraction formula (36.16).

EXAMPLE 6   A clock is at rest in the Earth frame. If a time Δt elapses as shown by this clock, how much time elapses according to the clocks of a spaceship moving at speed V relative to the Earth frame?

SOLUTION: Again, we begin by writing the Lorentz transformation equation (36.28) in terms of differences:

$$\Delta t' = \frac{\Delta t - V \Delta x/c^2}{\sqrt{1 - V^2/c^2}} \tag{36.31}$$

For the clock at rest in the Earth frame, $\Delta x = 0$, and therefore

$$\Delta t' = \frac{\Delta t}{\sqrt{1 - V^2/c^2}}$$

which is the expected time-dilation formula, Eq. (36.6).

COMMENT: Note that both the Lorentz transformation equations (36.26) and (36.28) have factors of $1/\sqrt{1 - V^2/c^2}$, even though the length contraction has a factor of $\sqrt{1 - V^2/c^2}$, not a factor of $1/\sqrt{1 - V^2/c^2}$. The reason becomes clear from inspection of these two examples: the Lorentz transformation equations (36.26) and (36.28) incorporate the time dilation for a clock at rest on Earth, but Eq. (36.26) provides the length of a rod at rest *in the spaceship frame*. If we want the length contraction for a rod at rest in the Earth frame, we would need to use the **inverse Lorentz transformation equations**, that is, the equations for x, t in terms of x', t'. These can be obtained by solving Eqs. (36.26) and (36.28) for x and t, that is, by pretending x and t are unknowns, to be evaluated by combining these equations to obtain each of them in terms of x' and t' (see Problem 33). The resulting equations have exactly the same form as Eqs. 36.26 and 36.28 with primed and unprimed space and time coordinates exchanged and with V replaced by $-V$.

inverse Lorentz transformations

Note that if the relative velocity between the two reference frames is small compared with the speed of light, then V/c in Eqs. (36.26) and (36.28) is small, and any term involving this quantity can be omitted in the equations. The Lorentz transformations then reduce to

$$x' \approx x - Vt$$

$$y' = y$$

$$t' \approx t$$

Thus, for low speeds, the Lorentz transformations reduce to the Galilean transformations (36.19)–(36.21).

The crucial feature of the Lorentz transformation equations is that they leave the speed of light unchanged. To verify this, we need to find the **relativistic combination rule for velocity**. If the x coordinate changes by dx in a time dt, then the Lorentz transformation equations tell us that

$$dx' = \frac{dx - Vdt}{\sqrt{1 - V^2/c^2}} \tag{36.32}$$

$$dt' = \frac{dt - Vdx/c^2}{\sqrt{1 - V^2/c^2}} \tag{36.33}$$

and dividing these two equations side by side, we find

$$\frac{dx'}{dt'} = \frac{dx - Vdt}{dt - Vdx/c^2} \tag{36.34}$$

On the right side we can divide both the numerator and the denominator by dt, with the result

$$\frac{dx'}{dt'} = \frac{dx/dt - V}{1 - V(dx/dt)/c^2} \tag{36.35}$$

In this expression, dx/dt is the x component of the velocity of a light signal or particle measured in the first reference frame and dx'/dt' is the x component of the velocity measured in the second reference frame. Hence Eq. (36.35) may be written

$$v_x' = \frac{v_x - V}{1 - v_x V/c^2} \tag{36.36}$$

relativistic velocity combination

This is the relativistic combination law for the x components of the velocity (there are somewhat different formulas for the combination of the y and z components of the velocity; see Problem 41).

It is instructive to compare the relativistic combination rule for velocities with the Galilean addition rule

$$v_x' = v_x - V \tag{36.37}$$

It is the denominator in Eq. (36.36) that makes all the difference. For instance, suppose that v_x is the velocity of a light signal propagating along the x axis of the first reference frame. Then $v_x = c$, and Eq. (36.36) yields

$$v_x' = \frac{c - V}{1 - cV/c^2} = \frac{c(1 - V/c)}{1 - V/c} = c \tag{36.38}$$

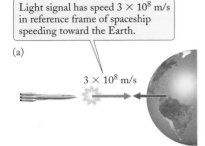

Light signal has speed 3×10^8 m/s in reference frame of spaceship speeding toward the Earth.

(a)

3×10^8 m/s

(b)

3×10^8 m/s

Light signal has the same speed of 3×10^8 m/s in reference frame of the Earth.

FIGURE 36.21 Addition of velocities according to the relativistic combination rule. A spaceship speeding toward the Earth emits a light signal toward the Earth. (a) Reference frame of the spaceship. (b) Reference frame of the Earth.

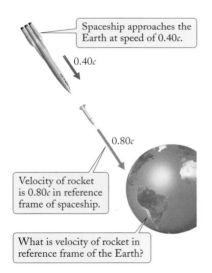

Spaceship approaches the Earth at speed of 0.40c.

$0.40c$

$0.80c$

Velocity of rocket is 0.80c in reference frame of spaceship.

What is velocity of rocket in reference frame of the Earth?

FIGURE 36.22 A spaceship launches a rocket toward Earth.

Thus, as required by the Principle of Universality of the Speed of Light, the velocity of the light signal in the second reference frame has exactly the same magnitude as in the first reference frame (see Fig. 36.21).

The relativistic combination rule for light velocities has been explicitly tested in an experiment at CERN, on the French-Swiss border; involving a beam of very fast pions. These particles decay spontaneously by a reaction that emits a flash of very intense, very energetic light (gamma rays). Hence, such a beam of pions can be regarded as a high-speed light source. In the experiment, the velocity of the pions relative to the laboratory was $V = 0.999\ 75c$. The Galilean addition law for velocity would have predicted laboratory velocities of $1.999\ 75c$ for light emitted in the forward direction and $0.000\ 25c$ for light emitted in the backward direction. But the experiments confirmed the relativistic combination rule—the laboratory velocity of the light had the same magnitude c in all directions.

EXAMPLE 7　An alien spaceship approaching the Earth at a speed of $0.40c$ fires a rocket at the Earth (see Fig. 36.22). If the velocity of the rocket is $0.80c$ in the reference frame of the spaceship, what is its velocity in the reference frame of the Earth?

SOLUTION: The equation for the combination of velocities is easiest to use if v_x is treated as known and v_x' as unknown. Accordingly, we take v_x to be the velocity in the reference frame of the spaceship, and we take v_x' to be the velocity in the reference frame of the Earth. The x axis is directed from the spaceship toward the Earth, and the velocity of the rocket in the reference frame of the spaceship is $v_x = 0.80c$. The velocity V must be taken to be that of the Earth relative to the spaceship; this velocity is negative, $V = -0.40c$. Then Eq. (36.36) gives

$$v_x' = \frac{v_x - V}{1 - v_x V/c^2} = \frac{0.80c - (-0.40c)}{1 - (0.80c)(-0.40c)/c^2} = 0.91c$$

✔ Checkup 36.5

QUESTION 1: Suppose that the new reference frame moves in the direction of the *negative x* axis of the first reference frame. What are the Lorentz transformation equations in this case?

QUESTION 2: How do we know that the Lorentz transformation equations are consistent with the requirement that the speed of light is left unchanged?

QUESTION 3: If the spaceship in Example 7 is moving away from the Earth instead of approaching the Earth, how does this change the answer for v_x'?

QUESTION 4: A radioactive nucleus approaches Earth at $v = c/2$ and emits an electron toward the Earth at $v = c/2$ relative to the nucleus. What is the electron's speed relative to the Earth?

(A) $c/4$　　　　(B) $4c/5$　　　　(C) $\sqrt{3/4}\,c$
(D) c　　　　(E) $4c/3$

36.6 RELATIVISTIC MOMENTUM AND ENERGY

The drastic revision that the theory of Special Relativity imposes on Newton's concepts of space and time implies a corresponding revision of the concepts of momentum and energy. The formulas for momentum and energy and the equations expressing their conservation are intimately tied to the transformation equations of the space and time coordinates. To see that this is so, we briefly examine the Newtonian (nonrelativistic) case.

In Newton's physics, the momentum of a particle of mass m and velocity $\mathbf{v}$ is

$$\mathbf{p} = m\mathbf{v} \tag{36.39}$$

To find the momentum of this particle in a new reference frame, we note that the Galilean transformation equation for the velocity vector is

$$\mathbf{v}' = \mathbf{v} - \mathbf{V} \tag{36.40}$$

Multiplying this by the mass, we find the transformation equation for the momentum:

$$\mathbf{p}' = m\mathbf{v}' = m\mathbf{v} - m\mathbf{V} \tag{36.41}$$

From this we see that the momentum $\mathbf{p}'$ in the new reference frame differs from the momentum $\mathbf{p}$ in the old reference frame by only the constant quantity $m\mathbf{V}$ (a quantity independent of the velocity $\mathbf{v}$ of the particle). Hence, if the total momentum of a system of colliding particles is conserved in one reference frame, it will also be conserved in the other reference frame—and the Law of Conservation of Momentum obeys the Principle of Relativity. This shows that the nonrelativistic formula for momentum and the nonrelativistic Galilean formula for the addition of velocities match in just the right way.

According to the relativistic physics of Einstein, we must replace the Galilean addition formula for velocities by the relativistic combination rule. If the Law of Conservation of Momentum is to obey the Principle of Relativity, we must then design a new relativistic formula for momentum that matches the new relativistic combination rule for velocities. The required relativistic formula for momentum is

$$\mathbf{p} = \frac{m\mathbf{v}}{\sqrt{1 - v^2/c^2}} \tag{36.42}$$

relativistic momentum

We will not give a proof of this formula.

If the speed of the particle is small compared with the speed of light, then $\sqrt{1 - v^2/c^2} \approx 1$ and Eq. (36.42) becomes approximately

$$\mathbf{p} \approx m\mathbf{v} \tag{36.43}$$

This shows that for low speeds, the relativistic and the Newtonian formulas for the momentum agree. We can therefore regard the Newtonian formula for the momentum as a simple and useful approximation for low speeds. This approximation is quite adequate for the description of all the phenomena we encounter in everyday life and (almost) all the phenomena we encounter in the realm of engineering, such as the phenomena we dealt with in the earlier chapters of this book. But at high speeds, the formulas differ drastically. We must then abandon the Newtonian formula, and rely entirely on the relativistic formula. Note that the relativistic momentum becomes infinite as the speed of the particle approaches the speed of light. Figure 36.23 is a plot of the magnitude of the momentum as a function of the speed.

EXAMPLE 8	An electron in the beam of a TV tube has a speed of 1.0×10^8 m/s. What is the magnitude of the momentum of this electron?

SOLUTION: For this electron, $v/c = (1.0 \times 10^8 \text{ m/s})/(3.0 \times 10^8 \text{ m/s}) = 0.33$.

According to Eq. (36.42), the magnitude of the momentum is then

$$p = \frac{mv}{\sqrt{1 - v^2/c^2}} = \frac{9.11 \times 10^{-31} \text{ kg} \times 1.0 \times 10^8 \text{ m/s}}{\sqrt{1 - (0.33)^2}} = 9.7 \times 10^{-23} \text{ kg·m/s}$$

COMMENTS: Note that if we had calculated the momentum according to the nonrelativistic formula $p = mv$, we would have obtained 9.1×10^{-23} kg·m/s, and we would have been in error by about 6%.

We also need a new relativistic formula for kinetic energy. This formula is

relativistic kinetic energy

$$K = \frac{mc^2}{\sqrt{1 - v^2/c^2}} - mc^2 \tag{36.44}$$

For low speeds, this relativistic formula for kinetic energy can be shown to agree approximately with the nonrelativistic formula $K = \frac{1}{2}mv^2$.

The relativistic kinetic energy becomes infinite as the speed of the particle approaches the speed of light. This indicates that, for any particle with mass (and for any body), the speed of light is unattainable, since it is impossible to supply a particle with an infinite amount of energy. Figure 36.24 is a plot of the kinetic energy vs. the speed.

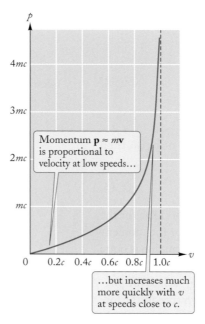

FIGURE 36.23 Momentum of a particle as a function of speed v.

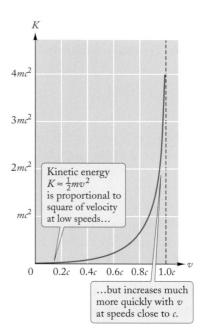

FIGURE 36.24 Kinetic energy of a particle as function of speed v.

EXAMPLE 9   The maximum speed that electrons achieve in the Stanford Linear Accelerator (SLAC) is 0.999 999 999 67c. What is the kinetic energy of an electron moving at this speed?

SOLUTION: The relativistic formula (36.44) contains a factor $\sqrt{1 - v^2/c^2}$. Since v/c for these electrons is extremely close to 1, most calculators are unable to evaluate $\sqrt{1 - v^2/c^2}$. To get around this difficulty, we write

$$\sqrt{1 - \frac{v^2}{c^2}} = \sqrt{1 + \frac{v}{c}}\sqrt{1 - \frac{v}{c}} \approx \sqrt{2}\sqrt{1 - \frac{v}{c}} \qquad (36.45)$$

and we evaluate $1 - v/c$ "by hand,"

$$1 - v/c = 1 - 0.999\ 999\ 999\ 67 = 3.3 \times 10^{-10}$$

The rest of the calculation is within the reach of our calculator:

$$K = mc^2 \left(\frac{1}{\sqrt{1 - v^2/c^2}} - 1 \right) \approx mc^2 \left(\frac{1}{\sqrt{2}\sqrt{1 - v/c}} - 1 \right)$$

$$= 9.11 \times 10^{-31}\ \text{kg} \times (3.00 \times 10^8\ \text{m/s})^2 \times \left(\frac{1}{\sqrt{2}\sqrt{3.3 \times 10^{-10}}} - 1 \right)$$

$$= 3.2 \times 10^{-9}\ \text{J}$$

Although the theory of Special Relativity requires a revision of the basic equations of mechanics, it does not require any revision of the basic equations of electricity and magnetism. Maxwell's equations are already relativistic, that is, they match the relativistic behavior of length and of time in just the right way. This concordance between Maxwell's equations and the requirements of relativity is no accident. Maxwell's equations incorporate a universal speed of light—they imply $c = 1/\sqrt{\mu_0 \epsilon_0}$ in every reference frame. Einstein's search for a theory of relativity was motivated by his faith in Maxwell's equations and his recognition that if Maxwell's equations were right then the Galilean coordinate transformations had to be wrong.

 Checkup 36.6

QUESTION 1: For a given speed, is the relativistic value of the momentum always larger than the Newtonian value? By what factor?

QUESTION 2: Is the relativistic momentum always in the direction of the velocity of the particle?

QUESTION 3: In science fiction stories, spaceships routinely reach speeds equal to or in excess of the speed of light. What is wrong with this?

QUESTION 4: According to the relativistic formula (36.44) for the kinetic energy, what is the kinetic energy of a particle of zero speed?

(A) 0   (B) mc^2 (C) $-mc^2$

(D) $2mc^2$ (E) Infinite

36.7 MASS AND ENERGY

One of the great discoveries that emerged from relativity is that energy can be transformed into mass, and mass can be transformed into energy. Thus, mass is a form of energy. The amount of energy contained in an amount m of mass at rest is given by Einstein's famous formula

rest mass energy

$$E = mc^2 \qquad (36.46)$$

The quantity mc^2 is called the **rest-mass energy**.[3] The formula (36.46) can be derived from the theory of relativity, but, as with some other equations in this chapter, we will not give the derivation.

The most spectacular demonstration of Einstein's mass–energy formula is found in the annihilation of matter and antimatter (see Chapter 41). If a proton collides with an antiproton, or an electron with an antielectron, the two colliding particles react violently and they annihilate each other in an explosion that generates an intense flash of very energetic light. According to Eq. (36.46), the annihilation of just 1000 kg of matter and antimatter (500 kg of each) would release an amount of energy

$$E = mc^2 = 1000 \text{ kg} \times (3.00 \times 10^8 \text{ m/s})^2 = 9.00 \times 10^{19} \text{ J} \qquad (36.47)$$

This is enough energy to satisfy the needs of the United States for a full year. Unfortunately, antimatter is not readily available in large amounts. On Earth, antiparticles can be obtained only from reactions induced by the impact of beams of high-energy particles on a target. These collisions occasionally result in the creation of a particle–antiparticle pair. Such pair creation is the reverse of pair annihilation. The creation process transforms some of the kinetic energy of the collision into mass, and a subsequent annihilation merely gives back the original energy.

But the relationship between energy and mass in Eq. (36.46) also has another aspect. *Energy has mass.* Whenever the internal energy stored in a body is changed, its rest mass (and weight) is changed. The change in rest mass that accompanies a given change of energy is

mass and energy changes

$$\Delta m = \Delta E / c^2 \qquad (36.48)$$

For instance, in the fission of uranium, the nuclear material loses energy, and correspondingly its mass (and weight) decreases. The complete fission of 1.0 kg of uranium releases an energy of 8.2×10^{13} J, and correspondingly the mass of the nuclear material decreases by $\Delta m = (8.2 \times 10^{13} \text{ J})/c^2 = 0.00091$ kg, or about 0.1%.

The fact that energy has mass indicates that energy is a form of mass. Conversely, as we have seen above, mass is a form of energy. Hence mass and energy must be regarded as different aspects of essentially the same thing. The laws of conservation of mass and conservation of energy are therefore not two independent laws—each implies the other. For example, consider the fission reaction of uranium inside the reactor vessel of a nuclear power plant (for details, see Chapter 40). The reaction conserves

---

[3] Throughout this section, *mass* means the mass that a body has when at rest or nearly at rest; to emphasize this, we use the term *rest mass*. The definition and measurement of mass for a body in motion at high (relativistic) speeds is rather tricky, because Newton's equation $m\mathbf{a} = \mathbf{F}$ fails and the direction of the acceleration is not necessarily the direction of the force. The only kind of mass that is unambiguously defined in Einstein's physics is the mass that the body has when at rest, and this is the only kind of mass we will consider.

energy—it merely transforms nuclear energy into heat, light, and kinetic energy but does not change the total amount of energy. The reaction also conserves mass—if the reactor vessel is hermetically sealed and thermally insulated from its environment, then the reaction does not change the mass of the contents of the vessel. However, if the vessel has an opening that lets some of the heat and light escape, then the mass of the residues will not match the mass of the original amount of uranium. The mass of the residues will be about 0.1% smaller than the original mass of the uranium. This mass defect represents the mass carried away by the energy that escapes. Thus, the nuclear fission reaction merely transforms energy into new forms of energy and mass into new forms of mass. In this regard, a nuclear reaction is not fundamentally different from a chemical reaction. The mass of the residues in an exothermic chemical reaction is slightly less than the original mass. The heat released in such a chemical reaction carries away some mass, but, in contrast to a nuclear reaction, this amount of mass is so small as to be quite immeasurable.

The total energy of a free particle in motion is the sum of its rest-mass energy (36.46) and its kinetic energy (36.44):

$$E = mc^2 + K = mc^2 + \frac{mc^2}{\sqrt{1 - v^2/c^2}} - mc^2 \qquad (36.49)$$

This leads to a simple formula for the **relativistic total energy** of the particle:

$$E = \frac{mc^2}{\sqrt{1 - v^2/c^2}} \qquad (36.50)$$

relativistic total energy

It is easy to verify (see Problem 66) that the relativistic energy can be expressed as follows in terms of the relativistic momentum:

$$E = \sqrt{c^2 p^2 + m^2 c^4} \qquad (36.51)$$

For an ultra-relativistic particle, moving at a speed close to that of light, the first term ($c^2 p^2$) within the square root is much larger than the second term ($m^2 c^4$). Hence, for such a particle we can ignore the second term, and we then obtain the simple result

$$E \approx \sqrt{c^2 p^2}$$

or

$$E \approx cp \quad \text{(ultra-relativistic particle)} \qquad (36.52)$$

Thus, the momentum and the energy of an ultra-relativistic particle are directly proportional.

EXAMPLE 10   Consider an electron of speed 0.999 999 999 67c, as in Example 9. What is the momentum of such an electron?

SOLUTION: Such as electron is ultra-relativistic. Its kinetic energy is much larger than its rest-mass energy, and the total energy is therefore approximately equal to the kinetic energy, which we have already calculated in Example 9:

$$E = mc^2 + K \approx K = 3.2 \times 10^{-9} \, \text{J}$$

Hence Eq. (36.47) yields

$$p \approx \frac{E}{c} = \frac{3.2 \times 10^{-9}\,\text{J}}{3.0 \times 10^{8}\,\text{m/s}} = 1.1 \times 10^{-17}\,\text{kg·m/s}$$

 Checkup 36.7

QUESTION 1: What is the rest-mass energy of a 1.0-kg piece of stone? Why can't we exploit this energy?

QUESTION 2: Does kinetic energy have mass? For instance, does the kinetic energy of the particles of a gas contribute to the overall mass of the gas?

QUESTION 3: For an ultra-relativistic particle (of speed near c), the momentum and the energy are proportional. Is this also true for a particle of lower speed, say, $0.9c$ or lower?

QUESTION 4: We must add energy to a hydrogen atom to ionize it and thus obtain a proton and an electron. A neutron, on the other hand, will spontaneously decay to provide a moving proton and a moving electron. From this information, which has a greater mass, the hydrogen atom or the neutron? Or do they have the same mass?

    (A) Hydrogen atom   (B) Neutron   (C) Both have same mass

SUMMARY

PRINCIPLE OF RELATIVITY   All the laws of physics are the same in all inertial reference frames.

PRINCIPLE OF UNIVERSALITY OF SPEED OF LIGHT   The speed of light is the same in all inertial reference frames.

TIME DILATION ($\Delta t'$ registered by clock in its own reference frame.)

$$\Delta t = \frac{\Delta t'}{\sqrt{1 - V^2/c^2}}$$

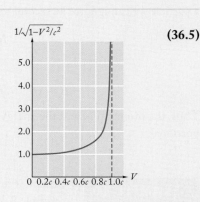

(36.5)

RELATIVISTIC DOPPLER SHIFT

$$f' = \sqrt{\frac{1 - V/c}{1 + V/c}}\, f \quad \text{for receding emitter}$$

$$f' = \sqrt{\frac{1 + V/c}{1 - V/c}}\, f \quad \text{for approaching emitter}$$

(36.13)

LENGTH CONTRACTION (L' is length of body in its own reference frame.)

$$L = \sqrt{1 - V^2/c^2}\, L'$$

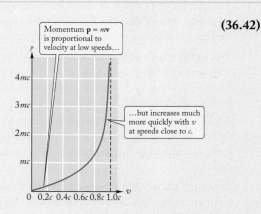

Lengths on spaceship along direction of motion measured by observers on Earth are short.

(36.16)

LORENTZ TRANSFORMATIONS

$$x' = \frac{x - Vt}{\sqrt{1 - V^2/c^2}}$$

(36.26)

$$y' = y$$

(36.27)

$$t' = \frac{t - Vx/c^2}{\sqrt{1 - V^2/c^2}}$$

(36.28)

RELATIVISTIC COMBINATION OF VELOCITIES

$$v_x' = \frac{v_x - V}{1 - v_x V/c^2}$$

(36.36)

RELATIVISTIC MOMENTUM

$$\mathbf{p} = \frac{m\mathbf{v}}{\sqrt{1 - v^2/c^2}}$$

Momentum $\mathbf{p} \approx m\mathbf{v}$ is proportional to velocity at low speeds…

…but increases much more quickly with v at speeds close to c.

(36.42)

RELATIVISTIC KINETIC ENERGY

$$K = \frac{mc^2}{\sqrt{1 - v^2/c^2}} - mc^2$$

Kinetic energy $K \approx \frac{1}{2}mv^2$ is proportional to square of velocity at low speeds…

…but increases much more quickly with v at speeds close to c.

(36.44)

RELATIVISTIC TOTAL ENERGY

$$E = \frac{mc^2}{\sqrt{1 - v^2/c^2}}$$

$$E = \sqrt{c^2 p^2 + m^2 c^4}$$

(36.50)

REST-MASS ENERGY

$$E = mc^2$$

(36.46)

MASS AND ENERGY CHANGES

$$\Delta m = \Delta E/c^2$$

(36.48)

QUESTIONS FOR DISCUSSION

1. An astronaut is inside a closed space capsule coasting through interstellar space. Is there any way the astronaut can measure the speed of the capsule without looking outside?

2. Why did Michelson and Morley use *two* light beams, rather than a single light beam, in their experiment?

3. When Einstein was a boy he wondered about the following question: A runner holds a mirror at arm's length in front of his face. Can he see himself in the mirror if he runs at (almost) the speed of light? Answer this question both according to the ether theory and according to the theory of Special Relativity.

4. Consider the piece of paper on which one page of this book is printed. Which of the following properties of the piece of paper are absolute, that is, which are independent of whether the paper is at rest or in motion relative to you? (a) The thickness of the paper, (b) the mass of the paper, (c) the volume of the paper, (d) the number of atoms in the paper, (e) the chemical composition of the paper, (f) the speed of light reflected by the paper, and (g) the color of the colored print on the paper.

5. Two streetlamps, one in Boston and the other in New York City, are turned on at exactly 6:00 P.M. Eastern Standard Time. Find a reference frame in which the streetlamp in New York was turned on late.

6. According to the theory of Special Relativity, the time order of events can be reversed under certain conditions. Does this mean that a sparrow might fall from the sky before it leaves the nest?

7. Because of the rotational motion of the Earth about its axis, a point on the equator moves with a speed of 460 m/s relative to a point on the North Pole. Does this mean that a clock placed on the equator runs more slowly than a similar clock placed on the pole?

8. According to Jacob Bronowski, author of *The Ascent of Man*, the explanation of time dilation is as follows: If you are moving away from a clock tower at a speed nearly equal to the speed of light, you keep pace with the light that the face of the clock sent out at, say, 11 o'clock. Hence, if you look toward the clock tower, you always see its hands at 11 o'clock. Is this explanation correct? If not, what is wrong with it?

9. Suppose you wanted to travel into the future and see what the twenty-fifth century is like. In principle, how could you do this? Could you ever return to the twenty-first century?

10. According to the arguments of Section 36.3, a light signal traveling along a track placed perpendicular to the direction of motion of the spaceship (see Fig. 36.11) takes a longer time to complete a round trip when measured by the clocks on the Earth than when measured by the clocks on the spaceship. Would the same be true for a light signal traveling along a track placed parallel to the direction of motion? Explain qualitatively.

11. A cannonball is perfectly round in its own reference frame. Describe the shape of this cannonball in a reference frame relative to which it has a speed of $0.95c$. Is the volume of the cannonball the same in both reference frames?

12. A rod at rest in the ground makes an angle of 30° with the x axis in the reference frame of the Earth. Will the angle be larger or smaller in the reference frame of a spaceship moving along the x axis?

13. In the charming tale "City Speed Limit" by George Gamow,[4] the protagonist, Mr. Tompkins, finds himself riding a bicycle in a city where the speed of light is very low, roughly 30 km/h. What weird effects must Mr. Tompkins have noticed under these circumstances?

14. A long spaceship is accelerating away from the Earth. In the reference frame of the Earth, are the instantaneous speeds of the nose and of the tail of the spaceship the same?

15. Suppose that a *very* fast runner holding a long horizontal pole runs through a barn open at both ends. The length of the pole (in its rest frame) is 6 m, and the length of the barn (in *its* rest frame) is 5 m. In the reference frame of the barn, the pole will suffer length contraction and, at one instant of time, all of the pole will be inside the barn. However, in the reference frame of the runner, the barn will suffer length contraction and all of the pole will never be inside the barn at one instant of time. Is this a contradiction?

16. Why can a spaceship not travel as fast as or faster than the speed of light?

17. If the beam from a revolving searchlight is intercepted by a distant cloud, the bright spot will move across the surface of the cloud very quickly, with a speed that can easily exceed the speed of light. Does this conflict with our conclusion of Section 36.6, that the speed of light is unattainable?

---

[4] George Gamow, *Mr. Tompkins in Wonderland.*

PROBLEMS

†36.1 The Speed of Light; the Ether

1. Consider the case where the Sun moves at a high speed v through the hypothetical ether. What are the minimum and maximum ether-wind speeds on Earth when the Sun moves through the ether at (a) 30 km/s and (b) 60 km/s? Assume the orbital speed of the Earth is 30 km/s.

*2. A Michelson–Morley interferometer determines the shift of two waves traveling in perpendicular directions (see also Section 35.2 and Fig. 35.9).

   (a) Assume that one wave travels a distance L along the ether wind with speed $c + V$ to a mirror, and back with speed $c - V$, as in Figs. 36.3a–b. Show that the round-trip time can be written

   $$t_{\parallel} = \frac{2L}{c}\left(1 - \frac{V^2}{c^2}\right)^{-1}$$

   (b) Assume the other wave travels the same distance perpendicular to the ether wind with speed $\sqrt{c^2 - V^2}$ (see Fig. 36.3c). Show that its round-trip time is

   $$t_{\perp} = \frac{2L}{c}\left(1 - \frac{V^2}{c^2}\right)^{-1/2}$$

   (c) Use the expansion $(1 - x)^n \approx 1 - nx$ for small x to show that the difference in arrival times is

   $$\Delta t = t_{\parallel} - t_{\perp} \approx \frac{LV^2}{c^3}$$

   (d) What fraction of a full period is this shift for light with $\lambda = 500$ nm? Use the values $L = 11$ m and $V = 30$ km/s.

*3. Ordinarily, the two arms of a Michelson–Morley interferometer cannot be set exactly equal, and instead have two values, L_1 and L_2. Insert these respective values into the results of Problem 2a and b and obtain a new expression for Δt (see Problem 2c). Note that this result alone cannot be used to determine the ether-wind speed, since the difference between L_1 and L_2 is not accurately known. In an actual experiment, the entire apparatus is rotated 90° (thus interchanging L_1 and L_2). Obtain an expression for the net shift by subtracting the differences in arrival times for the two orientations.

†,††36.2 Einstein's Principle of Relativity

4. A spaceship traveling at speed $\frac{1}{2}c$ relative to the Earth ejects a spacepod traveling in the forward direction at speed $\frac{1}{4}c$ relative to the spaceship. The spacepod emits a light signal toward the Earth at speed c relative to the spacepod. What is the speed of the light signal relative to the spaceship? What is the speed of the light signal relative to the Earth? Which observer (on the spaceship or on the spacepod) determines that the light strikes the Earth earlier?

††36.3 Time Dilation

5. If a moving clock is to have a time-dilation factor of 10, what must be its speed?

6. Neutrons have an average lifetime of 15 minutes when at rest in the laboratory. What is the average lifetime of neutrons of a speed of 25% of the speed of light? 50%? 90%?

7. Consider an unstable particle, such as a pion, which has a lifetime of only 2.6×10^{-8} s when at rest in the laboratory. What speed must you give such a particle to make its lifetime twice as long as when at rest in the laboratory?

8. The speed of the Sun around the center of our Galaxy is 200 km/s. Clocks in the Solar System will therefore run slow as compared with clocks at rest in the Galaxy. By what factor are the Solar System clocks slow?

9. The orbital speed of the Earth around the Sun is 30 km/s. In one year, how many seconds do the clocks on the Earth lose with respect to the clocks of an inertial reference frame at rest relative to the Sun? [Hint: If V/c is small, the approximation $\sqrt{1 - (V/c)^2} \approx 1 - \frac{1}{2}(V/c)^2$ is valid.]

10. In 1961, the cosmonaut G. S. Titov circled the Earth for 25 h at a speed of 7.8 km/s. According to Eq. (36.5), what was the time-dilation factor of his body clock relative to the clocks on Earth? By how many seconds did his body clock fall behind during the entire trip? (Hint: Use the approximation given in Problem 9.)

11. At a speed V, the time-dilation factor has some value. Suppose that at speed $2V$, the time-dilation factor has twice the previous value. What is the speed V?

12. An astronaut traveling at $V = 0.80c$ taps her foot 3.0 times per second. What is the frequency of taps determined by an observer on the Earth?

13. An atomic clock aboard a spaceship runs slow compared with an Earth-based atomic clock at a rate of 1.0 second per day. What is the speed of the spaceship?

14. A spaceship equipped with a chronometer is sent on a round-trip to Alpha Centauri, 4.4 light-years away. The spaceship travels at $0.10c$, and returns immediately.

   (a) According to clocks on the Earth, how long does this trip take?

   (b) According to the chronometer on the spaceship, how long does this trip take?

15. Consider the Doppler-shift formula for a receding source. By what factor does the frequency decrease for $V = 0.50c$? For $V = 0.70c$? For $V = 0.90c$?

---

† For help, see Online Concept Tutorial 41 at www.wwnorton.com/physics
†† For help, see Online Concept Tutorial 42 at www.wwnorton.com/physics

16. The frequencies of light received from distant galaxies and quasars are shifted due to the Doppler effect. Frequencies 5.0 times smaller than expected for a stationary source have been detected from receding quasars. What is V/c for such a quasar?

*17. In a test of the relativistic time-dilation effect, physicists compared the rates of vibration of nuclei of iron moving at different speeds. One sample of iron nuclei was placed on the rim of a high-speed rotor; another sample of similar nuclei was placed at the center. The radius of the rotor was 10 cm, and it rotated at 35 000 rev/min. Under these conditions, what was the speed of the rim of the rotor relative to the center? According to Eq. (36.5), what was the time-dilation factor of the sample at the rim compared with the sample at the center? (Hint: Use the approximation given in Problem 9.)

*18. If cosmonauts from the Earth wanted to travel to the Andromeda galaxy in a time of no more than 10 years as reckoned by clocks aboard their spaceship, at what (constant) speed would they have to travel? How much time would have elapsed on Earth after 10 years of time on the spaceship? The distance to the Andromeda galaxy is 2.2×10^6 light-years.

*19. Because of the rotation of the Earth, a point on the equator has a speed of 460 m/s relative to a point at the North Pole. According to the time-dilation effect of Special Relativity, by what factor do the rates of two clocks differ if one is located on the equator and the other at the North Pole? After 1.00 year has elapsed, by how many seconds will the clocks differ? Which clock will be ahead? (Although the special-relativistic time dilation slows one clock at the equator, there is an additional gravitational time dilation that slows the other clock. These two time-dilation effects balance, and the two clocks actually run at the same rate.)

**20. The star Alpha Centauri is 4.4 light-years away from us. Suppose that we send a spaceship on an expedition to this star. Relative to the Earth, the spaceship accelerates at a constant rate of 0.10g until it reaches the midpoint, 2.2 light-years from Earth. The spaceship then decelerates at a constant rate of 0.10g until it reaches Alpha Centauri. The spaceship performs the return trip in the same manner.

   (a) What is the time required for the complete trip according to the clocks on the Earth? Ignore the time that the spaceship spends at its destination.

   (b) What is the time required for the complete trip according to the clocks on the spaceship? Assume that the *instantaneous* time-dilation factor is still $\sqrt{1 - V^2/c^2}$ even though the speed V is a function of time.

36.4 Length Contraction

21. A meterstick is moving by an observer in a direction parallel to its length. The speed of the meterstick is 0.50c. What is its measured length in the reference frame of the observer?

22. According to the manufacturer's specifications, a spaceship has a length of 200 m. At what speed (relative to the Earth) will this spaceship have a length of 100 m in the reference frame of the Earth?

23. A cannonball flies through our laboratory at a speed of 0.30c. Measurement of the transverse diameter of the cannonball gives a result of 0.20 m. What can you predict for the measurement of the length, or the longitudinal diameter, of the cannonball?

24. What is the percent length contraction of an automobile traveling at 96 km/h? (Hint: Use the approximation given in Problem 9.)

25. A hangar for housing spaceships is 100 m long. How fast must a 200-m-long spaceship be traveling to (briefly) fit in the hangar?

26. A right triangle of sheet metal with two 45° angles lies in the x–y plane, with one of its sides along the x axis (see Fig. 36.25). The length of each side is 0.20 m, and the length of the hypotenuse is $\sqrt{2} \times 0.20$ m. Suppose that this triangle is observed from an x', y' reference frame moving at 0.80c along the x axis. What are the lengths of the sides and of the hypotenuse in this reference frame? What are the angles?

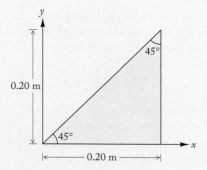

FIGURE 36.25 A triangle.

*27. Two identical spaceships are traveling in the same direction. An observer on Earth measures the first to have speed 0.80c and observes the second to be 1.50 times as long as the first one. What is the speed of the second spaceship?

*28. Suppose that a meterstick at rest in the reference frame of the Earth lies in the x–y plane and makes an angle of 30° with the x axis. Suppose that one end of the meterstick is at the origin. At a fixed time t, what are the x and y components of the displacement from this end of the meterstick to the other? At a fixed time t', what are the x' and y' components of the displacement from one end of the meterstick to the other in a new reference frame moving with velocity $V = 0.70c$ in the positive x direction? What is the angle the meterstick makes with the x' axis of this new reference frame?

*29. Electric charge is uniformly distributed throughout a sphere; the charge density is 2.0×10^{-6} C/m³. If this sphere is put in motion relative to the laboratory at a speed of 0.80c, what will be the charge density? Keep in mind that the total amount of electric charge is unchanged by the motion of the sphere.

*30. It can be shown that when a point charge moves at uniform velocity of relativistic magnitude, its pattern of electric field lines is contracted by the usual length-contraction factor $\sqrt{1 - V^2/c^2}$ in the longitudinal direction and is unchanged in the transverse direction. Figure 36.26 shows the resulting pattern of field lines for a speed $V = 0.60c$. Draw a similar picture for a speed of $0.80c$.

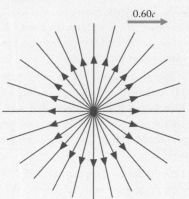

0.60c

FIGURE. 36.26 Electric field lines of a charge moving at $V = 0.60c$.

*31. A flexible drive belt runs over two flywheels whose axles are mounted on a rigid base (see Fig. 36.27). In the reference frame of the base, the horizontal portions of the belt have a speed v and therefore are subject to length contraction, which tightens the belt around the flywheels. However, in a reference frame moving to the right with the upper portion of the belt, the base is subject to length contraction, which ought to loosen the belt around the flywheels. Resolve this paradox by a qualitative argument. (Hint: Consider the lower portion of the belt as seen in the reference frame of the upper portion.)

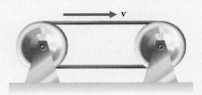

v

FIGURE 36.27 A drive belt and two flywheels.

36.5 The Lorentz Transformations and the Combination of Velocities

32. In the reference frame of the Earth, a firecracker is observed to explode at $x = 6.0 \times 10^8$ m, at $t = 4.0$ s. According to the Lorentz transformation equations, what are the x' and t' coordinates of this event as observed in the reference frame of a spaceship traveling in the x direction at a speed of $0.50c$? According to the Galilean transformation equations?

33. Obtain the **inverse Lorentz transformation equations** by solving Eqs. (36.26) and (36.28) for x and t, each in terms of x' and t'.

34. A spaceship has a length of 300 m, measured in its own reference frame. It is traveling in the positive x direction at a speed of $0.80c$ relative to the Earth. A strobe light at the nose of the spaceship sends a pulse of light toward the tail of the spaceship.

(a) As measured in the reference frame of the spaceship, how long does this light pulse take to reach the tail?

(b) As measured in the reference frame of the Earth, how long does this light pulse take to reach the tail?

35. A spaceship is moving at a speed of $0.60c$ toward the Earth. A second spaceship, following the first one, is moving at a speed of $0.90c$. What is the speed of the second spaceship as observed in the reference frame of the first?

36. Find the inverse of Eq. (36.36); that is, express v_x in terms of v_x'.

37. The captain of a spaceship traveling away from Earth in the x direction at $V = 0.80c$ observes that a nova explosion occurs at a point with spacetime coordinates $t' = -6.0 \times 10^8$ s, $x' = 1.9 \times 10^{17}$ m, $y' = 1.2 \times 10^{17}$ m as measured in the reference frame of the spaceship. He reports this event to the Earth via radio without delay.

(a) What are the spacetime coordinates of the explosion in the reference frame of the Earth? Assume that the master clock of the spaceship coincides with the master clock of the Earth at the instant $t = t' = 0$ when the midpoint of the spaceship passes by the Earth, and that the origin of the spaceship x', y' coordinates is at the midpoint of the spaceship.

(b) Will the Earth receive the captain's report before or after astronomers on the Earth see the nova explosion in their telescopes? No calculation is required for this question.

*38. Consider the situation described in Problem 37. Since light takes some time to travel from the nova to the spaceship, the space and time coordinates that the captain reports are not directly measured but, rather, deduced from the time of arrival and the direction of the nova light reaching the spaceship.

(a) At what time (t' time) did the nova light reach the spaceship?

(b) If the captain sends a report to Earth via radio as soon as he sees the nova, at what time (t time) does the Earth receive the report?

(c) At what time do Earth astronomers see the nova?

*39. At $11^h0^m0.0000^s$ A.M. a boiler explodes in the basement of the Museum of Modern Art in New York City. At $11^h0^m0.0003^s$ A.M. a similar boiler explodes in the basement of a soup factory in Camden, New Jersey, at a distance of 150 km from the first explosion. Show that, in the reference frame of a spaceship moving at a speed greater than $V = 0.60c$ from New York toward Camden, the first explosion occurs *after* the second.

*40. A radioactive atom in a beam produced by an accelerator has a speed $0.80c$ relative to the laboratory. The atom decays and ejects an electron of speed $0.50c$ relative to itself. What is the speed of the electron relative to the laboratory, if ejected in the forward direction? If ejected in the backward direction?

*41. In a manner similar to the procedure of Eqs. (36.32)–(36.36), show that relativistic combination formula for the y component of the velocity is

$$v_y' = \frac{v_y \sqrt{1 - V^2/c^2}}{1 - v_x V^2/c^2}$$

*42. Consider two speeds v_x and V, each of which is less than the speed of light. Show that if these speeds are combined by the relativistic combination formula, the result is always less than the speed of light.

*43. The speed of light with respect to a medium is $v_x' = c/n$, where n is the index of refraction. Suppose that the medium, say, flowing water, is moving past a stationary observer in the same direction as the light with speed V. Show that the observer measures the speed of light to be approximately

$$v_x = \frac{c}{n} + \left(1 - \frac{1}{n^2}\right)V$$

This effect was first observed by Fizeau in 1851.

**44. The acceleration of a particle in one reference frame is $a_x = dv_x/dt$, where the particle has instantaneous velocity v_x in that frame. Consider a reference frame moving with speed V parallel to the positive x axis of the first frame. Show that the acceleration in the second frame is given by

$$a_x' = \frac{dv_x'}{dt'} = a_x \frac{(1 - V^2/c^2)^{3/2}}{(1 - v_x V/c^2)^3}$$

36.6 Relativistic Momentum and Energy

45. Consider a particle of mass m moving at a speed of $0.10c$. What is its kinetic energy according to the relativistic formula? What is its kinetic energy according to the Newtonian formula? What is the percent deviation between these two results?

46. Suppose you want to give a rifle bullet of mass 0.010 kg a speed of 1.0% of the speed of light. What kinetic energy must you supply?

47. The yearly energy expenditure of the United Stated is about 8×10^{19} J. Suppose that all of this energy could be converted into kinetic energy of an automobile of mass 1000 kg. What would be the speed of this automobile?

48. The speed of an electron in a hydrogen atom is 2.6×10^6 m/s. For this speed, does your calculator show any difference between the kinetic energies calculated according to the relativistic formula and the Newtonian formula?

49. What is the speed of an electron if its kinetic energy is 1.6×10^{-13} J?

50. What is the momentum and what is the kinetic energy of an electron moving at a speed of one-half the speed of light?

51. Show that the momentum of a particle can be expressed in the concise form

$$\mathbf{p} = \frac{E\mathbf{v}}{c^2}$$

*52. What is the percent difference between the Newtonian and the relativistic values for the momentum of a meteoroid reaching the Earth at a speed of 72 km/s?

*53. Consider three accelerators that produce high-energy particles: the Large Hadron Collider, which will soon produce protons with kinetic energy 7 TeV; the Stanford Linear Accelerator, which produces electrons with kinetic energy 50 GeV; and the Relativistic Heavy Ion Collider, which produces gold nuclei (mass 197 u) with kinetic energy 20 TeV. In each case, calculate the difference $c - v$ between the speed of light and the speed of the particle.

*54. The most energetic cosmic-ray particles have energies of about 50 J. Assume that such a cosmic ray consists of a proton. By how much does the speed of such a proton differ from the speed of light? Express your answer in meters per second. [Hint: Use the approximation given in Eq. (36.45)].

*55. Consider the electrons of a speed of 0.999 999 999 67c produced by the Stanford Linear Accelerator. What is the magnitude of the momentum of such an electron?

*56. At the Fermilab accelerator, protons are given kinetic energies of 1.6×10^{-7} J. By how many meters per second does the speed of such a proton differ from the speed of light? What is the magnitude of the momentum of such a proton?

*57. A mass M at rest decays into two particles of masses m_1 and m_2. Use Eq. (36.51) to show that the magnitude of the momentum of each of the two particles is

$$p = \frac{\sqrt{M^2 - (m_1 + m_2)^2} \sqrt{M^2 - (m_1 - m_2)^2}\, c}{2M}$$

*58. A particle of mass m_1 is at rest. A second particle of mass m_2 and kinetic energy K strikes the first particle and sticks to it, a perfectly inelastic collision. Use Eq. (36.51) to show that the mass M of the composite particle is

$$M = \sqrt{(m_1 + m_2)^2 + \frac{2m_1 K}{c^2}}$$

*59. Show that the velocity of a relativistic particle can be expressed as follows:

$$\mathbf{v} = \frac{c\mathbf{p}}{\sqrt{m^2 c^2 + p^2}}$$

**60. At the Brookhaven AGS accelerator, protons of kinetic energy 5.3×10^{-9} J are made to collide with protons at rest.

(a) What is the speed of a moving proton in the laboratory reference frame?

(b) What is the speed of a reference frame in which the two colliding protons have the same speed (and are moving in opposite directions)?

(c) What is the total energy of each proton in the latter reference frame?

36.7 Mass and Energy

61. How much energy will be released by the annihilation of one electron and one antielectron (both initially at rest)? Express your answer in electron-volts.

62. The atomic bomb dropped on Hiroshima had an explosive energy equivalent to that of 20 000 tons of TNT, or 8.4×10^{13} J. How many kilograms of rest mass must have been converted into energy in this explosion?

63. The mass of the Sun is 2.0×10^{30} kg. The thermal energy in the Sun is about 2×10^{41} J. How much does the thermal energy contribute to the mass of the Sun? Express your answer in percent.

64. Combustion of one gallon of gasoline releases 1.3×10^8 J of energy. How much mass is converted to energy? Compare this with 2.8 kg, the mass of one gallon of gasoline.

65. The masses of the proton, electron, and neutron are $1.672\,623 \times 10^{-27}$ kg, 9.11×10^{-31} kg, and $1.674\,929 \times 10^{-27}$ kg, respectively. When a neutron decays into a proton and an electron, how much energy is released (other than the energy of the rest mass of the proton and electron)? Compare this extra energy with the energy of the rest mass of the electron.

66. From Eqs. (36.42) and (36.50) show that the relativistic energy and the relativistic momentum are related by

$$E^2 = c^2 p^2 + m^2 c^4$$

**67. A K^0 particle at rest decays spontaneously into a π^+ particle and a π^- particle. What will be the speed of each of the latter? The mass of the K^0 is 8.87×10^{-28} kg, and the masses of the π^+ and π^- particles are 2.49×10^{-28} kg each.

REVIEW PROBLEMS

*68. Muons are unstable particles which—if at rest in a laboratory—decay after a time of only 2.2×10^{-6} s. Suppose that a muon is created in a collision between a cosmic ray and an oxygen nucleus at the top of the Earth's atmosphere, at an altitude of 20 km above sea level.

   (a) If the muon has a downward speed of $v = 0.990c$ relative to the Earth, at what altitude will it decay? Ignore gravity in this calculation.

   (b) Without time dilation, at what altitude would the muon have decayed?

*69. Suppose that a special breed of cat (*Felis einsteinii*) lives for exactly 7.0 years according to its own body clock. When such a cat is born, we put it aboard a spaceship and send it off at $V = 0.80c$ toward the star Alpha Centauri. How far from the Earth (reckoned in the reference frame of the Earth) will the cat be when it dies? How long after the departure of the spaceship will a radio signal announcing the death of the cat reach us? The radio signal is sent out from the spaceship at the instant the cat dies.

*70. Suppose that a proton speeds by the Earth at $v = 0.80c$ along a line parallel to the axis of rotation of the Earth.

   (a) In the reference frame of the proton, what is the polar diameter of the Earth? The equatorial diameter?

   (b) In the reference frame of the proton, how long does the proton take to travel from the point of closest approach to the North Pole to the point of closest approach to the South Pole? In the reference frame of the Earth, how long does this take?

71. A spaceship travels in the positive x direction with speed $0.80c$. A man on Earth makes these observations:

   At $t = 0$, $x = 0$, a photon with $\lambda = 400$ nm is emitted at the rear of the ship moving toward the front.

   At $t = 1.00$ μs, $x = 960$ m, a photon with $\lambda = 600$ nm is emitted at the front of the spaceship moving toward the rear.

   A woman on the spaceship observes the same events.

   (a) What time interval does she measure between the two events? Which happens earlier?

   (b) What wavelengths does she measure for the two photons?

   (c) What is the length of the ship (as determined by the woman on it)?

*72. An observer on Earth sees one spaceship traveling away to the west at speed $0.40c$ and a second spaceship, also traveling away but to the east, at $0.70c$. Each spaceship emits a signal in its own reference frame at 2.00 GHz. What frequency does the Earth observer measure for each signal? What frequency does each spaceship measure for the signal from the other?

*73. Consider a cube measuring 1.0 m $\times$ 1.0 m $\times$ 1.0 m in its own reference frame. If this cube moves relative to the Earth at a speed of $0.60c$, what are its dimensions in the reference frame of the Earth? What are the areas of its faces? What is its volume? Assume that the cube moves in a direction perpendicular to one of its faces.

*74. A spaceship has a length of 200 m in its own reference frame. It is traveling at $0.95c$ relative to the Earth. Suppose that the tail of the spaceship emits a flash of light.

(a) In the reference frame of the spaceship, how long does the light take to reach the nose?

(b) In the reference frame of the Earth, how long does this take? Calculate the time directly from the motions of the spaceship and the flash of light; then compare it with the results calculated by applying the Lorentz transformations to the result obtained in (a).

75. Suppose that a spaceship is moving at a speed of $V = 0.20c$ relative to the Earth and a meteoroid is moving at a speed of $v_x = 0.10c$ relative to the Earth; in the same direction as the spaceship. What is the speed v'_x of the meteoroid relative to the spaceship according to Eq. (36.36)? What is the percent difference between this relativistic result and the Galilean result?

76. A collision between two gamma rays creates an electron and an antielectron that travel away from the point of creation in opposite directions, each with a speed of $0.95c$ in the laboratory. What is the speed of the antielectron in the reference frame of the electron, and vice versa?

*77. A spaceship traveling at $0.70c$ away from the Earth launches a projectile of muzzle speed $0.90c$ (relative to the spaceship). What is the speed of the projectile relative to the Earth if it is launched in the forward direction? In the backward direction?

*78. Three particles are moving along the positive x axis, in the positive direction. The first particle has a speed of $0.60c$ relative to the second, the second has a speed of $0.80c$ relative to the third, and the third has a speed of $0.50c$ relative to the laboratory. What is the speed of the first particle relative to the laboratory?

79. What is the kinetic energy of a spaceship of rest mass 50 metric tons moving at a speed of $0.50c$? How many metric tons of matter–antimatter mixture would have to be consumed to make this much energy available?

80. A particle has a kinetic energy equal to its rest-mass energy. What is the speed of this particle?

81. Suppose that a spaceship traveling at $0.80c$ through our Solar System suffers a totally inelastic collision with a small meteoroid of mass 2.0 kg.

(a) What is the kinetic energy of the meteoroid in the reference frame of the spaceship?

(b) In the collision all of this kinetic energy suddenly becomes available for inelastic processes that damage the spaceship. The effect on the spaceship is similar to an explosion. How many tons of TNT will release the same explosive energy? One ton of TNT releases 4.2×10^9 J.

*82. At the Brookhaven AGS accelerator, protons of kinetic energy 5.3×10^{-9} J are made to collide with protons at rest.

(a) What is the speed of one of these moving protons in the laboratory reference frame?

(b) What is the magnitude of the momentum?

*83. At the SSC accelerator that was to be built in the United States, protons would have been given kinetic energies of 3.2×10^{-6} J. What is the value of $c - v$ for such a proton, that is, by how many meters per second does the speed differ from the speed of light?

*84. Free neutrons decay spontaneously into a proton, an electron, and an antineutrino:

$$n \rightarrow p + e + \bar{\nu}$$

The neutron has a rest mass of 1.6749×10^{-27} kg; the proton, 1.6726×10^{-27} kg; the electron, 9.11×10^{-31} kg; and the antineutrino nearly zero. Assume that the neutron is at rest. Other than the rest-mass energy of the proton and electron, what is the energy released in this decay?

**85. A K^0 particle moving at a speed of $0.60c$ through the laboratory decays into a muon and an antimuon.

(a) In the reference frame of the K^0, what is the speed of each muon? The mass of the K^0 is 8.87×10^{-28} kg, and the masses of the muon and the antimuon are 1.88×10^{-28} kg each.

(b) Assume that the muon moves in a direction parallel to the original direction of motion of the K^0 and that the antimuon moves in the opposite direction. What are the speeds of the muon and the antimuon with respect to the laboratory?

Answers to Checkups

Checkup 36.1

1. Yes. Sound waves propagate in a medium (air), and thus the speed of sound waves relative to you depends on your speed relative to the air. The speed is largest for sound waves traveling opposite to your motion, from the front to the back of the convertible automobile.

2. An ether wind due only to rotation means that the Earth remains translationally at rest in the ether (the ether moves with the Earth). The original experiment had a sensitivity of about 5 km/s, and thus could not detect an ether wind due only to the rotational speed of 0.46 km/s.

3. Such an observation would have led to the conclusion that the Earth has an absolute motion of 30 km/s relative to the ether. Since this is the same as the velocity of the Earth relative to the Sun, such a result, if observed year round, would have also implied that the Sun is at rest with respect to the ether.

4. (C) At a different time of year. A single null result could have implied that the Earth was (coincidentally) nearly at rest with respect to the ether. Repeating the experiment when the Earth's velocity was in a different direction ensures that this was not the case.

Checkup 36.2

1. The speed of light in vacuum always has the same value, 3.00×10^8 m/s.

2. No; the clocks shown in Fig. 36.10 are as observed from the Earth. For the crew or anyone in the reference frame of the spaceship, the clocks are all synchronized.

3. Relative to a spaceship traveling westward, the Earth reference frame is traveling eastward. Thus, New York is at the leading edge of this reference frame, so the New York clocks, and the New York earthquake, are late.

4. (D) c. The speed of an electromagnetic wave is the speed of light; it does not depend on the speed of the emitter or receiver.

Checkup 36.3

1. Relativity of synchronization means that the times indicated by different clocks in a moving reference frame are different but all these clocks run at the same rate; relativity of rates means that the rate of the clocks in the moving reference frame differs from that of the clocks on the ground.

2. At low speeds, a factor of 2 increase in velocity has a negligible effect on the time-dilation factor, which remains nearly equal to unity. At higher speeds, the time-dilation factor increases more quickly; for example, an increase from $0.45c$ to $0.90c$ increases the time-dilation factor from 1.1 to 2.3, more than a factor of 2, and higher speeds result in larger-factor increases.

3. An approaching receiver is the same as an approaching emitter, since only relative motion matters; and the lower relation in Eq. (36.13) applies. Thus the frequency increases. So the wavelength decreases, and you detect blue light.

4. (C) Slow; slow. The time-dilation effect is symmetric, so observers in each reference frame measure a clock in the other reference frame to be running slow.

Checkup 36.4

1. Like the shape of the Earth moving at high speed relative to the reference frame of the proton in Example 4, the cannonball is an ellipsoid, flattened along the direction of motion, in a reference frame relative to which it is moving at $0.5c$.

2. No; contraction along the direction of motion does not affect which pipe fits inside which. Moreover, any apparent contradiction can be explained in terms of differences in simultaneity at any two different positions along the direction of motion.

3. Yes. Since the volume decreases with increasing speed, the density, or mass per unit volume, increases with increasing speed.

4. (B) 50 m. The 100-m track is contracted by the factor $\sqrt{1 - V^2/c^2} = \sqrt{0.25} = 0.50$, that is, to a length of 0.50×100 m = 50 m.

Checkup 36.5

1. The Lorentz transformation equations will be the same as Eqs. (36.26)–(36.28), but now with a negative value of V.

2. We know the Lorentz transformation equations are consistent with a speed of light that is unchanged because they lead directly to the relativistic velocity combination law, Eq. (36.36), which gives $v'_x = c$ when $v_x = c$.

3. This is the same as if the Earth is moving away from the spaceship, that is, V is now positive, and so the relativistic velocity combination rule gives $v'_x = (0.80c - 0.40c)/(1 - 0.80 \times 0.40) = 0.59c$.

4. (B) $4c/5$. As in Example 7, we can obtain the relative speed from the velocity combination rule, Eq. (36.36): $v'_x = [(c/2) - (-c/2)]/[1 - (1/2) \times (-1/2)] = c/(5/4) = (4/5)c$.

Checkup 36.6

1. Yes; other than for speed $v = 0$, the relativistic value of the momentum, given by Eq. (36.42), is always larger than the Newtonian value, by the factor $1/\sqrt{1 - v^2/c^2}$.

2. Yes, the momentum vector $\mathbf{p}$ is proportional to the velocity $\mathbf{v}$, and so $\mathbf{p}$ is always in the same direction as $\mathbf{v}$.

3. For any mass, attaining the speed of light would require infinite kinetic energy, and this is impossible.

4. (A) 0. With $v = 0$ in the first term of Eq. (36.44), the two terms in the realistic kinetic energy cancel.

Checkup 36.7

1. From $E = mc^2$ we have 1.0 kg $\times (3.00 \times 10^8$ m/s$)^2 = 9.0 \times 10^{16}$ J. This energy can't be converted to useful forms of energy, except by annihilation, which would require an equal amount of (unavailable) antimatter.

2. Yes. For example, the mass of a warm container of gas is greater than the mass of a cold container of gas, due to the additional kinetic energy

3. No. This is only true when $v \approx c$. For example, for $v \gtrsim 0.99c$, the energy becomes very nearly proportional to the momentum.

4. (B) Neutron. Since the neutron produces a proton, an electron, and some kinetic energy, its total energy (its mass) must be greater than the hydrogen atom, which requires added energy just to separate the proton and electron.

Appendix 1: Greek Alphabet

A	α	alpha	N	ν	nu
B	β	beta	Ξ	ξ	xi
Γ	γ	gamma	O	o	omicron
Δ	δ	delta	Π	π	pi
E	ϵ	epsilon	P	ρ	rho
Z	ζ	zeta	Σ	σ	sigma
H	η	eta	T	τ	tau
Θ	θ	theta	Υ	υ	upsilon
I	ι	iota	Φ	ϕ	phi
K	κ	kappa	X	χ	chi
Λ	λ	lambda	Ψ	ψ	psi
M	μ	mu	Ω	ω	omega

Appendix 2: Mathematics Review

A 2.1 Symbols

$a = b$ means a equals b

$a \neq b$ means a is not equal to b

$a > b$ means a is greater than b

$a < b$ means a is less than b

$a \geq b$ means a is not less than b

$a \leq b$ means a is not greater than b

$a \propto b$ means a is proportional to b

$a \approx b$ means a is approximately equal to b

$a \gg b$ means a is much greater than b

$a \ll b$ means a is much less than b

$\pi = 3.141\ 59 \ldots$

$e = 2.718\ 28 \ldots$

A 2.2 Powers and Roots

For any number a, the nth *power* of the number is the number multiplied by itself n times. This is written as a^n, and n is called the **exponent**. Thus,

$$a^1 = a \quad a^2 = a \cdot a \quad a^3 = a \cdot a \cdot a \quad a^4 = a \cdot a \cdot a \cdot a \quad \text{etc.}$$

For instance,

$$3^2 = 3 \times 3 = 9 \quad 3^3 = 3 \times 3 \times 3 = 27 \quad 3^4 = 3 \times 3 \times 3 \times 3 = 81 \quad \text{etc.}$$

A negative exponent indicates that the number is to be divided n times into 1; thus

$$a^{-1} = \frac{1}{a} \quad a^{-2} = \frac{1}{a^2} \quad a^{-3} = \frac{1}{a^3} \quad \text{etc.}$$

A zero exponent yields 1, regardless of the value of a:

$$a^0 = 1$$

The rules for the combination of exponents in products, in ratios, and in powers of powers are

$$a^n \cdot a^m = a^{m+n}$$

$$\frac{a^n}{a^m} = a^{n-m}$$

$$(a^n)^m = a^{nm}$$

For instance, it is easy to verify that

$$3^2 \times 3^3 = 3^5$$

$$\frac{3^2}{3^3} = 3^{-1} = \frac{1}{3}$$

$$(3^2)^3 = 3^{2 \times 3} = 3^6$$

Note that for any two numbers a and b

$$(a \cdot b)^n = a^n \cdot b^n$$

For instance,

$$(2 \times 3)^3 = 2^3 \times 3^3$$

The nth root of a is a number such that its nth power equals a. The nth root is written $a^{1/n}$. The second root $a^{1/2}$ is usually called the square root, and designated by $\sqrt{a}$:

$$a^{1/2} = \sqrt{a}$$

As suggested by the notation $a^{1/n}$, roots are fractional powers, and they obey the usual rules for the combination of exponents:

$$(a^{1/n})^n = a^{n/n} = a$$

$$(a^{1/n})^m = a^{m/n}$$

A 2.3 Arithmetic in Scientific Notation

The scientific notation for numbers (see the first page of the Prelude) is quite handy for the multiplication and the division of very large or very small numbers, because we can deal with the decimal parts and the power-of-10 parts in the numbers separately. For example, to multiply 4×10^{10} by 5×10^{12}, we multiply 4 by 5 and 10^{10} by 10^{12}, as follows:

$$(4 \times 10^{10}) \times (5 \times 10^{12}) = (4 \times 5) \times (10^{10} \times 10^{12})$$

$$= 20 \times 10^{10+12} = 20 \times 10^{22} = 2 \times 10^{23}$$

To divide these numbers, we proceed likewise:

$$\frac{4 \times 10^{10}}{5 \times 10^{12}} = \frac{4}{5} \times \frac{10^{10}}{10^{12}} = 0.8 \times 10^{10-12} = 0.8 \times 10^{-2} = 8 \times 10^{-3}$$

When performing additions or subtractions of numbers in scientific notation, we must be careful to begin by expressing the numbers with the same power of 10. For example, the sum of 1.5×10^9 and 3×10^8 is

$$1.5 \times 10^9 + 3 \times 10^8 = 1.5 \times 10^9 + 0.3 \times 10^9 = 1.8 \times 10^9$$

A 2.4 Algebra

An equation is a mathematical statement that tells us that one quantity or a combination of quantities is equal to another quantity or combination. We often have to solve for one of the quantities in the equation in terms of the other quantities. For instance, we may have to solve the equation

$$x + a = b$$

for x in terms of a and b. Here a and b are numerical constants or mathematical expressions which are regarded as known, and x is regarded as unknown.

The rules of algebra instruct us how to manipulate equations and accomplish their solution. The three most important rules are:

1. Any equation remains valid if equal terms are added or subtracted from its left side and its right side.

This rule is useful for solving the equation $x + a = b$. We simply subtract a from both sides of this equation and find

$$x + a - a = b - a$$

that is,

$$x = b - a$$

To see how this works in a concrete numerical example, consider the equation

$$x + 7 = 5$$

Subtracting 7 from both sides, we obtain

$$x = 5 - 7$$

or

$$x = -2$$

Note that given an equation of the form $x + a = b$, we may want to solve for a in terms of x and b, if x is already known from some other information but a is a mathematical quantity that is not yet known. If so, we must subtract x from both sides of the equation, and we obtain

$$a = b - x$$

Most equations in physics contain several mathematical quantities which sometimes play the role of known quantities, sometimes the role of unknown quantities, depending on circumstances. Correspondingly, we will sometimes want to solve the equation for one quantity (such as x), sometimes for another (such as a).

2. Any equation remains valid if the left and the right sides are multiplied or divided by the same factor.

This rule is useful for solving

$$ax = b$$

We simply divide both sides by a, which yields

$$\frac{ax}{a} = \frac{b}{a}$$

or

$$x = \frac{b}{a}$$

Often it will be necessary to combine both of the above rules. For instance, to solve the equation

$$2x + 10 = 16$$

we begin by subtracting 10 from both sides, obtaining

$$2x = 16 - 10$$

or

$$2x = 6$$

and then we divide both sides by 2, with the result

$$x = \frac{6}{2}$$

or

$$x = 3$$

3. Any equation remains valid if both sides are raised to the same power.

This rule permits us to solve the equation

$$x^3 = b$$

Raising both sides to the power $\frac{1}{3}$, we find

$$(x^3)^{1/3} = b^{1/3}$$

or

$$x = b^{1/3}$$

As a final example, let us consider the equation

$$x = -\tfrac{1}{2}gt^2 + x_0$$

(as established in Chapter 2, this equation describes the vertical position of a particle that starts at a height x_0 and falls for a time t; but the meaning of the equation need not concern us here). Suppose that we want to solve for t in terms of the other quantities in the equation. This will require the use of all our rules of algebra. First, subtract x from both sides and then add $\frac{1}{2}gt^2$ to both sides. This leads to

$$0 = -\tfrac{1}{2}gt^2 + x_0 - x$$

and then to

$$\tfrac{1}{2}gt^2 = x_0 - x$$

Next, multiply both sides by 2 and divide both sides by g; this yields

$$t^2 = \frac{2}{g}(x_0 - x)$$

Finally, raise both sides to the power $\frac{1}{2}$, or, equivalently, extract the square root of both sides. This gives us the final result

$$t = \sqrt{\frac{2}{g}(x_0 - x)}$$

A 2.5 Equations with Two Unknowns

If we seek to solve for two unknowns simultaneously, then we need two independent equations containing these two unknowns. The solution of such simultaneous equations can be carried out by the method of elimination: begin by using one equation to solve for the first unknown in terms of the second, then use this result to eliminate the first unknown from the other equation. An example will help to make this clear. Consider the following two simultaneous equations with two unknowns x and y:

$$4x + 2y = 8$$

$$2x - y = -2$$

To solve the first equation for x in terms of y, subtract $2y$ from both sides and then divide both sides by 4:

$$x = \frac{8 - 2y}{4}$$

Next, substitute this expression for x into the second equation:

$$2 \times \frac{8 - 2y}{4} - y = -2$$

To simplify this equation, multiply both sides by 4:

$$2 \times (8 - 2y) - 4y = -8$$

and combine the two terms containing y:

$$16 - 8y = -8$$

This is an ordinary equation for the single unknown y, and it can be solved by the methods we discussed in the preceding section, with the result

$$y = 3$$

It then follows from the above expression for x that

$$x = \frac{8 - 2y}{4} = \frac{8 - 2 \times 3}{4} = \frac{2}{4} = \frac{1}{2}$$

A 2.6 The Quadratic Formula

The quadratic equation $ax^2 + bx + c = 0$ has two solutions:

$$x = \frac{-b \pm \sqrt{b^2 - 4ac}}{2a}$$

A 2.7 Logarithms and the Exponential Function

The **base-10 logarithm** of a (positive) number is the power to which 10 must be raised to obtain this number. Thus, from $10 = 10^1$ and $100 = 10^2$ and $1000 = 10^3$ and $10\,000 = 10^4$ we immediately deduce that

$$\log 10 = 1$$

$$\log 100 = 2$$

$$\log 1000 = 3$$

$$\log 10\ 000 = 4, \text{ etc.}$$

Likewise

$$\log 1 = 0$$

$$\log 0.1 = -1$$

$$\log 0.01 = -2$$

$$\log 0.001 = -3, \text{ etc.}$$

Thus, the logarithm of a number between 1 and 10 is somewhere between 0 and 1, but to find the logarithm of such a number, we need the help of a computer program (many calculators have built-in computer programs that yield the value of the logarithm at the touch of a button). For some calculations, it is convenient to remember that $\log 2 = 0.301 \approx 0.3$ and $\log 5 = 0.699 \approx 0.7$.

The logarithm of the product of two numbers is the sum of the individual logarithms, and the logarithm of the ratio of two numbers is the difference of the individual logarithms. This rule makes it easy to find the logarithm of a number expressed in scientific notation. For example, the logarithm of 2×10^6 is

$$\log(2 \times 10^6) = \log 2 + \log 10^6 = 0.301 + 6 = 6.301$$

Note that the logarithm of any (positive) number smaller than 1 is negative. For example,

$$\log(5 \times 10^{-3}) = \log 5 + \log 10^{-3} = 0.699 - 3 = -2.301$$

The **exponential function** $\exp(x)$ is defined by the following infinite series:

$$\exp(x) = 1 + x + \frac{x^2}{2} + \frac{x^3}{3 \cdot 2} + \frac{x^4}{4 \cdot 3 \cdot 2} + \cdots$$

This function is equivalent to raising the constant $e = 2.718\ 28 \ldots$ to the power x:

$$\exp(x) = e^x$$

The **natural logarithm** $\ln x$ is the inverse of the exponential function, so

$$x = e^{\ln x}$$

and

$$x = \ln(e^x)$$

Natural logarithms obey the usual rules for logarithms,

$$\ln(x \cdot y) = \ln x + \ln y$$

$$\ln\left(\frac{x}{y}\right) = \ln x - \ln y$$

$$\ln(x^a) = a \ln x$$

Note that

$$\ln e = 1$$

and

$$\ln 10 = 2.3026$$

If we designate the base-10 logarithm, or **common logarithm**, by log x, then the relationship between the two kinds of logarithm is as follows:

$$\ln x = \ln(10^{\log x}) = (\log x)(\ln 10) = 2.3026 \log x$$

Appendix 3: Geometry and Trigonometry Review

A3.1 Perimeters, Areas, and Volumes

[perimeter of a circle of radius r] $= 2\pi r$
[area of a circle of radius r] $= \pi r^2$
[area of a triangle of base b, altitude h] $= hb/2$
[surface area of a sphere of radius r] $= 4\pi r^2$
[volume of a sphere of radius r] $= 4\pi r^3/3$
[area of curved surface of a cylinder of radius r, height h] $= 2\pi rh$
[volume of a cylinder of radius r, height h] $= \pi r^2 h$

A3.2 Angles

The angle between two intersecting straight lines is defined as the fraction of a complete circle included between these lines (Fig. A3.1). To express the angle in **degrees**, we assign an angular magnitude of 360° to the complete circle; any arbitrary angle is then an appropriate fraction of 360°. To express the angle in **radians**, we assign an angular magnitude of 2π radians to the complete circle; any arbitrary angle is then an appropriate fraction of 2π. For example, the angle shown in Fig. A3.1 is $\frac{1}{12}$ of a complete circle, that is, 30°, or $\pi/6$ radian. In view of the definition of angle, the length of arc included between the two intersecting straight lines is proportional to the angle θ between these lines; if the angle is expressed in radians, then the constant of proportionality is simply the radius:

$$s = r\theta \tag{1}$$

Since 2π radians $= 360°$, it follows that

$$1 \text{ radian} = \frac{360°}{2\pi} = \frac{360°}{2 \times 3.141\ 59} = 57.2958° \tag{2}$$

Each degree is divided into 60 minutes of arc (arcminutes), and each of these into 60 seconds of arc (arcseconds). In degrees, minutes of arc, and seconds of arc, the radian is

$$1 \text{ radian} = 57° \ 17' \ 44.8'' \tag{3}$$

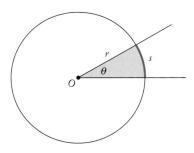

FIGURE A3.1 The angle θ in this diagram is $\theta = 30°$, or $\pi/6$ radian.

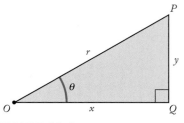

FIGURE A3.2 A right triangle.

A3.3 The Trigonometric Functions

The trigonometric functions of an angle are defined as ratios of the lengths of the sides of a right triangle erected on this angle. Figure A3.2 shows an acute angle θ and a right triangle, one of whose angles coincides with θ. The adjacent side OQ has a length x, the opposite side QP a length y, and the hypotenuse OP a length r. The **sine**, **cosine**, **tangent**, **cotangent**, **secant**, and **cosecant** of the angle θ are then defined as follows:

$$\text{sine} \qquad \sin\theta = y/r \tag{4}$$

$$\text{cosine} \qquad \cos\theta = x/r \tag{5}$$

$$\text{tangent} \qquad \tan\theta = y/x \tag{6}$$

$$\text{cotangent} \qquad \cot\theta = x/y \tag{7}$$

$$\text{secant} \qquad \sec\theta = r/x \tag{8}$$

$$\text{cosecant} \qquad \csc\theta = r/y \tag{9}$$

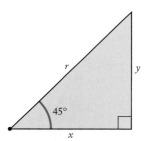

FIGURE A3.3 A right triangle with an angle of 45°.

> **EXAMPLE 1** Find the sine, cosine, and tangent for angles of 0°, 90°, and 45°.
>
> **SOLUTION:** For an angle of 0°, the opposite side is zero ($y = 0$), and the adjacent side coincides with the hypotenuse ($x = r$). Hence
>
> $$\sin 0° = 0 \quad \cos 0° = 1 \quad \tan 0° = 0 \tag{10}$$
>
> For an angle of 90°, the adjacent side is zero ($x = 0$), and the opposite side coincides with the hypotenuse ($y = r$). Hence
>
> $$\sin 90° = 1 \quad \cos 90° = 0 \quad \tan 90° = \infty \tag{11}$$
>
> Finally, for an angle of 45° (Fig. A3.3), the adjacent and the opposite sides have the same length ($x = y$) and the hypotenuse has a length of $\sqrt{2}$ times the length of either side ($r = \sqrt{2}x = \sqrt{2}y$). Hence
>
> $$\sin 45° = \frac{1}{\sqrt{2}} \quad \cos 45° = \frac{1}{\sqrt{2}} \quad \tan 45° = 1 \tag{12}$$

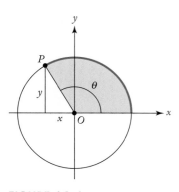

FIGURE A3.4 The angle θ in this diagram is larger than 90°.

The definitions (4)–(9) are also valid for angles greater than 90°, such as the angle shown in Fig. A3.4. In the general case, the quantities x and y must be interpreted as the rectangular coordinates of the point P. For any angle larger than 90°, one or both of the coordinates x and y are negative. Hence some of the trigonometric functions will also be negative. For instance,

$$\sin 135° = \frac{1}{\sqrt{2}} \quad \cos 135° = -\frac{1}{\sqrt{2}} \quad \tan 135° = -1 \tag{13}$$

Figure A3.5 shows plots of the sine, cosine, and tangent vs. θ.

(a)

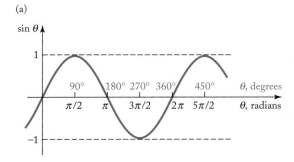

(b)

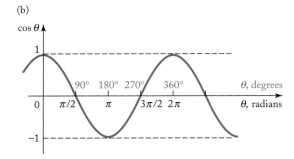

(c)

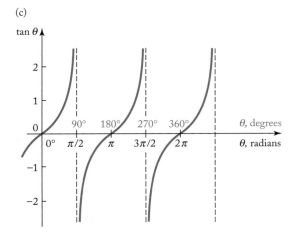

FIGURE A3.5 Plots of the sine, cosine, and tangent functions.

A3.4 Trigonometric Identities

From the definitions (4)–(9) we immediately find the following identities:

$$\tan \theta = \sin \theta / \cos \theta \tag{14}$$

$$\cot \theta = 1/\tan \theta \tag{15}$$

$$\sec \theta = 1/\cos \theta \tag{16}$$

$$\csc \theta = 1/\sin \theta \tag{17}$$

Figure A3.6 shows a right triangle with angles θ and $90° - \theta$. Since the adjacent side for the angle θ is the opposite side for the angle $90° - \theta$ and vice versa, we see that the trigonometric functions also obey the following identities:

$$\sin (90° - \theta) = \cos \theta \tag{18}$$

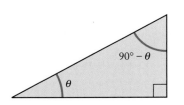

FIGURE A3.6 A right triangle with angles θ and $90° - \theta$.

$$\cos(90° - \theta) = \sin\theta \qquad (19)$$

$$\tan(90° - \theta) = \cot\theta = 1/\tan\theta \qquad (20)$$

According to the **Pythagorean theorem**, $x^2 + y^2 = r^2$. With $x = r\cos\theta$ and $y = r\sin\theta$, this becomes $r^2\cos^2\theta + r^2\sin^2\theta = r^2$, or

$$\cos^2\theta + \sin^2\theta = 1 \qquad (21)$$

The following are a few other trigonometric identities, which we state without proof:

$$\sec^2\theta = 1 + \tan^2\theta \qquad (22)$$

$$\csc^2\theta = 1 + \cot^2\theta \qquad (23)$$

$$\sin 2\theta = 2\sin\theta\,\cos\theta \qquad (24)$$

$$\cos 2\theta = 2\cos^2\theta - 1 \qquad (25)$$

$$\sin(\alpha + \beta) = \sin\alpha\cos\beta + \cos\alpha\sin\beta \qquad (26)$$

$$\cos(\alpha + \beta) = \cos\alpha\cos\beta - \sin\alpha\sin\beta \qquad (27)$$

A3.5 The Laws of Cosines and Sines

In an arbitrary triangle the lengths of the sides and the angles obey the laws of cosines and of sines. The **law of cosines** states that if the lengths of two sides are A and B and the angle between them is γ (Figure A3.7), then the length of the third side is given by

$$C^2 = A^2 + B^2 - 2AB\cos\gamma \qquad (28)$$

The **law of sines** states that the sines of the angles of the triangle are in the same ratio as the lengths of the opposite sides (Figure A3.7):

$$\frac{\sin\alpha}{A} = \frac{\sin\beta}{B} = \frac{\sin\gamma}{C} \qquad (29)$$

Both of these laws are very useful in the calculation of unknown lengths or angles of a triangle.

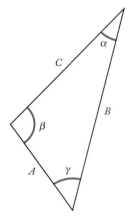

FIGURE A3.7 An arbitrary triangle.

Appendix 4: Calculus Review

A4.1 Derivatives

We saw in Section 2.3 that if the position of a particle is some function of time, say, $x = x(t)$, then the instantaneous velocity of the particle is the derivative of x with respect to t:

$$v = \frac{dx}{dt} \qquad (1)$$

This derivative is defined by first looking at a small increment Δx that results from a small increment Δt, and then evaluating the ratio $\Delta x/\Delta t$, in the limit when both Δx and Δt tend toward zero. Thus

$$\frac{dx}{dt} = \lim_{\Delta t \to 0} \frac{\Delta x}{\Delta t} \qquad (2)$$

Graphically, in a plot of position vs. time, the derivative dx/dt is the slope of the straight line tangent to the curved line at the time t (see Figure A4.1).

In general, if $f = f(u)$ is some given function of a variable u, the **derivative** of f with respect to u is defined by

$$\frac{df}{du} = \lim_{\Delta u \to 0} \frac{\Delta f}{\Delta u} \tag{3}$$

In a plot of f vs. u, this derivative is the slope of the straight line tangent to the curve representing $f(u)$.

Starting with the definition (3) we can find the derivative of any function (provided the function is sufficiently smooth so the derivative exists!). For example, consider the function $f(u) = u^2$. If we increase u to $u + \Delta u$, the function $f(u)$ increases to

$$f + \Delta f = (u + \Delta u)^2 \tag{4}$$

and therefore

$$\Delta f = (u + \Delta u)^2 - f = (u + \Delta u)^2 - u^2$$
$$= 2u\,\Delta u + (\Delta u)^2 \tag{5}$$

The derivative df/du is then

$$\frac{df}{du} = \lim_{\Delta u \to 0} \frac{\Delta f}{\Delta u} = \lim_{\Delta u \to 0} \frac{2u\,\Delta u + (\Delta u)^2}{\Delta u} \tag{6}$$

$$= \lim_{\Delta u \to 0} (2u) + \lim_{\Delta u \to 0} (\Delta u) \tag{7}$$

The second term on the right side vanishes in the limit $\Delta u \to 0$; the first term is simply $2u$. Hence

$$\frac{df}{du} = 2u \tag{8}$$

or

$$\frac{d}{du}(u^2) = 2u \tag{9}$$

This is one instance of the general rule for the differentiation of u^n:

$$\frac{d}{du}(u^n) = nu^{n-1} \tag{10}$$

This general rule is valid for any positive or negative number n, including zero. The proof of this rule can be constructed by an argument similar to that above. Table A4.1 lists the derivatives of the most common functions.

A4.2 Important Rules for Differentiation

1. Derivative of a constant times a function:

$$\frac{d}{du}(cf) = c\frac{df}{du} \tag{11}$$

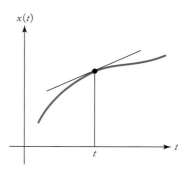

FIGURE A4.1 The derivative of $x(t)$ at t is the slope of the straight line tangent to the curve at t.

TABLE A4.1

SOME DERIVATIVES

$$\frac{d}{du}u^n = nu^{n-1}$$

$$\frac{d}{du}\ln u = \frac{1}{u}$$

$$\frac{d}{du}e^u = e^u$$

(In all the following formulas, u is in radian:)

$$\frac{d}{du}\sin u = \cos u$$

$$\frac{d}{du}\cos u = -\sin u$$

$$\frac{d}{du}\tan u = \sec^2 u$$

$$\frac{d}{du}\cot u = -\csc^2 u$$

$$\frac{d}{du}\sec u = \tan u \sec u$$

$$\frac{d}{du}\csc u = -\cot u \csc u$$

$$\frac{d}{du}\sin^{-1} u = 1/\sqrt{1 - u^2}$$

$$\frac{d}{du}\cos^{-1} u = -1/\sqrt{1 - u^2}$$

$$\frac{d}{du}\tan^{-1} u = \frac{1}{1 + u^2}$$

For instance,

$$\frac{d}{du}(6u^2) = 6\frac{d}{du}(u^2) = 6 \times 2u = 12u$$

2. Derivative of the sum of two functions:

$$\frac{d}{du}(f + g) = \frac{df}{du} + \frac{dg}{du} \tag{12}$$

For instance,

$$\frac{d}{du}(6u^2 + u) = \frac{d}{du}(6u^2) + \frac{d}{du}(u) = 12u + 1$$

3. Derivative of the product of two functions:

$$\frac{d}{du}(f \times g) = g\frac{df}{du} + f\frac{dg}{du} \tag{13}$$

For instance,

$$\frac{d}{du}(u^2 \sin u) = \sin u \frac{d}{du}u^2 + u^2\frac{d}{du}\sin u$$

$$= \sin u \times 2u + u^2 \times \cos u$$

4. Chain rule for derivatives: If f is a function of g and g is a function of u, then

$$\frac{d}{du}f(g) = \frac{df}{dg}\frac{dg}{du} \tag{14}$$

For instance, if $g = 2u$ and $f(g) = \sin g$, then

$$\frac{d}{du}\sin(2u) = \frac{d\sin(2u)}{d(2u)}\frac{d(2u)}{du}$$

$$= \cos(2u) \times 2$$

5. Partial derivatives: If f is a function of more than one variable, then the *partial derivative* of f with respect to one of the variables, say x, is denoted $\partial f/\partial x$, and is obtained by treating all the *other* variables as constants when differentiating. For instance, if $f = x^2y + y^2z$, then

$$\frac{\partial f}{\partial x} = 2xy, \quad \frac{\partial f}{\partial y} = x^2 + 2yz, \quad \text{and} \quad \frac{\partial f}{\partial z} = y^2$$

A4.3 Integrals

We have learned that if the position of a particle is known as a function of time, then we can find the instantaneous velocity by differentiation. What about the converse problem: if the instantaneous velocity is known as a function of time, how can we find the position? In Section 2.5 we learned how to deal with this problem in the special case of motion with constant acceleration. The velocity is then a fairly simple function of time [see Eq. (2.17)]

$$v = v_0 + at \tag{15}$$

and the position deduced from this velocity is [see Eq. (2.22)]

$$x = x_0 + v_0 t + \tfrac{1}{2} a t^2 \tag{16}$$

where x_0 and v_0 are the initial position and velocity at the initial time $t_0 = 0$. Now we want to deal with the general case of a velocity that is an arbitrary function of time,

$$v = v(t) \tag{17}$$

Figure A4.2 shows what a plot of v vs. t might look like. At the initial time t_0, the particle has an initial position x_0 (for the sake of generality we now assume that $t_0 \neq 0$). We want to find the position at some later time t. For this purpose, let us divide the time interval $t - t_0$ into a large number of small time intervals, each of the duration Δt. The total number of intervals is N, so $t - t_0 = N \Delta t$. The first of these intervals lasts from t_0 to $t_0 + \Delta t$; the second from $t_0 + \Delta t$ to $t_0 + 2\Delta t$; etc.

In Figure A4.3 the beginnings and the ends of these intervals have been marked t_0, t_1, t_2, etc., with $t_1 = t_0 + \Delta t, t_2 = t_0 + 2\Delta t$, etc. If Δt is sufficiently small, then during the first time interval the velocity is approximately $v(t_0)$; during the second, $v(t_1)$; etc. This amounts to replacing the smooth function $v(t)$ by a series of steps (see Fig. A4.3). Thus, during the first time interval, the displacement of the particle is approximately $v(t_0) \Delta t$; during the second interval, $v(t_1) \Delta t$; etc. The net displacement of the particle during the entire interval $t - t_0$ is the sum of all these small displacements:

$$x(t) - x_0 \approx v(t_0) \Delta t + v(t_1) \Delta t + v(t_2) \Delta t + \cdots \tag{18}$$

Using the standard mathematical notation for summation, we can write this as

$$x(t) - x_0 \approx \sum_{i=0}^{N-1} v(t_i) \Delta t \tag{19}$$

We can give this sum the following graphical interpretation: since $v(t_i) \Delta t$ is the area of the rectangle of height $v(t_i)$ and width Δt, the sum is the net area of all the rectangles shown in Figure A4.3, i.e., it is approximately the area under the velocity curve. Note that if the velocity is negative, the area must be reckoned as negative!

Of course, Eq. (19) is only an approximation. To find the exact displacement of the particle we must let the step size Δt tend to zero (while the number of steps N tends to infinity). In this limit, the steplike horizontal and vertical line segments in Fig. A4.3 approach the smooth curve. Thus,

$$x(t) - x_0 = \lim_{\substack{\Delta t \to 0 \\ N \to \infty}} \sum_{i=0}^{N-1} v(t_i) \Delta t \tag{20}$$

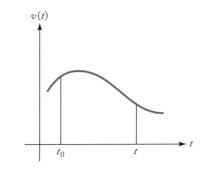

FIGURE A4.2 Plot of a function $v(t)$.

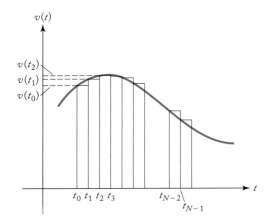

FIGURE A4.3 The interval $t - t_0$ has been divided into N equal intervals of duration Δt, so $t_1 = t_0 + \Delta t$, etc.

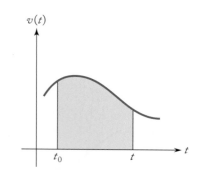

FIGURE A4.4 The area under the velocity curve.

In the notation of calculus, the right side of Eq. (20) is usually written in the following fashion:

$$x(t) - x_0 = \int_{t_0}^{t} v(t')\, dt' \tag{21}$$

The right side is called the **integral** of the function $v(t)$. The subscript and the superscript on the integration symbol $\int$ are called, respectively, the lower and the upper limit of integration; and t' is called the variable of integration (the prime on the variable of integration t' merely serves to distinguish that variable from the limit of integration t). Graphically, the integral is the exact area under the velocity curve between the limits t_0 and t in a plot of v vs. t (see Fig. A4.4). Areas below the t axis must be reckoned as negative.

In general, if $f(u)$ is a function of u, then the integral of this function is defined by a limiting procedure similar to that described above for the special case of the function $v(t)$. The integral over an interval from $u = a$ to $u = b$ is

$$\int_{a}^{b} f(u)\, du = \lim_{\substack{\Delta u \to 0 \\ N \to \infty}} \sum_{i=0}^{N-1} f(u_i)\, \Delta u \tag{22}$$

where $u_i = a + i\,\Delta u$. As in the case of the integral of $v(t)$, this integral can again be interpreted as an area: it is the area under the curve between the limits a and b in a plot of f vs. u.

For the explicit evaluation of integrals we can take advantage of the connection between integrals and antiderivatives. An **antiderivative** of a function $f(u)$ is simply a function $F(u)$ such that $dF/du = f$. For example, if $f(u) = u^n$ and $n \neq -1$, then an antiderivative of $f(u)$ is $F(u) = u^{n+1}/(n + 1)$. The fundamental theorem of calculus states that the integral of any function $f(u)$ can be expressed in terms of antiderivatives:

$$\int_{a}^{b} f(u)\, du = F(b) - F(a) \tag{23}$$

In essence, this means that integration is the inverse of differentiation. We will not prove this theorem here, but we remark that such an inverse relationship between integration and differentiation should not come as a surprise. We have already run across an obvious instance of such a relationship: we know that velocity is the derivative of the position, and we have seen above that the position is the integral of the velocity.

We will sometimes write Eq. (23) as

$$\int_{a}^{b} f(u)\, du = F(u)\Big|_{a}^{b} \tag{24}$$

where the notation $F(u)\big|_{a}^{b}$ means that the function $F(u)$ is to be evaluated at a and at b, and these values are to be subtracted. For example, if $n \neq -1$,

$$\int_{a}^{b} u^n\, du = \frac{u^{n+1}}{n + 1}\Big|_{a}^{b} = \frac{b^{n+1}}{n + 1} - \frac{a^{n+1}}{n + 1} \tag{25}$$

Table A4.2 lists some frequently used integrals. In this table, the limits of integration belonging with Eq. (24) have been omitted for the sake of brevity.

TABLE A4.2	SOME INTEGRALS

$$\int u^n du = \frac{u^{n+1}}{n+1} \qquad \text{for } n \neq -1$$

$$\int \frac{1}{u} du = \ln u \qquad \text{for } u > 0$$

$$\int e^{ku} du = \frac{e^{ku}}{k}$$

$$\int \ln u \, du = u \ln u - u$$

$$\int \sin(ku) \, du = -\frac{1}{k} \cos(ku) \qquad \text{(where } ku \text{ is in radians)}$$

$$\int \cos(ku) \, du = \frac{1}{k} \sin(ku) \qquad \text{(where } ku \text{ is in radians)}$$

$$\int \frac{du}{1+ku} = \frac{1}{k} \ln(1+ku)$$

$$\int \frac{du}{\sqrt{k^2 - u^2}} = \sin^{-1}\left(\frac{u}{k}\right)$$

$$\int \frac{du}{\sqrt{u^2 \pm k^2}} = \ln\left(u + \sqrt{u^2 \pm k^2}\right)$$

$$\int \sqrt{k^2 - u^2} \, du = \frac{1}{2}\left[u\sqrt{k^2 - u^2} + k^2 \sin^{-1}\left(\frac{u}{k}\right)\right]$$

$$\int \frac{du}{k^2 + u^2} = \frac{1}{k} \tan^{-1}\left(\frac{u}{k}\right)$$

$$\int \frac{du}{u\sqrt{k^2 \pm u^2}} = -\frac{1}{k} \ln\left(\frac{k + \sqrt{k^2 \pm u^2}}{u}\right)$$

$$\int \frac{du}{(u^2 + k^2)^{3/2}} = \frac{u}{k^2 \sqrt{u^2 + k^2}}$$

A4.4 Important Rules for Integration

1. Integral of a constant times a function:

$$\int_a^b cf(u) \, du = c\int_a^b f(u) \, du \tag{26}$$

For instance,

$$\int_a^b 5u^2 \, du = 5\int_a^b u^2 \, du = 5\left(\frac{b^3}{3} - \frac{a^3}{3}\right) \tag{27}$$

2. Integral of a sum of two functions:

$$\int_a^b [f(u) + g(u)] \, du = \int_a^b f(u) \, du + \int_a^b g(u) \, du \tag{28}$$

For instance,

$$\int_a^b (5u^2 + u)\, du = \int_a^b 5u^2\, du + \int_a^b u\, du = 5\left(\frac{b^3}{3} - \frac{a^3}{3}\right) + \left(\frac{b^2}{2} - \frac{a^2}{2}\right) \quad (29)$$

3. Change of limits of integration:

$$\int_a^b f(u)\, du = \int_a^c f(u)\, du + \int_c^b f(u)\, du \quad (30)$$

$$\int_a^b f(u)\, du = -\int_b^a f(u)\, du \quad (31)$$

4. Change of variable of integration: If u is a function of v, then

$$\int_a^b f(u)\, du = \int_{v(a)}^{v(b)} f(u) \frac{du}{dv}\, dv \quad (32)$$

For instance, with $u = v^2$,

$$\int_a^b u^3\, du = \int_a^b v^6\, du = \int_{\sqrt{a}}^{\sqrt{b}} v^6 (2v)\, dv \quad (33)$$

Finally, let us apply these general results to some specific examples of integration of the velocity.

EXAMPLE 1   A particle with constant acceleration has the following velocity as a function of time [compare Eq. (15)]:

$$v(t) = v_0 + at$$

where v_0 is the velocity at $t = 0$.

By integration, find the position as a function of time.

SOLUTION: According to Eq. (21), with $t_0 = 0$,

$$x(t) - x_0 = \int_0^t v(t')\, dt' = \int_0^t (v_0 + at')\, dt'$$

Using rule 2 and rule 1, we find that this equals

$$x(t) - x_0 = \int_0^t v_0\, dt' + \int_0^t at'\, dt' = v_0 \int_0^t dt' + a \int_0^t t'\, dt' \quad (34)$$

The first entry listed in Table A4.2 gives $\int dt' = t'$ (for $n = 0$) and $\int t'\, dt' = t'^2/2$ (for $n = 1$). Thus,

$$x(t) - x_0 = v_0\, t' \Big|_0^t + \tfrac{1}{2} a t'^2 \Big|_0^t$$

$$= v_0 t + \tfrac{1}{2} a t^2 \quad (35)$$

This, of course, agrees with Eq. (16).

EXAMPLE 2	The instantaneous velocity of a projectile traveling through air is the following function of time:

$$v(t) = 655.9 - 61.14t + 3.26t^2$$

where $v(t)$ is measured in meters per second and t is measured in seconds. Assuming that $x = 0$ at $t = 0$, what is the position as a function of time? What is the position at $t = 3.0$ s?

SOLUTION: With $x_0 = 0$ and $t_0 = 0$, Eq. (21) becomes

$$x(t) = \int_0^t (655.9 - 61.14t' + 3.26t'^2)\, dt'$$

$$= 655.9 \int_0^t dt' - 61.14 \int_0^t t'\, dt' + 3.26 \int_0^t t'^2\, dt'$$

$$= 655.9(t')\Big|_0^t - 61.14(t'^2/2)\Big|_0^t + 3.26(t'^3/3)\Big|_0^t$$

$$= 655.9t - 61.14t^2/2 + 3.26t^3/3$$

When evaluated at $t = 3.0$ s, this yields

$$x(3.0) = 655.9 \times 3.0 - 61.14 \times (3.0)^2/2 + 3.26 \times (3.0)^3/3$$
$$= 1722 \text{ m}$$

EXAMPLE 3	The acceleration of a mass pushed back and forth by an elastic spring is

$$a(t) = B \cos \omega t \tag{36}$$

where B and ω are constants. Find the position as a function of time. Assume $v = 0$ and $x = 0$ at $t = 0$.

SOLUTION: The calculation involves two steps: first we must integrate the acceleration to find the velocity, then we must integrate the velocity to find the position. For the first step we use an equation analogous to Eq. (21),

$$v(t) - v_0 = \int_{t_0}^t a(t')\, dt' \tag{37}$$

This equation becomes obvious if we remember that the relationship between acceleration and velocity is analogous to that between velocity and position. With $v_0 = 0$ and $t_0 = 0$, we obtain from Eq. (29)

$$v(t) = \int_0^t B \cos \omega t'\, dt' = B\frac{1}{\omega} \sin \omega t'\Big|_0^t$$

$$= \frac{B}{\omega} \sin \omega t \tag{38}$$

Next,

$$x(t) = \int_0^t v(t')\, dt' = \int_0^t \frac{B}{\omega} \sin \omega t'\, dt' = \frac{B}{\omega}\left(-\frac{1}{\omega} \cos \omega t'\right)\Big|_0^t$$

$$= -\frac{B}{\omega^2} \cos \omega t + \frac{B}{\omega^2} \tag{39}$$

A4.5 The Taylor Series

Suppose that $f(u)$ is a smooth function of u in some neighborhood of a given point $u = a$, so the function has continuous derivatives of all orders. Then the value of the function at an arbitrary point near a can be expressed in terms of the following infinite series, where all the derivatives are evaluated at the point a:

$$f(u) = f(a) + \frac{df}{du}(u - a) + \frac{1}{2}\frac{d^2f}{du^2}(u - a)^2 + \frac{1}{3 \cdot 2}\frac{d^3f}{du^3}(u - a)^3 + \cdots \quad (40)$$

This is called the **Taylor series** for the function $f(u)$ about the point a. The series converges, and is valid, provided u is sufficiently close to a. How close is "sufficiently close" depends on the function f and on the point a. Some functions, such as $\sin u$, $\cos u$, and e^u, are extremely well behaved, and their Taylor series converge for any choice of u and of a. The Taylor series gives us a convenient method for the approximate evaluation of a function.

EXAMPLE 4   Find the Taylor series for $\sin u$ about the point $u = 0$.

SOLUTION: The derivatives of $\sin u$ evaluated at $u = 0$ are

$$\frac{d}{du}\sin u = \cos u = 1$$

$$\frac{d^2}{du^2}\sin u = \frac{d}{du}\cos u = -\sin u = 0$$

$$\frac{d^3}{du^3}\sin u = \frac{d}{du}(-\sin u) = -\cos u = -1$$

$$\frac{d^4}{du^4}\sin u = \frac{d}{du}(-\cos u) = \sin u = 0, \quad \text{etc.}$$

Hence Eq. (32) gives

$$\sin u = 0 + 1 \times (u - 0) + \frac{1}{2} \times 0 \times (u - 0)^2 + \frac{1}{3 \cdot 2} \times (-1) \times (u - 0)^3$$

$$+ \frac{1}{4 \cdot 3 \cdot 2} \times 0 \times (u - 0)^4 + \cdots$$

$$= u - \frac{1}{6}u^3 + \cdots$$

Note that for very small values of u, we can neglect all higher powers of u, so $\sin u \approx u$, which is an approximation often used in this book.

A4.6 Some Approximations

By constructing Taylor series, we can obtain the following useful approximations, all of which are valid for small values of u. It is often sufficient to keep just the first one or two terms on the right side.

$$\sqrt{1 + u} = 1 + \frac{1}{2}u - \frac{1}{8}u^2 + \frac{1}{16}u^3 + \cdots \quad (41)$$

$$\frac{1}{1 + u} = 1 - u + u^2 - u^3 + \cdots \tag{42}$$

$$\frac{1}{\sqrt{1 + u}} = 1 - \frac{1}{2}u + \frac{3}{8}u^2 - \frac{5}{16}u^3 + \cdots \tag{43}$$

$$\frac{1}{(1 + u)^n} = 1 - nu + \frac{n(n + 1)}{2}u^2 - \frac{n(n + 1)(n + 2)}{2 \cdot 3}u^3 + \cdots \tag{44}$$

$$e^u = 1 + u + \frac{1}{2}u^2 + \frac{1}{2 \cdot 3}u^3 + \cdots \tag{45}$$

$$\ln(1 + u) = u - \frac{1}{2}u^2 + \frac{1}{3}u^3 + \cdots \tag{46}$$

In all the following formulas, u is in radians:

$$\sin u = u - \frac{1}{6}u^3 + \frac{1}{120}u^5 + \cdots \tag{47}$$

$$\cos u = 1 - \frac{1}{2}u^2 + \frac{1}{24}u^4 - \frac{1}{720}u^6 + \cdots \tag{48}$$

$$\tan u = u + \frac{1}{3}u^3 + \frac{2}{15}u^5 + \cdots \tag{49}$$

$$\sin^{-1} u = u + \frac{1}{6}u^3 + \frac{3}{40}u^5 + \cdots \tag{50}$$

$$\tan^{-1} u = u - \frac{1}{3}u^3 + \frac{1}{5}u^5 + \cdots \tag{51}$$

Appendix 5: Propagating Uncertainties

Experimentalists carefully work to measure physical quantities and to determine the uncertainty in each quantity. We must often calculate a new result from a measured quantity or from several quantities; we must therefore understand the propagation of uncertainties through functions and formulas.

To keep things simple, we will make the assumption that the uncertainties in each quantity are symmetrically distributed about its measured value and that the various measured quantities are independent of each other. This is not always true. But by ignoring correlations and assuming symmetry, we can reduce all the necessary propagation of uncertainties to some simple formulas.

Suppose we have a measured quantity and its uncertainty, $x \pm \Delta x$, where Δx is a positive quantity and has the same units as x, and is also known as the *absolute uncertainty* in x. What, then, is the uncertainty of some function, $f(x)$, of this data? Under the assumption that the uncertainty is small, we can obtain the uncertainty from the first terms of the Taylor series expansion of f: $f(x + \Delta x) = f(x) + (df(x)/dx)\Delta x + \cdots$ From this we find the uncertainty $\Delta f = |f(x + \Delta x) - f(x)|$ in the function value $f(x)$ is

$$\Delta f = \left| \frac{df}{dx} \Delta x \right| \tag{1}$$

with the derivative evaluated at the point x. We can generalize this result to functions of several variables as follows: given the data $x \pm \Delta x$, $y \pm \Delta y$, . . ., the function $f(x, y, \ldots)$ has the associated uncertainty

$$\Delta f = \left| \frac{\partial f}{\partial x} \Delta x \right| + \left| \frac{\partial f}{\partial y} \Delta y \right| + \cdots \tag{2}$$

where all the partial derivatives (see App. 4.2) are evaluated at the point $x, y, \ldots$. If we recall that we defined absolute uncertainties to be positive, we can write this as

$$\Delta f = \left| \frac{\partial f}{\partial x} \right| \Delta x + \left| \frac{\partial f}{\partial y} \right| \Delta y + \cdots \tag{3}$$

From this relationship, we can derive several simple results for uncertainty propagation.

EXAMPLE 1   **Addition and Subtraction.**

Given $f(x, y) = 3x + y - z + 5$, find Δf:

$$\Delta f = \left| \frac{\partial f}{\partial x} \right| \Delta x + \left| \frac{\partial f}{\partial y} \right| \Delta y + \left| \frac{\partial f}{\partial z} \right| \Delta z$$

$$= |3| \Delta x + |1| \Delta y + |-1| \Delta z$$

$$= 3 \Delta x + \Delta y + \Delta z$$

Thus in addition or subtraction, the uncertainties add, and in multiplication by a constant, the uncertainty is multiplied by the same constant.

EXAMPLE 2   **Multiplication, Division, and Exponentiation.**

Given $f(x, y) = x^2 y / (5z)$, find Δf:

$$\Delta f = \left| \frac{\partial f}{\partial x} \right| \Delta x + \left| \frac{\partial f}{\partial y} \right| \Delta y + \left| \frac{\partial f}{\partial z} \right| \Delta z$$

$$= |2xy/(5z)| \Delta x + |x^2/(5z)| \Delta y + |-x^2 y/(5z^2)| \Delta z$$

Equivalently, for multiplication and division, we add *relative uncertainties* (e.g., $\Delta x/x$), and for exponentiation, we multiply the relative uncertainty by the magnitude of the exponent, to get the relative uncertainty of the product, quotient, or power.

EXAMPLE 3   **Numerical Application to Ohm's Law, $V = IR$.**

Given $V = 1.5 \pm 0.1$ Volt and $I = 0.50 \pm 0.02$ A, find R and ΔR:

Rearranging we find $R = V/I = (1.5 \text{ Volt})/(0.50 \text{ A}) = 3.0 \ \Omega$, and

$$\Delta R = \left| \frac{\partial R}{\partial V} \right| \Delta V + \left| \frac{\partial R}{\partial I} \right| \Delta I$$

$$= \left| \frac{1}{I} \right| \Delta V + \left| \frac{-V}{I^2} \right| \Delta I$$

$$= \left| \frac{1}{0.50 \text{ A}} \right| (0.1 \text{ Volt}) + \left| \frac{-1.5 \text{ Volt}}{(0.50 \text{ A})^2} \right| (0.02 \text{A})$$

$$= 0.2 \ \Omega + 0.12 \ \Omega = 0.4 \ \Omega$$

Note in the last step that unlike an ordinary calculation, we have rounded this final result up; uncertainties should always be rounded up, never down.

Appendix 6: The International System of Units (SI)

A6.1 Base Units

The SI system of units is the modern version of the metric system. The SI system recognizes seven fundamental, or base, units for length, mass, time, electric current, thermodynamic temperature, amount of substance, and luminous intensity.[b] The following definitions of the base units were adopted by the Conférence Générale des Poids et Mesures in the years indicated:

meter (m) "The metre is the length of the path travelled by light in vacuum during a time interval of 1/299 792 458 of a second." (Adopted in 1983.)

kilogram (kg) "The kilogram is . . . the mass of the international prototype of the kilogram." (Adopted in 1889 and in 1901.)

second (s) "The second is the duration of 9 192 631 770 periods of the radiation corresponding to the transition between the two hyperfine levels of the ground state of the cesium-133 atom." (Adopted in 1967.)

ampere (A) "The ampere is that constant current which, if maintained in two straight parallel conductors of infinite length, of negligible circular cross section, and placed one meter apart in vacuum, would produce between these conductors a force equal to 2×10^{-7} newton per meter of length." (Adopted in 1948.)

kelvin (K) "The kelvin . . . is the fraction 1/273.16 of the thermodynamic temperature of the triple point of water." (Adopted in 1967.)

---

[b] At least two of the seven base units of the SI system are redundant. The mole is merely a certain number of atoms or molecules, in the same sense that a dozen is a number; there is no need to designate this number as a unit. The candela is equivalent to $\frac{1}{683}$ watt per steradian; it serves no purpose that is not served equally well by watt per steradian. Two other base units could be made redundant by adopting new definitions of the unit of temperature and of the unit of electric charge. Temperature could be measured in energy units because, according to the equipartition theorem, temperature is proportional to the energy per degree of freedom. Hence the kelvin could be defined as a derived unit, with $1 \text{ K} = \frac{1}{2} \times 1.38 \times 10^{-23}$ joule per degree of freedom. Electric charge could also be defined as a derived unit, to be measured with a suitable combination of the units of force and distance, as is done in the cgs system.

Furthermore, the definitions of the supplementary units—radian and steradian—are gratuitous. These definitions properly belong in the province of mathematics and there is no need to include them in a system of physical units.

TABLE A6.1	NAMES OF DERIVED UNITS		
QUANTITY	**DERIVED UNIT**	**NAME**	**SYMBOL**
frequency	$1/s$	hertz	Hz
force	$kg \cdot m/s^2$	newton	N
pressure	N/m^2	pascal	Pa
energy	$N \cdot m$	joule	J
power	J/s	watt	W
electric charge	$A \cdot s$	coulomb	C
electric potential	J/C	volt	V
electric capacitance	C/V	farad	F
electric resistance	V/A	ohm	Ω
conductance	A/V	siemen	S
magnetic flux	$V \cdot s$	weber	Wb
magnetic field	$V \cdot s/m^2$	tesla	T
inductance	$V \cdot s/A$	henry	H
temperature	K	degree Celsius	°C
luminous flux	$cd \cdot sr$	lumen	lm
illuminance	$cd \cdot sr/m^2$	lux	lx
radioactivity	$1/s$	becquerel	Bq
absorbed dose	J/kg	gray	Gy
dose equivalent	J/kg	sievert	Sv

TABLE A6.2	PREFIXES FOR UNITS	
FACTOR	**PREFIX**	**SYMBOL**
10^{24}	yotta	Y
10^{21}	zetta	Z
10^{18}	exa	E
10^{15}	peta	P
10^{12}	tera	T
10^{9}	giga	G
10^{6}	mega	M
10^{3}	kilo	k
10^{2}	hecto	h
10	deka	da
10^{-1}	deci	d
10^{-2}	centi	c
10^{-3}	milli	m
10^{-6}	micro	μ
10^{-9}	nano	n
10^{-12}	pico	p
10^{-15}	femto	f
10^{-18}	atto	a
10^{-21}	zepto	z
10^{-24}	yocto	y

mole   "The mole is the amount of substance of a system which contains as many elementary entities as there are atoms in 0.012 kilogram of carbon-12." (Adopted in 1967.)

candela (cd)   "The candela is the luminous intensity, in a given direction, of a source that emits monochromatic radiation of frequency 540×10^{12} Hz and that has a radiant intensity in that direction of $\frac{1}{683}$ watt per steradian." (Adopted in 1979.)

Besides these seven base units, the SI system also recognizes two supplementary units of angle and solid angle:

radian (rad)   "The radian is the plane angle between two radii of a circle which cut off on the circumference an arc equal in length to the radius."

steradian (sr)   "The steradian is the solid angle which, having its vertex in the center of a sphere, cuts off an area equal to that of a [flat] square with sides of length equal to the radius of the sphere."

A6.2 Derived Units

The derived units are formed out of products and ratios of the base units. Table A6.1 lists those derived units that have been glorified with special names. (Other derived units are listed in the tables of conversion factors in Appendix 8.)

A6.3 Prefixes

Multiples and submultiples of SI units are indicated by prefixes, such as the familiar *kilo, centi,* and *milli* used in *kilometer, centimeter,* and *millimeter,* etc. Table A6.2 lists all the accepted prefixes. Some enjoy more popularity than others; it is best to avoid the use of uncommon prefixes, such as *atto* and *exa,* since hardly anybody will recognize those.

Appendix 7: Best Values of Fundamental Constants

The values in the following table are the "2002 CODATA Recommended Values" by P. J. Mohr and B. N. Taylor Listed at the website physics.nist.gov/constants of the National Institute of Standards and Technology. The digits in parentheses are the one–standard deviation uncertainty in the last digits of the given value.

TABLE A7.1   BEST VALUES OF FUNDAMENTAL CONSTANTS

QUANTITY	SYMBOL	VALUE	UNITS	RELATIVE UNCERTAINTY (PARTS PER MILLION)
UNIVERSAL CONSTANTS				
speed of light in vacuum	c	299 792 458	$\text{m}\cdot\text{s}^{-1}$	(exact)
magnetic constant	μ_0	$4\pi \times 10^{-7}$	$\text{N}\cdot\text{A}^{-2}$	
		$= 12.566\,370\,614\ldots \times 10^{-7}$	$\text{N}\cdot\text{A}^{-2}$	(exact)
electric constant $1/\mu_0 c^2$	ϵ_0	$8.854\,187\,817\ldots \times 10^{-12}$	$\text{F}\cdot\text{m}^{-1}$	(exact)
gravitational constant	G	$6.6742(10) \times 10^{-11}$	$\text{m}^3\cdot\text{kg}^{-1}\cdot\text{s}^{-2}$	1.5×10^{-4}
Planck constant	h	$6.626\,0693(11) \times 10^{-34}$	$\text{J}\cdot\text{s}$	1.7×10^{-7}
in eV·s	$\hbar$	$4.135\,667\,43(35) \times 10^{-15}$	$\text{eV}\cdot\text{s}$	8.5×10^{-8}
$h/2\pi$		$1.054\,571\,68(18) \times 10^{-34}$	$\text{J}\cdot\text{s}$	1.7×10^{-7}
in eV·s		$6.582\,119\,15(56) \times 10^{-16}$	$\text{eV}\cdot\text{s}$	8.5×10^{-8}
ELECTROMAGNETIC CONSTANTS				
elementary charge	e	$1.602\,176\,53(14) \times 10^{-19}$	C	8.5×10^{-8}
magnetic flux quantum $h/2e$	Φ_0	$2.067\,833\,72(18) \times 10^{-15}$	Wb	8.5×10^{-8}
quantum	$2e^2/h$	$7.748\,091\,733(26) \times 10^{-5}$	S	3.3×10^{-9}
Josephson constant	$2e/h$	$483\,597.879(41) \times 10^{9}$	$\text{Hz}\cdot\text{V}^{-1}$	8.5×10^{-8}
Bohr magneton $e\hbar/2m_e$	μ_B	$927.400\,949(80) \times 10^{-26}$	$\text{J}\cdot\text{T}^{-1}$	8.6×10^{-8}
in eV·T^{-1}		$5.788\,381\,804(39) \times 10^{-5}$	$\text{eV}\cdot\text{T}^{-1}$	6.7×10^{-9}
nuclear magneton $e\hbar/2m_p$	μ_N	$5.050\,783\,43(43) \times 10^{-27}$	$\text{J}\cdot\text{T}^{-1}$	8.6×10^{-8}
in eV·T^{-1}		$3.152\,451\,259(21) \times 10^{-8}$	$\text{eV}\cdot\text{T}^{-1}$	6.7×10^{-9}

(continued)

QUANTITY	SYMBOL	VALUE	UNITS	RELATIVE UNCERTAINTY (PARTS PER MILLION)
ATOMIC AND NUCLEAR CONSTANTS				
General				
fine-structure constant $e^2/4\pi\epsilon_0\hbar c$	α	$7.297\ 352\ 568(24) \times 10^{-3}$		3.3×10^{-9}
inverse fine-structure constant	α^{-1}	$137.035\ 999\ 11(46)$		3.3×10^{-9}
Rydberg constant $\alpha^2 m_e c/2h$	R_∞	$10\ 973\ 731.568\ 525(73)$	m^{-1}	6.6×10^{-12}
Bohr radius $4\pi\epsilon_0\hbar^2/m_e e^2$	a_0	$0.529\ 177\ 2108(18) \times 10^{-10}$	m	3.3×10^{-9}
Electron				
electron mass	m_e	$9.109\ 3826(16) \times 10^{-31}$	kg	1.7×10^{-7}
in u		$5.485\ 799\ 0945(24) \times 10^{-4}$	u	4.4×10^{-10}
energy equivalent in MeV	$m_e c^2$	$0.510\ 998\ 918(44)$	MeV	8.6×10^{-8}
electron-proton mass ratio	m_e/m_p	$5.446\ 170\ 2173(25) \times 10^{-4}$		4.6×10^{-10}
electron charge to mass quotient	$-e/m_e$	$-1.758\ 820\ 12(15) \times 10^{11}$	$C \cdot kg^{-1}$	8.6×10^{-8}
Compton wavelength $h/m_e c$	λ_C	$2.426\ 310\ 238(16) \times 10^{-12}$	m	6.7×10^{-9}
classical electron radius $\alpha^2 a_0$	r_e	$2.817\ 940\ 325(28) \times 10^{-15}$	m	1.0×10^{-8}
Thomson cross section $(8\pi/3)r_e^2$	σ_e	$0.665\ 245\ 837(13) \times 10^{-28}$	m^2	2.0×10^{-8}
electron magnetic moment	μ_e	$-928.476\ 412(80) \times 10^{-26}$	$J \cdot T^{-1}$	8.6×10^{-8}
to Bohr magneton ratio	μ_e/μ_B	$-1.001\ 159\ 652\ 1859(38)$		3.8×10^{-12}
to nuclear magneton ratio	μ_e/μ_N	$-1838.281\ 971\ 07(85)$		4.6×10^{-10}
electron magnetic moment anomaly $|\mu_e|/\mu_B - 1$	a_e	$1.159\ 652\ 1859(38) \times 10^{-3}$		3.2×10^{-9}
electron g-factor $-2(1 + a_e)$	g_e	$-2.002\ 319\ 304\ 3718(75)$		3.8×10^{-12}
Muon				
muon mass	m_μ	$1.883\ 531\ 40(33) \times 10^{-28}$	kg	1.7×10^{-7}
in u				
energy equivalent in MeV	$m_\mu c^2$	$0.113\ 428\ 9264(30)$	u	2.6×10^{-8}
		$105.658\ 3692(94)$	MeV	8.9×10^{-8}
muon-electron mass ratio	m_μ/m_e	$206.768\ 2838(54)$		2.6×10^{-8}
muon Compton wavelength $h/m_\mu c$	$\lambda_{C,\mu}$	$11.734\ 441\ 05(30) \times 10^{-15}$	m	2.5×10^{-8}
muon magnetic moment	μ_μ	$-4.490\ 447\ 99(40) \times 10^{-26}$	$J \cdot T^{-1}$	8.9×10^{-8}
to Bohr magneton ratio	μ_μ/μ_B	$-4.841\ 970\ 45(13) \times 10^{-3}$		2.6×10^{-8}
muon magnetic moment anomaly $|\mu_\mu|/(e\hbar/2m_\mu) - 1$	a_μ	$1.165\ 919\ 81(62) \times 10^{-3}$		5.3×10^{-7}
muon g-factor $-2(1 + a_\mu)$	g_μ	$-2.002\ 331\ 8396(12)$		6.2×10^{-10}
Proton				
proton mass	m_p	$1.672\ 621\ 71(29) \times 10^{-27}$	kg	1.7×10^{-7}
in u		$1.007\ 276\ 466\ 88(13)$	u	1.3×10^{-10}
energy equivalent in MeV	$m_p c^2$	$938.272\ 029(80)$	MeV	8.6×10^{-8}
proton-electron mass ratio	m_p/m_e	$1836.152\ 672\ 61(85)$		4.6×10^{-10}
proton-neutron mass ratio	m_p/m_n	$0.998\ 623\ 478\ 72(58)$		5.8×10^{-10}
proton charge to mass quotient	e/m_p	$9.578\ 833\ 76(82) \times 10^7$	$C \cdot kg^{-1}$	8.6×10^{-8}
proton Compton wavelength $h/m_p c$	$\lambda_{C,p}$	$1.321\ 409\ 8555(88) \times 10^{-15}$	m	6.7×10^{-9}
proton magnetic moment	μ_p	$1.410\ 606\ 71(12) \times 10^{-26}$	$J \cdot T^{-1}$	8.7×10^{-8}
to Bohr magneton ratio	μ_p/μ_B	$1.521\ 032\ 206(15) \times 10^{-3}$		1.0×10^{-8}
to nuclear magneton ratio	μ_p/μ_N	$2.792\ 847\ 351(28)$		1.0×10^{-8}
Neutron				
neutron mass	m_n	$1.674\ 927\ 28(29) \times 10^{-27}$	kg	1.7×10^{-7}
in u		$1.008\ 664\ 915\ 60(55)$	u	5.5×10^{-10}
energy equivalent in MeV	$m_n c^2$	$939.565\ 360(81)$	MeV	8.6×10^{-8}
neutron-electron mass ratio	m_n/m_e	$1838.683\ 6598(13)$		7.0×10^{-10}
neutron-proton mass ratio	m_n/m_p	$1.001\ 378\ 418\ 70(58)$		5.8×10^{-10}

(continued)

QUANTITY	SYMBOL	VALUE	UNITS	RELATIVE UNCERTAINTY (PARTS PER MILLION)
neutron Compton wavelength $h/m_n c$	$\lambda_{C,n}$	$1.319\ 590\ 9067(88) \times 10^{-15}$	m	6.7×10^{-9}
neutron magnetic moment	μ_n	$-0.966\ 236\ 45(24) \times 10^{-26}$	$J \cdot T^{-1}$	2.5×10^{-7}
to Bohr magneton ratio	μ_n/μ_B	$-1.041\ 875\ 63(25) \times 10^{-3}$		2.4×10^{-7}
to nuclear magneton ratio	μ_n/μ_N	$-1.913\ 042\ 73(45)$		2.4×10^{-7}
Deuteron				
deuteron mass	m_d	$3.343\ 583\ 35(57) \times 10^{-27}$	kg	1.7×10^{-7}
in u		$2.013\ 553\ 212\ 70(35)$	u	1.7×10^{-10}
energy equivalent in MeV	$m_d c^2$	$1875.612\ 82(16)$	MeV	8.6×10^{-8}
deuteron-electron mass ratio	m_d/m_e	$3670.482\ 9652(18)$		4.8×10^{-10}
deuteron-proton mass ratio	m_d/m_p	$1.999\ 007\ 500\ 82(41)$		2.0×10^{-10}
deuteron magnetic moment	μ_d	$0.433\ 073\ 482(38) \times 10^{-26}$	$J \cdot T^{-1}$	8.7×10^{-8}
to Bohr magneton ratio	μ_d/μ_B	$0.466\ 975\ 4567(50) \times 10^{-3}$		1.1×10^{-8}
to nuclear magneton ratio	μ_d/μ_N	$0.857\ 438\ 2329(92)$		1.1×10^{-8}
Alpha Particle				
alpha particle mass	m_α	$6.644\ 6565(11) \times 10^{-27}$	kg	1.7×10^{-7}
in u		$4.001\ 506\ 179\ 149(56)$	u	1.4×10^{-11}
energy equivalent in MeV	$m_\alpha c^2$	$3727.379\ 17(32)$	MeV	8.6×10^{-8}
alpha particle to electron mass ratio	m_α/m_e	$7294.299\ 5363(32)$		4.4×10^{-10}
alpha particle to proton mass ratio	m_α/m_p	$3.972\ 599\ 689\ 07(52)$		1.3×10^{-10}
PHYSICO-CHEMICAL CONSTANTS				
Avogadro constant	N_A	$6.022\ 1415(10) \times 10^{23}$	$mole^{-1}$	1.7×10^{-7}
atomic mass constant $m_u = \frac{1}{12} m(^{12}C) = 1\,u$	m_u	$1.660\ 538\ 86(28) \times 10^{-27}$	kg	1.7×10^{-7}
energy equivalent in MeV	$m_u c^2$	$931.494\ 043(80)$	MeV	8.6×10^{-8}
Faraday constant $N_A e$	F	$96\ 485.3383(83)$	$C \cdot mole^{-1}$	8.6×10^{-8}
molar gas constant	R	$8.314\ 472\ (15)$	$J \cdot mole^{-1} \cdot K^{-1}$	1.7×10^{-6}
Boltzmann constant R/N_A	k	$1.380\ 6505(24) \times 10^{-23}$	$J \cdot K^{-1}$	1.8×10^{-6}
in $eV \cdot K^{-1}$		$8.617\ 343(15) \times 10^{-5}$	$eV \cdot K^{-1}$	1.8×10^{-6}
molar volume of ideal gas RT/p $T = 273.15$ K, $p = 101.325$ kPa	V_m	$22.413\ 996(39) \times 10^{-3}$	$m^3 \cdot mole^{-1}$	1.7×10^{-6}
Loschmidt constant N_A/V_m	n_0	$2.686\ 7773(47) \times 10^{25}$	m^{-3}	1.8×10^{-6}
Stefan-Boltzmann constant $(\pi^2/60)k^4/\hbar^3 c^2$	σ	$5.670\ 400(40) \times 10^{-8}$	$W \cdot m^{-2} \cdot K^{-4}$	7.0×10^{-6}
Wien displacement law constant $b = \lambda_{max} T$	b	$2.897\ 7685(51) \times 10^{-3}$	$m \cdot K$	1.7×10^{-6}

Appendix 8: Conversion Factors

The units for each quantity are listed alphabetically, except that the SI unit is always listed first. The numbers are based on "American National Standard; Metric Practice" published by the Institute of Electrical and Electronics Engineers, 1982.

Angle

1 radian $= 57.30° = 3.438 \times 10^3{}' = (1/2\pi)$ rev $= 2.063 \times 10^5{}''$
1 degree (°) $= 1.745 \times 10^{-2}$ radian $= 60' = 3600'' = \frac{1}{360}$ rev
1 minute of arc (') $= 2.909 \times 10^{-4}$ radian $= \frac{1}{60}° = 4.630 \times 10^{-5}$ rev $= 60''$
1 revolution (rev) $= 2\pi$ radians $= 360° = 2.160 \times 10^4{}' = 1.296 \times 10^6{}''$
1 second of arc ('') $= 4.848 \times 10^{-6}$ radian $= \frac{1}{3600}° = \frac{1}{60}' = 7.716 \times 10^{-7}$ rev

Length

1 meter (m) $= 1 \times 10^{-9}$ nm $= 1 \times 10^{10}$ Å $= 6.685 \times 10^{-12}$ AU $= 100$ cm $= 1 \times 10^{15}$ fm $= 3.281$ ft $= 39.37$ in. $= 1 \times 10^{-3}$ km $= 1.057 \times 10^{-16}$ light-year $= 1 \times 10^6\ \mu$m $= 5.400 \times 10^{-4}$ nmi $= 6.214 \times 10^{-4}$ mi $= 3.241 \times 10^{-17}$ pc $= 1.094$ yd
1 angstrom (Å) $= 1 \times 10^{-10}$ m $= 1 \times 10^{-8}$ cm $= 1 \times 10$ nm $= 1 \times 10^{-5}$ fm $= 3.281 \times 10^{-10}$ ft $= 1 \times 10^{-4}\ \mu$m
1 astronomical unit (AU) $= 1.496 \times 10^{11}$ m $= 1.496 \times 10^{13}$ cm $= 1.496 \times 10^8$ km $= 1.581 \times 10^{-5}$ light-year $= 4.848 \times 10^{-6}$ pc
1 centimeter (cm) $= 0.01$ m $= 1 \times 10^8$ Å $= 1 \times 10^{13}$ fm $= 3.281 \times 10^{-2}$ ft $= 0.3937$ in. $= 1 \times 10^{-5}$ km $= 1.057 \times 10^{-18}$ light-year $= 1 \times 10^4\ \mu$m
1 fermi, or **femtometer** (fm) $= 1 \times 10^{-15}$ m $= 1 \times 10^{-13}$ cm $= 1 \times 10^5$ Å
1 foot (ft) $= 0.3048$ m $= 30.48$ cm $= 12$ in. $= 3.048 \times 10^5\ \mu$m $= 1.894 \times 10^{-4}$ mi $= \frac{1}{3}$ yd
1 inch (in.) $= 2.540 \times 10^{-2}$ m $= 2.54$ cm $= \frac{1}{12}$ ft $= 2.54 \times 10^4\ \mu$m $= \frac{1}{36}$ yd
1 kilometer (km) $= 1 \times 10^3$ m $= 1 \times 10^5$ cm $= 3.281 \times 10^3$ ft $= 0.5400$ nmi $= 0.6214$ mi $= 1.094 \times 10^3$ yd
1 light-year $= 9.461 \times 10^{15}$ m $= 6.324 \times 10^4$ AU $= 9.461 \times 10^{17}$ cm $= 9.461 \times 10^{12}$ km $= 5.879 \times 10^{12}$ mi $= 0.3066$ pc
1 micron, or **micrometer** (μm) $= 1 \times 10^{-6}$ m $= 1 \times 10^4$ Å $= 1 \times 10^{-4}$ cm $= 3.281 \times 10^{-6}$ ft $= 3.937 \times 10^{-5}$ in.
1 nautical mile (nmi) $= 1.852 \times 10^3$ m $= 1.852 \times 10^5$ cm $= 6.076 \times 10^3$ ft $= 1.852$ km $= 1.151$ mi
1 statute mile (mi) $= 1.609 \times 10^3$ m $= 1.609 \times 10^5$ cm $= 5280$ ft $= 1.609$ km $= 0.8690$ nmi $= 1760$ yd
1 parsec (pc) $= 3.086 \times 10^{16}$ m $= 2.063 \times 10^5$ AU $= 3.086 \times 10^{18}$ cm $= 3.086 \times 10^{13}$ km $= 3.262$ light-years
1 yard (yd) $= 0.9144$ m $= 91.44$ cm $= 3$ ft $= 36$ in. $= \frac{1}{1760}$ mi

Time

1 second (s) $= 1.157 \times 10^{-5}$ day $= \frac{1}{3600}$ h $= \frac{1}{60}$ min $= 1.161 \times 10^{-5}$ sidereal day $= 3.169 \times 10^{-8}$ yr
1 day $= 8.640 \times 10^4$ s $= 24$ h $= 1440$ min $= 1.003$ sidereal days $= 2.738 \times 10^{-3}$ yr
1 hour (h) $= 3600$ s $= \frac{1}{24}$ day $= 60$ min $= 1.141 \times 10^{-4}$ yr

1 minute (min) = 60 s = 6.944×10^{-4} day = $\frac{1}{60}$ h = 1.901×10^{-6} yr

1 sidereal day = 8.616×10^4 s = 0.9973 day = 23.93 h = 1.436×10^3 min
   = 2.730×10^{-3} yr

1 year (yr) = 3.156×10^7 s = 365.24 days = 8.766×10^3 h =
   5.259×10^5 min = 366.24 sidereal days

Mass

1 kilogram (kg) = 6.024×10^{26} u = 5000 carats = 1.543×10^4 grains =
   1000 g = 1×10^{-3} t = 35.27 oz = 2.205 lb = 1.102×10^{-3} short
   ton = 6.852×10^{-2} slug

1 atomic mass unit (u) = 1.6605×10^{-27} kg = 1.6605×10^{-24} g

1 carat = 2×10^{-4} kg = 0.2 g = 7.055×10^{-3} oz = 4.409×10^{-4} lb

1 grain = 6.480×10^{-5} kg = 6.480×10^{-2} g = 2.286×10^{-3} oz = $\frac{1}{7000}$ lb

1 gram (g) = 1×10^{-3} kg = 6.024×10^{23} u = 5 carats = 15.43 grains =
   1×10^{-6} t = 3.527×10^{-2} oz = 2.205×10^{-3} lb = 1.102×10^{-6} short ton
   = 6.852×10^{-5} slug

1 metric ton, or **tonne** (t) = 1×10^3 kg = 1×10^6 g = 2.205×10^3 lb =
   1.102 short tons = 68.52 slugs

1 ounce (oz) = 2.835×10^{-2} kg = 141.7 carats = 437.5 grains = 28.35 g = $\frac{1}{16}$ lb

1 pound (lb)^c = 0.4536 kg = 453.6 g = 4.536×10^{-4} t = 16 oz =
   $\frac{1}{2000}$ short ton = 3.108×10^{-2} slug

1 short ton = 907.2 kg = 9.072×10^5 g = 0.9072 t = 2000 lb

1 slug = 14.59 kg = 1.459×10^4 g = 32.17 lb

Area

1 square meter (m^2) = 1×10^4 cm^2 = 10.76 ft^2 = 1.550×10^3 in.^{2} =
   1×10^{-6} km^2 = 3.861×10^{-7} mi^2 = 1.196 yd^2

1 barn = 1×10^{-28} m^2 = 1×10^{-24} cm^2

1 square centimeter (cm^2) = 1×10^{-4} m^2 = 1.076×10^{-3} ft^2 = 0.1550 in.^{2}
   = 1×10^{-10} km^2 = 3.861×10^{-11} mi^2

1 square foot (ft^2) = 9.290×10^{-2} m^2 = 929.0 cm^2 = 144 in.^{2} =
   3.587×10^{-8} mi^2 = $\frac{1}{9}$ yd^2

1 square inch (in.^{2}) = 6.452×10^{-4} m^2 = 6.452 cm^2 = $\frac{1}{144}$ ft^2

1 square kilometer (km^2) = 1×10^6 m^2 = 1×10^{10} cm^2
   = 1.076×10^7 ft^2 = 0.3861 mi^2

1 square statute mile (mi^2) = 2.590×10^6 m^2 = 2.590×10^{10} cm^2 =
   2.788×10^7 ft^2 = 2.590 km^2

1 square yard (yd^2) = 0.8361 m^2 = 8.361×10^3 cm^2 = 9 ft^2 = 1296 in.^{2}

Volume

1 cubic meter (m^3) = 1×10^6 cm^3 = 35.31 ft^3 = 264.2 gal =
   6.102×10^4 in.^{3} = 1×10^3 liters = 1.308 yd^3

1 cubic centimeter (cm^3) = 1×10^{-6} m^3 = 3.531×10^{-5} ft^3 =
   2.642×10^{-4} gal = 6.102×10^{-2} in.^{3} = 1×10^{-3} liter

1 cubic foot (ft^3) = 2.832×10^{-2} m^3 = 2.832×10^4 cm^3 = 7.481 gal =
   1728 in.^{3} = 28.32 liters = $\frac{1}{27}$ yd^3

^{c} This is the "avoirdupois" pound. The "troy" or "apothecary" pound is 0.3732 kg, or 0.8229 lb avoirdupois.

1 gallon $(\text{gal})^d = 3.785 \times 10^{-3} \text{ m}^3 = 0.1337 \text{ ft}^3$
1 cubic inch $(\text{in.}^3) = 1.639 \times 10^{-5} \text{ m}^3 = 16.39 \text{ cm}^3 = 5.787 \times 10^{-4} \text{ ft}^3$
1 liter $(\text{l}) = 1 \times 10^{-3} \text{ m}^3 = 1000 \text{ cm}^3 = 3.531 \times 10^{-2} \text{ ft}^3 = 0.2642 \text{ gal}$
1 cubic yard $(\text{yd}^3) = 0.7646 \text{ m}^3 = 7.646 \times 10^5 \text{ cm}^3 = 27 \text{ ft}^3 = 202.0 \text{ gal}$

Density

1 kilogram per cubic meter $(\text{kg/m}^3) = 1 \times 10^{-3} \text{ g/cm}^3 =$
     $6.243 \times 10^{-2} \text{ lb/ft}^3 = 8.345 \times 10^{-3} \text{ lb/gal} = 3.613 \times 10^{-5} \text{ lb/in.}^3 =$
     $8.428 \times 10^{-4} \text{ short ton/yd}^3 = 1.940 \times 10^{-3} \text{ slug/ft}^3$
1 gram per cubic centimeter $(\text{g/cm}^3) = 1 \times 10^3 \text{ kg/m}^3 = 62.43 \text{ lb/ft}^3 =$
     $8.345 \text{ lb/gal} = 3.613 \times 10^{-2} \text{ lb/in.}^3 = 0.8428 \text{ short ton/yd}^3 = 1.940 \text{ slugs/ft}^3$
1 lb per cubic foot $(\text{lb/ft}^3) = 16.02 \text{ kg/m}^3 = 1.602 \times 10^{-2} \text{ g/cm}^3 =$
     $0.1337 \text{ lb/gal} = 1.350 \times 10^{-2} \text{ short ton/yd}^3 = 3.108 \times 10^{-2} \text{ slug/ft}^3$
1 pound-per gallon $(1 \text{ lb/gal}) = 119.8 \text{ kg/m}^3 = 7.481 \text{ lb/ft}^3 = 0.2325 \text{ slug/ft}^3$
1 short ton per cubic yard $(\text{short ton/yd}^3) = 1.187 \times 10^3 \text{ kg/m}^3 = 74.07 \text{ lb/ft}^3$
1 slug per cubic foot $(\text{slug/ft}^3) = 515.4 \text{ kg/m}^3 = 0.5154 \text{ g/cm}^3 =$
     $32.17 \text{ lb/ft}^3 = 4.301 \text{ lb/gal}$

Speed

1 meter per second $(\text{m/s}) = 100 \text{ cm/s} = 3.281 \text{ ft/s} = 3.600 \text{ km/h} =$
     $1.944 \text{ knots} = 2.237 \text{ mi/h}$
1 centimeter per second $(\text{cm/s}) = 0.01 \text{ m/s} = 3.281 \times 10^{-2} \text{ ft/s} =$
     $3.600 \times 10^{-2} \text{ km/h} = 1.944 \times 10^{-2} \text{ knot} = 2.237 \times 10^{-2} \text{ mi/h}$
1 foot per second $(\text{ft/s}) = 0.3048 \text{ m/s} = 30.48 \text{ cm/s} = 1.097 \text{ km/h} =$
     $0.5925 \text{ knot} = 0.6818 \text{ mi/h}$
1 kilometer per hour $(\text{km/h}) = 0.2778 \text{ m/s} = 27.78 \text{ cm/s} = 0.9113 \text{ ft/s}$
     $= 0.5400 \text{ knot} = 0.6214 \text{ mi/h}$
1 knot, or **nautical mile per hour** $= 0.5144 \text{ m/s} = 51.44 \text{ cm/s} =$
     $1.688 \text{ ft/s} = 1.852 \text{ km/h} = 1.151 \text{ mi/h}$
1 mile per hour $(\text{mi/h}) = 0.4470 \text{ m/s} = 44.70 \text{ cm/s} = 1.467 \text{ ft/s} =$
     $1.609 \text{ km/h} = 0.8690 \text{ knot}$

Acceleration

1 meter per second squared $(\text{m/s}^2) = 100 \text{ cm/s}^2 = 3.281 \text{ ft/s}^2 = 0.1020\,g$
1 centimeter per second squared $(\text{cm/s}^2) = 0.01 \text{ m/s}^2 =$
     $3.281 \times 10^{-2} \text{ ft/s}^2 = 1.020 \times 10^{-3}\,g$
1 foot per second squared $(\text{ft/s}^2) = 0.3048 \text{ m/s}^2 = 30.48 \text{ cm/s}^2 = 3.108 \times 10^{-2}\,g$
1 $g = 9.807 \text{ m/s}^2 = 980.7 \text{ cm/s}^2 = 32.17 \text{ ft/s}^2$

Force

1 newton $(\text{N}) = 1 \times 10^5 \text{ dynes} = 0.2248 \text{ lb-f} = 1.124 \times 10^{-4} \text{ short ton-force}$
1 dyne $= 1 \times 10^{-5} \text{ N} = 2.248 \times 10^{-6} \text{ lb-f} = 1.124 \times 10^{-9} \text{ short ton-force}$
1 pound-force $(\text{lb-f}) = 4.448 \text{ N} = 4.448 \times 10^5 \text{ dynes} = \frac{1}{2000} = \text{short ton-force}$
1 short ton-force $= 8.896 \times 10^3 \text{ N} = 8.896 \times 10^8 \text{ dynes} = 2000 \text{ lb-f}$

---

^{d}This is the U.S. gallon; the U.K. and the Canadian gallon are $4.546 \times 10^{-3} \text{ m}^3$, or 1.201 U.S. gallons.

Energy

1 joule (J) $= 9.478 \times 10^{-4}$ Btu $= 0.2388$ cal $= 1 \times 10^{7}$ ergs $=$
6.242×10^{18} eV $= 0.7376$ ft·lb-f $= 2.778 \times 10^{-7}$ kW·h

1 British thermal unit (Btu)^{e} $= 1.055 \times 10^{3}$ J $= 252.0$ cal $=$
1.055×10^{10} ergs $= 778.2$ ft·lb-f $= 2.931 \times 10^{-4}$ kW·h

1 calorie (cal)^{f} $= 4.187$ J $= 3.968 \times 10^{-3}$ Btu $= 4.187 \times 10^{7}$ ergs $=$
3.088 ft·lb-f $= 1 \times 10^{-3}$ kcal $= 1.163 \times 10^{-6}$ kW·h

1 erg $= 1 \times 10^{-7}$ J $= 9.478 \times 10^{-7}$ Btu $= 2.388 \times 10^{-8}$ cal $=$
6.242×10^{11} eV $= 7.376 \times 10^{-8}$ ft·lb-f $= 2.778 \times 10^{-14}$ kW·h

1 electron-volt (eV) $= 1.602 \times 10^{-19}$ J $= 1.602 \times 10^{-12}$ erg $=$
1.182×10^{-19} ft·lb-f

1 foot-pound-force (ft·lb-f) $= 1.356$ J $= 1.285 \times 10^{-3}$ Btu $= 0.3239$ cal $=$
1.356×10^{7} ergs $= 8.464 \times 10^{18}$ eV $= 3.766 \times 10^{-7}$ kW·h

1 kilocalorie (kcal), or **large calorie** (Cal) $= 4.187 \times 10^{3}$ J $= 1 \times 10^{3}$ cal

1 kilowatt-hour (kW·h) $= 3.600 \times 10^{6}$ J $= 3412$ Btu $= 8.598 \times 10^{5}$ cal $=$
3.6×10^{13} ergs $= 2.655 \times 10^{6}$ ft·lb-f

Power

1 watt (W) $= 3.412$ Btu/h $= 0.2388$ cal/s $= 1 \times 10^{7}$ ergs/s $=$
0.7376 ft·lb-f/s $= 1.341 \times 10^{-3}$ hp

1 British thermal unit per hour (Btu/h) $= 0.2931$ W $=$
7.000×10^{-2} cal/s $= 0.2162$ ft·lb-f/s $= 3.930 \times 10^{-4}$ hp

1 calorie per second (cal/s) $= 4.187$ W $= 14.29$ Btu/h $=$
4.187×10^{7} ergs/s $= 3.088$ ft·lb-f/s $= 5.615 \times 10^{-3}$ hp

1 erg per second (erg/s) $= 1 \times 10^{-7}$ W $= 2.388 \times 10^{-8}$ cal/s $=$
7.376×10^{-8} ft·lb-f/s $= 1.341 \times 10^{-10}$ hp

1 foot-pound-force per second (ft·lb-f/s) $= 1.356$ W $= 0.3238$ cal/s $=$
4.626 Btu/h $= 1.356 \times 10^{7}$ ergs/s $= 1.818 \times 10^{-3}$ hp

1 horsepower (hp)^{g} $= 745.7$ W $= 2.544 \times 10^{3}$ Btu/h $= 178.1$ cal/s
$= 550$ ft·lb-f/s

1 kilowatt (kW) $= 1 \times 10^{3}$ W $= 3.412 \times 10^{3}$ Btu/h $= 238.8$ cal/s $=$
737.6 ft·lb-f/s $= 1.341$ hp

Pressure

1 newton per square meter (N/m^{2}), or **pascal** (Pa) $= 9.869 \times 10^{-6}$ atm $=$
1×10^{-5} bar $= 7.501 \times 10^{-3}$ mm-Hg $= 10$ dynes/cm^{2} $= 2.953 \times 10^{-4}$ in.-Hg
$= 2.089 \times 10^{-2}$ lb-f/ft^{2} $= 1.450 \times 10^{-4}$ lb-f/in.^{2} $= 7.501 \times 10^{-3}$ torr

1 atmosphere (atm) $= 1.013 \times 10^{5}$ N/m^{2} $= 760.0$ mm-Hg $=$
1.013×10^{6} dynes/cm^{2} $= 29.92$ in.-Hg $= 2.116 \times 10^{3}$ lb-f/ft^{2}
$= 14.70$ lb-f/in.^{2}

1 bar $= 1 \times 10^{5}$ N/m^{2} $= 0.9869$ atm $= 750.1$ mm-Hg

1 dyne per square centimeter (dyne/cm^{2}) $= 0.1$ N/m^{2} $=$
9.869×10^{-7} atm $= 7.501 \times 10^{-4}$ mm-Hg $= 2.089 \times 10^{-3}$ lb-f/ft^{2} $=$
1.450×10^{-5} lb-f/in.^{2}

---

^{e}This is the "International Table" Btu; there are several other Btus.

^{f}This is the "International Table" calorie, which equals exactly 4.1868 J. There are several other calories; for instance, the thermochemical calorie, which equals 4.184 J.

^{g}There are several other horsepowers; for instance, the metric horsepower, which equals 735.5 W.

1 inch of mercury (in.-Hg) $= 3.386 \times 10^3$ N/m$^2 = 3.342 \times 10^{-2}$ atm $=$ 25.40 mm-Hg $= 0.4912$ lb-f/in.^{2}

1 pound-force per square inch (lb-f/in.^{2}, or psi) $= 6.895 \times 10^3$ N/m$^2 =$ 6.805×10^{-2} atm $= 6.895 \times 10^4$ dynes/cm$^2 = 2.036$ in.-Hg $=$ 7.031×10^{-2} kp/cm^2

1 torr, or **millimeter of mercury** (mm-Hg) $= 1.333 \times 10^2$ N/m$^2 = 1/760$ atm $=$ 1.333×10^{-3} bar $= 1.333 \times 10^3$ dynes/cm$^2 = 0.03937$ in.-Hg $= 0.01934$ lb-f/in.^{2}

Electric Charge[h]

1 coulomb (C) $\Leftrightarrow 2.998 \times 10^9$ statcoulombs, or esu of charge $\Leftrightarrow$ 0.1 abcoulomb, or emu of charge

Electric Current

1 ampere (A) $\Leftrightarrow 2.998 \times 10^9$ statamperes, or esu of current $\Leftrightarrow$ 0.1 abampere, or emu of current

Electric Potential

1 volt (V) $\Leftrightarrow 3.336 \times 10^{-3}$ statvolt, or esu of potential $\Leftrightarrow 1 \times 10^8$ abvolts, or emu of potential

Electric Field

1 volt per meter (V/m) $\Leftrightarrow 3.336 \times 10^{-5}$ statvolt/cm $\Leftrightarrow 1 \times 10^6$ abvolts/cm

Magnetic Field

1 tesla (T), or **weber per square meter** (Wb/m^2) $= 1 \times 10^4$ gauss

Electric Resistance

1 ohm (Ω) $\Leftrightarrow 1.113 \times 10^{-12}$ statohm, or esu of resistance $\Leftrightarrow 1 \times 10^9$ abohms, or emu of resistance

Electric Resistivity

1 ohm-meter ($\Omega \cdot$m) $\Leftrightarrow 1.113 \times 10^{-10}$ statohm-cm $\Leftrightarrow 1 \times 10^{11}$ abohm-cm

Capacitance

1 farad (F) $\Leftrightarrow 8.988 \times 10^{11}$ statfarads, or esu of capacitance $\Leftrightarrow 1 \times 10^{-9}$ abfarad, or emu of capacitance

Inductance

1 henry (H) $\Leftrightarrow 1.113 \times 10^{-12}$ stathenry, or esu of inductance $\Leftrightarrow 1 \times 10^9$ abhenrys, or emu of inductance

[h] The dimensions of the electric quantities in SI units, electrostatic units (esu), and electromagnetic units (emu) are usually different; hence the relationships among most of these units are correspondences ($\Leftrightarrow$) rather than equalities (=).

Appendix 9: The Periodic Table and Chemical Elements

TABLE A9.1 **THE PERIODIC TABLE**

Group designation — Atomic number — Symbol for element — Atomic mass

IA (1)	IIA (2)	IIIB (3)	IVB (4)	VB (5)	VIB (6)	VIIB (7)	VIIIB (8)	VIIIB (9)	VIIIB (10)	IB (11)	IIB (12)	IIIA (13)	IVA (14)	VA (15)	VIA (16)	VIIA (17)	VIIIA (18)
1 **H** 1.00794																	2 **He** 4.002602
3 **Li** 6.941	4 **Be** 9.012182											5 **B** 10.811	6 **C** 12.0107	7 **N** 14.0067	8 **O** 15.9994	9 **F** 18.99840	10 **Ne** 20.1797
11 **Na** 22.98977	12 **Mg** 24.3050											13 **Al** 26.98154	14 **Si** 28.0855	15 **P** 30.97376	16 **S** 32.065	17 **Cl** 35.453	18 **Ar** 39.948
19 **K** 39.0983	20 **Ca** 40.078	21 **Sc** 44.955910	22 **Ti** 47.867	23 **V** 50.9415	24 **Cr** 51.9961	25 **Mn** 54.938049	26 **Fe** 55.845	27 **Co** 58.93320	28 **Ni** 58.6934	29 **Cu** 63.546	30 **Zn** 65.409	31 **Ga** 69.723	32 **Ge** 72.64	33 **As** 74.92160	34 **Se** 78.96	35 **Br** 79.904	36 **Kr** 83.798
37 **Rb** 85.4678	38 **Sr** 87.62	39 **Y** 88.90585	40 **Zr** 91.224	41 **Nb** 92.90638	42 **Mo** 95.94	43 **Tc** 98.9072	44 **Ru** 101.07	45 **Rh** 102.90550	46 **Pd** 106.42	47 **Ag** 107.8682	48 **Cd** 112.411	49 **In** 114.818	50 **Sn** 118.710	51 **Sb** 121.760	52 **Te** 127.60	53 **I** 126.90447	54 **Xe** 131.293
55 **Cs** 132.90545	56 **Ba** 137.327	57 ***La** 138.9055	72 **Hf** 178.49	73 **Ta** 180.9479	74 **W** 183.84	75 **Re** 186.207	76 **Os** 190.23	77 **Ir** 192.217	78 **Pt** 195.078	79 **Au** 196.96654	80 **Hg** 200.59	81 **Tl** 204.3833	82 **Pb** 207.2	83 **Bi** 208.98037	84 **Po** 208.9824	85 **At** 209.9871	86 **Rn** 222.0176
87 **Fr** 223.0197	88 **Ra** 226.0277	89 **†Ac** 227.0277	104 **Rf** 261.1089	105 **Db** 262.1144	106 **Sg** 263.118	107 **Bh** 262.12	108 **Hs** 265.1306	109 **Mt** (268)	110 **Ds** (271)	111 **Uuu** (272)	112 **Uub** (285)		114 **Uuq** (289)				

Periods 1–7 (row labels)

*Lanthanides	58 **Ce** 140.116	59 **Pr** 140.90765	60 **Nd** 144.24	61 **Pm** 144.9127	62 **Sm** 150.36	63 **Eu** 151.964	64 **Gd** 157.25	65 **Tb** 158.92534	66 **Dy** 162.50	67 **Ho** 164.93032	68 **Er** 167.26	69 **Tm** 168.93421	70 **Yb** 173.04	71 **Lu** 174.967
†Actinides	90 **Th** 232.0381	91 **Pa** 231.0359	92 **U** 238.0289	93 **Np** 237.0482	94 **Pu** 244.0642	95 **Am** 243.0614	96 **Cm** 247.07003	97 **Bk** 247.0703	98 **Cf** 251.0796	99 **Es** 252.083	100 **Fm** 257.0951	101 **Md** 258.0984	102 **No** 259.1011	103 **Lr** 262.110

TABLE A9.2   ATOMIC MASSES AND ATOMIC NUMBERS OF CHEMICAL ELEMENTS

Data were obtained from the National Institute for Standards and Technology; values are for the elements as they exist naturally on Earth or for the most stable isotope, with carbon-12 (the reference standard) having a mass of exactly 12 u. The estimated uncertainties in values between ± and ± 9 units in the last digit of an atomic mass are in parentheses after the atomic mass.
(Source: http://physics.nist.gov/PhysRefData/Compositions/index.html)

ELEMENT	SYMBOL	ATOMIC NUMBER	ATOMIC MASS (u)	ELEMENT	SYMBOL	ATOMIC NUMBER	ATOMIC MASS (u)
Actinium	Ac	89	227.027 7	Mercury	Hg	80	200.59 (2)
Aluminum	Al	13	26.981 538 (2)	Molybdenum	Mo	42	95.94 (1)
Americium	Am	95	243.061 4	Neodymium	Nd	60	144.24 (3)
Antimony	Sb	51	121.760 (1)	Neon	Ne	10	20.179 7 (6)
Argon	Ar	18	39.948 (1)	Neptunium	Np	93	237.048 2
Arsenic	As	33	74.921 60 (2)	Nickel	Ni	28	58.693 4 (2)
Astatine	At	85	209.987 1	Niobium	Nb	41	92.906 38 (2)
Barium	Ba	56	137.327 (7)	Nitrogen	N	7	14.006 7 (2)
Berkelium	Bk	97	247.070 3	Nobelium	No	102	259.101 1
Beryllium	Be	4	9.012 182 (3)	Osmium	Os	76	190.23 (3)
Bismuth	Bi	83	208.980 38 (2)	Oxygen	O	8	15.999 4 (3)
Bohrium	Bh	107	264.12	Palladium	Pd	46	106.42 (1)
Boron	B	5	10.811 (7)	Phosphorus	P	15	30.973 761 (2)
Bromine	Br	35	79.904 (1)	Platinum	Pt	78	195.078 (2)
Cadmium	Cd	48	112.411 (8)	Plutonium	Pu	94	244.064 2
Calcium	Ca	20	40.078 (4)	Polonium	Po	84	208.982 4
Californium	Cf	98	251.079 6	Potassium	K	19	39.098 3 (1)
Carbon	C	6	12.010 7 (8)	Praseodymium	Pr	59	140.907 65 (2)
Cerium	Ce	58	140.116 (1)	Promethium	Pm	61	144.912 7
Cesium	Cs	55	132.905 45 (2)	Protactinium	Pa	91	231.035 88 (2)
Chlorine	Cl	17	35.453 (9)	Radium	Ra	88	226.025 4
Chromium	Cr	24	51.996 1 (6)	Radon	Rn	86	222.017 6
Cobalt	Co	27	58.933 200 (9)	Rhenium	Re	75	186.207 (1)
Copper	Cu	29	63.546 (3)	Rhodium	Rh	45	102.905 50 (2)
Curium	Cm	96	247.070 3	Rubidium	Rb	37	85.467 8 (3)
Darmstadtium	Ds	110	271	Ruthenium	Ru	44	101.07 (2)
Dubnium	Db	105	262.114 4	Rutherfordium	Rf	104	261.108 9
Dysprosium	Dy	66	162.500 (1)	Samarium	Sm	62	150.36 (3)
Einsteinium	Es	99	252.083	Scandium	Sc	21	44.955 910 (8)
Erbium	Er	68	167.259 (3)	Seaborgium	Sg	106	263.118 6
Europium	Eu	63	151.964 (1)	Selenium	Se	34	78.96 (3)
Fermium	Fm	100	257.095 1	Silicon	Si	14	28.085 5 (3)
Fluorine	F	9	18.998 403 2 (5)	Silver	Ag	47	107.868 2 (2)
Francium	Fr	87	223.019 7	Sodium	Na	11	22.989 770 (2)
Gadolinium	Gd	64	157.25 (3)	Strontium	Sr	38	87.62 (1)
Gallium	Ga	31	69.723 (1)	Sulfur	S	16	32.065 (6)
Germanium	Ge	32	72.64 (1)	Tantalum	Ta	73	180.947 9 (1)
Gold	Au	79	196.966 55 (2)	Technetium	Tc	43	98.907 2
Hafnium	Hf	72	178.49 (2)	Tellurium	Te	52	127.60 (3)
Hassium	Hs	108	265.130 6	Terbium	Tb	65	158.925 34 (2)
Helium	He	2	4.002 602 (2)	Thallium	Tl	81	204.383 3 (2)
Holmium	Ho	67	164.930 32 (2)	Thorium	Th	90	232.038 1 (1)
Hydrogen	H	1	1.007 94 (7)	Thulium	Tm	69	168.934 21 (2)
Indium	In	49	114.818 (3)	Tin	Sn	50	118.710 (7)
Iodine	I	53	126.904 47 (3)	Titanium	Ti	22	47.867 (1)
Iridium	Ir	77	192.217 (3)	Tungsten	W	74	183.84 (1)
Iron	Fe	26	55.845 (2)	Ununbium	Uub	112	285
Krypton	Kr	36	83.798 (2)	Unununium	Uuu	111	272
Lanthanum	La	57	138.905 5 (2)	Ununquadium	Uuq	114	289
Lawrencium	Lr	103	262.110	Uranium	U	92	238.028 9 (1)
Lead	Pb	82	207.2 (1)	Vanadium	V	23	50.941 5 (1)
Lithium	Li	3	6.941 (2)	Xenon	Xe	54	131.293 (2)
Lutetium	Lu	71	174.967 (1)	Ytterbium	Yb	70	173.04 (3)
Magnesium	Mg	12	24.305 0 (6)	Yttrium	Y	39	88.905 85 (2)
Manganese	Mn	25	54.938 049 (9)	Zinc	Zn	30	65.409 (4)
Meitnerium	Mt	109	268	Zirconium	Zr	40	91.224 (2)
Mendelevium	Md	101	258.098 4				

Appendix 10: Formula Sheets

Chapters 1–21

$v = dx/dt$

$a = dv/dt = d^2x/dt^2$

$x = x_0 + v_0 t + \frac{1}{2}at^2$

$a(x - x_0) = \frac{1}{2}(v^2 - v_0^2)$

$A_x = A \cos \theta$

$A = \sqrt{A_x^2 + A_y^2 + A_z^2}$

$\mathbf{A} \cdot \mathbf{B} = AB \cos \phi$

$\qquad = A_x B_x + A_y B_y + A_z B_z$

$|\mathbf{A} \times \mathbf{B}| = AB \sin \phi$

$a = v^2/r$

$\mathbf{v}' = \mathbf{v} - \mathbf{V}_O$

$m\mathbf{a} = \mathbf{F}_{net}$

$w = mg$

$f_k = \mu_k N$

$f_s \leq \mu_s N$

$F = -kx$

$W = F_x \, \Delta x$

$W = \mathbf{F} \cdot \mathbf{s}$

$W = \int \mathbf{F} \cdot d\mathbf{s}$

$K = \frac{1}{2}mv^2$

$U = mgy$

$E = K + U = [\text{constant}]$

$U(x) = -\int_{x_0}^{x} F_x(x') \, dx'$

$F_x = -\dfrac{dU}{dx}$

$U = \frac{1}{2}kx^2$

$E = mc^2$

$P = dW/dt$

$P = \mathbf{F} \cdot \mathbf{v}$

$F = GMm/r^2$

$v^2 = GM_S/r$

$g = GM_E/R_E^2$

$U = -GMm/r$

$\mathbf{p} = m\mathbf{v}$

$\mathbf{r}_{CM} = \dfrac{1}{M} \displaystyle\int \mathbf{r} \, dm$

$\mathbf{I} = \displaystyle\int_0^{\Delta t} \mathbf{F} \, dt$

$v_1' = \dfrac{m_1 - m_2}{m_1 + m_2} v_1; \; v_2' = \dfrac{2m_1}{m_1 + m_2} v_1$

$\omega = d\phi/dt$

$\alpha = d\omega/dt = d^2\phi/dt^2$

$v = R\omega$

$K = \frac{1}{2}I\omega^2$

$I = \displaystyle\int R^2 \, dm$

$I_{CM} = MR^2 \text{ (hoop)}; \frac{1}{2}MR^2 \text{ (disk)};$

$\qquad \frac{2}{5}MR^2 \text{ (sphere)}; \frac{1}{12}ML^2 \text{ (rod)}.$

$I = I_{CM} + Md^2$

$\tau = FR \sin \theta$

$I\alpha = \tau$

$P = \tau\omega$

$L = I\omega$

$\mathbf{L} = \mathbf{r} \times \mathbf{p}$

$\dfrac{d\mathbf{L}}{dt} = \mathbf{r} \times \mathbf{F}$

$x = A \cos(\omega t + \delta)$

$T = 2\pi/\omega; \; f = 1/T = \omega/2\pi$

$m \, d^2x/dt^2 = -kx$

$\omega = \sqrt{k/m}$

$\omega = \sqrt{g/l}; \; T = 2\pi\sqrt{l/g}$

$\omega = \sqrt{mgd/I}$

$y = A \cos k(x \pm vt) = A \cos(kx \pm \omega t)$

$\lambda = 2\pi/k; \; f = v/\lambda; \; \omega = 2\pi f$

$v = \sqrt{F/(M/L)}$

$f_{beat} = f_1 - f_2$

$f' = f(1 \pm V_R/v)$

$f' = f/(1 \mp V_E/v)$

$\sin \theta = v/V_E$

$p - p_0 = -\rho g y$

$\frac{1}{2}\rho v^2 + \rho g y + p = [\text{constant}]$

$pV = NkT$

$T_C = T - 273.15$

$v_{rms} = \sqrt{3kT/m}$

$TV^{\gamma-1} = [\text{constant}];$

$\qquad pV^\gamma = [\text{constant}]; \; \gamma = C_p/C_V$

$\Delta E = Q - W$

$e = 1 - T_2/T_1$

$\Delta S = \displaystyle\int_A^B dQ/T$

$g = 9.81 \text{ m/s}^2$

$G = 6.67 \times 10^{-11} \text{ N·m}^2/\text{kg}^2$

$M_E = 5.98 \times 10^{24} \text{ kg}$

$R_E = 6.37 \times 10^6 \text{ m}$

$m_e = 9.11 \times 10^{-31} \text{ kg}$

$m_p = 1.67 \times 10^{-27} \text{ kg}$

$c = 3.00 \times 10^8 \text{ m/s}$

$N_A = 6.02 \times 10^{23}/\text{mole}$

$k = 1.38 \times 10^{-23} \text{ J/K}$

$1 \text{ cal} = 4.19 \text{ J}$

$$F = \frac{1}{4\pi\epsilon_0} \frac{qq'}{r^2}$$

$$E = \frac{1}{4\pi\epsilon_0} \frac{q'}{r^2}$$

$$E = \sigma/2\epsilon_0$$

$$p = lQ$$

$$\boldsymbol{\tau} = \mathbf{p} \times \mathbf{E}$$

$$U = -\mathbf{p} \cdot \mathbf{E}$$

$$\Phi_E = \int \mathbf{E} \cdot d\mathbf{A}$$

$$\oint \mathbf{E} \cdot d\mathbf{A} = \oint E_\perp \, dA = \frac{Q_{\text{inside}}}{\epsilon_0}$$

$$V = \frac{1}{4\pi\epsilon_0} \frac{q'}{r}$$

$$E_x = -\frac{\partial V}{\partial x}, \quad E_y = -\frac{\partial V}{\partial y}, \quad E_z = -\frac{\partial V}{\partial z}$$

$$U = \tfrac{1}{2}Q_1 V_1 + \tfrac{1}{2}Q_2 V_2 + \tfrac{1}{2}Q_3 V_3 + \cdots$$

$$u = \tfrac{1}{2}\epsilon_0 E^2$$

$$C = Q/\Delta V$$

$$C = \epsilon_0 A/d$$

$$E = E_{\text{free}}/\kappa$$

$$\oint \kappa E_\perp \, dA = \frac{Q_{\text{free, inside}}}{\epsilon_0}$$

$$\Delta Y = -\int_{P_0}^{P} \mathbf{F} \cdot d\mathbf{s}$$

$$u = \tfrac{1}{2}\kappa\epsilon_0 E^2$$

$$I = \Delta V/R$$

$$R = \rho l/A$$

$$P = I\mathcal{E} \, ; \quad P = I\,\Delta V$$

$$F = \frac{\mu_0}{2\pi} \frac{qvI}{r}$$

$$\mathbf{F} = q\mathbf{v} \times \mathbf{B}$$

$$d\mathbf{B} = \frac{\mu_0}{4\pi} \frac{I\,d\mathbf{s} \times \mathbf{r}}{r^3}$$

$$\oint \mathbf{B} \cdot d\mathbf{s} = \oint B_\parallel \, ds = \mu_0 I$$

$$B = \mu_0 n I$$

$$r = \frac{p}{qB}$$

$$d\mathbf{F} = I \, d\mathbf{l} \times \mathbf{B}$$

$$\boldsymbol{\mu} = I \times [\text{area of loop}]$$

$$\boldsymbol{\tau} = \boldsymbol{\mu} \times \mathbf{B}$$

$$U = -\boldsymbol{\mu} \cdot \mathbf{B}$$

$$\mathcal{E} = vBl$$

$$\mathcal{E} = -\frac{d\Phi_B}{dt}$$

$$\Phi_B = \int \mathbf{B} \cdot d\mathbf{A}$$

$$\oint \mathbf{E} \cdot d\mathbf{s} = \oint E_\parallel \, ds = -\frac{d\Phi_B}{dt}$$

$$\Phi_B = LI$$

$$\mathcal{E} = -L\frac{dI}{dt}$$

$$U = \tfrac{1}{2}LI^2$$

$$u = \frac{1}{2\mu_0} B^2$$

$$\omega_0 = 1/\sqrt{LC}$$

$$Z = \sqrt{R^2 + \left(\omega L - \frac{1}{\omega C}\right)^2}$$

$$\mathcal{E}_2 = \mathcal{E}_1 \frac{N_2}{N_1}$$

$$\oint \mathbf{B} \cdot d\mathbf{A} = \oint B_\perp \, dA = 0$$

$$\oint \mathbf{B} \cdot d\mathbf{s} = \oint B_\parallel \, ds = \mu_0 I + \mu_0\epsilon_0 \frac{d\Phi_E}{dt}$$

$$B = E/c$$

$$\mathbf{S} = \frac{1}{\mu_0} \mathbf{E} \times \mathbf{B}$$

$$[\text{pressure}] = S/c$$

$$\frac{\partial^2 E}{\partial x^2} = \mu_0\epsilon_0 \frac{\partial^2 E}{\partial t^2}$$

$$c = 1/\sqrt{\mu_0\epsilon_0}$$

$$v = c/n$$

$$n_1 \sin\theta_1 = n_2 \sin\theta_2$$

$$f = \pm\tfrac{1}{2}R$$

$$\frac{1}{s} + \frac{1}{s'} = \frac{1}{f}$$

Interference minima:
$$d \sin\theta = \tfrac{1}{2}\lambda, \tfrac{3}{2}\lambda, \tfrac{5}{2}\lambda, \ldots$$

Interference maxima:
$$d \sin\theta = 0, \lambda, 2\lambda, \ldots$$

Diffraction minima:
$$a \sin\theta = \lambda, 2\lambda, 3\lambda, \ldots$$

$$a \sin\theta = 1.22\lambda$$

$$f' = \sqrt{\frac{1 \mp v/c}{1 \pm v/c}} \, f$$

$$x' = \frac{x - Vt}{\sqrt{1 - V^2/c^2}}$$

$$y' = y$$

$$t' = \frac{t - Vx/c^2}{\sqrt{1 - V^2/c^2}}$$

$$\Delta t = \frac{\Delta t'}{\sqrt{1 - V^2/c^2}}$$

$$L = \sqrt{1 - V^2/c^2}\, L'$$

$$v'_x = \frac{v_x - V}{1 - v_x V/c^2}$$

$$\mathbf{p} = \frac{m\mathbf{v}}{\sqrt{1 - v^2/c^2}} \, ; \quad E = \frac{mc^2}{\sqrt{1 - v^2/c^2}}$$

$$E^2 = p^2 c^2 + m^2 c^4$$

$$E = hf$$

$$p = hf/c$$

$$\Delta y \, \Delta p_y \geq h/4\pi$$

$$L = n\hbar$$

$$E_n = -\frac{m_e e^4}{2(4\pi\epsilon_0)^2 \hbar^2} \frac{1}{n^2} = -\frac{13.6 \text{ eV}}{n^2}$$

$$\lambda = h/p$$

$$\mu_{\text{spin}} = \frac{e\hbar}{2m_e}$$

$$E = \frac{J(J+1)\hbar^2}{2I}$$

$$R = (1.2 \times 10^{-15} \text{ m}) \times A^{1/3}$$

$$n = n_0 e^{-t/\tau}; \quad \tau = t_{1/2}/0.693$$

$$e = 1.60 \times 10^{-19} \text{ C}$$

$$\epsilon_0 = 8.85 \times 10^{-12} \text{ F/m}$$

$$\mu_0 = 1.26 \times 10^{-6} \text{ H/m}$$

$$c = 3.00 \times 10^8 \text{ m/s}$$

$$h = 2\pi\hbar = 6.63 \times 10^{-34} \text{ J·s}$$

$$m_e = 9.11 \times 10^{-31} \text{ kg}$$

$$m_p = 1.67 \times 10^{-27} \text{ kg}$$

Appendix 11: Answers to Odd-Numbered Problems and Review Problems

Chapter 22

1. 5.8×10^5 N
3. 58 N, 3.5×10^{28} m/s^2
5. 51 N
7. 1.6×10^{20} electrons
9. 2.39×10^{-7} N
11. $F_g = 6.7 \times 10^{-13}$ N, $F_e = 2.3 \times 10^{-8}$ N
13. 9.63×10^4 C
15. 4.7×10^{13} electrons
17. 3.2×10^{19} N
19. 1.3
21. 99.9 %
23. 2.9×10^{-9} N/m
25. $(-2.3 \times 10^{-5}$ N$)\mathbf{i} + (-3.5 \times 10^{-5}$ N$)\mathbf{j}$, $(2.3 \times 10^{-5}$ N$)\mathbf{i}$ $+ (3.5 \times 10^{-5}$ N$)\mathbf{j}$
27. 6.81×10^{32} electrons on Earth and 1.9×10^{32} electrons on Moon
29. 1.0×10^{-9} at 1 m, 1.0×10^{-5} at 1×10^4 m
31. 3.8×10^{-39} C, ratio $= 2.8 \times 10^6$ (attractive)
33. $\dfrac{2kqxQ}{\left[\frac{d^2}{4} + x^2\right]^{3/2}}$, $+x$ direction
35. $-(1.9 \times 10^{-7}$ N$)\,\mathbf{i} - (1.7 \times 10^{-7}$ N$)\,\mathbf{j}$
37. 1.0×10^{-7} C
39. $-1.35k\,\dfrac{Q^2}{L^2}\mathbf{i} - 1.35k\,\dfrac{Q^2}{L^2}\mathbf{j} = -1.35k\,\dfrac{Q^2}{L^2}\,(\mathbf{i}-\mathbf{j})$
41. $-(3.1 \times 10^{-15}$ N$)\,\mathbf{i} + (6.9 \times 10^{-16}$ N$)\,\mathbf{j}$
43. 1.2×10^{-4} kg or 0.12 g
45. $2\sqrt{2}\left(\dfrac{\sqrt{3}}{2} + 1\right)^2 = 9.85$
47. $kQq\left[\dfrac{1}{x^2} - \dfrac{4}{(d + 2x)^2}\right]$
49. $k\dfrac{Q^2}{a^2}\,(1.116\,\mathbf{i} - 1.75\,\mathbf{j} - 0.5\mathbf{k})$
51. $0.35d$
53. $\mathrm{p} + \mathrm{p} \to \mathrm{n} + \mathrm{n} + \pi^+$, $\mathrm{p} + \mathrm{p} \to \mathrm{n} + \mathrm{p} + \pi^0$, $\mathrm{p} + \mathrm{p} \to \mathrm{n} +$ $\mathrm{p} + \pi^0 + \pi^-$
55. 1
57. 5.6×10^{21} electrons
59. 1.9×10^{-9} kg
61. any negative charge
63. 3.6×10^{-8} N, 6.9×10^{-2} m/s^2
65. 1.0×10^{-6} C, 6.5×10^{12} electrons
67. 0 C, $+e$, $+e$, 0 C

Chapter 23

1. $F = 5.4 \times 10^{-14}$ N, $a = 6.0 \times 10^{16}$ m/s^2
3. $\theta = -20°$
5. $\mathbf{E} = -2.1 \times 10^5$ N/C $\mathbf{j}$
7. $F_e = 0.6\,F_g$
9. 6.3×10^{-7} m
11. $E = 5.1 \times 10^{11}$ N/C
13. $E = 5.1 \times 10^{12}$ N/C
15. 28 N/C
17. $\mathbf{E}_A = \left(1.15 \times 10^{10}\,\dfrac{\mathrm{N\cdot m^2}}{\mathrm{C^2}}\right)\dfrac{Q}{L^2}\mathbf{i}$,

$\mathbf{E}_B = \left(1.15 \times 10^{10}\,\dfrac{\mathrm{N\cdot m^2}}{\mathrm{C^2}}\right)\dfrac{Q}{L^2}\mathbf{j}$,

$\mathbf{E}_C = -\left(1.15 \times 10^{10}\,\dfrac{\mathrm{N\cdot m^2}}{\mathrm{C^2}}\right)\dfrac{Q}{L^2}\mathbf{i}$,

$\mathbf{E}_D = -\left(1.15 \times 10^{10}\,\dfrac{\mathrm{N\cdot m^2}}{\mathrm{C^2}}\right)\dfrac{Q}{L^2}\,\mathbf{j}$
19. $E_P = 9.5 \times 10^3$ N/C $\mathbf{i} - 2.8 \times 10^4$ N/C $\mathbf{j}$
21. $E(x) = 2 \times \left(8.99 \times 10^9\,\dfrac{\mathrm{N\cdot m^2}}{\mathrm{C^2}}\right)$

$\left[\dfrac{(-40\ \mathrm{C})(10000\ \mathrm{m})}{\left[x^2 + (10000\ \mathrm{m})^2\right]^{3/2}} + \dfrac{(30\ \mathrm{C})(4000\ \mathrm{m})}{\left[x^2 + (4000\ \mathrm{m})^2\right]^{3/2}}\right]$
23. $E \approx \dfrac{Q}{2\pi\epsilon_0 x^2}$
25. a) E_{max} is at $y = \pm R\,\dfrac{\sqrt{2}}{2}$
   b) The field distribution for the ring is the same as that for two positive charges rotated around the y axis.
27. $E_{0.5} = 7.2 \times 10^{-4}$ N/m, $E_{1.0} = 3.6 \times 10^{-4}$ N/m, $E_{1.5} = 2.4 \times 10^{-4}$ N/m
29. $\mathbf{E} = \left(2.6 \times 10^{10}\,\dfrac{\mathrm{N\cdot m^2}}{\mathrm{C^2}}\right)\dfrac{\lambda}{L}\,(\mathbf{i}+\mathbf{j})$
31. $\mathbf{E} = \dfrac{1}{\pi\epsilon_0}\dfrac{\lambda}{2d}\,(\cos 30°)\,\mathbf{j}$
33. $E_x = \dfrac{1}{4\pi\epsilon_0}\left[\dfrac{-\lambda}{\sqrt{x^2 + y^2}}\right]$, $E_y = \dfrac{\lambda}{4\pi\epsilon_0 y}\left[1 + \dfrac{x}{\sqrt{x^2 + y^2}}\right]$
35. $\mathbf{E}_A = 1.13 \times 10^5$ N/C $\mathbf{j}$, $\mathbf{E}_B = -1.13 \times 10^5$ N/C $\mathbf{j}$, $\mathbf{E}_C = -3.39 \times 10^5$ N/C $\mathbf{j}$, $\mathbf{E}_D = -1.13 \times 10^5$ N/C $\mathbf{j}$
37. $E = 2.4 \times 10^7$ N/C directed at an angle of 45° with respect to each sheet
39. $E_z = -\dfrac{z\sigma}{2\epsilon_0\sqrt{z^2 + R^2}}$

41. a) $\mathbf{E}_P = \dfrac{1}{4\pi\epsilon_0}\dfrac{Q}{l}\left(\dfrac{1}{x} - \dfrac{1}{x+l}\right)\mathbf{i}$,

b) $\mathbf{E}_{P'} = \dfrac{1}{2\pi\varepsilon_0}\dfrac{Q}{y}\dfrac{1}{\sqrt{l^2 + 4y^2}}\mathbf{j}$

43. a) $\lambda_0 = \dfrac{2Q}{l}$,

b) $\mathbf{E}_p = \dfrac{1}{2\pi\epsilon_0}\dfrac{-Q}{l^2}\left[\dfrac{1}{2}\ln\left(\dfrac{x+l}{x}\right)^2 + \dfrac{x}{x+l} - 1\right]\mathbf{i}$,

c) When $x \gg l$, the electric field resembles that of a point charge and goes to zero as x goes to infinity.

45. $\mathbf{E} = \dfrac{\lambda\sqrt{2}}{\pi\epsilon_0 l}\mathbf{i} - \dfrac{\lambda\sqrt{2}}{\pi\epsilon_0 l}\mathbf{j}$

47. $\mathbf{E} = -\dfrac{\lambda}{4\pi\epsilon_0}\left\{\left[\dfrac{1}{x}\left(1 + \dfrac{y}{(x^2+y^2)^{1/2}}\right) + \dfrac{1}{(x^2+y^2)^{1/2}}\right]\mathbf{i} + \left[\dfrac{1}{y}\left(1 + \dfrac{x}{(x^2+y^2)^{1/2}}\right) + \dfrac{1}{(x^2+y^2)^{1/2}}\right]\mathbf{j}\right\}$

49. $E = 0.0609\dfrac{\lambda}{\epsilon_0 R}$

51. The solution is a sketch of the electric field.

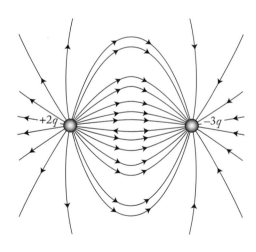

53. The solution is a sketch of the electric field.

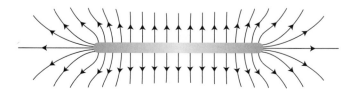

55. The solution is a sketch of the electric field.

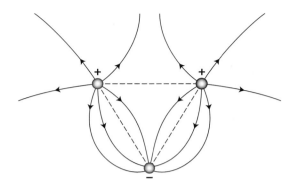

57. $E = 3.1 \times 10^5$ N/C

59. $v_0 = 1.4 \times 10^5$ m/s

61. $v_0 = 110$ m/s

63. 1085 electrons

65. $p = 2.0 \times 10^5$ C·m

67. a) $p = 1.6 \times 10^{-29}$ C·m, b) The dipole moment is reduced due to the motion of electrons.

69. $E = 520$ N/C

71. $E = 1.1 \times 10^7$ N/C, $\theta = 3°$ with respect to the y-axis

73. $E = 1.21 \times 10^4$ N/C, $\theta = 74°$ with respect to the y-axis

75. (a) $E = \dfrac{1}{4\pi\epsilon_0}\left(\dfrac{\lambda}{d}\right)$, (b) $F = \dfrac{q}{4\pi\epsilon_0}\left(\dfrac{\lambda}{d}\right)$,

(c) The magnitude of the force is the same as in (b), but its direction is opposite that of the rod on the charge.

77. $\mathbf{E} = \dfrac{Q}{2\pi\epsilon_0}\left(\dfrac{1}{y^2} - \dfrac{1}{(y^2 + d^2)^{\frac{3}{2}}}\right)\mathbf{j}$

79. $E = \dfrac{1}{4\pi\epsilon_0}\dfrac{Q}{y^2}$

Chapter 24

1. 1.1×10^{12} N·m^2/C

3. 0.16 N·m^2/C

5. $-\dfrac{\sqrt{3}}{8}\dfrac{\sigma}{\epsilon_0}a^2, \dfrac{\sqrt{3}}{24}\dfrac{\sigma}{\epsilon_0}a^2$

7. $\phi_{E,1} = \phi_{E,6} = 0, \phi_{E,2} = -\phi_{E,4} = 2.0$ N·m^2/C, $\phi_{E,3} = -\phi_{E,5} = -3.5$ N·m^2/C, $\phi_{total} = 0$

9. $\phi = \dfrac{q}{\epsilon_0}$

11. 1.4×10^3 N·m^2/C

13. 0.038 N·m^2/C

15. (a) $\dfrac{q}{2\epsilon_0}$ (b) $\dfrac{\sigma q}{2\epsilon_0}$ (c) $F = \dfrac{\sigma q}{2\epsilon_0}, E = \dfrac{\sigma}{2\epsilon_0}$

17. (a) $\dfrac{q}{4\epsilon_0}$ (b) $\dfrac{\sigma}{4\epsilon_0}$

19. $+45$ N·m²/C, 5.5×10^{-8} C/m

21. 2.3×10^2 N·m²/C

23. $\phi_G = 4\pi G m_{inside}$

25. Imagine a small cube of volume dV. If the cube itself contains no charge and it's located in a uniform electric field, the net flux through the six faces of the cube must be zero, no matter how the cube is oriented in the field. If the cube contains charge, then the flux through it cannot be zero and the field cannot be uniform. If the field is uniform, Gauss' law tells us that the charge density inside the cube, which is the charge inside divided by the volume dV, must be zero.

27. 160 N

29. $EA = \dfrac{q_{enc}}{\epsilon_0} \Rightarrow q_{enc} = 0 \therefore EA = 0$

since $A \neq 0, E = 0$

31. $\dfrac{\lambda r}{2\pi R^2 \epsilon_0}$ in direction perpendicular to axis

33. $r \le a\ E = 0, a \le r \le b\ E = \dfrac{\rho}{3\epsilon_0} \dfrac{(r^3 - a^3)}{r^2}$,

$r \ge b\ E = \dfrac{\rho}{3\epsilon_0} \dfrac{(b^3 - a^3)}{r^2}$

35. $r \le a\ E = 0, a \le r \le b\ E = \dfrac{Q}{4\pi\epsilon_0 r^2} \dfrac{(r^3 - a^3)}{(b^3 - a^3)}$,

$r \ge b\ E = \dfrac{Q}{4\pi\epsilon_0 r^2}$

37. (a) $F = qE$ where $q = -e$. $E = \dfrac{Q}{4\pi\epsilon_0 r^2} r$ where

$Q = +e\left(\dfrac{r^3}{R^3}\right)$. Substituting yields $\vec{F} = -\dfrac{e^2 r}{4\pi\epsilon_0 R^3} \hat{r}$

where the minus sign indicates this is a restoring force and

the magnitude of that force is $F = \dfrac{e^2 r}{4\pi\epsilon_0 R^3}$. (b) 7.2×10^{15} Hz

39. (a) $C = \dfrac{Q}{2\pi R^2}$ (b) $E = \dfrac{Q}{4\pi\epsilon_0 R^2}$ (c) $E = \dfrac{Q}{4\pi\epsilon_0 r^2}$

41. (a) $Q = \dfrac{4\pi k}{n + 3} r^{n+3}$ (b) $E = \dfrac{kr^{n+1}}{\epsilon_0(n + 3)}$ (c) $n = -1$, $E = k/2$ (d) if $n = -3, Q = \infty$

43. For $|x| \le \dfrac{d}{2}, E = \dfrac{Cx^3}{3\epsilon_0}$. For $|x| \ge \dfrac{d}{2}\ E = \dfrac{Cd^3}{24\epsilon_0}$.

45. $\phi = \displaystyle\int_{-\infty}^{\infty} E_y 2\pi R dx = 2\pi R \int_{-\infty}^{\infty} E_y dx = \dfrac{q'}{\epsilon_0}$.

$\displaystyle\int_{-\infty}^{\infty} E_y dx = \dfrac{q'}{2\pi R\epsilon_0} \quad \therefore \dfrac{q}{v} \int_{-\infty}^{\infty} E_y dx = \dfrac{q}{v} \dfrac{q'}{2\pi\epsilon_0 R}$

$\therefore \displaystyle\int_{-\infty}^{\infty} F_y dt = \dfrac{q}{v} \dfrac{q'}{2\pi\epsilon_0 R}$

47. $E = \dfrac{Q}{4\pi\epsilon_0}\left(\dfrac{1}{r^2} - \dfrac{1}{8\left(r - \dfrac{R}{2}\right)^2}\right) r$

49. $\mathbf{r} = \dfrac{1}{4}d\hat{i}\ \vec{E} = \dfrac{\rho d}{6\epsilon_0}\hat{i}, \quad \mathbf{r} = 2d\hat{i}\ \mathbf{E} = \dfrac{47\rho d}{96\epsilon_0}\hat{i},$

$\vec{r} = \dfrac{1}{2}d\hat{i} + d\hat{j}\ \vec{E} = \dfrac{\rho d}{\epsilon_0}\left[\left(\dfrac{1}{2} - \dfrac{1}{30\sqrt{5}}\right)\hat{i} - \dfrac{1}{15\sqrt{5}}\hat{j}\right]$

51. $\mathbf{E}_{bigsphere} = \dfrac{q\mathbf{r}}{4\pi\epsilon_0 R^3}$ and $\mathbf{E}_{smallsphere} = \dfrac{q\mathbf{r'}}{4\pi\epsilon_0 R^3}$

where $\mathbf{r'} = \mathbf{d} - \mathbf{r}\ \mathbf{E}_{total} = \mathbf{E}_{bigsphere} + \mathbf{E}_{smallsphere}$

$\therefore \mathbf{E}_{total} = \dfrac{q\mathbf{d}}{4\pi\epsilon_0 R^3} \quad \therefore E = \dfrac{qd}{4\pi\epsilon_0 R^3}$

53. 1.1×10^5 C, 0 C, 1.1×10^5 C

55. $E = \dfrac{\lambda}{2\pi\epsilon_0 x}, E = 0$

57. $r < a E = \dfrac{q}{4\pi\epsilon_0 r^2}, a < r < b\ E = 0, r > b$

$E = \dfrac{q}{4\pi\epsilon_0 r^2}, \sigma_a = -\dfrac{q}{4\pi a^2}, \sigma_b = \dfrac{q}{4\pi b^2}$

59. 8.85×10^{-10} C/m²

61. (a) $Q_{small,inner} = -Q_1, Q_{small,outer} = +Q_1$

(b) $E = \dfrac{Q_1}{4\pi\epsilon_0 r^2}$ (c) $Q_{large,inner} = -Q_1,$

$Q_{large,outer} = Q_1 + Q_2$ (d) $E = \dfrac{Q_1 + Q_2}{4\pi\epsilon_0 r^2}$

63. $\phi_{base} = \dfrac{q}{2\epsilon_0}, \phi_{curved} = \dfrac{q}{2\epsilon_0}$

65. (a) $\phi = \dfrac{q}{4\epsilon_0}$ (b) $\phi = \dfrac{3q}{4\epsilon_0}$

67. (a) 1.1×10^{12} N·m²/C (b) -1.1×10^{12} N·m²/C (c) 1.7×10^{12} N·m²/C (d) 0

69. $r \le R'\ E = -\dfrac{r\lambda}{3\pi R^2\epsilon_0}$ where the minus sign indicates the field points towards the center of the sphere. $r \ge R'$

$E = -\dfrac{R'^3\lambda}{3\pi\epsilon_0 R^2 r^2}$ where again, the minus sign indicates

the field points towards the center of the sphere.

71. $m = \dfrac{4\pi R^3 \rho \sigma}{6g\epsilon_0}$

73. $E = \dfrac{\lambda}{2\pi\epsilon_0 R}$, $\vec{E}_{total} = \dfrac{2\lambda}{5\pi\epsilon_o R}\vec{i}$

75. $r \le a\ E = -\dfrac{\lambda}{2\pi\epsilon_0 r}$, $a \le r \le b\ E = -\dfrac{\lambda}{2\pi\epsilon_0 r}\dfrac{b^2 - r^2}{b^2 - a^2}$,

$r \ge b\ E = 0$

77. 1.2×10^7 m/s, 2.3×10^{-13} J, 1.1×10^{-34} kg·m²/s,

6.6×10^{20} Hz

Chapter 25

1. 2.4×10^6 J

3. 60,000 V

5. 6.9×10^6 m/s

7. a) 2.1×10^6 m/s; b) 1.45×10^6 m/s

9. 2.7×10^7 m/s

11. 9.2×10^2 V

13. 2.05×10^5 m/s

15. 7.36×10^{-8} C

17. 3.86×10^{-12} J

19. 5.8×10^{-12} J

21. $\dfrac{Q}{a\pi\epsilon_0}\left(1 + \dfrac{2}{\sqrt{5}} + \dfrac{1}{3}\right)$

23. $\dfrac{\lambda\theta}{4\pi\epsilon_0}$

25. a) 5.5×10^{-12} J; b) 4.0×10^7 m/s

27. -23 V

29. $\dfrac{Q}{l}\dfrac{1}{4\pi\epsilon_0}\ln\left(\dfrac{2 + \sqrt{3}}{2 - \sqrt{3}}\right)$

31. $\dfrac{Q}{l}\dfrac{1}{4\pi\epsilon_0}\left[\ln\left(\dfrac{l+x}{x}\right) + \ln\left(\dfrac{1 + \sqrt{l^2 + (x+1)^2}}{1+x}\right)\right.$

$\left. + \ln\left(\dfrac{1 + \sqrt{x^2 + l^2}}{x}\right) + \ln\left(\dfrac{x + 1 + \sqrt{l^2 + (x+1)^2}}{x + \sqrt{x^2 + l^2}}\right)\right]$

33. $\dfrac{\lambda}{2\pi\epsilon_0}\left[\dfrac{a^2\ln(b/a)}{b^2 - a^2} - \dfrac{1}{2}\right]$

35. 1.16×10^4 V/m for the wire and the same for the cylinder.

37. $V(r) = \dfrac{4k}{\epsilon_0\sqrt{r}}$ Volt

Take any point (x,y,z) on the surface of the sphere. Then
$x^2 + y^2 + z^2 = R^2$

$V = \dfrac{1}{4\pi\epsilon_0}\left[\dfrac{Q}{r_1} + \dfrac{-Q(R/h)}{2}\right]$

r_1 and r_2 as shown

$r_1 = \sqrt{x^2 + y^2 + (h - z)^2}$

$r_2 = \sqrt{x^2 + y^2 + \left(\dfrac{R^2}{h} - z\right)^2}$

Therefore,

$V(x,y,z) = \dfrac{Q}{4\pi\epsilon_0}\left[\dfrac{1}{\sqrt{x^2 + y^2(h-z)^2}}\right.$

$\left. -\dfrac{R}{h}\dfrac{1}{\sqrt{x + y + \left(\dfrac{R^2}{h} - z\right)^2}}\right]$

$= \dfrac{Q}{4\pi\epsilon_0}\left[\dfrac{1}{(x^2 + y^2 + z^2 - 2zh + h^2)^{1/2}} = \dfrac{R}{h}\right.$

$\left.\dfrac{1}{(x^2 + y^2 + z^2 - 2R^2z/h + R^4/h^2)^{1/2}}\right]$

$= \dfrac{Q}{4\pi\epsilon_0}\left[\dfrac{1}{(R^2 - 2zh + h^2)^{1/2}} - \right.$

$\left.\dfrac{1}{h/r(R^2 - 2R^2z/h + R^4/h^2)^{1/2}}\right]$

$= \dfrac{Q}{4\pi\epsilon_0}\left[\dfrac{1}{(R^2 - 2zh + h^2)^{1/2}} - \right.$

$\left.\dfrac{1}{(h^2 - 2zh + R^2)^{1/2}}\right] = 0$

Therefore, potential constant is at 0 on entire surface.

39. (a) $r > R_2$, $V_a = 0$

(b) $R_1 < r < R_2$, $V_b = 0$

(c) $R < r < R_1$, $V_c = \dfrac{Q}{4\pi\epsilon_0 R_1} - \dfrac{Q}{4\pi\epsilon_0 r}$

(d) $R_0 < r < R$, $V_d = \dfrac{Q}{4\pi\epsilon_0 R_1} - \dfrac{Q}{4\pi\epsilon_0 r}$

41. $\dfrac{3Q}{4\pi\epsilon_0 r}; \dfrac{3Q}{4\pi\epsilon_0 c}; \dfrac{3Q}{4\pi\epsilon_0 c} + \dfrac{Q}{4\pi\epsilon_0 r} - \dfrac{Q}{4\pi\epsilon_0 b};$

$\dfrac{Q}{4\pi\epsilon_0}\left(\dfrac{3}{c} + \dfrac{1}{a} - \dfrac{1}{b} - \dfrac{r^2}{a^3} + \dfrac{1}{2a}\right)$

47. $E_x = -\dfrac{2\pi}{a}\sin\dfrac{2\pi x}{a}\cos\dfrac{2\pi y}{b}\cos\dfrac{2\pi z}{c};$

$E_y = -\dfrac{2\pi}{b}\cos\dfrac{2\pi x}{a}\sin\dfrac{2\pi y}{b}\cos\dfrac{2\pi z}{c};$

$$E_z = -\frac{2\pi}{c}\cos\frac{2\pi x}{a}\cos\frac{2\pi y}{b}\sin\frac{2\pi z}{c}$$

49. $E_x = \dfrac{p}{4\pi\epsilon_0}\dfrac{3xz}{(x^2+y^2+z^2)^{5/2}}$;

$E_y = \dfrac{p}{4\pi\epsilon_0}\dfrac{3yz}{(x^2+y^2+z^2)^{5/2}}$;

$E_z = \dfrac{p}{4\pi\epsilon_0}\left[\dfrac{3z^2}{(x^2+y^2+z^2)^{5/2}} - \dfrac{1}{(x^2+y^2+z^2)^{3/2}}\right]$;

On the z axis: $E_x = E_y = 0$, $E_z = \dfrac{p}{2\pi\epsilon_0 z^3}$; On the x axis:

$E_x = E_y = 0$, $E_z = -\dfrac{p}{2\pi\epsilon_0 x^3}$

51. a) $\dfrac{Q}{l}\dfrac{1}{4\pi\epsilon_0}\left[\ln\left(\dfrac{x+l/2+\sqrt{R^2+(x+l/2)^2}}{x-l/2+\sqrt{(x+l/2)^2+R^2}}\right)\right]$;

b) $-\dfrac{Q}{4\pi\epsilon_0}$

$$\dfrac{4l}{\sqrt{l^2+4lx+4(R^2+x^2)}\left(-l+2x+\sqrt{l^2+4lx+4(R^2+x^2)}\right)}$$

53. -8.1×10^{-18} J

55. 6.3×10^{-4} J

57. 3×10^{-2} J

59. a) 4.4×10^{-8} J/m³; b) 2.3×10^{11} J

61. 1.8×10^9 J/m³

63. $\dfrac{Q^2}{4\pi\epsilon_0 d}(4+\sqrt{2})$

65. 5.8×10^6 eV

67. -50 eV

69. $\dfrac{Q^2}{8\pi\epsilon_0}\left(\dfrac{7}{10R}\right)$

73. 8.6×10^5 eV

75. b) $U_{grav} = \dfrac{3}{5}\dfrac{GM^2}{R} = 1.24\times10^{29}$ J

$\dfrac{U_{grav}}{U_{rest\,mass}} = \dfrac{(3/5)(GM^2/R)}{Mc^2} = \dfrac{3}{5}\dfrac{GM}{Rc^2} = 1.9\times10^{11}$

77. 4.1×10^{-7} J.

79. a) $\dfrac{\lambda^2}{8\pi^2\epsilon_0 r^2}$; b) $\dfrac{\lambda^2}{4\pi\epsilon_0}\left(\dfrac{1}{4}+\ln\dfrac{b}{a}\right)$

81. 9.96×10^{-2} m; 2.5×10^5 V

83. 0 V

85. 3.1×10^7 m/s

87. a) $\dfrac{2Q}{3\pi\epsilon_0 R^2}\left(\sqrt{R^2+x^2} - \sqrt{\dfrac{R^2}{4}+x^2}\right)$;

b) $-\dfrac{2Q}{3\pi\epsilon_0 R^2}\left(\dfrac{x}{\sqrt{R^2+x^2}} - \dfrac{x}{\sqrt{\dfrac{R^2}{4}+x^2}}\right)\hat{i}$

89. $V = \dfrac{Q^2}{4\pi\epsilon_0 l^2}\left[x\ln\dfrac{(x+2l)x}{(x+l)^2} + 2l\ln\dfrac{(x+2l)}{(x+l)}\right]$

91. $U_{total} = \dfrac{Q^2}{8\pi\epsilon_0 b(b^3-a^3)^2}\left(\dfrac{6}{5}b^6 - 3a^3b^3 + \dfrac{9}{5}a^5b\right)$

Chapter 26

1. $Q_2 = 3Q_1$, $V_1 = 3V_2$

3. $C = 1.1\times10^{-11}$ F, $Q = 1.1\times10^{-6}$ C

5. $C = 1.1\times10^{-9}$ F, $Q = 1.3\times10^{-8}$ C

7. $n = 4.5\times10^{14}$ electrons

9. $A_{min} = 5.6\times10^{-4}$ m², $A_{max} = 6.8\times10^{-3}$ m²

11. $Q = 125$ C

13. $C = 1.8\times10^{-17}$ F, $Q = 1.8\times10^{-18}$ C

15. $C = 8.0\times10^{-12}$ F/m

17. $C = 9.9\ \mu$F

19. $Q = 1.7\times10^{-3}$ C

21. $C_{net} = \dfrac{2}{3}$ C

23. $Q = 12.0\ \mu$C, $\Delta V = 3.5$ V

25. $Q = 1.1\times10^{-5}$ C, $\Delta V_{2.5\mu F} = 4.5$ V, $\Delta V_{5.0\mu F} = 2.3$ V

27. The only arrangement of capacitors to give the same net capacitance is two pairs of two capacitors in series, connected in parallel.

29. $Q_1 = 3.0\times10^{-4}$ C, $Q_2 = 4.1\times10^{-4}$ C,

$Q_3 = 4.8\times10^{-4}$ C, $\Delta V_{PP'} = 68$ V

31. $C = 4\pi\epsilon_0\kappa R$

33. $k = 5100$

35. $k = 1.7$

37. $C = (\kappa_1 + \kappa_2)\left(\dfrac{\epsilon_0 A}{2d}\right)$

39. a) $E_{free} = \dfrac{2\Delta V}{d}\left(\dfrac{\kappa}{1+\kappa}\right)$, b) $E = \dfrac{2\Delta V}{d}\left(\dfrac{1}{1+\kappa}\right)$,

c) $\sigma_{bound} = \dfrac{2\epsilon_0\Delta V}{d}\left(\dfrac{1}{1+\kappa}\right)$

41. a) $C_0 = 2.0\times10^{-11}$ F, b) $C = 2.9\times10^{-11}$ F

43. $Q = 9.0\times10^{-6}$ C, $\Delta V_{filled} = 1.8$V, $\Delta V_{empty} = 4.5$ V

45. a) $\sigma_{bound,inside} = -2.6\times10^{-6}$ C/m²,

$\sigma_{bound,outside} = 1.7\times10^{-6}$ C/m²,

b) $E_{inner} = 1.6 \times 10^5$ V/m, $E_{outer} = 1.1 \times 10^5$ V/m,

c) $E_{free} = 3.0 \times 10^5$ V/m

47. $F = 0.050$ N

49. $C = 2\pi\epsilon_0 \left(\dfrac{1}{R_1} - \dfrac{1}{R_2} \right)^{-1} (\kappa_{top} + \kappa_{bottom})$

51. a) $C = 1.6 \times 10^{-10}$ F, b) $\Delta V = 380$ V,

    c) $E = 7.5 \times 10^4$ V/m, d) $u = 0.025$ J/m^3,

    e) $U = 1.1 \times 10^{-5}$ J

53. $Q = 0.2$ C, $U = 2000$ J

55. $\Delta V = 2000$ V

57. $Q_{super} = 17$ C, $Q_{supply} = 0.33$ C (52 times more charge in the supercapacitor), $U_{super} = 21$ J, $U_{supply} = 66$ J (3 times more energy in the supply capacitor)

59. a) $C = 7.1 \times 10^{-10}$ F, $Q = 8.5 \times 10^{-9}$ C, $\Delta V = 12$V,

    $U = 5.1 \times 10^{-8}$ J,

    b) $C = 2.1 \times 10^{-9}$ F, $Q = 2.5 \times 10^{-8}$ C, $\Delta V = 12$ V,

    $\Delta U = 1.0 \times 10^{-7}$ J,

    c) $C = 7.1 \times 10^{-10}$ F, $Q = 2.5 \times 10^{-8}$ C, $\Delta V = 36$ V,

    d) $C = 2.1 \times 10^{-9}$ F, $Q = 8.5 \times 10^{-9}$ C, $\Delta V = 4.0$ V

61. The solution is a proof.

63. $Q_1 = 3.5 \times 10^{-4}$ C, $Q_2 = 1.5 \times 10^{-4}$ C,

    $Q_3 = 2.0 \times 10^{-4}$ C, $U_1 = 3.1 \times 10^{-2}$ J,

    $U_2 = 1.9 \times 10^{-3}$ J, $U_3 = 2.5 \times 10^{-3}$ J

65. $F = 11$ N

67. $Q_1 = Q_2 = 9.6 \times 10^{-4}$ C, $Q_3 = 1.2 \times 10^{-3}$ C,

    $U_1 = 0.115$ J, $U_2 = 0.077$ J, $U_3 = 0.24$ J

69. a) $C_{net} = 8.0 \times 10^{-6}$ F, b) $C_{net} = 8.0 \times 10^{-7}$ F

71. $k = 2.0$

73. a) $U = 13$ J, b) $u = 4.8 \times 10^4$ J/m^3,

    Volume $= 2.7 \times 10^{-4}$ m^3

75. $C_2 = 8.0$ μF, $U_1 = 7.2 \times 10^{-5}$ J, $U_2 = 3.6 \times 10^{-5}$ J

77. $\dfrac{C}{l} = 2\pi\epsilon_0 \left(\dfrac{\kappa_1 \kappa_2}{\kappa_2 \ln\left(\dfrac{c}{a}\right) + \kappa_1 \ln\left(\dfrac{b}{c}\right)} \right)$

79. $\Delta V = 4.7 \times 10^{-2}$ V

81. a) $F = \dfrac{Q^2}{2A\epsilon_0}$, b) $W = -\dfrac{Q^2}{2A\epsilon_0}\Delta d$, c) $\Delta U = \dfrac{Q^2}{2\epsilon_0 A}\Delta d$,

    d) Compare the answers from parts a) and c).

Chapter 27

1. $Q = 1800$ C/h, $n = 1.1 \times 10^{22}$ electrons/h

3. $t = 1.2 \times 10^{-4}$ s

5. $I = 4.0$ A, $E = 6.0$ V/m

7. $I = 7.5$ A, $Q = 15$ C

9. $I_{(0.0s)} = 1.0$ A, $I_{(1.0s)} = 0.25$ A, $Q_{(2.0s)} = 0.67$ C

11. $\tau = 3.8 \times 10^{-14}$ s, $v_d = 0.054$ m/s

13. $R = 3.0$ Ω

15. $I = 1.3$ A, $n = 8.2 \times 10^{18}$ electrons/s

17. $R_{net} = R/32$

19. $R = 0.40$ Ω

21. $E = 0.069$ V/m

23. $R = 0.87$ Ω

25. $\rho = 5.7 \times 10^{-7}$ $\Omega\cdot$m

27. $\Delta R = 0.92$ Ω

29. $j = 5.7 \times 10^6$ A/m^2, $E = 0.097$ V/m

31. $\sigma = 5.9 \times 10^7 (\Omega\cdot\text{m})^{-1}$

33. $T = 22°$C

35. $\Delta V = 9.9$ V

37. $m_{Cu} = 380$ kg, $m_{Al} = 190$ kg

39. $d = 0.16$ cm

41. $I = 8.0 \times 10^{-5}$ A

43. $R = 4.4 \times 10^{10}$ Ω, $I = 6.8 \times 10^{-9}$ A

45. $R_{eq} = 2.2$ Ω, $I = 5.4$ A, $I_1 = 2.4$ A,

    $I_2 = 1.7$ A, $I_3 = 1.3$ A

47. $I_{iron} = 1.7$ A, $I_{brass} = 4.3$ A

49. In series: $R_{eq} = 9$ Ω, in parallel: $R_{eq} = 0.92$ Ω, in combinations of series and parallel: $R_{eq} = 3.7$ Ω, 4.3 Ω, 5.2 Ω, 1.6 Ω, 2.0 Ω, 2.2 Ω

51. $I = 27$ A

53. a) $I_{Cu} = 210$ A, b) $I_{rubber} = 2.7 \times 10^{-19}$ A

55. $d = 1.5$ km from point A

57. $R = 3.2 \times 10^{-3}$ Ω

59. $R = 5.4 \times 10^{-2}$ Ω/m

61. a) $R_{net} = 4.4$ Ω, b) $I = 1.8$ A,

    c) $I_1 = 1.8$ A, $I_2 = 1.1$ A, $I_3 = 0.7$ A,

    $\Delta V_1 = 3.6$ V, $\Delta V_2 = \Delta V_3 = 4.4$ V

63. $\Delta V = 36$ V

65. $R = 2.73$ Ω

67. $Q = 1.4 \times 10^5$ C, $n = 9.0 \times 10^{23}$ electrons

69. $E = 2300$ V/m

71. $\dfrac{\Delta R}{R_0} = 0.089$ or 8.9%

73. $R = 5.2 \times 10^{-4}$ Ω, $\Delta V = 0.31$ V

75. Connecting two sets of two resistors in parallel gives $R_{net} = 1.0$ Ω

77. $I = 0.30$ A

79. $j = 3.3 \times 10^6$ A/m^2, $v_d = 2.4 \times 10^{-4}$ m/s

81. a) $R_{eq} = 1.3 \times 10^6$ Ω, b) $I_1 = 9.2$ μA, harmless, $I_2 = 88$ μA, harmless, $I_3 = 0.18$ A, fatal

Chapter 28

1. smallest battery: 3.3×10^{-3} $\frac{\text{kW}\cdot\text{h}}{\text{kg}} = 1.2 \times 10^4$ $\frac{\text{J}}{\text{kg}}$;

    largest battery: 1.9×10^{-2} $\frac{\text{kW}\cdot\text{h}}{\text{kg}} = 6.7 \times 10^4$ $\frac{\text{J}}{\text{kg}}$

3. 6.9×10^6 J

5. 1.3×10^5 J

7. 0.38 A, 4.0×10^3 J

9. 6.0 A, $I_2 = I_1 = 3.0$ A

11. current through R_1 does not change, but it increases through the other two resistors

13. (a) 2.4×10^{-4} A; (b) 4.8×10^{-6} V

15. 62 Ω, 115 V, 4.8 V

17. (a) 1.7 A, (b) 0.86 A through resistors 2, 3, and 4

19. 0.49 V

21. 0.11 A

23. 31.0 A, out of the junction

25. 8.0 Ω

27. $I_1 = -15$ A, $I_2 = 25$ A, $I_3 = -5$ A, $I_4 = 15$ A, $I_5 = -20$ A, $R_5 = 4.0$ Ω

29. $I_1 = 1.25$ A, $I_2 = 1.75$ A, $I_3 = -0.50$ A

31. 16 V

33. The current through R_1 is 2.58 A, through R_2 is 1.71 A, through R_3 is 1.28 A, through R_4 is 2.15 A, through R_5 is 0.85 A. $V = 2.6$ V

35. 6.0×10^{-4} W

37. \$0.36

39. 3.7×10^{-3} A

41. 1100 W

43. 4.9 A

45. 6.3×10^{12} protons/s, 7.0×10^2 W

47. 6.7 h

49. 58 h

51. 1.2×10^6 W

53. 10%

55. $R_{115} = 13.23$ Ω, $R_{230} = 52.90$ Ω

57. 0.50 W

59. The solution is a proof.

61. (a) 3.3%, (b) 33%, lower current is more efficient

63. (a) 180 V, (b) 1.8×10^6 W

65. 19 liters/min,

67. 7.0×10^2 Ω

69. 3.958 A

71. $R_{Cu} = 5.0$ Ω, $R_{constantan} = 5.6$ Ω

73. 3.2×10^{-4} s

75. 4.0×10^{-4} s

77. 0.011 s

79. $Q = C\varepsilon\left(1 - e^{-t/(R_1 + R_2)C}\right)$,

$\Delta V = \dfrac{ER_1}{(R_1 + R_2)}\left(e^{-t/(R_1 + R_2)C}\right)$

81. (a) $A = E_2C$, $B = C(E_1 - E_2)$, $\tau = R_2C$, $Q(t) = E_2C + (E_1 - E_2)Ce^{-t/R_2C}$

(b) $A = E_1C$, $B = C(E_2 - E_1)$, $\tau = R_1CI$, $Q(t) = E_1C + (E_2 - E_1)Ce^{-t/R_1C}$

83. 0.020 s

85. 5.3 s, 5.0×10^{-6} A

87. (a) 1380 Ω, (b) 430 A, (c) 2670 A

89. 3.0×10^1 W, 1.7×10^1 W

91. $I_1 = \dfrac{ER'_i}{R_iR' + RR' - RR_i}$, $I_2 = \dfrac{E}{R'_i + R\left(\dfrac{R'}{R_i} - 1\right)}$,

93. $I_1 = 0.45$ A, $I_2 = 1.31$ A, $I_3 = 0.85$ A, $P_1 = 5.4$ W, $P_2 = 14$ W

95. (a) 0.024 m, (b) 3.4×10^6 W

97. (a) 47 A, 7.1×10^3 m

99. 2.2%

101. (a) 7.6×10^{-5} C, (b) 1.2×10^{-4} C, (c) 6.0×10^{-2} A

Chapter 29

1. $|F| = 1.07 \times 10^{-16}$ N, opposite the direction of the current

3. $|F| = 1.04 \times 10^{-16}$ N, $a = 1.14 \times 10^{14}$ m/s^2

5. $|F| = 1.38 \times 10^{-24}$ N

7. $\theta = 19.5°$

9. $B = 4.18$ T, pointing downward

11. $B = 1.44 \times 10^{-5}$ T, $B_{earth} = 4.2\,B$

13. $F = 8.2 \times 10^{-16}$ N

15. $\mathbf{F} = (-2.98 \times 10^{-18}\,\mathbf{i} + 6.8 \times 10^{-19}\,\mathbf{j} + 3.33 \times 10^{-18}\,\mathbf{k})$ N

17. $F = 9.98 \times 10^{-18}$ N and points in direction 16° above the horizontal in North direction.

19. a) $I = \lambda v$,

b) $B = \dfrac{\mu_0 I}{2\pi r} = \dfrac{\mu_0 \lambda v}{2\pi r} = \dfrac{\mu_0 v}{2\pi r}2\pi\epsilon_0 Er = \epsilon_0\mu_0 vE$

21. $B = 0.19$ T

23. $B = 4 \times 10^{-8}$ T, $F = 1.92 \times 10^{-18}$ N

25. $B = 5.03 \times 10^{-3}$ T

27. $\displaystyle\int B\,ds = \dfrac{\mu_0 I}{R}$

29. $a = 1.5 \times 10^{13}$ m/s^2

31. a) $B = 5 \times 10^{-6}$ T, b) $\theta = 16°$

33. $B = \dfrac{\mu_0}{2\pi}\dfrac{I}{d}\dfrac{3}{2}\sqrt{2}$

35. $B = \dfrac{\mu_0 I}{2R} - \dfrac{\mu_0 I}{2\pi R}$, pointing into the page.

37. $r < r_1: B = \dfrac{\mu_0 Ir}{2\pi r_1^2}$, $r_1 < r < r_2: B = \dfrac{\mu_0 I}{2\pi r}$, $r_2 < r < r_3:$

$B = \dfrac{\mu_0 I}{2\pi r}\dfrac{r_3^2 - r^2}{r_3^2 - r_2^2}$, $r < r_3: B = 0$.

39. For $y > 0$, $z > 0$, $-\infty < x < +\infty: \mathbf{B} = 0$,

for $y < 0$, $z > 0$, $-\infty < x < +\infty: \mathbf{B} = \mu_0\sigma\,\mathbf{i}$,

for $y < 0$, $z < 0$, $-\infty < x < +\infty: \mathbf{B} = 0$,

for $y > 0$, $z < 0$, $-\infty < x < +\infty: \mathbf{B} = -\mu_0\sigma\,\mathbf{i}$

41. $B = \dfrac{\mu_0}{\pi}\dfrac{I}{b}\tan^{-1}\left(\dfrac{b}{2z}\right)$

43. For $z > 2R$, $B_1 = \dfrac{\mu_0 IR}{\pi(z^2 - R^2)}$, for $R < z < 2R$,

$B_2 = \dfrac{\mu_0 I}{2\pi}\left(\dfrac{z - R}{R^2} - \dfrac{1}{z + R}\right)$, for $0 < z < R$,

$B_3 = -\dfrac{\mu_0 I}{2\pi}\left(\dfrac{R - z}{R^2} + \dfrac{1}{z + R}\right)$, for $-R < z < 0$,

$B_4 = -\dfrac{\mu_0 I}{2\pi}\left(\dfrac{1}{R - z} + \dfrac{R + z}{R^2}\right)$, for $-2R < z < -R$,

$B_5 = \dfrac{\mu_0 I}{2\pi}\left(\dfrac{1}{R - z} - \dfrac{z + R}{R^2}\right)$, for $z < -2R$,

$B_6 = \dfrac{\mu_0 IR}{\pi(z^2 - R^2)}$, $B_{max} = B_3(z = 0) = \dfrac{\mu_0 I}{\pi R}$

45. $I = 26.5$ A

47. $B = 1.26 \times 10^{-2}$ T

49. $B_{max} = 3.34$ T, $B_{min} = 1.43$ T

51. $B = \mu_0 nn'(r_2 - r_1)I$ for $r < r_1$;
    $B = \mu_0 nn'(r_2 - r)I$ for $r_1 < r < r_2$

53. $B = \dfrac{4\mu_0 I}{\sqrt{2}\pi L}$

55. $B = 7.9 \times 10^{-5}$ T, directed into the paper

57. $B = \dfrac{\mu_0 I}{2\pi R} + \dfrac{\mu_0 I}{8R}$, directed into the paper

59. $B = 0.11\dfrac{\mu_0 I}{L}$, directed into the paper

61. a) $B_P = \dfrac{8\mu_0 I}{\sqrt{125}R}$,

b) At any point z, the fields produced by the coils are

$B_{bottom} = \dfrac{\mu_0 I}{2}\dfrac{R^2}{(z^2 + R^2)^{3/2}}$

$B_{top} = \dfrac{\mu_0 I}{2}\dfrac{R^2}{[(R - z)^2 + R^2]^{3/2}}$

Their first derivatives are

$\dfrac{dB_{bottom}}{dz} = -\dfrac{3\mu_0 I}{2}\dfrac{zR^2}{(z^2 + R^2)^{5/2}}$

$\dfrac{dB_{top}}{dz} = \dfrac{3\mu_0 I}{2}\dfrac{(R - z)R^2}{[(R - z)^2 + R^2]^{5/2}}$

Both of these derivatives cancel each other at $z = R/2$. The second derivatives are

$\dfrac{d^2B_{bottom}}{dz^2} = -\dfrac{3\mu_0 IR^2(R^2 - 4z^2)}{2(z^2 + R^2)^{7/2}}$

$\dfrac{d^2B_{top}}{dz^2} = \dfrac{3\mu_0 IR^2\left[R^2 - 4(R - z)^2\right]}{2\left[(R - z)^2 + R^2\right]^{7/2}}$

Both of these derivatives cancel each other at $z = R/2$.

63. $F = 4.8 \times 10^{-17}$ N, $a = 2.87 \times 10^{10}$ m/s^2, both directed opposite the current.

65. For currents in the same direction, $B = 7.33 \times 10^{-5}$ T, for currents in the opposite direction, $B = 1.05 \times 10^{-5}$ T.

67. $|\mathbf{B}| = 1.05\dfrac{\mu_0 I}{R}$, the angle between $\mathbf{B}$ and the straight wire is $\theta = 17.7°$

69. For $r < R_1$, $B = 2\mu_0 nI$; for $R_1 < r < R_2$, $B = \mu_0 NI$; for $r > R_2$, $B = 0$.

71. $B = \dfrac{\mu_0 nI}{2\pi}$, directed in a plane parallel to the plane of the wires and perpendicular to the current, with the exception along the edges.

73. $B = \dfrac{\mu_0 I\sqrt{5}}{2\pi L}$, directed into the paper.

75. $B = \dfrac{\mu_0 I}{2R}\left(\dfrac{1}{\pi} + \dfrac{3}{8}\right)$, directed out of the paper.

Chapter 30

1. $p = 1.1 \times 10^{-17}$ kg·m/s

3. $p = 3.4 \times 10^{-17}$ kg·m/s

5. $B = 3.3$ T

7. $p = 3.8 \times 10^{-19}$ kg·m/s

9. $B = 0.036$ T

11. $\dfrac{r_{12}}{r_{14}} = 1$

13. $I = 0.39$ A

15. (a) The electron will follow a circular path that spirals in the direction of the magnetic field;
    (b) $f = 1.4 \times 10^7$ Hz, $T = 7.2 \times 10^{-8}$ s;
    (c) $x = 0.29$ m.

17. $B = 3.2$ T

19. (a) $v_p = 1.6 \times 10^3$ m/s; (b) $\theta_p = 3.3 \times 10^{-3}$ rad

21. $F = 6.7 \times 10^{-5}$ N

23. Consider an element $d\mathbf{l}$ of the long, straight wire. As the current I flows through the loop, the magnetic field $\mathbf{B}$ produced by the loop exerts a magnetic force $\mathbf{F}$ on the element $d\mathbf{l}$ given by $d\mathbf{F} = I\,d\mathbf{l}\,?\,\mathbf{B}$. This force is perpendicular to element $d\mathbf{l}$. Applying Newton's third law of motion, the element $d\mathbf{l}$ exerts a force that is directed oppositely to $d\mathbf{F}$, which is also perpendicular to element $d\mathbf{l}$.

25. $\mathbf{F} = -13$ N $\mathbf{j}$

27. $I = 1.0 \times 10^6$ A

29. $\tau = 2.4 \times 10^{-7}$ N·m

31. (a) $\tau = (1.1 \times 10^{-26} \sin \theta)$ N·m, sinusoidal behavior;
    (b) $W = 2.25 \times 10^{-26}$ J

33. $\mu = 0.023$ A·m^2

35. $\mu = \frac{1}{4}Q\omega R^2$ The magnetic and electric fields surrounding the spinning paper disk are illustrated as follows:

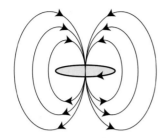

37. (a) $\tau = (1.71 \times 10^{-7} \sin \theta)$ N·m; (b) $f = 0.47$ Hz

39. (a) $M = 1.7 \times 10^6$ A/m; (b) $\dfrac{\mu}{V} = 2.0 \times 10^{-23}$ A·m^2

41. (a) $\chi = -1$; (b) $\mu = 0$

43. $\Delta V_H = 3.1$ μV

45. $B = 2.5 \times 10^{-2}$ T

47. (a) $v_d = 2.7 \times 10^{-3}$ m/s; (b) $\Delta V_H = 1.6 \times 10^{-5}$ V

49. $R_H = -4.7 \times 10^{-10}$ m^3/C

51. (a) $F = 0$ N; (b) $F = 0.18$ N due north; (c) $F = 0.18$ N due east; (b) $F = 0.16$ N due east.

53. $B = 0.010$ T, $f = 1.6 \times 10^5$ Hz

55. (a) $y = \frac{1}{2}a_y t^2 = \frac{1}{2}\left(\dfrac{F}{m}\right)\left(\dfrac{L}{v_{0x}}\right)^2 = \dfrac{eEL^2}{2mv^2}$,

(b) $y = \dfrac{eEL^2}{2mv}$, therefore $\dfrac{e}{m} = \dfrac{2yE}{B^2L^2}$

57. $B = \dfrac{4mv}{qd}$

59. $a = 3.1 \times 10^3$ rad/s^2

Chapter 31

1. 3.0×10^{-3} V; left side is positive and right side is negative

3. 0.5 m/s

5. 2.2×10^{-3} V

7. $v = 0.48$ m/s, current flows counterclockwise, $F = 10$ N to the left

9. 9.8×10^{-3} V

11. -5.5×10^{-5} T·m^2

13. 22 rev/sec

15. 9.7×10^{-3} V; patient does not need to be pushed more slowly

17. $|E| = 6.88$ V, $I = 0.606$ A, $W = 146$ J

19. 0.047 V

21. 3.5×10^{-7} T·m^2

23. (a) 6.6×10^{-6} V/m; (b) 7.4×10^{-6} V/m

25. $I = 0.565$ A, $\tau = 5.59 \times 10^{-3}$ N·m

27. 3.0×10^2 A

29. 0.63 A

31. At $r = 0.80$ m: $E_0 = 140$ V/m $\varepsilon_0 = 704$ V; at $r = 1.5$ m: $E_0 = 117$ V/m $\varepsilon_0 = 1100$ V

33. 0.010 C

35. 0.40 H

37. (a) 0.5 H; (b) 10 V

39. 1.1×10^4 A/s

41. -1.9 V

43. $M = 200\mu_0 n\pi R^2$; the shape of the coil wire does not matter

45. 1.1×10^{-7} H

49. 1.2×10^{-5} J

51. 1.0 J/m^3

53.

B-field (T)	u (J/m^3)
10^8	4.0×10^{21}
10^3	4.0×10^{11}
45	8.1×10^8
8	2.5×10^7
2	1.6×10^6
1.5	9.0×10^5

55. $U = 5.29 \times 10^4$ J; $V = 1.64 \times 10^{-3}$ m^3

57. 211 J/km

59. 1.0×10^{-3} J

61. $IR + L\dfrac{dI}{dt} = 0$; $I = \dfrac{E}{R}e^{-t/\tau}$ satisfies the differential equation if $= L/R$

63. (a) 3.75 Ω; (b) 1.58×10^4 W; (c) 9.51×10^4 J

65. (a) 2.0 A; (b) -3.0×10^3 V

67. (a) $\dfrac{L}{R_2}$; (b) $\dfrac{L}{R_1 + R_2}$

69. (a) $\dfrac{dI_1}{dt} = 0.75$ A/s, $\dfrac{dI_2}{dt} = 1.5$ A/s; (b) $I_{Resistor} = 0.50$ A, $I_{L_1} = 0.17$ A, $I_{L_2} = 0.33$ A

71. $R = 0.125$ Ω, $L = 0.0396$ H

73. 2.4×10^{-4} V

75. 0.014 J/m^3

77. 7.5×10^{-5} V

79. (a) 2.6×10^{-3} V; (b) 0; (c) 9.2×10^{-15} C/m^2

81. (a) 0.0188 T/s; (b) 6.4×10^{-3} V; (c) 6.4×10^{-3} V

83. 102 V

85. (a) $I_{max} = 48$ A, $t = 14$ s; (b) $U_{max} = 576$ J, percentage of energy that remains = 25%

Chapter 32

1. (a) $I_{rms} = 10.4$ A, $I_{max} = 14.8$ A; (b) $P_{max} = 2400$ W, $P_{min} = 0$ W

3. (a) $E_{max} = 3.25 \times 10^5$ V, $I_{max} = 1.05 \times 10^3$ A;

  (b) $P_{max} = 3.4 \times 10^8$ W, $P_{ave} = 1.7 \times 10^8$ W

5. $I_{rms} = 10.4$ A, $I_{max} = 14.8$ A, $R = 11.1$ Ω

7. $I_{rms} = 31.1$ A, $P_{ave} = 48.5$ kW

9. 0.032 A, 0.064 A

11. 2.0×10^6 Ω, 1.5×10^6 Ω

13. (a) 208 Ω; (b) 9.6×10^{-4} A;

  (c) $I(t = 0) = 0$ A, $I(t = 4\pi/\omega) = -6.8 \times 10^{-4}$ A

15. 1.0×10^{-9} F

17. For an amplitude of 1.00 V:

t (s)	$0.0012\cos^2(600t)$ (W)
0.001	8.17E-04
0.002	1.58E-04
0.003	6.19E-05
0.004	6.52E-04

19. 5.6×10^{-5} F

21. 1.0×10^5 A/s

23. $f = 1.6 \times 10^3$ Hz

25. 0.033 A, 0.017 A

27. 0.049 A

29. 0.343 A/s, 0.00218 Hz

31. 0.018 Ω at 60 Hz, 3.0×10^4 Ω at 100 MHz

33. 3.6×10^3 Hz

35. (a) $I_{max} = 5.0 \times 10^{-6}$ A,

  (b) $C = 1.66 \times 10^{-6}$ F, $L = 6.6 \times 10^{-2}$ H

37. 9.4×10^3 rad/s

39. 379 Hz

41. 663 Hz

43. 2.11 Hz, 259 H

45. 2.76×10^{-8} F, 7.5×10^{-8} H

47. 8.23×10^{-6} F, 6.25 W

49. (a) $\overline{P} = \dfrac{1}{2} I_{max} E_{max} \left(\dfrac{R}{Z} \right)$

  (b) From the impedance triangle, $\cos\phi = \dfrac{R}{Z}$

  (c) $\cos\phi = 0.642$, $\phi = \pm 50°$. No.

51. (a) 1.9×10^3 V; (b) 12 V; (c) 24 W; (d) 1.6×10^2

53. $0.995 E_{max}$; $E_{max}/10$; $\omega = \dfrac{R}{L}$

55.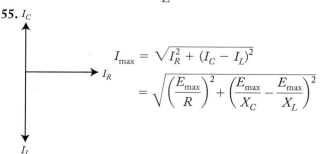

$$I_{max} = \sqrt{I_R^2 + (I_C - I_L)^2}$$
$$= \sqrt{\left(\dfrac{E_{max}}{R}\right)^2 + \left(\dfrac{E_{max}}{X_C} - \dfrac{E_{max}}{X_L}\right)^2}$$

57. 4348 turns

59. Transformer 1: $\dfrac{N_1}{N_2} = 0.044$, Transformer 2: $\dfrac{N_1}{N_2} = 7.6$,

  Transformer 3: $\dfrac{N_1}{N_2} = 16.5$, Transformer 4: $\dfrac{N_1}{N_2} = 34.8$

61. $I_{generator} = 9.09 \times 10^4$ A, $I_{line} = 5.0 \times 10^3$ A

63. 36 W

65. 0.139 A

67. $I_{max} = 1.63$ A. One-quarter cycle after the maximum $I = 0$. One-half cycle after maximum $I = -1.63$ A, three-quarters cycle after maximum $I = 0$.

69. (a) $I = (2.4 + 0.43\cos 360t)$ A; (b) 29.3 W

71. (a) $Q = 6.5 \times 10^{-8} \sin(120\,\pi t)$ F; (b) First max. at $t = 1/240$ sec, first min. at $t = 0$; (c) 5.9×10^{-8} J, 2.95×10^{-8} J

73. $I_{max} = 0.89$ A. One quarter cycle later, $I = 0$. One half cycle later, $I = -0.89$ A. Three-quarters of a cycle later, $I = 0$.

75. (a) $I_R(t) = 1.5 \times 10^{-3}$ A $+ (7.5 \times 10^{-4}$ A$)\cos(6000\pi t)$, $I_L(t) = 60t + (1.6 \times 10^{-3})\sin(6000\pi t)$ A.

  (b) $P_R = 4.5 \times 10^{-3} + 1.1 \times 10^{-3}\cos^2(6000\pi t)$ W,

$P_L = 180t + 90t\cos(6000\pi t) + (4.8 \times 10^{-3})\sin(6000\pi t)$
$+ (2.4 \times 10^{-3} \sin(6000\pi t \cos(6000\pi t)$ W

77. 1.4×10^{-3} J; 7.9×10^{-4} s for fully magnetic energy; 1.6×10^{-3} s for fully electric energy.

79. (a) 1.36×10^3 Hz, (b) 1.2×10^3 V, (c) 78

81. 0.026 A at the power plant, 0.0012 A in the transmission line

83. $E_2 = 15 \sin(3000\pi t)$ V

Chapter 33

1. (c) $I_{disp} = \epsilon_0 \dfrac{d\Phi_E}{dt} = \dfrac{dQ}{dt}$

3. (a) 4.0 A; (b) 4.5×10^{11} V·m/s

5. (a) 0.10 A; (b) 6.3×10^{-3} A

9. (a) 2.0×10^{-3} A; (b) 1.0 sec; (c) 8.9×10^{-4} A, 8.9×10^{-9} T, 1.8×10^{-8} T, 2.7×10^{-8} T

11. (a) $E = \dfrac{\rho Ct}{\pi R^2}$; (b) $B = \dfrac{\mu_0\epsilon_0\rho Cr}{2\pi R^2}$; (c) $B = \dfrac{\mu_0\epsilon_0\rho C}{2\pi r}$

13. $\oint \mathbf{E} \cdot d\mathbf{A} = \dfrac{Q}{\kappa\epsilon_0}$, $\oint \mathbf{B} \cdot d\mathbf{A} = 0$, $-\oint \mathbf{E} \cdot d\mathbf{s} = -\dfrac{d\Phi_B}{dt}$,

$\oint \mathbf{B} \cdot d\mathbf{s} = \mu_0 \left(I + \kappa\varepsilon_0 \dfrac{d\Phi_E}{dt} \right)$

15. **E** is ∥ to acceleration. **B** is ⊥ to acceleration and propagation.

17.

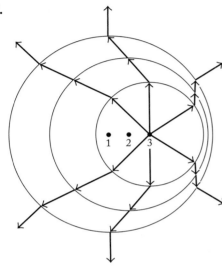

19. west

21. 1.1×10^{-5} A/m

23. 3.3×10^{-11} s

25. 5.00×10^{3} s (8.33 min)

27. 3.3×10^{-9} s, 0.30 GHz, no

29. North, 2.0×10^{-9} T

31. 1/4

33. 1.5%

35. 39°

37. (b) $\theta = 0°$, $\theta = 90°$, $\theta = 180°$, $\theta = 270°$; The field rotates clockwise with a period of $2\pi/\omega$

39. $\dfrac{I_0 \cos^2\phi}{2(\sin^2\phi \cos^2\alpha + \cos^2\phi)}$

41. 3.0×10^{11} Hz

43. AM: 566 m to 187 m; FM: 3.41 m to 2.78 m

45. (a) 1.5×10^{-10} m (X-ray); (b) 1.0 cm (microwave radio); (c) 5.0×10^{6} m ("electric" wave)

47. 0.027 m

49. 2.8×10^{-8} J/m^3

51. 9.5×10^{-6} W/m^2

53. 1.7×10^{-9} T down; 6.6×10^{-4} W/m^2 north

55. 2.8×10^{28} W

57. 4.0×10^{26} W

59. 6.42×10^{-5} W/m^2, 2.02×10^{4} W

61. 0.065 V/m, 0.043 V/m

63. 6.37×10^{3} W/m^2, 2.19×10^{3} V/m, 1.67×10^{-10} J

65. 1.49 W/m^2, 7.81×10^{-11} N

67. 8.0×10^{3} m^2

69. 51

71. 6.0×10^{8} N, 3.5×10^{22} N

73. (a) 1.77×10^{-17} J/m^3 each; (b) 4.0×10^{-3} V/m, and 0 T, or 0 V/m and 1.33×10^{-11} T; (c) one is 7.08×10^{-17} J/m^3, the other is zero.

75. 80.0 m, 0.0785 m^{-1}, 2.36×10^{7} s^{-1}, minus z direction, $E_y = cB_0 \cos(kx + \omega t)$

79. $V_0 \sin\omega t/R$, $\varepsilon_0 \dfrac{A}{d}\omega V_0 \cos\omega t$, $\dfrac{V_0 \sin\omega t}{R} + \varepsilon_0 \dfrac{A}{d} V_0 \omega \cos\omega t$, $\dfrac{\mu_0}{2\pi}\left(\dfrac{V_0 \sin\omega t}{rR} + \dfrac{\varepsilon_0 \pi r}{d} V_0 \omega \cos\omega t \right)$

81. 0.067 sec

83. 0.31 W/m^2

85. 5.00×10^{-7} T north

87. 9.60×10^{3} V/m, 3.2×10^{-5} T

89. 5.5×10^{-2} V/m, 1.8×10^{-10} T

91. 2.9×10^{-20} W, 3.6×10^{27} W

93. (a) 3.4×10^{-15} W; (b) 5.9×10^{17} m = 63 ly, 3600 stars

Chapter 34

1. 40

3. $90° - \theta$

5. 20°

7. 7

9. $H = 0.90$ m; $W = 0.31$ m

11. 4.74×10^{14} Hz, 416 nm, 1.97×10^{8} m/s

13. 1.09

15. 50°

17. 39.014°

19. Fused Quartz

21. 41.7°

23. 42°

25. 24.4°

27. 55.6°, 56.1°, 57.7°

29. 64.5°

31. 41.1°, 61.0°

33. 1.000 21

35. 36.9°

37. 0.77 mm

39. 53.1°

41. $1 \leq n \leq \sqrt{2}$

43. 48.8°, no dependence on n'

45. 1.4002, 40.25%.

47. 137°29′, 139°14′

49. −10.9 cm

51. $s > \frac{1}{2}R$, $s < \frac{1}{2}R$

53. −2.5 cm

55. 16.7 cm

57. 120.0 cm, concave

59. 60.0 cm, 30.0 cm

61. 8.3 mm

63. 21 cm

65. 12 cm

67. −8.6 cm, virtual, upright, smaller, 0.57

69. 14 cm, 14

71. 45 cm, inverted, enlarged

73. The solution is a proof.

75. −60 cm, −21 cm

77. The solution is a proof.

79. 475 cm behind the lens (also behind the mirror)

81. The solution is a proof.

83. 3.9 cm to the right of the center of the ball

85. 25.8 cm to infinity

87. 8.9×10^{-2} s

89. −22 cm, diverging lens, −0.72 cm

91. The solution is a proof.

93. 1/64

95. 1340

97. The solution is a proof.

99. 31°, 50°

101. 56°

103. 7.5 cm

105. $\alpha \geq 38.7°$

107.

mirror	object distance	image characteristics
concave	$s < f$	virtual, upright, and magnified
	$f < s < 2f$	real, inverted, and magnified
	$s > 2f$	real, inverted, and reduced
convex	all	virtual, upright, and reduced

109. −1.7 m, 0.24 m

111.

lens	object distance	image characteristics
concave	$s < f$	virtual, upright, and reduced
	$s > f$	virtual, upright, and reduced
convex	$s < f$	virtual, upright, and magnified
	$f < s < 2f$	real, inverted, magnified
	$s > 2f$	real, inverted, reduced

113. −30 cm

115. 1.9 mm, 108

Chapter 35

1. 3.2×10^8 m/s

3. 127 nm for both

5. 244 m

7. 78 nm

9. $2d = \dfrac{\lambda}{n_2}, \dfrac{2\lambda}{n_2}, \dfrac{3\lambda}{n_2}, \cdots$; yes

11. $2d = \dfrac{1}{2}\lambda, \dfrac{3}{2}\lambda, \dfrac{5}{2}\lambda, \ldots$

13. (a) Only one reflected ray suffers a phase reversal; (c) 1.22 mm

15. 0.257 mm

17. 5.76 mm/s

19. 1.000277

21. 7.4 mm

23. 1.9 mm

25. 0.994

27. measured angles: 5.5°, 20.5°, 35.5°, 51.5°; predicted angles: 6.1°, 18.6°, 32.2°, 48.2°

29. 0, ± 0.0074°, ± 0.0149°, ± 0.0223°, . . .

31.

θ (degrees)	θ (radians)	I/I_{max}
1.0	0.0175	0.626
2.0	0.0349	0.064
3.0	0.0524	0.154

33. 5.4×10^{-7} m

39. 19.0°, 40.5°, 77.2°

41. First order: $\theta_{400} = 13.7°$, $\theta_{700} = 24.4°$; Second order: $\theta_{400} = 28.2°$, $\theta_{700} = 55.7°$ Third order: $\theta_{400} = 45.1°$; no third order max for 700 nm Second and third orders overlap

43. 1.38×10^{-5} rad

45. 0.34 mm

47.

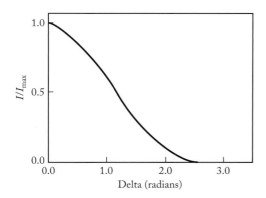

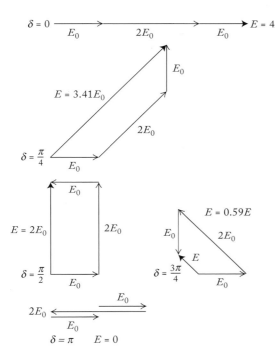

49. (b) $\phi = 0°$; (c) $\theta = \phi$

51. 2.5 cm

53. 0.042 mm

55. 47°

57. (a) not resolved; (b) resolved; (c) colors smeared, not resolved

59. $0.045I_{max}$

61. (b) m interference maxima between adjacent diffraction minima

63. $0.952(3\pi)$

65. about 0.22 mm in diameter

67. 94°

69. about 0.06 arcsec, about eight times better than from Earth

71. (a) 6.7×10^{-5} rad; (b) about 1 cm in diameter

73. minimum separation $\approx 6 \times 10^7$ m; it is possible to resolve the moons but they can only be seen as points of light

75. (a) 9.6×10^{-5} radian; (b) tremor is 9.7×10^{-5} radian to 1.5×10^{-4} radian, eliminating tremor would improve resolution somewhat

77. (a) 2.24×10^{-6} radian; (b) 0.34 m; (c) 0.0054 mm

79. cat barely distinguishes the mice

81. film is at least 120 nm thick

83. 454 nm

85. $\pm 14.5, \pm 48.6°$

87. (b) 95.9 MHz

89. 13 km

91. 4.2 cm, 2.1 cm

93. 1.63×10^{-3} radian; 5.7 km

95. (a) 1.95 km; (b) 5.02×10^{-4} W/m^2

Chapter 36

1. (a) 0, 60 km/s; (b) 30 km/s, 90 km/s

3. $\Delta t_1 = \dfrac{2}{c}\left[(L_1 - L_2) + \left(L_1 - \dfrac{L_2}{2}\right)\left(\dfrac{V}{c}\right)^2\right]$;

$\Delta t_2 = \dfrac{2}{c}\left[(L_2 - L_1) + \left(L_2 - \dfrac{L_1}{2}\right)\left(\dfrac{V}{c}\right)^2\right]$;

$\Delta t_1 - \Delta t_2 = \dfrac{2(L_1 - L_2)}{c}\left[2 - \dfrac{3}{2}\left(\dfrac{V}{c}\right)^2\right]$

5. $0.995c$

7. $0.866c$ (2.60×10^8 m/s)

9. 0.16 s.

11. $0.447c$

13. $4.8 \times 10^{-3}c$ (1.4×10^6 m/s)

15. $\dfrac{f'}{f} = 0.58; 0.42; 0.23$

17. 366 m/s; time dilation factor $= 1 + 7.4 \times 10^{-13}$

19. time dilation factor $= 1 + 1.17 \times 10^{-12}$; the clock at the North Pole will be ahead by 3.7×10^{-5} s in one year.

21. 0.866 m

23. 0.19 m

25. $0.866c$

27. $0.44c$

29. 3.3×10^{-6} C/m^3

31. Relative to the upper part of the belt, the lower belt segment is moving with a speed greater than V and appears shortened even more than the base. Therefore the belt is tightened in both reference frames.

35. $0.65c$

37. (a) $x = 7.67 \times 10^{16}$ m, $y = 1.2 \times 10^{17}$ m, $t = -1.56 \times 10^8$ s; (b) The light from the nova will arrive at earth before the radio message.

45. $K_{rel} = 5.04 \times 10^{-3} mc^2$, $K_N = 5.00 \times 10^{-3} mc^2$, deviation = 0.8%

47. $0.85c$

49. $0.94c$

53. 2.7 m/s; 0.016 m/s; 1.3×10^4 m/s.

55. 1.06×10^{-17} kg·m/s

61. 1.02×10^6 eV

63. $\dfrac{m_{thermal}}{m_{sun}} = 1.1 \times 10^{-4}$%

65. 1.26×10^{-13} J, 1.5 times the rest mass energy of the electron.

67. $0.828c$

69. 9.33 light years from earth, message arrives 21 years after departure

71. (a) To the ship observer, the 600 nm pulse is emitted 2.60 μs before the 400 nm pulse; (b) 133 nm instead of 400 nm, 1800 nm instead of 600 nm; (c) 1.2 km

73. Dimensions will be 1.0 m $\times$ 1.0 m $\times$ 0.8 m. The area of the two faces perpendicular to the direction of motion will be 1.0 m^2. The four faces whose planes are parallel to the direction of motion will have area = 0.80 m^2. The volume of the cube will be 0.80 m^3.

75. The deviation is 2%.

77. $0.98c$ if forward, $0.54c$ if backward

79. 6.96×10^{20} J; 7.74 metric tons

81. 1.2×10^{17} J; 2.9×10^7 tons of TNT

83. 0.33 m/s

85. (a) $0.906c$; (b) $|v_{muon}| = 0.98c$, $|v_{antimuon}| = 0.67c$

Photo credits

Part Opener 4: Andrew Syred/Photo Researchers, Inc.; **Part Opener 5**: NASA/Marshall Space Flight Center.

Chapter Opener 22: Canon Inc.; **fig. 22.3**: Departamento de Física Faculdade de Ciências e Tecnologia Universidade de Coimbra; **p. 698**: Art Resource, NY; **fig. 22.10a**: Corbis; **fig. 22.10c**: Department of Physics, University of Oslo, Norway; **fig. 2210d**: IBMRL/Visuals Unlimited; **fig. 22.9a**: Jun Yang, Ting-Jie Wang, Hong He, Fei Wei, Yong Jin. *Particle size distribution and morphology of in situ suspension polymerized toner*. Industrial & Engineering Chemistry Research. 2003, 42 (22): 5568-5575; **fig. 22.10b**: Wolfson Nanometrology Laboratory at the University of Strathclyde by Gregor Welsh; **fig. 22.13**: Chuck Doswell/Visuals Unlimited; **fig. 22.14**: H. David Seawell/Corbis; **p. 710**: North Carolina Museum of Art/Corbis; **fig. 22.15**: Lightingmaster, Streamer Delaying Air Terminal; **fig. 22.17**: Larry Stepanowicz/Visuals Unlimited.

Chapter Opener 23: Kent Wood/Photo Researchers, Inc.; **p. 735**: (Pip fig. 2): PPC Industries; **fig. 23.22**: Courtesy of Ionoptika Ltd., England.

Chapter Opener 24: Trustees of Princeton University; **p. 763**: American Institute of Physics; **fig. 24.22**: Trustees of Princeton University; **fig. 24.46**: Tom Pantages.

Chapter Opener 25: Hank Morgan/Photo Researchers, Inc.; **p. 793**: Bettmann/Corbis; **p. 805**: (Pip fig. 3) IT Stock Int'l/indexphoto.com; **p. 805**: (Pip fig. 4) Euclid Garmet/Georgia Power; **fig. 25.32**: Stanford University.

Chapter Opener 26: (left and right) National Ignition Facility; **fig. 26.18**: Edward Kinsman/Photo Researchers, Inc.; **fig. 26.23**: Edward Kinsman/Photo Researchers, Inc.

Chapter Opener 27: William Taufic/Corbis; **fig. 27.2**: American Journal of Physics 30, 19, 1963; **fig. 27.3**: American Journal of Physics 30, 19, 1963; **Table 27.1a**: National Center for Atmospheric Research; **Table 27.1b**: Patrick Bennett/Corbis; **Table 27.1c**: Brownie Harris/Corbis; **Table 27d**: Courtesy of Bosch; **Table 27e**: Corbis; **p. 866**: Bettmann/Corbis; **fig. 27.9**: © Crown copyright 1999. Reproduced by permission of the Controller of HMSO and the Queen's Printer for Scotland; **fig. 27.21**: John Wilkes Studio/Corbis; **fig. 27.26**: *Principals of Physics* by Hans Ohanian, 1994; **fig. 27.28**: Courtesy of Superpower, Inc.

Index